Statistics for Biology and Health

Statistics for Biology and Health (SBH) includes monographs and advanced textbooks on statistical topics relating to biostatistics, epidemiology, biology, and ecology.

Daniel Sorensen

Statistical Learning in Genetics

An Introduction Using R

Springer

Daniel Sorensen
Aarhus University
Aarhus, Denmark

ISSN 1431-8776 ISSN 2197-5671 (electronic)
Statistics for Biology and Health
ISBN 978-3-031-35850-0 ISBN 978-3-031-35851-7 (eBook)
https://doi.org/10.1007/978-3-031-35851-7

This Springer imprint is published by the registered company Springer Nature Switzerland AG
The registered company address is: Gewerbestrasse 11, 6330 Cham, Switzerland

Paper in this product is recyclable.

Til min elskede Pia

Preface

This book evolved from a set of notes written for a graduate course on Likelihood and Bayesian Computations held at Aarhus University in 2016 and 2018. The audience was life-science PhD students and post-docs with a background in either biology, agriculture, medicine or epidemiology, who wished to develop analytic skills to perform genomic research. This book is addressed to this audience of numerate biologists, who, despite an interest in quantitative methods, lack the formal mathematical background of the professional statistician. For this reason, I offer considerably more detail in explanations and derivations than may be needed for a more mathematically oriented audience. Nevertheless, some mathematical and statistical prerequisites are needed in order to extract maximum benefit from the book. These include introductory courses on calculus, linear algebra and mathematical statistics, as well as a grounding in linear and nonlinear regression and mixed models. Applied statistics and biostatistics students may also find the book useful, but may wish to browse hastily through the introductory chapters describing likelihood and Bayesian methods.

I have endeavoured to write in a style that appeals to the quantitative biologist, while remaining concise and using examples profusely. The intention is to cover ground at a good pace, facilitating learning by interconnecting theory with examples and providing exercises with their solutions. Many exercises involve programming with the open-source package `R`, a statistical software that can be downloaded and used with the free graphical user interface `RStudio`. Most of today's students are competent in `R` and there are many tutorials online for the uninitiated. The `R`-code needed to solve the exercises is provided in all cases and is written, with few exceptions, with the objective of being transparent rather than efficient. The reader has the opportunity to run the codes and to modify input parameters in an experimental fashion. This hands-on computing contributes to a better understanding of the underlying theory.

The first objective of this introduction is to provide readers with an understanding of the techniques used for analysis of data, with emphasis on genetic data. The second objective is to teach them to implement these techniques. Meeting these objectives is an initial step towards acquiring the skills needed to perform data-driven genetics/genomics research. Despite the focus on genetic applications, the mathematics of the statistical models and their implementation are relevant for many other branches of quantitative methods. An appendix in the opening chapter provides an overview of basic quantitative genomic concepts, making the book more accessible to an audience of "non-geneticists".

I have attempted to give a balanced account of frequentist/likelihood and Bayesian methods. Both approaches are used in classical quantitative genetic and modern genomic analyses and constitute essential ingredients in the toolkit of the well-trained quantitative biologist.

The book is organised in three parts. Part I (Chaps. 2–5) presents an overview of likelihood and Bayesian inference. Chapter 2 introduces the basic elements of the likelihood paradigm, including the likelihood function, the score and the maximum likelihood estimator. Properties of the maximum likelihood estimator are summarised and several examples illustrate the construction of simple likelihood models, the derivation of the maximum likelihood estimators and their properties. Chapter 3 provides a review of three computational methods for fitting likelihood models: Newton-Raphson, the EM (expectation-maximisation) algorithm and gradient descent. After a brief description of the methods and the essentials of their derivation, several examples (13 in all) are developed to illustrate their implementation. Chapter 4 covers the basics of the Bayesian approach, mostly through examples. The first set of examples illustrate the type of inferences that are possible (joint, conditional and marginal inferences), when the posterior distributions have known closed forms. In this case, inferences can be exact using analytical methods, or can be approximated using Monte Carlo draws from the posterior distribution. A number of options are available when the posterior distribution is only known up to proportionality. After a very brief account of Bayesian asymptotics, the chapter focuses on Markov chain Monte Carlo (McMC) methods. These are recipes for generating approximate draws from posterior distributions. Using these draws, one can obtain Monte Carlo estimates of the complete posterior distribution, or Monte Carlo estimates of summaries such as the mean, variance and posterior intervals. The chapter provides a description of the Gibbs sampling algorithm and of the joint and single-site updating of parameters based on the Metropolis-Hastings algorithm. An overview of the tools needed for analysis of the McMC output concludes the chapter. An appendix provides the mathematical details underlying the magic of McMC within the constraints imposed by the author's limited mathematics. Chapter 5 illustrates applications of McMC. Several of the examples discussed in connection with Newton-Raphson and the EM algorithm are revisited and implemented from a Bayesian McMC perspective.

Part II of the book has the heading *Prediction*. The boundaries between Parts I and II should not be construed as rigid. However, the heading emphasises the main thread of Chaps. 6–11, with an important detour in Chap. 8 that discusses multiple testing. Chapter 6 introduces many important ingredients of prediction: best predictor, best linear predictor, overfitting, bias-variance trade-off, cross-validation. Among the topics discussed is the accuracy with which future observations can be predicted, how is this accuracy measured, the factors affecting it and importantly, how a measure of uncertainty can be attached to accuracy. The body of the chapter deals with prediction from a classical/frequentist perspective. Bayesian prediction is illustrated in several examples throughout the book and particularly in Chap. 10. In Chap. 6, many important ideas related to prediction are illustrated using a simple least-squares setting, where the number of records n is larger than the number of parameters p of the model; this is the $n > p$ setup. However, in many modern genetic problems, the number of parameters greatly exceeds the number of records; the $p \gg n$ setup. This calls for some form of regularisation, a topic introduced in Chap. 7 under the heading *Shrinkage Methods*. After an introduction to ridge regression, the chapter provides a description of the lasso (least absolute shrinkage and selection operator) and of a Bayesian spike and slab model. The spike and slab model can be used for both prediction and for discovery of relevant covariates that have an effect on the records. In a genetic context, these covariates could be observed genetic markers and the challenge is how to find as many promising markers among the hundreds of thousands available, while incurring a low proportion of false positives. This leads to the topic reviewed in Chap. 8: False Discovery Rate. The subject is first presented from a frequentist perspective as introduced by Benjamini and Hochberg in their highly acclaimed work, and is also discussed using empirical Bayesian and fully Bayesian approaches. The latter is implemented within an McMC environment using the spike and slab model as driving engine. The complete marginal posterior distribution of the false discovery rate can be obtained as a by-product of the McMC algorithm. Chapter 9 describes some of the technical details associated with prediction for binary data. The topics discussed include logistic regression for the analysis of case-control studies, where the data are collected in a non-random fashion, penalised logistic regression, lasso and spike and slab models implemented for the analysis of binary records, area under the curve (AUC) and prediction of a genetic disease of an individual, given information on the disease status of its parents. The chapter concludes with an appendix providing technical details for an approximate analysis of binary traits. The approximation can be useful as a first step, before launching the full McMC machinery of a more formal approach. Chapter 10 deals with Bayesian prediction, where many of the ideas scattered in various parts of the book are brought into focus. The chapter discusses the sources of uncertainty of predictors from a Bayesian and frequentist perspective and how they affect accuracy of prediction as measured by the Bayesian and frequentist expectations of the sample mean squared error of prediction. The final part of the chapter introduces, via an example, how specific aspects of a Bayesian model can be tested using posterior predictive simulations, a topic that combines frequentist and Bayesian ideas. Chapter 11 completes Part II and provides an overview of selected nonparametric methods. After an introduction of traditional nonparametric models, such as the binned estimator and kernel smoothing methods, the chapter concentrates on four more recent approaches: kernel methods using basis expansions, neural networks, classification and regression trees, and bagging and random forests.

Part III of the book consists of exercises and their solutions. The exercises (Chap. 12) are designed to provide the reader with deeper insight of the subject discussed in the body of the book. A complete set of solutions, many involving programming, is available in Chap. 13.

The majority of the datasets used in the book are simulated and intend to illustrate important features of real-life data. The size of the simulated data is kept within the limits necessary to obtain solutions in reasonable CPU time, using straightforward `R`-code, although the reader may modify size by changing input parameters. Advanced computational techniques required for the analysis of very large datasets are not addressed. This subject requires a specialised treatment beyond the scope of this book.

The book has not had the benefit of having been used as material in repeated courses by a critical mass of students, who invariably stimulate new ideas, help with a deeper understanding of old ones and, not least, spot errors in the manuscript and in the problem sections. Despite these shortcomings, the book is completed and out of my hands. I hope the critical reader will make me aware of the errors. These will be corrected and listed on the web at https://github.com/SorensenD/SLGDS. The GitHub site also contains most of the `R`-codes used in the book, which can be downloaded, as well as notes that include comments, clarifications or additions of themes discussed in the book.

Aarhus, Denmark Daniel Sorensen
May 2023

Acknowledgements

Many friends and colleagues have assisted in a variety of ways. Bernt Guldbrandtsen (University of Copenhagen) has been a stable helping hand and helping mind. Bernt has generously shared his deep biological and statistical knowledge with me on many, many occasions, and provided also endless advice with LaTeX and MarkDown issues, with programming details, always with good spirits and patience. I owe much to him. Ole Fredslund Christensen (Aarhus University) read several chapters and wrote a meticulous list of corrections and suggestions. I am very grateful to him for this effort. Gustavo de los Campos (Michigan State University) has shared software codes and tricks and contributed with insight in many parts of the book, particularly in Prediction and Kernel Methods. I have learned much during the years of our collaboration. Parts of the book were read by Andres Legarra (INRA), Miguel Pérez Enciso (University of Barcelona), Bruce Walsh (University of Arizona), Rasmus Waagepetersen (Aalborg University), Peter Sørensen (Aarhus University), Kenneth Enevoldsen (Aarhus University), Agustín Blasco (Universidad Politécnica de Valencia), Jens Ledet Jensen (Aarhus University), Fabio Morgante (Clemson University), Doug Speed (Aarhus University), Bruce Weir (University of Washington), Rohan Fernando (retired from Iowa State University) and Daniel Gianola (retired from the University of Wisconsin-Madison). I received many helpful comments, suggestions and corrections from them. However, I am the only responsible for the errors that escaped scrutiny. I would be thankful if I could be made aware of these errors.

I acknowledge Eva Hiripi, Senior Editor, Statistics Books, Springer, for consistent support during this project.

I am the grateful recipient of many gifts from my wife Pia. One has been essential for concentrating on my task: happiness.

Contents

Chapter 1
Overview

1.1 Introduction

Suppose there is a set of data consisting of observations in humans on forced expiratory volume (FEV, a measure of lung function; lung function is a predictor of health and a low lung function is a risk factor for mortality), or on the presence or absence of heart disease and that there are questions that could be answered using these data. For example, a statistical geneticist may wish to know:

1. Is there a genetic component contributing to the total variance of these traits? A positive answer suggests that genetic factors are at play. The next step would be to investigate the following:
2. Is the genetic component of the traits driven by a few genes located on a particular chromosome, or are there many genes scattered across many chromosomes? How many genes are involved and is this a scientifically sensible question?
3. Are the genes detected protein-coding genes, or are there also noncoding genes involved in gene regulation?
4. How is the strength of the signals captured in a statistical analysis related to the two types of genes? What fraction of the total genetic variation is allocated to both types of genes?
5. What are the frequencies of the genes in the sample? Are the frequencies associated with the magnitude of their effects on the traits?
6. What is the mode of action of the genes?
7. What proportion of the genetic variance estimated in 1 can be explained by the discovered genes?
8. Given the information on the set of genes carried by an individual, will a genetic score constructed before observing the trait help with early diagnosis and prevention?
9. How should the predictive ability of the score be measured?

D. Sorensen, *Statistical Learning in Genetics*, Statistics for Biology and Health,
https://doi.org/10.1007/978-3-031-35851-7_1

10. Are there other non-genetic factors that affect the traits, such as smoking behaviour, alcohol consumption, blood pressure measurements, body mass index and level of physical exercise?
11. Could the predictive ability of the genetic score be improved by incorporation of these non-genetic sources of information, either additively or considering interactions? What is the relative contribution from the different sources of information?

The first question has been the focus of quantitative genetics during many years long before the so-called genomic revolution, that is, before breakthroughs in molecular biology made technically and economically possible the sequencing of whole genomes, resulting in hundreds of thousands or millions of genetic markers (single nucleotide polymorphisms (SNPs)) for each individual in the data set. Until the end of the twentieth century before dense genetic marker data were available, genetic variation of a given trait was inferred using resemblance between relatives. This requires equating the expected proportion of genotypes shared identical by descent, given a pedigree, with the observed phenotypic correlation between relatives. The fitted models also retrieve "estimates of random effects", the predicted genetic values that act as genetic scores and are used in selection programs of farm animals and plants.

Answers to questions 2 – 7 would provide insight into genetic architecture and thereby, into the roots of many complex traits and diseases. This has important practical implications for drug therapies targeted to particular metabolic pathways, for personalised medicine and for improved prediction. These questions could not be sensibly addressed before dense marker data became available (perhaps with the exception provided by complex segregation analysis that allowed searching for single genes).

Shortly after a timid start where use of low-density genetic marker information made its appearance, the first decade of the twenty-first century saw the construction of large biomedical databases that could be accessed for research purposes where health information was collected. One such database was the *British* 1958–*cohort study* including medical records from approximately 3000 individuals genotyped for one million SNPs. These data provided for the first time the opportunity to begin addressing questions 2 – 7. However, a problem had to be faced: how to fit and validate a model with one million unknowns to a few thousand records and how to find a few promising genetic markers from the million available avoiding a large proportion of false positives? This resulted in a burst of activity in the fields of computer science and statistics, leading to development of a methodology designed to meet the challenges posed by *Big Data*.

In recent years, the amount of information in modern data sets has grown and become formidable and the challenges have not diminished. One example is the *UK Biobank* that provides a wealth of health information from half a million UK participants. The database is regularly updated and a team of scientists recently reported that the complete exome sequence was completed (about 2% of the genome involved in coding for proteins and considered to be important for identifying disease-causing or rare genetic variants). The study involved more than 150,000 individuals genotyped for more than 500 million SNPs (Halldorsson et al 2022). These data are paired with detailed medical information and constitute an unparalleled resource for linking human genetic variation to human biology and disease.

An important task for the statistical geneticist is to adapt, develop and implement models that can extract information from these large-scale data and to contribute to finding answers to the 11 questions posed above. This is an exercise on inference (such as estimation of genetic variation), on gene detection (among the millions of genetic markers that may be included in a probability model, how to screen the "relevant" ones for further study?), on prediction (how does the quality of prediction of future records, for example, outcome of a disease, improve with this new knowledge about the trait?) and on how to fit the probability models. There are several areas of expertise that must be developed in order to fulfil this data-driven research task. An initial step is to understand the methodology that underlies the probability models and to learn the modern computer-intensive methods required for fitting these models. The objective of this book is to guide the reader to take this first step.

This opening chapter gives an overview of the book's content, omitting many technicalities that are revealed in later chapters, and is intended to give a flavour of the way ahead. The first part is about methodology and introduces, by means of an example, the concepts of probability distribution, likelihood and the maximum likelihood estimator. This is followed by a brief description of Bayesian methods indicating how prior knowledge can be incorporated in a probability model and how it can affect inferences. The second part of the chapter presents models for prediction and for detection of genes using parametric and nonparametric approaches. There is an appendix that offers a brief tour of the quantitative genetic/genomic model. The goal is to introduce the jargon and the basic quantitative genetic/genomic concepts used in the book.

1.2 The Sampling Distribution of a Random Variable

A useful starting point is to establish the distinction between a probability distribution and a likelihood function. For example, assume a random variable X that has a Bernoulli probability distribution. This random variable can take 1 or 0 as possible values (more generally, it can have two modalities) with probabilities θ and $1 - \theta$, respectively. The mean of the distribution is

$$E(X|\theta) = 0 \times \Pr(X = 0|\theta) + 1 \times \Pr(X = 1|\theta) = \theta$$

and the variance is

$$\begin{aligned}\mathrm{Var}(X|\theta) &= \mathrm{E}(X^2|\theta) - [\mathrm{E}(X|\theta)]^2 \\ &= \theta - \theta^2 = \theta\,(1-\theta)\,.\end{aligned}$$

A binomial distribution arises from the sum of n mutually independent Bernoulli random variables all having the same probability θ. Therefore, the expected value and the variance of a binomially distributed random variable are $n\theta$ and $n\theta\,(1-\theta)$, respectively.

With this background, imagine that a sample of size n of unrelated haploid individuals is obtained from some population with the objective of estimating allele frequency at a biallelic locus. The sample contains x copies of allele A and $n-x$ copies of allele a. The n data points are draws assumed to be identically and independently distributed, and in each draw, the probability of observing an A allele is θ. Since the random variable can take two modalities (A or a), the number of copies drawn, X, is binomially distributed with parameters n and θ and probability mass function equal to

$$\Pr(X = x|n,\theta) = \binom{n}{x}\theta^x\,(1-\theta)^{n-x}\,, \quad x = 0, 1, \ldots, n, \quad 0 < \theta < 1. \tag{1.1}$$

For fixed values of n and θ, one can plot (1.1) as a function of $x = 0, 1, \ldots, n$, and this defines the *probability distribution* of X. Figure 1.1 shows two different binomial distributions. Importantly in (1.1), the parameters n and θ are fixed and the random variable is X, the number of A alleles drawn (here I distinguish between

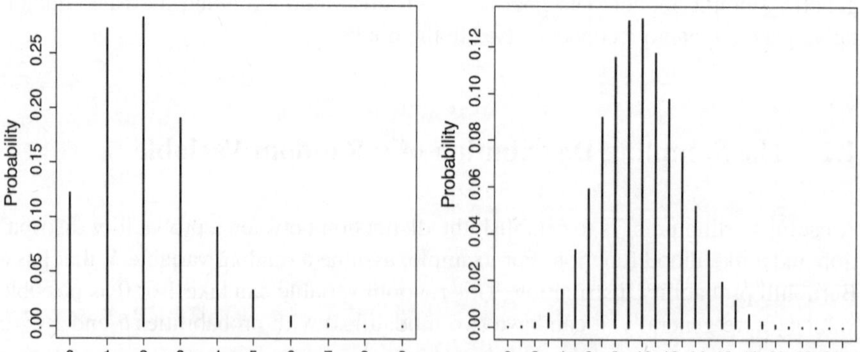

Fig. 1.1 Left: binomial probability distribution with parameters $n = 20$, $\theta = 0.1$; right: binomial probability distribution with parameters $n = 100$, $\theta = 0.1$

the random variable X and its realised value, x. This distinction is not necessarily followed throughout the book).

The probability distribution in the right panel of Fig. 1.1 is more symmetrical than in the left panel. This is due to the different sample sizes n. As sample size increases further, X will approach its limiting distribution which is the normal distribution by virtue of the central limit theorem.

1.3 The Likelihood and the Maximum Likelihood Estimator

Consider now viewing (1.1) in a different manner, whereby x and n are fixed and θ varies. To be specific, assume that the sample size is $n = 27$ and the number of copies of allele A in the sample is $x = 11$. One can plot the probability of obtaining $x = 11$ copies of A in a sample of size $n = 27$, for all permissible values of θ as in Fig. 1.2. For example, for $\theta = 0.1$, $\Pr(X = 11|n = 27, \theta = 0.1) = 0.242 \times 10^{-4}$ and for $\theta = 0.6$, $\Pr(X = 11|n = 27, \theta = 0.6) = 0.203 \times 10^{-1}$. This plot is the *likelihood function* for θ, $L(\theta|x, n)$, and the value of θ that maximises this function is known as the maximum likelihood estimate of θ (I will use MLE short for maximum likelihood estimator or maximum likelihood estimate and ML for maximum likelihood).

Fig. 1.2 Binomial model: likelihood function for θ, given data $n = 27$, $x = 11$

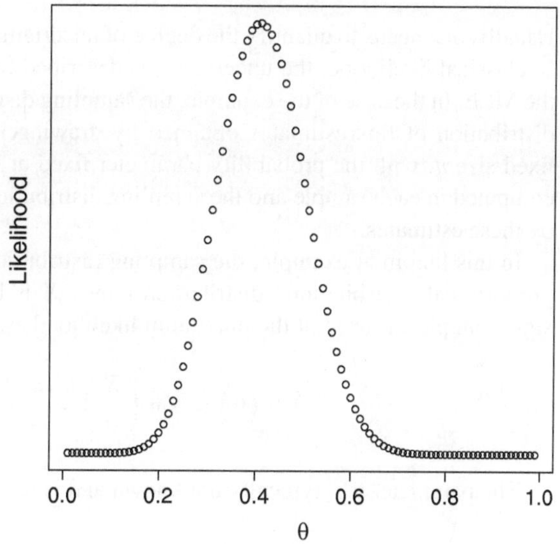

One way of finding the *maximum likelihood estimate* of θ is to differentiate (1.1) and find the maximiser. It is equivalent—but often easier—to maximise the logarithm of the likelihood function, the loglikelihood, denoted as $\ell(\theta|x,n)$:

$$\frac{\partial \ell(\theta|x,n)}{\partial \theta} = \frac{\partial}{\partial \theta}\left[\log\binom{n}{x} + x\log(\theta) + (n-x)\log(1-\theta)\right].$$

Carrying out the differentiation and setting the result equal to zero shows that the MLE of θ must satisfy

$$\frac{x}{\theta} - \frac{n-x}{1-\theta} = 0.$$

Solving for θ yields the MLE

$$\hat{\theta} = \frac{x}{n}, \tag{1.2}$$

that in the case of the example, with $x = 11$ and $n = 27$, gives $\hat{\theta} = 0.41$.

The Sampling Variance of the Maximum Likelihood Estimator

Usually, one needs to quantify the degree of uncertainty associated with an estimate. In classical likelihood, the uncertainty is described by the *sampling distribution* of the MLE. In the case of the example, the sampling distribution of $\hat{\theta}$ is the probability distribution of this estimator obtained by drawing repeated binomial samples of fixed size n, with the probability parameter fixed at its MLE, $\theta = \hat{\theta}$. The MLE is computed in each sample and the sampling distribution of the MLE is characterised by these estimates.

In this binomial example, the sampling distribution of $\hat{\theta}$ is known exactly; it is proportional to a binomial distribution (since X is binomial and n is fixed). The small sample variance of the maximum likelihood estimator is

$$\text{Var}\left(\hat{\theta}\right) = \text{Var}\left(\frac{X}{n}\right) = \frac{\theta(1-\theta)}{n}. \tag{1.3}$$

The parameter θ is typically not known and is replaced by the MLE $\hat{\theta}$. Then

$$\widehat{\text{Var}}\left(\hat{\theta}\right) = \frac{\hat{\theta}\left(1-\hat{\theta}\right)}{n}.$$

In many cases, the MLE does not have a closed form and the small sample variance is not known. One can then appeal to large sample properties of MLE; one

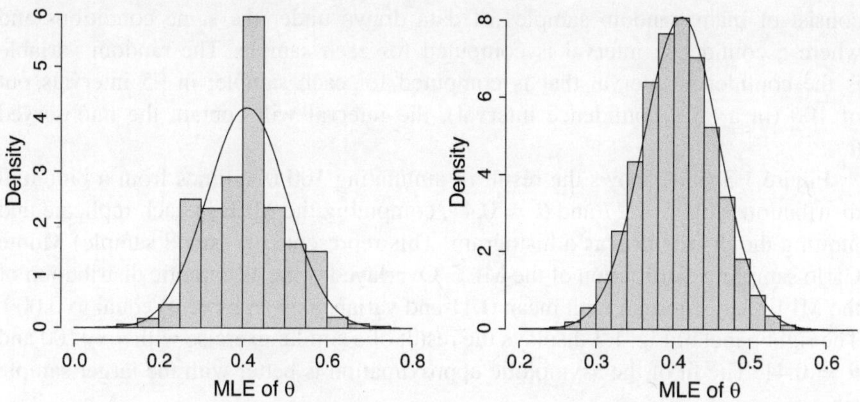

Fig. 1.3 Left: histogram of the Monte Carlo distribution of the MLE for the binomial model, with $n = 27, \theta = 0.41$. Right: histogram of the Monte Carlo distribution of the MLE for the binomial model, with $n = 100, \theta = 0.41$. The overlaid normal curves represent the asymptotic approximation of the distribution of the MLE

of these is that, asymptotically, the MLE is normally distributed, with mean equal to the parameter and variance given by minus the inverse of the second derivative of the loglikelihood evaluated at $\theta = \hat{\theta}$. The second derivative of the loglikelihood is

$$\frac{\partial^2 \ell(\theta|x,n)}{(\partial\theta)^2} = \frac{\partial^2}{(\partial\theta)^2}\left[\log\binom{n}{x} + x\log(\theta) + (n-x)\log(1-\theta)\right]$$
$$= -\frac{x}{\theta^2} - \frac{n-x}{(1-\theta)^2}.$$

In this expression, substituting θ with the MLE $\hat{\theta}$ and taking a reciprocal yields

$$-\left(\frac{\partial^2 \ell(\theta|x,n)}{(\partial\theta)^2}\right)^{-1}\Bigg|_{\theta=\hat{\theta}} = \widehat{\text{Var}}\left(\hat{\theta}\right) = \frac{\hat{\theta}\left(1-\hat{\theta}\right)}{n} \approx 0.009. \tag{1.4}$$

In this simple example, the asymptotic variance agrees with the small sample variance. An approximate 95% confidence interval for θ based on asymptotic theory is

$$0.41 \pm 1.96 \times 0.095 = (0.22, 0.60)\,. \tag{1.5}$$

This means that there is a 95% probability that this interval contains the true parameter θ. The “probability” is interpreted with respect to a set of hypothetical repetitions of the entire data collection and analysis procedure. These repetitions consist of many random samples of data drawn under the same conditions and where a confidence interval is computed for each sample. The random variable is the confidence interval that is computed for each sample; in 95 intervals out of 100 (in a 95% confidence interval), the interval will contain the unobserved θ.

Figure 1.3 (left) shows the result of simulating 100,000 times from a binomial distribution with $n = 27$ and $\theta = 0.41$, computing the MLE in each replicate and plotting the distribution as a histogram. This represents the (small sample) Monte Carlo sampling distribution of the MLE. Overlayed is the asymptotic distribution of the MLE that is normal with mean 0.41 and variance given by (1.4) equal to 0.009. The right panel of Fig. 1.3 displays the result of a similar exercise with $n = 100$ and $\theta = 0.41$. The fit of the asymptotic approximation is better with the larger sample size.

A glance at a standard calculus book reveals that the curvature of a function f at a point θ is given by

$$c(\theta) = \frac{f''(\theta)}{\left[1 + f'(\theta)^2\right]^{3/2}}.$$

In the present case, the function f is the loglikelihood ℓ whose first derivative evaluated at $\theta = \hat{\theta}$ is equal to zero. The curvature of the loglikelihood at $\theta = \hat{\theta}$ is

$$c\left(\hat{\theta}\right) = \ell''\left(\hat{\theta}\right).$$

(I use the standard notation $\ell''\left(\hat{\theta}\right)$ for $\left(\frac{\partial^2 \ell(\theta|x,n)}{(\partial\theta)^2}\right)\Big|_{\theta=\hat{\theta}}$).

Note As the loglikelihood increases or decreases, so does the likelihood; therefore, the value of the parameter that maximises one also maximises the other. Working with the loglikelihood is to be preferred to working with the likelihood function because it is easier to differentiate a sum than a product. The curvature of the loglikelihood at the MLE is related to the sample variance of the MLE. This last point is illustrated in Fig. 1.4. As n increases from 27 to 100, the likelihood function becomes sharper and more concentrated about the MLE.

1.4 Incorporating Prior Information

Imagine that there is prior information about the frequency θ of allele A from comparable populations. Bayesian methods provide a natural way of incorporating such prior information into the model. This requires eliciting a *prior distribution* for θ that captures what is known about θ before obtaining the data sample. This prior distribution is combined with the likelihood (which, given the model,

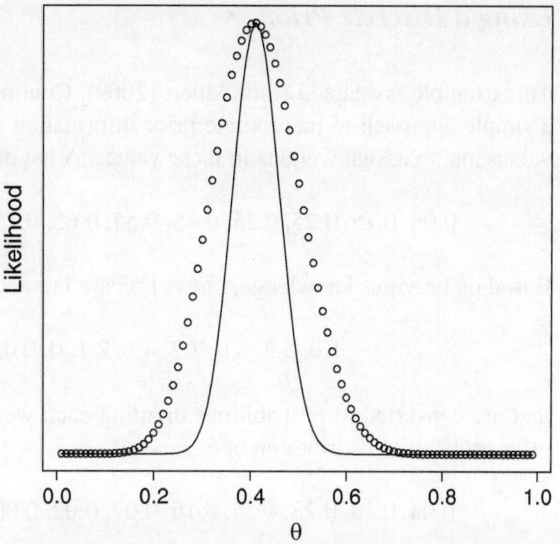

Fig. 1.4 Circled line: likelihood function for θ, given $n = 27$, $x = 11$. Full line: likelihood function for θ, given $n = 100$, $x = 41$

contains all the information arising from the data) to form the *posterior distribution* that is the basis for the Bayesian inference. Specifically using Bayes theorem:

$$\text{Posterior} \propto \text{Prior} \times \text{Likelihood}. \tag{1.6}$$

If the prior density of θ is labelled $g(\theta)$, (1.6) becomes

$$p(\theta|x,n) \propto g(\theta) L(\theta|x,n), \tag{1.7}$$

indicating that the posterior density is proportional to the prior density times the likelihood. Probability statements about θ require scaling (1.7). This involves dividing the right-hand side of (1.7) by

$$\sum_i g(\theta_i) L(\theta_i|x,n), \tag{1.8}$$

if θ is discrete (in which case g is a probability mass function), or by

$$\int g(\theta) L(\theta|x,n)\, d\theta, \tag{1.9}$$

if it is continuous (in which case g is a probability density function).

Using a Discrete Prior

This example is adapted from Albert (2009). Continuing with the binomial model, a simple approach to incorporate prior information on θ is to write down possible values and to assign weights to these values. A list of possible values of θ could be

$$0.05, 0.15, 0.25, 0.35, 0.45, 0.55, 0.65, 0.75, 0.85, 0.95. \tag{1.10}$$

Based on previous knowledge, one is prepared to assign the weights

$$1.0, 5.2, 8.0, 7.2, 4.6, 2.1, 0.7, 0.1, 0.0, 0.0$$

that are converted to probabilities dividing each weight by the sum; this gives the prior probability distribution of θ

$$0.04, 0.18, 0.28, 0.25, 0.16, 0.07, 0.02, 0.00, 0.00, 0.00. \tag{1.11}$$

The likelihood for θ is proportional to (1.1). The combinatorial term does not contain information about θ, so one can write

$$L(\theta|x = 11, n = 27) \propto \theta^{11} (1 - \theta)^{27-11}.$$

Evaluating this expression for all the possible values of θ in (1.10) yields a list of ten numbers (too small to be written down here). Label these ten numbers:

$$L(0.05), L(0.15), \ldots, L(0.95). \tag{1.12}$$

To obtain the posterior (1.7), the terms in (1.11) are multiplied by the corresponding term in (1.12). For example,

$$p(\theta = 0.05|x = 11, n = 27) \propto 0.04 \times L(0.05),$$
$$p(\theta = 0.15|x = 11, n = 27) \propto 0.18 \times L(0.15),$$

and so on with the remaining *eight* terms. After scaling with the sum

$$0.04 \times L(0.05) + 0.18 \times L(0.15) + \cdots + 0.00 \times L(0.95),$$

posterior probabilities can be assigned to the ten possible values of θ. These posterior probabilities are (rounded to two decimal places)

$$0.00, 0.00, 0.13, 0.48, 0.33, 0.06, 0.00, 0.00, 0.00, 0.00.$$

Based on these posterior probabilities, the posterior mean of θ is 0.38, and the probability that θ falls in the set $\{0.25, 0.35, 0.45\}$ is 0.94.

Using a Beta Prior: The Beta-Binomial Model

Another possible prior is to assign a beta distribution to θ with the appropriate parameters to reflect prior information. This is a continuous distribution with support between 0 and 1 and has two parameters denoted as a and b that determine the shape. When $a = b$, the distribution is symmetric.

One way of using a beta distribution that matches the prior probabilities (1.11) is as follows. Notice that the sum of the first three probabilities in (1.11) represents the probability that θ is smaller than or equal to 0.25. This probability is equal to 0.50. Similarly, the sum of the first five probabilities is the probability that θ is smaller than or equal to 0.45. This probability is equal to 0.91. The values of θ equal to 0.25 and 0.45 are two quantiles. Let $F(0.25; a, b)$ and $F(0.45; a, b)$ represent the cumulative distribution functions (cdf) of the beta distribution for $\theta = 0.25$ and for $\theta = 0.45$, respectively (the cdf is $F(x; a, b) = \Pr(X \leq x; a, b)$). Then the parameters a and b of the beta distribution that match the prior probabilities (1.11) can be found by minimising the function

$$(F(0.25; a, b) - 0.5)^2 + (F(0.45; a, b) - 0.91)^2$$

with respect to a and b. This can be achieved using the function OPTIM in R as indicated in the following code:

```
mod <- function(par){
  a <- par[1]
  b <- par[2]
  fct <- (pbeta(0.25,a,b)-0.5)^2 + (pbeta(0.45,a,b)-0.91)^2
  return(fct)
}
res <- optim(par=c(3,3),mod)
res$par
```

```
## [1] 2.89705 8.04717
```

The function returns $a = 2.90$, $b = 8.05$. As a check, one can compute the cumulative distribution functions:

```
pbeta(0.45,2.9,8.05)
```

```
## [1] 0.9099298
```

```
pbeta(0.25,2.9,8.05)
```

```
## [1] 0.4995856
```

Figure 1.5 displays the discrete prior defined by (1.10) and (1.11) and the prior based on $Be(2.90, 8.05)$.

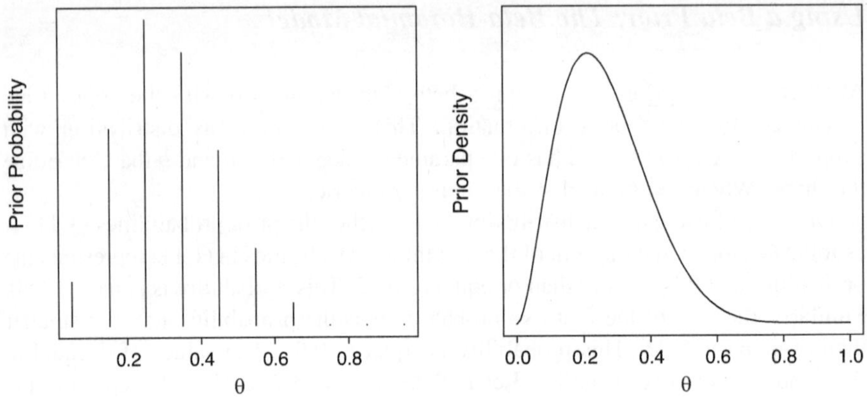

Fig. 1.5 Left: discrete prior distribution defined by (1.10) and (1.11). Right: beta prior $Be(2.90, 8.05)$

As mentioned above, the likelihood is proportional to (1.1); that is,

$$L(\theta|x, n) \propto \theta^x (1-\theta)^{n-x}. \tag{1.13}$$

Seen as a function of θ, this is the kernel of a beta distribution with $a = x + 1$ and $b = n - x + 1$. The pdf (probability density function, sometimes referred to as density function) of the beta distribution is

$$p(\theta) = \frac{\Gamma(a+b)}{\Gamma(a)\Gamma(b)} \theta^{a-1} (1-\theta)^{b-1}, \quad \theta \in [0, 1], \quad a, b > 0,$$

where Γ is the gamma function. The posterior distribution is obtained by combining this likelihood with the prior $Be(2.90, 8.05)$. This results in a posterior distribution that has the beta density with parameters $x+1+2.90-1$ and $n-x+1+8.05-1$. The posterior distribution has the form

$$p(\theta|x = 11, n = 27) = Be(13.90, 24.05). \tag{1.14}$$

Figure 1.6 displays plots of the prior, the likelihood and the posterior for the example. The posterior distribution is sharper than the prior distribution, and its probability mass is concentrated between that of the prior distribution and the likelihood. In Bayesian inference, the posterior distribution provides the necessary information for drawing conclusions about θ. For example, the mean, mode and median of (1.14) are 0.37, 0.36, 0.36. The 95% posterior interval for the posterior mean is (0.24, 0.50). The inference using the Bayesian approach is sharper than that based on the likelihood (1.5). Further, the frequentist confidence interval and the Bayesian posterior intervals have different interpretations. In the latter, the

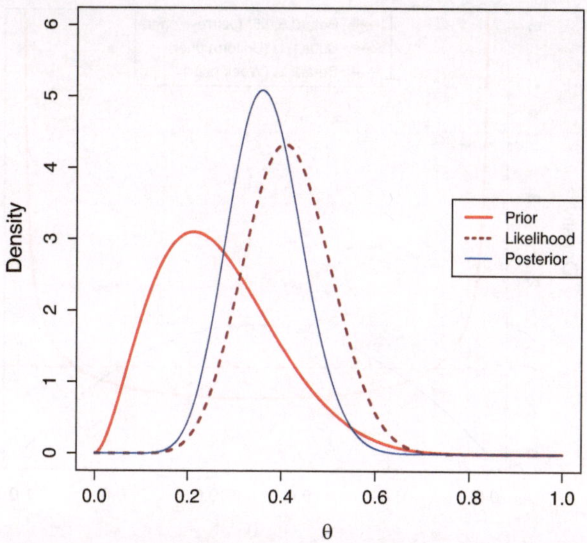

Fig. 1.6 The prior density $Be(2.90, 8.05)$, the likelihood $Be(12, 17)$ and the posterior density $Be(13.90, 24.05)$ of the probability θ

confidence interval is fixed and the associated probability is the probability that the true parameter falls in the interval.

Note If a prior for θ is $Be\,(a, b)$, then $p\,(\theta) \propto \theta^{a-1}\,(1-\theta)^{b-1}$. The likelihood of the binomial model is proportional to $\theta^{x}\,(1-\theta)^{n-x}$ which is the kernel of $Be\,(x+1, n-x+1)$. The posterior is then proportional to $\theta^{a+x-1}\,(1-\theta)^{b+n-x-1}$ that is the kernel of $Be\,(a+x, b+n-x)$.

Prior Influence on Inferences

A Bayesian analysis is seldom complete without investigating how prior information affects the conclusions. In this example, one can compare the inference about θ using either the discrete or the beta prior. With the former, the mean is 0.38 and the probability that θ falls in the set $\{0.25, 0.35, 0.45\}$ is 0.94. The numerical values arrived at with the beta prior are quite similar.

In a situation where no prior information is available about θ, one could resort either to maximum likelihood or to a Bayesian approach with a non-informative prior. The development of non-informative (or reference) priors can become quite technical, especially in complex model scenarios; a pragmatic approach is often chosen along the following lines. In the absence of information about θ, the investigator may consider three possible beta distributions as displayed in Fig. 1.7 (taken from Carlin and Louis 1996). The so-called Jeffrey's prior is transformation

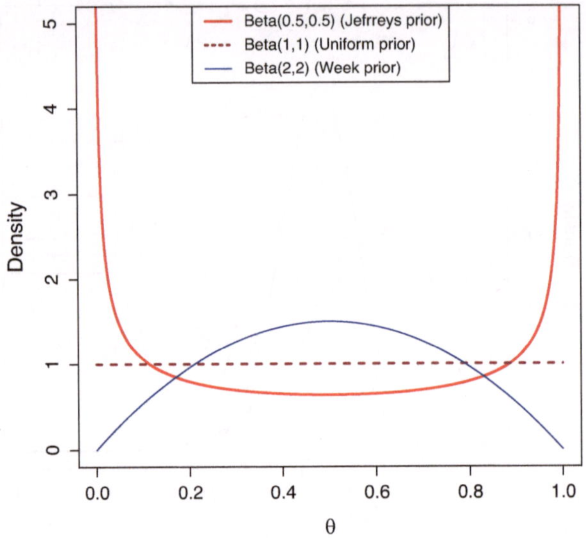

Fig. 1.7 The densities $Be(0.5, 0.5)$, $Be(1, 1)$ and $Be(2, 2)$ to model prior information about θ

invariant; the $Be(1, 1)$ is a special case that retrieves a uniform distribution (assigning equal probabilities to all values of θ); the $Be(2, 2)$ is mildly informative assigning larger probabilities to intermediate values of θ. The combination of these priors with the likelihood $\theta^{11}\,(1-\theta)^{16}$ gives rise to the three posterior distributions shown in Fig. 1.8.

In this particular example, three very different prior distributions give rise to very similar posterior distributions. This is often the case when the likelihood is very informative relative to the prior distribution. Using the uniform prior $Be(1, 1)$, the posterior is $Be(12, 17)$ with mean value (the mode and median are almost the same) and 95% posterior interval equal to 0.41 and (0.24, 0.59), respectively. The posterior interval is a little wider than that based on the sharper prior $Be\,(2.90, 8.05)$ and almost identical (numerically) to the one based on the normal approximation to the maximum likelihood estimator. The posterior probability that θ is less than or equal to 0.2 is

$$\int_{\theta=0}^{\theta=1} I\,(\theta \leq 0.2)\; p\,(\theta|y, n)\, d\theta = 0.00496. \tag{1.15}$$

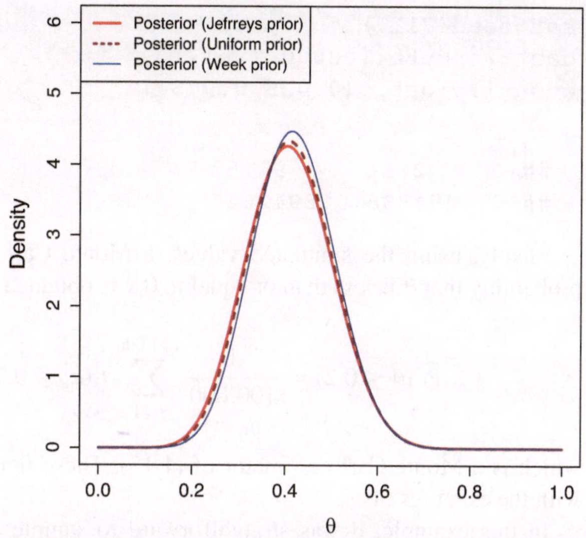

Fig. 1.8 The three posterior distributions corresponding to the three priors of Fig. 1.7 when $n = 27, x = 11$

Simulating from the Posterior Distribution

Inferences drawn in the examples above were exact. This is possible when the analytical form of the posterior distribution is known and features from it (such as probability intervals and moments) can be calculated. In principle, features from any posterior distribution can also be obtained using samples drawn from it, making use of standard theorems from the time series literature. For example, any function of the random variable X, $h(X)$ with finite expectation $\mathrm{E}(h(X))$ can be approximated by

$$\mathrm{E}(h(X)) \approx \frac{\sum_{i=1}^{N} h(x_i)}{N}, \tag{1.16}$$

where N is the sample size. Using R, a sample $(x_i)_{i=1}^{100{,}000}$ of size 100,000 from the posterior $Be(12, 17)$ results in a sample mean:

$$\mathrm{E}(X) \approx \sum_{i=1}^{100{,}000} \frac{x_i}{100{,}000} = 0.41.$$

A 95% posterior interval for θ can be estimated as the 2.5th to 97.5th percentiles of the empirical distribution of the draws x_i using the R function `quantile`. If the original dataset is denoted `dat`, the R code is:

```
set.seed(7117)
dat <- rbeta(100000,12,17)
quantile(dat,c(0.025,0.975))
```

```
##      2.5%     97.5%
## 0.2454898 0.5942647
```

Finally, using the simulated values, a Monte Carlo estimate of the posterior probability that θ is less than or equal to 0.2 is obtained using

$$\widehat{\Pr}(\theta \leq 0.2) = \frac{1}{100{,}000} \sum_{i=1}^{100{,}000} I(x_i \leq 0.2) = 0.00499,$$

which is a Monte Carlo estimator of (1.15). These figures are in good agreement with the exact results.

In this example, it was straightforward to sample directly from the posterior distribution because the normalising constant (1.9) is known, and therefore the form of the posterior is fully specified. Often the normalising constant cannot be obtained in closed form, particularly when θ contains many elements. Chapter 4 discusses how Monte Carlo draws from the approximate posterior distribution can still be obtained using Markov chain Monte Carlo (McMC) methods.

An important issue with inferences based on Monte Carlo samples from posterior distributions is the accuracy of posterior summaries. The latter are subject to sampling uncertainty that depends on the size of the Monte Carlo sample and on the degree of autocorrelation of the samples. Methods to quantify this uncertainty are reviewed in Chap. 4.

Estimating Moments Using (Correlated) Samples from Posterior Distributions

Result (1.16) is extremely useful and is applied routinely in an McMC environment to estimate features from posterior distributions. Typically, the elements that constitute the sample are correlated. However, despite this correlation structure, consistent estimators of features of posterior distributions can be obtained. For example, the sample mean

$$\widehat{\mu} = \frac{1}{N} \sum_{i=1}^{N} x_i, \tag{1.17}$$

the lag-k sample autocovariance

$$\widehat{\gamma}(k) = \frac{1}{N}\sum_{i=1}^{N}(x_i - \widehat{\mu})(x_{i+k} - \widehat{\mu}) \tag{1.18}$$

and the lag-k sample autocorrelation

$$\widehat{\rho}(k) = \frac{\widehat{\gamma}(k)}{\widehat{\gamma}(0)} \tag{1.19}$$

are consistent estimators of the respective population parameters. In (1.19), $\widehat{\gamma}(0)$ is the sample variance.

As an example, consider generating draws from the lag-1 autoregressive model

$$x_t = \rho x_{t-1} + e_t, \quad |\rho| < 1, \quad e_t \sim N\left(0, \sigma^2 = 1\right), \quad t = 1, \ldots, N, \tag{1.20}$$

where the $e's$ are iid (independently and identically distributed). Using `R`, I simulated $N = 1000$, 10,000 and 100,000 observations from (1.20) using $\rho = 0.8$ and $\sigma^2 = 1$, with the initial condition $x_1 \sim N(0, 1)$. This generates a strongly autocorrelated structure among the draws. The marginal mean and variance of this process are

$$\mathrm{E}(x_t) = 0,$$

$$\mathrm{Var}(x_t) = \frac{\sigma^2}{1 - \rho^2}.$$

The estimates of the mean (0.0), variance (2.778) and correlation (0.8) with samples of size $N = 1000$, $N = 10{,}000$ and $N = 100{,}000$ are $(-0.060, 2.731, 0.793)$, $(-0.046, 2.749, 0.797)$ and $(0.018, 2.774, 0.800)$, respectively. Despite the rather strong degree of autocorrelation, the estimates are quite acceptable and get closer to the true values as sample size increases.

1.5 Frequentist or Bayesian?

There has been much heated debate between frequentists and Bayesians about the advantages and shortcomings of both methods of inference. One can easily construct examples where one of the methods gives silly answers and the other performs fairly well. One such example is the following. Imagine that in the situation discussed above, rather than obtaining $x = 11$ copies of allele A in a sample of size $n = 27$, one obtained $x = 0$. This outcome is not unlikely when θ is small; for example, the probability of obtaining $x = 0$ when $n = 27$ and

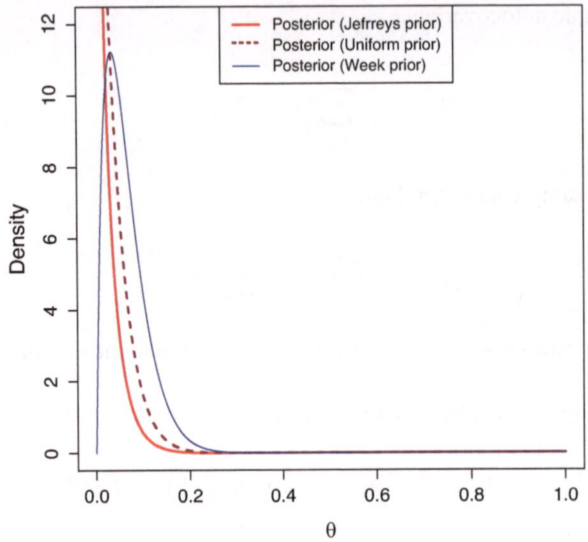

Fig. 1.9 The three posterior distributions corresponding to the three priors of Fig. 1.7 when $n = 27, x = 0$

$\theta = 0.04$ is $\Pr(x = 0|n = 27, \theta = 0.04) = 0.33$. In this situation, the maximum likelihood (ML) estimate (1.2) is 0 and its variance is also 0, which is clearly a silly result (classical maximum likelihood is problematic when the estimate lies on the boundary of the parameter space).

How does the Bayesian approach behave in such a situation? The three posterior distributions corresponding to the three priors of Fig. 1.7 are shown in Fig. 1.9. Jeffrey's prior and mathematical form as the likelihood, proportional to $((1-\theta)^{27})$, lead to posterior distributions $Be\,(0.5, 27.5)$ and $Be\,(1, 28)$, respectively; these have modal values of 0. The weakly informative prior $Be\,(2, 2)$ yields a posterior of the form $Be\,(2, 29)$, which has a mode at $\theta \approx 0.034$. The posterior means of these three distributions are 0.018, 0.034 and 0.065, respectively. The 95% posterior intervals are

$$\left(0.18 \times 10^{-4}, 0.88 \times 10^{-1}\right),$$
$$\left(0.90 \times 10^{-3}, 0.12\right),$$
$$\left(0.81 \times 10^{-2}, 0.17\right),$$

respectively. The posterior probabilities that θ is less than or equal to 0.05 for $Be\,(0.5, 27.5)$, $Be\,(1, 28)$ and $Be\,(2, 29)$ are 0.91, 0.76 and 0.45, respectively. In this extreme situation, prior information plays an important role (certainly, compared to the case $n = 27$, $x = 11$, displayed in Fig. 1.8).

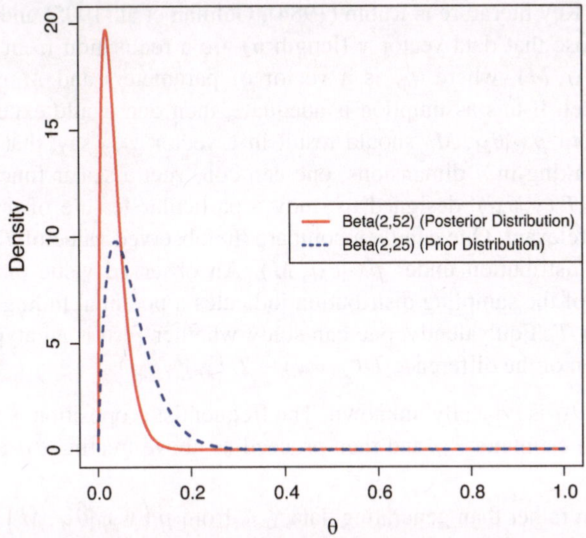

Fig. 1.10 A $Be(2, 25)$ prior distribution and a $Be(2, 52)$ posterior distribution when $n = 27, x = 0$

One may consider other priors from the beta family that put strong probability mass in the neighbourhood of zero. One possibility is to use a $Be\ (2, 25)$ that has a mode at $\theta = 0.04$. With $n = 27$ and $x = 0$, this results in a posterior $Be\ (2, 52)$. The prior and posterior distributions are plotted in Fig. 1.10. The mean and the mode of the posterior distribution are 0.037 and 0.019, respectively. The 95% posterior interval is now $(0.005, 0.101)$ and $\Pr(\theta \leq 0.05|n = 27, x = 0) = 0.75$.

The beta prior does not assign probability mass to $\theta = 0$, and this rules out the possibility that $\theta = 0$ in its posterior distribution. One way of including 0 as a possible value for θ is to use a two-component mixture prior, where one component is a point mass at zero and the other is a beta prior. Mixture distributions are discussed in Chaps. 3, 7 and 9.

There is flexibility associated with the Bayesian approach and a carefully chosen prior distribution will lead to stable inferences about θ. The cost is a result which is partly affected by prior input. With small amount of data and when parameters lie in the border of the parameter space, there is little else to choose from. In such a situation, the most important role of prior distributions may well be to obtain inferences about θ that are stable and that provide a fair picture of posterior uncertainty, *conditional on the model*.

Frequentist and Bayesian

There are situations where instead of choosing between *frequentist or Bayesian*, one could use *frequentist and Bayesian* tools in a meaningful way. This is the case with model checking where one is interested in studying the ability of a model to account for particular features of the data or to give reasonable predictions of future observations. Key literature is Rubin (1984), Gelman et al (1995) and Gelman et al (1996). Suppose that data vector y (length n) are a realisation from the sampling model $p(y|\theta_M, M)$, where θ_M is a vector of parameters and M represents the assumed model. If this assumption is adequate, then one would expect that a new realisation from $p(\cdot|\theta_M, M)$ should result in a vector y_{rep} say, that resembles y. Instead of working in n dimensions, one can construct a scalar function T of the data and θ_M, $T(y, \theta_M)$, designed to study a particular feature of the data that is scientifically relevant. One can then compare the observed value of $T(y, \theta_M)$ with its sampling distribution under $p(\cdot|\theta_M, M)$. An observed value that falls in the extreme tails of the sampling distribution indicates a potential failing of the model to account for T. Equivalently, one can study whether zero is an atypical value in the distribution of the difference $T(y, \theta_M) - T(y_{rep}, \theta_M)$.

Parameter θ_M is typically unknown. The frequentist proposition is to replace θ_M by some point estimator $\hat{\theta}_M$ and then proceed as above treating θ_M as known and equal to $\hat{\theta}_M$.

A Bayesian rather than generating data y_{rep} from $p\left(y_{rep}|\hat{\theta}_M, M\right)$ does so from $p(y_{rep}|y, M)$, the density of the posterior predictive distribution, given by

$$p(y_{rep}|y, M) = \int p(y_{rep}|\theta_M, y, M)\, p(\theta_M|y, M)\, d\theta_M \tag{1.21}$$

where often, $p(y_{rep}|\theta_M, y, M) = p(y_{rep}|\theta_M, M)$. One then observes whether zero is an extreme value in the posterior predictive distribution $T(y, \theta_M) - T(y_{rep}, \theta_M)$, where θ_M is generated from the posterior $[\theta_M|y, M]$ and given θ_M, y_{rep} is generated from $[y_{rep}|\theta_M, M]$. This "Bayesian frequentist", like the frequentist, accounts for the uncertainty in y_{rep} due to the sampling process from $p(y_{rep}|\theta_M, y, M)$. However, unlike the frequentist, account is also taken of the uncertainty about θ_M described by its posterior distribution $[\theta_M|y, M]$. Model checking applied in this manner, although embedded in the Bayesian paradigm, is frequentist in spirit because it decides whether the observed data look reasonable under the posterior predictive distribution based on repetitions of data that could have been generated by the model. All this is typically carried out using McMC methods that provide great flexibility to question models.

Model checking using posterior predictive distributions is discussed in Chap. 10.

1.6 Prediction

The second part of the book provides an introductory overview of prediction. A stylised setup is as follows. Data matrix z has the structure $z_i = (x_i, y_i)$, $i = 1, 2, \ldots, n$, where x_i is a p-dimensional vector of covariates (or predictor variables); y_i, a scalar, is a response variable; and the n vectors are independent and identically distributed realisations from some distribution. Examples of both parametric (frequentist and Bayesian) and nonparametric models are given here. In the case of parametric models where the response y is quantitative, a general form for the association between x and y is

$$y_i = f(x_i) + e_i, \tag{1.22}$$

where f, the conditional mean, is a fixed unknown function of x_i and of parameters and e_i is a random error term, assumed independent of x_i, with mean zero. Using the data z, an estimate of f labelled $\widehat{f}$ is obtained by some method that leads to $\hat{\mathrm{E}}(y_0|x_0) = \hat{y}_0$, a point prediction of the average value of y_0, evaluated at a new value of the covariate $x = x_0$ (in a frequentist setting, conditional on estimates of parameters that index f):

$$\widehat{y_0} = \widehat{f}(z, x_0). \tag{1.23}$$

The notation emphasises that the estimation procedure inputs data z and yields a prediction $\hat{y}_0$ for $x = x_0$. For example, in standard least squares regression $f(x_i) = \mathrm{E}(y_i|x_i) = x_i'b$, $\widehat{f}(z, x_0) = x_0'\widehat{b}$, where $\widehat{b} = (x'x)^{-1} x'y$ and x_i' is the ith row of matrix x.

With binary responses, one may fit a logistic regression. Here, the modelling is at the level of the probability. Specifically, letting $\Pr(y_i = 1|x_i) = \pi(x_i)$, the logistic model can be written as

$$\ln\left[\frac{\pi(x_i)}{1-\pi(x_i)}\right] = x_i'b, \quad i = 1, 2, \ldots, n.$$

A maximum likelihood estimate $\widehat{b}$ together with a new input x_0 results in a predicted probability $\widehat{\pi}(x_0)$ that can be transformed into a predicted value according to the rule:

$$\widehat{y_0} = \begin{cases} 1 \text{ if } \widehat{\pi}(x_0) \geq 0.5 \\ 0 \text{ if } \widehat{\pi}(x_0) < 0.5. \end{cases} \tag{1.24}$$

Measuring Prediction Performance

The performance of the predictions can be evaluated measuring how well they match observed data. One measure of predictive performance is the sample mean squared error (MSE) :

$$\mathrm{MSE}_t = \frac{1}{n_t} \sum_{i=1}^{n_t} (y_i - \widehat{y_i})^2 \tag{1.25}$$

where n_t is the number of records and $\widehat{y_i} = \widehat{f}(z, x_i)$. When MSE_t (1.25) is computed using the data that was used to fit the model, the *training data*, it is known as the sample *training mean squared error*. If the objective is to study how well the model predicts a yet-to-be-observed record, MSE_t can be misleading as it

Table 1.1 Training and validating mean squared errors for the prostate cancer data, as the number of covariates included in the linear predictor increases from 5 to 30. A standard logistic model is implemented, and the mean squared errors represent the proportion of misclassifications in the training and validating data

No. of covariates	5	10	15	20	25	30
MSE_t	0.29	0.31	0.16	0.12	0.06	0.00
MSE_v	0.27	0.37	0.25	0.33	0.35	0.39

overestimates predictive performance. In fact, it can be made arbitrarily small by including a large number of covariates.

A more reliable measure of the prediction ability of the model is to test how well predictions match observations from a new sample of data z_0 (or hold-out data), the *validating data*, drawn from the same distribution as the training data. The validating data is $z_{0i} = (y_{0i}, x_{0i})$, $i = 1, 2, \ldots, n_v$ and the sample *validating mean squared error* is

$$\text{MSE}_v = \frac{1}{n_v} \sum_{i=1}^{nv} (y_{0i} - \widehat{y}_{0i})^2 \,. \tag{1.26}$$

In (1.26) $\widehat{y}_{0i}$ is the ith prediction computed using the training data z evaluated at the value of the ith covariate x_{0i}. That is, $\widehat{y}_{0i} = \widehat{f}(z, x_{0i})$. With binary observations, (1.25) and (1.26) represent the proportion of cases where $\widehat{y}_{0i} \neq y_{0i}$, or misclassification error .

As an illustration of some of these concepts, I use data from a microarray study of prostate cancer from Singh et al (2002). The study includes 52 men with tumour, 50 healthy men and a total of $n = 102$ men. The genetic expression of a panel of $p = 6033$ genes was measured for each man. The level of gene expression is associated with the level of activity of the gene. A larger number implies a more active gene. The $n \times p$ matrix of covariates is then $x = \{x_{ij}\}, i = 1, 2, \ldots, n = 102; j = 1, 2, \ldots, p = 6033$, with $p \gg n$.

Fitting Traditional Logistic Regression

To illuminate some of the consequences of overfitting, a first analysis is undertaken with traditional logistic regression models involving $p < n$. Data are divided into training and validating sets with equal numbers in each. The logistic models are fitted to the training data using maximum likelihood, and the estimates of the parameters are used to predict the outcome (healthy/not healthy) in the validating data. The models differed in the number of covariates included. The change in MSE_t and MSE_v as the number of covariates (columns in x) increases in the different models from $p = 5$ to $p = 30$ in steps of 5 is shown in Table 1.1 for one training/validating split. The covariates were arbitrarily chosen as the first 5 columns of x, the first 10 and so on. The figures in the table show an increase in MSE_v as the number of covariates increases beyond 15 and a parallel increase of the overstatement of the model's predictive ability, as judged by the steady fall in MSE_t when the number of covariates is larger than 10.

The code below reads the data (singh2002), splits it into a training and a validating/testing set (y.test and y.train) and in the bottom part,

- fits a logistic regression to the training data using the R-function GLM
- using the ML estimates, computes the predicted liabilities in the training and validating data
- based on these liabilities, computes $\Pr(Y = 1|\widehat{b})$, where $\widehat{b}$ is the ML estimate
- transforms the probabilities into the 0/1 scale
- computes MSE (misclassification error) in the training and validating data

The figures in Table 1.1 were generated using this code. The code illustrates the case with 15 covariates (first 15 columns of matrix x). The output agrees with the figures in the third column of the table:

```
# CODE0101
# READING SINGH ET AL 2002 DATA
rm(list=ls()) # CLEAR WORKSPACE
# Lasso solutions using package glmnet
#install.packages("glmnet", .libPaths()[1])
#install.packages("sda")
library("sda")
```

```
library(glmnet)
```

```
data(singh2002)
X<-singh2002$x
y<-ifelse(singh2002$y=="cancer",1,0)
n<-nrow(X)
Xlasso<-X
set.seed(3037)
train=sample(1:nrow(X),nrow(X)/2)
test=(-train)
y.test=y[test]
y.train<-y[train]
#
# RESAMPLES TRAIN/TEST DATA AND COMPUTES MSE
# FOR EACH RESAMPLE/REPLICATE
t1 <- seq(1,15,1) # CHOOSE THE FIRST 15 COLUMNS OF x
X1 <-X[,t1]
n <- length(t1)
datarf <- data.frame(cbind(y,X))
nc <- 1 # EXAMPLE WITH 1 REPLICATE ONLY
res <- matrix(data=NA, nrow=nc,ncol=3)
for(i in 1:nc){
  if(i > 1){train <- sample(1:nrow(datarf),nrow(datarf)/2)}
  glm.fit <- glm(y[train] ~ X1[train,] ,
                 family=binomial(link="logit"))
```

```
# CALCULATE PREDICTED LIABILITY FOR THE TRAINING (liabT) AND THE
# VALIDATING DATA (liabV)
  liabV <- X1[-train,1:n]%*%glm.fit$coefficients[2:(n+1)]+
    glm.fit$coefficients[1]
  liabT <- X1[train,1:n]%*%glm.fit$coefficients[2:(n+1)]+
    glm.fit$coefficients[1]
# COMPUTE Pr(Y=1) BASED ON THESE LIABILITIES
  probT <- exp(liabT)/(1+exp(liabT))
  probV <- exp(liabV)/(1+exp(liabV))
# COMPUTE PREDICTED VALUES IN TRAINING AND VALIDATING DATA
# ON THE 0/1 SCALE
  predT <- ifelse(probT > 0.5, "1", "0")
  predV <- ifelse(probV > 0.5, "1", "0")
# COMPUTE MISCLASSIFICATION ERROR
  predclassT <- mean((as.numeric(predT) - y.train)^2)
  predclassV <- mean((as.numeric(predV) - y.test)^2)

 # IF CURIOUS COMPUTE LOG-LIKELIHOOD, DEVIANCE,
  # AIC USING TRAINING DATA
#  ll <- sum(y.train*liabT) - sum(log(1+exp(liabT)))
#  dev <- -2*ll
#  AIC <- dev + 2*(n+1)
# ***********************************

  res[i,] <- c(n,predclassT,predclassV)
}
res
```

```
##          [,1]       [,2]      [,3]
## [1,]    15 0.1568627 0.254902
```

Selection of Covariates for Prediction

An objective of the experiment above is to find genes that may have an effect on prostate cancer. The data are typical of a genomic setup where the number of variables p (e.g. genetic markers) measured for each individual is considerably larger than the number of individuals n; the classical $p \gg n$ scenario. In the context of prediction, these variables enter as covariates in linear regression or logistic regression models, but only a subset are likely to contribute meaningfully. Inclusion of redundant variables may improve model fit (reflected in small values of training mean squared error, MSE_t), but will definitely result in poor predictions. From the point of view of model implementation, when $p \gg n$, some form of regularisation or shrinkage is needed. This constitutes an important topic of the book. Three examples are provided in this overview. The first is a parametric model; the other two are nonparametric approaches specifically developed, but not restricted to, to deal with the $p \gg n$ situation. Like many of the tools used for the analysis of the type of data sets commonly found in genomic studies, these models can be useful as guidance in the choice of predictors for further study.

Fitting the Lasso

The parametric example is based on the *lasso* (Tibshirani 1996 , "least absolute shrinkage and selection operator") that is a regularisation method with a tuning parameter governing the amount of shrinkage of the regression parameters towards zero. Since the lasso solutions typically include many coefficients equal to zero when the tuning parameter is sufficiently large, it does model selection and shrinkage simultaneously.

The lasso logistic regression model is fitted using the public package **glmnet** (Friedman et al 2009) implemented in R. Documentation about **glmnet** can be found in Hastie and Qian (2016) and in Friedman et al (2010).

To obtain predictions, the code below executes first the function **cv.glmnet** on the training data in order to find the value of the tuning parameter (λ) that optimises prediction ability measured by MSE. In a second step, **glmnet** is executed again on the training data using this best λ to obtain the final values of the regression parameters. The code then constructs the predictions from the output of this second run. The model is finally tested on the training and on the validating data.

A more direct implementation of **glmnet** without the need to generate estimates of regression parameters is indicated at the bottom of the code.

The lasso logistic regression was run on the prostate data including all 6033 covariates representing the gene expression profiles. Lasso chooses 36 covariates and sets the remaining equal to zero. The model with these 36 covariates was used to classify the observations in the validating data and resulted in a MSE_v equal to 0.25. In other words, $51 \times 0.25 \approx 13$ out of the 51 observations in the validating data are incorrectly classified. At face value, the result for MSE_v matches that obtained in Table 1.1 when the first 15 columns of x were included in the linear predictor. The latter can be interpreted as a logistic regression model where 15 out of 6033 covariates are randomly chosen. The result based on the lasso is not encouraging:

```
# CODE0102
# READING SINGH ET AL 2002 DATA
rm(list=ls()) # CLEAR WORKSPACE
# Lasso solutions using package glmnet
#install.packages("glmnet", .libPaths()[1])
#install.packages("sda")
library("sda")
library(glmnet)
data(singh2002)
X<-singh2002$x
y<-ifelse(singh2002$y=="cancer",1,0)
n<-nrow(X)
Xlasso<-X
set.seed(3037)
train=sample(1:nrow(X),nrow(X)/2)
test=(-train)
y.test=y[test]
y.train<-y[train]
#
# **********  FOR PREDICTION USING LASSO   *****************
repl <- 1 # NUMBER OF REPLICATES
#          (RESAMPLES TRAINING / VALIDATING)
result <- matrix(data=NA, nrow=repl,ncol=4)
set.seed(3037)
for (i in 1:repl){
  if(i > 1){train <- sample(1:nrow(Xlasso),nrow(Xlasso)/2)}
  y.train <- y[train]
```

```
  y.test <- y[-train]
# STEP 1: cross-validation; find best value of lambda
# alpha=1: LASSO; alpha=0: RIDGE REGRESSION
  cv.out=cv.glmnet(Xlasso[train,],y[train],alpha=1,
          family="binomial",type = "class")
 #plot(cv.out)
  bestlam=cv.out$lambda.min
 #bestlam

# Using best lambda, fit model on training data
#  to obtain final parameter estimates

# STEP 2
  fm=glmnet(y=y[train],x=Xlasso[train,],alpha=1,lambda=bestlam,
           family="binomial",type.measure= "class")
  nzcf<-coef(fm)
  cf<-which(fm$beta[,1]!=0)
  if (length(cf) == 0){
    out <-c(i,length(cf))
    print(out)
    break
  }
#length(cf) # NO. REGRESSION PARAMETERS IN FINAL MODEL
# CONSTRUCT PREDICTIONS FROM OUTPUT OF fm
#     1. VALIDATING DATA
  predglmnet<-fm$a0+Xlasso[-train,cf]%*%fm$beta[cf]
  probs <- exp(predglmnet)/(1+exp(predglmnet))
  predclass_test <- as.numeric(ifelse(probs > 0.5, "1", "0"))
#     2. TRAINING DATA
  predglmnet<-fm$a0+Xlasso[train,cf]%*%fm$beta[cf]
  probs <- exp(predglmnet)/(1+exp(predglmnet))
  predclass_train <- as.numeric(ifelse(probs > 0.5, "1", "0"))
  result[i,] <- c(mean((predclass_train-y.train)^2),
           mean((predclass_test-y.test)^2),bestlam,length(cf))
}
result
```

```
##        [,1]      [,2]       [,3] [,4]
## [1,]      0 0.254902 0.01948886   36
```

```
#NOTE: for prediction, GLMNET can be implemented more directly,
# using in STEP2:
#################################################################
#  fm.predclass=predict(cv.out,s=bestlam,newx=Xlasso[test,],
#       family="binomial",type="class")
#  mean((as.numeric(fm.predclass)-y.test)^2) # VALIDATION ERROR
#       RATE (BASED ON CLASS LABELS)

#################################################################
```

The somewhat disappointing performance of the lasso was investigated further by random splitting the training/testing data set 50 times. This provides a picture of the sampling variation of the MSE over the joint distribution of training/testing data. The mean validating mean squared error, over the 50 replicates, was 0.30, with a minimum of 0.16 and a maximum of 0.53. The number of covariates not set equal to zero ranged from 2 to 41 with a median of 24. There is not a model that is consistently singled out as a good predictor over replications. This is a reflection of the ubiquitous multicollinearity in a multidimensional setting where covariates become highly correlated (a covariate can be expressed as a linear combination of others). Therefore, a different set of covariates is chosen in each replication.

Fitting a Classification Tree

The first of the two nonparametric models that are fitted to the data is a `classification tree` (Breiman et al 1984), that is described briefly via the example generated by the `R`-code below. The code executes the `R` function `tree`; this requires installation of the package `tree`. The data set `singh2002` includes 6033 covariates and the response variable is y: a binary classification variable with modalities "healthy" and "cancer":

```
# CODE0103
rm(list=ls()) # CLEAR WORKSPACE
set.seed(30331)
#install.packages("tree")
library(sda)
library(tree)
# library(glmnet)
data(singh2002)
d <- data.frame(singh2002$x)
d$y <- singh2002$y
nrep <- 1 # NUMBER OF REPLICATES
res <- matrix(data=NA,nrow=nrep,ncol=3)
ptm<-proc.time()
for ( i in 1:nrep ) {
    cat(i,"\n",sep="")
    train <- c(sample(1:50,25),sample(51:102,26))
# FIT THE TREE TO THE TRAINING DATA
    trees <- tree(y ~ . ,  data=d[train,])
# FIT FUNCTION PREDICT TO THE TRAINING AND VALIDATING DATA
    predtreev <- predict(trees,d[-train,],type="class")
    predtreet <- predict(trees,d[train,],type="class")
# CLASSIFICATION ERROR IN TRAINIMG AND VALIDATING DATA
    predv <- sum(predtreev==d$y[-train])/length(d$y[-train])
    predt <- sum(predtreet==d$y[train])/length(d$y[train])
# RECORD TRAINING / VALIDATING CLASSIFICATION ERROR AND
# NUMBER OF COVARIATES IN TREE
    res[i,]<-c((1-predt),(1-predv),length(summary(trees)$used))
}
```

```
## 1
```

```
proc.time()-ptm
```

```
##    user  system elapsed
##    2.95    0.17    3.12
```

```
res
```

```
##      [,1]      [,2] [,3]
## [1,]    0 0.1764706    2
```

```
tab <- table(predtreev,d$y[-train])
tab
```

```
##
## predtreev cancer healthy
##    cancer      17       0
##    healthy      9      25
```

```
# CHECK CLASSIFICATION ERROR
(tab[1,2]+tab[2,1])/(length(d$y[-train]))
```

```
## [1] 0.1764706
```

```
#summary(res)
#plot(trees)
#text(trees,pretty=0)
```

Figure 1.11 indicates that in the particular replicate, the algorithm isolated 2 of the 6033 covariates, X_{77} and X_{237}. Starting at the top of the tree, the 51 cases in the training data have been split into two groups: one, to the left, that shows expression profile for X_{77} less than a threshold $t_1 = -0.777$ and those to the right with $t_1 > -0.777$. The group on the left is not split further and constitutes a *terminal node*. On the right side, a second split based on the profile of X_{237} and a threshold $t_2 = -0.855$ gives rise to two terminal nodes. The result can be interpreted as an interaction between the two markers.

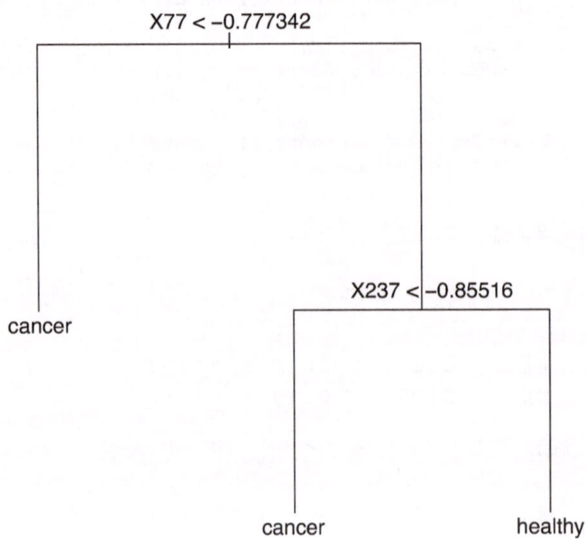

Fig. 1.11 Output from a classification tree fitted to data `singh2002` from Singh et al (2002) using the `R` function `tree` (`R`-code CODE0103). Results from one replicate

A predicted value is attached to each terminal node. For a new individual, a predicted value is obtained by starting at the top of the tree and following the splits downwards until the terminal node with its predicted value is reached. The new individual is assigned the prediction given by that terminal node. In the example, if the new individual were to show a value for $X_{77} = -0.3$ and for $X_{237} = -0.1$, following the tree from top to bottom would lead to a prediction taking the modality "healthy".

Typing the tree object (in this case, *trees*) gives further details associated with the figure. Terminal nodes are indicated with asterisks. R prints output from each branch of the tree in the form of the *node*, *split criterion t*, the number of observations in the branch *n*, the *deviance*, the classification for the branch ("cancer"/"healthy" in the present case) and the proportion of observations in the branch that take the values "cancer"/"healthy":

```
library(tree)
trees
```

```
## node), split, n, deviance, yval, (yprob)
##       * denotes terminal node
##
## 1) root 51 70.68 cancer ( 0.5098 0.4902 )
##   2) X77 < -0.777342 20   0.00 cancer ( 1.0000 0.0000 ) *
##   3) X77 > -0.777342 31 30.46 healthy ( 0.1935 0.8065 )
##     6) X237 < -0.85516 6   0.00 cancer ( 1.0000 0.0000 ) *
##     7) X237 > -0.85516 25   0.00 healthy ( 0.0000 1.0000 ) *
```

The output above indicates that at the top of the tree at X_{77} (which is the root since the tree is upside down), there are 51 records (the training data), and the proportion of "cancer" is 0.5098. After the first split, to the left, the split is $t_1 < -0.777$, and 20 observations are classified as "cancer" and 0 as "healthy" leading to proportions of (1.00, 0.00). To the right, the split is $t_1 > -0.777$ that gives rise to 31 records, with a proportion of "healthy" equal to 0.8065 (25 out of the 31 records are "healthy, those whose $t_2 > -0.855$, associated with covariate X_{237}).

Various algorithms are available to decide which variable to split and the splitting value t to use for the construction of the tree. Some of these topics are deferred to the chapter on nonparametric methods. Here, I concentrate on the predictive ability of the method. For the particular replicate, the classification error in the training and validating data is 0 and 0.18, respectively. Replicating the experiment 50 times gives a mean classification error in training and validating data equal to 0.016 and 0.197, respectively, with (minimum, maximum) values of (0.000, 0.078) and (0.098, 0.333), respectively. With the parameters for the utility function `tree.control` set at the default values, the number of covariates over the 50 replicates included in each tree fluctuates between 2 and 3, and these covariates vary over replicates. For these data, the classification tree performs considerably better than the lasso.

Interestingly, in all the cases, the classification trees capture what can be interpreted as an interaction involving two or three covariates. These are not the same covariates for the 50 trees. Perhaps, this must not come as a surprise: 6033 covariates give rise to more than 18 million different two-way interactions. Therefore, predictors based on interacting covariates are prone to be highly correlated. More generally and as noted with the lasso, in the high-dimensional setting, the multicollinearity of the covariate matrix is often extreme, and any of the p covariates in the $n \times p$ matrix can be written as a linear combination of the others. This means that there are likely many sets of pairs of covariates (other than X_{77} and X_{237}) that could predict just as well. It does not follow that the model cannot be trusted as a prediction tool, but rather that one must not overstate the importance of X_{77} and X_{237} as the only genes associated with the response variable. As with the lasso, the analysis with the classification tree provides inconclusive evidence of specific genes affecting prostate cancer.

Fitting a Random Forest

A problem often mentioned with trees is that they exhibit high variability. Small changes in the data can result in the construction of very different trees and their predictions can be impaired. However, they are an integral part of another method known as *random forest* (Breiman 2001) whose prediction performance benefits by the process of averaging. The random forest consists of many classification trees and each is created as follows:

- Create a sample of size n_v by drawing with replacement from the n_v data in the training data. Repeat this B times to generate B samples. (With random forests, there is an alternative way of estimating validating mean squared error using the entire data, without cross-validation. Details are discussed in Chap. 11).
- For each sample, generate a classification tree. Each time a split in the tree is considered, a random sample of m unique predictors is chosen as split candidates from the p predictors. For classification, it is customary to use $m \approx \sqrt{p}$. This step has the effect of *decorrelating* the ensemble of B trees (in classification trees, the construction of a split involves all the predictors).

For classification, once the trees are available, the final prediction is obtained by a majority vote. Thus, for $B = 10$, say, if for a particular observation in the validating data six or more trees classify it as "1", the predicted value for this observation is "1". The prediction obtained in this manner usually outperforms the prediction of classification trees. This improvement in performance arises from the fact that a prediction based on B predictors with very low correlation has smaller variance than a single prediction. The low correlation is ensured in the second step. The first step involving the bootstrapping of the training data is known as *bagging*, short for *bootstrap aggregating*, whereby the results of several bootstrap samples are averaged. (The *mean squared error* is a measure of the performance of a predictor, whose expectation includes the variance of the predictor, a squared bias term and a pure noise term associated with the variance of the predictand. Therefore, the performance of a predictor improves as its variance is reduced. This reduction

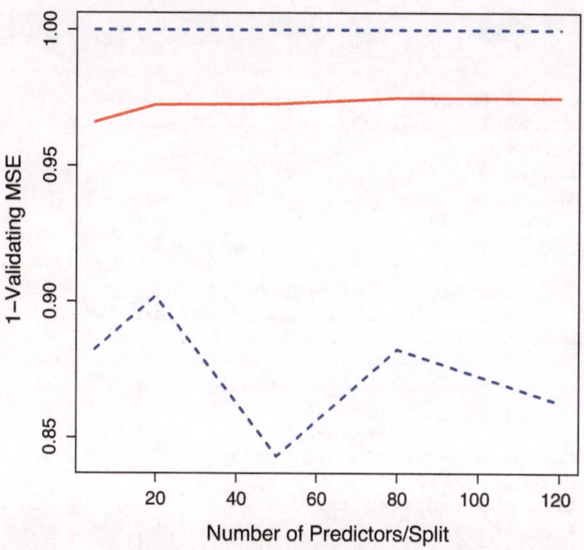

Fig. 1.12 Average proportion of correct classifications in the validating data (in red) of a *random forest* over 200 replicates against the number of covariates included in the ensemble of trees. Maximum and minimum over replicates in blue

is achieved by constructing a prediction averaged over several bootstrap samples whose variance is smaller than the variance of an estimate based on a single sample).

To study prediction ability, the random forest was implemented on the *singh2002* data set using the R function `RandomForest`. I executed 200 replicates (200 splits of training/testing data), and in each replicate, the number of covariates included in the split of a particular tree ranged from 5 to 120 as indicated in the code below in the variable `mtry <- c(5,20,50,80,120)`. The average proportion of correct classifications in the validating data ("cancer", "healthy"), as well as the minimum and maximum over the 200 replicates as a function of the number of covariates ($mtry$ in the $x-$ axis), is shown in Fig. 1.12.

The results are quite impressive. The average proportion of correct classifications in the validating data is of the order of 97% with a minimum −maximum in the range 84 % − 100 %.

The code used to implement the random forest is shown below:

```
# CODE0104
#install.packages("randomForest")
rm(list=ls()) # CLEAR WORKSPACE
library(sda)
library(randomForest)
data(singh2002)
d <- data.frame(X=singh2002$x)
d$y <- singh2002$y
n0 <- sum(d$y=="healthy")
n1 <- sum(d$y=="cancer")
```

```
set.seed(3037)
p <- .5
nrep <- 1
mtry <- c(5,20,50,80,120)
sumd <- data.frame()
res <- rep(NA,nrep)
ptm<-proc.time()
for ( m in mtry) {
  cat("mtry ",m,"\n",sep="")
  for ( rep in 1:nrep ) {
    cat("Replicate ",rep,"\n",sep="")
    train <- c(sample( 1:n0,floor(p*n0) ),
               sample( (n0+1):(n0+n1),floor(p*n1) ))
    rf.singh =randomForest(y ~.,
                           data=d,
                            subset =train,
                            mtry=m,
                            importance =TRUE)
    predict <- predict(rf.singh,d[-train,])
    observed <- d$y[-train]
    t <- table(observed,predict)
    print(t)
    res[rep] <- (t[1,1]+t[2,2])/sum(t)
  }
  sumd <- rbind(sumd,c(m,min(res),mean(res),median(res),
                     max(res),var(res)))
}
proc.time()-ptm
names(sumd) <- c("mtry","min","mean","median","max","var")

with(sumd,plot(mtry,mean,type="l",col="red",ylim=c(min(min),1),
        ylab="1 - Mean Squared Error",
        xlab="Number of Predictors Considered at each Split"))
with(sumd,lines(mtry,min,lty=2,col="blue"))
with(sumd,lines(mtry,max,lty=2,col="blue"))
```

While in this particular set of data the random forest was the clear winner among the prediction machines tested, it is important to mention that there is no uniformly best prediction machine. A different set of data may produce different results. Very marked differences among prediction methods ought to raise suspicion and warrant careful investigation of the data (Efron 2020). This is particularly important in this era of increasingly larger data sets where the consequence of bias due to non-random sampling is magnified. The point is elaborated in Meng (2018). Spurious results may be obtained by complex interactions between a prediction method and a particular structure in the training data at hand that may not be reproduced when the model is deployed using validating data.

1.7 Appendix: A Short Overview of Quantitative Genomics

I provide a brief and compact description of the quantitative genetics/genomics model and introduce terms used repeatedly in the book, such as *allele, locus, diploid, haploid, genotype, Hardy-Weinberg law, single nucleotide polymorphisms (SNPs), genomewide association study (GWAS), allele content, quantitative trait loci (QTL), linkage, linkage disequilibrium, phenotype, genotype, genetic value, genetic variance, additive genetic value (breeding value), additive genetic effect (additive effect of a gene substitution), additive genetic variance, heritability, expected additive genetic relationship, additive genetic relationship matrix, genomic relationship matrix, genomic model, genomic value and genomic variance.*

The Classical Quantitative Genetics Model

The starting point of the mathematical genetics model is the metaphor that describes chromosomes as strings of beads, with each bead representing a gene. Genes are the unit of inheritance. In mammals and many other groups, each cell carries two copies of each chromosome; they are said to be diploid. Most fungi, algae and human gametes have only one chromosome set and are haploid.

The complete set of chromosomes of an organism includes sex chromosomes and autosomes. For example, in humans, there are 23 pairs of chromosomes and of these, 22 pairs are non-sex chromosomes known as autosomes. The majority of genes are located on the autosomes and in this book I consider autosomal loci only.

In diploid organisms, at a specific location on the chromosome called the locus, each of the two copies of the chromosome carries a gene. The pair of genes constitute the genotype at the particular locus. Genes exist in different forms known as alleles. Here, I consider biallelic loci, so for a given locus in diploid individuals, if the two alleles are A and a, the three genotypes could be denoted, say AA, Aa and aa (no distinction is made between Aa and aA). For example, an individual with genotype Aa received one allele (say A) from the mother and the other from the father.

The standard quantitative genetic model assumes that the expression of a trait value y (the phenotype, here centred with zero mean) in diploid individuals is determined by the additive contributions of a genetic value G and an environmental value e,

$$y = G + e,$$

where $e \sim (0, \sigma^2)$ is often assumed independent of G. The genetic value is defined as the conditional mean of the phenotype, given genotype, $\mathrm{E}(y|G)$ and is the result of the joint action of a typically unknown number q of quantitative trait loci (QTL).

The Single Locus Model

Consider first a trait affected by a single biallelic locus with the three genotypes labelled AA, Aa and aa. Let p denote the frequency of allele A in the population (assumed to be the same in both sexes). In a large population, assuming random mating among parents and in the absence of random genetic drift, selection and mutation, in the offspring generation, the frequency of genotype AA is p^2, of genotype Aa is $2p(1-p)$ and of genotype aa is $(1-p)^2$. In this overview, gene frequencies p are treated as known constants that remain unchanged over repeated cycles of random mating.

Define the random variable z^* as

$$z^* = \begin{cases} 2, & \text{with probability } p^2, \\ 1, & \text{with probability } 2p\,(1-p)\,, \\ 0, & \text{with probability } (1-p)^2\,, \end{cases}$$

The random variable z^* is known as the allele content (here, allele A is arbitrarily taken as reference and $z^* = 2$ if the genotype has *two* copies of allele A). For this locus, $\mathrm{E}\,(z^*) = 2p$, $\mathrm{Var}\,(z^*) = 2p\,(1-p)$ and for individuals k and j, $\mathrm{Cov}\left(z_j^*, z_k^*\right) = a_{jk} 2p\,(1-p)$ where a_{jk} is the expected additive genetic relationship (given a pedigree) between k and j (e.g. $a_{jk} = 0.5$ if j and k are non-inbred full sibs or parent and an offspring), also interpreted as the expected proportion of alleles shared identical by descent between j and k (genes that are identical by descent (IBD) are copies of a specific gene carried by some ancestral individual). The note note0101.pdf at https://github.com/SorensenD/SLGDS has a derivation of these results.

In a large (idealised) random mating population, in the absence of selection, mutation or migration, the relationship between gene frequency p and genotype frequency p^2 remains constant from generation to generation. The property is derived from a theorem known as the Hardy-Weinberg law that provides one explanation for the maintenance of genetic variation in such idealised random mating population. In a population in Hardy-Weinberg equilibrium, genotype frequencies at a particular locus in the offspring generation depend on gene frequencies in the parent generation.

From now on, the codes for the three genotypes are centred as $z = z^* - \mathrm{E}\,(z^*)$ so that $\mathrm{E}\,(z) = 0$, $\mathrm{Var}\,(z) = 2p\,(1-p)$ and phenotypic values are also centred, so that $\mathrm{E}(y) = 0$.

The genetic value $G\,(z)$ at the locus can take three modalities corresponding to the three genotypes at the biallelic locus and can be decomposed into two terms:

$$G\,(z) = \alpha z + \delta, \tag{1.27}$$

where αz is the best linear predictor of genetic value. The best linear predictor αz is also known as the additive genetic value or breeding value: the best linear approximation describing the relationship between genetic value and allele content z (best linear prediction is discussed on page 259; see also the example on page 261 for more details on the concepts of additive genetic values and effects, where it is shown that α, the additive genetic effect of the locus or average substitution effect at the locus is also the regression of y on z). The residual term δ is orthogonal to z and includes deviations between $G(z)$ and αz.

The genetic variance contributed by the locus in the population (based on the law of total variance)

$$\text{Var}(G(z)) = \text{Var}_z(\text{E}[G(z)|z]) + \text{E}_z(\text{Var}[G(z)|z]) \tag{1.28}$$

is orthogonally decomposed into an additive genetic component of variance σ_a^2, the first term in the right-hand side and a residual or dominant component of genetic variance , $\text{Var}(\delta)$, the second term. The additive genetic variance (variance of the additive genetic values) in this single locus model, assuming Hardy-Weinberg equilibrium, is

$$\sigma_a^2 = \text{Var}_z(\text{E}[G(z)|z]) = \text{Var}(\alpha z|\alpha) = 2\alpha^2 p(1-p).$$

If the linear fit is perfect, the genetic variance is equal to the additive genetic variance. Importantly, additive genetic variation at the locus arises due to variation in allele content z among individuals at the locus. The substitution effect α is treated as a fixed albeit unknown parameter (this is stressed by conditioning on α).

The (narrow sense) heritability of the trait is defined as the ratio of the additive genetic variance to the phenotypic variance : $h^2 = \sigma_a^2 / \sigma_y^2$, where $\sigma_y^2 = \text{Var}(y)$, the marginal variance of the phenotype.

Models with Many Loci

The extension to q biallelic loci involves a random vector $z = (z_1, \ldots, z_q)'$ of allele contents of the q genotypes. Under random mating, $\text{Var}(z_k) = 2p_k(1-p_k)$, $k = 1, \ldots, q$ and $\text{Cov}(z_k, z_l) = 2D_{kl}$, where the linkage disequilibrium (LD) parameter D_{kl} between loci k and l is defined as follows. Choose the paternal (or maternal) gamete and let the random variable U take the value 1 if allele A_k is present in the paternal gamete at locus k and zero otherwise; let the random variable V take the value 1 if allele A_l is present in the paternal gamete at locus l and zero otherwise. Then D_{kl} is defined as the covariance between U and V:

$$D_{kl} = \text{Cov}(U, V),$$

and $\text{Cov}(z_k, z_l) = 2D_{kl}$ since in the diploid model, the genotype results from the random union of two gametes. Covariances involving alleles of different loci between gametes are zero. Linkage disequilibrium is created by evolutionary forces such as selection, mutation and drift and is broken down by random mating, as a function of time (measured in generations) and of the distance that separates the intervening loci. Generally, loci that are physically close together show stronger LD.

With random mating in a large population, the variance structure of vector z containing the q allele contents, a population parameter, is

$$\Sigma_z = \begin{bmatrix} 2p_1(1-p_1) & 2D_{12} & \cdots & 2D_{1q} \\ 2D_{21} & 2p_2(1-p_2) & \cdots & 2D_{2q} \\ \vdots & \vdots & \ddots & \vdots \\ 2D_{q1} & 2D_{q2} & \cdots & 2p_q(1-p_q) \end{bmatrix}. \tag{1.29}$$

The combined action of the q loci defines the genetic value $G(z)$. A useful starting point is to assume that the effects of the loci combine additively. A genetic value has the same structure as (1.27), $G(z) = \alpha' z + \delta$, where now α is the $q \times 1$ column vector of average substitution effects (or additive genetic effects) at the q loci. The additive genetic variance contributed by the q loci is

$$\sigma_a^2 = \text{Var}\left(\alpha' z | \alpha\right) = \alpha' \Sigma_z \alpha. \tag{1.30}$$

Expanding (1.30) reveals that the additive genetic variance takes the form

$$\sigma_a^2 = 2\sum_{i=1}^{q} \alpha_i^2 p_i (1-p_i) + 4\sum_{i=1}^{q-1}\sum_{j=i+1}^{q} \alpha_i \alpha_j D_{ij}. \tag{1.31}$$

The first term in the right-hand side is the *equilibrium additive genetic variance* , and the second term is the contribution due to disequilibrium (Bulmer 1971). If the loci are in linkage equilibrium ($D_{ij} = 0$ for all i, j), genotypes in the population are uncorrelated, (1.29) is a diagonal matrix and the additive genetic variance reduces to the first term. Among the loci of neutral traits, some of the disequilibrium terms are positive and others negative. In contrast, among loci of selected traits, depending on the type of selection, in large populations, the disequilibrium terms are either positive (disruptive selection) or negative (optimising and directional selection) (Bulmer 1971).

The phenotypic value y of an individual in the multilocus model can be written as

$$\begin{aligned} y &= G(z) + e \\ &= \alpha' z + (\delta + e) \\ &= \alpha' z + \epsilon, \end{aligned} \tag{1.32}$$

where the residual term ϵ includes environmental effects e and non-additive genetic effects δ. The genetic value is the conditional expected value of the phenotype given the genotype

$$G(z) = \text{E}[y|z].$$

Covariance Between Relatives

Consider two related individuals i and j whose causal genotypes at q loci are described by the random vectors of allele contents z_i and z_j, where $z_i = (z_{i1}, \ldots, z_{iq})'$. While (1.29) describes associations within individuals, here we seek associations between individuals. The covariance between i and j at the same locus k is $\text{Cov}(z_{ik}, z_{jk}) = 2a_{ij} p_k (1 - p_k)$, where a_{ij} is the coefficient of expected additive genetic relationship between i and j. The covariance involving two different loci k in i and l in j is $\text{Cov}(z_{ik}, z_{jl}) = 2a_{ij} D_{kl}$. The contribution to the covariance between the additive genetic values of i and j from loci k and l is then

$$\begin{aligned}
\text{Cov}\left(\alpha' z_i, z_j' \alpha | \alpha\right) &= \alpha' \,\text{Cov}\left(z_i, z_j'\right) \alpha \\
&= 2a_{ij} \sum_{m=k,l} \alpha_m^2\, p_m (1 - p_m) + 4a_{ij} \alpha_k \alpha_l D_{kl},
\end{aligned} \tag{1.33}$$

that has the same form as (1.31) except for the factor a_{ij}.

Expression (1.33) generalises to n individuals displaying a family structure given by a pedigree. Based on the pedigree, the $n \times n$ matrix A can be constructed. This matrix, known as the additive genetic relationship matrix, has elements a_{ij} that denote the expected additive genetic relationship between individuals i and j.

Let $Z\alpha$ represent the vector of n additive genetic values, where Z is the $n \times q$ matrix of genotypic codes and α is the $q \times 1$ vector of additive genetic effects for the q loci. Then the variance-covariance matrix of the n additive genetic values, treating Z as random and α as a fixed parameter, is

$$\begin{aligned}
\text{Var}(Z\alpha | \alpha) &= \begin{bmatrix} a_{11}\alpha' \Sigma_z \alpha & a_{12}\alpha' \Sigma_z \alpha & \ldots & a_{1n}\alpha' \Sigma_z \alpha \\ a_{21}\alpha' \Sigma_z \alpha & a_{22}\alpha' \Sigma_z \alpha & \ldots & a_{2n}\alpha' \Sigma_z \alpha \\ \vdots & \vdots & \ddots & \vdots \\ a_{n1}\alpha' \Sigma_z \alpha & \ldots & \ldots & a_{nn}\alpha' \Sigma_z \alpha \end{bmatrix} \\
&= A \left[\sum_{k=1}^{q} \alpha_k^2\, 2p_k (1 - p_k) + \sum_{k=1}^{q-1} \sum_{l=k+1}^{q} 4\alpha_k \alpha_l D_{kl} \right] \\
&= A\sigma_a^2,
\end{aligned} \tag{1.34}$$

where σ_a^2 is given by (1.31) and is equal to the term in square brackets.

Expressions (1.33) and (1.34) assume that the (large) population is maintained in a steady state of gene frequency and linkage disequilibrium from generation to generation resulting in constant σ_a^2 and that the computation of the expected additive genetic relationship between individuals i and j involving loci k and l is tracing i and j to the most recent common ancestor. The constancy of σ_a^2 is an approximation that can also be justified by considering cohorts spanning one or two generations as often found in human pedigrees.

The covariance between relatives in multilocus systems is part of a subject of difficult entry. An exact general treatment involving only pairs of loci constitutes a formidable challenge leading to unwieldy expressions, as shown by Weir and Cockerham (1977). The curious reader may wish to glance with awe at formula (6) for the genetic variance in their article that is almost two pages long! Results assuming lack of inbreeding, epistasis and assortative mating, but accounting for dominance, linkage and the dynamics of the linkage disequilibrium parameter over generations, lead to simpler expressions and are given by Weir et al (1980). Further details and an informal derivation of the covariance between allele contents of two individuals at different loci are in the note note0101.pdf at https://github.com/SorensenD/SLGDS.

Expected Value of the Genomic Relationship Matrix

In the population with n related individuals and q causal loci, the centred matrix Z contains the coded genotypes of the causal loci, z_{ij}, $i = 1, \ldots, n$; $j = 1, \ldots, q$. These are realisations from a random Mendelian process. The expectation $E\left(ZZ'\right)$ is proportional to the expected additive genetic relationship matrix, as sketched below.

The ijth element of ZZ' involving individuals i and j is computed using the inner product $z_i' z_j$, where z_i' is the ith row of Z, containing the causal genotypes of individual i. The expectation of the inner product $z_i' z_j$ involves only the diagonal elements of the matrix Σ_z in (1.29). Then conditional on gene frequencies,

$$E\left(z_i' z_j\right) = a_{ij} \sum_{k=1}^{q} 2p_k\left(1 - p_k\right)$$

and

$$E\left(ZZ'\right) = A \sum_{k=1}^{q} 2p_k\left(1 - p_k\right), \tag{1.35}$$

proportional to A, as in Habier et al (2007). This expectation, here derived for causal (QTL) genotypes, holds for any set of randomly drawn autosomal genotypes, including genetic marker loci.

The Genomic Model

The description so far is in essence the standard model of quantitative genetics. It is viewed as the model that generates the data, the *true model*. "True" must be interpreted in a statistical sense and not as a mechanistic description of the complex biological process that results in an observed phenotype. It is perhaps best understood relative to an *operational model* that by its very nature cannot be viewed as having generated the data. A *genomic model* is such an operational model. A genomic model incorporates genetic markers in a regression equation, often assuming a linear association between the phenotype and the genetic markers. This association arises as a result of the correlation (linkage disequilibrium) between the observed genetic markers and the unobserved causal loci (QTL). The genetic markers are DNA sequences with a known location on a chromosome. One type is single-base pairs (single nucleotide polymorphisms (SNPs)), that give rise to different alleles containing alternative bases at a given position. SNPs are most often phenotypically neutral, but sometimes they can affect the phenotype, for example, by causing a change in the amino acid sequence of a gene product, or by affecting gene regulation in noncoding sequences. In this case, the SNP becomes a causal locus and has a direct effect on phenotype.

The use of genetic markers as covariates instead of causal loci induces a new set of parameters that must be defined: the *genomic value*, the *genomic variance* and the *genomic heritability*.

At the population level, using p markers with centred marker genotype codes x_i in vector $x = \left(x_1, \ldots, x_p\right)'$, the genomic value (the part of the genetic value explained by the linear regression on markers) is defined as

$$\begin{aligned}
\mathrm{E}\left[G\left(z\right)|x\right] &= \mathrm{E}_{z|x}\left[\mathrm{E}\left(G\left(z\right)|z, x\right)\right] \\
&= \mathrm{E}_{z|x}\left[\mathrm{E}\left(G\left(z\right)|z\right)\right] \\
&= \alpha' \mathrm{E}\left(z|x\right) \\
&= \alpha' \Sigma_{zx} \Sigma_x^{-1} x = \alpha' \widehat{z} \\
&= \beta' x.
\end{aligned} \tag{1.36}$$

In this expression, the marker effects $\beta = \Sigma_x^{-1} \Sigma_{xz} \alpha$ are population parameters, defined as the vector of coefficients in a multiple linear regression of additive genetic values on marker genotypes. In the fourth line, $\widehat{z} = \Sigma_{zx} \Sigma_x^{-1} x$ is the best linear predictor of allele contents at the QTL, z, given allele content at markers x. The parameter Σ_{zx} is the covariance matrix involving marker and QTL genotypes, and the parameter Σ_x is the variance covariance matrix of marker genotypes that has a structure similar to (1.29).

The genomic variance or variance of genomic values $\beta' x$ is the part of the genetic variance explained by the linear regression on marker genotypes:

$$\mathrm{Var}_x\left(\mathrm{E}\left[G\left(z\right)|x\right]\right) = \beta' \Sigma_x \beta = \alpha' \Sigma_{zx} \Sigma_x^{-1} \Sigma_{xz} \alpha. \tag{1.37}$$

The *genomic heritability* is the ratio of genomic variance to phenotypic variance, also a population parameter. These are the parameters to be inferred when the statistical models are fitted to the data.

There are a couple of details to note in relation to (1.37). The first and obvious one is that the proportion of the additive genetic variance captured by the marker genotypes depends on the strength of the association between the markers and the QTL genotypes (defined by the degree of linkage disequilibrium between marker loci and causal loci). This is dictated by Σ_{zx}. When LD between marker genotypes and causal genotypes is perfect, $\Sigma_{zx} = \Sigma_z$, $\Sigma_x = \Sigma_z$ and the genomic variance is equal to the additive genetic variance. In this situation, the genomic heritability achieves its upper bound, equal to the heritability of the trait. The second is that when the QTL genotypes are part of the marker panel (e.g. with full genome sequence), the (population parameter) genomic variance (1.37) is equal to the (population parameter) additive genetic variance (1.30). In other words, $\mathrm{Var}(\beta' x) = \alpha' \Sigma_z \alpha$. The proof of this result involves a little manipulation with the inverse of partitioned matrices and can be found in de los Campos et al (2015). It does not follow that implementing a statistical model that incorporates full genome sequence will automatically uncover the additive genetic variance of the trait. This depends on the input data and on the properties of the estimator of the genomic variance of the statistical model.

In several examples used in the book, I generate a matrix of genotypic markers drawing independent samples from a binomial distribution and a proportion is assigned as QTL. Since the QTL are part of the marker panel, the variance of the independent marker genotypes fully accounts for the equilibrium additive genetic variance.

Further details on these subjects can be found in Gianola et al (2009) and de los Campos et al (2015) who lay the background for most of what is written here.

Fitting Models Incorporating Marker Genotypes

When marker genotypes are fitted one at a time, the analysis is known as *genomewide association study* (GWAS). GWAS exploits the association, at the population level, between observed genetic markers scattered across the genome and unobserved causal loci. The significance of markers is assessed by regressing the phenotype on each marker in turn. This is based on a simple t−test using often a Bonferroni adjustment to correct for multiple testing (see Chap. 8). GWAS have identified thousands of genetic variants associated with human complex traits and diseases. Despite this success, the significant SNPs explain a small proportion of the genetic variation estimated from pedigrees. This has been known as the "missing heritability" problem and indicates that most of the genetic variation uncovered by GWAS remains unexplained. Part of the explanation lies in the multiple testing procedure, in the method's attempt to avoid false discoveries, on the size of the effects and frequency of the causal genes affecting the complex trait and on the incomplete degree of association between marker and causal genotypes. Many loci contributing to genetic variation do not reach the stringent significance level imposed and are missed. Despite the limitations, the success of GWAS as a gene discovery tool cannot be denied, particularly when used together with a number of refinements designed to partially overcome some of its shortcomings (Yang et al 2012). In 2008, a GWAS study involving several thousand individuals detected approximately 40 genetic variants associated with height that account for about 5% of the heritability (Visscher 2008). In 2022, with larger size of testing populations (resulting in higher power) and larger number of genetic markers (leading to larger amounts of linkage disequilibrium), the number of discovered SNPs increased to approximately 12, 000, explaining 40% of the heritability of height (Yengo et al 2022).

An alternative to fitting one marker at a time is to use a genomic model that incorporates all the markers simultaneously in a multiple linear regression disregarding statistical significance and in contrast with traditional GWAS, accounting for the structure among marker genotypes. A specification could be

$$y|X \sim N\left(1\mu + Xb, I\sigma^2\right), \tag{1.38}$$

where 1 is an $n \times 1$ vector of $1'$s, μ is a scalar, y is $n \times 1$ vector of observed records, X is a centred matrix of observed marker genotypes of order $n \times p$ and b is an unobserved $p \times 1$ vector of marker effects at the level of the fitted model. When typically $p > n$, some additional structure must be incorporated to fit the p marker effects b to the n data points. One common approach is to treat b as the random variable

$$b|\sigma_b^2 \sim N\left(0, I\sigma_b^2\right), \tag{1.39}$$

where σ_b^2 can be interpreted as quantifying prior uncertainty about the genetic marker effects b. The specification of marker effects as random variables is in contrast with their definition in (1.36) and has a direct consequence on the relationship between the genomic variance defined in (1.37) and the genomic variance of the fitted model. To see this, define the vector of genomic values (at the level of this operational model) as $g = Xb$. The genomic variance of the fitted model is defined as the unconditional variance:

$$\begin{aligned}\sigma_g^2 &= \mathrm{Var}\left(b'X\right) = \mathrm{E}_b\left[\mathrm{Var}\left(b'X|b\right)\right] + \mathrm{Var}_b\left[\mathrm{E}\left(b'X|b\right)\right] \\ &= \mathrm{E}_b\left(b'\Sigma_X b\right) + 0 \\ &= \sigma_b^2 tr\left(\Sigma_X\right) = \sigma_b^2 \sum_{j=1}^{p} 2p_j\left(1-p_j\right). \end{aligned} \tag{1.40}$$

In this expression, $\mathrm{Var}_b\left[\mathrm{E}\left(b'X|b\right)\right] = 0$ because with centred X, $E(b'X|b) = b'E(X) = 0$. A glance at the genomic variance of the fitted model (1.40) and at the parameter defined in (1.37) reveals a lack of correspondence between the two. This is elaborated a little further in the first *Comment* on page 44 and a deeper discussion is in de los Campos et al (2015).

Let $k = \sum_{j=1}^{p} 2p_j\left(1-p_j\right)$. An alternative parametrisation in terms of g that is useful when $p > n$ is

$$y|X \sim N\left(1\mu + g, I\sigma^2\right), \tag{1.41a}$$

$$g|X \sim N\left(0, G_g\sigma_g^2\right), \tag{1.41b}$$

where the genomic relationship matrix $G_g = \frac{1}{k}XX'$ has rank at most $n-1$ if X is centred and if $p > n$. This is equivalent to constructing the genomic relationship matrix from $G_g = XX'$, when the elements of X are centred and scaled by $\sqrt(k)$. In this parametrisation, $\sigma_g^2 = \sigma_b^2$, independent of the number of markers. In general, when X is centred, the vector of genomic values g has a singular normal distribution.

The model of marker effects defined in (1.39) shrinks all marker effects homogeneously towards zero and is therefore not ideal for discovery. It is best deployed for inferences of genomic variance σ_g^2 and for prediction of genomic values g. An early application of a version of this genomic model in humans to infer genomic heritability (Yang et al 2010) incorporated 294, 831 SNPs genotyped on 3925 unrelated individuals (meaning distantly related) and reported that 45% of the variance of height could be accounted for by considering all the SNPs simultaneously. The increase in variance explained relative to the GWAS analysis is attributed to the many SNPs with small effects that do not pass the significant threshold in GWAS. This was still considerably less than the 70% – 80% explained by traditional methods using pedigrees. The authors conjectured that most of the unaccounted variance could be due to causal variants at low frequency, leading to very low correlations between markers and causal variants. Genetic markers are typically at intermediate frequencies, and the disparity of gene frequencies between markers and causal variants results in low correlations. Support for this conjecture was obtained years later using whole genome sequence data and larger population sizes, leading to an estimate of the proportion of variance of height attributed to SNPs (the genomic heritability) of 68% (Wainschtein et al 2022), quite close to the estimate based on pedigrees.

The picture that is emerging is interpreted in a paradigm in which complex traits are driven by a very large number of loci spread along the genome. However, complex traits are affected not only by protein-coding genes but also by noncoding variants, perhaps related to gene regulation that plays a direct role in the expression of the trait. Despite their relevance, the noncoding variants show weak signals in GWAS studies and are difficult to detect. Most of the genetic variation of complex trait seems to be driven by a very large number of peripheral genes of small effect (Boyle et al 2017).

The new insights of the genetics of complex traits stimulated further developments and refinements of the genomic model for estimation of genomic parameters and for prediction. For example, prior knowledge can be incorporated to partition genomic variance across groups defined by allele frequency of genetic markers, linkage disequilibrium and genotype certainty and accounting for SNP function and metabolic pathways (Speed et al 2017, 2021; Patxot et al 2021).

The genomic model (1.41) is revisited several times in the book. A likelihood model using Newton-Raphson and using the EM algorithm is described in Chap. 3 and in the Problems section on pages 553 and 609. A Bayesian model is presented in Chap. 5 and in the Problems section on pages 560 and 645.

The genomic model (1.41) can be extended for use in simultaneous inference, prediction and gene discovery. An early review is provided by de los Campos et al (2013a). One approach is based on modifying the prior distribution of SNP effects (1.39) and adopting instead a mixture prior of two densities: one with small variance (the spike) and one with large variance (the slab). These are known as spike and slab models. For example, if a two component mixture is chosen, these could be two Gaussian densities (George and McCulloch 1993) or one Gaussian and a point mass at zero (Mitchell and Beauchamp 1988; Meuwissen et al 2001; Habier et al 2011). If two Gaussian distributions are used, they are often parametrised such that one has a mean of zero and a very small user-tuned variance, generating very small SNP effects and the other a mean of zero and a larger variance, allowing for larger effects. For gene discovery, the idea is to obtain an estimate of the posterior probability that a genetic marker is drawn from either mixture component, for all genetic markers. An example of a spike and slab model with a point mass at zero applied to gene discovery can be found in Chap. 8. Chapter 7 provides a detailed Bayesian-McMC implementation of this model for the analysis of continuous traits, and Chap. 9 extends the algorithm for the analysis of binary data.

Comments on the Genomic Variance and the Genomic Relationship Matrix

- The genomic variance σ_g^2 (1.40) of the fitted model has a tenuous correspondence with the genomic variance of the genomic model defined in (1.37). An approach that results in an estimate of genomic variance that is better aligned with parameter (1.37) can be derived drawing from Sorensen et al (2001) as follows. Consider a group or cohort consisting of n individuals; the genomic value of individual i is g_i. This is a random variable that can take n possible values each with probability n^{-1}. By definition, the variance of g_i conditional on the statistical model adopted and on the particular group of the n individuals is

$$\sigma_{gen}^2 = E\left(g_i^2\right) - [E(g_i)]^2 = \frac{1}{n}\sum_{i=1}^{n} g_i^2 - \overline{g}^2, \tag{1.42}$$

where $\overline{g} = n^{-1}\sum_i g_i$. Let $g_i = \beta' x_i$, where x_i is the column vector of the p centred marker genotypes of individual i and $\beta = \Sigma_x^{-1}\Sigma_{xz}\alpha$ is the column vector of p marker effects (the parameters of a multiple regression of additive genetic values $\alpha' z$ on marker genotypes x). Then if β is known and the elements of x are centred, $\overline{g} = 0$ and the genomic variance in the cohort is

$$\begin{aligned}\sigma_{gen}^2 &= n^{-1}\sum_{i=1}^{n} \beta' x_i x_i' \beta \\ &= \beta' \left[n^{-1} X'X\right] \beta.\end{aligned} \tag{1.43}$$

This expression has the same form as (1.37) with Σ_x replaced by its method of moments estimator $n^{-1}X'X$. The marker effects β are not observed, and adopting a Bayesian approach can be inferred from their marginal posterior distribution $[\beta|y]$. The posterior mean of the distribution of the resulting genomic variance is then

$$\begin{aligned}\mathrm{E}\left(\sigma_{gen}^2|y\right) &= \mathrm{E}\left(\beta' Q \beta|y\right) \\ &= \mathrm{tr}\left(Q\,\mathrm{Var}(\beta|y)\right) + \mathrm{E}(\beta|y)'\, Q\, \mathrm{E}(\beta|y) \\ &= \mathrm{tr}\left[Q\left(\mathrm{E}\left(\beta\beta'|y\right) - \mathrm{E}(\beta|y)\,\mathrm{E}(\beta|y)'\right)\right] + \mathrm{tr}\left[\mathrm{E}(\beta|y)'\, Q\, \mathrm{E}(\beta|y)\right] \\ &= \mathrm{tr}\left[Q\,\mathrm{E}\left(\beta\beta'|y\right)\right],\end{aligned} \tag{1.44}$$

where $Q = n^{-1}X'X$ (the equality in the final line stems from applying the cyclic property of the trace operator in the second trace of the third line). All this is typically implemented using McMC. A Monte Carlo estimate of the posterior distribution of the genomic variance in the cohort σ_{gen}^2 is easily obtained by calculating the variance among the draws from $[g|y]$, without the need to infer the marker effects. A draw from this posterior distribution at round t of an McMC implementation is

$$\sigma_{gen}^{2[t]} = \frac{1}{n}\sum_{i=1}^{n}\left(g_i^{2[t]} - \overline{g}^{2[t]}\right),$$

where $g_i^{[t]}$ is sampled from $[g_i|y]$ belonging to the ith individual in the cohort. The posterior mean of the genomic variance under the parametrisation $g = X\beta$ takes the form

$$\mathrm{E}\left(\sigma_{gen}^2|y\right) = \frac{1}{n}\,\mathrm{tr}\left[\mathrm{E}\left(gg'|y\right)\right]. \tag{1.45}$$

The subject is elaborated a little further on page 233 where an McMC implementation is shown to involve computations using scalar quantities.

- The genomic relationship matrix constructed using genetic markers describes realised patterns of inheritance in a particular sample of individuals in the region of the genome defined by the marker loci. There are a number of ways of constructing the genomic relationship matrix and most use moment-based estimators (VanRaden 2008; de los Campos et al 2013a). Here, it was defined in connection with (1.41b)

$$G_g = \frac{1}{k}XX', \quad k = \sum_{i=1}^{p} 2p_i\left(1 - p_i\right), \tag{1.46}$$

where the elements of X (observed marker genotypes) are centred. Provided the model includes an intercept as in (1.41a), centring does not affect predictions or inferences of variances. However, standardising the elements of X may affect predictions, depending on the degree of similarity between the distributions of observed marker genotypes and of unobserved causal loci genotypes. Inferences, such as estimation of genomic heritability, are also sensitive to the way the genomic relationship matrix is constructed (Speed et al 2012). In a prediction context, the choice of a method can be evaluated by cross-validation.

- The genomic model builds on the incorporation of a large number of genetic marker loci spread along the genome of individuals from which G_g is constructed. These genetic marker loci are not necessarily causal, but are correlated with the causal loci. The latter are typically unobserved while the genetic markers are observed and can provide information about the genetics of a trait. The ability of the genomic model to separate true signal (the genetic value at the causal loci) from noise depends on how well the genomic relationship matrix G_g describes relationships realised at causal loci. Realised relationships in different regions of the genome are the result of a random process, with expected value dictated by the pedigree and variation due to Mendelian sampling. This creates variability in the patterns of genetic similarity across the genome; this pattern can be very different for marker genotypes and for causal genotypes. The variability is accentuated for data composed of nominally unrelated individuals (Hill and Weir 2011), often seen in human data and plays a smaller role when pedigrees display strong family relationships as in livestock. With distantly related individuals, fitting a large number of markers, many of which are potentially uncorrelated with genotypes at causal loci, can lead to an incorrect specification of the covariance structure, and this can affect inferences of variance parameters of the model (de los Campos et al 2015). This problem can be partly addressed fitting mixture models (spike and slab models) where inferences of variance parameters and model selection are performed simultaneously. These models are introduced in Chap. 7.

- Genomic models that incorporate dense marker genotypes, particularly in humans, typically include phenotypes on nominally unrelated individuals (to be understood as distantly related). These models are often used to estimate genomic heritability. Inference about genomic heritability is driven by quantifying the proportion of the phenotypic resemblance between distant relatives explained by the short genome segments they share (which is tagged by the observed marker loci via the genomic relationship matrix). In such a situation, there is a need to avoid bias due to non-genetic effects, such as common environment among close relatives, that inflates the phenotypic resemblance between relatives. Also, by excluding close relatives such as twins, full-sibs and parent and offspring that would explain most of the phenotypic resemblance, the estimate obtained reflects the proportion of the short genome sequences shared among the distantly related individuals. These short shared sequences transmitted from remote common ancestors generate linkage disequilibrium; this LD is exploited via dense marker genotyping by capturing contributions from unobserved causal loci. In so doing, a proportion of the additive genetic variation is unmasked: the genomic variation.
- A decomposition that has attractive computational properties introduced in Chap. 3 is the eigenvalue or spectral decomposition of XX'

$$\begin{aligned} XX' &= U \Delta U' \\ &= \sum_{i=1}^{n} \lambda_i U_i U_i', \end{aligned}$$

where X, here centred and scaled, has dimension $n \times m$, (m being the number of marker genotypes), $U = [U_1, U_2, \ldots, U_n]$, of dimension $n \times n$ (n being the number of individuals) is the matrix of eigenvectors of XX', U_j is the jth column (dimension $n \times 1$) and Δ is a diagonal matrix with elements equal to the eigenvalues $\lambda_1, \lambda_2, \ldots, \lambda_n$ associated with the n eigenvectors. Since XX' must be non-negative definite, the eigenvalues are $\lambda_i \geq 0$, $i = 1, 2, \ldots, n$. The eigenvectors satisfy $U'U = UU' = I$. Then an alternative representation of the genomic relationship matrix (when X is centred and scaled) is

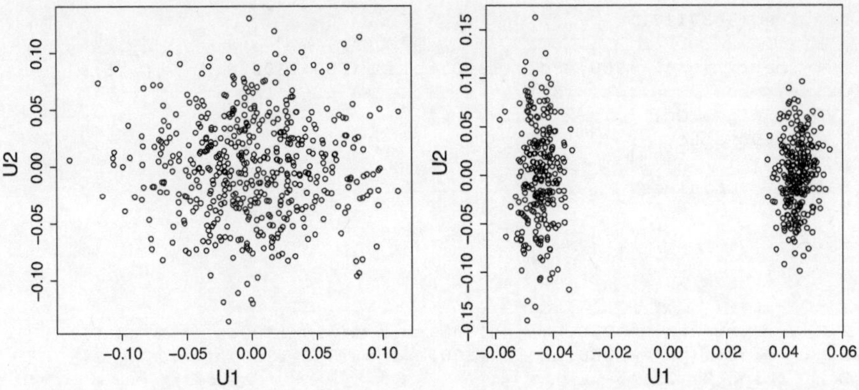

Fig. 1.13 Left: plot of the first two eigenvectors for homogeneous matrix composed of draws from $Bi\,(x; n = 2, q = 0.5)$; right: plot of the first two eigenvectors for heterogeneous matrix composed of draws from $Bi\,(x; n = 2, q = 0.5)$ and draws from $Bi\,(x; n = 2, q = 0.3)$

$$
\begin{aligned}
G_g &= \frac{1}{m} X X' \\
&= \frac{1}{m} U \Delta U'.
\end{aligned}
$$

This decomposition can be used to investigate the presence of unobserved substructure in X, which can cause spurious associations between marker genotypes and phenotypes and also artefact genomic variation. As an illustration, the `R`-code below constructs two genomic relationship matrices drawing independent samples from binomial distributions. In the first genomic relationship matrix, each x representing a marker genotype is a draw from $Bi\,(x; n = 2, q = 0.5)$, where x can take values 0, 1, 2 and q is the probability of drawing allele A. The draws are elements of X that in the example is of order 500×1300. The second relationship matrix is generated by appending two submatrices one on top of the other, where the first (order 250×1300) has elements drawn from $Bi\,(x; n = 2, q = 0.5)$ and the second (order 250×1300) has elements drawn from $Bi\,(x; n = 2, q = 0.3)$. This matrix displays fairly strong substructure caused by the difference of allele frequencies. The eigenvalue decomposition is performed on both matrices and Fig. 1.13 shows the plot of the first two eigenvectors. The left subfigure corresponding to the homogeneous matrix does not reveal any form of structure, but the right subfigure does. Estimation of marker effects without accounting for population structure can be a problem when they are used to predict records of a sample that has a different structure:

```
# CODE0105
# EIGEN DECOMPOSITION
rm(list=ls()) # CLEAR WORKSPACE
```

```
set.seed(30371171)
# SIMULATE Z FROM AN UNSTRUCTURED POPULATION
Z <- matrix(nrow= 500,ncol= 1300,rbinom(500*1300,size=2,p=.5))
Gz <- tcrossprod(scale(Z)) # CENTRING AND SCALING
EVD <- eigen(Gz)
U <- EVD$vector
tU<-t(U)
val <- EVD$values
#qr(Gz)$rank
#plot(U[,1],U[,2],xlab='U1',ylab='U2')

# SIMULATE Z FROM TWO POPULATIONS WITH
# DIFFERENT GENE FREQUENCIES
Z1 <- matrix(nrow= 250,ncol= 1300,rbinom(250*1300,size=2,p=.5))
Z2 <- matrix(nrow= 250,ncol= 1300,rbinom(250*1300,size=2,p=.3))
Z <- rbind(Z1,Z2)
Gz <- tcrossprod(scale(Z))
EVD <- eigen(Gz)
U <- EVD$vector
tU<-t(U)
val <- EVD$values
qr(Gz)$rank
#plot(U[,1],U[,2],xlab='U1',ylab='U2')
```

Part I
Fitting Likelihood and Bayesian Models

Chapter 2
Likelihood

A central problem in statistics is the estimation of parameters that index a probability model proposed to describe aspects of the state of nature. In the classical approach to inference, these parameters have a “true” but unknown value and given the model, can be estimated from a set of observations. A firmly entrenched inferential approach in statistics is the method of maximum likelihood proposed and termed by Fisher (1922), although, as is often the case in science, the subject had been in the air long before Fisher disguised in the terminology of inverse probability. An excellent account is in Hald (1998).

This chapter starts by providing an intuition for the concept of likelihood emphasising the conceptual difference between a likelihood function and a probability function. This is followed by a summary of the basic results needed for classical likelihood inference. Proofs of most of the results presented here are not difficult and can be found, for example, in Sorensen and Gianola (2002). The final part of the chapter consists of examples that illustrate the construction of simple likelihood models, the derivation of the maximum likelihood estimators and some of their properties.

Readers that require a quick brush up of concepts of basic probability distributions and random variables may consider browsing through the first two chapters of Sorensen and Gianola (2002) that cover what is necessary of the subject to follow comfortably the material in this book.

2.1 A Little Intuition

Imagine that you are given the following *iid* (independently and identically distributed) data y sampled from a normal distribution $N\left(\mu, \sigma^2 = 2\right)$ with unknown mean μ and known variance $\sigma^2 = 2$:

$$y = (y_1 = 11.8, \quad y_2 = 6.1, \quad y_3 = 7.1, \quad y_4 = 11.1, \quad y_5 = 5.8)\,.$$

D. Sorensen, *Statistical Learning in Genetics*, Statistics for Biology and Health,
https://doi.org/10.1007/978-3-031-35851-7_2

You will agree that 10 is a better candidate for μ than 20, given the sample y (in fact, the sample was drawn from $N\left(\mu=10, \sigma^2=2\right)$ using the `R` function `rnorm(5,10,sqrt(2))`). In a general setting, suppose that a sample y from the distribution of Y is drawn, given the true (typically unknown) value of the parameter μ and the true known $\sigma^2 = 2$. This sample drawn with a certain probability is to provide information about μ. For fixed $Y = y$, compute $p\left(Y = y|\mu = \mu_1, \sigma^2 = 2\right)$ and $p\left(Y = y|\mu = \mu_2, \sigma^2 = 2\right)$, in order to choose between μ_1 and μ_2 as two possible values of μ. Viewed as a function of μ, $p\left(Y = y|\mu, \sigma^2 = 2\right)$ is the *likelihood function*. If $p\left(Y = y|\mu_1, \sigma^2 = 2\right) > p\left(Y = y|\mu_2, \sigma^2 = 2\right)$ we would choose $\mu = \mu_1$ on the grounds that, given the sample y, this is a "more likely" value than $\mu = \mu_2$. In fact, it is possible that if the true value of μ is μ_2, we could still observe for the particular sample y, that $p\left(Y = y|\mu_1, \sigma^2 = 2\right) > p\left(Y = y|\mu_2, \sigma^2 = 2\right)$, but this occurs with small probability and this probability approaches zero as sample size tends to infinity (Lehmann and Casella 1998). This is what is meant by "more likely". The value of the likelihood function $p\left(Y = y|\mu, \sigma^2 = 2\right)$ evaluated at $\mu = \mu_1$ agrees with the *probability of observing* $Y = y$ only when the *true value* of the parameter μ is in fact μ_1 (this is so if Y is discrete; otherwise, for continuous Y, the likelihood evaluated at $\mu = \mu_1$ is proportional to the probability that Y takes values in a small set containing y). However, the likelihood function is not a probability distribution. To quote Fisher (Fisher 1922): "The likelihood that any parameter (or set of parameters) should have any assigned value (or set of values) is proportional to the probability that if this were so, the totality of observations should be that observed".

Irrespective of the true value of the parameter, the value of μ that maximises the likelihood function is the maximum likelihood estimator of μ, $\widehat{\mu}$. The great appeal of the method is mostly due to the desirable statistical properties of the maximum likelihood estimator, many of which hold asymptotically. To be specific, given that the true value of μ is 10, the probability of observing $Y = y_1 = 11.8$ is proportional to $N\left(11.8|\mu = 10, \sigma^2 = 2\right)$:

$$\Pr(Y = 11.8) \propto \texttt{dnorm(11.8, 10, sqrt(2))} = \texttt{0.1255}.$$

The probability of observing $Y = y$ is proportional to

$$\begin{aligned}\prod_{i=1}^{5} N\left(y_i|\mu = 10, \sigma^2 = 2\right) &= \texttt{prod(dnorm(y, 10, sqrt(2)))}\\ &= \texttt{1.9457} \times \texttt{10}^{-8}.\end{aligned}$$

On the other hand, the contribution to the likelihood function of $Y = y_1$ is

$$N\left(11.8|\mu, \sigma^2 = 2\right) = (2\pi 2)^{-\frac{1}{2}} \exp\left(-\frac{1}{2 \times 2}(11.8 - \mu)^2\right)$$

and in view of the *iid* nature of the data, the contribution from the whole data y to the likelihood function is

$$\prod_{i=1}^{5} N\left(y_i|\mu, \sigma^2 = 2\right) = (2\pi 2)^{-\frac{5}{2}} \exp\left(-\frac{1}{2 \times 2}\sum_{i=1}^{5}(y_i - \mu)^2\right). \tag{2.1}$$

The value of μ that maximises (2.1) is $\hat{\mu} = 8.38$. This is the maximum likelihood estimate.

Notation

- The shorthand used for probability density function is pdf, for probability mass function is pmf and for cumulative distribution function is cdf.
- The shorthand used for independent and identically distributed is iid.
- MLE is shorthand for maximum likelihood estimator or maximum likelihood estimate and ML for maximum likelihood
- Often I use $\mathrm{E}_X[T(X)]$, where $T(X)$ is some function of the random variable X with probability density function $p(x|\theta)$. By this I mean "integration over the distribution of X":

  $$\mathrm{E}_X[T(X)] = \int T(x)\, p(x|\theta)\, dx$$

  where θ is a parameter that indexes the probability density of the random variable X.
- The notation $X \sim N(\mu, \sigma^2)$ stands for X *is distributed* as $N(\mu, \sigma^2)$, a normal distribution with mean μ and variance σ^2.
- The notation $N(x|\mu, \sigma^2)$ is used to specify the probability density function of X at the value $X = x$, when$X \sim N(\mu, \sigma^2)$.
- If λ is a $k \times 1$ vector and θ is a $p \times 1$ vector, then the jth row of the $k \times p$ matrix $\partial\lambda/\partial\theta'$ is

  $$\left[\partial\lambda_j/\partial\theta_1\ \partial\lambda_j/\partial\theta_2 \cdots \partial\lambda_j/\partial\theta_p\right]$$

  and the jth row of the $p \times k$ matrix $\partial\lambda'/\partial\theta = \left(\partial\lambda/\partial\theta'\right)'$ is

  $$\left[\partial\lambda_1/\partial\theta_j\ \partial\lambda_2/\partial\theta_j \cdots \partial\lambda_k/\partial\theta_j\right].$$
- When f is a function of several variables $f(x) = f(x_1, x_2, \ldots, x_n)$, the *gradient* of f is the column vector of n partial derivatives $\left(\frac{\partial f}{\partial x_1}, \frac{\partial f}{\partial x_{12}}, \ldots, \frac{\partial f}{\partial x_n}\right)'$ and is denoted ∇f.

- I use $[\theta|y]$ (with or without square brackets) to represent the distribution of the random variable θ given y. This may be a posterior distribution.
- Random variables, say, Y, are not consistently distinguished in notation from their realised values y, unless the need to avoid confusion demands it. For example, I may write *the distribution of* y *given* x (symbolised as $[y|x]$), rather than *the distribution of* Y *given* $X = x$ (symbolised as $[Y|X = x]$) , or for the probability mass function, $\Pr(y = 1|x)$ rather than $\Pr(Y = 1|X = x)$.
- I do not use boldface to distinguish matrices or vectors from scalars. Depending on context, matrices may be denoted with capital letters X or with lower-case letters x. In both cases, I make clear the dimension of the quantity at hand.
- The transpose of a matrix X (or of matrix x) is denoted by X' (or by x').

2.2 Summary of Likelihood Results

Let $p(y|\theta)$ represent the sampling density or mass function of the data vector y indexed by a parameter vector θ with p elements. The likelihood of the parameter θ given the data y, $L(\theta|y)$ is equal to $p(y|\theta)$:

$$L(\theta|y) = p(y|\theta). \tag{2.2}$$

If the elements of $y = (y_1, \ldots, y_n)'$ are iid,

$$L(\theta|y) = \prod_{i=1}^{n} p(y_i|\theta). \tag{2.3}$$

The loglikelihood (also known as *support*) is

$$\ln L(\theta|y) = \ell(\theta|y) = \sum_{i=1}^{n} \ln p(y_i|\theta). \tag{2.4}$$

The *maximum likelihood estimator* $\hat{\theta}$ is the value of θ that maximises (2.4) (or (2.3)) over the parameter space of θ. Formally, this involves a constrained maximisation. Often an unconstrained maximisation is performed; therefore, one must check that the resulting maximum likelihood estimate is within the allowed parameter space.

One may always multiply a likelihood function by a constant c independent of θ. Therefore likelihood functions $L(\theta|y)$ and $cL(\theta|y)$ are equivalent in the sense that both lead to the same maximum likelihood estimate. Similarly, two loglikelihood functions that differ only by an additive constant lead to the same maximum likelihood estimate.

The first derivative of ℓ with respect to the vector θ is known as the *score function*, $S(\theta)$. Letting p be the number of components in θ and using ℓ for $\ell(\theta|y)$,

$$S(\theta) = \frac{\partial}{\partial\theta}\ell(\theta|y) = \begin{pmatrix} \partial\ell/\partial\theta_1 \\ \vdots \\ \partial\ell/\partial\theta_p \end{pmatrix}, \tag{2.5}$$

so the score is an $p \times 1$ vector. It can be shown (Sorensen and Gianola 2002 pages 132–133) that the score has zero mean and variance given by Fisher's expected information matrix (see (2.9)); that is,

$$S(\theta) \sim \left[0, \mathrm{E}_y\left[S(\theta)\, S(\theta)'\right]\right]. \tag{2.6}$$

Evaluated at the maximum $\hat{\theta}$

$$S\left(\hat{\theta}\right) = 0$$

when $\hat{\theta}$ is in the interior of the parameter space.

The matrix of second derivatives of the loglikelihood is known as the *Hessian*, $H(\theta|y)$. For example, if $p = 2$, $H(\theta|y)$ takes the form

$$H(\theta|y) = \frac{\partial^2}{\partial\theta\partial\theta'}\ell(\theta|y) = \begin{bmatrix} \partial^2\ell/(\partial\theta_1)^2 & \partial^2\ell/(\partial\theta_1\partial\theta_2) \\ \partial^2\ell/(\partial\theta_2\partial\theta_1) & \partial^2\ell/(\partial\theta_2)^2 \end{bmatrix}.$$

The *observed information matrix* (of dimension $p \times p$) is given by

$$I(\theta|y) = -\frac{\partial^2}{\partial\theta\partial\theta'}\ell(\theta|y) = -H(\theta|y) \tag{2.7}$$

which involves second derivatives and is a function of both θ and y. An approximation that uses first derivatives only is given by the outer product of the vector of first derivatives of the score:

$$\left(S(\theta)\, S(\theta)'\right). \tag{2.8}$$

For example, if $p = 2$,

$$\begin{aligned} S(\theta)\, S(\theta)' &= \begin{bmatrix} \partial\ell/\partial\theta_1 \\ \partial\ell/\partial\theta_2 \end{bmatrix} \begin{bmatrix} \partial\ell/\partial\theta_1 & \partial\ell/\partial\theta_2 \end{bmatrix} \\ &= \begin{bmatrix} (\partial\ell/\partial\theta_1)^2 & (\partial\ell/\partial\theta_1)(\partial\ell/\partial\theta_2) \\ (\partial\ell/\partial\theta_2)(\partial\ell/\partial\theta_1) & (\partial\ell/\partial\theta_2)^2 \end{bmatrix}. \end{aligned}$$

Fisher's *expected information matrix* is defined to be the average of (2.7) or (2.8) over conceptual repetitions of y:

$$i(\theta) = \mathrm{E}_y\left[\left(\frac{\partial}{\partial\theta}\ell(\theta|y)\right)\left(\frac{\partial}{\partial\theta}\ell(\theta|y)\right)'\right] = -\mathrm{E}_y\left(\frac{\partial^2}{\partial\theta\partial\theta'}\ell(\theta|y)\right) \tag{2.9}$$

and is a function of θ only. Typically, these information matrices are evaluated at $\theta = \widehat{\theta}$.

Let θ_0 denote the true value of θ. A standard result says that the sampling distribution of the MLE has a limiting normal distribution

$$\widehat{\theta} \to N\left(\theta_0, i(\theta_0)^{-1}\right) \tag{2.10}$$

as sample size goes to infinity. This sampling distribution may be approximated using

$$N\left(\widehat{\theta}, i\left(\widehat{\theta}\right)^{-1}\right) \text{ or } N\left(\widehat{\theta}, I\left(\widehat{\theta}\right)^{-1}\right). \tag{2.11}$$

Estimates of standard errors are obtained from

$$\sqrt{i\left(\widehat{\theta}\right)^{-1}_{jj}} \text{ or } \sqrt{I\left(\widehat{\theta}\right)^{-1}_{jj}}. \tag{2.12}$$

Cramér-Rao Theorem Given certain regularity conditions, the variance of any unbiased estimator of a parameter θ (scalar or vector) must be at least as large as

$$\mathrm{Var}\left(\widehat{\theta}\right) \geq i(\theta)^{-1}. \tag{2.13}$$

When θ is a vector, the above inequality implies that

$$\mathrm{Var}\left(\widehat{\theta}\right) - i(\theta)^{-1}$$

is positive semidefinite. Any unbiased estimator that achieves this lower bound is *efficient* and no better unbiased estimator can be found. It follows from (2.10) that the ML estimator satisfies the Cramér-Rao lower bound, asymptotically; therefore, the asymptotic variance is the smallest possible.

An important property of ML is *functional invariance*. If $\hat{\theta}$ is the MLE of θ and if $\lambda = g(\theta)$ is a one-to-one transformation such that $\theta = g^{-1}(\lambda)$ exists, then the MLE of $g(\theta)$ is $g\left(\widehat{\theta}\right)$.

If θ is a scalar, under the new parametrisation, the score can be written as

$$\frac{d\ell\left(g^{-1}(\lambda)|y\right)}{d\lambda} = \frac{d\ell(\theta|y)}{d\theta}\frac{d\theta}{d\lambda} \tag{2.14}$$

and if θ and λ are vectors

$$\frac{\partial \ell\left(g^{-1}(\lambda)\,|y\right)}{\partial \lambda} = \frac{\partial \theta'}{\partial \lambda}\frac{\partial \ell(\theta|y)}{\partial \theta}. \tag{2.15}$$

If θ is a scalar, the expected information about the new parameter λ contained in the sample is

$$i(\lambda) = i(\theta)\left(\frac{d\theta}{d\lambda}\right)^2 = i\left(g^{-1}(\lambda)\right)\left[\frac{dg^{-1}(\lambda)}{d\lambda}\right]^2 \tag{2.16}$$

and the asymptotic variance is

$$\text{Var}\left(\widehat{\lambda}\right) = [i(\lambda)]^{-1} = i(\theta)^{-1}\left(\frac{d\lambda}{d\theta}\right)^2 = \text{Var}\left(\widehat{\theta}\right)\left(\frac{d\lambda}{d\theta}\right)^2. \tag{2.17}$$

When θ and λ are $p \times 1$ vectors, the expected information matrix is

$$i(\lambda) = \frac{\partial \theta'}{\partial \lambda} i(\theta)\frac{\partial \theta}{\partial \lambda'} = \frac{\partial \left[g^{-1}(\lambda)\right]'}{\partial \lambda} i\left(g^{-1}(\lambda)\right)\frac{\partial \left[g^{-1}(\lambda)\right]}{\partial \lambda'}, \tag{2.18}$$

where $\frac{\partial \theta'}{\partial \lambda}$ and $i(\theta)$ are $p \times p$ matrices. The asymptotic variance evaluated at $\lambda = \hat{\lambda}$ is

$$\begin{aligned}\text{Var}\left(g\left(\hat{\theta}\right)\right) &= \left[i\left(\hat{\lambda}\right)\right]^{-1} \\ &= \left.\frac{\partial \lambda}{\partial \theta'}\right|_{\theta=\widehat{\theta}} \text{Var}\left(\widehat{\theta}\right)\left.\frac{\partial \lambda'}{\partial \theta}\right|_{\theta=\widehat{\theta}}\end{aligned} \tag{2.19}$$

where $\frac{\partial \lambda}{\partial \theta'}$ and $\text{Var}\left(\widehat{\theta}\right)$ are $p \times p$ matrices. Expression (2.19) is the multiparameter extension of (2.17) that can also be derived using the "delta method" (Sorensen and Gianola 2002, page 95). The derivation based on the delta method does not require that vectors θ and λ have the same number of elements. An application of (2.19) is in the Exercises on page 612.

Also asymptotically, for scalar θ and $g(\theta)$,

$$g\left(\hat{\theta}\right) \sim N\left(g(\theta), i(\theta)^{-1}\left(\frac{dg(\theta)}{d\theta}\right)^2\right). \tag{2.20}$$

If $g\left(\hat{\theta}\right)$ is a $p \times 1$ vector, a similar results holds, except that the asymptotic variance in (2.20) is replaced by the $p \times p$ matrix given by (2.19). In applications θ is replaced by $\hat{\theta}$.

The MLE is also invariant to transformations of the data that do not depend on the parameter θ. If data X that have density function $p_X(x|\theta)$ are transformed to $Y = f(X)$ so that the inverse transformation is $f^{-1}(Y) = X$ exists, then the density of the transformed data Y is

$$\begin{aligned} p_Y(y|\theta) &= p_X\left(f^{-1}(y)|\theta\right)\left|\frac{df^{-1}(y)}{dy}\right| \\ &= p_X(x|\theta)\left|\frac{dx}{dy}\right| \end{aligned} \tag{2.21}$$

and therefore the likelihood function based on X and Y differ only by a factor (the Jacobian of the transformation) that does not depend on the parameter θ.

Note
The equality in the second line of (2.19) is based on the following two results. Firstly, for full-rank square matrices A, B, C,

$$[ABC]^{-1} = C^{-1}B^{-1}A^{-1}.$$

Secondly,

$$\left[\frac{\partial\theta}{\partial\lambda'}\right]^{-1} = \left[\frac{\partial\lambda}{\partial\theta'}\right].$$

The derivation holds when the transformation $\lambda = g(\theta)$ is one-to-one.

When the dimension of λ is $k \times 1$ and of θ is $p \times 1$, direct application of the delta method yields

$$\text{Var}\left(g\left(\widehat{\theta}\right)\right) = \left.\frac{\partial\lambda}{\partial\theta'}\right|_{\theta=\widehat{\theta}} \text{Var}\left(\widehat{\theta}\right) \left.\frac{\partial\lambda'}{\partial\theta}\right|_{\theta=\widehat{\theta}}$$

as in (2.19), but now $\frac{\partial\lambda}{\partial\theta'}$ is a $k \times p$ matrix. The stronger assumption of normality of the asymptotic distribution of $g\left(\widehat{\theta}\right)$ arrived at using the delta method requires that $k \le p$ and that the rank of $\frac{\partial\lambda}{\partial\theta'}$ is equal to k.

Summary of Properties of Maximum Likelihood Estimators

Little is known about small sample properties of MLE. In many situations, it is known that MLE are not unbiased in small samples.

In large samples, MLE

(i) are consistent
(ii) are asymptotically normal (i.e. (2.10))
(iii) in a well-specified model, the MLE achieves the Cramér-Rao lower bound

Things to be aware of:

(i) The MLE may not be a turning point (the likelihood increases continuously).
(ii) The MLE may not be unique (the maximum is found over a flat region spanning a wide interval comprising the parameter—"flat" likelihood functions).
(iii) If iterative procedures are used to find a maximum, rate of convergence may be extremely slow, particularly in complex multiparameter models.
(iv) An iterative procedure may converge to a local maximum, not necessarily to a global maximum.
(v) Asymptotic normality of the MLE is compromised if the parameter lies on the border of the parameter space.
(vi) The zeroes of the first derivatives only locate extreme points in the interior of the domain of a function. If extrema only occur on the boundaries or corners, first derivatives may not be zero at those points.

2.3 Example: The Likelihood Function of Transformed Data

An example of (2.21) is the following. Let X have density

$$p(x) = \frac{1}{\sqrt{2\pi}} \exp\left[-\frac{1}{2}(x-\mu)^2\right] \tag{2.22}$$

which is the density of the normal process $N(\mu, 1)$. Assume that X is transformed to $Y = f(X) = \exp(X)$. The inverse transformation is $X = f^{-1}(Y) = \ln(Y)$. The density function of Y is then, using (2.21),

$$\begin{aligned} p(y) &= p\left(f^{-1}(y)\right)\left|\frac{d}{dy} f^{-1}(y)\right| \\ &= \frac{1}{y\sqrt{2\pi}} \exp\left[-\frac{1}{2}(\ln y-\mu)^2\right], \end{aligned} \tag{2.23}$$

where $1/y$ enters through the Jacobian of the transformation. Then $Y = \exp(X)$ is said to have a lognormal distribution with density as given above. The likelihood based on Y is proportional to this density. Inspection of (2.22) and (2.23) reveals that the two likelihoods are maximised when the exponential term is equal to 1. Therefore, the MLE of μ based on X is x and based on Y is $\ln y$. In other words, the MLE of the lognormal likelihood is the same as that of the normal likelihood fitted to the logarithm of the data.

2.4 Example: Linear Regression

Linear regression is one of the most widely adopted statistical methods and its use in genetics probably dates back to Galton (1885). For example, a simple approach for estimating heritability of quantitative traits is based on regressing offspring records on the records of their fathers and mothers, or regressing offspring records on mid-parental records.

This example, taken from Sorensen and Gianola (2002), is based on such a linear regression model where observations (e.g. the offspring records) are postulated to be linked to an intercept β_0 and to a slope parameter β_1 via the relationship

$$y_i = \beta_0 + \beta_1 x_i + e_i .$$

In this expression, the scalars x_i $(i = 1, 2, ..., n)$ are known values of an explanatory variable (e.g. the mid-parental values), and $e_i \sim N\left(0, \sigma^2\right)$ is an unobserved random residual term. The n residuals are assumed to be iid. The parameter vector is $[\beta_0, \beta_1, \sigma^2]$. In a genetic context, β_1 (the regression of offspring on mid-parents) is interpreted as the heritability of the quantitative trait (Falconer and Mackay 1996).

The likelihood function can be written as

$$L\left(\beta_0, \beta_1, \sigma^2|y\right) \propto \left(\sigma^2\right)^{-\frac{n}{2}} \exp\left[-\frac{1}{2\sigma^2}\sum_{i=1}^{n}(y_i - \beta_0 - \beta_1 x_i)^2\right]$$

and the corresponding loglikelihood, apart from an additive constant, is

$$l\left(\beta_0, \beta_1, \sigma^2|y\right) = -\frac{n}{2}\ln\left(\sigma^2\right) - \frac{1}{2\sigma^2}\sum_{i=1}^{n}(y_i - \beta_0 - \beta_1 x_i)^2 .$$

The score vector is given by

$$\begin{bmatrix} \frac{\partial l}{\partial \beta_0} \\ \frac{\partial l}{\partial \beta_1} \\ \frac{\partial l}{\partial \sigma^2} \end{bmatrix} = \begin{bmatrix} \frac{1}{\sigma^2}\sum_{i=1}^{n}(y_i - \beta_0 - \beta_1 x_i) \\ \frac{1}{\sigma^2}\sum_{i=1}^{n} x_i (y_i - \beta_0 - \beta_1 x_i) \\ -\frac{n}{2\sigma^2} + \frac{1}{2\sigma^4}\sum_{i=1}^{n}(y_i - \beta_0 - \beta_1 x_i)^2 \end{bmatrix}.$$

Setting the score vector equal to zero and solving simultaneously for the unknown parameters gives explicit solutions to the ML equations. The MLEs are

$$\widehat{\sigma}^2 = \frac{\sum_{i=1}^{n}(y_i - \hat{\beta}_0 - \hat{\beta}_1 x_i)^2}{n},$$

where $\hat{\beta}_0$ and $\hat{\beta}_1$ are the solutions to the matrix equation:

$$\begin{bmatrix} n & \sum_{i=1}^n x_i \\ \sum_{i=1}^n x_i & \sum_{i=1}^n x_i^2 \end{bmatrix} \begin{bmatrix} \hat{\beta}_0 \\ \hat{\beta}_1 \end{bmatrix} = \begin{bmatrix} \sum_{i=1}^n y_i \\ \sum_{i=1}^n x_i y_i \end{bmatrix}.$$

Explicitly,

$$\hat{\beta}_0 = \overline{y} - \hat{\beta}_1 \overline{x}$$

and

$$\widehat{\beta}_1 = \frac{\sum_{i=1}^n x_i y_i - \frac{1}{n} \sum_{i=1}^n x_i \sum_{i=1}^n y_i}{\sum_{i=1}^n x_i^2 - \frac{1}{n} \left(\sum_{i=1}^n x_i\right)^2}.$$

The first two columns of the 3×3 matrix of the negative of second derivatives of the loglikelihood with respect to the parameters, or observed information matrix, are

$$\begin{bmatrix} \sigma^{-2} n & \sigma^{-2} \sum_{i=1}^n x_i \\ \sigma^{-2} \sum_{i=1}^n x_i & \sigma^{-2} \sum_{i=1}^n x_i^2 \\ \sigma^{-4} \sum_{i=1}^n (y_i - \beta_0 - \beta_1 x_i) & \sigma^{-4} \sum_{i=1}^n x_i (y_i - \beta_0 - \beta_1 x_i) \end{bmatrix}$$

and the last column is

$$\begin{bmatrix} \sigma^{-4} \sum_{i=1}^n (y_i - \beta_0 - \beta_1 x_i) \\ \sigma^{-4} \sum_{i=1}^n x_i (y_i - \beta_0 - \beta_1 x_i) \\ -\left(2\sigma^4\right)^{-1} n + \sigma^{-6} \sum_{i=1}^n (y_i - \beta_0 - \beta_1 x_i)^2 \end{bmatrix}.$$

It is easy to verify that the expected value of each of the elements of the score vector is null. For example,

$$\mathrm{E}\left(\frac{\partial l}{\partial \beta_1}\right) = \frac{1}{\sigma^2} \sum_{i=1}^n x_i \, \mathrm{E}\,(y_i - \beta_0 - \beta_1 x_i) = 0$$

and

$$\begin{aligned} \mathrm{E}\left(\frac{\partial l}{\partial \sigma^2}\right) &= -\frac{n}{2\sigma^2} + \frac{1}{2\sigma^4} \sum_{i=1}^n \mathrm{E}\,(y_i - \beta_0 - \beta_1 x_i)^2 \\ &= -\frac{n}{2\sigma^2} + \frac{1}{2\sigma^4} n\sigma^2 = 0. \end{aligned}$$

Further, the expected information matrix is given by

$$i(\theta) = \sigma^{-2} \begin{bmatrix} n & \sum_{i=1}^n x_i & 0 \\ \sum_{i=1}^n x_i & \sum_{i=1}^n x_i^2 & 0 \\ 0 & 0 & \left(2\sigma^2\right)^{-1} n \end{bmatrix}.$$

Note that the elements $(1, 2)$ and $(2, 1)$ of this matrix are not null. This illustrates that in a multiparameter model, it is often more sensible to speak about joint information on a set of parameters rather than about information on individual parameters themselves.

Invariance

The original parametrisation was in terms of $\theta = [\beta_0, \beta_1, \sigma^2]'$. Imagine that there is interest in a new parametrisation consisting of the vector

$$\eta = \begin{bmatrix} \eta_1 \\ \eta_2 \\ \eta_3 \end{bmatrix} = \begin{bmatrix} \beta_0/\beta_1 \\ \beta_1 \\ \sigma^2 \end{bmatrix}$$

with inverse

$$\theta = f^{-1}(\eta) = \begin{bmatrix} \eta_1\eta_2 \\ \eta_2 \\ \eta_3 \end{bmatrix}.$$

The 3×3 matrix defined after (2.15) would be

$$\frac{\partial \theta'}{\partial \eta} = \begin{bmatrix} \partial\theta_1/\partial\eta_1 & \partial\theta_2/\partial\eta_1 & \partial\theta_3/\partial\eta_1 \\ \partial\theta_1/\partial\eta_2 & \partial\theta_2/\partial\eta_2 & \partial\theta_3/\partial\eta_2 \\ \partial\theta_1/\partial\eta_3 & \partial\theta_2/\partial\eta_3 & \partial\theta_3/\partial\eta_3 \end{bmatrix} = \begin{bmatrix} \eta_2 & 0 & 0 \\ \eta_1 & 1 & 0 \\ 0 & 0 & 1 \end{bmatrix}. \tag{2.24}$$

Differentiating the loglikelihood twice under the new parametrisation, multiplying by -1 and taking expectations yields the information matrix:

$$I(\eta) = \frac{1}{\eta_3} \begin{bmatrix} n\eta_2^2 & \eta_2\left(n\eta_1 + \sum_{i=1}^n x_i\right) & 0 \\ \eta_2\left(n\eta_1 + \sum_{i=1}^n x_i\right) & \sum_{i=1}^n (\eta_1 + x_i)^2 & 0 \\ 0 & 0 & \dfrac{n}{2\eta_3} \end{bmatrix}.$$

It can be verified that the same result is obtained from the matrix product given in (2.18), employing the matrix (2.24).

2.5 Example: Bivariate Normal Model with Missing Records

Consider a data set consisting of two rows; the first has phenotypic records on fathers (x) and the second on sons (y), one son per father. Some of the records of the sons are missing. The ordered data can be represented as

$$\begin{aligned} &x_1, x_2, \ldots, x_m, x_{m+1}, \ldots, x_n, \\ &y_1, y_2, \ldots, y_m. \end{aligned} \tag{2.25}$$

There are m complete bivariate observations $(x_i, y_i), i = 1, 2, \ldots, m$, and $(n - m)$ univariate records from fathers without sons. Each father that left offspring is assumed to have mated with a randomly chosen unknown female and produced one son. At this stage, it is assumed that the observed pattern of missing records may depend on the observed x's but not on the missing values themselves. In other words, the missing data $(y_{m+1}, \ldots, y_n)$ are *missing at random* (Rubin 1976, 2002), and the missing data mechanism is *ignorable*, in the sense that this mechanism does not have to be incorporated as part of the likelihood model for correct inferences.

The data could represent LDL cholesterol levels in humans collected with the objective of estimating the parameters of a probability model entertained by the data analyst. The parameters could be mean levels of LDL cholesterol and variance components and functions thereof, such as heritability. Information on the latter indicates whether genetic factors may affect LDL cholesterol.

Assume that a record of a father and his son is a draw from the bivariate normal model:

$$\left[(y_i, x_i)\,|\mu_y, \mu_x, \sigma_{xy}, \sigma_{yy}, \sigma_{xx}\right] \sim N\left(\begin{bmatrix}\mu_y \\ \mu_x\end{bmatrix}, \begin{bmatrix}\sigma_{yy} & \sigma_{xy} \\ \sigma_{xy} & \sigma_{xx}\end{bmatrix}\right), \quad i = 1, \ldots, n, \tag{2.26}$$

and assume that the pairs (y_i, x_i) are iid. The covariance term is usually interpreted as $\sigma_{xy} = \sigma_{xx}h^2/2$, where h^2 is the heritability (twice the regression of sons on fathers) in the population from which fathers were sampled (with assumptions, such as absence of common environmental effects between parents and offspring). The correlation coefficient is $\rho = \sigma_{xy}/\sqrt{\sigma_{yy}\sigma_{xx}}$ and is equal to $h^2/2$ when $\sigma_{xx} = \sigma_{yy}$.

In the absence of missing records (if the $(n - m)$ $y's$ were not missing), the joint density of the n pairs (y_i, x_i) is

$$\begin{aligned} &p\left(y, x|\mu_y, \mu_x, \rho, \sigma_{yy}, \sigma_{xx}\right) = \\ &\left(2\pi\sigma_{yy}\sigma_{xx}\left(1-\rho^2\right)\right)^{-\frac{n}{2}} \exp\left[-\frac{1}{2\left(1-\rho^2\right)}\left(\frac{\sum_{i=1}^n (x_i - \mu_x)^2}{\sigma_{xx}}\right.\right. \\ &\left.\left.+\frac{\sum_{i=1}^n \left(y_i - \mu_y\right)^2}{\sigma_{yy}} - 2\rho\frac{\sum_{i=1}^n (x_i - \mu_x)\left(y_i - \mu_y\right)}{\left(\sigma_{xx}\sigma_{yy}\right)^{\frac{1}{2}}}\right)\right], \end{aligned} \tag{2.27}$$

and the likelihood is proportional to (2.27). In this situation (no missing records), the ML estimators are in closed form and are given by

$$\widehat{\mu}_x = \frac{1}{n}\sum_{i=1}^{n} x_i, \qquad \widehat{\mu}_y = \frac{1}{n}\sum_{i=1}^{n} y_i,$$
$$\widehat{\sigma}_{xx} = \frac{1}{n}\sum_{i=1}^{n} (x_i - \widehat{\mu}_x)^2, \qquad \widehat{\sigma}_{yy} = \frac{1}{n}\sum_{i=1}^{n} \left(y_i - \widehat{\mu}_y\right)^2, \tag{2.28}$$
$$\widehat{\sigma}_{xy} = \frac{1}{n}\sum_{i=1}^{n} (x_i - \widehat{\mu}_x)\left(y_i - \widehat{\mu}_y\right),$$
$$\widehat{\rho} = \frac{\widehat{\sigma}_{xy}}{\left(\widehat{\sigma}_{xx}\widehat{\sigma}_{yy}\right)^{\frac{1}{2}}}.$$

However, with the present pattern of missing records, the likelihood is proportional to

$$p\left(y, x|\mu_y, \mu_x, \rho, \sigma_{yy}, \sigma_{xx}\right) =$$
$$\left(2\pi\sigma_{yy}\sigma_{xx}\left(1-\rho^2\right)\right)^{-\frac{m}{2}} \exp\left[-\frac{1}{2\left(1-\rho^2\right)}\left(\frac{\sum_{i=1}^{m}(x_i-\mu_x)^2}{\sigma_{xx}}\right.\right.$$
$$\left.\left. + \frac{\sum_{i=1}^{m}\left(y_i-\mu_y\right)^2}{\sigma_{yy}} - 2\rho\frac{\sum_{i=1}^{m}(x_i-\mu_x)\left(y_i-\mu_y\right)}{\left(\sigma_{xx}\sigma_{yy}\right)^{\frac{1}{2}}}\right)\right] (2\pi\sigma_{xx})^{-\frac{(n-m)}{2}}$$
$$\exp\left[-\frac{\sum_{i=m+1}^{n}(x_i-\mu_x)^2}{2\sigma_{xx}}\right]. \tag{2.29}$$

The likelihood equations derived from this likelihood function are difficult to solve analytically.

Rather than expressing the likelihood as in (2.29), an alternative strategy can be followed. The normal density of the joint distribution (2.26) can be factorised as the product of two normal densities:

$$p\left(y_i, x_i|\mu_y, \mu_x, \sigma_{xy}, \sigma_{yy}, \sigma_{xx}\right) = p\left(x_i|\mu_x, \sigma_{xx}\right) p\left(y_i|x_i, \mu_y, \mu_x, \sigma_{xy}, \sigma_{yy}, \sigma_{xx}\right). \tag{2.30}$$

The parameters of the two terms in the right-hand side of (2.30) are not distinct because (μ_x, σ_{xx}) are common to both factors. Therefore, a global maximisation of the likelihood requires that the two terms of the right-hand side of (2.30) are maximised jointly. The resulting algebra is quite messy.

Considering all the records, the first term in the right hand side of (2.30) is

$$(2\pi\sigma_{xx})^{-\frac{n}{2}} \exp\left[-\frac{\sum_{i=1}^{n}(x_i-\mu_x)^2}{2\sigma_{xx}}\right] \tag{2.31}$$

and the second term is (2.30) and

$$\left(2\pi\sigma_{y.x}\right)^{-\frac{m}{2}} \exp\left[-\frac{\sum_{i=1}^{m}\left(y_i-\mu_y-\frac{\sigma_{xy}}{\sigma_{xx}}(x_i-\mu_x)\right)^2}{2\sigma_{y.x}}\right], \tag{2.32}$$

where

$$\sigma_{y.x} = \sigma_{yy} - \frac{\left(\sigma_{xy}\right)^2}{\sigma_{xx}},$$

is the variance of the conditional distribution of y given x. The mean and variance of the distribution $[x_i|\mu_x, \sigma_{xx}]$ are then μ_x and σ_{xx}, respectively, and the corresponding parameters of $\left[y_i|x_i, \mu_y, \mu_x, \sigma_{xy}, \sigma_{yy}, \sigma_{xx}\right]$ are

$$\mathrm{E}\left(y_i|x_i, \mu_y, \mu_x, \sigma_{xy}, \sigma_{yy}, \sigma_{xx}\right) = \mu_y + \frac{\sigma_{xy}}{\sigma_{xx}}(x_i-\mu_x),$$

$$\mathrm{Var}\left(y_i|x_i, \mu_y, \mu_x, \sigma_{xy}, \sigma_{yy}, \sigma_{xx}\right) = \sigma_{yy} - \frac{\left(\sigma_{xy}\right)^2}{\sigma_{xx}}.$$

Consider using the factorised likelihood (2.30) but rather than parametrising with $\theta = \left(\mu_y, \mu_x, \sigma_{xy}, \sigma_{yy}, \sigma_{xx}\right)$, use instead $\phi = \left(\mu_x, \sigma_{xx}, \beta_0, \beta_1, \sigma_{y.x}\right)$, where (μ_x, σ_{xx}) are common to both parametrisations and the other components of ϕ are given by the following functions of the components of θ:

$$\beta_0 = \mu_y - \beta_1\mu_x,$$

$$\beta_1 = \frac{\sigma_{xy}}{\sigma_{xx}},$$

$$\sigma_{y.x} = \sigma_{yy} - \frac{\left(\sigma_{xy}\right)^2}{\sigma_{xx}}.$$

Under this parametrisation, the density of the pair (y_i, x_i) becomes

$$p\left(y_i, x_i|\mu_y, \mu_x, \sigma_{xy}, \sigma_{yy}, \sigma_{xx}\right) = p\left(x_i|\mu_x, \sigma_{xx}\right) p\left(y_i|x_i, \beta_0, \beta_1, \sigma_{y.x}\right). \tag{2.33}$$

The form of the first term in the right-hand side of (2.33) is the same as in (2.31) and the form of the second term is

$$p\left(y_i|x_i,\beta_0,\beta_1,\sigma_{y.x}\right)=\left(2\pi\sigma_{y.x}\right)^{-\frac{m}{2}}\exp\left[-\frac{\sum_{i=1}^{m}\left(y_i-\beta_0-\beta_1x_i\right)^2}{2\sigma_{y.x}}\right]. \tag{2.34}$$

The mean and variance of the distribution corresponding to (2.34) are

$$\begin{aligned}\mathrm{E}\left(y_i|x_i,\beta_0,\beta_1\right)&=\beta_0+\beta_1x_i \qquad \left(=\mu_y+\beta_1\left(x_i-\mu_x\right)\right),\\ \mathrm{Var}\left(y_i|x_i,\sigma_{y.x}\right)&=\sigma_{y.x}.\end{aligned}$$

The parameters in the two terms of the right-hand side of (2.33) are distinct; the global maximisation of the likelihood can proceed by maximising the likelihood of each term separately. Each of the two terms corresponds to the likelihood of a straightforward problem: $p\left(x_i|\mu_x,\sigma_{xx}\right)$ is the likelihood of $N\left(\mu_x,\sigma_{xx}\right)$, the marginal distribution of x_i, and $p\left(y_i|x_i,\beta_0,\beta_1,\sigma_{y.x}\right)$ is the likelihood of $N\left(\beta_0+\beta_1x_i,\sigma_{y.x}\right)$, the likelihood of the regression of father on son. This development was apparently first given by Anderson (1957).

2.6 Example: Likelihood Inferences Using Selected Records

Consider the following problem discussed in Lush and Shrode (1950) and in Henderson et al (1959). One is to obtain an estimate of the difference in milk production between years. The data on which inferences are to be based are selected: second year records are obtained only from those cows that had the highest first year records. The lowest producers of first year records are not allowed to produce in year 2 and are culled from the herd. The pattern of missing records in this example is identical to the pattern displayed in (2.25): the missing data mechanism is only a function of the observed data (the first lactation records in this example) and therefore the data are *missing at random* (Rubin 2002) and the selection process does not have to be incorporated as part of the likelihood model; the selection mechanism is ignorable. A heuristic conditional argument provides an alternative explanation. Given a data structure of the form in (2.25), the likelihood can be factorised as

$$\begin{aligned}&p\left(x_1,x_2,\ldots,x_m,x_{m+1},\ldots,x_n,y_1,y_2,\ldots,y_m;\theta\right)\\ &=p\left(y_1,y_2,\ldots,y_m|x_1,x_2,\ldots,x_m;\theta_1\right)p\left(x_1,x_2,\ldots,x_m,x_{m+1},\ldots,x_n;\theta_2\right),\end{aligned}$$

where the θ's are parameters of the respective densities. The conditional likelihood, given x (the first term on the right-hand side), is unaffected by selection on x and the second term includes all the parental data. Therefore, selection is ignorable.

Let y_{i1} and y_{i2} be age-adjusted records of the ith cow in years 1 and 2, respectively, with

$$y_{i1} = \mu + c_i + e_{i1}, \quad i = 1, 2, \ldots, N \tag{2.35}$$

and

$$y_{i2} = \mu + c_i + \delta + e_{i2}, \quad i = 1, 2, \ldots, n, \tag{2.36}$$

where c_i is an effect of cow assumed $N\left(0, \sigma_c^2\right)$, e_{i1} and $e_{i2} \sim N\left(0, \sigma_e^2\right)$ and δ is the difference between second and first year records that one wishes to estimate. There are N cows that have a first lactation record and n have a second lactation record. In the absence of culling, $n = N$. With culling based on first lactation information, $n < N$.

From the models,

$$\begin{aligned}
\mathrm{E}\left(y_{i1}\right) &= \mu \\
\mathrm{E}\left(y_{i2}\right) &= \mu + \delta \\
\mathrm{E}\left(y_{i2}\right) - \mathrm{E}\left(y_{i1}\right) &= \delta \\
\mathrm{Var}\left(y_{i1}\right) = \mathrm{Var}\left(y_{i2}\right) &= \sigma^2 = \sigma_c^2 + \sigma_e^2 \\
\mathrm{Cov}\left(y_{i1}, y_{i2}\right) &= \sigma_c^2 = \rho\sigma^2 \\
\mathrm{Corr}\left(y_{i1}, y_{i2}\right) = \rho &= \frac{\mathrm{Cov}\left(y_{i1}, y_{i2}\right)}{\sqrt{\mathrm{Var}\left(y_{i1}\right)}\sqrt{\mathrm{Var}\left(y_{i2}\right)}} = \frac{\sigma_c^2}{\sigma^2}.
\end{aligned}$$

Above, ρ is the repeatability or correlation between y_{i1} and y_{i2}. Assume that, in the absence of selection, y_{i1} and y_{i2} are bivariate normally distributed. Then,

$$\begin{bmatrix} y_{i1} \\ y_{i2} \end{bmatrix} \sim N\left[\begin{bmatrix} \mu \\ \mu + \delta \end{bmatrix}, \begin{bmatrix} \sigma^2 & \rho\sigma^2 \\ \rho\sigma^2 & \sigma^2 \end{bmatrix}\right]. \tag{2.37}$$

Therefore,

$$y_{i2}|y_{i1} \sim N\left[\mu + \delta + \rho\left(y_{i1} - \mu\right), \sigma^2\left(1 - \rho^2\right)\right] \tag{2.38}$$

and

$$y_{i1} \sim N\left(\mu, \sigma^2\right). \tag{2.39}$$

If the conditional expectation of a second year record, given the first year record

$$\mathrm{E}\left(y_{i2}|y_{i1}\right) = \mu + \delta + \rho\left(y_{i1} - \mu\right), \tag{2.40}$$

is averaged over all cows that have a year 1 record

$$\begin{aligned}\mathrm{E}_{y_{i1}}\left[\mathrm{E}\left(y_{i2}|y_{i1}\right)\right] &= \mathrm{E}\left(y_{i2}\right) = \mu + \delta + \rho\, \mathrm{E}_{y_{i1}}\left[\left(y_{i1} - \mu\right)\right] \\ &= \mu + \delta,\end{aligned} \tag{2.41}$$

as expected from the model. However, only the best cows in year 1 are allowed to produce in year 2. This means that, with selection, the expected value of year 1 records is not μ but

$$\mathrm{E}\left(y_{i1}|y_{i1} \in S\right) = \mu_s \neq \mu, \tag{2.42}$$

where S is the set of cows with records in year 1 that are allowed to produce a second record in year 2 and μ_s is the mean of the selected records of year 1. Then we have

$$\begin{aligned}\mathrm{E}_{y_{i1}}\left[\mathrm{E}\left(y_{i2}|y_{i1} \in S\right)\right] &= \mu + \delta + \rho\, \mathrm{E}_{y_{i1}}\left[\left(y_{i1} - \mu\right)|y_{i1} \in S\right] \\ &= \mu + \delta + \rho\left(\mu_s - \mu\right).\end{aligned} \tag{2.43}$$

Estimation by Least Squares

This section presents properties of the least squares estimator of δ, the difference between milk production in year 1 and 2. A least squares estimator of this difference, including only cows that have a record on both years, is

$$\widehat{d_1} = \overline{y}_{.2} - \overline{y}_{.1}, \quad y_{i1} \in S.$$

The expected value of $\widehat{d_1}$ is

$$\begin{aligned}\mathrm{E}\left(\widehat{d_1}\right) &= \mathrm{E}_{y_{i1}}\left[\mathrm{E}\left(\overline{y}_{.2} - \overline{y}_{.1}|y_{i1} \in S\right)\right] = \left(\mu + \delta + \rho\left(\mu_s - \mu\right)\right) - \mu_s \\ &= \delta - (1 - \rho)\left(\mu_s - \mu\right).\end{aligned} \tag{2.44}$$

Since $\mu_s > \mu$, the estimator of the difference is biased downwards. With selected data, least squares estimates of year effects based on (2.44) give the impression of negative environmental trend with time.

Another possibility would be to estimate δ by taking the average of all year 2 records minus all year 1 records. The expected value of this second estimator is

$$\begin{aligned}\mathrm{E}\left(\widehat{d_2}\right) &= \mathrm{E}_{y_{i1}}\left[\mathrm{E}\left(\overline{y}_{.2}|y_{i1} \in S\right)\right] - \mathrm{E}\left(\overline{y}_{.1}\right) = \left(\mu + \delta + \rho\left(\mu_s - \mu\right)\right) - \mu \\ &= \delta + \rho\left(\mu_s - \mu\right),\end{aligned} \tag{2.45}$$

which is biased unless $\rho = 0$, in which case selection would not be effective in changing milk production.

The conclusion is that with this type of selection (culling type), the least squares estimate of the difference in milk production is biased. On the other hand, BLUP (best linear unbiased predictor) yields unbiased estimates of genetic and environmental trends under certain types of selection. The paper where this is discussed at length is Henderson (1975). The following subsection describes in a simplified manner estimation of year effects using maximum likelihood. An important paper on the topic is the one by Curnow (1961).

Estimation by Maximum Likelihood

The parameters of the model defined by (2.38) and (2.39) are

$$\theta' = \left(\mu, \delta, \sigma^2, \rho\right).$$

However, in this simplified problem, it will be assumed that ρ and σ^2 are known. In view of (2.38) and (2.39), the likelihood is

$$L(\theta|\mathbf{y}) \propto \left(2\pi\sigma^2\right)^{-\frac{N}{2}} \exp\left[-\frac{\sum_{i=1}^{N}(y_{i1}-\mu)^2}{2\sigma^2}\right]$$
$$\times \left(2\pi\left(1-\rho^2\right)\sigma^2\right)^{-\frac{n}{2}} \exp\left[-\frac{\sum_{i=1}^{n}(y_{i2}-\mu-\delta-\rho(y_{i1}-\mu))^2}{2\left(1-\rho^2\right)\sigma^2}\right]. \quad (2.46)$$

With ρ and σ^2 assumed known, the loglikelihood, apart from an additive constant, is

$$l(\mu,\delta|\mathbf{y}) = -\frac{\left(1-\rho^2\right)\sum_{i=1}^{N}(y_{i1}-\mu)^2 + \sum_{i=1}^{n}(y_{i2}-\mu-\delta-\rho(y_{i1}-\mu))^2}{2\left(1-\rho^2\right)\sigma^2}. \quad (2.47)$$

This loglikelihood can be written as

$$l(\mu,\delta|\mathbf{y}) = k - \frac{N(\overline{y}-\mu)^2}{2\sigma^2}$$
$$-\frac{n\left(\overline{y}_2-\mu-\delta-\rho\left(\overline{y}_1-\mu\right)\right)^2}{2\left(1-\rho^2\right)\sigma^2}, \quad (2.48)$$

where k is a constant that does not depend on (μ, δ) and

$$\overline{y} = \frac{\sum_{i=1}^{N} y_{i1}}{N}$$
$$\overline{y}_1 = \frac{\sum_{i=1}^{n} y_{i1}}{n}$$
$$\overline{y}_2 = \frac{\sum_{i=1}^{n} y_{i2}}{n}.$$

With ρ and σ^2 assumed known, the loglikelihood (2.48) is a function of μ and δ only. To obtain the maximum likelihood estimates of μ and δ, (2.48) must be maximised with respect to μ and δ. One way of doing this is to take partial derivatives of (2.48) with respect to μ and δ, to set the resulting two equations equal to zero and to solve for μ and δ. The resulting maximum likelihood estimators of μ and δ are

$$\widehat{\mu} = \overline{y}, \tag{2.49}$$

and

$$\widehat{\delta} = \overline{y}_2 - \overline{y} - \rho\left(\overline{y}_1 - \overline{y}\right). \tag{2.50}$$

If all first lactation cows produce a second lactation (i.e. no selection), then $\overline{y}_1 = \overline{y}$ and the maximum likelihood estimator of δ reduces to $\overline{y}_2 - \overline{y}$, the difference between mean production in years 2 and 1.

To study whether (2.49) and (2.50) are biased, I compute their expectations. From (2.39),

$$\mathrm{E}\left(\widehat{\mu}\right) = \mathrm{E}\left(\overline{y}\right) = \mu. \tag{2.51}$$

Using (2.42), (2.51) and (2.43), the expected value of (2.50) is

$$\begin{aligned} \mathrm{E}\left(\widehat{\delta}\right) &= \mathrm{E}_{y_{i1}}\left[\mathrm{E}\left(\overline{y}_2|y_{i1} \in S\right)\right] - \mathrm{E}\left(\overline{y}\right) - \rho\left[\mathrm{E}\left(\overline{y}_1|y_{i1} \in S\right) - \mathrm{E}\left(\overline{y}\right)\right] \\ &= \mu + \delta + \rho\left(\mu_s - \mu\right) - \mu - \rho\left(\mu_s - \mu\right) \\ &= \delta. \end{aligned} \tag{2.52}$$

Therefore, the maximum likelihood estimators of μ and δ are unbiased by selection of the records, despite the fact that the selection mechanism is not accounted for in the computation of the likelihood (2.47). This holds in the present example because the data are *missing at random* in the sense defined by Rubin (1976) and therefore the selection mechanism is *ignorable*. If this is not the case, inferences based on the likelihood require incorporating the selection process as part of the model. This is illustrated in the next example.

2.7 Example: The Likelihood Function with Truncated Data

Consider data collected on n individuals with the objective of inferring the mean and variance of some trait. For some reason beyond the control of the analyst, only those individuals that exceed a known threshold become available for analysis. Can the mean and variance of the (unselected) trait be correctly inferred using these data? The example is presented in a stylised fashion, but the data structure is not uncommon for field data in animal breeding, where records of culled individuals are typically excluded.

As a reminder, let X have probability density function $p(x)$ and cumulative distribution function $F(x) = \Pr(X \leq x)$. Let a and b be constants lying within the support of the distribution of X (the support of the distribution is the set of values of x where the pdf (or pmf) is positive). Then,

$$p(x|a < X \leq b) = \frac{p(x)}{F(b) - F(a)}, \tag{2.53a}$$

$$p(x|X > a) = \frac{p(x)}{1 - F(a)}, \tag{2.53b}$$

$$p(x|X \leq b) = \frac{p(x)}{F(b)}. \tag{2.53c}$$

In each case, the unconditional density is adjusted by a scaling constant so that the conditional density still integrates to unity. In other words, the selection mechanism is taken into account.

With this detail in place, consider data vector $Y = (Y_o, Y_m)$ of length n from a normal distribution with mean μ and variance σ^2 that consist of r observed records Y_o and $n-r$ missing records Y_m. The $n-r$ missing records were discarded because they were smaller than an observed threshold T. In this particular case, the missing data mechanism involves the complete data (the observed and the missing data). Therefore, the missing data mechanism must be incorporated in the likelihood for correct inferences.

The contribution to the likelihood function from each element of Y_o is

$$L_{Y_o}\left(\mu, \sigma^2 | y_i > T\right) \propto \frac{p\left(y_i|\mu, \sigma^2\right)}{\int_T^\infty p\left(y_i|\mu, \sigma^2\right) dy_i} = \frac{p\left(y_i|\mu, \sigma^2\right)}{1 - \Phi(z)}, \quad i = 1, \ldots, r;$$

$$z = \frac{T - \mu}{\sigma},$$

where $p\left(y_i|\mu, \sigma^2\right)$, in this example, is the density of $N\left(\mu, \sigma^2\right)$ and $\Phi(\cdot)$ is the distribution function (or cumulative distribution function) of the standard normal.

Concerning the $n-r$ missing records, the only information available is that they are smaller than T. Therefore, the contribution to the likelihood function from each missing record is

$$L_{Y_m}\left(\mu,\sigma^2|y_i<T\right)=\Pr\left(y_i<T|\mu,\sigma^2\right)=\Phi\left(z\right),\quad i=r+1,\ldots,n.$$

Assuming the records are *iid*, after ordering, the likelihood becomes

$$\prod_{i=1}^{r} L_{Y_o}\left(\mu,\sigma^2|y_i\geq T\right)\prod_{i=r+1}^{n} L_{Y_m}\left(\mu,\sigma^2|y_i<T\right)\propto$$
$$\frac{\prod_{i=1}^{r} p\left(y_i|\mu,\sigma^2\right)}{\left(1-\Phi\left(z\right)\right)^r}\left(\Phi\left(z\right)\right)^{n-r}, \tag{2.54}$$

where z is a function of μ and σ. Maximisation over μ and σ^2 provides inferences of the base (untruncated) population parameters. This would not be the case if the selection mechanism had been ignored and if inferences of μ and σ^2 had instead been based on the likelihood constructed assuming $Y_o \sim N\left(\mu,\sigma^2\right)$. The resulting estimators would be biased.

Likelihood (2.54) is a nonlinear function of μ and σ^2 and no analytic form for the estimators exists. However, solutions using the method of moments, the EM (*expectation maximisation*) algorithm or McMC are easily available as shown in the chapters ahead.

A slightly different situation arises if information is restricted to the r observed records only, known to have been selected because each was larger than T. There is no information about how many records were discarded. In this case, the likelihood is equal to (2.54) with the second term $\left(\Phi\left(z\right)\right)^{n-r}$ omitted.

The Exercises section, pages 547 and 592, introduces a censored model where data are drawn from an exponential distribution.

2.8 Example: The Likelihood Function of a Genomic Model

So far no distinction has been made between the model assumed to generate the data, the *true model*, and the model used for analysis, the *operational model*. When the likelihood based on an operational model differs meaningfully from that based on the true model, the likelihood is *misspecified* and **inferences** can be affected. I underline **inferences** because in the context of prediction, this is not necessarily a problem. The true model may constitute a poorer prediction machinery than an operational alternative.

I revisit (see page 39) a model that is used repeatedly in this book: the *genomic model* that has its origins in the influential work of Meuwissen et al (2001). Advances in molecular genetic techniques have allowed genotyping vast numbers of markers spread across the whole genome of individuals. These genetic marker loci are not necessarily causal but are correlated (they are in *linkage disequilibrium*) with the causal loci, the so-called quantitative trait loci (QTL). The latter are typically unobserved, and the genetic markers are observed and can provide information about the genetics of a trait.

Genetic marker information on a large number of loci allows the assessment of kinship (additive genetic relationships) among nominally unrelated individuals (meaning distantly related). However, due to Mendelian sampling, the patterns of allele sharing at markers and at causal loci may be very different (Hill and Weir 2011). As a consequence, the variance structure specified by the operational model can differ from that of the true model, leading to misspecification. This disparity between the true model and the operational model may affect inferences but is not necessarily a problem in the context of prediction. The topic is elaborated in de los Campos et al (2015).

With this warning in mind, I introduce the (operational) likelihood function of a genomic model. Consider, first in general terms, the following mixed model:

$$y|\alpha, g, \sigma_e^2 \sim N\left(Z\alpha + g, I\sigma_e^2\right), \tag{2.55a}$$

$$g|G, \sigma_g^2 \sim N\left(0, G\sigma_g^2\right). \tag{2.55b}$$

The first line specifies that the conditional distribution of the data y (vector of order $n \times 1$), given all the parameters, is normal with conditional mean $Z\alpha + g$ and conditional variance $I\sigma_e^2$. The scalar σ_e^2 is the residual variance, Z is an observed matrix and α is an unobserved vector of systematic effects. The second line describes the model for the random variable g given parameters. This is again normal with mean 0 and variance $G\sigma_g^2$. The system described by (2.55) is an example of a hierarchical model. Vector g could represent genomic or additive genetic values, G could represent a genomic or additive genetic relationship matrix and σ_g^2, a scalar, can be a genomic or an additive genetic variance component. When g is a vector of additive genetic values, taken to be equal to the sum of the contributions from a very large number of unobserved loci, the model is known as the *infinitesimal model* (Bulmer 1980). This has been the standard model for analysis of many quantitative traits in the pre-genomics era. Matrix G specifies the conditional expected value of allele sharing between individuals, given an observed pedigree, also known as the additive genetic relationship matrix.

In the genomic model, where g is a vector of genomic values, it takes the form $g = Wb$, where b is a vector of p unobserved genetic marker effects, often assigned the normal distribution:

$$b|\sigma_b^2 \sim N\left(0, I\sigma_b^2\right).$$

The scalar parameter σ_b^2 can be interpreted as the prior variance associated with each element of the $p \times 1$ vector b. The observed matrix W, of order $n \times p$, consists of marker labels X_{ij} (often centred and scaled, in which case the rank of W is at most $n-1$ if $p > n$):

$$W_{ij} = \frac{X_{ij} - \mathrm{E}\left(X_{ij}\right)}{\sqrt{\mathrm{Var}\left(X_{ij}\right)}}.$$

When the rank of W is smaller than n, g in (2.55b) has a singular normal distribution. The random variable X_{ij} can take values $0, 1, 2$ according to the number of the arbitrarily chosen allele of marker locus j of individual i. Therefore, $\mathrm{E}\left(W_{ij}\right) = 0$, $\mathrm{Var}\left(W_{ij}\right) = 1$. Matrix W is a random variable that takes a particular realised value in the data at hand. Inferences based on genomic models are typically conditional on W. Parametrising in terms of g rather than in terms of b is useful when $p > n$.

The vector g of genomic values is a proxy for the true additive genetic values determined by the causal QTL. The conditional variance of g given W is

$$\begin{aligned}\mathrm{Var}\left(g|W\right) &= WW'\sigma_b^2 \\ &= \frac{1}{m}WW'\sigma_g^2,\end{aligned}$$

where the genomic relationship matrix $G_g = (1/m)\, WW'$ is the average (over marker loci) realised observed matrix of genetic relationships among the n individuals, and $\sigma_g^2 = m\sigma_b^2$ is the unconditional (with respect to W) variance of an element of g, interpreted as the amount of additive genetic variance captured by marker loci, or *genomic variance* at the level of the operational model. The equality $\sigma_g^2 = m\sigma_b^2$ stems from the fact that

$$\begin{aligned}\mathrm{Var}\left(g_i\right) &= \mathrm{E}\left[\mathrm{Var}\left(g_i|W_i\right)\right] + \mathrm{Var}\left[\mathrm{E}\left(g_i|W_i\right)\right] \\ &= \mathrm{E}\left[\mathrm{Var}\left(g_i|W_i\right)\right]\end{aligned}$$

because $\mathrm{E}\left(g_i|W_i\right) = 0$. Labelling W_i' as the ith row of matrix W, the ith diagonal term WW' is $W_i'W_i = \sum_{j=1}^{p} W_{ij}^2$. Then

$$\mathrm{Var}\left(g_i\right) = \mathrm{E}\left[\sum_{j=1}^{p} W_{ij}^2\right] = m\sigma_b^2 = \sigma_g^2$$

because $\mathrm{E}\left(W_{ij}^2\right) = 1$. A *genomic heritability* or proportion of variance accounted for by the genetic markers is defined as

$$h_g^2 = \frac{\sigma_g^2}{\sigma_g^2 + \sigma_e^2}. \tag{2.56}$$

The construction of the likelihood involves integration over g:

$$L\left(\alpha, \sigma_g^2, \sigma_g^2 | y, G_g\right) = \int_g p\left(y|\alpha, g, \sigma_e^2\right) p\left(g|G_g, \sigma_g^2\right) dg, \tag{2.57}$$

which in view of the normality assumptions is the likelihood of the normal distribution:

$$N\left(Z\alpha, G_g\sigma_g^2 + I\sigma_e^2\right).$$

In a classical likelihood setup, the g's are not "parameters" but latent variables that do not feature in the likelihood. The only parameters are α, σ_g^2 and σ_e^2.

In this example, the likelihood is explicit but the maximum likelihood estimators do not have a closed form. Numerical methods are needed to obtain solutions. This is the subject of the next chapter. In most non-Gaussian setups, the integration in (2.57) cannot be written in closed form, and therefore neither the likelihood function nor the maximum likelihood estimators can be obtained explicitly.

Chapter 3
Computing the Likelihood

Estimation using the likelihood function proceeds by solving for θ the equation $S(\theta) = 0$ where $S(\theta)$ is the score. In many cases, there may be no explicit solution, either because the system of equations is not linear or because the likelihood cannot be written explicitly. Therefore, numerical methods must be employed to obtain the maximum likelihood estimates. There is a large number of optimisation algorithms for nonlinear problems (see, e.g. Dahlquist and Björck 1974). The chapter provides an outline of two classical approaches that are used for fitting models using maximum likelihood and a third one that is often used in high-dimensional settings such as neural networks. The first two are Newton-Raphson and the the EM algorithm and the third is gradient descent. Examples illustrate implementation of the methods.

3.1 Newton-Raphson and the Method of Scoring

Newton-Raphson is a general procedure to solve $g(x) = 0$. Newton-Raphson finds the root of a function and therefore it is applied to the derivative of the loglikelihood. To illustrate, assume that x is a scalar. The starting point is a Taylor expansion around an initial estimate of x labelled x_0

$$g(x) \approx g(x_0) + g'(x_0)(x - x_0) = 0, \tag{3.1}$$

where

$$g'(x_0) = \left. \frac{dg(x)}{dx} \right|_{x=x_0}.$$

D. Sorensen, *Statistical Learning in Genetics*, Statistics for Biology and Health,
https://doi.org/10.1007/978-3-031-35851-7_3

From (3.1),

$$x = x_0 - \frac{g(x_0)}{g'(x_0)}$$

and the iterative process is

$$x_{t+1} = x_t - \frac{g(x_t)}{g'(x_t)}. \tag{3.2}$$

ML estimation involves finding the value of θ that satisfies $S(\theta) = 0$, where θ is a vector of p parameters and $S(\theta)$ is the score function, the first derivative of the loglikelihood with respect to the parameters. The vectorial expression equivalent to (3.1) is

$$S(\theta) = S(\theta_0) + S'(\theta_0)(\theta - \theta_0) \tag{3.3}$$

where

$$S'(\theta_0) = \left.\frac{\partial^2 \ell(\theta|y)}{\partial\theta\partial\theta'}\right|_{\theta=\theta_0}.$$

The component ij of the $p \times p$ matrix of second derivatives $S'(\theta)$ (the *Hessian*) is

$$\frac{\partial^2 \ell(\theta|y)}{\partial\theta_i\partial\theta_j}$$

and $-S'(\theta_0) = I(\theta_0|y)$ is the *observed information*. After setting (3.3) equal to zero and solving for θ, the iterative process is

$$\theta_{t+1} = \theta_t - \left[S'(\theta_t)\right]^{-1} S(\theta_t). \tag{3.4}$$

The *method of scoring* replaces the inverse of the observed information in (3.4) by the *expected information* matrix and the iterative process is now

$$\theta_{t+1} = \theta_t - \left[\mathrm{E}_y\left(S'(\theta_t)\right)\right]^{-1} S(\theta_t). \tag{3.5}$$

The expected information is

$$\mathrm{E}_y\left(S'(\theta_t)\right) = -\left.\left[\mathrm{E}_y\left(\frac{\partial^2}{\partial\theta\partial\theta'}\ell(\theta|y)\right)\right]\right|_{\theta=\theta_t} = -\int \frac{\partial^2 \ell(\theta|y)}{\partial\theta\partial\theta'} p(y|\theta)\,dy.$$

Typically, the method of scoring requires fewer calculations in each iteration because many expressions vanish or simplify in the process of taking expectations.

However, it may converge at a slower rate. The two methods may not converge at all and, even if they do, there is no assurance that a global maximum would be reached. This is not surprising, as the methods search for stationarity without reference to the possible existence of multiple maxima. This is a potential problem in models having many parameters, and it is expected to occur more frequently when sample sizes are small, as the likelihood may have several "peaks and valleys". Both methods involve calculating the matrix of second derivatives of the loglikelihood.

A third variant (and there are others) of this iterative procedure avoids the use of second derivatives and is implemented with first derivatives only as

$$\theta_{t+1} = \theta_t - \left[\left(S(\theta)\, S(\theta)'\right)\right]^{-1} S(\theta_t). \tag{3.6}$$

This is known as the BHHH algorithm and was proposed by Berndt, Hall, Hall and Hausman (Berndt et al, 1974). The performance of this method can be more erratic than the previous two.

All these methods are based on a linear approximation to the score function that is equivalent to a quadratic approximation to the likelihood.

Example: Estimation of Gene Frequencies from ABO Blood Group Phenotypes

The ABO blood group antigens in man are encoded by one genetic locus on chromosome 9, the ABO locus that has three allelic forms: A, B and O. A child receives one of the three alleles from each parent, giving rise to six possible genotypes but only four phenotypic classes (blood types) can be observed.

Consider the blood group data in Table 3.1. The expected frequency of each genotype in the last column is derived assuming Hardy-Weinberg equilibrium. The problem at hand is to infer p_A, p_B and p_O, the frequency of alleles A, B and O, respectively, subject to the constraint $p_A + p_B + p_O = 1$. The observed data are (n_A, n_{AB}, n_B, n_O).

Before writing the likelihood corresponding to this problem, imagine that all six phenotypes were observed (one for each genotype). This would lead to the classical multinomial likelihood. To illustrate, assume that the complete data were instead

Table 3.1 Frequency of genotypes and phenotypes of ABO blood group data

Genotype	Phenotype	Observed counts	Frequency
AA	A	n_A	p_A^2
AO	A		$2p_Ap_O$
AB	AB	n_{AB}	$2p_Ap_B$
BB	B	n_B	p_B^2
BO	B		$2p_Bp_O$
OO	O	n_O	p_O^2

$(n_{AA}, n_{AO}, n_{AB}, n_{BB}, n_{BO}, n_{OO})$ associated with the six phenotypes and the total number of observations is

$$N = n_{AA} + n_{AO} + n_{AB} + n_{BB} + n_{BO} + n_{OO}.$$

The four phenotypic classes (blood types) are $n_A = n_{AA} + n_{AO}$, $n_{AB} = n_{AB}, n_B = n_{BB} + n_{BO}$ and $n_O = n_{OO}$. Under multinomial sampling, the joint probability function of the complete data is

$$\begin{aligned} &\Pr(n_{AA}, n_{AO}, n_{AB}, n_{BB}, n_{BO}, n_{OO} | p_A, p_B) \\ &= N! \frac{(p_A^2)^{n_{AA}}}{n_{AA}!} \frac{(2p_A p_O)^{n_{AO}}}{n_{AO}!} \frac{(2p_A p_B)^{n_{AB}}}{n_{AB}!} \frac{(p_B^2)^{n_{BB}}}{n_{BB}!} \frac{(2p_B p_O)^{n_{BO}}}{n_{BO}!} \frac{(p_O^2)^{n_{OO}}}{n_{OO}!}. \end{aligned}$$

The likelihood function is

$$\begin{aligned} &L(p_A, p_B | n_{AA}, n_{AO}, n_{AB}, n_{BB}, n_{BO}, n_{OO}) \\ &\propto \left(p_A^2\right)^{n_{AA}} (2p_A p_O)^{n_{AO}} (2p_A p_B)^{n_{AB}} \left(p_B^2\right)^{n_{BB}} (2p_B p_O)^{n_{BO}} \left(p_O^2\right)^{n_{OO}}, \end{aligned} \tag{3.7}$$

and the loglikelihood, up to an additive constant, is

$$\begin{aligned} &l(p_A, p_B | n_{AA}, n_{AO}, n_{AB}, n_{BB}, n_{BO}, n_{OO}) \\ &= 2n_{AA} \ln(p_A) + n_{AO} \ln(2p_A p_O) + n_{AB} \ln(2p_A p_B) + 2n_{BB} \ln(p_B) \\ &\quad + n_{BO} \ln(2p_B p_O) + 2n_{OO} \ln(p_O). \end{aligned} \tag{3.8}$$

Substituting $p_O = 1 - p_A - p_B$, taking partial derivatives with respect to p_A and p_B and setting these equal to zero leads to the equations:

$$\frac{2n_{AA} + n_{AB} + n_{AO}}{p_A} - \frac{2n_{OO} + n_{AO} + n_{BO}}{1 - p_A - p_B} = 0,$$

$$\frac{2n_{BB} + n_{AB} + n_{BO}}{p_B} - \frac{2n_{OO} + n_{AO} + n_{BO}}{1 - p_A - p_B} = 0.$$

The unique solutions are the closed-form ML estimators:

$$\widehat{p}_A = \frac{2n_{AA} + n_{AB} + n_{AO}}{2N}, \tag{3.9a}$$

$$\widehat{p}_B = \frac{2n_{BB} + n_{AB} + n_{BO}}{2N}. \tag{3.9b}$$

In the case of the data in Table 3.1, n_{AO} and n_{BO} are not observed and the likelihood becomes nonlinear. Given the data in Table 3.1, the loglikelihood is

$$l(p_A, p_B|n_A, n_{AB}, n_B, n_O) \propto n_A \ln[p_A(2 - p_A - 2p_B)] + n_{AB} \ln[2p_A p_B] + n_B \ln[p_B(2 - p_B - 2p_A)] + 2n_O \ln[(1 - p_A - p_B)]. \tag{3.10}$$

Differentiating with respect to p_A and p_B yields the nonlinear system of equations:

$$\frac{\partial l(p_A, p_B|n_A, n_{AB}, n_B, n_O)}{\partial p_A} = \frac{n_{AB}}{p_A} + \frac{n_A(2 - 2p_A - 2p_B)}{p_A(2 - p_A - 2p_B)} - \frac{2n_B}{2 - 2p_A - p_B} - \frac{2n_O}{1 - p_A - p_B}, \tag{3.11}$$

$$\frac{\partial l(p_A, p_B|n_A, n_{AB}, n_B, n_O)}{\partial p_B} = \frac{n_{AB}}{p_B} + \frac{n_B(2 - 2p_A - 2p_B)}{p_B(2 - 2p_A - p_B)} - \frac{2n_A}{2 - p_A - 2p_B} - \frac{2n_O}{1 - p_A - p_B}. \tag{3.12}$$

A solution can be obtained using Newton-Raphson. This requires the following second derivatives:

$$\frac{\partial^2 l(p_A, p_B|n_A, n_{AB}, n_B, n_O)}{(\partial p_A)^2} = \frac{n_A(2 - 2p_A - 2p_B)}{p_A(2 - p_A - 2p_B)^2} - \frac{2n_A}{p_A(2 - p_A - 2p_B)} - \frac{n_A(2 - 2p_A - 2p_B)}{p_A^2(2 - p_A - 2p_B)} - \frac{n_{AB}}{p_A^2} - \frac{2n_O}{(1 - p_A - p_B)^2} - \frac{4n_B}{(2 - 2p_A - p_B)^2},$$

$$\frac{\partial^2 l(p_A, p_B|n_A, n_{AB}, n_B, n_O)}{(\partial p_B)^2} = \frac{n_B(2 - 2p_A - 2p_B)}{p_B(2 - 2p_A - p_B)^2} - \frac{2n_B}{p_B(2 - 2p_A - p_B)} - \frac{n_B(2 - 2p_A - 2p_B)}{p_B^2(2 - 2p_A - p_B)} - \frac{n_{AB}}{p_B^2} - \frac{2n_O}{(1 - p_A - p_B)^2} - \frac{4n_A}{(2 - p_A - 2p_B)^2},$$

$$\frac{\partial^2 l(p_A, p_B|n_A, n_{AB}, n_B, n_O)}{\partial p_A \partial p_B} = -\frac{2n_A}{(2 - p_A - 2p_B)^2} - \frac{2n_B}{(2 - 2p_A - p_B)^2} - \frac{2n_O}{(1 - p_A - p_B)^2}.$$

Suppose the data are $n_A = 725$, $n_{AB} = 72$, $n_B = 258$, $n_O = 1073$. This example is discussed in Weir (1996) where references to the original source of the data can be found. Using these expressions in (3.4) yields, at convergence, the ML estimates: $\widehat{p}_A = 0.2091$ and $\widehat{p}_B = 0.0808$. The observed information matrix evaluated at the ML estimates is

$$I\left(\widehat{p}_A, \widehat{p}_B | n_A, n_{AB}, n_B, n_O\right) = \begin{bmatrix} 23,210.6 & 5031.56 \\ 5031.56 & 56,008.53 \end{bmatrix},$$

resulting in an estimate of the asymptotic covariance matrix equal to

$$\begin{aligned} \text{Var}\left(\widehat{p}_A, \widehat{p}_B | n_A, n_{AB}, n_B, n_O\right) &= \left[I\left(\widehat{p}_A, \widehat{p}_B | n_A, n_{AB}, n_B, n_O\right)\right]^{-1} \\ &= 10^{-6} \begin{bmatrix} 43.939 & -3.947 \\ -3.947 & 18.209 \end{bmatrix}. \end{aligned}$$

The R-code for the Newton-Raphson computations, spelled out line by line, is as follows:

```
# CODE0301
rm(list=ls())
set.seed(30371)
fd<-matrix(data=NA,nrow=2,ncol=1)
sd<-matrix(data=NA,nrow=2,ncol=2)
freq<-matrix(data=NA,nrow=2,ncol=1)
niter<-20
# DATA
n_A<-725
n_AB<-72
n_B<-258
n_O<-1073
# INITIALISE GENE FREQ
freq[1,1]<-0.3
freq[2,1]<-0.2
p_A<-freq[1,1]
p_B<-freq[2,1]
# ITERATION LOOP
for (i in 1:niter){
  fd[1,1] <- n_AB/p_A+n_A*(2-2*p_A-2*p_B)/(p_A*(2-p_A-2*p_B))-
    2*n_B/(2-2*p_A-p_B)-2*n_O/(1-p_A-p_B)
  fd[2,1] <- n_AB/p_B+n_B*(2-2*p_A-2*p_B)/(p_B*(2-2*p_A-p_B))-
    2*n_A/(2-p_A-2*p_B) - 2*n_O/(1-p_A-p_B)
  s11a <- -n_AB/((p_A)^2)
  s11b <- (n_A*(2-2*p_A-2*p_B))/((p_A*(2-p_A-2*p_B)^2))
  s11c <- - (2*n_A)/((p_A*(2-p_A-2*p_B)))
  s11d <-  - (n_A*(2-2*p_A-2*p_B))/((p_A^2*(2-p_A-2*p_B)))
  s11e <- - (4*n_B)/((2-2*p_A-p_B)^2)
  s11f <- - (2*n_O)/((1-p_A-p_B)^2)
  sd[1,1] <- s11a + s11b + s11c + s11d + s11e + s11f
  s22a <- - n_AB/((p_B)^2)
  s22b <-  n_B*(2-2*p_A-2*p_B)/(((2-2*p_A-p_B)^2) * p_B)
  s22c <- - (2*n_B)/(p_B*(2-2*p_A-p_B))
  s22d <- - (n_B*(2-2*p_A-2*p_B))/((p_B^2*(2-2*p_A-p_B)))
  s22e <- -(4*n_A)/((2-p_A-2*p_B)^2)
```

```
  s22f <- -(2*n_O)/((1-p_A-p_B)^2)
  sd[2,2] <- s22a + s22b + s22c + s22d + s22e + s22f
  sd[1,2] <- -2*n_A/((2-p_A-2*p_B)^2) - 2*n_O/((1-p_A-p_B)^2) -
    2*n_B/((2-2*p_A-p_B)^2)
  sd[2,1] <- sd[1,2]
  freq<-freq-solve(sd)%*%fd
  p_A<-freq[1,1]
  p_B<-freq[2,1]
}
# ML ESTIMATES ARE
freq
```

```
##            [,1]
## [1,] 0.20913065
## [2,] 0.08080101
```

```
# OBSERVED INFORMATION IS
-sd
```

```
##           [,1]      [,2]
## [1,] 23210.614  5031.559
## [2,]  5031.559 56008.529
```

```
# ASYMPTOTIC VAR-COVAR MATRIX IS
-solve(sd)
```

```
##               [,1]          [,2]
## [1,]  4.393943e-05 -3.947325e-06
## [2,] -3.947325e-06  1.820903e-05
```

A computationally simpler alternative is to use the R function OPTIM to solve minimisation problems (to maximise a function such as the loglikelihood (3.10), the function multiplied by -1 must be supplied). The code is as follows:

```
# CODE0302
rm(list=ls()) # CLEAR WORKSPACE
fr<-function(par){
  p_A <- par[1]
  p_B <- par[2]
  -(725*log(p_A*(2 - p_A - 2* p_B)) + 72*log(2*p_A*p_B) +
  258*log(p_B*(2 - p_B - 2*p_A)) + 2*1073*log(1 - p_A - p_B))}
result <- optim(par=c(0.3,0.2),fr,hessian=TRUE)
result$par
```

```
## [1] 0.20913767 0.08081225
```

```
solve(result$hessian)
```

```
##                [,1]          [,2]
## [1,]  4.393974e-05 -3.947440e-06
## [2,] -3.947440e-06  1.820851e-05
```

```
-result$value
```

```
## [1] -2303.55
```

The value of the loglikelihood at convergence can be extracted from `OPTIM` and is equal to -2303.55.

Example: A Regression Model for Binary Data

Categorical data are ubiquitous in genetics and arise when the outcome is an assignment into one of several mutually exclusive classes. A distinction is made based on whether the classes are unordered or ordered. The ABO blood groups illustrate the former. Another example is hair colour. In these cases, there is no clear ordering from lowest to highest. In contrast, the degree of severity of a disease can be meaningfully classified into several categories, ranging from very low severity to very high severity.

Only two categories are assumed here, so the response is said to be binary. If y_i denotes the response, without loss of generality, the two possible values may be coded as 1 and 0. Therefore,

$$\mathrm{E}(y_i) = 1\,\Pr(y_i = 1) + 0\,\Pr(y_i = 0) = \Pr(y_i = 1). \tag{3.13}$$

This section describes methods to study how covariates or explanatory variables influence $\Pr(y_i = 1)$. Later sections will consider joint inferences of explanatory variables and of components of variance of random effects as factors related to $\Pr(y_i = 1)$. The variance of the random effects in such a mixed model could, for example, inform on the existence of genetic variation underlying disease data.

Assume that binary records are available on $i = 1, 2, \dots, N$ individuals $(y_1, x_1), (y_2, x_2), \dots, (y_N, x_N)$, where $y_i = 0, 1$, and x_i is a vector with p elements representing covariates. The objective is to construct a model to study the influence of the covariates on the observations y. A possible choice is the regression model

$$y_i = x_i'\beta + e_i \tag{3.14}$$

and to estimate the $p \times 1$ vector of regression coefficients β using least squares. The problem with this approach is as follows. Taking expected values of (3.14)

$$\mathrm{E}\,(y_i|x_i) = x_i'\beta.$$

From (3.13) this expectation is equal to $\Pr(y_i = 1|x_i)$. This probability must satisfy

$$0 \leq \Pr(y_i = 1|x_i) \leq 1 \tag{3.15}$$

but the prediction using $\mathrm{E}(y_i|x_i) = x_i'\beta$ may yield outcomes outside the constraint (3.15).

A way of building a model in which the constraint (3.15) is automatically satisfied is to use a transformation of the output y, generating a nonlinear relationship between it and the covariate. This is achieved by defining the probability that $y_i = 1$ as a nonlinear function of x_i of the form

$$\Pr(y_i = 1|x_i) = F\left(x_i'\beta\right), \tag{3.16}$$

where F is any distribution function (cumulative distribution function). Two common choices for F are the normal distribution, leading to the probit model, and the logistic distribution, leading to the logistic model.

For the probit model, $F = \Phi$ and

$$\Pr(y_i = 1|x_i) = \Phi\left(x_i'\beta\right), \tag{3.17}$$

where Φ is the standard normal integral

$$\Phi(t) = \frac{1}{\sqrt{2\pi}} \int_{-\infty}^{t} \exp\left(-\frac{1}{2}u^2\right) du.$$

The relationship (3.17) is linearised by the inverse normal transformation:

$$\Phi^{-1}\left(x_i'\beta\right) = x_i'\beta. \tag{3.18}$$

For the probit model,

$$\begin{aligned} F(t) &= \int_{-\infty}^{t} p(u)\, du = \int_{-\infty}^{t} \frac{\exp(u)}{\left[1 + \exp(u)\right]^2} du \\ &= \frac{\exp(t)}{1 + \exp(t)}, \end{aligned} \tag{3.19}$$

where $p(u)$ is the pdf of the standard logistic distribution, with $\mathrm{E}(u) = 0$ and $\mathrm{Var}(u) = \pi^2/3$. Therefore, for the logistic model,

$$\Pr(y_i = 1|x_i) = \frac{\exp\left(x_i'\beta\right)}{1 + \exp\left(x_i'\beta\right)}, \tag{3.20a}$$

$$\Pr(y_i = 0|x_i) = \frac{1}{1 + \exp\left(x_i'\beta\right)}. \tag{3.20b}$$

The relationship is linearised by the logit transformation:

$$\ln\left[\frac{\Pr(y_i = 1|x_i)}{\Pr(y_i = 0|x_i)}\right] = x_i'\beta. \tag{3.21}$$

The Liability Model

Ordered categorical data and binary data are often analysed using a threshold model, first used by Wright (1934) to study the number of digits in guinea pigs and introduced in human genetics in a more modern version by Falconer (1965). This model assumes that there is an underlying or latent unobservable variable, u, in genetics of disease often called the liability after Falconer (1965). The categories of response result from the value of u relative to a fixed unobserved threshold t. The liability u is often assumed to follow a normal distribution or a logistic distribution. The objective is to show that in the first case, the liability model is equivalent to the probit model (3.17) and in the second case to the logistic model (3.20).

Let the dichotomy be say "survival" versus "death". If $u > t$ then the individual survives and the binary variable (the observed datum) takes the value $y = 1$. If $u \le t$ the individual dies and the observed datum is $y = 0$. Denote the liability associated with datum i as u_i and suppose that it is related to an unknown parameter vector β of order $p \times 1$ via the linear structure:

$$u_i = x_i'\beta + e_i, \quad i = 1, 2, ..., N, \tag{3.22}$$

where x_i' is the ith row of a known $N \times p$ matrix x of explanatory variables (covariates) and e_i is a random residual with pdf $p(e_i)$. Assume that the residuals are independent and identically distributed. The probability of survival of individual i (which is the pmf of the random variable y_i) is

$$\begin{aligned}
\Pr(y_i = 1|\beta, x) &= \Pr(u_i > t|\beta, x) = \Pr\left(u_i - x_i'\beta > t - x_i'\beta|\beta, x\right) \\
&= \Pr\left(e_i > t - x_i'\beta|\beta, x\right) = \int_{t-x_i'\beta}^{\infty} p(e_i)\, de_i \\
&= \int_{-\infty}^{-(t-x_i'\beta)} p(e_i)\, de_i = \int_{-\infty}^{x_i'\beta - t} p(e_i)\, de_i.
\end{aligned} \tag{3.23}$$

The equality in the third line requires e_i to be symmetrically distributed around 0. The liabilities cannot be observed and a convenient origin is to set the value of the threshold t equal to 0. Hence, the scale is one of deviations from the threshold. This constraint makes the likelihood model identifiable and the Hessian becomes negative definite. Then,

$$\Pr(y_i = 1|\beta, x) = \int_{-\infty}^{x_i'\beta} p(e_i)\, de_i = F\left(x_i'\beta\right), \tag{3.24}$$

as in (3.16), where F is the cdf of the random variable e_i, indicating the equivalence between the liability model and the original formulation of the model (3.16).

The use of an underlying liability is a computational trick. However, the liability can have an intrinsic mechanistic significance. For example, in the case of twinning in cattle or humans, the liability can be thought of as levels of hormones, which play a central role in factors determining monozygotic twinning. The liability formulation has also advantages in Gibbs sampling computations and provides a simple framework for extensions of the model to include hierarchical structures and for analysis of ordered categorical traits.

A Digression on Parameter Interpretation

The liability u_i changes with x_i at a constant rate, but this is not so at the level of the probabilities. This is verified by noting that

$$\frac{\partial u_i}{\partial x_i} = \beta,$$

whereas from expression (3.24)

$$\begin{aligned}\frac{\partial \Pr(y=1|\beta,x)}{\partial x_i} &= \frac{\partial}{\partial x_i}\left[\int_{-\infty}^{x_i'\beta} p(e_i)\,de_i\right] \\ &= \frac{\partial}{\partial x_i'\beta}\left[\int_{-\infty}^{x_i'\beta} p(e_i)\,de_i\right]\frac{\partial x_i'\beta}{\partial x_i} \\ &= F\left(x_i'\beta\right)\beta. \qquad (3.25)\end{aligned}$$

The change is not constant and depends on the value of the explanatory vector x_i.

In the Bernoulli distribution, $\mathrm{E}(y|\beta,x) = \Pr(y=1|\beta,x)$. The model at the level of the expectation of y is nonlinear, in contrast with the standard linear regression model, where $\mathrm{E}(y|\beta,x) = \mu + \beta x$. According to (3.25), a unit change in the covariate x leads to a nonlinear change in the probability $\mathrm{E}(y|\beta,x) = \Pr(y=1|\beta,x)$.

To get a little insight into the meaning of β in the case of binary responses, consider the *odds ratio*

$$\frac{\Pr(y=1|\beta,x)}{\Pr(y=0|\beta,x)} = \exp\left(x'\beta\right) \qquad (3.26)$$

or the *logit* or *logodds*

$$\ln\left[\frac{\Pr(y=1|\beta,x)}{\Pr(y=0|\beta,x)}\right] = x'\beta, \qquad (3.27)$$

showing that β describes the linear change per unit change of x at the level of the logit. To be specific, imagine that y denotes disease or absence of disease and (x_1, x_2) denotes level of exposure to two conditions. Let $x_i'\beta = \mu + \beta_1 x_{1i} + \beta_2 x_{2i}$. Then in terms of (3.26),

$$\frac{\Pr(y_i = 1|\beta, x)}{\Pr(y_i = 0|\beta, x)} = \exp\left[(\mu + \beta_1 x_{1i} + \beta_2 x_{2i})\right]. \tag{3.28}$$

The parameter $\exp(\mu)$ is the odds of disease for an unexposed individual ($x = 0$).

How does the *odds ratio* change when one of the conditions, say x_2, changes by one unit?

$$\begin{aligned}\frac{\Pr(y_i = 1|\beta, x_i)}{\Pr(y_i = 0|\beta, x_i)} &= \exp\left[(\mu + \beta_1 x_{1i} + \beta_2 (x_{2i} + 1))\right] \\ &= \exp\left[(\mu + \beta_1 x_{1i} + \beta_2 x_{2i})\right] \exp(\beta_2).\end{aligned}$$

The increase in x_2 by one unit to $x_2 + 1$ while keeping x_1 fixed multiplies the odds ratio by $\exp(\beta_2)$. In terms of the logit we have

$$\ln\left[\frac{\Pr(y_i = 1|\beta, x_i)}{\Pr(y_i = 0|\beta, x_i)}\right] = (\mu + \beta_1 x_{1i} + \beta_2 x_{2i}) + \beta_2,$$

indicating that increasing x_2 by one unit to $x_2 + 1$ while keeping x_1 constant increases the logit by β_2.

Likelihood Function

Assuming the logistic model, the pmf of a datum using (3.20) is

$$\Pr(Y_i = y_i|x_i) = \left[\frac{\exp(x_i'\beta)}{1 + \exp(x_i'\beta)}\right]^{y_i} \left[\frac{1}{1 + \exp(x_i'\beta)}\right]^{1-y_i}, \quad y_i = 1, 0.$$

For N independent binary observations collected in the vector y and the covariates in matrix x, the pmf is

$$\Pr(Y = y|x) = \prod_{i=1}^{N} \left[\frac{\exp(x_i'\beta)}{1 + \exp(x_i'\beta)}\right]^{y_i} \left[\frac{1}{1 + \exp(x_i'\beta)}\right]^{1-y_i}. \tag{3.29}$$

This is the likelihood function when viewed as a function of β. The loglikelihood is obtained taking natural logarithms:

$$l(\beta|y, x) = \sum_{i=1}^{N} \left\{y_i x_i'\beta - \ln\left[1 + \exp(x_i'\beta)\right]\right\}. \tag{3.30}$$

The Iterative System

To obtain ML estimators using Newton-Raphson, first and second derivatives of the loglikelihood are needed. The score vector is

$$
\begin{aligned}
l'(\beta|y) &= \sum_{i=1}^{N} \frac{\partial}{\partial\beta} \left\{(1 - y_i)\, x_i'\beta - \ln\left[1 + \exp\left(x_i'\beta\right)\right]\right\} \\
&= \sum_{i=1}^{N} \left\{ y_i x_i - \frac{\exp\left(x_i'\beta\right)}{1 + \exp\left(x_i'\beta\right)} x_i \right\} \\
&= \sum_{i=1}^{N} \left[y_i - \pi\left(x_i'\beta\right)\right] x_i,
\end{aligned}
\tag{3.31}
$$

where

$$
\Pr(y_i = 1|x_i) = \frac{\exp\left(x_i'\beta\right)}{1 + \exp\left(x_i'\beta\right)} = \pi\left(x_i'\beta\right).
$$

Let the $N \times 1$ vector of probabilities of survival for the N individuals be

$$
\pi(X\beta) = \left[\pi\left(x_1'\beta\right), \ldots, \pi\left(x_N'\beta\right)\right]'
$$

and observe that the score vector can be written as

$$
\begin{aligned}
\sum_{i=1}^{N} \left[y_i - \pi\left(x_i'\beta\right)\right] x_i &= \left\{x_1\left[y_1 - \pi\left(x_1'\beta\right)\right] + \ldots + x_N\left[y_N - \pi\left(x_N'\beta\right)\right]\right\} \\
&= X'\left[y - \pi(X\beta)\right].
\end{aligned}
$$

The vector $y - \pi(X\beta)$ consists of deviations of the observations from their expectations. Using this representation in (3.31), it can be seen that the first-order condition for a maximum is satisfied if

$$
X'\pi\left(X\widehat{\beta}\right) = X'y,
\tag{3.32}
$$

where $\pi\left(X\widehat{\beta}\right)$ is the vector of probabilities of survival for the N individuals evaluated at the ML estimator $\widehat{\beta}$, if this exists. The estimating equations (3.32) are not explicit in $\widehat{\beta}$ and must be solved iteratively. To obtain second derivatives, an additional differentiation of the loglikelihood with respect to the parameters gives

$$
\begin{aligned}
l''(\beta|y) &= \frac{\partial^2 l(\beta|y)}{\partial\beta\,\partial\beta'} = \frac{\partial}{\partial\beta'}\left\{\sum_{i=1}^{N}\left[y_i - \pi\left(x_i'\beta\right)\right]x_i\right\} \\
&= -\sum_{i=1}^{N} x_i \frac{\partial}{\partial\beta'}\pi\left(x_i'\beta\right). \qquad (3.33)
\end{aligned}
$$

Now,

$$
\begin{aligned}
\frac{\partial}{\partial\beta'}\pi\left(x_i'\beta\right) &= \frac{\partial}{\partial\beta'}\left[1+\exp\left(-x_i'\beta\right)\right]^{-1} \\
&= \left[1+\exp\left(x_i'\beta\right)\right]^{-2}\exp\left(-x_i'\beta\right)x_i' \\
&= \pi\left(x_i'\beta\right)\left[1-\pi\left(x_i'\beta\right)\right]x_i'.
\end{aligned}
$$

Using this in (3.33),

$$
l''(\beta|y) = -\sum_{i=1}^{N} x_i\pi\left(x_i'\beta\right)\left[1-\pi\left(x_i'\beta\right)\right]x_i' = -X'D(\beta)X, \qquad (3.34)
$$

where $D(\beta)$ is an $N \times N$ diagonal matrix with the ith diagonal element $\pi\left(x_i'\beta\right)\left[1-\pi\left(x_i'\beta\right)\right]$. Because the second derivatives do not depend on the observations, the expected information is equal to the observed information in this case. Hence, the Newton-Raphson and the scoring algorithms are identical. From (3.4), multiplying by $\left[X'D\left(\beta^{[t]}\right)X\right]$, the iteration can be represented as

$$
\left[X'D\left(\beta^{[t]}\right)X\right]\beta^{[t+1]} = \left[X'D\left(\beta^{[t]}\right)X\right]\beta^{[t]} + X'v\left(\beta^{[t]}\right), \qquad (3.35)
$$

where the vector $v\left(\beta^{[t]}\right) = y - \pi\left(X\beta^{[t]}\right)$. Now let

$$
y^*\left(\beta^{[t]}\right) = X\beta^{[t]} + D^{-1}\left(\beta^{[t]}\right)v\left(\beta^{[t]}\right)
$$

be a pseudo-data vector evaluated at iteration $[t]$. Then the system (3.35) can be written as

$$
\left[X'D\left(\beta^{[t]}\right)X\right]\beta^{[t+1]} = X'D\left(\beta^{[t]}\right)y^*\left(\beta^{[t]}\right). \qquad (3.36)
$$

This is an iterative reweighted least squares system with the matrix of weights equal to $D\left(\beta^{[t]}\right)$, with ith diagonal element

$$\pi\left(x_i'\beta^{[t]}\right)\left[1-\pi\left(x_i'\beta^{[t]}\right)\right].$$

The Newton-Raphson algorithm is iterated until the change in successive rounds is negligible. If convergence is to a global maximum, $\widehat{\beta}$ is the ML estimate. The asymptotic variance-covariance matrix is estimated as

$$\widehat{\text{Var}}\left(\widehat{\beta}\right)=\left[X'D\left(\widehat{\beta}\right)X\right]^{-1}. \tag{3.37}$$

The variance of the maximum likelihood estimator is larger at extreme probabilities of survival.

Example: A Genomic Model

The genomic model was introduced on page 39 and discussed briefly on page 72. In the present likelihood setting, one may be interested in estimating the proportion of trait variance explained by marker information. This would give a first indication that genetic factors are influencing the trait. The justification for this conjecture is that marker genotypes are correlated (in LD) with unobserved causal genotypes and/or that causal genotypes are part of the marker panel.

In this example, the genomic model is implemented with Newton-Raphson using a decomposition of the genomic relationship matrix that simplifies computations. This decomposition is used repeatedly in the book.

For simplicity, here it is assumed that the data are centred and have zero mean. The genomic model is

$$y|g,\sigma_e^2 \sim N\left(g, I\sigma_e^2\right), \tag{3.38a}$$

$$g|W,\sigma_g^2 \sim SN\left(0, G\sigma_g^2\right), \tag{3.38b}$$

$$G=\frac{1}{m}WW', \tag{3.38c}$$

$$W=\left\{W_{ij}\right\}. \tag{3.38d}$$

In these expressions, g is the vector of genomic values, σ_g^2 is the genomic variance at the level of this operational model, SN is a shorthand for singular normal, m is the number of SNPs (single nucleotide polymorphisms) and W_{ij} is a label for the jth marker in individual i, $(i=1,\dots n;\ j=1,\dots,m;\ m>n)$:

$$W_{ij}=\frac{X_{ij}-\text{E}\left(X_{ij}\right)}{SD\left(X_{ij}\right)},\quad X_{ij}=0,1,2,$$

where SD stands for standard deviation. Due to the centring, W_{ij} has rank $(n-1)$, matrix G is singular and $(g|W, \sigma_g^2)$ is singular normally distributed. In practice the expectation and the standard deviation are replaced by their sample estimates.

Background

The following results are useful for deriving the likelihood for the genomic model. The eigenvalue decomposition of WW' is

$$\begin{aligned} WW' &= U \Delta U' \\ &= \sum_{i=1}^{n} \lambda_i U_i U_i', \end{aligned}$$

where $U = [U_1, U_2, \ldots, U_n]$, of order $n \times n$ is the matrix of eigenvectors of WW', U_j is the jth column (dimension $n \times 1$) and Δ is a diagonal matrix with elements equal to the eigenvalues $\lambda_1, \lambda_2, \ldots, \lambda_n$ associated with the n eigenvectors. Since WW' is non-negative definite, the eigenvalues are $\lambda_i \geq 0$, $i = 1, 2, \ldots, n$. The eigenvectors satisfy $U'U = UU' = I$.

In general, I work with the genomic relationship matrix G, defined as

$$\begin{aligned} G &= \frac{1}{m} WW' \\ &= \frac{1}{m} U \Delta U' \\ &= UDU' \end{aligned}$$

where

$$D = \frac{1}{m} \Delta.$$

The following result on determinants will be used:

$$\begin{aligned} \left| UDU'k + I \right| &= \left| U \left(Dk + I \right) U' \right| \\ &= |U| \, |Dk + I| \, \left| U' \right| \\ &= |Dk + I| \\ &= \prod_{i=1}^{n} \left(\lambda_i k + 1 \right), \end{aligned}$$

where $k = \frac{\sigma_g^2}{\sigma_e^2}$ and $|U| = \pm 1$, a property of orthogonal matrices ($UU' = I$ implies $|U|^2 = 1$).

Due to centring, the rank of G is typically $n-1$ and the singular normal density is

$$p\left(g|W, \sigma_g^2\right) = \frac{1}{(2\pi)^{\frac{n-1}{2}}\left(\lambda_1\sigma_g^2 \ldots \lambda_{n-1}\sigma_g^2\right)^{\frac{1}{2}}} \exp\left(-\frac{g'G^-g}{2\sigma_g^2}\right) \tag{3.39}$$

where the $\lambda' s$ are the non-zero eigenvalues of G and G^- is any generalised inverse of G (Mardia et al, 1979). One choice of generalised inverse of G is

$$G^- = UD^-U', \tag{3.40}$$

where

$$D^- = \begin{bmatrix} \frac{1}{\lambda_1} & 0 & \ldots & 0 \\ 0 & \ddots & \ldots & 0 \\ \vdots & \vdots & \frac{1}{\lambda_{n-1}} & 0 \\ 0 & 0 & 0 & 0 \end{bmatrix} = \begin{bmatrix} D_1^{-1} & 0_c \\ 0_c' & 0 \end{bmatrix}. \tag{3.41}$$

Above, $D_1 = diag\,(\lambda_i)_{i=1}^{n-1}$, a diagonal matrix of dimension $n-1 \times n-1$ that contains non-zero eigenvalues, after removing the last row and column of D, 0_c is a column vector of zeroes of size $(n-1 \times 1)$ and the element in the last row and column of D is the scalar 0.

A Probabilistically Equivalent Reparametrisation

Define a random variable α with distribution:

$$\alpha|U, \sigma_g^2 \sim SN\left(0, D\sigma_g^2\right). \tag{3.42}$$

Since the last diagonal element of D is 0, the last element of vector α is equal to zero (with probability 1), and therefore α has the singular normal distribution with density:

$$\begin{aligned} &\left(\lambda_1\sigma_g^2 \ldots \lambda_{n-1}\sigma_g^2\right)^{-\frac{1}{2}} \exp\left[-\frac{1}{2\sigma_g^2}\alpha' D^-\alpha\right] \\ &\propto \left(\sigma_g^2\right)^{-\left(\frac{n-1}{2}\right)} \exp\left[-\frac{1}{2\sigma_g^2}\left[(\alpha_1, \alpha_0)' D^- (\alpha_1, \alpha_0)\right]\right], \end{aligned} \tag{3.43}$$

In this expression, $\alpha' = (\alpha_1', \alpha_0')$, where α_1 is a column vector with $n-1$ elements and the scalar $\alpha_0 = 0$ with probability 1. It is simpler to work with vector α_1 that has probability density function:

$$p\left(\alpha_1|U,\sigma_g^2\right) \propto \left(\sigma_g^2\right)^{-\left(\frac{n-1}{2}\right)} \exp\left[-\frac{1}{2\sigma_g^2}\alpha_1' D_1^{-1}\alpha_1\right]. \tag{3.44}$$

This is the density of the $n-1$ dimensional multivariate normal distribution $N\left(0, D_1\sigma_g^2\right)$. Notice that $\alpha' D^{-}\alpha = \alpha_1' D_1^{-1}\alpha_1$.

Writing the Likelihood

The vector of genomic values $g = U\alpha$ and the model defined in (3.38) can be written as

$$y|\alpha,\sigma_e^2 \sim N\left(U\alpha, I\sigma_e^2\right), \tag{3.45a}$$

$$\alpha|U,\sigma_g^2 \sim SN\left(0, D\sigma_g^2\right), \tag{3.45b}$$

$$G = UDU', \tag{3.45c}$$

$$y|\sigma_g^2,\sigma_e^2 \sim N\left(0, UDU'\sigma_g^2 + I\sigma_e^2\right). \tag{3.45d}$$

Then the likelihood takes the form

$$p\left(\sigma_g^2,\sigma_e^2|y,W\right) \propto \left|UDU'\sigma_g^2 + I\sigma_e^2\right|^{-\frac{1}{2}} \exp\left(-\frac{1}{2}y'\left(UDU'\sigma_g^2 + I\sigma_e^2\right)^{-1}y\right). \tag{3.46}$$

Write

$$UDU'\sigma_g^2 + I\sigma_e^2 = \sigma_e^2\left(UDU'k + I\right), \quad k = \frac{\sigma_g^2}{\sigma_e^2}.$$

Then

$$\begin{aligned}\left|UDU'\sigma_g^2 + I\sigma_e^2\right| &= \left|\sigma_e^2\left(UDU'k + I\right)\right| \\ &= \left|\sigma_e^2 U\left(Dk + I\right)U'\right| \\ &= \left(\sigma_e^2\right)^n \left|U\left(Dk + I\right)U'\right|\end{aligned}$$

$$= \left(\sigma_e^2\right)^n |Dk + I|$$

$$= \left(\sigma_e^2\right)^n \prod_{i=1}^{n} (\lambda_i k + 1).$$

Also,

$$\left(UDU'\sigma_g^2 + I\sigma_e^2\right)^{-1} = \frac{1}{\sigma_e^2} U (Dk + I)^{-1} U',$$

using $U^{-1} = U'$. The likelihood is

$$\begin{aligned} p\left(\sigma_g^2, \sigma_e^2 | y, W\right) &\propto \left(\sigma_e^2\right)^{-\frac{n}{2}} \prod_{i=1}^{n} (\lambda_i k + 1) \exp\left[-\frac{1}{2\sigma_e^2} y' U (Dk + I)^{-1} U' y\right] \\ &= \left(\sigma_e^2\right)^{-\frac{n}{2}} \prod_{i=1}^{n} (\lambda_i k + 1)^{-\frac{1}{2}} \exp\left[-\frac{1}{2\sigma_e^2} \widetilde{y}' (Dk + I)^{-1} \widetilde{y}\right] \\ &= \left(\sigma_e^2\right)^{-\frac{n}{2}} \prod_{i=1}^{n} (\lambda_i k + 1)^{-\frac{1}{2}} \exp\left[-\frac{1}{2\sigma_e^2} \sum_{i=1}^{n} \frac{\widetilde{y}_i^2}{\lambda_i k + 1}\right], \end{aligned} \quad (3.47)$$

where $\widetilde{y} = U'y$, a column vector of length n whose ith element is $\widetilde{y}_i$.

The loglikelihood, up to an additive constant, is

$$\ln p\left(k, \sigma_e^2 | y, W\right) = -\frac{1}{2} \left\{ n \ln \sigma_e^2 + \sum_{i=1}^{n} \ln (\lambda_i k + 1) + \frac{1}{\sigma_e^2} \sum_{i=1}^{n} \frac{\widetilde{y}_i^2}{\lambda_i k + 1} \right\}. \quad (3.48)$$

It can be informative to plot contours of (3.48) as a function of the two variance components.

***NOTE: it may be numerically useful to parametrise in terms of the one-to-one transformation

$$\begin{aligned} \nu_e &= \ln \sigma_e^2, \\ \nu_g &= \ln \sigma_g^2, \end{aligned}$$

with inverse function

$$\begin{aligned} \sigma_e^2 &= \exp(\nu_e), \\ \sigma_g^2 &= \exp(\nu_g), \\ k &= \frac{\sigma_g^2}{\sigma_e^2} = \exp\left(\nu_g - \nu_e\right). \end{aligned}$$

Then the loglikelihood (3.48) takes the form

$$\ln p\left(\nu_g, \nu_e | y, W\right) =$$
$$-\frac{1}{2}\left\{n\nu_e + \sum_{i=1}^{n} \ln\left(\lambda_i \exp\left(\nu_g - \nu_e\right) + 1\right) + \frac{1}{\exp(\nu_e)} \sum_{i=1}^{n} \frac{\widetilde{y}_i^2}{\lambda_i \exp\left(\nu_g - \nu_e\right) + 1}\right\}. \tag{3.49}$$

This avoids negative values of the variance components under unconstrained maximisation.

Implementation Using Newton-Raphson

To fit the model using Newton-Raphson, first and second derivatives are needed. It is easier to work with (3.48).

First Derivatives

$$\frac{\partial}{\partial \sigma_e^2} \ln p\left(k, \sigma_e^2 | y, W\right) = -\frac{1}{2}\left(\frac{n}{\sigma_e^2} - \frac{1}{\left(\sigma_e^2\right)^2} \sum_i \frac{\widetilde{y}_i^2}{1 + k\lambda_i}\right). \tag{3.50}$$

$$\frac{\partial}{\partial k} \ln p\left(k, \sigma_e^2 | y, W\right) = -\frac{1}{2}\left(\sum_i \frac{\lambda_i}{1 + k\lambda_i} - \frac{1}{\sigma_e^2} \sum_i \frac{\lambda_i \widetilde{y}_i^2}{(1 + k\lambda_i)^2}\right). \tag{3.51}$$

Second Derivatives

$$\frac{\partial^2}{\left(\partial \sigma_e^2\right)^2} \ln p\left(k, \sigma_e^2 | y, W\right) = -\frac{1}{2}\left(\frac{2}{\left(\sigma_e^2\right)^3} \sum_i \frac{\widetilde{y}_i^2}{1 + k\lambda_i} - \frac{n}{\left(\sigma_e^2\right)^2}\right). \tag{3.52}$$

$$\frac{\partial^2}{(\partial k)^2} \ln p\left(k, \sigma_e^2 | y, W\right) = -\frac{1}{2}\left(\frac{1}{\sigma_e^2} \sum_i \frac{2\lambda_i^2 \widetilde{y}_i^2}{(1 + k\lambda_i)^3} - \sum_i \frac{\lambda_i^2}{(1 + k\lambda_i)^2}\right). \tag{3.53}$$

$$\frac{\partial^2}{\partial k \, \partial \sigma_e^2} \ln p\left(k, \sigma_e^2 | y, W\right) = -\frac{1}{2\left(\sigma_e^2\right)^2} \sum_i \frac{\widetilde{y}_i^2 \lambda_i}{(1 + k\lambda_i)^2}. \tag{3.54}$$

The iterative system is

$$\begin{bmatrix} \sigma_e^2 \\ k \end{bmatrix}_{t+1} = \begin{bmatrix} \sigma_e^2 \\ k \end{bmatrix}_t -$$

$$\begin{bmatrix} \frac{\partial^2}{(\partial\sigma_e^2)^2} \ln p\left(k, \sigma_e^2|y, W\right) & \frac{\partial^2}{\partial k\, \partial\sigma_e^2} \ln p\left(k, \sigma_e^2|y, W\right) \\ \frac{\partial^2}{\partial k\, \partial\sigma_e^2} \ln p\left(k, \sigma_e^2|y, W\right) & \frac{\partial^2}{(\partial k)^2} \ln p\left(k, \sigma_e^2|y, W\right) \end{bmatrix}^{-1} \begin{bmatrix} \frac{\partial}{\partial\sigma_e^2} \ln p\left(k, \sigma_e^2|y, W\right) \\ \frac{\partial}{\partial k} \ln p\left(k, \sigma_e^2|y, W\right) \end{bmatrix}. \tag{3.55}$$

The partial derivatives in (3.55) are evaluated at $\sigma_e^2 = \sigma_e^{2[t]}, k = k^{[t]}$.

An R-code That Performs the Eigenvalue Decomposition

As an example, an R-code that performs the eigenvalue decomposition of the genomic relationship matrix is shown below. First, matrix X of dimension equal to the number of individuals times the number of markers ($nindiv \times nmark$) is generated by allocating marker codes drawn from a binomial distribution $Bi\,(2, p = 0.5)$. The matrix of standardised marker genotypes W is obtained by centring and scaling X. Using W the code generates the genomic relationship matrix G ($nindiv \times nindiv$), and the eigen decomposition is performed on G. This decomposition gives rise to matrix U ($nindiv \times nindiv$) and to the vector of eigenvalues val, from which the diagonal matrix D ($nindiv \times nindiv$) is produced.

The last line of the code checks whether G and UDU' are equal: they are within the limits of numerical precision:

```
# CODE0303
# DATA BASED ON GENOMIC MODEL; OBTAIN THE SVD OF WW' (1/m)
rm(list=ls()) # CLEAR WORKSPACE
set.seed(1953)
nindiv<-10
nmark<-20
X<-matrix(nrow= nindiv,ncol= nmark,
          rbinom(n=nindiv*nmark,size=2,p=.5))
W <- matrix(data=NA,nrow= nindiv,ncol=nmark)
U <- matrix(data=NA,nrow= nindiv,ncol= nindiv)
G<-matrix(data=NA,nrow= nindiv,ncol= nindiv)
cm <- colMeans(X)
# CREATE MATRIX OF STANDARDISED MARKER GENOTYPE CODES
for (i in 1:nmark)
  {
W[,i] <-( X[,i]-cm[i]) / sd(X[,i])
}
# THIS IS MORE EFFICIENT THAN THE LOOP:
# W <- scale(X, center=TRUE, scale=TRUE)
qr(X)$rank
```

```
## [1] 10
```

```
qr(W)$rank
```

```
## [1] 9
```

```
# GENOMIC RELATIONSHIP MATRIX G
G <- (1/nmark)*W%*%t(W)
# THIS IS MORE EFFICIENT THAN THE LINE ABOVE:
# G <- (1/nmark)*tcrossprod(W)
# SVD OF G
EVD <- eigen(G)
names(EVD)
```

```
## [1] "values"  "vectors"
```

```
head(EVD$values[1:5])
```

```
## [1] 2.3596077 1.8733148 1.3984700 1.0492670 0.7181604
```

```
U <- EVD$vector
val <- EVD$values
val[nindiv] <-0
D <- diag(val,nrow=nindiv)
# CHECK THAT G = UDU':
identical(G, U%*%D%*%t(U))
```

```
## [1] FALSE
```

```
max(abs(G - U%*%D%*%t(U)))
```

```
## [1] 2.88658e-15
```

3.2 Gradient Descent and Stochastic Gradient Descent

Gradient descent is a first-order iterative optimisation algorithm for finding a local minimum of a differentiable function. This is achieved by taking repeated steps in the opposite sign of the gradient of the function at the current point, because this is the direction of greatest rate of decrease of the function. The method does not rely on matrix inversions and is often used in high-dimensional settings as encountered in machine learning.

Let θ_t be the value of a parameter $\theta \in \mathbb{R}^d$ at step t and let the column vector $\Delta\theta \in \mathbb{R}^d$ be the move taken at step t. Consider the first-order Taylor expansion of the function f to be minimised around θ_t:

$$f(\theta_t + \Delta\theta) \approx f(\theta_t) + \Delta\theta' \, \nabla f|_{\theta_t}$$

where $\nabla f|_{\theta_t}$ is the gradient of f at θ_t, i.e. a column vector in $\mathbb{R}^d$ whose ith coordinate is the partial derivative of f with respect to element i of θ evaluated at $\theta = \theta_t$. Since the objective is to find a minimum of the function f, the move must result in

$$f(\theta_t + \Delta\theta) \leq f(\theta_t)$$

and therefore $\Delta\theta' \, \nabla f|_{\theta_t} \leq 0$. This suggests setting

$$\Delta\theta = -\alpha \, \nabla f|_{\theta_t}, \quad \alpha > 0 \tag{3.56}$$

where α is a small positive number known as the *learning rate*. If

$$\theta_{t+1} = \theta_t - \alpha \, \nabla f|_{\theta_t} \tag{3.57}$$

then

$$f(\theta_{t+1}) = f(\theta_t) - \alpha \left(\left(\nabla f|_{\theta_t} \right)' \nabla f|_{\theta_t} \right) < f(\theta_t) \tag{3.58}$$

since $\left(\nabla f|_{\theta_t} \right)' \nabla f|_{\theta_t} > 0$.

In the case of Newton's method, $\alpha = \left[S'(\theta_t) \right]^{-1}$, which involves the inverse of second derivatives. The attraction of gradient descent is that it only uses first derivatives and does not require matrix inversions. The downside is that it can be slow to converge and the appropriate choice of α can be quite challenging, especially in high-dimensional settings where the gradients differ markedly among the $\theta's$. In such cases, one may consider using a *diagonally scaled gradient descent* where α is replaced by a diagonal matrix with elements tuned to each gradient. Alternatively, one can scale the features x_i so that they fall approximately in the range ± 1. Centring and scaling the features by their standard deviation is yet another option. It can also help to change α as a function of t, with larger values at the start of the iterative process.

A popular modification, especially with very large data sets, is *stochastic gradient descent*. At each iteration, the actual gradient is replaced by an estimate obtained from a randomly selected subset of the data. An extreme version uses a single data point at a time (see example below). These versions of gradient descent are also known as *mini batch gradient descent* as opposed to *batch gradient descent* when the complete set of data is used. In common with many optimisation algorithms, when data sets are very large, stochastic gradient descent converges faster in terms of total computation by rapidly calculating approximate values rather than exact values of the gradient. Stochastic gradient descent does not converge exactly to a local minimum of the cost function (negative of the loglikelihood) as classical gradient descent in theory does. It rather may oscillate around the local minimum. Recent theoretical and empirical results indicate that this does not seem to be an issue in large networks (LeCun et al, 2015). The noise associated with stochastic gradient descent can be beneficial to escape local minima, particularly in complex networks with non-convex loss functions.

There are several refinements to gradient descent as well as rules guiding the choice of α. A useful reference is Bottou (2012) and an overview of numerical computations for machine learning algorithms can be found in Goodfellow et al (2016).

If the cost function is concave such as many likelihood functions, the sign in (3.57) becomes positive in order to move towards a maximum, and this turns into a *gradient ascent* algorithm.

Note

The maximum likelihood estimator of a parameter θ involves the maximisation of the likelihood function with respect to θ, or the minimisation of the negative of the likelihood function with respect to θ. Both are tasks on optimisation. When minimisation is involved, the function can be referred to as a *cost function*. The cost function may include a penalty term, as, for example, in penalised logistic regression discussed on page 377.

A Toy Example

As an illustration, the R-code below applies classical gradient descent to find the minimum of a cost function (negative loglikelihood) of a simple linear model $y = Xb$ where b includes an intercept. The result is compared to the exact least squares solution $\hat{b} = (X'X)^{-1}X'y$ at the bottom of the code. The code immediately below implements stochastic gradient descent:

```
# CODE0304
# GRADIENT DESCENT WITH A LINEAR NODEL
rm(list=ls()) # Clear the workspace
set.seed(195021)
N<-100
x<-seq(from=0,to=5,length=N)
signal<-10 + 0.2*x
error<-rnorm(N)
y<-signal+error
one <- rep(1,N)
X <- cbind(one,x)
LHS <- crossprod(X) # LEFT HAND SIDE
RHS <- crossprod(X,y) # RIGHT HAND SIDE
bhat<- solve(LHS,RHS) # SOLUTION TO THE LEAST SQUARES EQUATIONS
```

```
#################################
nit <- 200
alfa <- 0.002
miu <- matrix(data=NA, nrow=nit+1,ncol=1)
b <- matrix(data=NA, nrow=nit+1,ncol=1)
c <- matrix(data=NA, nrow=nit+1,ncol=1)
cost <- function(miu,b){sum(y-miu-b*x)^2}

miu[1] <- 5
b[1] <- 1
c[1] <- cost(miu[1],b[1])
for(i in 1:nit) {
  fdmiu <- -sum(y - miu[i] - b[i] * x)
  fdbeta <- -sum((y - miu[i] - b[i] * x) * x)
  fd <- matrix(c(fdmiu, fdbeta), nrow = 2, ncol = 1)
  sol0 <- matrix(c(miu[i], b[i]), nrow = 2, ncol = 1)
  alfa <- 0.002
  sol1 <- sol0 - alfa * fd
  miu[i + 1] <- sol1[1, 1]
  b[i + 1] <- sol1[2, 1]
}
## CHECK ######################
beta <- c(miu[i],b[i])
beta # GRADIENT DESCENT SOLUTION
```

```
## [1] 10.1511233  0.1727193
```

```
bhat # LEAST SQUARES SOLUTION
```

```
##             [,1]
## one 10.1515028
## x    0.1726028
```

```
cost(miu[i],b[i]) # COST FUNCTION AFTER nit ITERATIONS
```

```
## [1] 7.781174e-05
```

The following R-code implements stochastic gradient descent with a decaying learning rate α. The implementation is based on two loops. The internal loop computes the gradient and updates parameters one record at a time, for all records. The external loop repeats the procedure `nit` times. Each internal loop that defines a pass through the data is known as an *epoch* in the machine learning literature.

As implemented below, the algorithm does not reshuffle the data at the start. Some versions of the mini batch algorithm do this to achieve better behaviour and to break down structures in the data that may affect the results:

```
# STOCHASTIC GRADIENT DESCENT WITH A LINEAR NODEL
# CODE0305
rm(list=ls()) # Clear the workspace
set.seed(195021)
N <- 1000
```

```
x<-seq(from=0,to=5,length=N)
signal<-10 + 0.2*x
error<-rnorm(N)
y<-signal+error
one <- rep(1,N)
X <- cbind(one,x)
LHS <- crossprod(X) # LEFT HAND SIDE
RHS <- crossprod(X,y) # RIGHT HAND SIDE
bh<- solve(LHS,RHS) # SOLUTION TO THE LEAST SQUARES EQUATIONS

#################################
nit <- 10
alfa_0 <- 0.0038
c <- matrix(data=NA, nrow=nit+1,ncol=1)
cost <- function(miu,b){sum(y-miu-b*x)^2}

miu <- 5
b <- 1
c[1] <- cost(miu,b)
for(j in 1:nit){
  alfa <- (1-(j/nit))*alfa_0 + ((j/nit)*alfa_0*0.01)
  for(i in 1:length(y)) {
#  cat(i, "\n",sep="")
  fdmiu <- -2*(y[i] - miu - b * x[i])
  fdbeta <- -2*((y[i] - miu - b * x[i]) * x[i])
  fd <- matrix(c(fdmiu, fdbeta), nrow = 2, ncol = 1)
  sol0 <- matrix(c(miu, b), nrow = 2, ncol = 1)
  sol1 <- sol0 - alfa * fd
  miu <- sol1[1, 1]
  b <- sol1[2, 1]
}
}
## CHECK ######################
beta <- c(miu,b)
beta # STOCHASTIC GRADIENT DESCENT SOLUTION
```

```
## [1] 10.0067857  0.1833919
```

```
bh # LEAST SQUARES SOLUTION
```

```
##           [,1]
## one 10.0305748
## x    0.1859502
```

Gradient descent is revisited in the chapter on nonparametric methods where it is applied to solve more demanding problems.

3.3 The EM Algorithm

An overview is provided of one of the most versatile iterative algorithms for computing maximum likelihood and posterior modes: the expectation-maximisation, or EM, algorithm. More details can be found in Sorensen and Gianola (2002) and considerably more in McLachlan and Krishnan (1997) and of course a tour de force is the celebrated paper by Dempster et al (1977). The algorithm is conceptually simple, at least in its basic form, and brings considerable insight into the statistical structure of a maximum likelihood or posterior mode problem, contrary to Newton-Raphson or scoring that are based primarily on numerical considerations. The basic idea behind the method is to transform an incomplete into a complete data problem for which the required maximisation is computationally more tractable. Also, the algorithm is numerically stable: each iteration increases the likelihood or posterior density and convergence is nearly always to a local maximum.

The concept of missing data is fairly broad. It includes, for example, missing data in an unbalanced layout, but it extends to observations from truncated distributions, censored data and latent variables. In these cases, one can view the complete data x as consisting of the vectors (y, z), where y is the observed data or incomplete data and z is the missing data. More generally, many statistical problems that at first glance do not appear to involve missing data can be reformulated into missing data problems by judicious augmentation of the data set with unobserved values. As such, one can view the observations at hand and the parameters of the posed model as data: part of these data is observed (the records) and another part is missing (the parameters). Mixed effects models, hierarchical models and models with latent variables, such as the threshold model, are typically amenable to an EM formulation. An example is an additive genetic model where inference may focus on $\theta = \left(\beta', \sigma_a^2, \sigma_e^2\right)'$, where β is a vector of location parameters and $\left(\sigma_a^2, \sigma_e^2\right)$ are variance components. Here, one may augment the observed data y, with the missing data a, the unobserved vector of additive genetic values. As shown later, this simplifies the computations involved in finding the ML estimates of θ, or the maximum of $p\left(\beta, \sigma_a^2, \sigma_e^2|y\right)$, or mode of the posterior distribution $\left[\beta, \sigma_a^2, \sigma_e^2|y\right]$. On the other hand, if one wishes to find the mode of the distribution with density $p\left(\sigma_a^2, \sigma_e^2|y\right)$, an EM strategy is to consider (β, a) as the missing data.

Derivation

The EM algorithm is derived here from a likelihood (as opposed to a posterior distribution) perspective. Conceptually, there is no difference. Consider the identity

$$p(y|\theta) = \frac{p(y, z|\theta)}{p(z|y, \theta)},$$

where y is the observed data and z is the missing data. Taking logarithms on both sides leads to

$$\ln p(y|\theta) = \ln p(y, z|\theta) - \ln p(z|y, \theta), \tag{3.59}$$

where the first term on the right-hand side is known as the *complete data loglikelihood* (there is an equivalent complete data logposterior). The next step is to take expectations of both sides with respect to $[z|\theta^{[t]}, y]$, where $\theta^{[t]}$ is the current guess of θ. The left-hand side of (3.59) does not depend on z, so averaging over z, providing the integrals exist, gives

$$\ln p(y|\theta) = \int \ln p(y, z|\theta)\, p\left(z|\theta^{[t]}, y\right) dz - \int \ln p(z|y, \theta)\, p\left(z|\theta^{[t]}, y\right) dz. \tag{3.60}$$

The first term on the right-hand side of (3.60) is a function of θ for fixed y and fixed $\theta^{[t]}$ and is denoted as $Q(\theta|\theta^{[t]})$ in the EM literature. The second term is denoted $H(\theta|\theta^{[t]})$. Thus,

$$\ln p(y|\theta) = Q\left(\theta|\theta^{[t]}\right) - H\left(\theta|\theta^{[t]}\right). \tag{3.61}$$

The EM algorithm involves working with the first term only, $Q(\theta|\theta^{[t]})$, disregarding $H(\theta|\theta^{[t]})$. The two steps are:

1. E-step: calculation of $Q(\theta|\theta^{[t]})$, the expectation of the complete data loglikelihood (logposterior) with respect to the conditional distribution of the missing data, given the observed data and the current guess for θ.
2. M-step: maximisation of $Q(\theta|\theta^{[t]})$ with respect to θ, solving for θ and setting the result equal to $\theta^{[t+1]}$, the new value of the parameter. If $\theta^{[t+1]}$ maximises $Q(\theta|\theta^{[t]})$, the M-step is such that

$$Q\left(\theta^{[t+1]}|\theta^{[t]}\right) \geq Q\left(\theta|\theta^{[t]}\right), \quad \text{for all} \quad \theta \in \Omega, \tag{3.62}$$

which implies that $\theta^{[t+1]}$ is a solution to the equation:

$$\frac{\partial Q\left(\theta|\theta^{[t]}\right)}{\partial \theta} = 0. \tag{3.63}$$

The two steps are repeated iteratively until convergence is reached.

An important property of EM (not of Newton-Raphson) is that the iterative sequence leads to a monotonic increase of the loglikelihood $\ln p(y|\theta)$. Therefore,

$$\ln p\left(y|\theta^{[t+1]}\right) \geqslant \ln p\left(y|\theta^{[t]}\right). \tag{3.64}$$

Since the loglikelihood increases in each step, the EM algorithm, with few exceptions, converges to a local mode.

In some models, the calculation of $Q(\theta|\theta^{[t]})$ in the E-step may be difficult. Wei and Tanner (1990) propose a Monte Carlo approach for overcoming this difficulty.

This consists of simulating $z_1, z_2, \ldots, z_m$ from $p(z|\theta^{[t]}, y)$ and then forming the simulation consistent estimator:

$$\hat{Q}(\theta|\theta^{[t]}) \approx \frac{1}{m} \sum_{i=1}^{m} \ln p(y, z_i|\theta).$$

In its original formulation, the EM algorithm does not yield estimates of asymptotic variances. However, several approaches have been suggested to remedy this and some are described and illustrated in Sorensen and Gianola (2002).

A Digression on a Multivariate Transformation for Discrete Random Variables

Before going through the examples, I present a result of the theory of transformations that is relevant for the next example. A multinomial distribution with three distinctive classes is considered.

Let n_i be the number of observations falling into the ith class and let p_i be the probability that an observation falls in the ith class, for $i = 1, 2, 3$, with $p_1 + p_2 + p_3 = 1$. Then $n = n_1 + n_2 + n_3$ and the joint probability mass function of (n_1, n_2) is (the dependence on parameters p_1, p_2 and n is omitted)

$$p(n_1, n_2) = \frac{n!}{n_1!\, n_2!\, (n - n_1 - n_2)!} p_1^{n_1} p_2^{n_2} (1 - p_1 - p_2)^{n-n_1-n_2}. \tag{3.65}$$

Let $(n_1, n_2) = (X, Y)$ and $(n_1, n_1 + n_2) = (U, V)$. Suppose that one needs to find the conditional probability distribution of n_1, given $n_1 + n_2$:

$$p(n_1|n_1 + n_2) = \frac{p(n_1, n_1 + n_2)}{p(n_1 + n_2)} = \frac{p(U, V)}{p(V)}. \tag{3.66}$$

To derive the numerator in (3.66), $p(n_1, n_1 + n_2) = p(U, V)$, note that the transformation $(X, Y) \rightarrow (U, V)$ can be written as

$$\begin{bmatrix} U \\ V \end{bmatrix} = f(X, Y) = \begin{bmatrix} 1 & 0 \\ 1 & 1 \end{bmatrix} \begin{bmatrix} X \\ Y \end{bmatrix} = \begin{bmatrix} X \\ X + Y \end{bmatrix} \tag{3.67}$$

with inverse transformation

$$\begin{bmatrix} X \\ Y \end{bmatrix} = f^{-1}(U, V) = \begin{bmatrix} 1 & 0 \\ -1 & 1 \end{bmatrix} \begin{bmatrix} U \\ V \end{bmatrix} = \begin{bmatrix} U \\ V - U \end{bmatrix}.$$

Therefore,

$$p_{U,V}(U,V) = \frac{n!}{U!\,(V-U)!\,(n-V)!} p_1^U p_2^{V-U} (1-p_1-p_2)^{n-V}. \tag{3.68}$$

To obtain $p(n_1|n_1+n_2)$, (3.68) must be divided by $p(V)$. The random variable $V = n_1 + n_2$ is binomially distributed:

$$V \sim Bi(p_1 + p_2, n).$$

This is so because the three classes can be regrouped into two "wider" categories, one where the counts $(n_1 + n_2)$ are observed to fall and the other involving the third original category with counts n_3. In view of the independence of the draws, it follows that $(n_1 + n_2)$ is binomially distributed. Dividing $p_{U,V}(U,V)$ in (3.68) by the marginal probability mass function $p_V(V)$ yields

$$\begin{aligned}
p(n_1|n_1+n_2) &= p(U|V) \\
&= \frac{p(U,V)}{p(V)} \\
&= \frac{V!}{U!\,(V-U)!} \frac{p_1^U p_2^{V-U}}{(p_1+p_2)^V} \\
&= \frac{(n_1+n_2)!}{n_1!\,n_2!} \frac{p_1^{n_1} p_2^{n_2}}{(p_1+p_2)^{n_1+n_2}} \\
&= \frac{(n_1+n_2)!}{n_1!\,n_2!} \left(\frac{p_1}{p_1+p_2}\right)^{n_1} \left(\frac{p_2}{p_1+p_2}\right)^{n_2}.
\end{aligned}$$

This implies that

$$[n_1|n_1+n_2] \sim Bi\left(\frac{p_1}{p_1+p_2}, n_1+n_2\right). \tag{3.69}$$

Hence, the conditional distribution $[n_1|n_1+n_2]$ has mean,

$$E(n_1|n_1+n_2) = (n_1+n_2)\frac{p_1}{p_1+p_2}$$

and variance

$$Var(n_1|n_1+n_2) = (n_1+n_2)\frac{p_1}{p_1+p_2}\frac{p_2}{p_1+p_2}.$$

Example: Estimation of Gene Frequencies from ABO Blood Group Phenotypes

The *ABO* blood groups' problem introduced on page 79 is studied using the EM algorithm. Let $n = (n_A, n_{AB}, n_B, n_O)'$ be the observed data, with $n_A = 725$, $n_{AB} = 72$, $n_B = 258$ and $n_O = 1073$. It is sensible to treat the unobserved counts n_{AO}, n_{AA}, n_{BB} and n_{BO} as missing data. The resulting complete data vector is

$$n_c = (n_{AA}, n_{AO}, n_{AB}, n_{BB}, n_{BO}, n_O)' .$$

The complete data loglikelihood, excluding an additive constant, is

$$\begin{aligned}\ln f\,(p_A, p_B|n_c) &= 2n_{AA}\ln(p_A) + n_{AO}\ln(2p_Ap_O) + n_{AB}\ln(2p_Ap_B)\\ &\quad + 2n_{BB}\ln(p_B) + n_{BO}\ln(2p_Bp_O) + 2n_O\ln(p_O),\end{aligned}$$

where $p_O = (1 - p_A - p_B)$. The E-step consists of computing the expected value of the complete data loglikelihood, conditional on the observed counts n and on the value of the parameters at iteration t, $\left(p_A^{[t]}, p_B^{[t]}\right)$. Explicitly, this is

$$\begin{aligned}Q\left(p_A, p_B|p_A^{[t]}, p_B^{[t]}\right) &= \mathrm{E}\left[\{2n_{AA}\ln(p_A) + n_{AO}\ln(2p_Ap_O) + n_{AB}\ln(2p_Ap_B)\right.\\ &\quad \left. + 2n_{BB}\ln(p_B) + n_{BO}\ln(2p_Bp_O) + 2n_O\ln(p_O)\}\,|p_A^{[t]}, p_B^{[t]}, n\right]\\ &= 2\widetilde{n}_{AA}\ln(p_A) + \widetilde{n}_{AO}\ln(2p_Ap_O) + n_{AB}\ln(2p_Ap_B) + 2\widetilde{n}_{BB}\ln(p_B)\\ &\quad + \widetilde{n}_{BO}\ln(2p_Bp_O) + 2n_O\ln(p_O),\end{aligned} \tag{3.70}$$

where

$$\widetilde{n}_{AA} = \mathrm{E}\left(n_{AA}|p_A^{[t]}, p_B^{[t]}, n\right),$$

$$\widetilde{n}_{AO} = \mathrm{E}\left(n_{AO}|p_A^{[t]}, p_B^{[t]}, n\right),$$

$$\widetilde{n}_{BB} = \mathrm{E}\left(n_{BB}|p_A^{[t]}, p_B^{[t]}, n\right),$$

$$\widetilde{n}_{BO} = \mathrm{E}\left(n_{BO}|p_A^{[t]}, p_B^{[t]}, n\right).$$

The M-step consists of maximizing (3.70) with respect to p_A and p_B. This yields the following closed-form solution for p_A and p_B at a round $(t + 1)$:

$$p_A^{[t+1]} = \frac{2\widetilde{n}_{AA} + n_{AB} + \widetilde{n}_{AO}}{2\,(n_A + n_{AB} + n_B + n_O)}, \tag{3.71}$$

$$p_B^{[t+1]} = \frac{2\widetilde{n}_{BB} + n_{AB} + \widetilde{n}_{BO}}{2\left(n_A + n_{AB} + n_B + n_O\right)}. \tag{3.72}$$

The unobserved counts at iteration t are imputed via their expected values, given n and $\left(p_A^{[t]}, p_B^{[t]}\right)$. Using (3.69), the unobserved counts are distributed binomially as follows:

$$n_{AA} \sim Bi\left(\frac{p_A^2}{p_A^2 + 2p_A\left(1 - p_A - p_B\right)}, n_A\right),$$

$$n_{AO} \sim Bi\left(\frac{2p_A\left(1 - p_A - p_B\right)}{p_A^2 + 2p_A\left(1 - p_A - p_B\right)}, n_A\right),$$

$$n_{BB} \sim Bi\left(\frac{p_B^2}{p_B^2 + 2p_B\left(1 - p_A - p_B\right)}, n_B\right),$$

and

$$n_{BO} \sim Bi\left(\frac{2p_B\left(1 - p_A - p_B\right)}{p_B^2 + 2p_B\left(1 - p_A - p_B\right)}, n_B\right).$$

Hence, expectations can be computed immediately. For example,

$$\widetilde{n}_{AA} = n_A \frac{p_A^{2[t]}}{p_A^{2[t]} + 2p_A^{[t]}\left(1 - p_A^{[t]} - p_B^{[t]}\right)},$$

and similarly for the other components of the missing data. Using starting values for the gene frequencies, the missing counts $\widetilde{n}_{ij}$ are imputed, and the next round of gene frequency values are computed from (3.71) and (3.72). In the case of the present example, starting with $p_A^{[0]} = p_B^{[0]} = 0.2$, *nine* EM iterations result in $\hat{p}_A = 0.2091$ and $\hat{p}_B = 0.0808$ (with $\widehat{p}_0 = 1 - \widehat{p}_A - \widehat{p}_B$).

The R-code below fits the EM algorithm to the *ABO* data. The third column of the output below displays the increasing values of the loglikelihood function with each EM iteration. This compares well with the output from OPTIM on page 84:

```
# CODE0306
#### EM algorithm ##########################
rm(list=ls())
set.seed(30371)
niter<-9
result <- matrix(data=NA,nrow=niter,ncol=3)

# DATA
n_A<-725
```

```
n_AB<-72
n_B<-258
n_0<-1073
# START VALUES FOR P_A and p_B
p_A <- 0.2
p_B <- 0.2
for (i in 1:niter){
# E-step
n_AA <- n_A * p_A^2/(p_A^2 + 2*p_A*(1-p_A-p_B))
n_A0 <- n_A *  (2*p_A*(1-p_A-p_B))/(p_A^2 + 2*p_A*(1-p_A-p_B))
n_BB <- n_B * p_B^2/(p_B^2 + 2*p_B*(1-p_A-p_B))
n_B0 <- n_B *  (2*p_B*(1-p_A-p_B))/(p_B^2 + 2*p_B*(1-p_A-p_B))
# M-step
p_A <-  (2*n_AA + n_AB + n_A0)/(2*(n_A + n_AB + n_B + n_0))
p_B <-  (2*n_BB + n_AB + n_B0)/(2*(n_A + n_AB + n_B + n_0))
loglik <-  (725*log(p_A*(2 - p_A - 2* p_B)) + 72*log(2*p_A*p_B) +
    258*log(p_B*(2 - p_B - 2*p_A)) + 2*1073*log(1 - p_A - p_B))
result[i,] <- c(p_A,p_B,loglik)

}
result
```

```
##              [,1]        [,2]      [,3]
##  [1,] 0.2116004 0.08619764 -2304.473
##  [2,] 0.2095705 0.08104311 -2303.555
##  [3,] 0.2091891 0.08081323 -2303.551
##  [4,] 0.2091379 0.08080178 -2303.550
##  [5,] 0.2091315 0.08080107 -2303.550
##  [6,] 0.2091308 0.08080101 -2303.550
##  [7,] 0.2091307 0.08080101 -2303.550
##  [8,] 0.2091307 0.08080101 -2303.550
##  [9,] 0.2091307 0.08080101 -2303.550
```

Example: A Regression Model for Binary Data

I return to the likelihood of the logit model fitted using Newton-Raphson. Here, a probit model is implemented with the EM algorithm.

The binary data y is interpreted as arising from the following process involving the unobserved liability (or latent variable) u:

$$y_i = \begin{cases} 1 \text{ if } u_i < 0, \\ 0 \text{ if } u_i > 0. \end{cases} \tag{3.73}$$

The liability is modelled as

$$u_i = x_i'\beta + e_i, \quad \text{i} = 1, 2, \ldots, \text{N}, \tag{3.74}$$

where in the probit model, the error terms are iid $\mathcal{N}(0, 1)$ and therefore $[u_i|x_i, \beta] \sim \mathcal{N}\left(x_i'\beta, 1\right)$. The row vector x_i' contains observed covariates with p elements and β is an $p \times 1$ vector of unknown regression coefficients. In the probit model specified by (3.73),

$$\begin{aligned}\Pr(y_i = 1|\beta, x_i) &= \Pr(u_i < 0|\beta, x_i)\\ &= \Pr\left(x_i'\beta + e_i < 0|\beta, x_i\right)\\ &= \Pr\left(e_i < -x_i'\beta\,|\beta, x_i\right)\\ &= 1 - \Phi\left(x_i'\beta\right)\end{aligned}$$

and

$$\begin{aligned}\Pr(y_i = 0|\beta, x_i) &= \Pr(u_i > 0|\beta, x_i)\\ &= \Phi\left(x_i'\beta\right).\end{aligned}$$

The probit likelihood is

$$L(\beta|x, y) \propto \prod_{i=1}^{N}\left[\left(1 - \Phi\left(x_i'\beta\right)\right)^{y_i}\left(\Phi\left(x_i'\beta\right)\right)^{1-y_i}\right]. \tag{3.75}$$

Augmenting the observed data y with the missing data u, the complete data for the ith observation is (y_i, u_i). The joint density of the complete data can be written as

$$\Pr(Y_i = y_i|u_i)\, p(u_i|\beta, x_i) \propto \begin{cases}\mathcal{N}\left(u_i|x_i'\beta, 1\right) I(u_i < 0), & \text{for } y_i = 1,\\ \mathcal{N}\left(u_i|x_i'\beta, 1\right) I(u_i > 0), & \text{for } y_i = 0,\end{cases} \tag{3.76}$$

where the results follow from (3.73) and (3.74). The term $\Pr(Y_i = y_i|u_i)$ is a degenerate probability mass function: given the model specified by (3.73), once u_i is observed, Y_i is not stochastic; it is known with certainty. This translates into the indicator functions in the right-hand side of (3.76). However, these indicator functions will be omitted because they become additive constants of the complete data loglikelihood; they are not a function of β and vanish in the computation of the M-step (the form of the complete data likelihood including indicator functions is given in Eq. (5.26) on page 217 that is an alternative way of expressing the right-hand side of (3.76)). Therefore, the complete data likelihood for the ith observation is proportional to $p(u_i|\beta, x_i)$.

The *complete data likelihood* is proportional to the joint distribution of the liabilities. Due to independence, the density of this joint distribution is

$$p(u|\beta, x) = \prod\nolimits_{i=1}^{N} p(u_i|\beta, x_i) \tag{3.77}$$

where each term in (3.77) is

$$p\left(u_i|\beta, x_i\right) = \mathcal{N}\left(u_i|x_i'\beta, 1\right). \tag{3.78}$$

The complete data loglikelihood (excluding additive terms that do not include β) is

$$\ell\left(\beta|u, x\right) = \sum_{i=1}^{N} \ln\left(p\left(u_i|\beta, x_i\right)\right). \tag{3.79}$$

The E-step consists of averaging (3.79) over the conditional distribution $\left[u|x, \beta^{[t]}, y\right]$; that is,

$$\begin{aligned} Q\left(\beta, \beta^{[t]}\right) &= \int \ell\left(\beta|u, x\right) p\left(u|x, \beta^{[t]}, y\right) du \\ &= \int \sum_{i=1}^{N} \ln\left(p\left(u_i|\beta, x_i\right)\right) p\left(u_i|x_i, \beta^{[t]}, y_i\right) du_i \\ &= \sum_{i=1}^{N} \int \ln\left(p\left(u_i|\beta, x_i\right)\right) p\left(u_i|x_i, \beta^{[t]}, y_i\right) du_i. \end{aligned} \tag{3.80}$$

E-step

The calculation of (3.80) needs the following results. For $y_i = 0$:

$$\begin{aligned} \mathrm{E}\left(u_i|\beta, x_i, y_i = 0\right) &= x_i'\beta + \mathrm{E}\left(e_i|y_i = 0\right) \\ &= x_i'\beta + \mathrm{E}\left(e_i|x_i'\beta + e_i > 0\right) \\ &= x_i'\beta + \mathrm{E}\left(e_i|e_i > -x_i'\beta\right) \\ &= x_i'\beta + \frac{1}{\Phi\left(x_i'\beta\right)} \int_{-x_i'\beta}^{\infty} e_i \left(2\pi\right)^{-\frac{1}{2}} \exp\left(-\frac{e_i^2}{2}\right) de_i \\ &= x_i'\beta + \frac{\phi\left(x_i'\beta\right)}{\Phi\left(x_i'\beta\right)}. \end{aligned} \tag{3.81}$$

For $y_i = 1$:

$$\begin{aligned} \mathrm{E}\left(u_i|\beta, x_i, y_i = 1\right) &= x_i'\beta + \mathrm{E}\left(e_i|y_i = 1\right) \\ &= x_i'\beta + \mathrm{E}\left(e_i|x_i'\beta + e_i < 0\right) \\ &= x_i'\beta + \mathrm{E}\left(e_i|e_i < -x_i'\beta\right) \end{aligned}$$

$$
\begin{aligned}
&= x_i'\beta + \frac{1}{\Phi(-x_i'\beta)} \int_{-\infty}^{-x_i'\beta} e_i (2\pi)^{-\frac{1}{2}} \exp\left(-\frac{e_i^2}{2}\right) de_i \\
&= x_i'\beta - \frac{\phi(x_i'\beta)}{1-\Phi(x_i'\beta)}. \qquad (3.82)
\end{aligned}
$$

In these expressions, $\phi(z)$ is the density of $\mathcal{N}(0,1)$ at $Z = z$ and $\Phi(t) = \Pr(U \le t)$, the cumulative distribution function of the standard normal distribution. The ratios $\phi(x_i'\beta)/\Phi(x_i'\beta)$ and $-\phi(x_i'\beta)/(1 - \Phi(x_i'\beta))$ are known as the *intensity of selection* in the quantitative genetics literature (for normally distributed characters). To arrive at the last line in (3.81) and (3.82), use

$$
\int_{-x_i'\beta}^{\infty} e_i (2\pi)^{-\frac{1}{2}} \exp\left(-\frac{e_i^2}{2}\right) de_i = -\int_{-\infty}^{-x_i'\beta} e_i (2\pi)^{-\frac{1}{2}} \exp\left(-\frac{e_i^2}{2}\right) de_i = \phi(x_i'\beta).
$$

With these results, one proceeds with the evaluation of (3.80). Each term $\ln(p(u_i|\beta, x_i))$ is equal to (excluding an additive constant)

$$
\ln(p(u_i|\beta, x_i)) = -\frac{(u_i - x_i'\beta)^2}{2} = -\frac{u_i^2 - 2u_i x_i'\beta + \beta' x_i x_i'\beta}{2}.
$$

For the ith term of (3.80), the expectation over the distribution $\left[u_i|x_i'\beta^{[t]}, y_i\right]$ is

$$
\begin{aligned}
Q_i\left(\beta, \beta^{[t]}\right) &= \mathrm{E}\left[\ln(p(u_i|\beta, x_i))\right] \\
&= -\frac{1}{2}\left[\mathrm{E}\left(u_i^2|x_i, \beta^{[t]}, y_i\right) - 2x_i'\beta\,\mathrm{E}\left(u_i|x_i, \beta^{[t]}, y_i\right) + \beta' x_i x_i'\beta\right]. \qquad (3.83)
\end{aligned}
$$

M-step

The M-step involves a differentiation of (3.80) with respect to β, setting the result equal to zero and solving for β. The solution leads to $\beta^{[t+1]}$.

Since the first term in (3.83) $\mathrm{E}\left(u_i^2|x_i'\beta^{[t]}, y_i\right)$ does not involve β, the partial derivative is

$$
\begin{aligned}
&\frac{\partial}{\partial\beta}\left\{-\frac{1}{2}\left[\mathrm{E}\left(u_i^2|x_i, \beta^{[t]}, y_i\right) - 2x_i'\beta\,\mathrm{E}\left(u_i|x_i, \beta^{[t]}, y_i\right) + \beta' x_i x_i'\beta\right]\right\} \\
&= -\frac{1}{2}\left[-2x_i\,\mathrm{E}\left(u_i|x_i, \beta^{[t]}, y_i\right) + 2x_i x_i'\beta\right],
\end{aligned}
$$

where $\mathrm{E}\left(u_i|x_i, \beta^{[t]}, y_i\right)$ is given by (3.81) or by (3.82). Setting equal to zero yields

$$x_i\left[\mathrm{E}\left(u_i|x_i, \beta^{[t]}, y_i\right) - x_i'\beta\right] = 0.$$

For the N records, we obtain

$$\sum_{i=1}^{N} x_i x_i'\beta = \sum_{i=1}^{N} x_i\,\mathrm{E}\left(u_i|x_i, \beta^{[t]}, y_i\right).$$

The resulting iterative system is

$$\beta^{[t+1]} = \left[\sum_{i=1}^{N}\left(x_i x_i'\right)\right]^{-1}\sum_{i=1}^{N} x_i\,\mathrm{E}\left(u_i|x_i, \beta^{[t]}, y_i\right). \tag{3.84}$$

This solution is used back again in (3.80) to start a new iteration round that requires calculation of $\mathrm{E}\left(u_i|x_i, \beta^{[t+1]}, y_i\right)$ using (3.81) or (3.82) depending on whether $y_i = 1$ or $y_i = 0$. Therefore, the implementation of EM for the probit model is as follows:

1. Start with a guess value $\beta^{[0]}$.
2. Compute $\mathrm{E}\left(u_i|x_i, \beta^{[0]}, y_i\right)$ using (3.81) or (3.82).
3. Solve for β using (3.84) and obtain $\beta^{[1]}$.
4. Go back to 2, update using $\beta^{[1]}$ and continue iterating until convergence.

Notes

If u denotes the $N \times 1$ vector of liabilities, then the model for the missing data can be written as

$$u = X\beta + e, \tag{3.85}$$

where X is an $N \times p$ observed matrix of covariates whose ith row is x_i'. The ith row of this linear system is given by (3.74). With this formulation, in (3.84),

$$\sum_{i=1}^{N}\left(x_i x_i'\right) = X'X.$$

Let $\tilde{E}^{[t]}$ denote the $N \times 1$ vector whose ith element is $\mathrm{E}\left(u_i|x_i, \beta^{[t]}, y_i\right)$. Then (3.84) can be written as

$$\beta^{[t+1]} = \left(X'X\right)^{-1} X'\tilde{E}^{[t]}.$$

Example: A Binomial Regression Model

An extension of the binary probit regression model is the binomial probit regression model. The unobserved original observations y_{ij} are independent binary (0, 1) as before, but now the observed records n_i are counts of one class and $(N_i - n_i)$ counts of the other class associated with covariate x_i, where N_i is the total number of counts in covariate x_i. There are $i = 1, 2, \ldots, C$ levels of the covariate and the total number of counts is $\sum_{i=1}^{C} N_i = N$. In other words, the observed counts n_i are the result of adding the unobserved original records y_{ij}:

$$n_i = \sum_{j=1}^{N_i} y_{ij}.$$

The binomial likelihood is

$$L(\beta|x, y) \propto \prod_{i=1}^{C} \left[\left(1 - \Phi\left(x_i'\beta\right)\right)^{n_i} \left(\Phi\left(x_i'\beta\right)\right)^{N_i - n_i} \right] \tag{3.86}$$

and the loglikelihood, up to an additive constant, is

$$\ell(\beta|x, y) = \sum_{i=1}^{C} \left[n_i \ln\left(1 - \phi(x_i'\beta)\right) + (N_i - n_i) \ln\left(\Phi\left(x_i'\beta\right)\right) \right]. \tag{3.87}$$

The extension of the EM algorithm to accommodate the binomial model consists of associating the liability defined in (3.74) to each of the $\sum_{i=1}^{C} N_i = N$ binary outcomes. The complete data loglikelihood takes the form

$$\ell(\beta|u, x) = \sum_{i=1}^{C} \sum_{j=1}^{N_i} \ln\left(p\left(u_{ij}|\beta, x_i\right)\right), \tag{3.88}$$

where

$$p\left(u_{ij}|\beta, x_i\right) = N\left(u_{ij}|x_i'\beta, 1\right), \quad j = 1, \ldots, N_i. \tag{3.89}$$

For example, x_i can represent the level i of a drug administered to N_i mice, n_i is the number of mice that are observed dead and $N_i - n_i$ are the number alive. The liability for all the N_i mice has the same form given by (3.89). The E-step of the algorithm is now

$$Q\left(\beta, \beta^{[t]}\right) = \int \ell(\beta|u, x)\, p\left(u|x, \beta^{[t]}, y\right) du$$

$$= \int \sum_{i=1}^{C} \sum_{j=1}^{N_i} \ln\left(p\left(u_{ij}|\beta, x_i\right)\right) \left[I\left(y_{ij} = 1\right) p\left(u_{ij}|\beta^{[t]}, x_i, y_{ij}\right)\right.$$

$$+I\left(y_{ij}=0\right)p\left(u_{ij}|\beta^{[t]},x_i,y_{ij}\right)\Big]du_{ij}$$

$$=\sum_{i=1}^{C}\left\{n_i\,\mathrm{E}_{[y=1]}\ln\left(p\left(u_{ij}|\beta,x_i\right)\right)+(N_i-n_i)\,\mathrm{E}_{[y=0]}\ln\left(p\left(u_{ij}|\beta,x_i\right)\right)\right\} \tag{3.90}$$

where $\mathrm{E}_{[y=1]}$ indicates expectation over $\left[u_{ij}|\beta^{[t]},x_i,y_{ij}=1\right]$ and $\mathrm{E}_{[y=0]}$ indicates expectation over $\left[u_{ij}|\beta^{[t]},x_i,y_{ij}=0\right]$. These two expectations are given by (3.81) and (3.82); that is,

$$\mathrm{E}\left(u_{ij}|\beta,x_i,y_{ij}=1\right)=x_i'\beta-\frac{\phi\left(x_i'\beta\right)}{1-\Phi\left(x_i'\beta\right)}, \tag{3.91a}$$

$$\mathrm{E}\left(u_{ij}|\beta,x_i,y_{ij}=0\right)=x_i'\beta+\frac{\phi\left(x_i'\beta\right)}{\Phi\left(x_i'\beta\right)}. \tag{3.91b}$$

The term $\ln\left(p\left(u_{ij}|\beta,x_i\right)\right)$, up to an additive constant, is equal to

$$-\frac{\left(u_{ij}-x_i'\beta\right)^2}{2}=-\frac{u_{ij}^2-2u_{ij}x_i'\beta+\beta'x_ix_i'\beta}{2}$$

and

$$\mathrm{E}_{[y=1]}\ln\left(p\left(u_{ij}|\beta,x_i\right)\right)=-\frac{1}{2}\Big[\mathrm{E}_{[y=1]}\left(u_{ij}^2|x_i,\beta^{[t]},y_{ij}\right)$$
$$-2x_i'\beta\,\mathrm{E}_{[y=1]}\left(u_{ij}|x_i,\beta^{[t]},y_{ij}\right)+\beta'x_ix_i'\beta\Big]$$

with a similar structure for $\mathrm{E}_{[y=0]}$.

The M-step involves differentiation of (3.90) with respect to β. For the ith term in (3.90) this gives, after a little simplification,

$$n_ix_i\,\mathrm{E}_{[y=1]}\left(u_{ij}|x_i,\beta^{[t]},y_{ij}\right)-n_ix_ix_i'\beta+(N_i-n_i)x_i$$
$$\mathrm{E}_{[y=0]}\left(u_{ij}|x_i,\beta^{[t]},y_{ij}\right)-(N_i-n_i)x_ix_i'\beta$$
$$=n_ix_i\,\mathrm{E}_{[y=1]}\left(u_{ij}|x_i,\beta^{[t]},y_{ij}\right)+(N_i-n_i)x_i$$
$$\mathrm{E}_{[y=0]}\left(u_{ij}|x_i,\beta^{[t]},y_{ij}\right)-N_ix_ix_i'\beta.$$

Therefore differentiation of (3.90) with respect to β is equal to

$$\frac{\partial Q\left(\beta, \beta^{[t]}\right)}{\partial \beta} = \sum_{i=1}^{C} n_i x_i \,\mathrm{E}_{[y=1]}\left(u_{ij}|x_i, \beta^{[t]}, y_{ij}\right)$$

$$+\sum_{i=1}^{C}(N_i - n_i)x_i \,\mathrm{E}_{[y=0]}\left(u_{ij}|x_i, \beta^{[t]}, y_{ij}\right) - \sum_{i=1}^{C} N_i x_i x_i' \beta$$

Setting equal to zero yields

$$\sum_{i=1}^{C} N_i x_i x_i' \beta = \sum_{i=1}^{C} n_i x_i \,\mathrm{E}_{[y=1]}\left(u_{ij}|x_i, \beta^{[t]}, y_{ij}\right)$$

$$+\sum_{i=1}^{C}(N_i - n_i)x_i \,\mathrm{E}_{[y=0]}\left(u_{ij}|x_i, \beta^{[t]}, y_{ij}\right)$$

and the iterative system becomes

$$\beta^{[t+1]} = \left[\sum_{i=1}^{C} N_i x_i x_i'\right]^{-1} \left\{\sum_{i=1}^{C} n_i x_i \,\mathrm{E}_{[y=1]}\left(u_{ij}|x_i, \beta^{[t]}, y_{ij}\right)\right.$$

$$\left.+\sum_{i=1}^{C}(N_i - n_i)x_i \,\mathrm{E}_{[y=0]}\left(u_{ij}|x_i, \beta^{[t]}, y_{ij}\right)\right\}. \tag{3.92}$$

Let D_N, D_n and D_δ denote $C \times C$ diagonal matrices with diagonal elements N_i, n_i and $(N_i - n_i)$, respectively. Let $\tilde{E}^{[t]}_{[y=1]}$ be the $C \times 1$ vector whose ith element $(i = 1, \ldots, C)$ is $\mathrm{E}\left(u_{ij}|x_i, \beta^{[t]}, y_{ij} = 1\right)$ and let $\tilde{E}^{[t]}_{[y=0]}$ be the $C \times 1$ vector whose ith element is $\mathrm{E}\left(u_{ij}|x_i, \beta^{[t]}, y_{ij} = 0\right)$. Then (3.92) can be written more compactly as

$$\beta^{[t+1]} = \left(X' D_N X\right)^{-1} \left[X' D_n \tilde{E}^{[t]}_{[y=1]} + X' D_\delta \tilde{E}^{[t]}_{[y=0]}\right]. \tag{3.93}$$

A Digression on Some Matrix Algebra Results

I outline results that are useful for the examples that follow.

Assume that the joint distribution of the vector of data y and the vector of additive genetic values a is multivariate normal

$$\left.\begin{matrix} y \\ a \end{matrix}\right| \theta \sim N\left(\begin{bmatrix} X\beta \\ 0 \end{bmatrix}, \begin{bmatrix} V & ZA\sigma_a^2 \\ AZ'\sigma_a^2 & A\sigma_a^2 \end{bmatrix}\right),$$

with $V = ZAZ'\sigma_a^2 + I\sigma_e^2$, $\theta = (\beta, \sigma_a^2, \sigma_e^2)$, β is a location vector of "fixed" effects, A is the additive genetic relationship matrix, X and Z are observed incidence matrices and σ_a^2, σ_e^2 are components of variance.

- *Conditional mean*. From properties of the multivariate normal distribution

$$a|\theta, y \sim N\left(\mathrm{E}(a|\theta, y), \mathrm{Var}(a|\theta, y)\right),$$

where

$$\mathrm{E}(a|\theta, y) = AZ'V^{-1}(y - X\beta)\sigma_a^2.$$

Let $k = \sigma_e^2/\sigma_a^2$. Substituting V^{-1} with

$$V^{-1} = \frac{1}{\sigma_e^2}I - \frac{1}{\sigma_e^2}Z\left(Z'Z + A^{-1}k\right)^{-1}Z'$$

leads to the following expression for the conditional mean:

$$\begin{aligned}
&\mathrm{E}(a|\theta, y) \\
&= AZ'\left[\frac{1}{\sigma_e^2}I - \frac{1}{\sigma_e^2}Z\left(Z'Z + A^{-1}k\right)^{-1}Z'\right](y - X\beta)\sigma_a^2 \\
&= \left[A\frac{1}{k} - \frac{1}{k}AZ'Z\left(Z'Z + A^{-1}k\right)^{-1}\right]Z'(y - X\beta) \\
&= \left(Z'Z + A^{-1}k\right)^{-1}Z'(y - X\beta). \qquad (3.94)
\end{aligned}$$

The last line follows because

$$A\frac{1}{k} - \frac{1}{k}AZ'Z\left(Z'Z + A^{-1}k\right)^{-1} = \left(Z'Z + A^{-1}k\right)^{-1}.$$

This can be verified by post-multiplying the left-hand side by the inverse of the right-hand side, which recovers I.

- *Conditional variance*. The conditional variance is

$$\mathrm{Var}(a|\theta, y) = \left(Z'Z + A^{-1}k\right)^{-1}\sigma_e^2. \qquad (3.95)$$

This is obtained using properties of the multivariate normal distribution:

$$\begin{aligned}
\mathrm{Var}(a|\theta, y) &= A\sigma_a^2 - AZ'\sigma_a^2V^{-1}ZA\sigma_a^2 \\
&= A\sigma_a^2 - \left[A\frac{1}{k} - \frac{1}{k}AZ'Z\left(Z'Z + A^{-1}k\right)^{-1}\right]Z'ZA\sigma_a^2
\end{aligned}$$

$$= A\sigma_a^2 - \left(Z'Z + A^{-1}k\right)^{-1} Z'ZA\sigma_a^2$$

$$= \left(Z'Z + A^{-1}k\right)^{-1} \sigma_e^2.$$

The last line can be verified by premultiplying the third line by

$$\left(Z'Z + A^{-1}k\right),$$

that recovers $I\sigma_e^2$.

An alternative derivation from a Bayesian perspective (that implies assigning a uniform, improper prior distribution to the location vector β) is to start from

$$\left.\begin{matrix}\beta \\ a\end{matrix}\right| \sigma_a^2, \sigma_e^2, y \sim N\left(\begin{bmatrix}\widehat{\beta} \\ \widehat{a}\end{bmatrix}, \begin{bmatrix}C^{11} & C^{12} \\ C^{21} & C^{22}\end{bmatrix}\sigma_e^2\right) \tag{3.96}$$

where

$$\begin{bmatrix}C^{11} & C^{12} \\ C^{21} & C^{22}\end{bmatrix}$$

is the inverse of the coefficient matrix of the mixed model equations and $(\widehat{\beta}, \widehat{a})$ is the posterior mean vector:

$$\begin{bmatrix}\widehat{\beta} \\ \widehat{a}\end{bmatrix} = \begin{bmatrix}C^{11} & C^{12} \\ C^{21} & C^{22}\end{bmatrix}\begin{bmatrix}X'y \\ Z'y\end{bmatrix}. \tag{3.97}$$

Using results outlined on page 158 or in Example 1.18 of Chapter 1 in Sorensen and Gianola (2002), one can show that the mean vector of $\left[a|\beta, \sigma_a^2, \sigma_e^2, y\right]$ is equal to

$$\left(Z'Z + A^{-1}k\right)^{-1} Z'(y - X\beta),$$

and its covariance matrix is equal to (3.95).

Example: ML Estimation in the Mixed Linear Model

This example illustrates the derivation of the iterative EM equations for ML estimation of fixed effects and of variance components in the univariate Gaussian mixed linear model with *two* variance components. A classical quantitative genetics setup is the mixed linear model that includes fixed and random effects entering linearly into the conditional (given the random effects) expectation of the observations. Fixed effects could, for example, represent systematic sex differences or breed differences, and the random effects could represent additive genetic values. In this case, the two variance components involve the residual variance and the component of the variance-covariance structure of the vector of additive genetic values. The implicit assumption of the model is that variance components are the same in the two sexes and breeds.

Specifically, assume that the data y (vector of dimension $n \times 1$) is a realisation from

$$y|\beta, a, \sigma_e^2 \sim N\left(X\beta + Za, I\sigma_e^2\right),$$

and the unobserved $q \times 1$ vector of the additive genetic values is multivariate normally distributed:

$$a|A\sigma_a^2 \sim N\left(0, A\sigma_a^2\right).$$

The vector of fixed effects β has order $p \times 1$; X and Z are known incidence matrices and the unknown variance components are the scalars σ_a^2 and σ_e^2. The matrix A is known; it describes expected additive genetic relationships among individuals, given a known pedigree. The focus of inference is $\theta = \left(\beta', \sigma_a^2, \sigma_e^2\right)'$. The observed data likelihood is

$$\begin{aligned} L(\theta|y) &= \int p\left(y|\beta, a, \sigma_e^2\right) p\left(a|A\sigma_a^2\right) da \\ &\propto |V|^{-\frac{1}{2}} \exp\left[-\frac{1}{2}(y - X\beta)' V^{-1} (y - X\beta)\right], \end{aligned} \tag{3.98}$$

where $V = ZAZ'\sigma_a^2 + I\sigma_e^2$ is the unconditional variance-covariance matrix of the observed data y. Rather than working with (3.98), the ML estimate of θ is obtained using the EM algorithm.

Treating the random effects a as the missing data, the complete data likelihood is

$$\begin{aligned} L(\theta, a|y) = &\left|I\sigma_e^2\right|^{-\frac{1}{2}} \exp\left[-\frac{1}{2\sigma_e^2}(y - X\beta - Za)'(y - X\beta - Za)\right] \\ &\times \left|A\sigma_a^2\right|^{-\frac{1}{2}} \exp\left[-\frac{1}{2\sigma_a^2} a' A^{-1} a\right], \end{aligned} \tag{3.99}$$

and the corresponding loglikelihood is

$$\begin{aligned} \ln p(\theta, a|y) = &\text{constant} - \frac{n}{2}\ln\sigma_e^2 - \frac{q}{2}\ln\sigma_a^2 - \frac{1}{2\sigma_a^2} a' A^{-1} a \\ &- \frac{1}{2\sigma_e^2}(y - X\beta - Za)'(y - X\beta - Za). \end{aligned} \tag{3.100}$$

The E-step is

$$\begin{aligned}Q\left(\theta|\theta^{[t]}\right) &= \int \ln p\left(\theta, a|y\right) p\left(a|\theta^{[t]}, y\right) da \\ &= -\frac{n}{2}\ln\sigma_e^2 - \frac{q}{2}\ln\sigma_a^2 - \frac{1}{2\sigma_a^2}\mathrm{E}_{a|\theta^{[t]},y}\left[a'A^{-1}a\right] \\ &-\frac{1}{2\sigma_e^2}\mathrm{E}_{a|\theta^{[t]},y}\left(y - X\beta - Za\right)'\left(y - X\beta - Za\right).\end{aligned}$$

Let

$$\mathrm{E}_{a|\theta^{[t]},y}\left[a|\theta^{[t]}, y\right] = \widetilde{a}^{[t]}, \tag{3.101}$$

and

$$\mathrm{Var}_{a|\theta^{[t]},y}\left[a|\theta^{[t]}, y\right] = \widetilde{V}_a^{[t]}. \tag{3.102}$$

Using results for expectation of quadratic forms (Searle, 1971) (see also NOTE on page 134),

$$\begin{aligned}Q\left(\theta|\theta^{[t]}\right) &= -\frac{n}{2}\ln\sigma_e^2 - \frac{q}{2}\ln\sigma_a^2 - \frac{1}{2\sigma_a^2}\left[\widetilde{a}^{[t]\prime}A^{-1}\widetilde{a}^{[t]} + \mathrm{tr}\left(A^{-1}\widetilde{V}_a^{[t]}\right)\right] \\ &-\frac{1}{2\sigma_e^2}\left[\left(y - X\beta - Z\widetilde{a}^{[t]}\right)'\left(y - X\beta - Z\widetilde{a}^{[t]}\right) + \mathrm{tr}\left(Z'Z\widetilde{V}_a^{[t]}\right)\right].\end{aligned}$$

The M-step consists of setting the following equations equal to zero:

$$\frac{\partial Q\left(\theta|\theta^{[t]}\right)}{\partial\beta} = \frac{1}{\sigma_e^2}X'\left(y - X\beta - Z\widetilde{a}^{[t]}\right),$$

$$\frac{\partial Q\left(\theta|\theta^{[t]}\right)}{\partial\sigma_a^2} = -\frac{q}{2\sigma_a^2} + \frac{1}{2\left(\sigma_a^2\right)^2}\left[\widetilde{a}^{[t]\prime}A^{-1}\widetilde{a}^{[t]} + \mathrm{tr}\left(A^{-1}\widetilde{V}_a^{[t]}\right)\right],$$

and

$$\begin{aligned}\frac{\partial Q\left(\theta|\theta^{[t]}\right)}{\partial\sigma_e^2} &= -\frac{n}{2\sigma_e^2} \\ &+\frac{1}{2\left(\sigma_e^2\right)^2}\left[\left(y - X\beta - Z\widetilde{a}^{[t]}\right)'\left(y - X\beta - Z\widetilde{a}^{[t]}\right) + \mathrm{tr}\left(Z'Z\widetilde{V}_a^{[t]}\right)\right].\end{aligned}$$

Solving for θ, one obtains the iterative system

$$\beta^{[t+1]} = (X'X)^{-1} X' \left(y - Z\widetilde{a}^{[t]}\right), \tag{3.103}$$

$$\sigma_a^{2[t+1]} = \frac{1}{q}\left[\widetilde{a}^{[t]\prime} A^{-1}\widetilde{a}^{[t]} + \mathrm{tr}\left(A^{-1}\widetilde{V}_a^{[t]}\right)\right], \tag{3.104}$$

and

$$\sigma_e^{2[t+1]} = \frac{1}{n}\left[\left(y - X\beta^{[t+1]} - Z\widetilde{a}^{[t]}\right)'\left(y - X\beta^{[t+1]} - Z\widetilde{a}^{[t]}\right) + \mathrm{tr}\left(Z'Z\widetilde{V}_a^{[t]}\right)\right]. \tag{3.105}$$

Explicit expressions for $\widetilde{a}^{[t]}$ and $\widetilde{V}_a^{[t]}$ were given in (3.94) and (3.95).

Example: REML (Restricted Maximum Likelihood) Estimation in the Mixed Linear Model

The model is as in the preceding example, but now the focus of inference is $\theta = \left(\sigma_a^2, \sigma_e^2\right)$, with β ans a viewed as nuisance parameters. The restricted maximum likelihood (REML) equations are derived using a Bayesian perspective, and the mode of the posterior distribution with density $p\left(\sigma_a^2, \sigma_e^2|y\right)$ is chosen as the "REML" point estimator (Patterson and Thompson, 1971). You may want to return to this example after reading Chapter 4.

Assigning improper uniform prior distributions to each of $\left(\sigma_a^2, \sigma_e^2\right)$ and to β, then

$$p\left(\sigma_a^2, \sigma_e^2|y\right) \propto \int p\left(y|\beta, a, \sigma_e^2\right) p\left(a|A, \sigma_a^2\right) da\, d\beta.$$

In this setting, the mode of the posterior distribution of the variance components is identical to the REML estimator (Harville, 1977). Joint maximisation of this expression is difficult. However, it is relatively easy to structure an EM algorithm, where the missing data are now $z = \left(\beta', a'\right)'$. The complete data posterior distribution $p\left(\sigma_a^2, \sigma_e^2, z|y\right)$ is identical to (3.100) and the E-step is now

$$Q\left(\theta|\theta^{[t]}\right) = \int \ln p\left(\theta, \beta, a|y\right) p\left(\beta, a|\theta^{[t]}, y\right) da\, d\beta$$

$$= -\frac{n}{2}\ln\sigma_e^2 - \frac{q}{2}\ln\sigma_a^2 - \frac{1}{2\sigma_a^2}\mathrm{E}_{\beta,a|\theta^{[t]},y}\left[a'A^{-1}a\right]$$

$$-\frac{1}{2\sigma_e^2}\mathrm{E}_{\beta,a|\theta^{[t]},y}\left(y - X\beta - Za\right)'\left(y - X\beta - Za\right). \tag{3.106}$$

Taking expectations over the quadratic forms leads to

$$Q\left(\theta|\theta^{[t]}\right) = -\frac{n}{2}\ln\sigma_e^2 - \frac{q}{2}\ln\sigma_a^2$$
$$-\frac{1}{2\sigma_a^2}\left[\widehat{a}^{[t]\prime}A^{-1}\widehat{a}^{[t]} + \text{tr}\left(A^{-1}C^{22[t]}\right)\sigma_e^{2[t]}\right]$$
$$-\frac{1}{2\sigma_e^2}\left[\widehat{e}^{[t]\prime}\widehat{e}^{[t]} + \text{tr}\left[[X,Z]\,C^{-1[t]}\,[X,Z]'\right]\sigma_e^{2[t]}\right],$$

where $\hat{e}^{[t]} = \left(y - X\hat{\beta}^{[t]} - Z\hat{a}^{[t]}\right)$, $\widehat{a}$ is defined in (3.97) and

$$C^{-1} = \begin{bmatrix} C^{11} & C^{12} \\ C^{21} & C^{22} \end{bmatrix}.$$

Since the missing data are now $z = \left(\beta', a'\right)'$, expectations in (3.106) are taken with respect to $\left[\beta, a|\theta^{[t]}, y\right]$, displayed in (3.96). The M-step is

$$\frac{\partial Q\left(\theta|\theta^{[t]}\right)}{\partial\sigma_a^2} = -\frac{q}{2\sigma_a^2} + \frac{1}{2\left(\sigma_a^2\right)^2}\left[\widehat{a}^{[t]\prime}A^{-1}\widehat{a}^{[t]} + \text{tr}\left(A^{-1}C^{22[t]}\right)\sigma_e^{2[t]}\right],$$

and

$$\frac{\partial Q\left(\theta|\theta^{[t]}\right)}{\partial\sigma_e^2} = -\frac{n}{2\sigma_e^2}$$
$$+\frac{1}{2\left(\sigma_e^2\right)^2}\left[\widehat{e}^{[t]\prime}\widehat{e}^{[t]} + \text{tr}\left[[X,Z]\,C^{-1[t]}\,[X,Z]'\right]\sigma_e^{2[t]}\right].$$

Setting to zero yields the iterative system:

$$\sigma_a^{2[t+1]} = \frac{\widehat{a}^{[t]\prime}A^{-1}\widehat{a}^{[t]} + \text{tr}\left(A^{-1}C^{22[t]}\right)\sigma_e^{2[t]}}{q}, \tag{3.107}$$

$$\sigma_e^{2[t+1]} = \frac{\widehat{e}^{[t]\prime}\widehat{e}^{[t]} + \text{tr}\left[[X,Z]\,C^{-1[t]}\,[X,Z]'\right]\sigma_e^{2[t]}}{n}. \tag{3.108}$$

Contrary to ML estimation, REML estimation or inference via the posterior mode of $\left[\sigma_a^2, \sigma_e^2|y\right]$ requires inverting the entire coefficient matrix C (in each iteration).

Example: Bivariate Normal Model with Missing Records

The example on page 63 is revisited and maximum likelihood inferences are drawn using the EM algorithm. The data consist of records of fathers and of their sons, with a number of the latter missing. As was discussed in connection with the example, the construction of the likelihood does not require incorporation of the missing data mechanism for correct inferences. The ordered data are represented as

$$x_1, x_2, \ldots, x_m, x_{m+1}, \ldots, x_n,$$
$$y_1, y_2, \ldots, y_m.$$

To derive the EM equations, an obvious choice for the missing data is $y_* = y_{m+1}, \ldots, y_n$. Then the complete data loglikelihood, ignoring an additive constant is

$$l\left(\mu_y, \mu_x, \rho, \sigma_{yy}, \sigma_{xx}|x, y, y_*\right) = -\frac{n}{2}\left[\ln\sigma_{yy} + \ln\sigma_{xx} + \ln\left(1-\rho^2\right)\right]$$
$$-\frac{1}{2\left(1-\rho^2\right)}\left[\frac{\sum_{i=1}^n (x_i-\mu_x)^2}{\sigma_{xx}} + \frac{\sum_{i=1}^n \left(y_i-\mu_y\right)^2}{\sigma_{yy}}\right.$$
$$\left.-2\rho\frac{\sum_{i=1}^n (x_i-\mu_x)\left(y_i-\mu_y\right)}{(\sigma_{xx})^{\frac{1}{2}}\left(\sigma_{yy}\right)^{\frac{1}{2}}}\right] \tag{3.109}$$

As mentioned before, if $y_{m+1}, \ldots, y_n$ were observed, the ML estimators have a simple closed form.

Expanding the quadratic forms in (3.109),

$$l\left(\mu_y, \mu_x, \rho, \sigma_{yy}, \sigma_{xx}|x, y, y_*\right) = -\frac{n}{2}\left[\ln\sigma_{yy} + \ln\sigma_{xx} + \ln\left(1-\rho^2\right)\right]$$
$$-\frac{1}{2\left(1-\rho^2\right)}\left(\frac{\sum_{i=1}^n x_i^2 + n\mu_x^2 - 2\mu_x\sum_{i=1}^n x_i}{\sigma_{xx}} + \frac{\sum_{i=1}^n y_i^2 + n\mu_y^2 - 2\mu_y\sum_{i=1}^n y_i}{\sigma_{yy}}\right.$$
$$\left.-2\rho\frac{\sum_{i=1}^n x_i y_i - \mu_x\sum_{i=1}^n y_i - \mu_y\sum_{i=1}^n x_i + n\mu_x\mu_y}{(\sigma_{xx})^{\frac{1}{2}}\left(\sigma_{yy}\right)^{\frac{1}{2}}}\right). \tag{3.110}$$

In this expression, the following equalities hold:

$$\sum_{i=1}^n y_i^2 = \sum_{i=1}^m y_i^2 + \sum_{i=m+1}^n y_{*,i}^2,$$
$$\sum_{i=1}^n x_i y_i = \sum_{i=1}^m x_i y_i + \sum_{i=m+1}^n x_i y_{*,i}, \tag{3.111}$$
$$\sum_{i=1}^n y_i = \sum_{i=1}^m y_i + \sum_{i=m+1}^n y_{*,i},$$

where $y_{*,i}$ is the ith missing observation. The computation of $Q\left(\theta|\theta^{[t]}\right)$ requires the expectations of (3.110) with respect to the distribution of the missing data, given the parameters evaluated at round $[t]$ and given the observed data (x, y). Thus,

$$Q\left(\theta|\theta^{[t]}\right) = \int l\left(\mu_y, \mu_x, \rho, \sigma_{yy}, \sigma_{xx}|x, y, y_*\right) p\left(y_*|\theta^{[t]}, x, y\right) dy_*.$$

This calculation involves expectations of (3.111):

$$\mathrm{E}\left[\sum_{i=1}^{n} y_i^2|\theta^{[t]}, x, y\right] = \sum_{i=1}^{m} y_i^2 + \mathrm{E}\left[\sum_{i=m+1}^{n} y_{*,i}^2|\theta^{[t]}, x, y\right],$$

$$\mathrm{E}\left[\sum_{i=1}^{n} x_i y_i|\theta^{[t]}, x, y\right] = \sum_{i=1}^{m} x_i y_i + \mathrm{E}\left[\sum_{i=m+1}^{n} x_i y_{*,i}|\theta^{[t]}, x, y\right],$$

$$\mathrm{E}\left[\sum_{i=1}^{n} y_i|\theta^{[t]}, x, y\right] = \sum_{i=1}^{m} y_i + \mathrm{E}\left[\sum_{i=m+1}^{n} y_{*,i}|\theta^{[t]}, x, y\right].$$

Using properties of the bivariate normal distribution, the following results are easily derived:

$$\begin{aligned}
\mathrm{E}\left(y_{*,i}|\theta^{[t]}, x, y\right) &= \mathrm{E}\left(y_{*,i}|\theta^{[t]}, x_i\right) \\
&= \mu_y^{[t]} + \frac{\sigma_{xy}^{[t]}}{\sigma_{xx}^{[t]}}\left(x_i - \mu_x^{[t]}\right) \\
&= \mu_y^{[t]} + \rho^{[t]} \frac{\left[\sigma_{yy}^{[t]}\right]^{\frac{1}{2}}}{\left[\sigma_{xx}^{[t]}\right]^{\frac{1}{2}}}\left(x_i - \mu_x^{[t]}\right),
\end{aligned}$$

$$\mathrm{E}\left(x_i y_{*,i}|\theta^{[t]}, x, y\right) = x_i \, \mathrm{E}\left(y_{*,i}|\theta^{[t]}, x, y\right),$$

and

$$\begin{aligned}
\mathrm{E}\left(y_{*,i}^2|\theta^{[t]}, x, y\right) &= \sigma_{yy}^{[t]} - \frac{\left(\sigma_{xy}^{[t]}\right)^2}{\sigma_{xx}^{[t]}} + \left[\mathrm{E}\left(y_{*,i}|\theta^{[t]}, x, y\right)\right]^2 \\
&= \sigma_{yy}^{[t]}\left(1 - \left(\rho^{[t]}\right)^2\right) + \left[\mathrm{E}\left(y_{*,i}|\theta^{[t]}, x, y\right)\right]^2.
\end{aligned}$$

Then the E-step is

$$Q\left(\theta|\theta^{[t]}\right) = -\frac{n}{2}\left[\ln\sigma_{yy} + \ln\sigma_{xx} + \ln\left(1-\rho^2\right)\right]$$

$$-\frac{1}{2\left(1-\rho^2\right)}\left(\frac{\sum_{i=1}^n x_i^2 + n\mu_x^2 - 2\mu_x\sum_{i=1}^n x_i}{\sigma_{xx}}\right.$$

$$+\frac{\mathrm{E}\left[\sum_{i=1}^n y_i^2|\theta^{[t]}, x, y\right] + n\mu_y^2 - 2\mu_y\,\mathrm{E}\left[\sum_{i=1}^n y_i|\theta^{[t]}, x, y\right]}{\sigma_{yy}}$$

$$\left.-2\rho\frac{\mathrm{E}\left[\sum_{i=1}^n x_i y_i|\theta^{[t]}, x, y\right] - \mu_x\,\mathrm{E}\left[\sum_{i=1}^n y_i|\theta^{[t]}, x, y\right] - \mu_y\sum_{i=1}^n x_i + n\mu_x\mu_y}{(\sigma_{xx})^{\frac{1}{2}}\left(\sigma_{yy}\right)^{\frac{1}{2}}}\right), \tag{3.112}$$

and the M-step consists of finding the value of $\theta = \left(\mu_y, \mu_x, \rho, \sigma_{yy}, \sigma_{xx}\right)$ that maximises (3.112); this maximiser becomes $\theta^{[t+1]}$. To find this maximiser, first derivatives are needed:

$$\frac{\partial Q\left(\theta|\theta^{[t]}\right)}{\partial\mu_x} = \frac{n}{(\sigma_{xx})^{\frac{1}{2}}\left(1-\rho^2\right)}\left[\frac{\frac{\sum_{i=1}^n x_i}{n} - \mu_x}{(\sigma_{xx})^{\frac{1}{2}}} - \rho\frac{\frac{\mathrm{E}\left[\sum_{i=1}^n y_i|\theta^{[t]},x,y\right]}{n} - \mu_y}{\left(\sigma_{yy}\right)^{\frac{1}{2}}}\right], \tag{3.113}$$

$$\frac{\partial Q\left(\theta|\theta^{[t]}\right)}{\partial\mu_y} = \frac{n}{\left(\sigma_{yy}\right)^{\frac{1}{2}}\left(1-\rho^2\right)}\left[\frac{\frac{\mathrm{E}\left[\sum_{i=1}^n y_i|\theta^{[t]},x,y\right]}{n} - \mu_y}{\left(\sigma_{yy}\right)^{\frac{1}{2}}} - \rho\frac{\frac{\sum_{i=1}^n x_i}{n} - \mu_x}{(\sigma_{xx})^{\frac{1}{2}}}\right], \tag{3.114}$$

$$\frac{\partial Q\left(\theta|\theta^{[t]}\right)}{\partial\sigma_{xx}} = -\frac{1}{2\sigma_{xx}\left(1-\rho^2\right)}\left[n\left(1-\rho^2\right) - \frac{\sum_{i=1}^n (x_i-\mu_x)^2}{\sigma_{xx}}\right.$$

$$\left.+\rho\frac{\mathrm{E}\left[\sum_{i=1}^n x_i y_i|\theta^{[t]}, x, y\right] - \mu_x\,\mathrm{E}\left[\sum_{i=1}^n y_i|\theta^{[t]}, x, y\right] - \mu_y\sum_{i=1}^n x_i + n\mu_x\mu_y}{(\sigma_{xx})^{\frac{1}{2}}\left(\sigma_{yy}\right)^{\frac{1}{2}}}\right], \tag{3.115}$$

$$\frac{\partial Q\left(\theta|\theta^{[t]}\right)}{\partial\sigma_{yy}} = -\frac{1}{2\sigma_{yy}\left(1-\rho^2\right)}\left[n\left(1-\rho^2\right)\right.$$

$$-\frac{\mathrm{E}\left[\sum_{i=1}^n y_i^2|\theta^{[t]}, x, y\right] + n\mu_y^2 - 2\mu_y\,\mathrm{E}\left[\sum_{i=1}^n y_i|\theta^{[t]}, x, y\right]}{\sigma_{yy}}$$

$$\left.+\rho\frac{\mathrm{E}\left[\sum_{i=1}^n x_i y_i|\theta^{[t]}, x, y\right] - \mu_x\,\mathrm{E}\left[\sum_{i=1}^n y_i|\theta^{[t]}, x, y\right] - \mu_y\sum_{i=1}^n x_i + n\mu_x\mu_y}{(\sigma_{xx})^{\frac{1}{2}}\left(\sigma_{yy}\right)^{\frac{1}{2}}}\right] \tag{3.116}$$

and

$$\frac{\partial Q\left(\theta|\theta^{[t]}\right)}{\partial \rho}=\frac{1}{\left(1-\rho^{2}\right)}\left\{n\rho-\frac{1}{\left(1-\rho^{2}\right)}\left[\rho\left(\frac{\sum_{i=1}^{n}\left(x_{i}-\mu_{x}\right)^{2}}{\sigma_{xx}}\right.\right.\right.$$
$$\left.+\frac{\mathrm{E}\left[\sum_{i=1}^{n} y_{i}^{2}|\theta^{[t]}, x, y\right]+n\mu_{y}^{2}-2\mu_{y}\,\mathrm{E}\left[\sum_{i=1}^{n} y_{i}|\theta^{[t]}, x, y\right]}{\sigma_{yy}}\right)-(1+\rho)^{2}$$
$$\left.\left.\frac{\mathrm{E}\left[\sum_{i=1}^{n} x_{i} y_{i}|\theta^{[t]}, x, y\right]-\mu_{x}\,\mathrm{E}\left[\sum_{i=1}^{n} y_{i}|\theta^{[t]}, x, y\right]-\mu_{y}\sum_{i=1}^{n} x_{i}+n\mu_{x}\mu_{y}}{\left(\sigma_{xx}\right)^{\frac{1}{2}}\left(\sigma_{yy}\right)^{\frac{1}{2}}}\right]\right\}. \tag{3.117}$$

These equations must be solved simultaneously. After setting derivatives equal to zero, Eqs. (3.113) and (3.114) reduce to

$$\frac{\frac{\sum_{i=1}^{n} x_{i}}{n}-\mu_{x}}{\left(\sigma_{xx}\right)^{\frac{1}{2}}}=\rho\frac{\frac{\mathrm{E}\left[\sum_{i=1}^{n} y_{i}|\theta^{[t]}, x, y\right]}{n}-\mu_{y}}{\left(\sigma_{yy}\right)^{\frac{1}{2}}}$$

and

$$\frac{\frac{\mathrm{E}\left[\sum_{i=1}^{n} y_{i}|\theta^{[t]}, x, y\right]}{n}-\mu_{y}}{\left(\sigma_{yy}\right)^{\frac{1}{2}}}=\rho\frac{\frac{\sum_{i=1}^{n} x_{i}}{n}-\mu_{x}}{\left(\sigma_{xx}\right)^{\frac{1}{2}}}.$$

The only solution is

$$\widehat{\mu}_{x}^{[t+1]}=\widehat{\mu}_{x}=\frac{\sum_{i=1}^{n} x_{i}}{n}, \tag{3.118}$$

$$\widehat{\mu}_{y}^{[t+1]}=\frac{\mathrm{E}\left[\sum_{i=1}^{n} y_{i}|\theta^{[t]}, x, y\right]}{n}. \tag{3.119}$$

Since x is completely observed, the solution for $\hat{\mu}_x$ is explicit.

Expressions for σ_{xx}, σ_{yy} and ρ involve (3.115), (3.116) and (3.117). A little algebra results in

$$\widehat{\sigma}_{xx}=\frac{1}{n}\sum_{i=1}^{n}\left(x_{i}-\widehat{\mu}_{x}\right)^{2}, \tag{3.120}$$

$$\widehat{\sigma}_{yy}^{[t+1]} = \frac{1}{n}\left\{ \mathrm{E}\left[\sum_{i=1}^{n} y_i^2 | \theta^{[t]}, x, y \right] + n\left(\widehat{\mu}_y^{[t+1]}\right)^2 - 2\widehat{\mu}_y^{[t+1]} \mathrm{E}\left[\sum_{i=1}^{n} y_i | \theta^{[t]}, x, y \right] \right\}$$

$$= \frac{1}{n}\mathrm{E}\left[\sum_{i=1}^{n} y_i^2 | \theta^{[t]}, x, y \right] - \left(\widehat{\mu}_y^{[t+1]}\right)^2. \tag{3.121}$$

and

$$\widehat{\rho}^{[t+1]} = \frac{\frac{1}{n}\mathrm{E}\left[\sum_{i=1}^{n} x_i y_i | \theta^{[t]}, x, y\right] - \widehat{\mu}_x \widehat{\mu}_y^{[t+1]}}{\left(\widehat{\sigma}_{xx}\right)^{\frac{1}{2}} \left(\widehat{\sigma}_{yy}^{[t+1]}\right)^{\frac{1}{2}}}. \tag{3.122}$$

The EM iterations involve computation of $\hat{\mu}_x$ using (3.118) and of $\hat{\sigma}_{xx}$ using (3.120) and then looping over (3.119), (3.121) and (3.122) to update $\hat{\mu}_y^{[t+1]}$, $\hat{\sigma}_{yy}^{[t+1]}$ and $\hat{\rho}^{[t+1]}$.

Example: A Two-Component Mixture Model

Finite mixtures constitute a very flexible modelling tool that arise in practice when measurements are assumed to be drawn from several subpopulations or mixture components, where the component to which an observation belongs is not identified. Mixture models have a long history; perhaps the first major analysis involving the use of mixtures was undertaken in 1894 by Karl Pearson (Pearson, 1894) that fitted a mixture of two normal probability densities with different means and variances to data on crabs.

This example also assumes that the data are realisations from two subpopulations. A classical scenario in genomics is to decide which of the thousands or millions of genetic markers scattered along the genome has or has not an effect on a particular trait. The trait could have a binary expression, such as presence or absence of a disease, or it could be continuously distributed, such as height in humans. The model poses that each genetic marker effect (an unobserved quantity) is drawn from either a component that generates very small values or from the component that allows for larger values. The objective is to allocate each marker to one of the two components. In the example discussed here, the mixture operates at the level of the observed data.

Let z_i be the (unobserved) binary variable that assigns an observation y_i to a specific mixture component. Assume the marginal probability that Z_i is a draw from the mixture component j is

$$\Pr\left(Z_i = j | \pi_j\right) = \pi_j.$$

Conditional on $z_i = j,\ j = 1, 0$, assume y_i has density $p_j\left(y_i|\theta_j\right)$. That is,

$$\begin{aligned} p\left(y_i, z_i|\theta, \pi\right) &= p_j\left(y_i|z_i = j, \theta_j\right) \Pr\left(Z_i = j|\pi_j\right) \\ &= p_j\left(y_i|z_i = j, \theta_j\right) \pi_j. \end{aligned}$$

Treating z as missing data, let $x = (z, y)$ denote the complete data. The complete data loglikelihood is

$$\ell\left(\theta|x\right) = \sum_i \ell\left(\theta|x_i\right),$$

where $\ell\left(\theta|x_i\right)$ is the loglikelihood of the ith complete datum that takes the form

$$\ell\left(\theta|x_i\right) = \sum_{j=0}^{1}\left[I\left(z_i = j\right)\ln p_j\left(y_i|\theta_j\right) + I\left(z_i = j\right)\ln \pi_j\right]. \tag{3.123}$$

E-step

The E-step for the ith observation consists of averaging (3.123) over the conditional distribution $\left[z_i|\pi^{[t]}, \theta^{[t]}, y\right]$. The contribution from the ith complete datum is

$$Q_i\left(\pi, \theta, \pi^{[t]}, \theta^{[t]}\right) = \mathrm{E}\left\{\sum_{j=0}^{1}\left[I\left(z_i = j\right)\ln p_j\left(y_i|\theta_j\right) + I\left(z_i = j\right)\ln \pi_j\right]\right\} \tag{3.124}$$

and the contribution from the entire data,

$$\begin{aligned} Q\left(\pi, \theta, \pi^{[t]}, \theta^{[t]}\right) &= \mathrm{E}\left\{\sum_{i=1}^{n}\sum_{j=0}^{1}\left\{I\left(z_i = j\right)\left[\ln p_j\left(y_i|\theta_j\right) + \ln \pi_j\right]\right\}\right\} \\ &= \sum_{i=1}^{n}\sum_{j=0}^{1}\left\{\mathrm{E}\left[I\left(z_i = j\right)|\pi^{[t]}, \theta^{[t]}, y\right]\right. \\ &\quad \left.\times\left[\ln p_j\left(y_i|\theta_j\right) + \ln \pi_j\right]\right\} \\ &= \sum_{i=1}^{n}\sum_{j=0}^{1}\left\{\widehat{p}_{ij}\left[\ln p_j\left(y_i|\theta_j\right) + \ln \pi_j\right]\right\}, \end{aligned} \tag{3.125}$$

where the term in squared brackets $\left[\ln p_j\left(y_i|\theta_j\right) + \ln \pi_j\right]$ is a constant with respect to $\left[z_i|\pi^{[t]}, \theta^{[t]}, y\right]$ and

$$\begin{aligned} \widehat{p}_{ij} &= \mathrm{E}\left[I\left(Z_i = j\right)|\pi^{[t]}, \theta^{[t]}, y\right] \\ &= \Pr\left(Z_i = j|\pi^{[t]}, \theta^{[t]}, y\right) \end{aligned} \tag{3.126}$$

to be derived shortly. At this point, note that $\sum_j \pi_j = 1$, $\sum_j \widehat{p}_{ij} = 1$ and therefore $\sum_{i=1}^n \sum_j \widehat{p}_{ij} = \sum_{i=1}^n 1 = n$.

M-step

The Q function must be maximised with respect to the two sets of parameters, the $\theta's$ and π_j. The derivation of π_j requires a constrained maximisation using Lagrange multipliers. Maximising (3.125) with respect to π_j subject to $\sum_j \pi_j = 1$ gives

$$\frac{\partial}{\partial \pi_j}\left[\sum_{i=1}^n \sum_{j=0}^1 \widehat{p}_{ij}\left[\ln p_j\left(y_i|\theta_j\right) + \ln \pi_j\right] + \lambda\left(\sum_j \pi_j - 1\right)\right] = 0$$

or

$$\frac{1}{\pi_j}\sum_{i=1}^n \widehat{p}_{ij} + \lambda = 0.$$

Therefore,

$$-\frac{1}{\lambda}\sum_{i=1}^n \widehat{p}_{ij} = \pi_j. \tag{3.127}$$

Summing over j on both sides

$$-\frac{1}{\lambda}\sum_{i=1}^n \sum_{j=0}^1 \widehat{p}_{ij} = \sum_{j=0}^1 \pi_j$$

leading to

$$-\frac{1}{\lambda}\sum_{i=1}^n 1 = 1$$

which gives $\lambda = -n$. Substituting in (3.127) leads to the iterate

$$\pi_j^{[t+1]} = \frac{1}{n}\sum_{i=1}^n \widehat{p}_{ij}. \tag{3.128}$$

The derivation of θ_j requires maximisation of (3.125) with respect to θ_j:

$$\frac{\partial}{\partial \theta_j}\left[\sum_{i=1}^n \sum_{j=0}^1 \widehat{p}_{ij}\left[\ln p_j\left(y_i|\theta_j\right) + \ln \pi_j\right]\right] = 0$$

or

$$\sum_{i=1}^{n} \widehat{p}_{ij} \frac{\partial}{\partial \theta_j} \ln p_j \left(y_i | \theta_j \right) = 0. \tag{3.129}$$

To be specific, assume that $p_j \left(y_i | \theta_j \right) = N \left(y_i | \theta_j, \sigma^2 \right)$ and that observations are identically and independently distributed. Then ignoring terms that do not contain θ_j,

$$\begin{aligned}\sum_{i=1}^{n} \widehat{p}_{ij} \frac{\partial}{\partial \theta_j} \ln p_j \left(y_i | \theta_j \right) &= \sum_{i=1}^{n} \widehat{p}_{ij} \frac{\partial}{\partial \theta_j} \left[-\frac{\left(y_i - \theta_j \right)^2}{2\sigma^2} \right] \\ &= \sum_{i=1}^{n} \widehat{p}_{ij} \frac{\left(y_i - \theta_j \right)}{\sigma^2} = 0.\end{aligned}$$

Multiplying out by σ^2 yields

$$\theta_j^{[t+1]} = \frac{\sum_{i=1}^{n} \widehat{p}_{ij}\, y_i}{\sum_{i=1}^{n} \widehat{p}_{ij}}. \tag{3.130}$$

Similarly, for σ^2

$$\begin{aligned}\frac{\partial Q}{\partial \sigma^2} &= \sum_{i=1}^{n} \sum_{j=0}^{1} \widehat{p}_{ij} \frac{\partial}{\partial \sigma^2} \left[-\frac{1}{2} \ln \sigma^2 - \frac{\left(y_i - \theta_j \right)^2}{2\sigma^2} \right] \\ &= \sum_{i=1}^{n} \sum_{j=0}^{1} \widehat{p}_{ij} \left[-\frac{1}{2\sigma^2} + \frac{\left(y_i - \theta_j \right)^2}{2\left(\sigma^2 \right)^2} \right] = 0.\end{aligned}$$

Multiplying out by $2\sigma^2$,

$$\sum_{i=1}^{n} \sum_{j=0}^{1} \left[-\widehat{p}_{ij} + \frac{\widehat{p}_{ij} \left(y_i - \theta_j \right)^2}{\sigma^2} \right] = 0.$$

Then

$$\sum_{i=1}^{n} \sum_{j=0}^{1} \frac{\widehat{p}_{ij} \left(y_i - \theta_j \right)^2}{\sigma^2} = \sum_{i=1}^{n} \sum_{j=0}^{1} \widehat{p}_{ij} = \sum_{i=1}^{n} 1 = n$$

and finally

$$\sigma^{2[t+1]} = \frac{1}{n} \sum_{i=1}^{n} \sum_{j=0}^{1} \widehat{p}_{ij} \left(y_i - \theta_j \right)^2. \tag{3.131}$$

An expression for $\hat{p}_{ij}$ defined in (3.126) is obtained as follows:

$$\begin{aligned}\widehat{p}_{ij}^{[t+1]} &= \Pr\left(Z_i = j|\pi^{[t]}, \theta_j^{[t]}, y_i\right) \\ &= \frac{p_j\left(y_i, Z_i = j|\theta_j^{[t]}, \pi^{[t]}\right)}{\sum_j p_j\left(y_i, Z_i = j|\theta_j^{[t]}, \pi^{[t]}\right)} \\ &= \frac{p_j\left(y_i|\theta_j^{[t]}\right)\Pr\left(Z_i = j|\pi^{[t]}\right)}{\sum_j p_j\left(y_i|\theta_j^{[t]}\right)\Pr\left(Z_i = j|\pi^{[t]}\right)} \\ &= \frac{p_j\left(y_i|\theta_j^{[t]}\right)\pi_j^{[t]}}{\sum_j p_j\left(y_i|\theta_j^{[t]}\right)\pi_j^{[t]}}. \end{aligned} \tag{3.132}$$

A Bayesian implementation of this mixture model is on page 234, and examples using likelihood and Bayesian models are illustrated on pages 352 and 356.

Example: Genomic Model

This model was implemented using Newton-Raphson and here the EM algorithm is used instead. The focus of inference is the variance components with a view to learning how much of the total variance of the trait is explained by the regression on marker information. Full details of the hierarchical model are specified in (3.38).

If

$$\begin{aligned} y|U, \alpha, \sigma_e^2 &\sim N\left(U\alpha, I\sigma_e^2\right), \\ \alpha|U, \sigma_g^2 &\sim SN\left(0, D\sigma_g^2\right), \end{aligned}$$

then the distribution $[\alpha|y, \sigma_e^2, \sigma_g^2]$ is singular normal, with conditional mean

$$\begin{aligned} \mathrm{E}\left(\alpha|y, \sigma_e^2, \sigma_g^2\right) = \widetilde{\alpha} &= \sigma_g^2 DU'\left(UDU'\sigma_g^2 + I\sigma_e^2\right)^{-1} y \\ &= kD\left(Dk + I\right)^{-1}\widetilde{y} \end{aligned} \tag{3.133}$$

where $\tilde{y} = U'y$ and $k = \sigma_g^2/\sigma_e^2$. The conditional variance is

$$\begin{aligned} \mathrm{Var}\left(\alpha|y, \sigma_e^2, \sigma_g^2\right) = \widetilde{V}_\alpha &= D\sigma_g^2 - \sigma_g^2 DU'\left(UDU'\sigma_g^2 + I\sigma_e^2\right)^{-1} UD\sigma_g^2 \\ &= D\sigma_g^2 - \sigma_g^2 kD\left(Dk + I\right)^{-1} D. \end{aligned} \tag{3.134}$$

Both (3.133) and (3.134) have a diagonal structure. Simple operations on these expressions reveal that for the ith element,

$$\mathrm{E}\left(\alpha_i|y,\sigma_e^2,\sigma_g^2\right)=\widetilde{\alpha}_i=\frac{\lambda_i k}{\lambda_i k+1}\widetilde{y}_i, \tag{3.135a}$$

$$\mathrm{Var}\left(\alpha_i|y,\sigma_e^2,\sigma_g^2\right)=\widetilde{V}_{\alpha_i}=\sigma_g^2\frac{\lambda_i}{\lambda_i k+1},\quad i=1,\ldots,n-1, \tag{3.135b}$$

and when $i=n$ both terms are equal to zero.

The joint density of y and α, given the variance components, is

$$p\left(y|\alpha,\sigma_e^2\right)p\left(\alpha|\sigma_g^2\right)=\left|I\sigma_e^2\right|^{-\frac{1}{2}}\exp\left[-\frac{1}{2\sigma_e^2}\left(y-U\alpha\right)'\left(y-U\alpha\right)\right]\left(\sigma_g^2\right)^{-\left(\frac{n-1}{2}\right)}$$
$$\exp\left[-\frac{1}{2\sigma_g^2}\alpha' D^-\alpha\right]. \tag{3.136}$$

The full data likelihood (where α acts as the missing data) is proportional to (3.136), and the full data loglikelihood, up to an additive constant, is given by

$$\ell\left(\sigma_e^2,\sigma_g^2,\alpha|y\right)=-\frac{n}{2}\ln\sigma_e^2-\left(\frac{n-1}{2}\right)\sigma_g^2-\frac{1}{2\sigma_e^2}\left(y-U\alpha\right)'\left(y-U\alpha\right)-$$
$$-\frac{1}{2\sigma_g^2}\alpha' D^-\alpha. \tag{3.137}$$

The Q function that defines the E-step of the algorithm is

$$Q\left(\sigma_e^2,\sigma_g^2|\sigma_e^{2[t]},\sigma_g^{2[t]}\right)=$$
$$-\frac{n}{2}\ln\sigma_e^2-\left(\frac{n-1}{2}\right)\sigma_g^2-\frac{1}{2\sigma_e^2}\mathrm{E}\left[\left(y-U\alpha\right)'\left(y-U\alpha\right)\right]-\frac{1}{2\sigma_g^2}\mathrm{E}\left[\alpha' D^-\alpha\right] \tag{3.138}$$

Using results from expectations of quadratic forms (see NOTE at the end of the example) and the notation defined in (3.133) and (3.134) yields

$$Q\left(\sigma_e^2,\sigma_g^2|\sigma_e^{2[t]},\sigma_g^{2[t]}\right)=-\frac{n}{2}\ln\sigma_e^2-\left(\frac{n-1}{2}\right)\sigma_g^2-$$
$$-\frac{1}{2\sigma_e^2}\left[\left(y-U\widetilde{\alpha}^{[t]}\right)'\left(y-U\widetilde{\alpha}^{[t]}\right)+\mathrm{tr}\left(\widetilde{V}_\alpha^{[t]}\right)\right]$$
$$-\frac{1}{2\sigma_g^2}\left[\widetilde{\alpha}^{[t]\prime}D^-\widetilde{\alpha}^{[t]}+\mathrm{tr}\left(D^-\widetilde{V}_\alpha^{[t]}\right)\right]. \tag{3.139}$$

Differentiation with respect to σ_e^2 and σ_g^2 constitutes the M-step. This gives

$$\frac{\partial}{\partial \sigma_e^2} Q\left(\sigma_e^2, \sigma_g^2 | \sigma_e^{2[t]}, \sigma_g^{2[t]}\right) = -\frac{n}{2\sigma_e^2} + \frac{1}{2\left(\sigma_e^2\right)^2}\left[\left(y - U\widetilde{\alpha}^{[t]}\right)'\left(y - U\alpha^{[t]}\right) + \mathrm{tr}\left(\widetilde{V}_\alpha^{[t]}\right)\right],$$

$$\frac{\partial}{\partial \sigma_g^2} Q\left(\sigma_e^2, \sigma_g^2 | \sigma_e^{2[t]}, \sigma_g^{2[t]}\right) = -\frac{n-1}{2\sigma_g^2} + \frac{1}{2\left(\sigma_g^2\right)^2}\left[\widetilde{\alpha}^{[t]\prime} D^- \widetilde{\alpha}^{[t]} + \mathrm{tr}\left(D^- \widetilde{V}_\alpha^{[t]}\right)\right].$$

Setting these two equations equal to zero leads to the EM iterative system:

$$\sigma_g^{2[t+1]} = \frac{\widetilde{\alpha}^{[t]\prime} D^- \widetilde{\alpha}^{[t]} + \mathrm{tr}\left(D^- \widetilde{V}_\alpha^{[t]}\right)}{n-1}, \tag{3.140a}$$

$$\sigma_e^{2[t+1]} = \frac{\left(y - U\widetilde{\alpha}^{[t]}\right)'\left(y - U\widetilde{\alpha}^{[t]}\right) + \mathrm{tr}\left(\widetilde{V}_\alpha^{[t]}\right)}{n}. \tag{3.140b}$$

This system is computationally easy to implement because all operations involve diagonal matrices. Specifically,

$$\widetilde{\alpha}^{[t]\prime} D^- \widetilde{\alpha}^{[t]} = k^{2[t]} \sum_{i=1}^{n} \widetilde{y}_i^2 \frac{\lambda_i}{\left(\lambda_i k^{[t]} + 1\right)^2},$$

$$\mathrm{tr}\left(D^- \widetilde{V}_\alpha^{[t]}\right) = \sigma_g^{2[t]} \sum_{i=1}^{n} \frac{1}{\lambda_i k^{[t]} + 1} I\left(\lambda_i \neq 0\right),$$

$$\left(y - U\widetilde{\alpha}^{[t]}\right)'\left(y - U\widetilde{\alpha}^{[t]}\right) = \sum_{i=1}^{n} y_i^2 - 2\sum_{i=1}^{n} \widetilde{\alpha}_i^{[t]} \widetilde{y}_i + \sum_{i=1}^{n} \widetilde{\alpha}_i^{2[t]},$$

$$\mathrm{tr}\left(\widetilde{V}_\alpha^{[t]}\right) = \sigma_g^{2[t]} \sum_{i=1}^{n} \frac{\lambda_i}{\lambda_i k^{[t]} + 1},$$

where $\tilde{\alpha}_i^{[t]}$ is computed using (3.135a). The iterative system consists of cycling through the loop:

1. Compute expectations (3.135).
2. Compute (3.140) and return to 1.

Note

The step from (3.138) to (3.139) requires knowledge of expectations of quadratic forms. The standard result is that if the vector x has mean μ and variance V, then

$$\mathrm{E}\left(x'Ax\right) = \mu' A\mu + \mathrm{tr}\,(AV). \tag{3.141}$$

We need

$$\mathrm{E}\left[(y - U\alpha)'\,(y - U\alpha)\,|y, \sigma_e^{2[t]}, \sigma_g^{2[t]}\right].$$

Now,

$$\mathrm{E}\left((y - U\alpha)\,|y, \sigma_e^{2[t]}, \sigma_g^{2[t]}\right) = y - U\widetilde{\alpha}$$

and

$$\mathrm{Var}\left((y - U\alpha)\,|y, \sigma_e^{2[t]}, \sigma_g^{2[t]}\right) = U\widetilde{V}_\alpha^{[t]}U'.$$

Then using (3.141) results in

$$\mathrm{E}\left[(y - U\alpha)'\,(y - U\alpha)\,|y, \sigma_e^{2[t]}, \sigma_g^{2[t]}\right] = \left(y - U\widetilde{\alpha}^{[t]}\right)'\left(y - U\widetilde{\alpha}^{[t]}\right) + \mathrm{tr}\left(\widetilde{V}_\alpha^{[t]}\right),$$

where, for the second term in the right-hand side, I used

$$\mathrm{tr}\left(U\widetilde{V}_\alpha^{[t]}U'\right) = \mathrm{tr}\left(U'U\widetilde{V}_\alpha^{[t]}\right) = \mathrm{tr}\left(\widetilde{V}_\alpha^{[t]}\right).$$

Finally, direct application of (3.141) gives

$$\mathrm{E}\left[\left(\alpha' D^{-}\alpha\right)|y, \sigma_e^{2[t]}, \sigma_g^{2[t]}\right] = \widetilde{\alpha}^{[t]\prime} D^{-}\widetilde{\alpha}^{[t]} + \mathrm{tr}\left(D^{-}\widetilde{V}_\alpha^{[t]}\right).$$

Example: Likelihood Inferences with Truncated Data and Using the Method of Moments

The likelihood function for truncated data was introduced in the previous chapter on page 71. Here, I indicate how to derive ML estimates of parameters using the EM algorithm. The example concludes with a short outline of estimation using the method of moments.

The focus of inference is the parameters of the original, untruncated distribution, assumed in this example to be the Gaussian distribution $\mathcal{N}\left(\mu, \sigma^2\right)$. The information available is as follows: from the original sample of size N (drawn randomly from

$\mathcal{N}\left(\mu, \sigma^2\right)$), m observations were discarded because they were smaller than an observed threshold C. The observed data consist of n records larger than the threshold C. The complete data z consist of $N = m + n$ records, $z = (y^*, y)$ where y^* are the m unobserved missing records and y are the n observed records.

Let $\theta = \left(\mu, \sigma^2\right)$. The complete data loglikelihood excluding an additive constant is

$$\ln p\left(y^*, y|\theta\right) = -\frac{1}{2}(m+n)\ln\left(\sigma^2\right) - \frac{\sum_{i=1}^{m}\left(y_i^* - \mu\right)^2}{2\sigma^2} - \frac{\sum_{i=1}^{n}(y_i - \mu)^2}{2\sigma^2}. \tag{3.142}$$

The E$-$step consists of computing expectations of this complete data loglikelihood, conditional on the current value of θ, $\theta^{[t]}$ (remembering that the random variable is the unobserved missing data):

$$Q\left(\theta|\theta^{[t]}\right) = -\frac{1}{2}(m+n)\ln\left(\sigma^2\right) - \frac{\sum_{i=1}^{m}\mathrm{E}\left[\left(y_i^* - \mu\right)^2|\theta^{[t]}\right]}{2\sigma^2} - \frac{\sum_{i=1}^{n}(y_i - \mu)^2}{2\sigma^2}. \tag{3.143}$$

The M$-$step needs $\partial Q\left(\theta|\theta^{[t]}\right)/\partial\mu$ and $\partial Q\left(\theta|\theta^{[t]}\right)/\partial\sigma^2$. A little algebra shows that

$$\frac{\partial Q\left(\theta|\theta^{[t]}\right)}{\partial\mu} = \frac{1}{\sigma^2}\left[m\,\mathrm{E}\left(y_i^*|\theta^{[t]}\right) - m\mu + \sum_{i=1}^{n}(y_i - \mu)\right],$$

$$\frac{\partial Q\left(\theta|\theta^{[t]}\right)}{\partial\sigma^2} = \frac{1}{2}\left(\sigma^2\right)^{-2}\left[\sum_{i=1}^{m}\mathrm{E}\left[\left(y_i^* - \mu\right)^2|\theta^{[t]}\right] + \sum_{i=1}^{n}(y_i - \mu)^2 - (m+n)\right].$$

This leads to the iterative system:

$$\mu^{[t+1]} = \frac{m\,\mathrm{E}\left(y_i^*|\theta^{[t]}\right) + \sum_{i=1}^{n} y_i}{m+n}, \tag{3.144a}$$

$$\sigma^{2[t+1]} = \frac{\sum_{i=1}^{m}\mathrm{E}\left[\left(y_i^* - \mu^{[t+1]}\right)^2|\theta^{[t]}\right] + \sum_{i=1}^{n}\left(y_i - \mu^{[t+1]}\right)^2}{m+n}. \tag{3.144b}$$

Using results from the conditional expectations of truncated normal variables spelled out on page 407, one finds

$$\mathrm{E}\left(y_i^*|\theta^{[t]}\right) = \mu^{[t]} - \sigma^{[t]}\frac{\phi\left(c^{[t]}\right)}{\Phi\left(c^{[t]}\right)},$$

$$\mathrm{E}\left[\left(y_i^* - \mu\right)^2|\theta^{[t]}\right] = m\sigma^{2[t]}\left(1 - c^{[t]}\frac{\phi\left(c^{[t]}\right)}{\Phi\left(c^{[t]}\right)}\right),$$

where $\phi(\cdot)$ is the pdf of $\mathcal{N}(0, 1)$ and $c = (C - \mu)/\sigma$. The two expressions above replace the expectations in (3.144).

The R-code below illustrates details of the EM computations. The first part generates $N = 50,000$ records from $\mathcal{N}(10, 3)$ (the complete data), and those larger than $C = \mu + 1.5\sigma$ are kept, and the remaining are discarded. Using these truncated records, one has to draw inferences about the parameters of the original distribution, $\mu = 10$ and $\sigma^2 = 3$.

```
# CODE0307
# EM FOR TRUNCATED DATA; ESTIMATE MEAN OF UNTRUNCATED
# GENERATE Y ~ N(MEAN, VAR)
# TRUNCATE AT T SO THAT Z = Y > T ARE OBSERVED
# Y < T ARE MISSING (KNOWN INFORMATION)
rm(list=ls()) # CLEAR WORKSPACE
set.seed(12371)
nindiv<-50000
mean <- 10
var <- 3
T <- mean + 1.5*sqrt(var)
# CREATE DATA
y <- rnorm(nindiv,mean,sqrt(var))
z <- y[y>T]
length(z)
```

```
## [1] 3462
```

```
m <- length(y)-length(z)
mean(y)
```

```
## [1] 10.00188
```

```
mean(z)
```

```
## [1] 13.37054
```

```
var(z)
```

```
## [1] 0.4728943
```

```
# START VALUES FOR MEAN (mu) AND VARIANCE (sigmasq)
mu <- 0
sigmasq <- 2
sigma <- sqrt(sigmasq)
iter <- 1000
res <- matrix(data=NA, nrow=iter,ncol=2)
```

```
### EM LOOP ############################
for (i in 1:iter){
  T_star <- (T-mu)/sigma
  expymiss <- mu - (sigma * dnorm(T_star))/(pnorm((T_star)))
  mu <- (sum(z)+m*expymiss)/length(y)
  e <- z-mu
  sigmasq <-(m*sigmasq*(1-T_star*dnorm(T_star)/pnorm(T_star))+
             sum(e*e))/length(y)
  sigma <- sqrt(sigmasq)
  res[i,] <- c(mu,sigmasq)
}
tail(res)
```

```
##              [,1]     [,2]
##   [995,] 9.998747 3.077332
##   [996,] 9.998747 3.077332
##   [997,] 9.998747 3.077332
##   [998,] 9.998747 3.077332
##   [999,] 9.998747 3.077332
## [1000,] 9.998747 3.077332
```

```
emmu <- res[iter,1]
emmu
```

```
## [1] 9.998747
```

```
emsigmasq <- res[iter,2]
emsigmasq
```

```
## [1] 3.077332
```

```
emsel <- mu + sigma*dnorm(T_star)/(1-pnorm(T_star))
emsel
```

```
## [1] 13.37243
```

```
i <- dnorm(T_star)/(1-pnorm(T_star))
varsel <- sigmasq*(1-i*(i-T_star))
varsel
```

```
## [1] 0.4649044
```

The ML estimates are $\hat{\mu} = 9.999$ and $\hat{\sigma^2} = 3.077$, in good agreement with the parameters of the unselected population (10 and 3).

Ignoring the selection process and basing inferences on the likelihood of $Y_i \sim N\left(\mu, \sigma^2\right), \quad i = 1, \ldots, n$, one obtains

$$\hat{\mu} = \frac{1}{n}\sum_{i=1}^{n} y_i = 13.4, \tag{3.145a}$$

$$\hat{\sigma}^2 = \frac{1}{n}\sum_{i=1}^{n} \left(y_i - \hat{\mu}\right)^2 = 0.47, \tag{3.145b}$$

quite off the mark from the unselected population parameters. Notwithstanding, in this simple setting (no systematic effects other than a common mean, absence of clustering and variance heterogeneity), (3.145) are consistent estimators of the parameters of the selected population, of which the present data are a random draw. As indicated on page 407,

$$\mathrm{E}\left(y|y > C\right) = \mu + \sigma\frac{\phi\left(c\right)}{1-\Phi\left(c\right)},$$

$$\mathrm{Var}\left(y|y > C\right) = \sigma^2\left[1 - \frac{\phi\left(c\right)}{1-\Phi\left(c\right)}\left(\frac{\phi\left(c\right)}{1-\Phi\left(c\right)} - c\right)\right].$$

Likelihood properties sketched in the previous chapter indicate that MLE of these parameters are

$$\widehat{\mathrm{E}}\left(y|y > C\right) = \widehat{\mu} + \widehat{\sigma}\frac{\phi\left(\widehat{c}\right)}{1-\Phi\left(\widehat{c}\right)},$$

$$\widehat{\mathrm{Var}}\left(y|y > C\right) = \widehat{\sigma}^2\left[1 - \frac{\phi\left(\widehat{c}\right)}{1-\Phi\left(\widehat{c}\right)}\left(\frac{\phi\left(\widehat{c}\right)}{1-\Phi\left(\widehat{c}\right)} - \widehat{c}\right)\right].$$

Substituting in these expressions the converged values $\hat{\mu} = 10.000$ and $\widehat{\sigma}^2 = 3.077$ yields the ML estimates

$$\widehat{\mathrm{E}}\left(y|y > C\right) = 13.37,$$

$$\widehat{\mathrm{Var}}\left(y|y > C\right) = 0.46,$$

in good agreement with (3.145).

Estimation Using the Method of Moments

There is a simple alternative to the EM algorithm using the *method of moments* and properties of the normal distribution. A known truncation point $C = 12.59808$ leads to a "proportion selected" `P=length(y[y>C])/50,000=0.06924` (see `R`-code CODE0307 that uses T for the known truncation point). Then the "$z-$value" corresponding to P is `qnorm(1-P)=1.481475`; the pdf of the standard normal at this z−value is $\phi(z)$ =`dnorm(z)=0.1331442` and the “intensity of selection” is (the difference between the mean of the truncated group and the mean of the original population, measured in units of standard deviation) i =`dnorm(z)/P = 1.922937`.

The method of moments consists of equating sample moments with population moments. The moment estimators of μ and σ^2 are obtained by solving the two linear equations (the right-hand sides represent the mean and variance of truncated normal variables derived on page 407):

$$\frac{1}{n}\sum_{i=1}^{n} y_i = \mu + i\sigma,$$

$$\frac{1}{n}\sum_{i=1}^{n} (y_i - \overline{y})^2 = \sigma^2 \left(1 - i\left(i - z\right)\right).$$

This yields $\hat{\mu} \approx 9.97$ and $\hat{\sigma}^2 \approx 3.13$.

Chapter 4
Bayesian Methods

In classical likelihood, an important goal is to learn about a parameter θ regarded as a fixed unknown quantity. This is accomplished by collecting data y assumed to be a realisation from a probability model $p(y|\theta)$ indexed by θ. This probability model gives rise to the likelihood, a function of θ conditional on the realised y, from which the maximum likelihood estimate $\hat{\theta}$ is obtained. The ML estimator is a random variable (a function of y), whose distribution is characterised by conceptual replications of y. This distribution describes the (sampling) uncertainty of $\hat{\theta}$ and is typically unknown. It can be approximated using resampling techniques or one can appeal to asymptotic results so, informally,

$$\widehat{\theta} \sim N\left(\theta, I(\theta)^{-1}\right), \tag{4.1}$$

as explained before. Probability statements involving $\hat{\theta}$ (such as confidence intervals for θ) can be retrieved from this sampling distribution.

The Bayesian approach is radically different. First of all, the concept of probability is linked to a person's degree of belief. The goal is again to learn about θ, but Bayesians regard θ as a random variable, in the sense that there is uncertainty about the range of values it may take. This uncertainty is characterised by the probability distribution of θ. Before data are available and based on previous knowledge, mechanistic considerations or mathematical convenience, a *prior* probability distribution is elicited. The prior probability distribution has density $p(\theta)$. During the learning process, data are collected; these data are regarded as a realisation from a probability model $p(y|\theta)$, with θ random and unknown and y fixed and observed. This is often labelled as the likelihood. The final stage of the learning process consists of constructing the posterior probability distribution of θ, given data y, with density $p(\theta|y)$, using $p(\theta)$ and $p(y|\theta)$. This posterior probability distribution of θ is a description of the new (posterior) uncertainty about θ in the light of y. It mirrors how prior beliefs have been modified after observing the evidence: data y. During this process, the data have remained fixed and the inference

D. Sorensen, *Statistical Learning in Genetics*, Statistics for Biology and Health,
https://doi.org/10.1007/978-3-031-35851-7_4

is conditional on the particular realisation of these data. Specifically, Bayes theorem takes the form

$$p(\theta|y) = \frac{p(y|\theta)\,p(\theta)}{p(y)} \propto p(y|\theta)\,p(\theta) \tag{4.2}$$

The contentious part of the Bayesian approach is the elicitation of the prior probability $p(\theta)$ and how it may affect the conclusions of the analysis.

This chapter provides an introduction to Bayesian analysis mostly through examples. The objective of the first examples is to show the type of inferences that can be drawn (joint inferences, conditional inferences, marginal inferences) when the posterior distribution is known. In this case, inferences can be exact using analytical methods or can be estimated using Monte Carlo draws from the appropriate distribution. The Gaussian distribution $N(\mu, \sigma^2)$ is chosen as a classical example where all the necessary distributions can be obtained in closed form.

The normal linear model is heavily used for the analysis of continuous data. The chapter includes an example where the conditional distribution of location parameters is derived, given dispersion parameters. The fully conditional posterior distributions of location parameters are also derived; these are an essential ingredient for a Gibbs sampling implementation. A further example based on a classical regression model shows how the variance component can be inferred from its marginal posterior distribution. The modal value coincides with the restricted maximum likelihood estimator and the marginal posterior distribution has the form of a restricted likelihood.

The concept of *Bayesian learning* is the subject of two examples taken from Sorensen and Gianola (2002). This is a process by which prior information (with uncertainty described by the prior distribution) is modified by new data (generated with uncertainty under a sampling model), to become a posterior distribution, reflecting posterior uncertainty. If data accrue sequentially, at a particular stage i of the learning process, this posterior distribution acts as prior for stage $i + 1$. A single Bayesian analysis carried out at the end leads to the same inferences as one carried out sequentially. The two examples illustrate this probability-driven inductive process for discrete and for continuous data.

Most often the exact form of the posterior distribution is not known or only known up to proportionality. A large number of options are available to arrive at approximate results. After a very brief account of Bayesian asymptotics, the chapter focuses on Markov chain Monte Carlo (McMC) methods. These are recipes that make use of Markov chain theory to generate approximate draws from posterior distributions. Using these draws, one can obtain Monte Carlo estimates of expectations to construct summaries of the posterior distribution such as the posterior mean, variance and posterior intervals, or one can compute a posterior mode. Joint and single site updating of parameters using Metropolis-Hastings are explained and illustrated in the simple setting of the $N(\mu, \sigma^2)$ model. The same model is used to introduce the Gibbs sampler. The chapter concludes with an overview of tools needed for the analysis of the output from a Markov chain Monte Carlo.

An appendix provides the mathematical details of the McMC machinery.

4.1 Example: Estimating the Mean and Variance of a Normal Distribution

Consider the scalar random variable y that is normally distributed with probability density function:

$$p\left(y_i|\mu,\sigma^2\right) = \left(2\pi\sigma^2\right)^{-\frac{1}{2}} \exp\left(-\frac{(y_i-\mu)^2}{2\sigma^2}\right),$$

$$\sigma^2 > 0, -\infty < \mu < \infty, \ -\infty < y_i < \infty.$$

The central limit theorem provides a justification for the wide use of the normal distribution. It states that under very general conditions, the sum or the mean of a set of random variables is approximately normally distributed. Data that can be approximated using a continuous function and that result from the sum of the additive effects of a large number of factors fall into this category. The distribution has two parameters, the mean μ and the variance σ^2, and is denoted by $N\left(\mu,\sigma^2\right)$.

This example illustrates the type of posterior inferences that can be drawn when the forms of the posterior distributions are known.

Posterior Distribution of μ and σ^2

Consider now data vector $y = \{y_i\}_{i=1}^n$. Direct application of Bayes theorem leads to the posterior density of μ and σ^2:

$$p\left(\mu,\sigma^2|y\right) \propto p\left(\mu,\sigma^2\right) p\left(y|\mu,\sigma^2\right). \tag{4.3}$$

The second term in (4.3), the contribution from the n data points or likelihood, is

$$\begin{aligned}
p\left(y|\mu,\sigma^2\right) &= \left(2\pi\sigma^2\right)^{-\frac{n}{2}} \exp\left(-\frac{\sum_{i=1}^n (y_i-\mu)^2}{2\sigma^2}\right) \\
&= \left(2\pi\sigma^2\right)^{-\frac{n}{2}} \exp\left(-\frac{1}{2\sigma^2}\left[\sum_{i=1}^n (y_i-\overline{y})^2 + n\left(\overline{y}-\mu\right)^2\right]\right) \\
&= \left(2\pi\sigma^2\right)^{-\frac{n}{2}} \exp\left(-\frac{1}{2\sigma^2}\left[(n-1)\,S^2 + n\left(\overline{y}-\mu\right)^2\right]\right),
\end{aligned} \tag{4.4}$$

where

$$S^2 = \frac{\sum_{i=1}^{n} (y_i - \overline{y})^2}{n-1},$$

$$\overline{y} = \frac{\sum_{i=1}^{n} y_i}{n},$$

are the sample variance and mean, respectively.

The term $p\left(\mu, \sigma^2\right) = p\left(\mu|\sigma^2\right) p\left(\sigma^2\right)$ is the joint prior density. The choice for the form of this density is dictated by prior knowledge, by the nature of the scientific problem and/or by mathematical/computational convenience. The latter is less of an issue if the model is fit with Markov chain Monte Carlo (McMC) methods. For the time being, I settle for mathematical convenience and choose a set of conjugate priors. As in Gelman et al (1995), one possibility is to assume prior distributions:

$$\left[\mu|\sigma^2\right] \sim N\left(\mu_0, \sigma^2/n_0\right), \tag{4.5a}$$

$$\left[\sigma^2\right] \sim \chi^{-2}\left(\upsilon_0, S_0^2\right). \tag{4.5b}$$

The form of (4.5a) indicates an a priori dependence between μ and σ^2. The parameter n_0 can be regarded as prior sample size. The density is

$$p\left(\mu|\sigma^2\right) = \left(2\pi\frac{\sigma^2}{n_0}\right)^{-\frac{1}{2}} \exp\left(-\frac{1}{2}\frac{n_0\left(\mu-\mu_0\right)^2}{\sigma^2}\right). \tag{4.6}$$

The prior (4.5b) is a scaled inverted chi-square distribution with scale S_0^2 and degrees of freedom υ_0. The density is

$$p\left(\sigma^2\right) = \left(\frac{\upsilon_0 S_0^2}{2}\right)^{\frac{\upsilon_0}{2}} \left(\Gamma\left(\frac{\upsilon_0}{2}\right)\right)^{-1} \left(\sigma^2\right)^{-\left(\frac{\upsilon_0}{2}+1\right)} \exp\left(-\frac{\upsilon_0 S_0^2}{2\sigma^2}\right), \; \upsilon_0, S_0 > 0. \tag{4.7}$$

The term S_0^2 can be regarded as a prior variance component, so that $\upsilon_0 S_0^2$ is a prior sum of average squared deviations. The density of the joint prior distribution is

$$p\left(\mu, \sigma^2\right) \propto \left(\sigma^2\right)^{-\frac{1}{2}} \left(\sigma^2\right)^{-\left(\frac{\upsilon_0}{2}+1\right)} \exp\left(-\frac{1}{2\sigma^2}\left[\upsilon_0 S_0^2 + n_0\left(\mu-\mu_0\right)^2\right]\right), \tag{4.8}$$

proportional to a normal-inverse-chi-square density. As shown below, the assignment of a normal-inverse-chi-square prior distribution combined with a normal likelihood results in a normal-inverse-chi-square posterior. The property that the posterior distribution follows the same parametric form as the prior distribution is called *conjugacy*.

The joint posterior distribution is obtained replacing (4.4) and (4.8) in (4.3). This gives

$$\begin{aligned} p\left(\mu,\sigma^2|y\right) &\propto \left(\sigma^2\right)^{-\frac{n}{2}} \exp\left(-\frac{1}{2\sigma^2}\left[(n-1)\,S^2 + n\,(\overline{y}-\mu)^2\right]\right) \times \\ &\times \left(\sigma^2\right)^{-\frac{1}{2}} \left(\sigma^2\right)^{-\left(\frac{\upsilon_0}{2}+1\right)} \exp\left(-\frac{1}{2\sigma^2}\left[\upsilon_0 S_0^2 + n_0\,(\mu-\mu_0)^2\right]\right) \\ &= \left(\sigma^2\right)^{-\left(\frac{n+\upsilon_0+3}{2}\right)} \exp\left(-\frac{1}{2\sigma^2}\left[\upsilon_0 S_0^2 + n_0\,(\mu-\mu_0)^2 + (n-1)\,S^2 + n\,(\overline{y}-\mu)^2\right]\right). \end{aligned} \tag{4.9}$$

This expression is proportional to a normal-inverse-chi-square density confirming the conjugacy of the prior distribution.

Conditional Posterior Distribution of μ Given σ^2

The joint posterior distribution $\left[\mu,\sigma^2|y\right]$ can be factorised as follows:

$$p\left(\mu,\sigma^2|y\right) = p\left(\mu|\sigma^2,y\right) p\left(\sigma^2|y\right).$$

The densities of these two components are derived in turn. The distribution $\left[\mu|y,\sigma^2\right]$ can be obtained either by extracting from (4.9) those terms that are function of μ or by direct use of (4.2):

$$p\left(\mu|\sigma^2,y\right) \propto p\left(\mu|\sigma^2\right) p\left(y|\mu,\sigma^2\right), \tag{4.10}$$

where $p\left(\mu|\sigma^2\right)$ is the prior density of the conditional distribution $\left[\mu|\sigma^2\right]$. In both cases, this leads to (including terms in μ only)

$$p\left(\mu|\sigma^2,y\right) \propto \exp\left(-\frac{1}{2\sigma^2}\left[n_0\,(\mu-\mu_0)^2 + n\,(\overline{y}-\mu)^2\right]\right).$$

There is an identity for combining quadratic forms (Box and Tiao 1973, page 74 and page 418) stating that

$$\begin{aligned} A\,(z-a)^2 + B\,(z-b)^2 &= (A+B)\,(z-c)^2 + \frac{AB}{A+B}\,(a-b)^2, \\ c &= \frac{1}{A+B}\,(Aa+Bb). \end{aligned} \tag{4.11}$$

With

$$\begin{aligned}
A &= n_0, \\
B &= n, \\
z &= \mu, \\
a &= \mu_0, \\
b &= \overline{y}, \\
c &= \left(\frac{n_0\mu_0 + n\overline{y}}{n_0 + n}\right),
\end{aligned}$$

then

$$\begin{aligned}
&\exp\left(-\frac{1}{2\sigma^2}\left[n_0(\mu-\mu_0)^2 + n(\overline{y}-\mu)^2\right]\right) = \\
&= \exp\left(-\frac{1}{2\sigma^2}\left[(n_0+n)\left(\mu - \frac{n_0\mu_0 + n\overline{y}}{n_0+n}\right)^2 + \frac{n_0 n}{n_0+n}(\overline{y}-\mu_0)^2\right]\right) \\
&\propto \exp\left(-\frac{n_0+n}{2\sigma^2}\left[\left(\mu - \frac{n_0\mu_0 + n\overline{y}}{n_0+n}\right)^2\right]\right),
\end{aligned} \tag{4.12}$$

where the third line includes the only terms that are function of μ. This is the kernel of the normal distribution:

$$\left[\mu|\sigma^2, y\right] \sim N\left(\mu_1, \sigma_1^2\right). \tag{4.13}$$

where

$$\mathrm{E}\left(\mu|\sigma^2, y\right) = \mu_1 = \frac{n_0\mu_0 + n\overline{y}}{n_0+n} \tag{4.14}$$

is the posterior mean and

$$\mathrm{Var}\left(\mu|\sigma^2, y\right) = \sigma_1^2 = \frac{\sigma^2}{n_0+n} \tag{4.15}$$

is the posterior variance. Dividing numerator and denominator of (4.14) by σ^2, defining $\pi_0 = n_0/\sigma^2$ as the prior precision, $\pi_n = n/\sigma^2$ as the precision arising from the data, then (4.14) can also be written in the form of the shrunk estimator:

$$\mu_1 = \overline{y} - (\overline{y} - \mu_0)\frac{\pi_0}{\pi_0 + \pi_n}. \tag{4.16}$$

Expressions (4.14) and (4.16) express the posterior mean as a compromise between the prior mean μ_0 and the mean of the data $\overline{y}$.

Marginal Posterior Distribution of σ^2

There are two ways to obtain $p\left(\sigma^2|y\right)$. The **first** is to compute the integral

$$p\left(\sigma^2|y\right) = \int p\left(\mu, \sigma^2|y\right) d\mu$$

where $p\left(\mu, \sigma^2|y\right)$ is given up to proportionality in (4.9). Then

$$p\left(\sigma^2|y\right) \propto \int \left(\sigma^2\right)^{-\left(\frac{n+v_0+3}{2}\right)} \exp\left(-\frac{1}{2\sigma^2}\right.$$
$$\left.\left[v_0 S_0^2 + (n-1)\, S^2 + n_0 \left(\mu - \mu_0\right)^2 + n\left(\overline{y} - \mu\right)^2\right]\right) d\mu =$$
$$= \left(\sigma^2\right)^{-\left(\frac{n+v_0+3}{2}\right)} \exp\left(-\frac{1}{2\sigma^2}\left[v_0 S_0^2 + (n-1)\, S^2\right]\right)$$
$$\int \exp\left(-\frac{1}{2\sigma^2}\left[n_0 \left(\mu - \mu_0\right)^2 + n\left(\overline{y} - \mu\right)^2\right]\right) d\mu =$$
$$\left(\sigma^2\right)^{-\left(\frac{n+v_0+3}{2}\right)} \exp\left(-\frac{1}{2\sigma^2}\left[v_0 S_0^2 + (n-1)\, S^2 + \frac{n_0 n}{n_0+n}\left(\overline{y} - \mu_0\right)^2\right]\right)$$
$$\int \exp\left(-\frac{n_0+n}{2\sigma^2}\left[\left(\mu - \frac{n_0\mu_0 + n\overline{y}}{n_0+n}\right)^2\right]\right) d\mu. \tag{4.17}$$

The results in the last two lines are derived using equality (4.12), and the expression in the last line has a simple solution since it involves integration over a normal distribution. In general,

$$\int \exp\left(-\frac{\left(\mu - \mu_1\right)^2}{2\sigma_1^2}\right) d\mu = \left(2\pi\sigma_1^2\right)^{\frac{1}{2}} \propto \left(\sigma^2\right)^{\frac{1}{2}}.$$

Substituting in (4.17),

$$p\left(\sigma^2|y\right) \propto \left(\sigma^2\right)^{-\left(\frac{n+v_0}{2}+1\right)} \exp\left(-\frac{\left[v_0 S_0^2 + (n-1)\, S^2 + \frac{n_0 n}{n_0+n}\left(\overline{y} - \mu_0\right)^2\right]}{2\sigma^2}\right)$$
$$= \left(\sigma^2\right)^{-\left(\frac{\tilde{v}}{2}+1\right)} \exp\left(-\frac{\tilde{v}\tilde{S}^2}{2\sigma^2}\right), \tag{4.18}$$

where

$$\widetilde{S}^2 = \frac{\upsilon_0 S_0^2 + (n-1)\,S^2 + \frac{n_0 n}{n_0+n}\left(\overline{y}-\mu_0\right)^2}{\widetilde{\upsilon}},$$
$$\widetilde{\upsilon} = n + \upsilon_0.$$

This is proportional to the pdf of a scaled inverted chi-square distribution, with $\widetilde{\upsilon}$ degrees of freedom and scale parameter $\tilde{S}^2$. This is symbolised as

$$\left[\sigma^2|y\right] \sim \chi^{-2}\left(\tilde{\upsilon}, \tilde{S}^2\right). \tag{4.19}$$

For example, to draw samples from (4.19), first draw from a chi-square distribution with $\tilde{\upsilon}$ degrees of freedom, invert this number and multiply by

$$\widetilde{\upsilon}\tilde{S}^2 = \upsilon_0 S_0^2 + (n-1)\,S^2 + \frac{n_0 n}{n_0+n}\left(\overline{y}-\mu_0\right)^2. \tag{4.20}$$

A **second** way of arriving at $p\left(\sigma^2|y\right)$ is to use the identity

$$p\left(\sigma^2|y\right) = \frac{p\left(\mu,\sigma^2|y\right)}{p\left(\mu|\sigma^2,y\right)}. \tag{4.21}$$

Using (4.9)

$$p\left(\mu,\sigma^2|y\right) \propto \left(\sigma^2\right)^{-\left(\frac{n+\upsilon_0+3}{2}\right)} \exp\left(-\frac{1}{2\sigma^2}\right.$$
$$\left.\left[\upsilon_0 S_0^2 + (n-1)\,S^2 + n_0\left(\mu-\mu_0\right)^2 + n\left(\overline{y}-\mu\right)^2\right]\right) =$$
$$= \left(\sigma^2\right)^{-\left(\frac{n+\upsilon_0+3}{2}\right)} \exp\left(-\frac{1}{2\sigma^2}\right.$$
$$\left.\left[\upsilon_0 S_0^2 + (n-1)\,S^2 + (n_0+n)\left(\mu-\mu_1\right)^2 + \frac{n_0 n}{n_0+n}\left(\overline{y}-\mu_0\right)^2\right]\right) \tag{4.22}$$

and

$$p\left(\mu|\sigma^2,y\right) \propto \left(\sigma^2\right)^{-\frac{1}{2}} \exp\left(-\frac{n_0+n}{2\sigma^2}\left[\left(\mu-\mu_1\right)^2\right]\right).$$

Substituting in (4.21) yields (4.19).

Marginal Posterior Distribution of μ

The distribution $[\mu|y]$ is easily obtained regarding the t-distribution as a mixture of normal distributions with a common mean and variances distributed as scaled inverse chi-square random variables. For example, if $[y_i|V_i] \sim N(\mu, V_i)$ and $V_i \sim \chi^{-2}(v, \sigma^2)$, then $y_i \sim t(v, \mu, \sigma^2)$, a t-distribution with v degrees of freedom, mean μ and scale σ^2.

In the present case,

$$\left[\mu|\sigma^2, y\right] \sim N\left(\mu_1, \frac{\sigma^2}{n_0+n}\right) \tag{4.23}$$

and

$$\left[\sigma^2|y\right] \sim \chi^{-2}\left(\widetilde{v}, \widetilde{S}^2\right). \tag{4.24}$$

Therefore,

$$p(\mu|y) = \int p\left(\mu|\sigma^2, y\right) p\left(\sigma^2|y\right) d\sigma^2 \sim t\left(\widetilde{v}, \mu_1, \frac{\widetilde{S}^2}{n_0+n}\right), \tag{4.25}$$

a t-distribution with $\tilde{v}$ degrees of freedom, mean μ_1 and scale $\tilde{S}^2/(n_0+n)$. The density is

$$p(\mu|y) = \frac{\Gamma\left(\frac{\tilde{v}+1}{2}\right)}{\Gamma\left(\frac{\tilde{v}}{2}\right)\sqrt{\pi\tilde{v}}} \left(\frac{\tilde{S}^2}{n_0+n}\right)^{-\frac{1}{2}} \left(1 + \frac{n_0+n}{\tilde{v}\tilde{S}^2}(\mu-\mu_1)^2\right)^{-\left(\frac{\tilde{v}+1}{2}\right)}. \tag{4.26}$$

The mean and mode are given by

$$\mathrm{E}(\mu|y) = \mu_1,$$

and the variance is

$$\mathrm{Var}(\mu|y) = \left(\frac{\widetilde{S}^2}{n_0+n}\right)\left(\frac{\tilde{v}}{\tilde{v}-2}\right), \quad \widetilde{v} > 2.$$

The proof of this result is as follows. Using (4.22) and (4.20), the joint posterior of μ and σ^2 is

$$p\left(\mu, \sigma^2|y\right) =$$

$$= K\left(\sigma^2\right)^{-\left(\frac{n+v_0+3}{2}\right)} \exp\left(-\frac{1}{2\sigma^2}\right.$$

$$\left.\left[v_0 S_0^2 + (n-1)S^2 + (n_0+n)(\mu-\mu_1)^2 + \frac{n_0 n}{n_0+n}(\overline{y}-\mu_0)^2\right]\right)$$

$$= K\left(\sigma^2\right)^{-\left(\frac{\widetilde{v}+1}{2}+1\right)} \exp\left(-\frac{1}{2\sigma^2}\left[\widetilde{v}\widetilde{S}^2 + (n_0+n)(\mu-\mu_1)^2\right]\right)$$
$$= K\left(\sigma^2\right)^{-\alpha-1} \exp\left(-\frac{1}{2\sigma^2}\left[\widetilde{v}\widetilde{S}^2 + (n_0+n)(\mu-\mu_1)^2\right]\right)$$

where K is the constant of integration and $\alpha = \frac{\widetilde{v}+1}{2}$. Then

$$p(\mu|y) = \int_0^\infty K\left(\sigma^2\right)^{-\alpha-1} \exp\left(-\frac{1}{2\sigma^2}\left[\widetilde{v}\widetilde{S}^2 + (n_0+n)(\mu-\mu_1)^2\right]\right) d\sigma^2. \tag{4.27}$$

Let

$$A = \widetilde{v}\widetilde{S}^2 + (n_0+n)(\mu-\mu_1)^2,$$
$$x = \frac{A}{2\sigma^2},$$
$$d\sigma^2 = -\frac{A}{2}x^{-2}dx$$

Therefore, (4.27) can be written as

$$p(\mu|y) = -\left(\frac{A}{2}\right)\int_0^\infty K\left(\frac{A}{2x}\right)^{-\alpha-1} \exp(-x)\,x^{-2}dx$$
$$= -\left(\frac{A}{2}\right)\left(\frac{A}{2}\right)^{-\alpha-1}\int_0^\infty K x^{\alpha-1}\exp(-x)\,dx$$
$$\propto A^{-\alpha}\int_0^\infty x^{\alpha-1}\exp(-x)\,dx$$
$$\propto A^{-\alpha},$$

where the last line results from the fact that the integral, known as the Gamma function, is

$$\int_0^\infty x^{\alpha-1}\exp(-x)\,dx = \Gamma(\alpha), \quad \alpha > 0,$$

a constant with respect to μ. Then

$$p(\mu|y) \propto \left(\widetilde{v}\widetilde{S}^2 + (n_0+n)(\mu-\mu_1)^2\right)^{-\left(\frac{\widetilde{v}+1}{2}\right)}$$

$$\propto \left(1+\frac{(n_0+n)(\mu-\mu_1)^2}{\widetilde{v}\widetilde{S}^2}\right)^{-\left(\frac{\widetilde{v}+1}{2}\right)},$$

which is the kernel of (4.26).

4.2 Posterior Predictive Distribution for a New Observation

Anticipating a topic that will be discussed later in the book, I outline the derivation of the posterior predictive distribution for a new to-be-observed observation $\widetilde{y}_i$. A natural application of this distribution is in Bayesian prediction problems, discussed in Chapter 10. In the context of the above Gaussian setup, $\widetilde{y}_i$, given μ and σ^2, is assumed to be a draw from $N(\mu,\sigma^2)$. The density of the posterior predictive distribution of $\widetilde{y}_i$ is

$$\begin{aligned} p(\widetilde{y}|y) &= \iint p\left(\widetilde{y}|\mu,\sigma^2,y\right) p\left(\mu,\sigma^2|y\right) d\mu d\sigma^2 \\ &= \iint p\left(\widetilde{y}|\mu,\sigma^2\right) p\left(\mu|\sigma^2,y\right) p\left(\sigma^2|y\right) d\mu d\sigma^2. \end{aligned} \tag{4.28}$$

The first term in the integrand of the second line follows from the fact that, given $\left(\mu,\sigma^2\right)$, $\widetilde{y}$ is independent of y.

A first integration with respect to μ results in

$$p\left(\widetilde{y}|\sigma^2,y\right) = \int p\left(\widetilde{y}|\mu,\sigma^2\right) p\left(\mu|\sigma^2,y\right) d\mu. \tag{4.29}$$

Before deriving the density of $\left[\widetilde{y}|\sigma^2,y\right]$, one can easily derive the conditional mean and variance. First note that a future observation and a data point have the same sampling distribution. That is,

$$\widetilde{y}|\mu,\sigma^2 \sim N\left(\mu,\sigma^2\right).$$

Then the desired mean and variance are

$$\begin{aligned} E\left[\widetilde{y}|\sigma^2,y\right] &= E_{\mu|\sigma^2,y}\left[E\left(\widetilde{y}|\mu,\sigma^2\right)\right] \\ &= E_{\mu|\sigma^2,y}(\mu) \\ &= \frac{n_0\mu_0+n\overline{y}}{n_0+n} = \mu_1 \end{aligned} \tag{4.30}$$

using (4.14) and

$$\begin{aligned}\mathrm{Var}\left(\widetilde{y}|\sigma^2, y\right) &= \mathrm{Var}_{\mu|\sigma^2,y}\left[\mathrm{E}\left(\widetilde{y}|\mu, \sigma^2\right)\right] + \mathrm{E}_{\mu|\sigma^2,y}\left[\mathrm{Var}\left(\widetilde{y}|\mu, \sigma^2\right)\right] \\ &= \mathrm{Var}_{\mu|\sigma^2,y}\left(\mu\right) + \sigma^2 \\ &= \frac{\sigma^2}{n_0 + n} + \sigma^2 = \sigma_2^2 \end{aligned} \tag{4.31}$$

using (4.15). There are two sources of uncertainty in the posterior predictive distribution. The term σ^2 is the variance of the sampling model. The second represents uncertainty about μ as a result of the finiteness of the sample y. This term vanishes with large n but the first source of predictive uncertainty remains.

The form of the distribution $\left[\tilde{y}|\sigma^2, y\right]$ is readily obtained noting that the integrand in (4.29) involves two Gaussian distributions. That is,

$$p\left(\widetilde{y}|\sigma^2, y\right) = \int N\left(\mu, \sigma^2\right) N\left(\mu_1, \sigma_1^2\right) d\mu$$

Using the identity (4.11), it follows that

$$\widetilde{y}|\sigma^2, y \sim N\left(\mu_1, \sigma_2^2\right). \tag{4.32}$$

To arrive at the desired result, we need a second integration:

$$p\left(\widetilde{y}|y\right) = \int p\left(\widetilde{y}|\sigma^2, y\right) p\left(\sigma^2|y\right) d\sigma^2.$$

Using (4.24) and (4.32), this involves

$$p\left(\widetilde{y}|y\right) = \int N\left(\mu_1, \sigma_2^2\right) \chi^{-2}\left(\widetilde{v}, \widetilde{S}^2\right) d\sigma^2,$$

the integration of a normal distribution and a scaled inverse chi-square distribution. It follows from the result spelled out in (4.23) that $\left[\tilde{y}|y\right]$ has the density of a t-distribution with $\tilde{v}$ degrees of freedom, mean μ_1 and scale $\widetilde{S}^2\left(1 + \frac{1}{n_0+n}\right)$. That is,

$$\widetilde{y}|y \sim t\left(\widetilde{v}, \mu_1, \widetilde{S}^2\left(1 + \frac{1}{n_0 + n}\right)\right). \tag{4.33}$$

4.3 Example: Monte Carlo Inferences of the Joint Posterior Distribution of Mean and Variance

In the previous example, the forms of the joint posterior distribution $\left[\mu, \sigma^2|y\right]$ and the marginal posterior distributions of $[\mu|y]$ and $\left[\sigma^2|y\right]$ are known. Therefore, inferences can be obtained directly using these distributions. These "exact" inferences are compared with those based on Monte Carlo draws from the posterior distribution $\left[\mu, \sigma^2|y\right]$.

A simple way of drawing samples from $\left[\mu, \sigma^2|y\right]$ is based on the method of composition (Tanner 1996):

1. First draw σ_i^{2*} from $\left[\sigma^2|y\right]$.
2. Then draw μ_i^* from $\left[\mu|\sigma_i^{2*}, y\right]$.

Repeating this loop n times generates $\left(\mu_1^*, \sigma_1^{2*}\right), \ldots, \left(\mu_n^*, \sigma_n^{2*}\right)$ which are n realisations from $\left[\mu, \sigma^2|y\right]$. Further, $\left(\mu_1^*\right), \ldots, \left(\mu_n^*\right)$ and $\left(\sigma_1^{2*}\right), \ldots, \left(\sigma_n^{2*}\right)$ are n realisations from the marginal distributions $[\mu|y]$ and $\left[\sigma^2|y\right]$, respectively. There are no issues of convergence with the method of composition. The first extraction $\left(\mu_1^*, \sigma_1^{2*}\right)$ is a draw from $\left[\mu, \sigma^2|y\right]$.

The following data y mimic $n = 9$ bristle number measurements of male *Drosophila* flies which are assumed to be a realisation from $N\left(\mu, \sigma^2\right)$.

$$y = (35.14,\ 37.01,\ 30.97,\ 34.42,\ 38.22,\ 34.04,\ 30.79,\ 42.31,\ 33.78)'$$

The sample mean and variance are $\overline{y} = 35.19$ and $S^2 = 13.03$. As a first approximation, one may wish to compute a 95% confidence interval for μ based on $\left[\mu|\sigma^2, y\right]$. From previous knowledge about the trait you are willing to assign, a value to $\mu_0 = 34$ and you choose $n_0 = 1$. Using $\sigma^2 = S^2 = 13.03$, from (4.14) and (4.15),

$$\mu_1 = \frac{n_0\mu_0 + n\overline{y}}{n_0 + n} = \frac{1 \times 34 + 9 * 35.19}{1 + 9} = 35.07,$$

$$\sigma_1^2 = \frac{\sigma^2}{n_0 + n} = \frac{13.03}{10} = 1.303.$$

Then

$$\left[\mu|\sigma^2 = 13.03, y\right] \sim N\,(35.07, 1.303)\,.$$

A 95% quantile-based confidence interval for μ based on this conditional posterior distribution (given σ^2) gives (32.88; 37.31). This is obtained using the `R-code`

```
ci <- qnorm(c(0.0275,0.975),35.07,sqrt(1.303))
```

that yields (32.88, 37.31). This interval ignores the uncertainty about σ^2 since it assumes $\sigma^2 = S^2$, exactly.

To account properly for uncertainty, the method of composition is implemented drawing samples from $\left[\mu, \sigma^2|y\right]$. This requires to specify (4.23) and (4.24). Based on previous knowledge of the trait, one may assign $S_0^2 = 11.22$ and the parameters of (4.23) and (4.24) are

$$
\begin{aligned}
\mu_1 &= 35.07, \\
\frac{\sigma^2}{n_0 + n} &= \frac{\sigma^2}{1+9}, \\
\widetilde{v} &= v_0 + n = 1 + 9 = 10, \\
\widetilde{S}^2 &= \frac{v_0 S_0^2 + (n-1)\, S^2 + \frac{n_0 n}{n_0+n}\, (\overline{y} - \mu_0)^2}{\widetilde{v}} = \\
&= \frac{1 \times 11.22 + (9-1)\, 13.03 + \frac{1\times 9}{1+9}\, (35.19 - 34)^2}{10} = 11.67
\end{aligned}
$$

With these values, one can repeatedly draw

1. σ_i^{2*} from $\chi^{-2}\,(10, 11.67)$
2. μ_i^* from $N\left(35.07, \frac{\sigma_i^{2*}}{1+9}\right)$

The R-code to perform these calculations is

```
set.seed(377)
s <-(11.67*10)/rchisq(10000,10)
m <- rnorm(10000,35.07,sqrt(s/10))
quantile(m,c(0.025,0.975))
```

```
##      2.5%    97.5%
## 32.64573 37.45952
```

```
quantile(s,c(0.05,0.95))
```

```
##        5%       95%
##  6.403362 29.220816
```

The 95% quantile-based confidence interval for μ is now a little broader than before because the joint inference accounts for uncertainty about σ^2.

4.4 Approximating a Marginal Distribution

The conditional posterior distribution $\left[\mu|\sigma^2 = 13.03, y\right]$ can be regarded as an approximation to

$$p(\mu|y) = \int p\left(\mu|\sigma^2, y\right) p\left(\sigma^2|y\right) d\sigma^2.$$

The explanation lies in the following result. Assume a model that has two parameters, θ_1 and θ_2. Then

$$p(\theta_1|y) = \int p(\theta_1|\theta_2, y)\, p(\theta_2|y)\, d\theta_2. \tag{4.34}$$

Define $\hat{\theta}_2$ as the mode of $[\theta_2|y]$ and expand $p(\theta_1|\theta_2, y)$ in a Taylor series about $\widehat{\theta}_2$:

$$p(\theta_1|\theta_2, y) \approx p\left(\theta_1|\widehat{\theta}_2, y\right) + \left(\theta_2 - \widehat{\theta}_2\right) \left.\frac{\partial p(\theta_1|\theta_2, y)}{\partial \theta_2}\right|_{\theta_2 = \widehat{\theta}_2}.$$

Let

$$f\left(\widehat{\theta}_2\right) = \left.\frac{\partial p(\theta_1|\theta_2, y)}{\partial \theta_2}\right|_{\theta_2 = \widehat{\theta}_2}$$

and substitute in (4.34)

$$p(\theta_1|y) \approx \int \left[p\left(\theta_1|\widehat{\theta}_2, y\right) + \left(\theta_2 - \widehat{\theta}_2\right) f\left(\widehat{\theta}_2\right)\right] p(\theta_2|y)\, d\theta_2$$

$$= p\left(\theta_1|\hat{\theta}_2, y\right) \int p(\theta_2|y)\, d\theta_2 + f\left(\hat{\theta}_2\right) \int \theta_2\, p(\theta_2|y)\, d\theta_2 - f\left(\hat{\theta}_2\right) \hat{\theta}_2 \int p(\theta_2|y)\, d\theta_2$$

$$= p\left(\theta_1|\hat{\theta}_2, y\right) + f\left(\hat{\theta}_2\right)\left[\mathrm{E}(\theta|y) - \widehat{\theta}_2\right]. \tag{4.35}$$

When the posterior mode $\hat{\theta}_2$ is equal to the posterior mean $\mathrm{E}(\theta|y)$, the second term drops out and

$$p(\theta_1|y) \approx p\left(\theta_1|\hat{\theta}_2, y\right) \tag{4.36}$$

In the example above, the mean and the mode are markedly different and both are different from the value conditioned upon $\sigma^2 = 13.03$. The mean is

$$\begin{aligned} \mathrm{E}\left(\sigma^2|y\right) &= \frac{\widetilde{\upsilon}}{\widetilde{\upsilon} - 2} \widetilde{S}^2 \\ &= \frac{10}{8} 11.67 = 14.59 \end{aligned}$$

and the mode is

$$\begin{aligned}\arg\max p\left(\sigma^2|y\right) &= \frac{\widetilde{v}}{\widetilde{v}+2}\widetilde{S}^2 \\ &= \frac{10}{12}11.67 = 9.73,\end{aligned}$$

so in this case, the approximation is rather coarse.

4.5 Example: The Normal Linear Mixed Model

The mixed linear model, or (co)variance components model (Henderson 1953), includes fixed and random effects entering linearly into the conditional expectation of the observations (given the random effects). These effects are the location parameters of the model. Typically, random effects and model residuals are assigned Gaussian distributions which depend on components of variance or covariance. A classical example in animal genetics is modelling continuous data, where explanatory variables are "fixed" effects such as sex and breed and random effects are additive genetic values. In this setup, variance components could include the additive genetic variance and the residual variance. This example describes the computation of the posterior distribution of the location parameters, conditional on the variance components, and indicates the connection with the classic mixed model equations (Henderson et al 1959).

The second part of the example provides the derivation of the so-called fully conditional posterior distribution of location parameters; these are essential ingredients of Gibbs sampling algorithms.

Assume that the variance components are known and let b, of dimension $p \times 1$, and a, of dimension $q \times 1$, represent vectors of "fixed" and random effects, respectively. The conditional distribution (or sampling model) of data vector y, of dimension, $n \times 1$ is

$$y|b, a, \sigma^2 \sim N\left(Xb + Za, I\sigma^2\right),$$

where X, of dimension $n \times p$, and Z, of dimension $n \times q$, are observed incidence matrices. Assume $p(b) \propto$ *constant* and invoking an infinitesimal model, the prior distribution of a is

$$a|\sigma_a^2 \sim N\left(0, A\sigma_a^2\right),$$

where A is the $q \times q$ additive genetic relationship matrix. The scalars σ^2 and σ_a^2 are the residual and the additive genetic variances, respectively.

- This section provides the derivation of the posterior distribution of the location parameters (b, a). A benchmark paper is by Lindley and Smith (1972).

  Using Bayes theorem, the density of the posterior distribution of the parameters is

$$
\begin{aligned}
p\left(b, a|y, \sigma^2, \sigma_a^2\right) &\propto p\left(y|b, a, \sigma^2\right) p\left(a|\sigma_a^2\right) \\
&\propto \exp\left[-\frac{1}{2\sigma^2}(y - Xb - Za)'(y - Xb - Za)\right] \exp\left[-\frac{1}{2\sigma_a^2} a' A^{-1} a\right] \\
&= \exp\left[-\frac{1}{2\sigma^2}\left((y - Xb - Za)'(y - Xb - Za) + a' A^{-1} a\, k\right)\right]
\end{aligned}
$$

with $k = \sigma^2 / \sigma_a^2$. Let $W = (X\ Z)$, $\theta' = \left(b', a'\right)$ and write the mixed model equations as

$$
\left(W'W + \Sigma\right)\widehat{\theta} = W'y \tag{4.37}
$$

where

$$
\Sigma = \begin{bmatrix} 0 & 0 \\ 0 & A^{-1}k \end{bmatrix}.
$$

The two quadratic forms in the exponential term can be combined as follows:

$$
\begin{aligned}
&(y - Xb - Za)'(y - Xb - Za) + a' A^{-1} a\, k \\
&= (y - W\theta)'(y - W\theta) + \theta'\Sigma\theta \\
&= y'y - 2\theta' W' y + \theta'\left(W'W + \Sigma\right)\theta \\
&= y'y - 2\theta'\left(W'W + \Sigma\right)\widehat{\theta} + \theta'\left(W'W + \Sigma\right)\theta.
\end{aligned}
$$

Adding and subtracting $\hat{\theta}'\left(W'W + \Sigma\right)\hat{\theta}$ and rearranging yields

$$
\begin{aligned}
&(y - Xb - Za)'(y - Xb - Za) + a' A^{-1} a\, k \\
&= y'y - \widehat{\theta}'\left(W'W + \Sigma\right)\widehat{\theta} + \left(\theta - \widehat{\theta}\right)'\left(W'W + \Sigma\right)\left(\theta - \widehat{\theta}\right).
\end{aligned}
$$

This trick that I use here for combining the two quadratic forms is used repeatedly in the book, as well as formulas given in Box and Tiao (1973), page 418 (reproduced on page 145).

Since the first two terms do not depend on θ,

$$
p\left(\theta|y, \sigma^2, \sigma_a^2\right) \propto \exp\left[-\frac{1}{2\sigma^2}\left(\theta - \widehat{\theta}\right)'\left(W'W + \Sigma\right)\left(\theta - \widehat{\theta}\right)\right],
$$

which is the kernel of the normal distribution:

$$\left[\theta|y,\sigma^2,\sigma_a^2\right] \sim N\left(\widehat{\theta},\left(W'W+\Sigma\right)^{-1}\sigma^2\right). \tag{4.38}$$

In a normal linear model with known variances, the vector element $\hat{b}$ in $\hat{\theta}$ can be shown to be the maximum likelihood estimator of the vector of fixed effects, and the vector element $\hat{a}$ is the best linear unbiased predictor (BLUP) of a (Henderson et al 1959).

In the classical frequentist setting with b "fixed" and a "random", the inverse of the coefficient matrix of the mixed model equations (4.37) times σ^2 corresponds to

$$\text{Var}\begin{bmatrix}\widehat{b}\\ \widehat{a}-a\end{bmatrix} = \left(W'W+\Sigma\right)^{-1}\sigma^2$$

where $\text{Var}\left(\hat{b}\right)$ is the variance (in repeated sampling of y) of BLUE(b) (best linear unbiased estimator of b) and Var $(\widehat{a}-a)$ is the prediction error variance of the predicted additive genetic values.

- Let $\left(W'W+\Sigma\right)=C$, $W'y=r$ and write the mixed model equations (4.37) as

$$C\widehat{\theta}=r.$$

Partition $\theta'=(\theta_1,\theta_2)$ arbitrarily, such that θ_1 can be scalar or vector. *What is the conditional posterior distribution of* $\left(\theta_1|\theta_2,\sigma^2,\sigma_a^2,y\right)$? From multivariate normal theory,

$$\theta_1,\theta_2|y,\sigma^2,\sigma_a^2 \sim N\left[\begin{pmatrix}\widehat{\theta}_1\\ \widehat{\theta}_2\end{pmatrix},\begin{pmatrix}C^{11} & C^{12}\\ C^{21} & C^{22}\end{pmatrix}\sigma^2\right]$$

where

$$C^{-1}=\begin{pmatrix}C_{11} & C_{12}\\ C_{21} & C_{22}\end{pmatrix}^{-1}=\begin{pmatrix}C^{11} & C^{12}\\ C^{21} & C^{22}\end{pmatrix}.$$

Since $(\theta_1,\theta_2|y,\sigma^2,\sigma_a^2)$ is multivariate normal, so is the conditional distribution $\left(\theta_1|\theta_2,y,\sigma^2,\sigma_a^2\right)$.

Define $r'=\left(r_1',r_2'\right)$. The conditional expected value of $\left[\theta_1|\theta_2,y,\sigma^2,\sigma_a^2\right]$ is

$$\text{E}\left(\theta_1|\theta_2,y,\sigma^2,\sigma_a^2\right)=\widehat{\theta}_1+C^{12}\left(C^{22}\right)^{-1}\left(\theta_2-\widehat{\theta}_2\right).$$

Using the mixed model equations, write

$$\widehat{\theta}_1 = C^{11} r_1 + C^{12} r_2,$$
$$\widehat{\theta}_2 = C^{21} r_1 + C^{22} r_2.$$

Substituting above

$$\begin{aligned}
E\left(\theta_1|\theta_2, y, \sigma^2, \sigma_a^2\right) &= C^{11} r_1 + C^{12} r_2 + C^{12}\left(C^{22}\right)^{-1}\left(\theta_2 - C^{21} r_1 - C^{22} r_2\right) \\
&= \left(C^{11} - C^{12}\left(C^{22}\right)^{-1} C^{21}\right) r_1 + C^{12}\left(C^{22}\right)^{-1}\theta_2 \\
&= C_{11}^{-1}\left(r_1 + C_{11} C^{12}\left(C^{22}\right)^{-1}\theta_2\right) \\
&= C_{11}^{-1}\left(r_1 - C_{12}\theta_2\right). \qquad (4.39)
\end{aligned}$$

The derivation above uses the partition matrix results:

$$C^{11} - C^{12}\left(C^{22}\right)^{-1} C^{21} = C_{11}^{-1},$$
$$C_{11} C^{12}\left(C^{22}\right)^{-1} = -C_{12}.$$

The conditional variance is

$$\begin{aligned}
\mathrm{Var}\left(\theta_1|\theta_2, y, \sigma^2, \sigma_a^2\right) &= \left(C^{11} - C^{12}\left(C^{22}\right)^{-1} C^{21}\right)\sigma^2 \\
&= C_{11}^{-1}\sigma^2.
\end{aligned}$$

Therefore, the conditional posterior distribution is

$$\left[\theta_1|\theta_2, y, \sigma^2, \sigma_a^2\right] \sim N\left[C_{11}^{-1}\left(r_1 - C_{12}\theta_2\right), C_{11}^{-1}\sigma^2\right]. \qquad (4.40)$$

These results play a key role in the implementation of McMC methods that rely on conditional posterior distributions such as the Gibbs sampler. Notice that even if vector θ contains many parameters and C is a very large matrix, the conditional posterior distribution $\left(\theta_1|\theta_2, y, \sigma^2, \sigma_a^2\right)$ involves inverse of matrices of the order of the number of elements in θ_1. For example, if θ_1 is a scalar, only the inverse of the appropriate scalar element is needed.

4.6 Example: Inferring a Variance Component from a Marginal Posterior Distribution

Let the sampling distribution of the $n \times 1$ data vector y be

$$y|b, \sigma^2 \sim N\left(Xb, I\sigma^2\right), \tag{4.41}$$

where X is an observed $n \times p$ matrix of full rank, b is a $p \times 1$ vector of unobserved regressions, I is the $n \times n$ identity matrix and the unobserved scalar σ^2 is the residual variance. On assigning independent improper prior distributions to the parameters of the model (b, σ^2), the posterior distribution is proportional to the likelihood (4.41):

$$p\left(b, \sigma^2|y\right) \propto p\left(y|b, \sigma^2\right). \tag{4.42}$$

This example differs from the example on page 143 by extending the location parameter of the normal distribution to a vector b, by assuming improper prior distributions for (b, σ^2) and by focusing the inference on σ^2. The vector b is regarded as a nuisance parameter that must be integrated out of the joint posterior distribution $p(b, \sigma^2|y)$. This leads to the marginal posterior distribution $\left[\sigma^2|y\right]$. It will be shown that the modal value has the same form as the restricted maximum likelihood (REML) estimator.

The marginal posterior distribution of σ^2 requires computation of

$$\begin{aligned} p\left(\sigma^2|y\right) &= \int p\left(b, \sigma^2|y\right) db \\ &= \int \left(2\pi\sigma^2\right)^{-\frac{n}{2}} \exp\left[-\frac{1}{2\sigma^2}(y - Xb)'(y - Xb)\right] db. \end{aligned} \tag{4.43}$$

Defining $\widehat{b} = \left(X'X\right)^{-1} X'y$, adding and subtracting $X\widehat{b}$ in each term of the quadratic form and carrying out the expansion leads to

$$(y - Xb)'(y - Xb) = \left(y - X\widehat{b}\right)'\left(y - X\widehat{b}\right) + \left(b - \widehat{b}\right)' X'X\left(b - \widehat{b}\right).$$

This is so because the term $\left(b - \widehat{b}\right)' X'\left(y - X\widehat{b}\right) = 0$. Then

$$\begin{aligned} p\left(\sigma^2|y\right) &= \left(2\pi\sigma^2\right)^{-\frac{n}{2}} \exp\left[-\frac{1}{2\sigma^2}\left(y - X\widehat{b}\right)'\left(y - X\widehat{b}\right)\right] \\ &\int \exp\left[-\frac{1}{2\sigma^2}\left(b - \widehat{b}\right)' X'X\left(b - \widehat{b}\right)\right] db. \end{aligned} \tag{4.44}$$

The integration over the kernel of the normal distribution recovers the integration constant:

$$\left|2\pi \left(X'X\right)^{-1} \sigma^2\right|^{\frac{1}{2}} = \left|\left(X'X\right)^{-1}\right|^{\frac{1}{2}} \left(2\pi\sigma^2\right)^{\frac{p}{2}}.$$

Substituting in (4.44) yields

$$p\left(\sigma^2|y\right) \propto \left(\sigma^2\right)^{-\frac{(n-p)}{2}} \exp\left[-\frac{1}{2\sigma^2}\left(y - X\widehat{b}\right)'\left(y - X\widehat{b}\right)\right] \tag{4.45}$$

that is recognised (up to proportionality) as a scaled inverse chi-square distribution with $v = n - p - 2$ degrees of freedom and scale parameter $S^2 = \frac{1}{v}\left(y - X\widehat{b}\right)'\left(y - X\widehat{b}\right)$. That is,

$$p\left(\sigma^2|y\right) \propto \left(\sigma^2\right)^{-\left(\frac{v}{2}+1\right)} \exp\left(-\frac{vS^2}{2\sigma^2}\right). \tag{4.46}$$

Inferences about σ^2 or of functions thereof can be performed analytically from this marginal posterior distribution. It is also possible to draw samples from the posterior distribution of σ^2 and to use the samples to construct a Monte Carlo estimate of the posterior distribution of any function. To obtain a sample from this distribution, draw a chi-square variable with v degrees of freedom, invert this number and multiply by $\left(y - X\widehat{b}\right)'\left(y - X\widehat{b}\right)$.

The modal value of this marginal posterior distribution is obtained by maximising (4.45) or (4.46) over σ^2 that leads to

$$\arg\max_{\sigma^2} p\left(\sigma^2|y\right) = \frac{v}{v+2}S^2 = \frac{\left(y - X\widehat{b}\right)'\left(y - X\widehat{b}\right)}{n-p}. \tag{4.47}$$

In this particular example where (b, σ^2) have a priori an improper uniform distribution, (4.47) has the same exact form as the REML estimator of σ^2 and the marginal distribution (4.45) has the form of a restricted likelihood, here, arrived at by marginalising over the "fixed" effects b. The REML estimator for unbalanced data was originally derived from a frequentist perspective by maximising a likelihood that is invariant to the "fixed effects" in the highly cited work of Patterson and Thompson (1971).

It is revealing to compare the marginal posterior distribution $\left[\sigma^2|y\right]$ with the conditional posterior distribution $\left[\sigma^2|b, y\right]$. The latter takes the form

$$p\left(\sigma^2|b, y\right) \propto \left(\sigma^2\right)^{-\frac{n}{2}} \exp\left[-\frac{1}{2\sigma^2}(y - Xb)'(y - Xb)\right], \tag{4.48}$$

proportional to a scaled inverse chi square distribution with $\upsilon = n - 2$ degrees of freedom. Maximising over σ^2 leads to

$$\arg\max_{\sigma^2} p\left(\sigma^2|b, y\right) = \frac{(y - Xb)'(y - Xb)}{n}. \tag{4.49}$$

A comparison with (4.47) shows how marginalisation "accounts for the loss of degrees of freedom" associated with the p unknowns in b.

4.7 Example: Bayesian Learning—Inheritance of Haemophilia

This and the example below (both taken from Sorensen and Gianola 2002) illustrate the sequential nature of Bayesian inference. In this first example, data are discrete; continuous data involving additive genetic values (assumed to be multivariate normally distributed) are discussed in the next example.

The following is adapted from Gelman et al (1995). Haemophilia is a genetic disease in humans. The locus responsible for its expression is located on the sex chromosomes (these are denoted as XX in women and XY in men). The condition is observed in women only in double recessive individuals (aa) and in men that are carriers of the a allele on the X chromosome. Suppose there is a nonhaemophiliac woman whose father and mother are not affected by the disease, but her brother is known to be haemophiliac. This implies that her nonhaemophiliac mother must be heterozygote, a carrier of a. What is the probability that the propositus woman is also a carrier of the gene? Let θ be a random variable taking one of two mutually exclusive and exhaustive values. Either $\theta = 1$ if the woman is a carrier or $\theta = 0$ otherwise. Since it is known that the mother of the woman must be a carrier (this constitutes part of the system within which probabilities are assigned), the prior distribution of θ is

$$\Pr(\theta = 1) = \Pr(\theta = 0) = \frac{1}{2}.$$

In the absence of additional information, it is not possible to make a very sharp probability assignment. Suppose now that the woman has two sons, none of which is affected. Let Y_i be a binary random variable taking the value 0 if son i is not affected, or 1 if he has the disease. The values taken by Y_1 and Y_2 constitute the evidence. Given that $\theta = 1$, the probability of the observed data is

$$\begin{aligned}&\Pr(Y_1 = 0, Y_2 = 0|\theta = 1)\\ &= \Pr(Y_1 = 0|\theta = 1)\Pr(Y_2 = 0|\theta = 1) = \left(\frac{1}{2}\right)^2 = \frac{1}{4}.\end{aligned}$$

This follows because:

(a) the observations are assumed to be independent, conditionally on θ and
(b) if the woman is a carrier ($\theta = 1$), there is a 50% probability that she will not transmit the allele.

On the other hand, if she is not a carrier ($\theta = 0$):

$$\begin{aligned}&\Pr(Y_1 = 0, Y_2 = 0|\theta = 0)\\ &= \Pr(Y_1 = 0|\theta = 0)\Pr(Y_2 = 0|\theta = 0) = 1 \times 1 = 1,\end{aligned}$$

this being so because it is impossible for a son to have the disease unless the mother is a carrier (ignoring mutation). Hence, the data confer four times more likelihood to the hypothesis that the mother is not a carrier. Using the information that none of the sons is diseased, the posterior distribution of θ is then

$$\begin{aligned}\Pr(\theta = 1|Y_1 = 0, Y_2 = 0) &= \frac{\Pr(\theta = 1)\Pr(Y_1 = 0, Y_2 = 0|\theta = 1)}{\Pr(Y_1 = 0, Y_2 = 0)}\\ &= \frac{\Pr(\theta = 1)\Pr(Y_1 = 0, Y_2 = 0|\theta = 1)}{\sum_{i=0}^{1}\Pr(\theta = i)\Pr(Y_1 = 0, Y_2 = 0|\theta = i)}\\ &= \frac{\frac{1}{2}\frac{1}{4}}{\frac{1}{2}1 + \frac{1}{2}\frac{1}{4}} = \frac{1}{5}\end{aligned}$$

and

$$\Pr(\theta = 0|Y_1 = 0, Y_2 = 0) = 1 - \frac{1}{5} = \frac{4}{5}.$$

A sharper probability assignment can be made now, and the combination of prior information with the evidence suggests that the mother is probably not a carrier. The latter possibility cannot be ruled out, however, as there is a 20% probability that the mother is heterozygote. The posterior odds in favour of the hypothesis that the mother is not a carrier is given by the ratio

$$\begin{aligned}\frac{\Pr(\theta = 0|Y_1 = 0, Y_2 = 0)}{\Pr(\theta = 1|Y_1 = 0, Y_2 = 0)} &= \frac{\Pr(Y_1 = 0, Y_2 = 0|\theta = 0)}{\Pr(Y_1 = 0, Y_2 = 0|\theta = 1)}\frac{\Pr(\theta = 0)}{\Pr(\theta = 1)}\\ &= \frac{1}{\frac{1}{4}}\frac{\frac{1}{2}}{\frac{1}{2}} = 4,\end{aligned}$$

where the ratio

$$\frac{\Pr(\theta = 0)}{\Pr(\theta = 1)} = 1$$

is called the prior odds in favour of the hypothesis. Further,

$$B_{01} = \frac{\Pr(Y_1 = 0, Y_2 = 0|\theta = 0)}{\Pr(Y_1 = 0, Y_2 = 0|\theta = 1)} = 4$$

is called the *Bayes factor*, the factor by which the prior odds about the hypotheses are modified by the evidence and converted into posterior odds. In this example, the odds in favour of the hypothesis that $\theta = 0$ relative to $\theta = 1$ increase by a factor of 4 after observing two sons that are not affected by the disease. Suppose that the woman suspected of being a carrier has n children. The posterior distribution of θ can be represented as

$$\Pr(\theta = i|y) = \frac{\Pr(\theta = i)\Pr(y|\theta = i)}{\Pr(\theta = i)\Pr(y|\theta = i) + \Pr(\theta \neq i)\Pr(y|\theta \neq i)}, \qquad i = 0, 1,$$

where $y = [Y_1, Y_2, ..., Y_n]'$. Partition the data as $y = \left[y_A' y_B'\right]'$, with y_A being the records on the presence or absence of the disease for the first m progeny and with y_B containing data on the last $n - m$ children. The posterior distribution is

$$\Pr(\theta = i|y) = \frac{\Pr(\theta = i)\, p(y_A|\theta = i)\, p(y_B|y_A, \theta = i)}{\sum_{i=0}^{1} \Pr(\theta = i)\, p(y_A|\theta = i)\, p(y_B|y_A, \theta = i)}.$$

Dividing the numerator and denominator by the marginal probability of observing y_A, $p(y_A)$, gives

$$\Pr(\theta = i|y) = \frac{\dfrac{\Pr(\theta = i)\, p(y_A|\theta = i)}{p(y_A)} p(y_B|y_A, \theta = i)}{\sum_{i=0}^{1} \dfrac{\Pr(\theta = i)\, p(y_A|\theta = i)}{p(y_A)} p(y_B|y_A, \theta = i)}.$$

Note, however, that

$$\frac{\Pr(\theta = i)\, p(y_A|\theta = i)}{p(y_A)} = \Pr(\theta = i|y_A)$$

is the posterior probability after observing y_A, which acts as a prior before observing y_B. Then, it follows that

$$\Pr(\theta = i|y) = \frac{\Pr(\theta = i|y_A)\, p(y_B|y_A, \theta = i)}{\sum_{i=0}^{1} \Pr(\theta = i|y_A)\, p(y_B|y_A, \theta = i)}$$

illustrating the "memory" property of Bayes theorem. If the observations are conditionally independent, as assumed in this example, then

$$p\left(y_B|y_A, \theta = i\right) = p\left(y_B|\theta = i\right).$$

Suppose now that the woman has a third, unaffected son. The prior distribution now assigns probabilities of $\frac{4}{5}$ and $\frac{1}{5}$ to the events "not being a carrier" and "carrying the allele", respectively. The posterior probability of the woman being a carrier, after observing a third, unaffected child, is

$$\begin{aligned}
&\Pr\left(\theta = 1|Y_1 = 0, Y_2 = 0, Y_3 = 0\right) \\
&= \frac{\frac{1}{5}\Pr\left(Y_3 = 0|\theta = 1\right)}{\frac{1}{5}\Pr\left(Y_3 = 0|\theta = 1\right) + \frac{4}{5}\Pr\left(Y_3 = 0|\theta = 0\right)} \\
&= \frac{\frac{1}{5}\frac{1}{2}}{\frac{1}{5}\frac{1}{2} + \frac{4}{5}1} = \frac{1}{9}.
\end{aligned}$$

The same result is obtained starting from the prior before observing any children:

$$\begin{aligned}
\Pr\left(\theta = 1|Y_1 = 0, Y_2 = 0, Y_3 = 0\right) &= \frac{\frac{1}{2}.\left(\frac{1}{2}\right)^3}{\frac{1}{2}.\left(\frac{1}{2}\right)^3 + \frac{1}{2}.\left(1\right)^3} \\
&= \frac{1}{9}.
\end{aligned}$$

4.8 Example: Bayesian Learning—Updating Additive Genetic Values

The setting is in the same spirit as in the preceding example. Here, the vector of data $y = \left(y_1', y_2'\right)'$ is collected first at stage 1, y_1, and then at stage 2, y_2. Suppose that at stage 1 (2), measurements y_1 (y_2) are taken on n_1 (n_2) different individuals (so that an individual measured at any stage is not recorded at the other stage) and that the objective is to infer their additive genetic effects a_1 (a_2). The protocol for the evaluation in stages is as follows:

- Using data at stage 1, y_1, infer a_1 and a_2 and derive the marginal posterior distributions $[a_i|y_1]$, $i = 1, 2$. From $[a_2|y_1]$ obtain the mean and variance:

$$\tilde{a}_2 = \mathrm{E}\left(a_2|y_1\right), \quad \tilde{C}_2 = \mathrm{Var}\left(a_2|y_1\right).$$

(dropping here the conditioning on the location parameters and variances to avoid cluttering the notation).

- At stage 2, using as prior for a_2 the posterior from stage 1, obtain

$$p\left(a_2|y_2, y_2\right) \propto p\left(y_2|a_2\right) p\left(a_2|\tilde{a}_2, \tilde{C}_2\right)$$

and finally, with knowledge of the posterior density, obtain $\mathrm{E}\left(a_2|y_2, y_2\right)$ and $\mathrm{Var}\left(a_2|y_2, y_2\right)$.

Consider the following linear model:

$$\begin{bmatrix} y_1 \\ y_2 \end{bmatrix} = \begin{bmatrix} 1_1\mu_1 \\ 1_2\mu_2 \end{bmatrix} + \begin{bmatrix} a_1 \\ a_2 \end{bmatrix} + \begin{bmatrix} e_1 \\ e_2 \end{bmatrix}, \tag{4.50}$$

where $\mu_i, i = 1, 2$, is a known location parameter common to records collected in stages i, $1_i, i = 1, 2$, is a vector of $1'$s for stage i records and

$$\begin{bmatrix} e_1 \\ e_2 \end{bmatrix} \Bigg| \sigma_e^2 \sim N\left(\begin{bmatrix} 0 \\ 0 \end{bmatrix}, \begin{bmatrix} I_1 & 0 \\ 0 & I_2 \end{bmatrix} \sigma_e^2\right)$$

is a vector of independently distributed residual effects, where σ_e^2 is the (known) residual variance; I_i is an identity matrix of order $n_i \times n_i$, $(i = 1, 2)$.

In the classical infinitesimal model of inheritance, the additive genetic effects are assumed to follow the multivariate normal distribution (acting as a prior in the Bayesian sense):

$$\begin{bmatrix} a_1 \\ a_2 \end{bmatrix} \Bigg| \sigma_a^2 \sim N\left(\begin{bmatrix} 0 \\ 0 \end{bmatrix}, \begin{bmatrix} A_{11} & A_{12} \\ A_{21} & A_{22} \end{bmatrix} \sigma_a^2\right) \tag{4.51}$$

where σ_a^2 is the additive genetic variance in the population (also assumed known) and

$$A = \begin{bmatrix} A_{11} & A_{12} \\ A_{21} & A_{22} \end{bmatrix}$$

is the matrix of additive genetic relationships between individuals, or twice the matrix of coefficients of coancestry. This matrix is assumed to have full rank; clones or identical twins are not encountered.

The standard treatment using all the data y is presented first, where data collection at stage 2 has been completed and additive genetic values are inferred for all individuals. The posterior density of the additive genetic values can be obtained using results for the linear model derived in the example of page 156 and is given by

$$p\left(a_1, a_2|y_1, y_2, \mu_1, \mu_2, \sigma_e^2, \sigma_a^2\right)$$
$$\propto \exp\left\{-\frac{1}{2\sigma_e^2}\left[(a-\widehat{a})'\left(I + A^{-1}\frac{\sigma_e^2}{\sigma_a^2}\right)(a-\widehat{a})\right]\right\}, \tag{4.52}$$

which is the kernel of a normal process with mean

$$\widehat{a} = \left(I + A^{-1}\frac{\sigma_e^2}{\sigma_a^2}\right)^{-1}\begin{bmatrix} y_1 - 1_1\mu_1 \\ y_2 - 1_2\mu_2 \end{bmatrix} = \left(I + A^{-1}\frac{\sigma_e^2}{\sigma_a^2}\right)^{-1} w, \tag{4.53}$$

and variance-covariance matrix:

$$\text{Var}\left(a_1, a_2|y_1, y_2, \mu_1, \mu_2, \sigma_e^2, \sigma_a^2\right) = \left(I + A^{-1}\frac{\sigma_e^2}{\sigma_a^2}\right)^{-1} \sigma_e^2. \tag{4.54}$$

Using (4.53) and (4.54), the posterior density of a_1 of the "first-stage" distribution

$$\left[a_1|y_1, \mu_1, \sigma_e^2, \sigma_a^2\right]$$

is immediately found to be

$$p\left(a_1|y_1, \mu_1, \sigma_e^2, \sigma_a^2\right)$$
$$\propto \exp\left\{-\frac{1}{2\sigma_e^2}\left[(a_1 - \widetilde{a}_1)'\left(I_1 + A_{11}^{-1}\frac{\sigma_e^2}{\sigma_a^2}\right)(a_1 - \widetilde{a}_1)\right]\right\}. \tag{4.55}$$

The posterior mean at stage 1 is then

$$\widetilde{a}_1 = \left(I_1 + A_{11}^{-1}\frac{\sigma_e^2}{\sigma_a^2}\right)^{-1}(y_1 - 1_1\mu_1) = \left(I_1 + A_{11}^{-1}\frac{\sigma_e^2}{\sigma_a^2}\right)^{-1} w_1 \tag{4.56}$$

and the posterior covariance is

$$\text{Var}\left(a_1|y_1, \mu_1, \sigma_e^2, \sigma_a^2\right) = \left(I_1 + A_{11}^{-1}\frac{\sigma_e^2}{\sigma_a^2}\right)^{-1} \sigma_e^2 = \widetilde{C}_1. \tag{4.57}$$

The analysis in stages proceeds as follows. What can be said about all additive genetic effects at stage 1? The joint posterior at stage 1 is

$$p\left(a_1, a_2|y_1, \mu_1, \sigma_e^2, \sigma_a^2\right) \propto \left[p\left(y_1|\mu_1, a_1, \sigma_e^2\right) p\left(a_1|\sigma_a^2\right)\right] p\left(a_2|a_1, \sigma_a^2\right).$$

Noting that the expression in square brackets is the posterior of a_1 after stage 1, $p\left(a_1|y_1,\mu_1,\sigma_e^2,\sigma_a^2\right)$, one can write

$$p\left(a_1,a_2|y_1,\mu_1,\sigma_e^2,\sigma_a^2\right) \propto p\left(a_2|a_1,\sigma_a^2\right)p\left(a_1|y_1,\mu_1,\sigma_e^2,\sigma_a^2\right)$$

and this is the density of a normal process because the two intervening densities are in normal forms. Hence, the marginal distribution of a_2 at stage 1 is normal as well, with marginal density:

$$p\left(a_2|y_1,\mu_1,\sigma_e^2,\sigma_a^2\right) = \int p\left(a_2|a_1,\sigma_a^2\right)p\left(a_1|y_1,\mu_1,\sigma_e^2,\sigma_a^2\right)da_1.$$

Therefore, the mean of the posterior distribution of a_2 at stage 1 is

$$\begin{aligned}
\mathrm{E}\left(a_2|y_1,\mu_1,\sigma_e^2,\sigma_a^2\right) &= \int\left[\int a_2 p\left(a_2|a_1,\sigma_a^2\right)da_2\right]p\left(a_1|y_1,\mu_1,\sigma_e^2,\sigma_a^2\right)da_1 \\
&= \mathrm{E}_{a_1|y_1}\left[\mathrm{E}\left(a_2|a_1\right)\right] \\
&= A_{21}A_{11}^{-1}\widetilde{a}_1 \\
&= A_{21}A_{11}^{-1}\left(I_1 + A_{11}^{-1}\frac{\sigma_e^2}{\sigma_a^2}\right)^{-1} w_1.
\end{aligned} \tag{4.58}$$

This has the form of

$$\begin{aligned}
\mathrm{E}\left(a_2|a_1\right) &= \mathrm{E}\left(a_2\right) + A_{21}A_{11}^{-1}\left[a_1 - \mathrm{E}\left(a_1\right)\right] \\
&= A_{21}A_{11}^{-1}a_1
\end{aligned}$$

but with a_1 replaced by its posterior expectation $\widetilde{a}_1$. Likewise,

$$\begin{aligned}
\widetilde{C}_2 &= \mathrm{Var}\left(a_2|y_1,\mu_1,\sigma_e^2,\sigma_a^2\right) \\
&= \mathrm{E}_{a_1|y_1}\left[\mathrm{Var}\left(a_2|a_1\right)\right] + \mathrm{Var}_{a_1|y_1}\left[\mathrm{E}\left(a_2|a_1\right)\right] \\
&= \mathrm{E}_{a_1|y_1}\left[A_{22}\sigma_a^2 - A_{21}A_{11}^{-1}A_{12}\sigma_a^2\right] + \mathrm{Var}_{a_1|y_1}\left(A_{21}A_{11}^{-1}a_1\right) \\
&= A_{22}\sigma_a^2 - A_{21}A_{11}^{-1}A_{12}\sigma_a^2 + A_{21}A_{11}^{-1}\widetilde{C}_1A_{11}^{-1}A_{12}.
\end{aligned} \tag{4.59}$$

The first term represents the variance of a_2 before observing anything; the second term is the reduction in variance that would be obtained if a_1 were known and the third term is a penalty that results from having to infer a_1 from y_1. Thus, the prior distribution of a_2 at stage 2 (which is the posterior distribution of a_2 at stage 1) is a normal process with mean vector (4.58) and covariance matrix (4.59). Label this distribution $\left[a_2|\tilde{a}_2, \tilde{C}_2\right]$. Finally, at stage 2, the posterior density of a_2 is

$$p\left(a_2|y_1, y_2, \mu_1, \mu_2, \sigma_e^2, \sigma_a^2\right) \propto p\left(y_2|\mu_2, a_2, \sigma_e^2\right) p\left(a_2|\widetilde{a}_2, \widetilde{C}_2\right)$$

$$\propto \exp\left[-\frac{(y_2 - 1_2\mu_2 - a_2)'(y_2 - 1_2\mu_2 - a_2)}{2\sigma_e^2} - \frac{(a_2 - \widetilde{a}_2)'\,\widetilde{C}_2^{-1}\sigma_e^2\,(a_2 - \widetilde{a}_2)}{2\sigma_e^2}\right]. \quad (4.60)$$

We know that the density is in a normal form, so the quadratics on a_2 can be combined in the usual manner, to arrive at the mean vector and covariance matrix of the distribution. Alternatively, noting that a normal distribution is unimodal (so the mean is identical to the mode), the posterior mean at stage 2 can be found by maximizing the logarithm of (4.60). Let

$$F(a_2) = -\left[\frac{(y_2 - 1_2\mu_2 - a_2)'(y_2 - 1_2\mu_2 - a_2)}{2\sigma_e^2} + \frac{(a_2 - \widetilde{a}_2)'\,\widetilde{C}_2^{-1}\sigma_e^2\,(a_2 - \widetilde{a}_2)}{2\sigma_e^2}\right]$$

so

$$\frac{\partial F(a_2)}{\partial a_2} = (-1)\,\frac{-2\,(y_2 - 1_2\mu_2 - a_2) + 2\widetilde{C}_2^{-1}\sigma_e^2\,(a_2 - \widetilde{a}_2)}{2\sigma_e^2}.$$

Setting to 0 and solving for a_2 yields

$$\widehat{a}_2 = \left(I_2 + \widetilde{C}_2^{-1}\sigma_e^2\right)^{-1}\left(y_2 - 1_2\mu_2 + \widetilde{C}_2^{-1}\sigma_e^2\widetilde{a}_2\right) \quad (4.61)$$

as mean of the posterior distribution of a_2, after stages 1 and 2. This is a matrix weighted average of $\widetilde{a}_2$ and of $y_2 - 1_2\mu_2$. The variance-covariance matrix of the distribution is given by

$$\mathrm{Var}\left(a_2|y_1, y_2, \mu_1, \mu_2, \sigma_e^2, \sigma_a^2\right) = \left(I_2 + \widetilde{C}_2^{-1}\sigma_e^2\right)^{-1}\sigma_e^2. \quad (4.62)$$

It can be verified that this is equal to the inverse of minus the matrix of second derivatives of $F(a_2)$ with respect to a_2. Using tedious algebra, one can also verify that (4.61) is identical to the a_2-component of the solution to (4.53).

4.9 A Brief Account of Bayesian Asymptotics

When the form of the posterior distribution is unknown or when limiting results are of interest, a simple alternative is to obtain approximate Bayesian inferences appealing to asymptotic arguments. A result from Bayesian asymptotics (first-order approximations, and assuming a number of regularity conditions) says that as sample size n increases, the posterior distribution of a parameter θ converges to a normal distribution with mean equal to the mode of $[\theta|y]$ and asymptotic variance equal to the inverse of minus the second derivative of the logposterior (inverse of the observed information) evaluated at the mode θ,

$$[\theta|y] \sim N\left(\widehat{\theta}, \left[I\left(\widehat{\theta}\right)\right]^{-1}\right), \tag{4.63}$$

where the observed information evaluated at the mode of scalar θ is

$$I\left(\widehat{\theta}\right) = -\left.\frac{\partial^2 \log\left(p\left(\theta|y\right)\right)}{\left(\partial\theta\right)^2}\right|_{\theta=\widehat{\theta}}.$$

For large n, the amount of prior information will tend to be small compared with the information provided by the data and is ignored in this development. The posterior mode $\widehat{\theta}$ is now the ML estimator and the variance can be replaced by the inverse of either $I\left(\theta|y\right)$ or $i\left(\theta\right)$ defined in (2.9) and (2.7).

A heuristic proof of (4.63) is as follows. Write $p\left(\theta|y\right) = \exp\left[\log\left(p\left(\theta|y\right)\right)\right]$ and expand in a Taylor series about the posterior mode of $[\theta|y]$, $\hat{\theta}$. Consider first the case where θ is a scalar. This gives

$$\begin{aligned} p(\theta|y) = \exp&\left[\log\left(p\left(\hat{\theta}|y\right)\right) + \left(\theta-\hat{\theta}\right)\left[\frac{\partial \log\left(p\left(\theta|y\right)\right)}{\partial\theta}\right]_{\theta=\widehat{\theta}}\right. \\ &\left. + \frac{1}{2}\left(\theta-\widehat{\theta}\right)^2\left[\frac{\partial^2 \log\left(p\left(\theta|y\right)\right)}{\left(\partial\theta\right)^2}\right]_{\theta=\widehat{\theta}} + R\right], \end{aligned}$$

where R is of smaller order than $\left(\theta-\widehat{\theta}\right)^2$. At the mode, the first derivative is zero, the term $\log\left(p\left(\widehat{\theta}|y\right)\right)$ does not depend of θ and is absorbed into the normalising constant. Then

$$p\left(\theta|y\right) \propto \exp\left[-\frac{1}{2}\left(\theta-\widehat{\theta}\right)^2\left[\frac{-\partial^2 \log\left(p\left(\theta|y\right)\right)}{\left(\partial\theta\right)^2}\right]_{\theta=\widehat{\theta}} + R\right]$$

In the neighbourhood of $\widehat{\theta}$ as $n \to \infty$, R becomes negligible and $p\left(\theta|y\right)$ is proportional to the kernel of the normal distribution (4.63). Alternatively, Fisher's expected information $i\left(\theta\right)$ defined in (2.9) could be used in place of the observed information $I\left(\theta|y\right)$ in (4.63).

The quality of the approximation can be improved by applying a transformation to θ so that the transformed parameter looks more normal (see Likelihood exercises I, problem 2). The approximation can be accurate when the number of parameters is small relative to the number of data points, when there are no boundary problems associated with θ, and of course it requires the posterior distribution to be proper.

When θ is a vector, the same result (4.63) holds. Now $\hat{\theta}$ is the vector of posterior modes and the variance is the inverse of the matrix:

$$-\frac{\partial^2}{\partial\theta\partial\theta'} p\,(\theta|y)\,.$$

The reader is referred to Bernardo and Smith (1994) and references therein for a discussion of this delicate subject.

4.10 An Overview of Markov Chain Monte Carlo

The examples discussed so far do not pose computational challenges because the posterior distributions have standard forms. As a consequence, one can draw inferences analytically or using Monte Carlo methods by simulating directly from these posterior distributions. The posterior distributions of more complex models have seldom standard forms and so alternative methods are required. A very general method to obtain approximate samples from posterior distributions that does not require knowledge of their integration constant is Markov chain Monte Carlo (McMC). McMC has been successfully applied in many branches of science since its introduction to the statistical community in the early 1990s.

McMC algorithms are recipes for constructing a Markov chain that has the posterior distribution of interest as its *stationary distribution*. The goal is to obtain draws from this posterior distribution. Since doing this directly may be very complicated or impossible, a Markov chain X_i, $i = 1, 2, \ldots$, is constructed so that its stationary distribution is the posterior distribution. To achieve this, the variables X_i are drawn from a *proposal distribution*, but because this is not the same distribution as the posterior distribution, the drawn value is accepted in a stochastic manner by means of an *acceptance probability*. The derivation of this acceptance probability guarantees that the X_i's are an approximate Monte Carlo sample from the stationary distribution. To meet this, the Markov chain must satisfy a number of conditions which hold in most cases (but not always!). The samples can be used for Monte Carlo estimation of various expectations with respect to the stationary distribution. Thus, Monte Carlo estimates of the mean, variance or posterior intervals can be obtained using the draws from the Markov chain.

In what follows, two algorithms are described that are special cases of a third very general McMC algorithm. The focus is on examples that illustrate how the algorithms are applied. Technical details, including the derivation of the acceptance probability of the standard and general McMC algorithms, can be found in the appendix at the end of the chapter.

The algorithms are motivated with a simple example based on the normal linear model $y_i \overset{iid}{\sim} N\left(\mu, \sigma^2\right)$, where the vector of observations $y = \{y_i\}$ has n elements. The likelihood is proportional to

$$y|\mu, \sigma^2 \sim N\left(1\mu, I\sigma^2\right), \tag{4.64}$$

where 1 is a vector of ones with n elements and I is an $n \times n$ identity matrix. Is is assumed that the prior distribution of the parameters factorises into two independent prior distributions of the form

$$\mu \propto constant, \tag{4.65a}$$

$$\sigma^2 \propto \frac{1}{\sigma^2}. \tag{4.65b}$$

The prior for the variance is equivalent to assuming an improper uniform prior for $\ln \sigma$. The resulting joint posterior is

$$p\left(\mu, \sigma^2|y\right) \propto p\left(y|\mu, \sigma^2\right) p\left(\sigma^2\right) p\left(\mu\right)$$

$$\propto \exp\left[-\frac{\sum_{i=1}^{n}(y_i-\mu)^2}{2\sigma^2}\right]\left(\sigma^2\right)^{-\left(\frac{n}{2}+1\right)}. \tag{4.66}$$

For this particular model, there is no need to use McMC because explicit solutions for $p\left(\mu|y\right)$ and for $p\left(\sigma^2|y\right)$ are available. However, an McMC approach will be followed here.

4.11 The Metropolis-Hastings Algorithm

Two strategies are described. In the first, all the parameters of the model are updated simultaneously, while in the other, parameters are updated one at a time.

Joint Updating

Denote the target distribution by π (this is typically the posterior distribution) and let x be the current realisation of the chain. In the Metropolis-Hastings algorithm, a proposal Y is extracted from an arbitrary density $q\left(\cdot|x\right)$ that may depend on x.

Then the proposal Y is accepted with probability $\alpha(x, Y)$ given by (4.67). More specifically, given that at cycle t, $X^{(t)} = x = \left(x_1^{(t)}, x_2^{(t)}, \ldots, x_m^{(t)}\right)$, one cycle of the algorithm is as follows:

1. Generate Y from the proposal density $q(\cdot|x)$.
2. Generate a uniform number U on [0, 1].
3. If

$$U < \alpha(x, Y) = \min\left\{1, \frac{\pi(Y)\, q(x|Y)}{\pi(x)\, q(Y|x)}\right\} \tag{4.67}$$

then $X^{[t+1]} = Y$. Otherwise $X^{[t+1]} = x$.

This is repeated until enough samples from the posterior distribution have been collected. I return shortly to evaluate what is "enough".

In this formulation of the Metropolis-Hastings algorithm, all the components of $X^{(t)}$ are updated if the proposal is accepted.

Joint Updating: Example Using the Normal Model (4.64) and (4.65)

Let Y_μ and Y_{σ^2} denote the proposed values for μ and for σ^2, so that

$$Y = \left(Y_\mu, Y_{\sigma^2}\right),$$

and the state of the chain at stage t is

$$X^{(t)} = \left(\mu^{(t)}, \sigma^{2(t)}\right) = x.$$

Then,

$$\pi(Y) = \exp\left[-\frac{\sum_{i=1}^{n}\left(y_i - Y_\mu\right)^2}{2Y_{\sigma^2}}\right]\left(Y_{\sigma^2}\right)^{-\left(\frac{n}{2}+1\right)},$$

$$\pi(x) = \exp\left[-\frac{\sum_{i=1}^{n}\left(y_i - \mu^{(t)}\right)^2}{2\sigma^{2(t)}}\right]\left(\sigma^{2(t)}\right)^{-\left(\frac{n}{2}+1\right)}.$$

Assume that a $N(m, D)$ random walk proposal is used for generating

$$\left(\ln Y_{\sigma^2}, Y_\mu\right),$$

with

$$m = \left(\ln \sigma^2, \mu\right),$$

$$D = \begin{bmatrix} k_{\sigma^2} & 0 \\ 0 & k_\mu \end{bmatrix},$$

and where k_{σ^2} and k_μ are user-tuned variance parameters for σ^2 and for μ. Then (see **NOTE** below: if σ^2 (a variance) is a variable that is restricted to be positive and $\ln \sigma^2$ is normally distributed, then σ^2 lognormally distributed),

$$q(x|Y) = q\left(\mu^{(t)}, \sigma^{2(t)}|Y_\mu, Y_{\sigma^2}\right) = q\left(\mu^{(t)}|Y_u\right) q\left(\sigma^{2(t)}|Y_{\sigma^2}\right),$$

where

$$q\left(\sigma^{2(t)}|Y_{\sigma^2}\right) = \left(2\pi k_{\sigma^2}\right)^{-\frac{1}{2}} \exp\left[-\frac{\left(\ln \sigma^{2(t)} - \ln Y_{\sigma^2}\right)^2}{2k_{\sigma^2}}\right],$$

$$q\left(\mu^{(t)}|Y_\mu\right) = \left(2\pi k_\mu\right)^{-\frac{1}{2}} \exp\left[-\frac{\left(\mu^{(t)} - Y_\mu\right)^2}{2k_\mu}\right].$$

Similarly,

$$q(Y|x) = q\left(Y_\mu, Y_{\sigma^2}|\mu^{(t)}, \sigma^{2(t)}\right) = q\left(Y_u|\mu^{(t)}\right) q\left(Y_{\sigma^2}|\sigma^{2(t)}\right),$$

where

$$q\left(Y_{\sigma^2}|\sigma^{2(t)}\right) = \left(2\pi k_{\sigma^2}\right)^{-\frac{1}{2}} \exp\left[-\frac{\left(\ln Y_{\sigma^2} - \ln \sigma^{2(t)}\right)^2}{2k_{\sigma^2}}\right] \frac{1}{Y_{\sigma^2}}, \tag{4.68}$$

$$q\left(Y_u|\mu^{(t)}\right) = \left(2\pi k_\mu\right)^{-\frac{1}{2}} \exp\left[-\frac{\left(Y_\mu - \mu^{(t)}\right)^2}{2k_\mu}\right]. \tag{4.69}$$

Manipulating these expressions shows that the proposal $Y = \left(Y_\mu, Y_{\sigma^2}\right)$ is accepted with probability given by (4.67), equal to

$$\frac{\pi(Y)\, q(x|Y)}{\pi(x)\, q(Y|x)} =$$

$$= \exp\left[\frac{\sum_{i=1}^n \left(y_i - \mu^{(t)}\right)^2}{2\sigma^{2(t)}} - \frac{\sum_{i=1}^n \left(y_i - Y_\mu\right)^2}{2Y_{\sigma^2}}\right] \left(\frac{Y_{\sigma^2}}{\sigma^{2(t)}}\right)^{-\frac{n}{2}}. \tag{4.70}$$

In a programming environment in order to avoid under- or overflows, one computes

$$\ln \frac{\pi(Y)\, q(Y; x)}{\pi(x)\, q(x; Y)} =$$

$$= \left[\frac{\sum_{i=1}^{n} \left(y_i - \mu^{(t)}\right)^2}{2\sigma^{2(t)}} - \frac{\sum_{i=1}^{n} \left(y_i - Y_{\mu}\right)^2}{2Y_{\sigma^2}} \right] + \frac{n}{2} \left(\ln \sigma^{2(t)} - \ln Y_{\sigma^2} \right)$$

and in the final stage

$$\frac{\pi(Y)\, q(x|Y)}{\pi(x)\, q(Y|x)}$$

is calculated.

The user must choose the tuning parameters judiciously because this has a strong influence on the behaviour of the chain.

NOTE on the Theory of Transformations of Random Variables

Let $X = \ln \sigma^2$. Assume

$$X \sim N(\mu, v). \tag{4.71}$$

Then $Y = \sigma^2 = \exp X$ follows a lognormal distribution

$$Y = \sigma^2 = \exp X \sim \ln N(\mu, v), \quad 0 < Y < \infty, \quad -\infty < \mu < \infty, \quad v > 0, \tag{4.72}$$

with density

$$p(y) = (2\pi v)^{-\frac{1}{2}} \exp\left(-\frac{1}{2v} (\ln y - \mu)^2 \right) \frac{1}{y}. \tag{4.73}$$

To show this, I use the theory of transformation of random variables. If X has probability density function $p_X(x)$, that is

$$X \sim p_X(x),$$

and $Y = f(X)$ such that the inverse function f^{-1} exists, from which X can be retrieved

$$X = f^{-1}(Y),$$

then the probability density function of Y, $p_Y(y)$ is

$$p_Y(y) = p_X\left(f^{-1}(Y)\right)\left|\frac{df^{-1}(Y)}{dY}\right|, \tag{4.74}$$

where $\left|df^{-1}(Y)/dY\right|$ is the absolute value of the Jacobian of the transformation. The notation $p_X\left(f^{-1}(Y)\right)$ implies that one uses the form of the density of X (in the example, a $N(\mu, v)$) and replaces in this density, x by $f^{-1}(y)$. In the present case, the objective is to derive the distribution of the random variable $Y = \exp X = \sigma^2$. Then,

$$Y = f(X) = \exp(X) \Longrightarrow X = f^{-1}(Y) = \ln Y, \quad \left|\frac{df^{-1}(Y)}{dY}\right| = \frac{1}{Y}.$$

Using (4.74)

$$p_Y(y) = (2\pi v)^{-\frac{1}{2}} \exp\left[-\frac{(\ln y - \mu)^2}{2v}\right]\frac{1}{y}, \quad 0 < Y < \infty, \quad -\infty < \mu < \infty,$$
$$v > 0, \tag{4.75}$$

which is the density of the lognormal distribution with parameters μ and v.

Single-Site Updating

For a single-site updating Metropolis-Hastings algorithm, the components $x_1^{(t)}, x_2^{(t)}, \ldots, x_m^{(t)}$ are updated individually in a random or systematic order. Typically, a systematic order is chosen starting with $x_1^{(t)}$, then $x_2^{(t)}$ and so on, until the last component $x_m^{(t)}$ is updated.

Assume that the current state is $X^{(t)} = \left(x_1^{(t)}, x_2^{(t)}, \ldots, x_m^{(t)}\right) = x$ and that the jth component is to be updated. Then the next stage of the chain $X^{(t+1)}$ only differs from $X^{(t)}$ on the jth component that is generated as follows:

1. Generate a proposal Y_j from the proposal density $q_j(\cdot|x)$. Let

$$Y = \left(x_1^{(t)}, x_2^{(t)}, \ldots, x_{j-1}^{(t)}, Y_j, x_{j+1}^{(t)}, \ldots, x_m^{(t)}\right).$$

2. Generate a uniform number U on $[0, 1]$.

3. If

$$U < \alpha_j (x, Y) = \min \left\{ 1, \frac{\pi (Y) q_j (x_j|Y)}{\pi (x) q_j (Y_j|x)} \right\}, \tag{4.76}$$

then $X^{(t+1)} = Y$. Otherwise $X^{(t+1)} = x$.

Single-Site Updating: Example Using the Normal Model (4.64), (4.65)

1. Updating σ^2.
As before, the state of the chain at stage t is

$$X^{(t)} = \left(\mu^{(t)}, \sigma^{2(t)}\right) = x,$$

but now

$$Y = \left(\mu^{(t)}, Y_{\sigma^2}\right).$$

Then,

$$\pi (Y) \propto \exp \left(-\frac{\sum_{i=1}^{n} \left(y_i - \mu^{(t)}\right)^2}{2Y_{\sigma^2}} \right) \left(Y_{\sigma^2}\right)^{-\left(\frac{n}{2}+1\right)},$$

and

$$\pi (x) \propto \exp \left(-\frac{\sum_{i=1}^{n} \left(y_i - \mu^{(t)}\right)^2}{2\sigma^{2(t)}} \right) \left(\sigma^{2(t)}\right)^{-\left(\frac{n}{2}+1\right)}.$$

The proposal density of Y_{σ^2} is

$$q_j (Y_j|x)$$
$$= q_{\sigma^2} \left(Y_{\sigma^2}|\sigma^{2(t)}\right)$$
$$= \left(2\pi k_{\sigma^2}\right)^{-\frac{1}{2}} \exp \left[-\frac{\left(\ln Y_{\sigma^2} - \ln \sigma^{2(t)}\right)^2}{2k_{\sigma^2}} \right] \frac{1}{Y_{\sigma^2}}.$$

Likewise,

$$q_j\left(x_j|Y\right) =$$

$$q_{\sigma^2}\left(\sigma^{2(t)}|Y_{\sigma^2}\right) = \left(2\pi k_{\sigma^2}\right)^{-\frac{1}{2}} \exp\left[-\frac{\left(\ln \sigma^{2(t)} - \ln Y_{\sigma^2}\right)^2}{2k_{\sigma^2}}\right]\frac{1}{\sigma^{2(t)}}.$$

Manipulating these expressions, it is easy to show that the proposal is accepted with probability (4.76), where

$$\frac{\pi\left(Y\right)q_j\left(x_j|Y\right)}{\pi\left(x\right)q_j\left(Y_j|x\right)}$$

$$= \exp\left[\frac{\sum_{i=1}^{n}\left(y_i - \mu^{(t)}\right)^2}{2\sigma^{2(t)}} - \frac{\sum_{i=1}^{n}\left(y_i - \mu^{(t)}\right)^2}{2Y_{\sigma^2}}\right]\left(\frac{Y_{\sigma^2}}{\sigma^{2(t)}}\right)^{-\frac{n}{2}}. \tag{4.77}$$

In practice and in order to avoid under- or overflows, one computes

$$\ln\frac{\pi\left(Y\right)q_j\left(x_j|Y\right)}{\pi\left(x\right)q_j\left(Y_j|x\right)}$$

$$= \left[\frac{\sum_{i=1}^{n}\left(y_i - \mu^{(t)}\right)^2}{2\sigma^{2(t)}} - \frac{\sum_{i=1}^{n}\left(y_i - \mu^{(t)}\right)^2}{2Y_{\sigma^2}}\right] + \frac{n}{2}\left(\ln \sigma^{2(t)} - \ln Y_{\sigma^2}\right)$$

and at the final stage

$$\frac{\pi\left(Y\right)q_j\left(x_j|Y\right)}{\pi\left(x\right)q_j\left(Y_j|x\right)}$$

is calculated.

2. Updating μ.

   The state of the chain at stage t is again

$$X^{(t)} = \left(\mu^{(t)}, \sigma^{2(t)}\right) = x,$$

but now,

$$Y = \left(Y_\mu, \sigma^{2(t)}\right),$$

$$\pi\left(Y\right) \propto \exp\left[-\frac{\sum_{i=1}^{n}\left(y_i - Y_\mu\right)^2}{2\sigma^{2(t)}}\right]\left(\sigma^{2(t)}\right)^{-\left(\frac{n}{2}+1\right)},$$

$$\pi(x) \propto \exp\left[-\frac{\sum_{i=1}^{n}\left(y_i-\mu^{(t)}\right)^2}{2\sigma^{2(t)}}\right]\left(\sigma^{2(t)}\right)^{-\left(\frac{n}{2}+1\right)}.$$

The proposal densities are

$$q_j\left(Y_j|x\right)$$

$$= q_\mu\left(Y_\mu|\mu^{(t)}\right) = \left(2\pi k_\mu\right)^{-\frac{1}{2}}\exp\left[-\frac{\left(Y_\mu-\mu^{(t)}\right)^2}{2k_\mu}\right],$$

and

$$q_j\left(x_j|Y\right)$$

$$= q_\mu\left(\mu^{(t)}|Y_\mu\right) = \left(2\pi k_\mu\right)^{-\frac{1}{2}}\exp\left[-\frac{\left(\mu^{(t)}-Y_\mu\right)^2}{2k_\mu}\right].$$

These densities are symmetric and therefore cancel each other in the ratio. The proposal is accepted with probability (4.76), where

$$\frac{\pi(Y)\,q_j\left(x_j|Y\right)}{\pi(x)\,q_j\left(Y_j|x\right)}$$

$$= \exp\left[\frac{\sum_{i=1}^{n}\left(y_i-\mu^{(t)}\right)^2}{2\sigma^{2(t)}}-\frac{\sum_{i=1}^{n}\left(y_i-Y_\mu\right)^2}{2\sigma^{2(t)}}\right].$$

NOTE: A slightly different implementation of the algorithm is to parametrise the model in terms of $\ln\sigma^2$. Then the posterior is now proportional to

$$\exp\left[-\frac{\sum_{i=1}^{n}\left(y_i-\mu^{(t)}\right)^2}{2\sigma^2}\right]\left(\sigma^2\right)^{-\frac{n}{2}},$$

$q_{\sigma^2}\left(Y_{\sigma^2}|\ln\sigma^{2(t)}\right)$ is given by

$$\left(2\pi k_{\sigma^2}\right)^{-\frac{1}{2}}\exp\left[-\frac{\left(\ln Y_{\sigma^2}-\ln\sigma^{2(t)}\right)^2}{2k_{\sigma^2}}\right],$$

and one can verify that the acceptance probability for Y_{σ^2} is the same as in (4.77).

4.12 The Gibbs Sampling Algorithm

The Gibbs sampler is a very popular McMC algorithm because of its computational simplicity. The mechanics is as follows. Consider the vector of parameters of a model $(\theta_1, \theta_2, \ldots, \theta_p)$, with posterior density $p(\theta_1, \theta_2, \ldots, \theta_p|y)$ known up to proportionality. Assume that the user supplies "legal" starting values

$$\left(\theta_1^{(0)}, \theta_2^{(0)}, \ldots, \theta_p^{(0)}\right),$$

in the sense that $p\left(\theta_1^{(0)}, \theta_2^{(0)}, \ldots, \theta_p^{(0)}|y\right) > 0$. The implementation of the Gibbs sampler consists of iterating through the loop:

$$\text{draw } \theta_1^{(1)} \text{ from } p\left(\theta_1|\theta_2^{(0)}, \ldots, \theta_p^{(0)}, y\right),$$

$$\text{draw } \theta_2^{(1)} \text{ from } p\left(\theta_2|\theta_1^{(1)}, \theta_3^{(0)}, \ldots, \theta_p^{(0)}, y\right),$$

$$\text{draw } \theta_3^{(1)} \text{ from } p\left(\theta_3|\theta_1^{(1)}, \theta_2^{(1)}, \theta_4^{(0)}, \ldots, \theta_p^{(0)}, y\right),$$

$$\vdots$$

$$\text{draw } \theta_p^{(1)} \text{ from } p\left(\theta_p|\theta_1^{(1)}, \ldots, \theta_{p-1}^{(1)}, y\right),$$

$$\text{draw } \theta_1^{(2)} \text{ from } p\left(\theta_1|\theta_2^{(1)}, \ldots, \theta_p^{(1)}, y\right),$$

$$\vdots$$

and so on.

After an initial period during which samples are dependent on the starting value (burn-in period), the draws $\theta_1^{(i)}, \theta_2^{(i)}, \ldots, \theta_p^{(i)}$, for sufficiently large i, when the sampler converges, are regarded as samples from the normalised posterior distribution with density

$$p(\theta_1, \theta_2, \ldots, \theta_p|y) \Big/ \int p(\theta_1, \theta_2, \ldots, \theta_p|y)\, d\theta_1 \ldots d\theta_p.$$

The coordinate $\theta_j^{(i)}$ is regarded as a draw from its marginal posterior distribution with density

$$p(\theta_j|y) \Big/ \int p(\theta_j|y)\, d\theta_j.$$

In more general terms, let

$$\theta_{-i} = \left(\theta_1, \ldots, \theta_{i-1}, \theta_{i+1}, \ldots, \theta_p\right)$$

be the vector of dimension $(p - r)$, $p > r$, $r \geq 1$, which is equal to θ with its ith component, θ_i, deleted and where r is the number of elements in θ_i. The density of the fully conditional posterior distribution (fcpd) of θ_i is

$$\begin{aligned} p\left(\theta_i | \theta_{-i}, y\right) &= \frac{p\left(\theta_1, \ldots, \theta_{i-1}, \theta_i, \theta_{i+1}, \ldots, \theta_p | y\right)}{\int p\left(\theta_1, \ldots, \theta_{i-1}, \theta_i, \theta_{i+1}, \ldots, \theta_p | y\right) d\theta_i} \\ &\propto p\left(\theta_1, \ldots, \theta_{i-1}, \theta_i, \theta_{i+1}, \ldots, \theta_p | y\right). \end{aligned} \tag{4.78}$$

In many applications, $r = 1$ and parameters are updated one at a time. In general, single-site updating leads to moves along each coordinate, whereas updating several components in a block allows for more general moves. Joint updating, which incorporates information on the correlation structure among the components in the joint conditional posterior distribution, can result in faster convergence when correlations are strong (Liu et al 1994). A computational strategy that makes it feasible to draw the entire vector θ for some Gaussian linear models is described in Sorensen and Gianola (2002), page 584, originally proposed by García-Cortés and Sorensen (1996).

The Gibbs sampler is a special case of the more general Metropolis-Hastings algorithm. In the Gibbs sampler, the proposals (drawn from the fully conditional posterior distributions) are always accepted (see the **NOTE** below). It is simpler to implement than the Metropolis-Hastings algorithm because it does not require tuning parameters. However, it requires knowledge of the form of the fully conditional posterior distributions.

Example Using the Normal Model

Consider the normal model defined by the Eqs. (4.64) and (4.65). The fully conditional posterior distributions with densities $p\left(\mu|\sigma^2, y\right)$ and $p\left(\sigma^2|\mu, y\right)$ are derived using the posterior distribution (4.66) as the starting point. To obtain $\left[\mu|\sigma^2, y\right]$, one extracts from the posterior distribution (4.66) those terms that include μ. This results in

$$\begin{aligned} p\left(\mu|\sigma^2, y\right) &\propto \exp\left(-\frac{\sum_{i=1}^{n}\left(y_i - \mu\right)^2}{2\sigma^2}\right) \\ &= \exp\left(-\frac{1}{2\sigma^2}\left[n\left(\overline{y} - \mu\right)^2\right]\right) \end{aligned}$$

which is the kernel of a normal distribution with mean $\overline{y}$ and variance σ^2/n. Therefore,

$$\mu|\sigma^2, y \sim N\left(\overline{y}, \frac{\sigma^2}{n}\right). \tag{4.79}$$

To obtain $[\sigma^2|\mu, y]$, one extracts from the posterior distribution (4.66) those terms that include σ^2. This results in

$$p\left(\sigma^2|\mu, y\right) \propto \exp\left(-\frac{\sum_{i=1}^{n}(y_i-\mu)^2}{2\sigma^2}\right)\left(\sigma^2\right)^{-\left(\frac{n}{2}+1\right)}$$

which seen as a function of σ^2 is the kernel of a scale-inverted chi-square distribution with n degrees of freedom and scale

$$s^2 = \frac{\sum_{i=1}^{n}(y_i-\mu)^2}{n}.$$

Therefore,

$$\sigma^2|\mu, y \sim ns^2\chi_n^{-2}. \tag{4.80}$$

NOTE: The Acceptance Probability of the Gibbs Sampler Is Equal to 1
Let the state of the chain at stage t be

$$X_t = \left(\mu^{(t)}, \sigma^{2(t)}\right) = x$$

and let

$$Y = \left(\mu^{(t)}, Y_{\sigma^2}\right).$$

Then

$$\pi(Y) = \pi\left(\mu^{(t)}, Y_{\sigma^2}|y\right) \propto \exp\left(-\frac{\sum_{i=1}^{n}\left(y_i-\mu^{(t)}\right)^2}{2Y_{\sigma^2}}\right)\left(Y_{\sigma^2}\right)^{-\left(\frac{n}{2}+1\right)},$$

and

$$\pi(x) = \pi\left(\mu^{(t)}, \sigma^{2(t)}|y\right) \propto \exp\left(-\frac{\sum_{i=1}^{n}\left(y_i-\mu^{(t)}\right)^2}{2\sigma^{2(t)}}\right)\left(\sigma^{2(t)}\right)^{-\left(\frac{n}{2}+1\right)}.$$

The proposal density of Y_{σ^2} is

$$q\left(Y_{\sigma^2}|x\right) = \pi\left(Y_{\sigma^2}|\mu^{(t)}, y\right)$$

and

$$q\left(x|Y_{\sigma^2}\right) = \pi\left(\sigma^{2(t)}|\mu^{(t)}, y\right).$$

Then the acceptance probability for updating σ^2 is

$$\begin{aligned}
\frac{\pi(Y)\, q\left(x|Y_{\sigma^2}\right)}{\pi(x)\, q\left(Y_{\sigma^2}|x\right)} &= \frac{\pi(Y)}{\pi(x)} \frac{\pi\left(\sigma^{(t)}, \mu^{(t)}|y\right)}{\pi\left(\mu^{(t)}|y\right)} \frac{\pi\left(\mu^{(t)}|y\right)}{\pi\left(Y_{\sigma^2}, \mu^{(t)}|y\right)} \\
&= \frac{\pi(Y)}{\pi(x)} \frac{\pi(x)}{\pi\left(\mu^{(t)}|y\right)} \frac{\pi\left(\mu^{(t)}|y\right)}{\pi(Y)} = 1.
\end{aligned}$$

A similar result is obtained for the acceptance probability for updating μ.

4.13 Output Analysis

McMC is used to obtain approximate samples from posterior distributions. The information about unknowns contained in these samples depends on their size (length of the chain) and on the degree of autocorrelation among sampled values. When a particular feature from a posterior distribution is estimated from the samples, the uncertainty associated with the McMC estimate is the Monte Carlo error of estimation. This is a classical frequentist sampling uncertainty that depends on the length of the chain and on the degree of autocorrelation among the samples. In contrast, the posterior uncertainty, given the model, depends on the data. It is important to distinguish between these two sources of uncertainty.

There is a large literature on McMC convergence diagnosis and output analysis. A useful practical reference is Kass et al (1998). Here, the focus is on a few important issues of output analysis.

The first is to determine the length of the so-called burn-in period. This is related to the question of convergence of the sampled values to the target distribution. The chain is initialised with values of parameters that are not drawn from the posterior distribution. Since convergence, if reached, to the posterior distribution is gradual, one must decide how many of the initial values must be discarded in order to include only representative ones to draw Monte Carlo inferences. Figure 4.1 provides an illustration. The figure displays traceplots from the simulated values from $[\mu|y]$ (left panel) and from $\left[\sigma^2|y\right]$ (right panel) obtained with the single-site Metropolis-Hastings algorithm applied to model (4.64), (4.65a), (4.65b). Such a traceplot discloses information about the burn-in period and at the same time

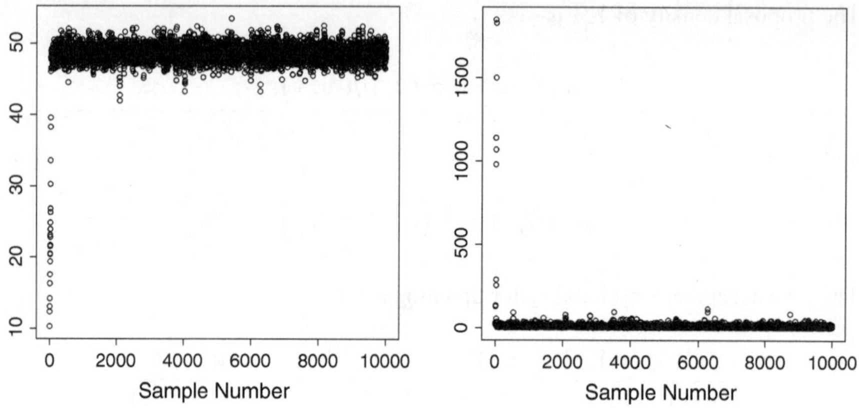

Fig. 4.1 Draws from the posterior distribution $[\mu|y]$ (left panel) and from $\left[\sigma^2|y\right]$ (right panel) obtained from the single-site Metropolis-Hastings algorithm

provides an informal check for convergence. The values of μ and σ^2 start far away from the support of the respective marginal posterior distributions but seem to reach convergence after a few iterations (burn-in). There are a number of methods that attempt to provide more formal checks of convergence than the "eye-balling" approach displayed by traceplots. None can guarantee convergence but can flag lack of convergence. A useful reference is Cowles and Carlin (1996).

A second concern is to determine the *degree of autocorrelation* of the sampled values and the so-called *effective chain length or effective sample size*. These determine the *McMC-sample variance or MC variance* of estimates of features of posterior distributions. Because the number of sampled values from the estimated posterior distribution is finite, there is always sampling uncertainty associated with an estimator of features of posterior distributions. In principle, the sampling uncertainty can be made as small as desired by taking a sufficiently large number of samples.

The MC variance can be estimated by running several independent chains and then calculating the empirical, between-chain variance of the estimates obtained for each chain. Since this is often computationally expensive, one resorts to theoretical estimators of MC variance. These estimators account for the autocorrelation among the samples taken from the target distribution. Useful references are Ripley (1987), Geyer (1992) and Chen et al (2000).

Consider the output of a Markov chain consisting of m samples $X^{(1)}, X^{(2)}, \ldots,$ $X^{(m)}$, where the $X^{(i)}$ are approximate draws from the posterior distribution $[X|y]$, where X is a vector of parameters, y is a vector of data and m is the length of the chain. The goal is to estimate the mean of some function of X, $h(X)$, over the posterior distribution of X, with density $p(x|y)$:

$$\mathrm{E}\left[h\left(X\right)|y\right] = \int h\left(x\right) p\left(x|y\right) dx$$

using

$$\widehat{\mu}_m = \frac{1}{m}\sum_{i=1}^{m} h\left(X^{(i)}\right). \tag{4.81}$$

Here, $h(X)$ is any function of X with finite expectation. If h is the identity function, (4.81) retrieves an estimator of the mean.

A central limit theorem asserts

$$\sqrt{m}\left(\widehat{\mu}_m - \mathrm{E}\left[h\left(X\right)|y\right]\right) \xrightarrow{\mathcal{D}} N\left(0, V_{\mathtt{asymp}}\right), \tag{4.82}$$

where $V_{\mathtt{asymp}}$ is the asymptotic variance of X. A little more informally, this can be written as

$$\widehat{\mu}_m \sim N\left(\mathrm{E}\left[h\left(X\right)|y\right], V_{\mathtt{asymp}}/m\right).$$

From (4.82) an approximate 95% confidence interval for the unknown true value $\mathrm{E}\left[h\left(X\right)|y\right]$ is $\hat{\mu}_m \pm 1.96\sqrt{V_{\mathtt{asymp}}}/\sqrt{m}$. With independent sampling, measures of sampling uncertainty are obtained by replacing the unknown $V_{\mathtt{asymp}}$ by the sample variance of X.

The sampling variance of the estimator (4.81) is $V_{\mathtt{asymp}}/m$ where

$$\begin{aligned} V_{\mathtt{asymp}} &= \lim_{m\to\infty} \mathrm{Var}\left(\sqrt{m}\widehat{\mu}_m\right) \\ &= \mathrm{Var}\left[h\left(X\right)|y\right]\left(1 + 2\sum_{j=1}^{\infty}\rho_j\right). \end{aligned} \tag{4.83}$$

(see **NOTE 2** below) and $\mathrm{Var}\left[h\left(X\right)|y\right]$ is the posterior variance of $h\left(X\right)$ under the limiting distribution. Above,

$$\begin{aligned} \rho_j &= \frac{\mathrm{Cov}\left[\left(h\left(X^{(i)}\right), h\left(X^{(i+j)}\right)\right)|y\right]}{\mathrm{Var}\left[h\left(X^{(i)}\right)|y\right]} \\ &= \frac{\gamma\left(j\right)}{\gamma\left(0\right)}, \quad j = 1, 2, \ldots \end{aligned}$$

for all i, where i refers to the ith draw and ρ_j is the lag$-j$ autocorrelation of $X^{(1)}, X^{(2)}, \ldots, X^{(m)}$. The above assumes that, under stationarity, the lag-covariance is constant for constant lag j (e.g. for $j = 2$, $\gamma\left(2\right) = \mathrm{Cov}\left[\left(h\left(X^{(1)}\right), h\left(X^{(3)}\right)\right)|y\right] = \mathrm{Cov}\left[\left(h\left(X^{(2)}\right), h\left(X^{(4)}\right)\right)|y\right]$) and that $\mathrm{Cov}\left[\left(h\left(X^{(i)}\right), h\left(X^{(i)}\right)\right)|y\right] = \mathrm{Var}\left[h\left(X^{(i)}\right)|y\right] = \gamma\left(0\right)$.

NOTE 1

The important property defined by expression (4.81) is used repeatedly in the book. For example, (4.81) is an estimator of

- the posterior mean if $h(X) = X$;
- the posterior variance if $h(X) = [X - E(X|y)]^2$. The estimator of the posterior variance is

$$\frac{1}{m}\sum_{i=1}^{m}\left[X^{(i)2} - \left[\widehat{E}(X|y)\right]^2\right],$$

  where $\widehat{E}(X|y) = \frac{1}{m}\sum_i X^{(i)}$;
- the posterior probability that $X \in A$, if $h(X) = I(X \in A)$, such that

$$\Pr(X \in A|) = \int I(X \in A)\, p(X|y)\, dX.$$

  Here, the estimator is $\sum_{i=1}^{m} I\left(X^{(i)} \in A\right)/m$. As a special case, the cumulative distribution function is estimated:

$$\widehat{F}(t) = \frac{1}{m}\sum_{i=1}^{n} I\left(X^{(i)} < t\right).$$

  Therefore, the estimator of the posterior probability that $t_1 < X < t_2$ is

$$\widehat{\Pr}(t_1 < X < t_2|y) = \frac{1}{m}\left[\sum_{i=1}^{m} I\left(t_1 < X^{(i)} < t_2\right)\right];$$

- the posterior predictive density:

$$p(z|y) = \int p(z|X, y)\, p(X|y)\, dX,$$

  where, usually, the form of the problem is such that $p(z|X, y) = p(z|X)$. In this setting, $h(X) = p(z|X)$, and the estimator of the predictive density is $\sum_{i=1}^{m} p\left(z|X^{(i)}\right)/m$.

  This important property of McMC is used to obtain Monte Carlo estimates of the marginal posterior distribution of mean squared errors and of false discovery rate.

NOTE 2

To arrive at (4.83), one writes

$$m\,\mathrm{Var}\left(\widehat{\mu}_m\right) = m^{-1}\sum_{i,j}\mathrm{Cov}\left[\left(h\left(X^{(i)}\right), h\left(X^{(j)}\right)\right)|y\right]$$

$$= \mathrm{Var}\left[h\left(X\right)|y\right]\left[1+2\sum_{j=1}^{m-1}\left(1-\frac{j}{m}\right)\rho_j\right]. \quad (4.84)$$

When $\rho_j = 0$ for all j,

$$\mathrm{Var}\left(\widehat{\mu}_m\right) = \frac{\mathrm{Var}\left[h\left(X\right)|y\right]}{m}$$

the familiar equation for the variance of the sample mean (assuming independent samples). The term in square brackets in (4.84) converges to $\left[1+2\sum_{j=1}^{\infty}\rho_j\right]$ as $m \to \infty$, in which case

$$\underset{\lim m\to\infty}{m\,\mathrm{Var}\left(\widehat{\mu}_m\right)} = V_{\mathtt{asymp}}$$

$$= \mathrm{Var}\left[h\left(X\right)|y\right]\left[1+2\sum_{j=1}^{\infty}\rho_j\right]$$

$$= \mathrm{Var}\left[h\left(X\right)|y\right]\tau,$$

where $\mathrm{Var}\left[h\left(X^{(i)}\right)|y\right] = \mathrm{Var}\left[h\left(X\right)|y\right]$ for all i, the posterior variance that, given the model, depends on data y and $\tau = 1+2\sum_{j=1}^{\infty}\rho_j$ is the *integrated autocorrelation* that is modulated by the McMC method used. The *effective sample size* is

$$m_{\mathtt{eff}} = \frac{m}{\tau},$$

which is equal to m if the draws from $[X|y]$ are uncorrelated. It is important to distinguish between $\mathrm{Var}\left[h\left(X\right)|y\right]$, the variance of $h\left(X\right)$ under the limiting distribution $[X|y]$ and $\mathrm{Var}\left(\sqrt{m}\widehat{\mu}_m\right)$ the limiting variance of $\sqrt{m}\widehat{\mu}_m$, equal to $\mathrm{Var}\left[h\left(X\right)|y\right]\tau$. The former is associated with the posterior uncertainty of $h(X)$, while the latter is associated with the sampling scheme from the posterior distribution. Under independent sampling, $\tau = 1$ and therefore both are equal to $\mathrm{Var}\left[h\left(X\right)|y\right]$.

There are a number of methods that can be used to estimate the Monte Carlo variance of features from posterior distributions, two of which are sketched below.

Geyer's Estimator of the Monte Carlo Variance

Geyer (1992) suggests the following estimator of $\text{Var}\left(\hat{\mu}_m\right)$ based on time-series theory (a classical reference is Priestley 1981). The lag$-t$ autocovariance is

$$\gamma(t) = \text{Cov}\left[\left(h\left(X^{(i)}\right), h\left(X^{(i+t)}\right)\right)|y\right],$$

which under stationarity is the same for all i. The estimator is

$$\widehat{\gamma}(t) = \frac{1}{m}\sum_{i=1}^{m-t}\left\{\left[h\left(X^{(i)}\right) - \widehat{\mu}_m\right]\left[h\left(X^{(i+t)}\right) - \widehat{\mu}_m\right]\right\}.$$

A Monte Carlo estimator of the asymptotic variance is

$$\widehat{V}_{\texttt{asymp}} = \widehat{\gamma}(0) + 2\sum_{i=1}^{i=2\delta+1}\widehat{\gamma}(i) \tag{4.85}$$

and a Monte Carlo estimator of the variance of (4.81) is

$$\widehat{\text{Var}}\left(\widehat{\mu}_m\right) = \frac{1}{m}\left[\widehat{\gamma}(0) + 2\sum_{i=1}^{i=2\delta+1}\widehat{\gamma}(i)\right] \tag{4.86}$$

where δ is chosen such that it is the largest integer satisfying

$$\widehat{\gamma}\left(2\delta'\right) + \widehat{\gamma}\left(2\delta' + 1\right) > 0, \quad \delta' = 0, 1, \ldots, \delta.$$

Above,

$$\begin{aligned}\widehat{\gamma}(0) &= \widehat{\text{Var}}\left[h\left(X^{(i)}|y\right)\right] \\ &= \widehat{\text{Var}}\left[h(X)|y\right], \quad \text{for all } i.\end{aligned}$$

An estimator of the integrated autocorrelation is obtained from (4.86) as follows:

$$\begin{aligned}\widehat{\text{Var}}\left(\widehat{\mu}_m\right) &= \frac{\widehat{\gamma}(0)}{m}\left[1 + 2\sum_{i=1}^{i=2\delta+1}\frac{\widehat{\gamma}(i)}{\widehat{\gamma}(0)}\right] \\ &= \frac{\widehat{\gamma}(0)}{m}\left[1 + 2\sum_{i=1}^{i=2\delta+1}\widehat{\rho}(i)\right].\end{aligned} \tag{4.87}$$

Therefore,

$$\begin{aligned}\widehat{\tau} &= 1 + 2 \sum_{i=1}^{i=2\delta+1} \widehat{\rho}(i) \\ &= \frac{m \widehat{\text{Var}}(\widehat{\mu}_m)}{\widehat{\gamma}(0)}. \end{aligned} \tag{4.88}$$

The estimator of effective sample size is

$$\begin{aligned}\widehat{m}_{\texttt{eff}} &= \frac{m}{\widehat{\tau}} \\ &= \frac{\widehat{\gamma}(0)}{\widehat{\text{Var}}(\widehat{\mu}_m)}. \end{aligned} \tag{4.89}$$

The Method of Batching

A popular method of estimating Monte Carlo variances that is easy to implement is known as "batching" (Hastings 1970). It is based on the idea that if individual draws $X^{(j)}$ are correlated, grouping successive draws into b batches or groups of size s each and computing the raw averages will lead to b batch means that are less strongly inter-correlated than the original draws. This can be so, provided that s is chosen appropriately. The larger the autocorrelation among samples, the larger s must be. Suppose that a chain of total length m is divided into b batches each of size s. Let the average of the ith batch be

$$\overline{x}_i = \frac{1}{s} \sum_{j=1}^{s} X^{(j)}, \quad i = 1, 2, \ldots, b.$$

Here, $h\left(X^{(j)}\right)$ is some feature of the posterior distribution evaluated at the sampled value $X^{(j)}$. The batch estimator of the variance of (4.81), assuming that s is large enough so that the $\overline{x}_i'$s are uncorrelated, is equal to

$$\widehat{\text{Var}}_b(\widehat{\mu}_m) = \frac{\sum_{i=1}^{b} (\overline{x}_i - \widehat{\mu}_m)^2}{b(b-1)}. \tag{4.90}$$

An estimate of the batch-effective chain length can be obtained as

$$\widehat{m}_{\texttt{eff\_b}} = \frac{\sum_{i=1}^{m} \left[X^{(i)} - \widehat{\mu}_m\right]^2}{(m-1)\widehat{\text{Var}}_b(\widehat{\mu}_m)}. \tag{4.91}$$

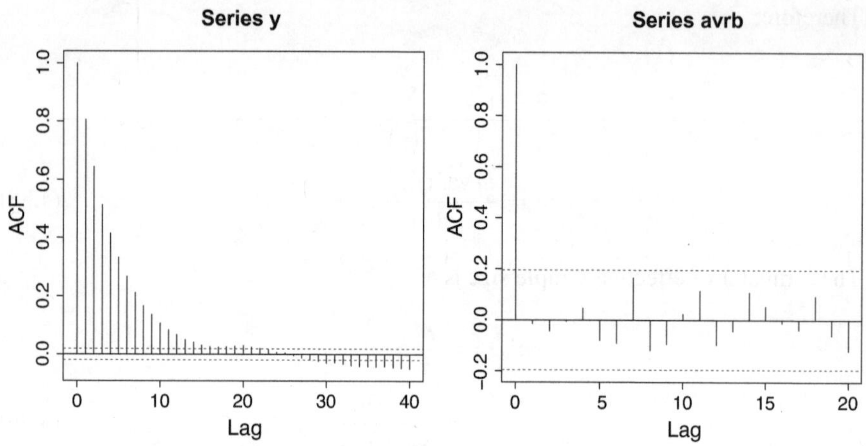

Fig. 4.2 Left: autocorrelations between samples of y_i versus lag; right: autocorrelations between batch means

When the samples are independent, $s = 1$, $b = m$, $\overline{x}_i = X^{(i)}$ for all i and $\hat{m}_{\texttt{eff\_b}} = m$.

If the autocorrelation among the samples of the chain is very high (> 0.95), estimator (4.86) seems to be preferred over (4.90).

Example: A Simulated Autoregressive Process

These concepts are illustrated with an autoregressive model that generates draws (vector y) of length 10,000 mimicking the correlated structure of a Monte Carlo Markov chain. A plot of the autocorrelations among the elements of y versus lag gives a visual impression of the output. The R function `acf(y)` (see the bottom of the second `R-code` below) is used to generate Fig. 4.2 (left panel). The figure indicates that at a lag of approximately 20, the autocorrelation among the $y's$ vanishes.

The `R-code` below has *two* sections. The first generates the 10,000 correlated samples y using an autocorrelation equal to 0.8. The second section computes the Monte Carlo variance of the mean of y, based on (4.86), the effective chain length based on (4.89) and the integrated autocorrelation based on (4.88).

The `R-code` is as follows:

```
# CODE0401
# CODE FOR THE MC VARIANCE TESTED ON
#A FIRST-ORDER AUTOREGRESSIVE PROCESS WITH CORRELATION RHO
rm(list=ls()) # Clear the workspace
set.seed(1237)
#GENERATE DATA: AUTOREGRESSIVE PROCESS WITH CORRELATION RHO
```

```
ns <- 10000
y <- rep(NA,ns)
rho<-0.8
sum <- 0
y[1] <- rnorm(1,0,1)
for(i in 2:ns)
{
  y[i] <- rho*y[i-1] + rnorm(1,0,1)
  sum <- sum + y[i]*y[i-1]
}
cov <- sum/ns
rhohat <- cov/var(y)
muhat <- mean(y)
gama0 <- var(y)
rhohat
```

```
## [1] 0.7971166
```

```
# CODE FOR THE MC VARIANCE BASED ON GEYER
svar<-var(y)*(ns-1)/ns
tau<-1
tausum<-0
ptm <- proc.time()
for(i in 0:ns)
{
  gamaj<-0.0
  gamak<-0.0
  j<-2*i
  k<-(2*i)+1
# FASTER CODE: JUMP THE LOOP AND USE FUNCTION ACF
#  for (ii in 1:(ns-j))
#  {
#   cov<-(y[ii]*y[ii+j]-mean(y)*(y[ii]+y[ii+j])+mean(y)*mean(y))
#   gamaj<-gamaj+cov
#  }
#  for(ii in 1:(ns-k))
#  {
#    gamak<-gamak+(y[ii]*y[ii+k]-mean(y)*(y[ii]+y[ii+k])+
  #          mean(y)*mean(y))
#  }
#  gamaj<-gamaj/ns
#  gamak<-gamak/ns

  lag1<-j
  lag2<-k
  #USE THE R-FUNCTION ACF TO COMPUTE AUTOCORRELATIONS
  cov1<-acf(y,type="covariance",lag.max=lag1,plot=FALSE)
  cov2<-acf(y,type="covariance",lag.max=lag2,plot=FALSE)
  gamaj<-cov1$acf[lag1+1]  gamak<-cov2$acf[lag2+1]
  tau<-gamaj+gamak
  if(tau<0)
  {
    break
  }
  tausum<-tausum+tau
}
proc.time()-ptm
```

```
##     user  system elapsed
##     0.06    0.02    0.08
```

```
muhat
```

```
## [1] -0.003605249
```

```
gama0
```

```
## [1] 2.723108
```

```
varch<- -svar+2*tausum
varch
```

```
## [1] 26.93673
```

```
mcvar<-varch/ns
mcvar
```

```
## [1] 0.002693673
```

```
efchsize<-svar/mcvar
efchsize
```

```
## [1] 1010.826
```

```
integrautoc<-varch/svar
integrautoc
```

```
## [1] 9.892896
```

The estimated feature is the mean of the distribution:

$$\hat{\mu}_m = \frac{1}{m}\sum_{i=1}^{m} X^{(i)} = -0.0036$$

where $m = 10{,}000$ is the length of the Markov chain. Using the method of Geyer,

$$\widehat{V}_{\texttt{asymp}} = \widehat{\text{Var}}\left(X^{(i)}|y\right)\widehat{\tau} = 2.72 \times 9.89 = 26.94,$$

$$\widehat{\text{Var}}\left(\hat{\mu}_m\right) = 0.0027,$$
$$\widehat{m}_{\texttt{eff}} = 1011,$$
$$\widehat{\tau} = 9.89.$$

where $\widehat{\text{Var}}\left(X^{(i)}|y\right)$ is the Monte Carlo estimate of the variance of the posterior distribution $[X|y]$.

The `R-code` below uses the method of batching to compute the Monte Carlo variance of the mean of y, based on (4.90) and the effective chain length based on (4.91). The batch size was set equal to 100. Figure 4.2, right panel, is generated using `acf(avrb)` and shows that batch means are effectively uncorrelated. The output is printed at the bottom of the code and compares well with the previous results:

```
#CODE FOR THE MC VARIANCE BASED ON BATCHING
#choose size of batch s
s<-100
x<-matrix(y,ncol=s, byrow=FALSE)
avrb<-apply(x,2,mean)
mcvarb<-var(avrb)/length(avrb)
efchsizebatch<-var(y)/mcvarb
mcvarb
```

```
## [1] 0.002656311
```

```
efchsizebatch
```

```
## [1] 1025.147
```

```
#PLOT THE AUTOCORRELATION VS LAG USING R-FUNCTION acf
#require(graphics)
#acf(y)
#acf(avrb)
```

The top code was also executed simulating again 10,000 samples with an autocorrelation now of $\rho = 0.5$. The Geyer statistics for this run are

$$\widehat{V}_{\texttt{asymp}} = 3.97$$
$$\widehat{\tau} = 3.03$$
$$\widehat{\text{Var}}\left(\hat{\mu}_m\right) = 0.0004,$$
$$\widehat{m}_{\texttt{eff}} = 3294,$$
$$\widehat{\tau} = 3.03.$$

4.14 Appendix: A Closer Look at the McMC Machinery

The appendix provides a brief description of the Metropolis-Hastings algorithm in its standard and general form. The intention is to provide an intuition for the rationale behind the form of the algorithm. The material may seem a little daunting at first sight, particularly due to the nature of the ideas than the mathematics. In fact, the appendix uses nuts and bolts mathematics and basic theory of transformation of random variables. Much of what is written below is taken from Waagepetersen and Sorensen (2001), where more details and an example can be found. A benchmark paper for the general Metropolis-Hastings algorithm is Green (1995); the paper is technical and its detailed understanding requires command of measure theory. An accessible overview of discrete and continuous Markov chains can be found in Chapters 10 and 11 of Sorensen and Gianola (2002).

The Standard Metropolis-Hastings Ratio

Consider a finite-state discrete Markov chain with state space S. The Markov chain is defined by a sequence of discrete random variables $X_i, i = 1, 2, \ldots$ each of which can take one of N values in the finite set S. The subscript i in X_i can be interpreted as stages or time periods, and the initial stage is $i = 0$. These random variables X_i satisfy the Markov property:

$$\Pr(X_i|X_{i-1}, X_{i-2}, \ldots, X_0) = \Pr(X_i|X_{i-1}). \tag{4.92}$$

Stationarity

A probability density π is *stationary* or *invariant*, if $X_i \sim \pi$ implies $X_{i+1} \sim \pi$, $i \geq 0$. Notationally, I often use x for the current value of the Markov chain and x' for the value at the following stage (not to be confused with the same symbol "′" used elsewhere to denote the transpose of a matrix or vector). A definition of stationarity is

$$\sum_x \pi(x) \Pr\left(X' = x'|X = x\right) = \pi\left(x'\right). \tag{4.93}$$

The idea behind stationarity is as follows. Consider the relationship

$$\Pr\left(X' = x'|X_0 = x_0\right) = \sum_x \Pr\left(X' = x'|X = x\right) \Pr(X = x|X_0 = x_0). \tag{4.94}$$

If the chain converges to a stationary distribution π that is independent of the starting point x_0, then

$$\Pr\left(X' = x'|X_0 = x_0\right) = \pi\left(x'\right),$$

$$\Pr\left(X = x|X_0 = x_0\right) = \pi\left(x\right)$$

and (4.94) reduces to (4.93).

The Metropolis-Hastings (MH) algorithm is a recipe for constructing a Markov chain that has π as its *stationary or invariant distribution*. The goal is to obtain draws from π which could be a posterior distribution. Since doing this directly may be very complicated or impossible, a draw from a *proposal distribution* (denoted here as $q(\cdot|x)$ that may depend on x) is taken instead, and because this is not the same distribution as π, the drawn value is accepted in a stochastic manner by means of an *acceptance probability*. This acceptance probability is derived in a way ensuring that the sequence $X_0, X_1, \ldots$, is an approximate Monte Carlo sample from the stationary distribution π. In order for this to hold, the Markov chain must be aperiodic and irreducible and hence, ergodic (an aperiodic chain does not return to the same state at regular time intervals; in an irreducible chain, every state is reachable from every other state in a finite number of transitions). An ergodic chain is one that converges to the stationary distribution, regardless of the starting value. This Markov chain can be used for Monte Carlo estimation of various expectations with respect to the stationary distribution. Before showing how the algorithm works, I define the concept of detailed balance on which the MH algorithm builds.

Detailed Balance

Let $T\left(x'|x\right) = \Pr\left(X' = x'|X = x\right)$ be the conditional probability that $X' = x'$, given that $X = x$. This conditional pmf is known as a *transition probability* that has the standard property:

$$\sum_{x'} T\left(x'|x\right) = 1. \tag{4.95}$$

In the case of finite-state discrete Markov chains, $T\left(x'|x\right)$ is an element in an $N \times N$ stochastic matrix (transition probability matrix) and (4.95) indicates that the sum of the elements of the rows of the stochastic matrix adds to one. Then π satisfies detailed balance with respect to T, if

$$\pi\left(x\right) T\left(x'|x\right) = \pi\left(x'\right) T\left(x|x'\right). \tag{4.96}$$

Note that when $x = x'$, (4.96) holds trivially because

$$\pi\left(x'\right) T\left(x'|x'\right) = \pi\left(x'\right) T\left(x'|x'\right). \tag{4.97}$$

The detailed balance equation can also be written as

$$\Pr\left(X = x, X' = x'\right) = \Pr\left(X = x', X' = x\right), \tag{4.98}$$

provided $X \sim \pi$.

A Markov chain that has stationary distribution π that satisfies detailed balance with respect to T is said to be a reversible Markov chain.

An intuition for expression (4.96) is that the total probability mass in the move from x to x' is equal to that of the reverse move, from x' to x. In the left-hand side, the probability mass at x is $\pi(x)$ and only a proportion $T\left(x'|x\right)$ moves to the right-hand side. The total probability mass in the move from x to x' is the product of these two quantities. Likewise, the probability mass on the right-hand side at $X' = x'$ is $\pi\left(x'\right)$ that may be different from $\pi(x)$. A proportion of $\pi\left(x'\right)$ equal to $T\left(x|x'\right)$ moves from x' to x. The total probability mass for this pair of states $\left(x, x'\right)$ is the same. This holds for all possible pairs x and x' that belong in S.

An important consequence of imposing the strong condition of detailed balance (4.96) is that the ergodic Markov chain is reversible and when it converges, has π as its stationary distribution. To show this, start with (4.96) and sum over x on both sides (if (4.96) holds, the equality still holds if we sum both sides over x):

$$\begin{aligned}\sum_x \pi(x)\, T\left(x'|x\right) &= \sum_x \pi\left(x'\right) T\left(x|x'\right) \\ &= \pi\left(x'\right) \sum_x T\left(x|x'\right) \\ &= \pi\left(x'\right),\end{aligned} \tag{4.99}$$

that is the definition of a stationary distribution. In the case of a Markov chain in continuous space with transition kernel p, the equivalent to (4.99) is obtained by integrating both sides with respect to x:

$$\begin{aligned}\int \pi(x)\, p\left(x'|x\right) dx &= \int \pi\left(x'\right) p\left(x|x'\right) dx \\ &= \pi\left(x'\right) \int p\left(x|x'\right) dx \\ &= \pi\left(x'\right)\end{aligned}$$

that again is the definition of a stationary distribution. This means that if (X_i, X_{i+1}) has stationary distribution π, then the time-reversed subchain (X_{i+1}, X_i) has the same stationary distribution, whenever X_i has density π.

The acceptance probability of the MH algorithm is derived assuming that the Markov chain satisfies detailed balance. If the Markov chain is ergodic, when it converges, detailed balance guarantees that π is the stationary distribution.

The Acceptance Probability of the Metropolis-Hastings Algorithm

Consider a move from x to x'. In the MH algorithm, a proposed value x^{p} for x' is drawn from the proposal distribution $q\left(x^{\mathrm{p}}|x\right)$. If the proposed value is accepted, $x' = x^{\mathrm{p}}$ and if its is rejected, $x' = x$, the value at the previous stage of the chain. There are two ways in which the state in the next stage is equal to x'. One is to draw the proposal x^{p} and to accept it with probability $a\left(x'|x\right)$. The other way in which the state at the next stage can take the value x' is to reject the proposal, to set $x' = x$ but X was already equal to x'. Let

$$s\left(x\right) = \Pr\left(x^{\mathrm{p}}\text{ rejected }|X = x\right) \tag{4.100}$$

Then

$$\begin{aligned} T\left(x'|x\right) &= \Pr\left(X' = x'|X = x\right) \\ &= q\left(x'|x\right) a\left(x'|x\right) + s\left(x\right) I\left(x = x'\right). \end{aligned}$$

The left-hand side of (4.96) can now be written as

$$\pi\left(x\right) T\left(x'|x\right) = \pi\left(x\right) q\left(x'|x\right) a\left(x'|x\right) + \pi\left(x\right) s\left(x\right) I\left(x = x'\right), \tag{4.101}$$

and by symmetry, replacing x by x', the right-hand side is equal to

$$\pi\left(x'\right) T\left(x|x'\right) = \pi\left(x'\right) q\left(x|x'\right) a\left(x|x'\right) + \pi\left(x'\right) s\left(x'\right) I\left(x' = x\right). \tag{4.102}$$

The second terms in the right-hand side of (4.101) and (4.102) are equal, both in the case when $x \neq x'$ in which case they are zero because the indicator function is zero, or trivially when $x = x'$. Therefore, detailed balance is satisfied if

$$\pi\left(x\right) q\left(x'|x\right) a\left(x'|x\right) = \pi\left(x'\right) q\left(x|x'\right) a\left(x|x'\right). \tag{4.103}$$

Then,

$$\frac{a\left(x'|x\right)}{a\left(x|x'\right)} = \frac{\pi\left(x'\right) q\left(x|x'\right)}{\pi\left(x\right) q\left(x'|x\right)}. \tag{4.104}$$

Relationship (4.104) is satisfied if $a(x'|x) = \pi\left(x'\right) q\left(x|x'\right) / b\left(x'|x\right)$, for some $b\left(x'|x\right) \geq \pi\left(x'\right) q\left(x|x'\right)$, to ensure that the acceptance probability $a\left(x'|x\right) \leq 1$ (a similar argument holds for $a\left(x|x'\right)$ due to symmetry). A valid $b(x'|x)$ is

$$b(x'|x) = \max(\pi\left(x'\right) q\left(x|x'\right), \pi\left(x\right) q\left(x'|z\right)) \tag{4.105}$$

but other choices are possible and these have an impact on the properties of the Markov chain. The subject is rather technical and is discussed by Hastings

(1970) and Peskun (1973), where it is shown that (4.105) leads to a Markov chain with the largest possible acceptance probabilities resulting in minimum asymptotic variances of moment estimates. This choice translates into the most commonly cited expression for the Metropolis-Hastings acceptance probability, given by

$$a\left(x'|x\right) = \min\left(1, \frac{\pi\left(x'\right) q\left(x|x'\right)}{\pi\left(x\right) q\left(x'|x\right)}\right), \tag{4.106}$$

that gives rise to the following algorithm:

- Initialise setting $X = x$, so that $\pi(x) > 0$.
- Choose the proposal distribution q
  Given current state $X_i = x$, go through the loop.
- Draw x^{p} from the proposal $q(\cdot|x)$.
- Draw u from $Un(0, 1)$.
- If $u < a(x^{\mathrm{p}}|x)$, accept x^{p} and set $X_{i+1} = x^{\mathrm{p}}$; otherwise, $X_{i+1} = x$.

The original idea of using Markov chain simulation of probability distributions is often attributed to Metropolis et al (1953).

The General Metropolis-Hastings Ratio

In the standard Metropolis-Hastings algorithm, the next state x' is obtained by drawing the candidate x^{p} from q. A more general mechanism to generate the move to x' is to construct a proposal y by applying a deterministic mapping to x and to a random component u that has density $q_U(u|x)$, which may depend on x (Green 1995). The proposal y (if the proposal is accepted, $x' = y$) is

$$y = g_1(x, u).$$

To ensure dimension matching in the move from (x, u) to x', if x is of size n_x and u of size n_u, it may be necessary to include a random variable u' of size $n_{u'}$ with density $q_{U'}\left(u'|x'\right)$ so that

$$n_x + n_u = n_{x'} + n_{u'}. \tag{4.107}$$

Then the vectors of Markov chain states and proposal random variables (x, u) and $\left(x', u'\right)$ are of equal dimension, and the densities $\pi(x) q_U(u|x)$ in the move from (x, u) to $\left(x', u'\right)$ and $\pi\left(x'\right) q_{U'}\left(u'|x'\right)$ in the move from $\left(x', u'\right)$ to (x, u) are joint densities on spaces of equal dimension. The mapping is then

$$\left(x', u'\right) = g(x, u) = \left(g_1(x, u), g_2(x, u)\right) \tag{4.108}$$

and the move in the opposite direction is

$$(x, u) = g^{-1}\left(x', u'\right) = \left(g_1^{-1}\left(x', u'\right), g_2^{-1}\left(x', u'\right)\right). \tag{4.109}$$

A necessary condition for the existence of the one-to-one mapping is that (4.107) is satisfied.

An **important detail** that becomes obvious in the derivation (more below) is that there is a constraint in the form of the deterministic function. The constraint is that

$$g = g^{-1}. \tag{4.110}$$

Therefore, we need

$$\begin{aligned} \left(x', u'\right) &= g\,(x, u) = \left(g_1\,(x, u)\,, g_2\,(x, u)\right), \\ (x, u) &= g^{-1}\left(x', u'\right) = g\left(x', u'\right) = \left(g_1\left(x', u'\right), g_2\left(x', u'\right)\right). \end{aligned} \tag{4.111}$$

The general Metropolis-Hastings acceptance ratio takes the form

$$a\left(x'|x\right) = \min\left\{1, \frac{\pi\left(x'\right) q_{U'}\left(u'|x'\right)}{\pi\,(x)\, q_U\,(u|x)}|J|\right\} \tag{4.112}$$

where $|J| = \left|\det \frac{\partial g(x,u)}{\partial(x,u)}\right|$ is the absolute value of the Jacobian of the transformation g. A detailed derivation is given below.

Stationarity

The Markov chain in continuous space is specified in terms of the distribution for the initial state X_0 and the transition kernel $P\,(\cdot, \cdot)$ which specifies the conditional distribution of X_{t+1} given the previous state X_t. If the current state is $X_t = x$, then the probability that X_{t+1} is in a set $B \subseteq \mathbb{R}^d$ is given by

$$P\,(x, B) = \Pr\left(X_{t+1} \in B|X_t = x\right). \tag{4.113}$$

Assume that π is a complex target distribution for a stochastic vector Z. Since expectations with respect to π cannot be evaluated analytically or by using techniques for numerical integration, a Markov chain X_i, $i = 1, 2, \ldots$, is constructed whose stationary distribution is π. If the chain is ergodic, then it can be used for Monte Carlo estimation of expectations $\mathrm{E}\,(h\,(Z))$ for any function h, with respect to the invariant density π. That is,

$$\mathrm{E}\,(h\,(Z)) = \int h\,(z)\,\pi\,(z)\,dz \approx \frac{1}{n}\sum\nolimits_{i=1}^{n} h\,(X_i) \tag{4.114}$$

as n tends to infinity. Thus, $\mathrm{E}\left(h\left(Z\right)\right)$ can be approximated by the sample average for large n, the length of the chain. The autocorrelation among the draws of the chain implies that the size of the Markov chain must be larger than when the draws are independent, in order to achieve a given level of accuracy.

For a continuous space Markov chain, the definition of stationarity is

$$\int P\left(x, B\right)\pi\left(x\right)dx = \int_B \pi\left(x\right)dx = \Pr\left(X \in B\right) \tag{4.115}$$

The distribution π is invariant (or stationary) for the Markov chain, if the transition kernel $P\left(\cdot,\cdot\right)$ of the Markov chain preserves π, so that $X_t \sim \pi$ implies $X_{t+1} \sim \pi$. In order to verify that π is the invariant density, using (4.115) is an infeasible task, since this involves integration with respect to π. The difficulty of doing this was the reason for using McMC in the first place. However, choosing a kernel that imposes the stronger condition of reversibility with respect to π is sufficient to guarantee that π is the invariant density of the ergodic Markov chain.

Reversibility

For a continuous state space Markov chain, the reversibility condition

$$P_{t,t+1}\left(A, B\right) = P_{t,t+1}\left(B, A\right) \tag{4.116}$$

requires that the equilibrium probability that the state of the chain is in a general set A and moves to a general set B to be the same with A and B reversed. In other words, expression (4.116) states that the joint probability that $X_t \in A$ and $X_{t+1} \in B$ (left-hand side) is the same as the joint probability that $X_t \in B$ and $X_{t+1} \in A$ (right-hand side). The left-hand side describes the move from, say, X to X', and the left-hand side the opposite move. The above can be written as

$$\iint I\left(x \in A, x' \in B\right) p\left(x, x'\right) dx dx' = \iint I\left(x \in B, x' \in A\right) p\left(x, x'\right) dx dx'. \tag{4.117}$$

The right-hand side is equal to the left-hand side with A and B reversed.

Since $\left(x, x'\right)$ in the integrals are dummy variables, one could relabel arbitrarily; for example, set $x = v$, $x' = w$ and write the right-hand side as

$$\iint I\left(v \in B, w' \in A\right) p\left(v, w\right) dv dw.$$

The reversibility condition (4.117) can also be written in terms of the transition kernel:

$$\int_A P(x, B)\,\pi(x)\,dx = \int_B P(x, A)\,\pi(x)\,dx. \tag{4.118}$$

Reversibility (4.118) implies (4.115) by taking $A = \mathbb{R}^d$. Then $P(x, A) = 1$ and (4.118) reduces to

$$\int P(x, B)\,\pi(x)\,dx = \int_B \pi(x)\,dx$$

which is equal to (4.115). Therefore, an ergodic Markov chain that satisfies (4.118) has stationary distribution π.

The Acceptance Probability for a General Metropolis-Hastings Algorithm

In a general setting, instead of generating X' from $q(\cdot|x)$ as is practised in the standard Metropolis-Hastings algorithm, X' can be defined in terms of a stochastic component $U \sim q(u|x)$ and a deterministic mapping g. As explained in connection with (4.108) and (4.109), in the move from X to X', the mapping is

$$\left(x', u'\right) = g(x, u) = \left(g_1(x, u), g_2(x, u)\right) \tag{4.119}$$

and in the reverse move

$$(x, u) = g^{-1}\left(x', u'\right) = \left(g_1^{-1}\left(x', u'\right), g_2^{-1}\left(x', u'\right)\right). \tag{4.120}$$

The transition kernel in the move from X to X'

$$P(x, B) = \Pr\left(X' \in B | X = x\right) = \int I\left(x' \in B\right) p\left(x'|x\right) dx'$$

is now constructed in three steps. According to the Metropolis-Hastings protocol, a transition from x to x' is accomplished by first drawing u from $q(\cdot|x)$. Secondly, constructing $x' = g_1(x, u)$ and thirdly accepting it with probability $a(g_1(x, u)|x)$. Then

$$\Pr\left(X' \in B | X = x\right) = \int I\left(g_1(x, u) \in B\right) q(u|x)\, a\left(g_1(x, u)|x\right) dx du$$
$$+ I(x \in B) \int q(u|x)\left[1 - a\left(g_1(x, u)|x_t\right)\right] du, \tag{4.121}$$

where the second term in the right-hand side accounts for the probability of rejection, but the current state already belongs in B. The left-hand side of (4.117) takes the form

$$\int\int I\left(x \in A, g_1(x,u) \in B\right)\pi(x)\, q(u|x)\, a\left(g_1(x,u)|x\right) dxdu$$
$$+\int I\left(x \in A \cap B\right)\int \pi(x)\, q(u|x)\left[1-a\left(g_1(x,u)|x\right)\right] dxdu. \quad (4.122)$$

The right-hand side of (4.117) can be expressed by interchanging A and B in (4.122). That is,

$$\int\int I\left(x \in B, g_1(x,u) \in A\right)\pi(x)\, q(u|x)\, a\left(g_1(x,u)|x\right) dxdu$$
$$+\int I\left(x \in B \cap A\right)\int \pi(x)\, q(u|x)\left[1-a\left(g_1(x,u)|x\right)\right] dxdu. \quad (4.123)$$

The second terms in (4.122) and (4.123) are equal and therefore a sufficient condition for (4.117) to hold is

$$\int\int I\left(x \in A, g_1(x,u) \in B\right)\pi(x)\, q(u|x)\, a\left(g_1(x,u)|x\right) dxdu$$
$$=\int\int I\left(x \in B, g_1(x,u) \in A\right)\pi(x)\, q(u|x)\, a\left(g_1(x,u)|x\right) dxdu. \quad (4.124)$$

The final step is to find a way to equalise the indicator functions of both sides of Eq. (4.124). This is accomplished in two steps. First note that since (x, u) are dummy variables of integration, they can be relabelled arbitrarily. Setting $x = x'$ and $u = u'$, the right-hand side of (4.124) becomes

$$\int\int I\left(x' \in B, g_1\left(x',u'\right) \in A\right)\pi\left(x'\right) q\left(u'|x'\right) a\left(g_1\left(x',u'\right)|x'\right) dx'du'. \quad (4.125)$$

Secondly, note that by applying the deterministic mapping

$$x = g_1\left(x',u'\right),$$
$$x' = g_1(x,u)$$

and more generally

$$g^{-1}\left(x',u'\right) = g\left(x',u'\right), \quad (4.126)$$

and substituting in the argument of the indicator function in (4.125), both indicator functions are equalised. It is at this point of the derivation that the constraint (4.110) or (4.126), in the form of the deterministic mapping, becomes relevant (see also (4.111)). Using this transformation and the change-of-variable formula (from (x', u') to (x, u)), then $dx'du' = |J|\, dxdu$, and (4.125) takes the form

$$\iint I\left(g_1(x,u) \in B, x \in A\right) \pi\left(g_1(x,u)\right)$$
$$q\left(g_2(x,u) | g_1(x,u)\right) a\left(x | g_1(x,u)\right) |J|\, dxdu \qquad (4.127)$$

where

$$J = \frac{\partial g(x,u)}{\partial (x,u)} = \frac{\partial (x', u')}{\partial (x,u)}.$$

Examination of the left-hand side of (4.124) and of (4.127) shows that the reversibility condition (4.117) is satisfied if

$$\pi(x)\, q(u|x)\, a\left(g_1(x,u)|x\right) = \pi\left(g_1(x,u)\right) q\left(g_2(x,u)|g_1(x,u)\right) a\left(x|g_1(x,u)\right) |J|.$$

In view of (4.111), $g_1(x,u) = x'$, $g_2(x,u) = u'$ and a valid choice for $a(x'|x)$ is

$$a(x'|x) = \min\left[1, \frac{\pi(x')\, q(u'|x')}{\pi(x)\, q(u|x)} |J|\right]. \qquad (4.128)$$

A Toy Example

The model is $[y|\mu, \lambda] \sim N(\mu, \lambda)$. Assume μ is known and a Metropolis-Hastings algorithm is constructed to update the variance λ.

Strategy 1

This is accomplished generating $u \sim Un(a, b)$ and letting $\lambda' = \lambda u$. The current state of the Markov chain is $(z, u) = (\lambda, u)$ and the move is to

$$\begin{aligned}(z', u') &= g(\lambda, u) \\ &= ((\lambda u), 1/u)\end{aligned}$$

where $u = 1/u'$. The inverse function that makes the move in the opposite direction possible is

$$\begin{aligned}(z,u) &= g^{-1}\left(z',u'\right)\\ &= g\left(\lambda',u'\right)\\ &= \left(\left(\lambda' u'\right),1/u'\right)\end{aligned}$$

where $u' = 1/u$ is also generated from $Un\,(a,b)$. The Jacobian of the transformation $g\,(\lambda,u) = (\lambda u, 1/u)$ is

$$\begin{aligned}J &= \left|\det\left[\frac{\partial g\,(\lambda,u)}{\partial\,(\lambda,u)}\right]\right|\\ &= \left|\det\begin{bmatrix} u & \lambda\\ 0 & -u^{-2}\end{bmatrix}\right| = u^{-1}.\end{aligned}$$

Since u and u' are drawn from $Un\,(a,b)$, $1/(b-a)$ cancels in the ratio $q_{U'}\left(u'|z'\right)/q_U\,(u|z)$ in (4.112) and the Metropolis-Hastings acceptance probability is

$$\min\left\{1, \frac{p\,(\mu,\lambda'|y)}{p\,(\mu,\lambda|y)}u^{-1}\right\}, \qquad u \in (a,b)\,. \tag{4.129}$$

NOTE

The above strategy satisfies the constraint (4.110) which in the example takes the form

$$\left(z',u'\right) = \left((zu)\,,\frac{1}{u}\right)$$

$$(z,u) = \left(\left(z'u'\right),\frac{1}{u'}\right)$$

with $u = 1/u'$. This is so because (recall, $g_1\,(a,b) = a \times b$, is a function that multiplies its arguments)

$$\begin{aligned}z' &= g_1\,(z,u)\\ &= g_1\left(g_1\left(z',u'\right),u\right)\\ &= g_1\left(z',u'\right)\times u\\ &= \left(z'\times u'\right)\times u\end{aligned}$$

and therefore $u = 1/u'$.

Strategy 2

An alternative way of arriving at (4.129) is to use as the Metropolis-Hastings ratio

$$\frac{p(\mu, \lambda u|y)}{p(\mu, \lambda|y)} \frac{p(\lambda|\lambda')}{p(\lambda'|\lambda)}. \tag{4.130}$$

This is the standard form of the Metropolis-Hastings ratio, using the proposal for the parameter rather than the auxiliary variables (u, u'). Then with $\lambda' = u\lambda$,

$$p(\lambda'|\lambda) = q_U(u) \left| \frac{du}{d\lambda'} \right| = \frac{1}{b-a} \frac{1}{\lambda}, \qquad \lambda' \in (\lambda a, \lambda b).$$

By symmetry,

$$p(\lambda|\lambda') = \frac{1}{b-a} \frac{1}{\lambda'}, \qquad \lambda \in (\lambda' a, \lambda b).$$

The bounds $\lambda \in (\lambda' a, \lambda' b)$ imply

$$\lambda u' a < \lambda < \lambda u' a,$$

$$a < \frac{1}{u'} < b.$$

The Metropolis-Hastings ratio is now

$$\begin{aligned} \frac{p(\mu, \lambda u|y)}{p(\mu, \lambda|y)} \frac{p(\lambda|\lambda')}{p(\lambda'|\lambda)} &= \frac{p(\mu, \lambda u|y)}{p(\mu, \lambda|y)} \frac{\lambda}{\lambda'} \\ &= \frac{p(\mu, \lambda u|y)}{p(\mu, \lambda|y)} \frac{1}{u}, \qquad u \in (a, b). \end{aligned} \tag{4.131}$$

and the resulting acceptance probability is

$$\min\left\{1, \frac{p(\mu, \lambda'|y)}{p(\mu, \lambda|y)} \frac{1}{u}\right\}, \qquad u \in (a, b). \tag{4.132}$$

Strategy 3

The third strategy consists of updating the variance using a random walk proposal density on the logvariance. That is,

$$\ln \lambda' \sim N(\ln \lambda, k),$$

a normal distribution with mean equal to the natural logarithm of the previous realisation of λ and variance given by k, a user-tuned parameter.

As a reminder, if X has density $p(x) = N(m, k)$ and $Y = f(X) = \exp(X)$, such that the inverse function f^{-1} exists and results in $X = f^{-1}(Y) = \ln Y$, then the Jacobian of the transformation from X to Y is $1/y$ and $p(y) = p\left(f^{-1}(y)\right)\frac{1}{y}$. In this particular case, in the move from λ to λ', we have $X = \ln \lambda'$; $Y = \exp(\ln \lambda') = \lambda'$; $f^{-1}(Y) = \ln \lambda'$. Therefore, if

$$q\left(\ln \lambda' | \ln \lambda, k\right) = N(\ln \lambda, k),$$

then

$$\begin{aligned} q\left(\lambda' | \ln \lambda, k\right) &= q\left(\ln \lambda' | \ln \lambda, k\right) \frac{1}{\lambda'} \\ &= N(\ln \lambda, k) \frac{1}{\lambda'}, \end{aligned} \tag{4.133}$$

which is the density of the lognormal distribution with parameters $(\ln \lambda, k)$. In these expressions, the variance of the normal distribution k is a user-tuned parameter. Then (4.133) is the proposal density evaluated at λ' and by symmetry,

$$q(\lambda | \ln \lambda, k) = q\left(\ln \lambda | \ln \lambda', k\right) \frac{1}{\lambda}.$$

Since $q\left(\ln \lambda' | \ln \lambda, k\right) = q\left(\ln \lambda | \ln \lambda', k\right)$, the Metropolis-Hastings ratio is

$$\frac{p\left(\mu, \lambda' | y\right)}{p(\mu, \lambda | y)} \frac{q(\lambda | \ln \lambda, k)}{q\left(\lambda' | \ln \lambda, k\right)} = \frac{p\left(\mu, \lambda' | y\right)}{p(\mu, \lambda | y)} \frac{\lambda'}{\lambda},$$

different from (4.131).

Chapter 5
McMC in Practice

This chapter illustrates applications of McMC in a Bayesian context. The treatment is mostly schematic; the objective is to present the mechanics of McMC in different modelling scenarios. Many of the examples, discussed in connection with the implementation of maximum likelihood (using Newton-Raphson and EM), are revisited from a Bayesian McMC perspective. These include the analysis of ABO blood group data, the binary regression, the genomic model, the two-component mixture model, and the Bayesian analysis of truncated data. Further examples are discussed in the second part of the book on Prediction and in the Exercise sections, including their solutions, at the end of the book.

5.1 Example: Estimation of Gene Frequencies from ABO Blood Group Phenotypes

The ABO blood group data discussed in Chap. 3 is reproduced in Table 5.1. The problem is to infer p_A, p_B and p_0, the frequency of the three alleles, A, B and 0, respectively, subject to $p_A + p_B + p_0 = 1$. Three alleles give rise to six genotypes but only four phenotypic classes are observed. The observed data are $n = (n_A, n_{AB}, n_B, n_0)$, and the multinomial likelihood is proportional to the pmf $f(n|p_A, p_B, p_0)$:

$$\begin{aligned} L(p_A, p_B|n) &\propto f(n|p_A, p_B, p_0) \\ &= \left(p_A^2 + 2p_A p_B\right)^{n_A} (2p_A p_B)^{n_{AB}} \left(p_B^2 + 2p_B p_0\right)^{n_B} \left(p_0^2\right)^{n_0}. \end{aligned} \tag{5.1}$$

A Dirichlet prior will be chosen as distribution for the gene frequencies $p_A, p_B, and p_0$. The Dirichlet is a multivariate generalisation of the univariate beta distribution and is symbolised $Di(\alpha)$ where α is a vector of positive real numbers.

D. Sorensen, *Statistical Learning in Genetics*, Statistics for Biology and Health,
https://doi.org/10.1007/978-3-031-35851-7_5

Table 5.1 Frequency of genotypes and phenotypes of ABO blood group data

Genotype	Phenotype	Observed counts	Frequency
AA	A	n_A	p_A^2
AO	A		$2p_A p_O$
AB	AB	n_{AB}	$2p_A p_B$
BB	B	n_B	p_B^2
BO	B		$2p_B p_O$
OO	O	n_O	p_O^2

The pdf is

$$f(x_1, \ldots, x_k | \alpha_1, \ldots, \alpha_k) = \frac{1}{B(\alpha)} \prod_{i=1}^{k} x_i^{\alpha_i - 1},$$
$$x_1, \ldots, x_{k-1} > 0,$$
$$x_1 + \ldots + x_{k-1} < 1,$$
$$x_k = 1 - x_1 - \cdots - x_{k-1}$$

and the normalising constant

$$B(\alpha) = \frac{\prod_{i=1}^{k} \Gamma(\alpha_i)}{\Gamma\left(\sum_{i=1}^{k} \alpha_i\right)}$$

is the multinomial beta function. A common special case is the symmetric Dirichlet distribution where all the elements of vector α have the same value. The pdf simplifies to

$$f(x_1, \ldots, x_k | \alpha) = \frac{\Gamma(\alpha k)}{\Gamma(\alpha)^k} \prod_{i=1}^{k} x_i^{\alpha - 1}. \tag{5.2}$$

When $\alpha = 1$, this becomes a uniform distribution in $k - 1$ dimensions. In this example, the Dirichlet prior is

$$f(p_A, p_B, p_O) \propto (p_A)^{\alpha_A - 1} (p_B)^{\alpha_B - 1} (p_O)^{\alpha_B - 1}. \tag{5.3}$$

The likelihood (5.1) and the Dirichlet prior give rise to the posterior density

$$f(p_A, p_B, p_O | n) \propto \left(p_A^2 + 2p_A p_B\right)^{n_A} (2p_A p_B)^{n_{AB}} \left(p_B^2 + 2p_B p_O\right)^{n_B} \left(p_O^2\right)^{n_O}$$
$$(p_A)^{\alpha_A - 1} (p_B)^{\alpha_B - 1} (p_O)^{\alpha_O - 1}. \tag{5.4}$$

The fully conditional posterior distributions do not have a closed form. An McMC approach requires a Metropolis-Hastings algorithm or using *data augmentation* that may lead to standard conditional posterior distributions. In the present example, data augmentation facilitates a computationally simple Gibbs sampling implementation, and I explore this alternative. First, data augmentation is described in general terms.

Data Augmentation

Imagine that there is an interest in obtaining the posterior distribution of a parameter θ. Due to analytical intractability, one chooses to approximate $p(\theta|y)$ using McMC. Often, the fully conditional posterior distribution $p(\theta|y)$ does not have a standard form, and the McMC algorithm can be difficult to implement. The idea of data augmentation is to augment with the so-called latent data or missing data φ, in order to exploit the simplicity of the resulting conditional posterior distribution $p(\theta|\varphi, y)$. This is in the same spirit as in the EM algorithm: by increasing the dimensionality of the problem, possibly at the expense of extra computing time, although this is not always the case (Swendsen and Wang 1987), the problem is simplified algorithmically. The focus of inference is

$$p(\theta|y) = \int p(\theta|\varphi, y)p(\varphi|y)d\varphi,$$

and this marginalisation is carried out via McMC. A key paper is Tanner and Wong (1987).

The mechanics works as follows: Start with an initial legal value of $\varphi^{[0]}$. Given $\varphi^{[t]}$ at iteration t,

1. Draw $\theta^{[t+1]}$ from $\left[\theta|\varphi^{[t]}, y\right]$.
2. Draw $\varphi^{[t+1]}$ from $\left[\varphi|\theta^{[t+1]}, y\right]$.

At convergence, the iterative procedure results in draws from the posterior distribution $[\theta, \varphi|y]$, and the margins correspond to draws from $[\theta|y]$ and $[\varphi|y]$.

To apply data augmentation to the ABO blood group data, one can define $n_A = n_{AA} + n_{A0}$ and $n_B = n_{BB} + n_{B0}$. Then the *missing data* is

$$n_m = (n_{A0}, n_{AA}, n_{BB}, n_{B0}).$$

If the missing data were observed,

$$\begin{aligned} &f(n, n_m|p_A, p_B, p_0) \\ &= \left(p_A^2\right)^{n_{AA}} (2p_A p_0)^{n_{A0}} (2p_A p_B)^{n_{AB}} \left(p_B^2\right)^{n_{BB}} (2p_B p_0)^{n_{B0}} \left(p_0^2\right)^{n_0}. \end{aligned}$$

This is the pmf of the standard multinomial distribution. The augmented posterior can be written

$$
\begin{aligned}
f(n_m, p_A, p_B, p_0|n) &\propto f(n_m, p_A, p_B, p_0)\, f(n|n_m, p_A, p_B, p_0) \\
&= f(n, n_m, p_A, p_B, p_0) \\
&= f(n, n_m|p_A, p_B, p_0)\, f(p_A, p_B, p_0) \\
&= \left(p_A^2\right)^{n_{AA}} (2p_A p_0)^{n_{A0}} (2p_A p_B)^{n_{AB}} \left(p_B^2\right)^{n_{BB}} \\
&\quad (2p_B p_0)^{n_{B0}} \left(p_0^2\right)^{n_0} (p_A)^{\alpha_A-1} (p_B)^{\alpha_B-1} (p_0)^{\alpha_0-1}. \qquad (5.5)
\end{aligned}
$$

This is proportional to

$$
\begin{aligned}
f(n_m, p_A, p_B, p_0|n) &\propto (p_A)^{2n_{AA}} (p_A)^{n_{A0}} (p_A)^{n_{AB}} (p_A)^{\alpha_A-1} \\
&\quad (p_B)^{2n_{BB}} (p_B)^{n_{B0}} (p_B)^{n_{AB}} (p_B)^{\alpha_B-1} \\
&\quad (p_0)^{2n_0} (p_0)^{n_{B0}} (p_0)^{n_{A0}} (p_0)^{\alpha_0-1} \\
&= (p_A)^{2n_{AA}+n_{A0}+n_{AB}+\alpha_A-1} (p_B)^{2n_{BB}+n_{B0}+n_{AB}+\alpha_B-1} \\
&\quad (p_0)^{2n_0+n_{A0}+n_{B0}+\alpha_0-1}. \qquad (5.6)
\end{aligned}
$$

The Gibbs sampling algorithm requires extracting the fully conditional posterior distributions

$$f(p_A, p_B, p_0|n, n_m), \qquad (5.7a)$$

$$f(n_{AA}|p_A, p_B, p_0, n_A, n), \qquad (5.7b)$$

$$f(n_{BB}|p_A, p_B, p_0, n_B, n) \qquad (5.7c)$$

from (5.5) or (5.6). The derivation of (5.7a) is straightforward: direct inspection of (5.6) indicates that the required fully conditional is

$$[p_A, p_B, p_0|n, n_m] \sim Di(a, b, c), \qquad (5.8)$$

a Dirichlet distribution with parameters

$$
\begin{aligned}
a &= 2n_{AA} + n_{A0} + n_{AB} + \alpha_A, \\
b &= 2n_{BB} + n_{B0} + n_{AB} + \alpha_B, \\
c &= 2n_0 + n_{A0} + n_{B0} + \alpha_0.
\end{aligned}
$$

To derive (5.7b) first replace in (5.5), $n_{A0} = n_A - n_{AA}$. Extracting terms in n_{AA} only yields

$$f(n_{AA}|p_A, p_B, p_0, n_A, n) \propto \left(p_A^2\right)^{n_{AA}} (2p_A p_0)^{n_A - n_{AA}},$$

which is the kernel of the binomial distribution

$$[n_{AA}|p_A, p_B, p_0, n_A, n] \sim Bi\left(\frac{p_A^2}{p_A^2 + 2p_A p_0}, n_A\right). \tag{5.9}$$

Similarly, after replacing $n_{BO} = n_B - n_{BB}$,

$$f(n_{BB}|p_A, p_B, p_0, n_B, n) \propto \left(p_B^2\right)^{n_{BB}} (2p_B p_0)^{n_B - n_{BB}}$$

which is the kernel of the binomial distribution

$$[n_{BB}|p_A, p_B, p_0, n_B, n] \sim Bi\left(\frac{p_B^2}{p_B^2 + 2p_B p_0}, n_B\right). \tag{5.10}$$

The Gibbs Sampling Implementation

The data in Weir (1996) used in Chap. 3 is $n_A = 725$, $n_{AB} = 72$, $n_B = 258$, and $n_0 = 1073$. The R-code below implements the Bayesian model with a Gibbs sampler, assuming $\alpha_A = \alpha_B = \alpha_0 = \alpha = 2$, start values $p_A = 0.33$, $p_B = 0.33$:

```
# CODE0501
# ABO BLOOD GROUP DATA - GIBBS SAMPLING
rm(list=ls()) # Clear the workspace
set.seed(1237)
#install.packages("MCMCpack", .libPaths()[1])
# to access function rdirichlet
library(MCMCpack)
#CHOOSE LENGTH OF GIBBS CHAIN rep
rep<-3000
result<-matrix(data=NA,nrow=rep,ncol=4)
# INITIALISE PARAMETERS
p_A<-0.33
p_B<-0.33
p_0<-1-p_A-p_B
alfa<-2
# DATA
```

```
n_A<-725
n_AB<-72
n_B<-258
n_0<-1073
#START WITH THE GIBBS LOOP
for (i in 1:rep)
{
  # SAMPLE n_AA AND n_BB
  n_AA<-rbinom(1,n_A,p_A^2/(p_A^2+2*p_A*p_0))
  n_BB<-rbinom(1,n_B,p_B^2/(p_B^2+2*p_B*p_0))
  n_A0<-n_A-n_AA
  n_B0<-n_B-n_BB
  # SAMPLE p_A,p_B,p_0
  a<-2*n_AA+n_A0+n_AB+alfa
  b<-2*n_BB+n_B0+n_AB+alfa
  c<-2*n_0+n_A0+n_B0+alfa
  draws<- rdirichlet(1,c(a,b,c))
  p_A<-draws[1,1]
  p_B<-draws[1,2]
  p_0<-draws[1,3]
  result[i, ]<-c(i,p_A,p_B,p_0)
}
# END OF GIBBS LOOP
meanp_A<-mean(result[,2])
meanp_A
```

```
## [1] 0.2092435
```

```
varp_A<-var(result[,2])
cip_A<- quantile(result[,2],c(0.025,0.975))
cip_A
```

```
##      2.5%     97.5%
## 0.1967966 0.2220510
```

```
meanp_B<-mean(result[,3])
meanp_B
```

```
## [1] 0.0811496
```

```
varp_B<-var(result[,3])
cip_B<- quantile(result[,3],c(0.025,0.975))
cip_B
```

```
##         2.5%      97.5%
## 0.07306027 0.08997164
```

```
covp_Ap_B<-cov(result[,2],result[,3])
```

These results are in good agreement with those from previous analyses.

5.2 Example: A Regression Model for Binary Data

The binary regression discussed in Chap. 3 was parametrised as

$$\begin{aligned}\Pr\left(y_i = 1|x_i'\beta\right) &= \Pr\left(u_i < 0|x_i, \beta\right),\\ \Pr\left(y_i = 0|x_i'\beta\right) &= \Pr\left(u_i > 0|x_i, \beta\right),\end{aligned}$$

where u_i is the unobserved liability. The logistic loglikelihood shown in (3.30) is

$$l\left(\beta|y, x\right) = \sum_{i=1}^{N}\left\{(1-y_i)\,x_i'\beta - \ln\left[1+\exp\left(x_i'\beta\right)\right]\right\}. \tag{5.11}$$

At the level of the liability u, the linear model for the ith datum is

$$u_i = x_i'\beta + e_i, \quad i = 1, 2, ..., N, \tag{5.12}$$

where x_i' is the ith row of the known non-stochastic $N \times p$ matrix of explanatory variables (covariates), β is an unknown parameter vector (of order $p \times 1$), and the residuals e_i are iid variables drawn from a standard logistic distribution with mean 0 and variance $\pi^2/3$.

A Bayesian implementation requires assigning a prior specification for β. One possibility is to assume a uniform prior. In this case, the posterior $p\left(\beta|y, x\right)$ is proportional to the likelihood, and therefore the logposterior takes the same form as the loglikelihood (5.11).

Note: The notation for the posterior density $p\left(\beta|y, x\right)$ indicates that conditioning is on the observations ($1's$ and $0's$) and the covariates $(x_1, \ldots, x_N)$ where N is the number of individuals. Hyperparameters are not included, and often, covariates are also omitted from the notation. When this is the case, the posterior density is written as $p\left(\beta|y\right)$.

Metropolis-Hastings Algorithm

An implementation with the Metropolis-Hastings algorithm requires a proposal density for β. A possibility is a normal distribution centred at the current value $\beta^{[t]}$ and covariance matrix λC. Let the proposed value be $Y = Y_\beta$, and let the current state of the chain be $X^{[t]} = x = \beta^{[t]}$. Then the algorithm is as follows:

1. Set $t = 0$ and choose a starting value for β equal to $\beta^{[0]}$; choose C and the tuning parameter λ.
2. Let $t = t + 1$.
3. Draw $Y_\beta \sim N\left(\beta^{[t]}, \lambda C\right)$. The density of this normal distribution evaluated at the proposed value Y_β is $q\left(Y_\beta|\beta^{[t]}, \lambda C\right)$.
4. Draw $u \sim Un\,(0, 1)$.
5. If $u < \alpha$, $\beta^{[t+1]} = Y_\beta$. Otherwise $\beta^{[t+1]} = \beta^{[t]}$.
6. Go to 2.

This is repeated until a chain of sufficient length is obtained. The decision is based on computation of effective chain length and Monte Carlo standard errors of estimates of features from the posterior distribution of β. The tuning parameter λ is used to obtain acceptance ratios in the range (0.25; 0.50).

The acceptance probability in step 5 is

$$\alpha = \frac{p\left(Y_\beta|y, x\right)}{p\left(\beta^{[t]}|y, x\right)} \frac{q\left(\beta^{[t]}|Y_\beta, \lambda C\right)}{q\left(Y_\beta|\beta^{[t]}, \lambda C\right)}. \tag{5.13}$$

In contrast with the general notation used in (4.67), here, a notation specific to the model is used. Comparing with (4.67),

$$\pi\,(Y) = p\left(Y_\beta|y, x\right),$$

$$\pi\,(x) = p\left(\beta^{[t]}|y, x\right),$$

$$q\,(x|Y) = q\left(\beta^{[t]}|Y_\beta, \lambda C\right),$$

$$q\,(Y|x) = q\left(Y_\beta|\beta^{[t]}, \lambda C\right).$$

The notation $q\left(\beta^{[t]}|Y_\beta, \lambda C\right)$ agrees with that used for $q\,(\cdot|Y)$ in (4.67). The proposal density $q\,(\cdot|Y)$ is evaluated at $\beta^{[t]}$, and $Y_\beta and \lambda C$ are, respectively, the mean and variance of the distribution. Due to the symmetry properties of the normal distribution, the second ratio in the right-hand side of (5.13) is equal to 1, and the acceptance probability reduces to the ratio of the posterior distributions $p\,(\cdot|y, x)$ evaluated at $\beta = Y_\beta$ and at $\beta = \beta^{[t]}$, respectively. The acceptance probability is

$$\alpha = \frac{p\left(Y_\beta|y, x\right)}{p\left(\beta^{[t]}|y, x\right)}. \tag{5.14}$$

An alternative approach is to use a *probit model*. The probit likelihood was given in (3.75):

$$L(\beta|x, y) \propto \prod_{i=1}^{N} \left[\left(1 - \Phi\left(x_i'\beta\right)\right)^{y_i} \left(\Phi\left(x_i'\beta\right)\right)^{1-y_i} \right]. \tag{5.15}$$

Assuming a uniform prior for β results in a posterior distribution proportional to the likelihood. The logposterior is obtained by taking logarithms of (5.15). This gives, up to an additive constant,

$$\ln p(\beta|x, y) = \sum_{i=1}^{N} \left[y_i \ln\left(1 - \Phi\left(x_i'\beta\right)\right) + (1 - y_i) \ln\left(\Phi\left(x_i'\beta\right)\right) \right]. \tag{5.16}$$

The Metropolis-Hastings algorithm based on the probit model follows along the same lines as with the logistic model. In the problems section Bayes Exercises II, you are asked to fit a logistic and a probit Bayesian model to binary data.

Gibbs Sampling Algorithm

The Gibbs sampling algorithm requires knowledge of the fully conditional posterior distributions (fcpd). The logposteriors (5.11) and (5.16) do not lead to fcpd that have a standard form. An alternative strategy is to augment the posterior distribution with the unobserved liability u. This is illustrated using the probit model. At the level of the liability, the linear model for the ith record is

$$u_i = x_i'\beta + e_i, \; e_i \overset{iid}{\sim} N(0, 1), \; i = 1, \ldots, N. \tag{5.17}$$

As before, assume

$$\Pr(y_i = 1|\beta, x_i) = \Pr(u_i < 0|\beta, x_i) \tag{5.18}$$

and

$$\Pr(y_i = 0|\beta, x_i) = \Pr(u_i > 0|\beta, x_i). \tag{5.19}$$

The density of the augmented posterior is

$$\begin{aligned} p(u, \beta|y) &\propto p(u, \beta)\, p(y|u, \beta) \quad \text{Bayes theorem} \\ &= p(u|\beta)\, p(\beta)\, p(y|u) \quad \text{given } u, y \text{ is independent of } \beta \\ &\propto p(u|\beta)\, p(y|u) \quad \text{assuming a uniform prior for } \beta. \end{aligned} \tag{5.20}$$

In this expression, $p(y|u)$ is a degenerate density: if u is observed, y is not stochastic because its value is known unambiguously as indicated in (5.18) and (5.19). That is,

$$\Pr(y_i = 1|u_i > 0) = \Pr(y_i = 0|u_i < 0) = 0,$$

$$\Pr(y_i = 1|u_i < 0) = \Pr(y_i = 0|u_i > 0) = 1.$$

Therefore, the joint density of vector y, given u, is

$$p(y|u) = \prod_{i=1}^{N} \left[I(u_i < 0)^{y_i} + I(u_i > 0)^{1-y_i} \right]. \tag{5.21}$$

It is important to realise that the focus of inference here is

$$p(\beta|y) = \int p(u, \beta|y)\, du$$

and not necessarily the augmented posterior $[u, \beta|y]$. When the system converges to the stationary distribution, the Gibbs sampler, in general, McMC, generates Monte Carlo draws from the joint posterior $[u, \beta|y]$; the margins are draws from $[\beta|y]$ and from $[u|y]$. In this sense, McMC is an automatic algorithm that performs the desired marginalisations. The draws from $[u, \beta|y]$ are obtained by sampling repeatedly from $[\beta|u, y]$ and $[u|\beta, y]$.

The fully conditional posterior distributions $[\beta|u, y]$ and $[u|\beta, y]$ are derived from (5.20) as follows:

- The fcpd $[\beta|u, y]$ is obtained by extracting from (5.20) the terms containing β. This yields

$$\begin{aligned} p(\beta|u, y) &\propto p(u|\beta) \\ &\propto \exp\left(-\frac{1}{2}(u - X\beta)'(u - X\beta)\right). \end{aligned} \tag{5.22}$$

  As a function of β, this density does not have a recognisable form. A little more work is needed. The quadratic form can be written as

$$(u - X\beta)'(u - X\beta) = u'u - 2u'X\beta + \beta'X'X\beta.$$

  Define

$$X'X\widehat{\beta} = X'u. \tag{5.23}$$

Then,

$$u'X = \widehat{\beta}'X'X,$$
$$u'X\beta = \widehat{\beta}'X'X\beta.$$

Now, replace $-2u'X\beta$ by $-2\widehat{\beta}'X'X\beta$ in the quadratic form

$$(u - X\beta)'(u - X\beta) = u'u - 2\widehat{\beta}'X'X\beta + \beta'X'X\beta.$$

Adding and subtracting $\widehat{\beta}'X'X\widehat{\beta}$ and keeping terms containing β yield

$$(u - X\beta)'(u - X\beta) = \left(\beta - \widehat{\beta}\right)' X'X\left(\beta - \widehat{\beta}\right) + k,$$

where k is a constant that does not depend on β. Substituting in (5.22) gives

$$p(\beta|u, y) \propto \exp\left(-\frac{1}{2}\left(\beta - \widehat{\beta}\right)' X'X\left(\beta - \widehat{\beta}\right)\right)$$

which is the kernel of a normal distribution with mean $\widehat{\beta}$ and variance $\left(X'X\right)^{-1}$. Therefore,

$$[\beta|u, y] \sim N\left(\widehat{\beta}, \left(X'X\right)^{-1}\right). \tag{5.24}$$

- The fcpd $[u|\beta, y]$ is obtained by extracting from (5.20) those terms that include u. This results in

$$p(u|\beta, y) \propto p(u|\beta)\, p(y|u).$$

The first term, seen as a function of u, is obtained directly from (5.17). This gives

$$[u|\beta] \sim N(X\beta, I). \tag{5.25}$$

The term $p(y|u)$ is given in (5.21). Therefore, the density of the fcpd $[u_i|\beta, y]$ can be written

$$p(u_i|\beta, y) \propto N\left(u_i|x_i'\beta, 1\right)\left[I(u_i < 0)^{y_i} + I(u_i > 0)^{1-y_i}\right]. \tag{5.26}$$

This means that if $y_i = 1$, $u_i < 0$, and the full conditional posterior distribution of u_i is a truncated normal, with mean $x_i'\beta$, variance 1, and support $(-\infty, 0)$. If $y_i = 0$, then $u_i > 0$ and the fully conditional posterior distribution of u_i is a truncated normal, with mean $x_i'\beta$, variance 1 and support $(0, \infty)$.

In Bayes Exercises II, problem 3, you are asked to analyse binary data with a Bayesian probit model implemented with a Gibbs sampler using the data augmentation algorithm.

Drawing Samples from Truncated Distributions

An efficient algorithm to sample from univariate truncated distributions is as follows: Let Y be a random variable from a normal distribution truncated between a (lower bound) and b (upper bound). To obtain a draw from the truncated normal $TN_{(a,b)}\left(\mu, \sigma^2\right)$, where μ and σ^2 are the mean and variance before truncation,

- Simulate U from a uniform distribution $Un\left(p_1, p_2\right)$, where $p_1 = \Phi\left[(a-\mu)/\sigma\right]$ and $p_2 = \Phi\left[(b-\mu)/\sigma\right]$.
- The truncated normal is

$$y = \mu + \sigma\Phi^{-1}(U),$$

where $\Phi(\cdot)$ is the cdf and $\Phi^{-1}(\cdot)$ is the inverse cdf of the standard normal distribution.

A more general method to sample from any univariate distribution truncated in the interval $[a, b]$ is as follows: If the cdf of the untruncated variate is F, then a draw from the truncated distribution is (Devroye 1986, page 38 Example 10)

$$y = F^{-1}\left\{F(a) + U\left[F(b) - F(a)\right]\right\}, \tag{5.27}$$

where U is a draw from a uniform distribution in the interval $[0, 1]$. If Φ is the distribution function of the standard normal, then to draw from $TN_{(a,b)}\left(\mu, \sigma^2\right)$, application of (5.27) yields

$$y = \mu + \left\{\sigma\Phi^{-1}\left[\Phi(\ell) + U\left(\Phi(u) - \Phi(\ell)\right)\right]\right\} \tag{5.28}$$

where the lower bound is $\ell = (a-\mu)/\sigma$, the upper bound is $u = (b-\mu)/\sigma$, and U is a draw from a uniform distribution in the interval $[0, 1]$. With binary data and $u \sim N(\mu, 1)$, if $y = 1$, $a = -\infty$, $b = 0$, and

$$\left.\begin{array}{r}\Phi(\ell) = 0, \\ \Phi(u) = \Phi(-\mu)\end{array}\right\} \text{ if } y = 1.$$

If $y = 0$, $a = 0$, $b = \infty$, and

$$\left.\begin{array}{r}\Phi(\ell) = \Phi(-\mu), \\ \Phi(u) = 1\end{array}\right\} \text{ if } y = 0.$$

An R-code below generates N samples from (5.26) using algorithm (5.28):

```
# CODE0502
rm(list=ls()) # CLEAR WORKSPACE
########## DRAWING ALL U's IN ONE GO
### REQUIRES TO GENERATE BINARY RECORDS y
##  USING PARAMETRISATION A OR B BELOW
# A
##### y = 0 -> TN(mean,1)(0,Infinity)
##### y = 1 -> TN(mean,1)(-Infinity,0)
######### OR ################
# B
##### y = 1 -> TN(mean,1)(0,Infinity)
##### y = 0 -> TN(mean,1)(-Infinity,0)
####################
set.seed(237777)
nrow <- 10
ncol <- 5
mu <- 0
# GENERATE X MATRIX
X<-matrix(nrow= nrow,ncol= ncol,rbinom(nrow*ncol,size=2,p=.5))
# GENERATE VECTOR b
b <- rnorm(ncol,0.5,1)
xb<-X%*%b
# LOGIT MODEL
#p1<-exp(mu+xb)/(1+exp(mu+xb)) # IF B
#p1 <- 1 - exp(mu+xb)/(1+exp(mu+xb)) # IF A
# PROBIT MODEL- GENERAte VECTOR OF DATA y
#p1 <- pnorm(mu+xb) # IF B
p1 <- 1 - pnorm(mu+xb) # IF A
y <- rbinom(nrow,1,p1)
mean <- mu+xb
sd <- 1

interm<-(1-y)*pnorm(0,mean=mean,sd=sd)+runif(length(y))*
    (pnorm(0,mean=mean,sd=sd)*(y)+
    (1-pnorm(0,mean=mean,sd=sd))*(1-y))

u <- qnorm(interm,mean=mean,sd=sd)
p1[1:5]
```

```
## [1] 0.70255823 0.82055114 0.05872622 0.52202695 0.37241559
```

```
p1[6:nrow]
```

```
## [1] 0.04796642 0.56227502 0.37241559 0.05222701 0.89662931
```

```
y[1:5]
```

```
## [1] 1 1 0 0 0
```

```
y[6:nrow]
```

```
## [1] 0 1 1 0 1
```

```
u[1:5]
```

```
## [1] -0.4976322 -0.6461563  3.8849389  0.4510179  1.1993790
```

```
u[6:nrow]
```

```
## [1]  1.962061 -1.652975 -1.563791  1.138155 -2.003726
```

As a more transparent case, consider drawing a sample y of size 1 from a normal distribution, with mean 10 and variance 4, that is truncated in the interval [11.0, 11.5]. Then application of (5.28) leads to the following algorithm:

```
rm(list=ls()) # CLEAR WORKSPACE
set.seed(1237)
l <- (11-10)/2
u <- (11.5-10)/2
y <- 10 + (2*qnorm(pnorm(l) + runif(1)*(pnorm(u) - pnorm(l))))
y
```

```
## [1] 11.15587
```

5.3 Example: A Regression Model for Correlated Binary Data

Very often data cluster in groups where members of the same group are correlated and groups can be uncorrelated. An example consists of independent full-sib families. In this setup, the model must be extended to account for the correlated structure of the observations (see Likelihood Exercises II, Problem 3 iii). This section illustrates a probit Bayesian model that can handle this situation.

The first step is to include a term in (5.17) that accounts for the correlated structure. The standard approach is to use a random variable f that represents a family effect:

$$u_{ij} = x'_{ij}\beta + f_i + e_{ij}, \tag{5.29}$$

$$f_i|\sigma_f^2 \overset{iid}{\sim} N\left(0, \sigma_f^2\right), \quad \sigma_f^2 \in [0, 1],$$

$$e_{ij} \overset{iid}{\sim} N(0, 1),$$

$$i = 1, \ldots, n_f, \; j = 1, \ldots, n.$$

Assume that

$$\begin{aligned}
\Pr(y_i = 1|\beta, x_i, f_i) &= \Pr(u_i < 0|\beta, x_i, f_i) \\
&= \Pr\left(x'_{ij}\beta + f_i + e_{ij} < 0|\beta, x_i, f_i\right) \\
&= \Pr\left(e_{ij} < -x'_{ij}\beta - f_i|\beta, x_i, f_i\right) \\
&= 1 - \Phi\left(x'_{ij}\beta + f_i\right)
\end{aligned}$$

and

$$\begin{aligned}
\Pr(y_i = 0|\beta, x_i, f_i) &= \Pr(u_i > 0|\beta, x_i) \\
&= \Phi\left(x'_{ij}\beta + f_i\right).
\end{aligned}$$

In this specification, x'_{ij} is a row vector with p columns (number of elements in β) corresponding to individual j of family i, f_i is the effect of the ith full-sib family, σ_f^2 is the variance between families (or covariance between full-sibs), n_f is the number of full-sib families, n is the number of full-sibs per family (here assumed to be the same for all families), and the total number of records is $N = n_f \times n$. Since according to quantitative genetic theory, in the absence of common environmental effects and assuming an additive genetic model, the heritability at the level of the liability is equal to

$$h^2 = \frac{2\sigma_f^2}{\sigma_f^2 + 1}, \quad 0 < h^2 < 1,$$

then it follows that $\sigma_f^2 \in [0, 1]$. The heritability on the liability scale does not depend on the prevalence of the disease in the population.

Gibbs Sampling Implementation

Using the data augmentation strategy, the parameters of the model are $\left[u, \beta, f, \sigma_f^2\right]$. The density of the prior distribution is assumed to admit the following factorisation:

$$\begin{aligned}
p\left(u, \beta, f, \sigma_f^2\right) &\propto p(u|\beta, f)\, p\left(f|\sigma_f^2\right) p\left(\sigma_f^2\right) \\
&\propto p(u|\beta, f)\, p\left(f|\sigma_f^2\right).
\end{aligned}$$

It is assumed that $p(\beta)$ and $p\left(\sigma_f^2\right)$ are proportional to constants. The augmented posterior is

$$p\left(u, \beta, f, \sigma_f^2 | y\right) \propto p\left(u, \beta, f, \sigma_f^2\right) p\left(y | u, \beta, f, \sigma_f^2\right)$$
$$\propto p\left(f | \sigma_f^2\right) \prod_{ij} p\left(u_{ij} | \beta, f_i\right) p\left(y_{ij} | u_{ij}\right), \quad \sigma_f^2 \in [0, 1]. \quad (5.30)$$

The Bayesian model can be implemented with a variety of McMC algorithms. One possibility that is explored here is to use data augmentation and update the parameters one at a time. The required fcpd are

$$[\beta | D]; [u | D]; [f | D]; \left[\sigma_f^2 | D\right],$$

where D is a vector containing the observations y and all the parameters of the model except the one to be updated.

Deriving $[\beta | D]$

Extracting from (5.30) the terms that depend on β yields

$$p(\beta | D) \propto \prod_{ij} p\left(u_{ij} | \beta, f_i\right)$$
$$\propto \exp\left(-\frac{1}{2}(u - X\beta - Zf)'(u - X\beta - Zf)\right)$$
$$= \exp\left(-\frac{1}{2}(\widetilde{u} - X\beta)'(\widetilde{u} - X\beta)\right), \quad (5.31)$$

where $\widetilde{u} = u - Zf$. This has the same form as (5.22). Then it is easy to see that

$$[\beta | D] \sim N\left(\widehat{\beta}, \left(X'X\right)^{-1}\right), \quad (5.32)$$

where here

$$\widehat{\beta} = \left(X'X\right)^{-1} X'(u - Zf).$$

Deriving $[u|D]$

Inspection of (5.30) and (5.20) reveals that $[u|D] = [u|y]$. Therefore, since the u_{ij} are conditionally independent, given f_i,

$$\left[u_{ij}|\beta, f, y\right] \sim N\left(x'_{ij}\beta + f_i, 1\right)\left[I\left(u_{ij} < 0\right)^{y_{ij}} + I\left(u_{ij} > 0\right)^{1-y_{ij}}\right]. \tag{5.33}$$

Deriving $[f|D]$

The fcpd $[f|D]$ is obtained by extracting from (5.30) those terms that depend on f. This gives

$$\begin{aligned} p(f|D) &\propto p\left(f|\sigma_f^2\right)\prod_{ij} p\left(u_{ij}|\beta, f_i\right) \\ &\propto \exp\left(-\frac{1}{2\sigma_f^2}f'f\right)\exp\left(-\frac{1}{2}\left(u - X\beta - Zf\right)'\left(u - X\beta - Zf\right)\right) \\ &= \exp\left(-\frac{1}{2}\left[\left(\tilde{u} - Zf\right)'\left(\tilde{u} - Zf\right) + kf'f\right]\right) \end{aligned} \tag{5.34}$$

where $\tilde{u} = u - X\beta$ and $k = \sigma_f^{-2}$. The quadratic form can be written as

$$\begin{aligned} \left(\tilde{u} - Zf\right)'\left(\tilde{u} - Zf\right) + kf'f &= \tilde{u}'\tilde{u} - 2f'Z'\tilde{u} + f'Z'Zf + kf'f \\ &= \text{constant} - 2f'Z'\tilde{u} + f'\left(Z'Z + Ik\right)f. \end{aligned}$$

Define

$$\left(Z'Z + Ik\right)\widehat{f} = Z'\tilde{u} \tag{5.35}$$

and substitute $Z'\tilde{u}$ by $\left(Z'Z + Ik\right)\widehat{f}$. Then

$$\left(\tilde{u} - Zf\right)'\left(\tilde{u} - Zf\right) + kf'f = \text{constant} - 2f'\left(Z'Z + Ik\right)\widehat{f} + f'\left(Z'Z + Ik\right)f.$$

Adding and subtracting $\hat{f}'\left(Z'Z + Ik\right)\hat{f}$ and keeping terms in f only yield

$$\left(\tilde{u} - Zf\right)'\left(\tilde{u} - Zf\right) + kf'f = \text{constant} + \left(f - \widehat{f}\right)'\left(Z'Z + Ik\right)\left(f - \hat{f}\right).$$

Substituting the above in the exponential of (5.34) reveals that the fully conditional is

$$\left[f|\sigma_f^2, D\right] \sim N\left(\widehat{f}, \left(Z'Z + Ik\right)^{-1}\right) \tag{5.36}$$

where $\hat{f}$ is given in (5.35). In this simple model with unrelated families, the term $Z'Z$ is diagonal, and computations are straightforward.

Deriving $\left[\sigma_f^2|D\right]$

Finally, the term $\left[\sigma_f^2|D\right]$ is obtained from

$$\begin{aligned} p\left(\sigma_f^2|D\right) &\propto p\left(f\,|\sigma_f^2\right) \\ &\propto \left(\sigma_f^2\right)^{-\frac{n_f}{2}} \exp\left(-\frac{1}{2\sigma_f^2} f' f\right) \\ &= \left(\sigma_f^2\right)^{-\left(\frac{v_f}{2}+1\right)} \exp\left(-\frac{v_f S_f}{2\sigma_f^2}\right), \quad \sigma_f^2 \in [0,1], \end{aligned} \tag{5.37}$$

which is the kernel of the density of a truncated scaled inverted chi-square distribution with scale $S_f = f'f/v_f$ and $v_f = n_f - 2$ degrees of freedom. A draw σ_f^{2*} from this distribution is

$$\sigma_f^{2*} \sim \frac{f'f}{\chi^2\left(v_f\right)}, \quad \sigma_f^{2*} \in [0,1].$$

where $\chi^2\left(v_f\right)$ is a draw from a chi-square distribution with v_f degrees of freedom.

One must confirm that the draw is within the constraint $\sigma_f^{2*} \in [0, 1]$.

A Metropolis Within Gibbs Implementation: Case 1

In the Gibbs sampling implementation, the bound on h^2 imposes the constraint $\sigma_f^2 \in [0, 1]$, but this was not incorporated in the prior distribution of σ_f^2. An unbounded uniform prior for σ_f^2 was chosen instead, even though the posterior is defined within the constraint $\sigma_f^2 \in [0, 1]$. The code of the McMC implementation must account for this constraint. In certain cases, this may require the need to draw from truncated distributions.

One may wish to incorporate the constraint on σ_f^2 assuming a beta prior. However, depending on how the algorithm is tailored, there may still be a need to confirm that the McMC samples are within the bounds imposed by the model.

A detail to observe is that the beta density does not assign a probability to $\sigma_f^2 = 0$. This implies that a value of exactly 0 is excluded in the posterior distribution of σ_f^2.

A mixture prior with a component that has a probability mass of 1 for $\sigma_f^2 = 0$ would remedy this shortcoming.

With this reservation in mind, one may explore using a beta prior distribution that may lead to more stable inferences when the likelihood is not very informative. Assuming a beta prior for σ_f^2 causes the density of the posterior distribution to change from (5.30) to

$$p\left(u, \beta, f, \sigma_f^2|y\right) \propto p\left(u, \beta, f, \sigma_f^2\right) p\left(y|u, \beta, f, \sigma_f^2\right)$$
$$\propto p\left(f|\sigma_f^2\right) p\left(\sigma_f^2\right) \prod_{ij} p\left(u_{ij}|\beta, f_i\right) p\left(y_{ij}|u_{ij}\right), \tag{5.38}$$
$$\sigma_f^2 \in [0, 1]$$

where

$$p\left(\sigma_f^2\right) \propto \left(\sigma_f^2\right)^{a-1} \left(1 - \sigma_f^2\right)^{b-1}, \quad \sigma_f^2 \in [0, 1], a > 0, b > 0. \tag{5.39}$$

As in Scenario 1, the fcpd are

$$[\beta|D]; [u|D]; [f|D]; \left[\sigma_f^2|D\right]$$

which, with the exception of $\left[\sigma_f^2|D\right]$, are unchanged. The fcpd of σ_f^2 is obtained by extracting from (5.38) the terms that are a function of σ_f^2. This gives

$$p\left(\sigma_f^2|D\right) \propto p\left(f|\sigma_f^2\right) p\left(\sigma_f^2\right)$$
$$\propto \left(\sigma_f^2\right)^{-\frac{n_f}{2}} \exp\left(-\frac{1}{2\sigma_f^2} f'f\right) \left(\sigma_f^2\right)^{a-1} \left(1 - \sigma_f^2\right)^{b-1}$$
$$= \left(\sigma_f^2\right)^{-\left(\frac{v_f}{2}+1\right)} \exp\left(-\frac{v_f S_f}{2\sigma_f^2}\right) \left(\sigma_f^2\right)^{a-1} \left(1 - \sigma_f^2\right)^{b-1}. \tag{5.40}$$

Above, $v_f = n_f - 2$ and $S_f = f'f/v_f$ where f is the vector with the sampled values of family effects. This density does not have a distinguishable closed form. Therefore, a Metropolis-Hastings step can be used to update σ_f^2.

Let $Y_{\sigma_f^2}$ denote the proposed value of σ_f^2 drawn from the proposal distribution with density $q\left(\cdot|\sigma_f^2\right)$ that may depend on σ_f^2 (it does not have to). Then the Metropolis-Hastings step at iteration t is as follows:

1. At iterate $t = 0$, choose a start value for σ_f^2; label it $\sigma_f^{2[0]}$.
2. Set $t = t + 1$.

3. Draw $Y_{\sigma_f^2}$ from proposal distribution. If $Y_{\sigma_f^2} > 0.999$, then set $Y_{\sigma_f^2} = \sigma_f^{2[t]}$, and go to 7.
4. Draw $u \sim Un(0, 1)$.
5. Compute acceptance probability α.
6. If $u < \alpha$, $\sigma_f^{2[t+1]} = Y_{\sigma_f^2}$. Otherwise, $\sigma_f^{2[t+1]} = \sigma_f^{2[t]}$
7. Continue with the Gibbs sampling step, and when completed, go to 2

where

$$\alpha = \frac{p\left(Y_{\sigma_f^2}|y, x\right)}{p\left(\sigma_f^{2[t]}|y, x\right)} \frac{q\left(\sigma_f^{2[t]}|Y_{\sigma_f^2}\right)}{q\left(Y_{\sigma_f^2}|\sigma_f^{2[t]}\right)}$$

$$= \frac{\left(Y_{\sigma_f^2}\right)^{-\left(\frac{v_f}{2}+1\right)} \exp\left(-\frac{v_f S_f}{2Y_{\sigma_f^2}}\right)\left(Y_{\sigma_f^2}\right)^{a-1}\left(1-Y_{\sigma_f^2}\right)^{b-1}}{\left(\sigma_f^{2[t]}\right)^{-\left(\frac{v_f}{2}+1\right)} \exp\left(-\frac{v_f S_f}{2\sigma_f^{2[t]}}\right)\left(\sigma_f^{2[t]}\right)^{a-1}\left(1-\sigma_f^{2[t]}\right)^{b-1}} \frac{q\left(\sigma_f^{2[t]}|Y_{\sigma_f^2}\right)}{q\left(Y_{\sigma_f^2}|\sigma_f^{2[t]}\right)} \tag{5.41}$$

The behaviour of the algorithm is influenced by the choice of the proposal distribution q. One possibility is to use a scaled inverted chi-square distribution (that does not depend on σ_f^2) of the form

$$q\left(Y_{\sigma_f^2}\right) \propto \left(Y_{\sigma_f^2}\right)^{-\left(\frac{v_f}{2}+1\right)} \exp\left(-\frac{v_f S_f}{2Y_{\sigma_f^2}}\right).$$

The ratio of the proposal terms is

$$\frac{q\left(\sigma_f^{2[t]}|\sigma_f^{2[p]}\right)}{q\left(Y_{\sigma_f^2}|\sigma_f^{2[t]}\right)} = \frac{\left(\sigma_f^{2[t]}\right)^{-\left(\frac{v_f}{2}+1\right)} \exp\left(-\frac{v_f S_f}{2\sigma_f^{2[t]}}\right)}{\left(Y_{\sigma_f^2}\right)^{-\left(\frac{v_f}{2}+1\right)} \exp\left(-\frac{v_f S_f}{2Y_{\sigma_f^2}}\right)}$$

and the Metropolis-Hastings acceptance ratio (5.41) reduces to

$$\alpha = \frac{\left(Y_{\sigma_f^2}\right)^{a-1}\left(1-Y_{\sigma_f^2}\right)^{b-1}}{\left(\sigma_f^{2[t]}\right)^{a-1}\left(1-\sigma_f^{2[t]}\right)^{b-1}}. \tag{5.42}$$

This acceptance probability must be set equal to 0 if the proposed value is outside the permissible interval.

A Metropolis Within Gibbs Implementation: Case 2

Another possibility is to draw at cycle t the proposed value $Y_{\sigma_f^2}$ from a lognormal distribution centred at $\ln \sigma_f^{2[t]}$. The lognormal density is

$$q\left(Y_{\sigma_f^2}|\sigma_f^{2[t]}\right) = (2\pi k)^{-\frac{1}{2}} \exp\left[-\frac{\left(\ln Y_{\sigma_f^2} - \ln \sigma_f^{2[t]}\right)^2}{2k}\right]\frac{1}{Y_{\sigma_f^2}},$$

where k is a tuning parameter chosen by the user. The ratio of the proposal terms is

$$\frac{q\left(\sigma_f^{2[t]}|Y_{\sigma_f^2}\right)}{q\left(Y_{\sigma_f^2}|\sigma_f^{2[t]}\right)} = \frac{Y_{\sigma_f^2}}{\sigma_f^{2[t]}}$$

and the Metropolis-Hastings acceptance ratio (5.41) is given by

$$\alpha = \frac{\left(Y_{\sigma_f^2}\right)^{-\left(\frac{v_f}{2}+1\right)} \exp\left(-\frac{v_f S_f}{2Y_{\sigma_f^2}}\right)\left(Y_{\sigma_f^2}\right)^{a-1}\left(1 - Y_{\sigma_f^2}\right)^{b-1}}{\left(\sigma_f^{2[t]}\right)^{-\left(\frac{v_f}{2}+1\right)} \exp\left(-\frac{v_f S_f}{2\sigma_f^{2[t]}}\right)\left(\sigma_f^{2[t]}\right)^{a-1}\left(1 - \sigma_f^{2[t]}\right)^{b-1}} \frac{Y_{\sigma_f^2}}{\sigma_f^{2[t]}}. \tag{5.43}$$

This acceptance probability must be set equal to 0 if the proposed value is outside the permissible interval.

5.4 Example: A Genomic Model

The genomic model was implemented in a likelihood setting with Newton-Raphson and with the EM algorithm. Here, the model is fitted using a Bayesian McMC approach.

To recapitulate, in its original parametrisation, the initial hierarchy of the genomic model is defined by the expressions

$$y|\mu, b, \sigma_e^2 \sim N\left(1\mu + Wb, I\sigma_e^2\right),$$

$$b|\sigma_b^2 \sim N\left(0, I\sigma_b^2\right),$$

where the matrix of marker genotype codes W of dimension $n \times m$ is centred and scaled and the vector of marker effects b has m elements. The genomic values are defined as $g = Wb$.

The reparametrised model uses the decomposition

$$\begin{aligned} G &= \frac{1}{m} W W' \\ &= \frac{1}{m} U \Delta U' \\ &= U \Lambda U', \quad \Lambda = \frac{1}{m} \Delta . \end{aligned}$$

The diagonal matrix Λ contains the (scaled) eigenvalues λ_i of U, $i = 1, \ldots, n$. The last eigenvalue is $\lambda_n = 0$.

On defining $\alpha \sim N\left(0, \Lambda \sigma_g^2\right)$, where $\sigma_g^2 = m\sigma_b^2$, the alternative parametrisation as defined on page 93 is

$$\begin{aligned} y|\mu, \alpha, \sigma_e^2 &\sim N\left(1\mu + U\alpha, I\sigma_e^2\right), \\ \alpha|U, \sigma_g^2 &\sim SN\left(0, \Lambda\sigma_g^2\right), \\ G &= U\Lambda U', \\ y|\mu, \sigma_g^2, \sigma_e^2 &\sim N\left(1\mu, U\Lambda U'\sigma_g^2 + I\sigma_e^2\right). \end{aligned}$$

In contrast with (3.38), here, a scalar mean μ is included. In the reparametrised model, the column vector of genomic values is $g = U\alpha$, where α is the regression of genomic values on eigenvectors.

The vector

$$\begin{aligned} \alpha' &= (\alpha_1, \ldots, \alpha_{n-1}, \alpha_n)' \\ &= \left(\alpha'_{-n} \alpha_n\right) \end{aligned}$$

of dimension n has its first $n-1$ elements different from zero (the vector α_{-n} above) and the last element $\alpha_n = 0$. As mentioned before, the marginal density of α_n is a point mass at zero, because $\lambda_n = 0$ and therefore $\left[\alpha|U, \sigma_g^2\right]$ has the form of a singular normal distribution.

The parameters of the Bayesian model are $\left(\mu, \alpha, \sigma_g^2, \sigma_e^2\right)$, and the density of the posterior distribution is

$$\begin{aligned} p\left(\mu, \alpha, \sigma_g^2, \sigma_e^2|y\right) &\propto p\left(\mu, \alpha, \sigma_g^2, \sigma_e^2\right) p\left(y|\mu, \alpha, \sigma_g^2, \sigma_e^2\right) \\ &\propto p\left(\alpha|\sigma_g^2\right) p\left(y|\mu, \alpha, \sigma_e^2\right) \end{aligned} \tag{5.44}$$

where it is assumed that

$$p\left(\mu, \sigma_g^2, \sigma_e^2\right) \propto \text{constant}.$$

A Gibbs sampling algorithm requires to update the parameters from the fcpd's

$$[\mu|D]; [\alpha|D]; \left[\sigma_g^2|D\right]; \left[\sigma_e^2|D\right].$$

Deriving $[\mu|D]$

Extracting from (5.44) those terms that include μ gives

$$\begin{aligned} p(\mu|D) &\propto p\left(y|\mu, \alpha, \sigma_e^2\right) \\ &\propto \exp\left[-\frac{1}{2\sigma_e^2}(y - 1\mu - U\alpha)'(y - 1\mu - U\alpha)\right] \\ &= \exp\left[-\frac{1}{2\sigma_e^2}(y^* - 1\mu)'(y^* - 1\mu)\right], \end{aligned}$$

where $y^* = y - U\alpha$. This has the same form as (5.31) and therefore

$$[\mu|D] \sim N\left(\widehat{\mu}, \left(1'1\right)^{-1}\sigma_e^2\right), \tag{5.45a}$$

where

$$\widehat{\mu} = \left(1'1\right)^{-1} 1'(y - U\alpha).$$

Deriving $[\alpha|D]$

Extracting from (5.44) those terms that include α gives

$$\begin{aligned}p(\alpha|D) &\propto p\left(\alpha|\sigma_g^2\right) p\left(y|\mu,\alpha,\sigma_e^2\right) \\ &\propto \exp\left[-\frac{1}{2\sigma_g^2}\alpha'\Lambda^-\alpha\right]\exp\left[-\frac{1}{2\sigma_e^2}(y-1\mu-U\alpha)'(y-1\mu-U\alpha)\right].\end{aligned}$$

In this expression,

$$\Lambda^- = \begin{bmatrix} (\Lambda_{-n})^{-1}_{(n-1)\times(n-1)} & 0_{(n-1)\times 1} \\ 0_{1\times(n-1)} & 0_{1\times 1} \end{bmatrix}_{n\times n}$$

and $(\Lambda_{-n})^{-1}$ is the $(n-1)\times(n-1)$ diagonal matrix whose diagonal elements are the inverse of the non-zero eigenvalues $\lambda_1,\ldots,\lambda_{n-1}$. The two terms in the exponentials can be brought together to obtain

$$p(\alpha|D) \propto \exp\left[-\frac{1}{2\sigma_e^2}\left\{(y-1\mu-U\alpha)'(y-1\mu-U\alpha)+k\alpha'\Lambda^-\alpha\right\}\right], \tag{5.46}$$

where $k=\sigma_e^2/\sigma_g^2$. The term in curly brackets that only contains terms in α is

$$\begin{aligned}&\left\{(y-1\mu-U\alpha)'(y-1\mu-U\alpha)+k\alpha'\Lambda^-\alpha\right\} = \\ &= -2\alpha'U'y+2\alpha'U'1\mu+\alpha'U'U\alpha+k\alpha'\Lambda^-\alpha \\ &= -2\alpha'U'(y-1\mu)+\alpha'\alpha+k\alpha'\Lambda^-\alpha \\ &= -2\alpha'U'(y-1\mu)+\alpha'\left(I+\Lambda^-k\right)\alpha,\end{aligned}$$

excluding an additive constant. Defining

$$\left(I+\Lambda^-k\right)\widehat{\alpha} = U'(y-1\mu) \tag{5.47}$$

and replacing $-2U'(y-1\mu)$ by the left-hand side of this expression gives

$$\begin{aligned}&\left\{(y-1\mu-U\alpha)'(y-1\mu-U\alpha)+k\alpha'\Lambda^-\alpha\right\} \propto \\ &\propto -2\alpha'\left(I+\Lambda^-k\right)\widehat{\alpha}+\alpha'\left(I+\Lambda^-k\right)\alpha.\end{aligned}$$

excluding an additive constant. Adding and subtracting $\widehat{\alpha}'\left(I+\Lambda^{-}k\right)\widehat{\alpha}$ and keeping only terms that contain α lead to (excluding an additive constant)

$$\left\{(y-1\mu-U\alpha)'(y-1\mu-U\alpha)+k\alpha'\Lambda^{-}\alpha\right\}=$$
$$=(\alpha-\widehat{\alpha})'\left(I+\Lambda^{-}k\right)(\alpha-\widehat{\alpha}).$$

Substituting in (5.46),

$$p(\alpha|D)\propto\exp\left[-\frac{1}{2\sigma_e^2}(\alpha-\widehat{\alpha})'\left(I+\Lambda^{-}k\right)(\alpha-\widehat{\alpha})\right]$$

which is the kernel of the normal distribution

$$[\alpha|D]\sim N\left(\widehat{\alpha},\left(I+\Lambda^{-}k\right)^{-1}\sigma_e^2\right). \tag{5.48}$$

In practice, one implements the Gibbs sampler updating the non-zero $n-1$ elements α_{-n} from the $(n-1)$ dimensional normal distribution

$$\alpha_{-n}|D\sim N\left(\widehat{\alpha}_{-n},\left(I+\Lambda_{-n}^{-1}k\right)^{-1}\sigma_e^2\right) \tag{5.49}$$

where $\widehat{\alpha}_{-n}$ contains the first $(n-1)$ non-zero elements of $\widehat{\alpha}$ defined in (5.47). The the nth element of $\widehat{\alpha}$ is equal to zero.

The update is computationally straightforward since the $\alpha's$ are conditionally independent. A little manipulation with (5.48) or (5.49) shows that

$$E(\alpha_i|D)=\widehat{\alpha}_i=\frac{\lambda_i}{\lambda_i+k}U_i'(y-1\mu),$$

$$Var(\alpha_i|D)=\frac{\lambda_i\sigma_e^2}{\lambda_i+k},\quad i=1,\ldots,n-1$$

and therefore each α_i is updated from $N(E(\alpha_i|D), Var(\alpha_i|D))$. These expressions make it clear that when $\lambda_n=0$, the mean and variance are zero, and the density collapses to a point mass at zero.

Deriving $\left[\sigma_g^2|D\right]$

The fcpd of the genomic variance component defined at the level of this operational model is obtained by extracting from (5.44) those terms that include σ_g^2. This results in

$$\begin{aligned} p\left(\sigma_g^2|D\right) &\propto p\left(\alpha|\sigma_g^2\right) \\ &\propto \left(\sigma_g^2\right)^{-\left(\frac{n-1}{2}\right)} \exp\left(-\frac{\alpha'\Lambda^-\alpha}{2\sigma_g^2}\right), \end{aligned} \tag{5.50}$$

that is proportional to the kernel of a scaled inverse chi-square distribution with $v_\alpha = n - 3$ degrees of freedom and scale $S_\alpha = \left(\alpha'\Lambda^-\alpha/v_\alpha\right)$. The density up to proportionality is equal to

$$p\left(\sigma_g^2|D\right) \propto \left(\sigma_g^2\right)^{-\left(\frac{v_\alpha}{2}+1\right)} \exp\left(-\frac{v_\alpha S_\alpha}{2\sigma_g^2}\right). \tag{5.51}$$

To extract a sample from (5.51), draw from a chi-square distribution with v_α degrees of freedom

$$\chi^2\left(v_\alpha\right)$$

and then compute

$$\frac{\alpha'_{-n}(\Lambda_{-n})^{-1}\alpha_{-n}}{\chi^2\left(v_\alpha\right)}.$$

Deriving $\left[\sigma_e^2|D\right]$

Extracting from (5.44) those terms that include σ_e^2 gives

$$\begin{aligned} p\left(\sigma_e^2|D\right) &\propto p\left(y|\mu,\alpha,\sigma_e^2\right) \\ &\propto \left(\sigma_e^2\right)^{-\frac{n}{2}} \exp\left[-\frac{1}{2\sigma_e^2}\left(y - 1\mu - U\alpha\right)'\left(y - 1\mu - U\alpha\right)\right]. \end{aligned}$$

Regarded as a function of σ_e^2, this is proportional to a scaled inverted chi-square density, with scale

$$S_e = \frac{(y - 1\mu - U\alpha)' (y - 1\mu - U\alpha)}{v_e}$$

and with $v_e = n - 2$ degrees of freedom. That is,

$$p\left(\sigma_e^2|D\right) \propto \left(\sigma_e^2\right)^{-\left(\frac{v_e}{2}+1\right)} \exp\left(-\frac{v_e S_e}{2\sigma_e^2}\right). \tag{5.52}$$

To extract a sample from (5.52), draw from a chi-square distribution with v_e degrees of freedom

$$\chi^2(v_e)$$

and compute

$$\frac{(y - 1\mu - U\alpha)' (y - 1\mu - U\alpha)}{\chi^2(v_e)}.$$

An Alternative Definition of Genomic Variance

The Appendix of Chap. 1 discusses the distinction between the parameters defined at the level of the quantitative genetic model, regarded as the data-generating mechanism and the parameters defined at the level of the operational model used to analyse the data. The genomic model is such an operational model. It was noted that the genomic variance of the genomic operational model (1.40), that is a parameter corresponding to σ_g^2 in the present example (here, the matrix of marker genotypes W is centred and scaled; in (1.40), it is centred), has an unclear connection with the parameter (1.37), that is argued, is the real focus of inference.

An alternative estimator of genomic variance (1.42) was defined on page 44, which is better aligned with (1.37). This subsection shows that estimator (1.42) can be obtained with one line of extra code, when the Bayesian model is implemented in an McMC computing environment.

The $n \times 1$ vector of genomic values is $g = Wb = U\alpha$, where W is the $n \times p$ centred and scaled matrix of marker genotypes, b is the $p \times 1$ vector of genetic marker effects, and α is the regression of genomic value on eigenvectors. The genomic variance or variance of genomic values was defined in (1.42) and in terms of the reparametrised model can be written as

$$\begin{aligned}\sigma_G^2 &= \frac{1}{n}g'g - \left(\frac{1}{n}1'g\right)^2 \\ &= \frac{1}{n}\alpha'U'U\alpha \\ &= \frac{1}{n}\sum_{i=1}^{n}\alpha_i^2. \end{aligned} \tag{5.53}$$

The equality in the second line follows because when matrix W is column-centred, $1'g = 1'Wb = 0$. Vector α is unknown and is therefore inferred from its marginal posterior distribution. When the Markov chain converges to its stationary distribution, extractions from this distribution are given by (5.49). For example, if $\alpha^{[t]}$ is the vector of draws from (5.49) at round t of the McMC sampler, an extraction of σ_G^2 from its marginal posterior distribution is $\sigma_G^{2[t]} = \frac{1}{n}\alpha^{[t]\prime}\alpha^{[t]}$. The `R`-code is `mean(alpha*alpha)`, where `alpha` is the vector $\alpha^{[t]}$.

5.5 Example: A Mixture Model of Two Gaussian Components

An EM algorithm to obtain maximum likelihood estimates for the mixture model of two Gaussian components was presented on page 127. Here, the mixture model is implemented with McMC in a Bayesian setup. Let z_i, $i = 1, \ldots, n$, be the (unobserved) binary random variable that indicates which of the two mixture components observation i comes from. Assume that the marginal probability of Z_i is Bernoulli $Br(\pi_1)$

$$\Pr(Z_i = z_i|\pi_1) = \pi_1^{z_i}(1-\pi_1)^{1-z_i}, \quad z_i = 0, 1,$$

with

$$\begin{aligned}\Pr(Z_i = 1|\pi_1) &= \pi_1, \\ \Pr(Z_i = 0|\pi_0) &= \pi_0 = 1 - \pi_1.\end{aligned}$$

An McMC data augmentation strategy is to use the unobserved Z to generate an augmented posterior distribution. Conditional on $z_i = j$, $j = 1, 0$, assume y_i has density $p_j(y_i|\theta_j, z_j)$. The pairs (Y_i, Z_i) are assumed to be iid with joint density

$$\begin{aligned}p(y_i, z_i|\theta, \pi) &= p_j(y_i|z_i = j, \theta_j)\Pr(Z_i = j|\pi_j) \\ &= p_j(y_i|z_i = j, \theta_j)\pi_j.\end{aligned}$$

To be specific, let $p_j\left(y_i|\theta_j,\sigma^2\right)$ be the density of the normal distribution $N\left(\theta_j,\sigma^2\right)$. The parameters of the model are $\theta_0, \theta_1, \sigma^2, \pi_0, z_i, i = 1, \ldots, n$.

Assume that the priors for the $\theta's$ are improper uniforms and that the density of the prior distribution of the remaining parameters factorises as follows:

$$p\left(\theta_j, \sigma^2, \pi_1, z\right) \propto p\left(\sigma^2\right) p\left(z|\pi_1\right) p\left(\pi_1\right),$$

where $p\left(z|\pi_1\right) = \prod_{i=1}^{n} \Pr\left(Z_i = z_i|\pi_1\right)$. The probability π_1 is assigned a beta prior distribution with user-tuned hyperparameters α and β,

$$\pi_1|\alpha, \beta \sim Be\left(\alpha, \beta\right) \tag{5.54}$$

and the conditional variance of the data is a scaled inverted chi-square prior distribution

$$\sigma^2|S, v \sim S\chi^{-2}\left(S, v\right), \tag{5.55}$$

where S and the v are user-tuned hyperparameters. The density of the augmented posterior distribution takes the form

$$p\left(\theta_j, \sigma^2, \pi_1, z|y\right) \propto p\left(y|z, \theta_j, \sigma^2\right) p\left(\sigma^2\right) p\left(z|\pi_1\right) p\left(\pi_1\right). \tag{5.56}$$

The McMC Gibbs sampling algorithm for this model is straightforward and can be implemented extracting the fully conditional posterior distributions (fcpd) from (5.56).

Deriving $[\theta_1|D]$

From (5.56), the fcpd of θ_j is

$$\begin{aligned} p\left(\theta_j|D\right) &\propto p\left(y|z, \theta_j, \sigma^2\right) \\ &\propto \exp\left[-\frac{1}{2\sigma^2}\sum_{i=1}^{n}\left\{I\left(z_i = j\right)\left(y_{ij} - \theta_j\right)^2\right\}\right]. \end{aligned}$$

A little algebra leads to

$$\left[\theta_j|D\right] \sim N\left(\widehat{\theta}_j, \frac{\sigma^2}{n_j}\right). \tag{5.57}$$

In this expression,

$$\widehat{\theta}_j = \frac{1}{n_j} \sum\nolimits_{i=1}^{n} y_i I\left(z_i = j\right),$$

$$n_j = \sum\nolimits_{i=1}^{n} I\left(z_i = j\right).$$

Deriving $[z_i | D]$

From (5.56), the fcpd of z_i is

$$\Pr\left(Z_i = j|D\right) \propto p\left(y_i|z_i, \theta_j, \sigma^2\right) \Pr\left(z_i = j|\pi_j\right).$$

When $j = 0$,

$$p\left(y_i|z_i = 0, \theta_j, \sigma^2\right) \propto \exp\left[-\frac{1}{2\sigma^2}\left(y_i - \theta_0\right)^2\right]$$

and

$$\Pr\left(Z_i = 0|D\right) \propto \exp\left[-\frac{1}{2\sigma^2}\left(y_i - \theta_0\right)^2\right]\left(1 - \pi_1\right).$$

When $j = 1$,

$$p\left(y_i|z_i = 1, \theta_j, \sigma^2\right) \propto \exp\left[-\frac{1}{2\sigma^2}\left(y_i - \theta_1\right)^2\right]$$

and

$$\Pr\left(Z_i = 1|D\right) \propto \exp\left[-\frac{1}{2\sigma^2}\left(y_i - \theta_1\right)^2\right]\pi_1.$$

The fcpd is then the Bernoulli process

$$\Pr\left(Z_i = 1|D\right) = \frac{\exp\left[-\frac{1}{2\sigma^2}\left(y_i - \theta_1\right)^2\right]\pi_1}{\exp\left[-\frac{1}{2\sigma^2}\left(y_i - \theta_1\right)^2\right]\pi_1 + \exp\left[-\frac{1}{2\sigma^2}\left(y_i - \theta_0\right)^2\right]\left(1 - \pi_1\right)} \tag{5.58}$$

and the logodds are given by

$$\ln\left[\frac{\Pr(Z_i=1|D)}{\Pr(Z_i=0|D)}\right]=\frac{1}{2\sigma^2}\left[(y_i-\theta_0)^2-(y_i-\theta_1)^2\right]-[\ln(1-\pi_1)-\ln\pi_1]=K_i. \tag{5.59}$$

In a computing environment, an extraction from (5.58) can be obtained efficiently as follows: Let un be a realisation from

$$un \sim Un(0,1).$$

Then

$$z_i=\begin{cases}1, \text{ if } \ln\left(\frac{un}{1-un}\right)\le\ln\left[\frac{\Pr(Z_i=1|D)}{\Pr(Z_i=0|D)}\right],\\ 0, \text{ otherwise.}\end{cases} \tag{5.60}$$

There are two ways of characterising $\Pr(Z_i=1|y)$. One is using the McMC draws $z_i^{[j]}$, the jth sample from $\Pr(Z_i=1|D)$. When the system converges, these are McMC draws from marginal posterior distribution $[Z_i=1|y]$, and their average is a Monte Carlo point estimate of the marginal posterior distribution

$$\varphi_i=\Pr(Z_i=1|y).$$

The estimator is

$$\widehat{\varphi}_i=\widehat{\Pr}(Z_i=1|y)=\frac{1}{l}\sum_{j=1}^{l}z_i^{[j]}, \tag{5.61}$$

where l is the length of the Gibbs chain.

The second way of characterising $\Pr(Z_i=1|y)$ is by constructing φ_i at each round of the McMC chain. Using the logodds (5.59),

$$\varphi_i=\frac{\exp(K_i)}{1+\exp(K_i)}. \tag{5.62}$$

When the McMC algorithm converges the value constructed from (5.62) at iteration j, $\varphi_i^{[j]}$ is an extraction from the marginal posterior distribution $[\varphi_i|y]$. The draws $\varphi_i^{[j]}$, $j=1,\ldots,l$, provide a Monte Carlo description of the complete marginal posterior distribution $[\varphi_i|y]$.

Deriving $\left[\sigma^2|D\right]$

From (5.56), the fcpd σ^2 is

$$p\left(\sigma^2|D\right) \propto p\left(\sigma^2\right) p\left(y|z,\theta_j,\sigma^2\right)$$

$$\propto \left(\sigma^2\right)^{-\left(1+\frac{v}{2}\right)} \exp\left[-\frac{vS}{2\sigma^2}\right] \left(\sigma^2\right)^{-\frac{n}{2}} \exp\left[-\frac{\sum_{i=1}^{n}\sum_{j=0}^{1} I\left(z_i=j\right)\left\{\left(y_i-\theta_j\right)^2\right\}}{2\sigma^2}\right]$$

$$=\left(\sigma^2\right)^{-\left(1+\frac{\widetilde{v}}{2}\right)} \exp\left[-\frac{\widetilde{v}\widetilde{S}}{2\sigma^2}\right],$$

where $\widetilde{v} = v+n$ and $\widetilde{S} = \left[\left\{\sum_{i=1}^{n}\sum_{j=0}^{1}\left[I\left(z_i=j\right)\left(y_i-\theta_j\right)^2\right]\right\} + vS\right] \Big/ \widetilde{v}$. This is the kernel of a scaled inverted chi-square distribution with parameters $\widetilde{v}$ and $\widetilde{S}$:

$$\left[\sigma^2|D\right] \sim \widetilde{v}\widetilde{S}\chi^{-2}\left(\widetilde{v}\right). \tag{5.63}$$

To obtain a sample from (5.63), draw a chi-square distribution with $\widetilde{v}$ degrees of freedom, and the reciprocal of this number is multiplied by $\widetilde{v}\widetilde{S}$.

Deriving $[\pi_1|D]$

From (5.56), the fcpd of π_1 is

$$\begin{aligned} p\left(\pi_1|D\right) &\propto p\left(z|\pi_1\right) p\left(\pi_1\right) \\ &= \pi_1^{\alpha-1}\left(1-\pi_1\right)^{\beta-1}\prod_{i=1}^{n} \Pr\left(Z_i=1|\pi_1\right)^{z_i}\left(1-\Pr\left(Z_i=1|\pi_1\right)\right)^{1-z_i} \\ &= \pi_1^{\sum_{i=1}^{n} z_i+\alpha-1}\left(1-\pi_1\right)^{n-\sum_{i=1}^{n} z_i+\beta-1} \end{aligned} \tag{5.64}$$

which is the kernel of a beta distribution with parameters $\sum_{i=1}^{n} z_i + \alpha$ and $n - \sum_{i=1}^{n} z_i + \beta$. Thus,

$$[\pi_1|D] \sim Be\left(\sum_{i=1}^{n} z_i + \alpha, n - \sum_{i=1}^{n} z_i + \beta\right). \tag{5.65}$$

An implementation of this Bayesian model is discussed on page 356.

5.6 Example: An Application of the EM Algorithm in a Bayesian Context—Estimation of SNP Effects

A classical parametrisation of the genomic model is based on the multiple linear regression

$$y = Z\alpha + X\beta + e, \tag{5.66}$$

where Z is an observed incidence matrix of dimension $n \times p$ that associates fixed effects α with the data y (a column vector of length n) and matrix X, of dimension $n \times m$, is an observed matrix of SNP genotypes where each element takes the value 0, 1 or 2. Row i of X contains the marker genotypes of the m SNPs of individual i. Column vector β (of length m) represents unobserved SNP effects, and e is a vector of residuals with iid elements $N\left(0, \sigma_e^2\right)$.

Sun et al (2012) present an interesting application of the EM algorithm for estimation of SNP effects in a Bayesian framework. In contrast with model (1.39) on page 41 that poses a single variance parameter for all SNP effects, Sun et al (2012) assume that the vector of SNP effects can be represented with the normal structure

$$\beta|\sigma^2 \sim N(0, D), \quad \sigma^2 = \left(\sigma_1^2, \sigma_2^2, \ldots, \sigma_m^2\right), \tag{5.67}$$

where

$$D = \text{diag}\left\{\sigma_j^2\right\}, \quad j = 1, \ldots, m, \tag{5.68}$$

and

$$\sigma_j^2 \overset{iid}{\sim} \upsilon_\beta S_\beta \chi^{-2}(\upsilon_\beta), \text{ for all } j. \tag{5.69}$$

The SNP model assumes that a priori, SNP effects β are conditionally (given D) independent and normally distributed and that each SNP has its own variance σ_j^2. The variances of the m SNPs are iid realisations from a common scaled inverse chi-square distribution $\upsilon_\beta S_\beta \chi^{-2}(\upsilon_\beta)$ with hyperparameters υ_β (the degrees of freedom) and scale S_β. It follows that the density of the marginal distribution of an SNP effect is

$$p\left(\beta_j|\upsilon_\beta, S_\beta\right) = \int p\left(\beta_j|\sigma_j^2\right) p\left(\sigma_j^2|\upsilon_\beta, S_\beta\right) d\sigma_j^2,$$

the density of a t−distribution with υ_β degrees of freedom and scale S_β (see page 149 for a characterisation of a t−distributed random variable as a mixture).

The EM algorithm proposed by Sun et al (2012) treats the variance of each SNP effect σ_j^2 as missing data and maximises the resulting Q function with respect to α, β and σ_e^2. The objective is to find the joint modal value of α, β and σ_e^2.

Preliminaries

Before deriving the EM equations, a little background is needed. The Q function involves expectation of $\ln p(y, \beta, \sigma, \sigma_e) = \ln\left[p\left(y|\alpha, \beta, \sigma_e^2\right) p\left(\beta|\sigma^2\right) p\left(\sigma^2\right)\right]$ with respect to the distribution $\left[\sigma^2|\beta^{[t]}, y\right]$. First, the form of the conditional posterior distribution $\left[\sigma_j^2|\beta^{[t]}, y\right]$ is needed. The posterior density is proportional to $p(y|\alpha, \beta, \sigma_e) p\left(\beta|\sigma^2\right) p\left(\sigma^2\right)$, and extracting terms including σ_j^2 yields

$$
\begin{aligned}
p\left(\sigma_j^2|\beta, y\right) &\propto p\left(\beta_j|\sigma_j^2\right) p\left(\sigma_j^2\right) \\
&= \left(\sigma_j^2\right)^{-\frac{1}{2}} \exp\left(-\frac{\beta_j^2}{2\sigma_j^2}\right) \left(\sigma_j^2\right)^{-\left(\frac{v_\beta}{2}+1\right)} \exp\left(-\frac{v_\beta S_\beta}{2\sigma_j^2}\right) \\
&= \left(\sigma_j^2\right)^{-\left(\frac{v_\beta}{2}+\frac{3}{2}\right)} \exp\left(-\frac{\beta_j^2 + v_\beta S_\beta}{2\sigma_j^2}\right) \\
&= \left(\sigma_j^2\right)^{-\left(\frac{v_\beta^*}{2}+1\right)} \exp\left(-\frac{v_\beta^* S_\beta^*}{2\sigma_j^2}\right)
\end{aligned}
\tag{5.70}
$$

that is the kernel of a scaled inverse chi-square density with v_β^* degrees of freedom and scale parameter S_β^* where

$$
\begin{aligned}
v_\beta^* &= v + 1, \\
S_\beta^* &= \frac{\beta_j^2 + v_\beta S_\beta}{v_\beta^*}.
\end{aligned}
$$

As shown below, evaluation of the Q function involves the expected value of $\left(\sigma_j^2\right)^{-1}$. Using the theory of transformation of random variables, the pdf of $w = \frac{1}{\sigma_j^2}$, from (5.70), is

$$
\begin{aligned}
p(w|\beta, y) &\propto \left(w^{-1}\right)^{-\left(\frac{v_\beta^*}{2}+1\right)} \exp\left(-\frac{v_\beta^* S_\beta^* w}{2}\right) w^{-2} \\
&= w^{\frac{v_\beta^*}{2}-1} \exp\left(-\frac{v_\beta^* S_\beta^* w}{2}\right).
\end{aligned}
\tag{5.71}
$$

Defining $a = v_\beta^*/2$ and $b = v_\beta^* S_\beta^*/2$, (5.71) can be written as

$$w^{a-1} \exp(-bw), \tag{5.72}$$

which is the kernel of a Gamma distribution whose mean is

$$E(w|\beta, y) = \frac{a}{b} = \frac{1}{S_\beta^*} = \left(\frac{\beta_j^2 + v_\beta S_\beta}{v_\beta^*}\right)^{-1}. \tag{5.73}$$

E-Step

The Q function involved in the E step is

$$Q = E\left(\ln\left[p(y|\alpha, \beta, \sigma_e) p\left(\beta|\sigma^2\right) p\left(\sigma^2\right)\right]\right)$$

where integration is with respect to $\left[\sigma^2|\beta^{[t]}, y\right]$. Taking logarithms and noting that the first term is a constant with respect to $\left[\sigma^2|\beta^{[t]}, y\right]$ yield

$$Q = \ln p(y|\alpha, \beta, \sigma_e) + E\left[\ln p\left(\beta|\sigma^2\right)\right] + E\left[\ln p\left(\sigma^2\right)\right]. \tag{5.74}$$

The last term in the right-hand side does not involve $\alpha, \beta, \sigma_e^2$ and therefore does not need to be evaluated (it drops out in the $M-$step). Working with the middle term of the right-hand side, due to independence, the expectation can be computed element-wise yielding

$$\begin{aligned}
\mathrm{E}\left[\ln p\left(\beta|\sigma^2\right)\right] &= \sum_{i=1}^{m} \mathrm{E}\left(\ln\left[\left(2\pi\sigma_i^2\right)^{-\frac{1}{2}} \exp\left(-\frac{\beta_i^2}{2\sigma_i^2}\right)\right]\right) \\
&= -\frac{1}{2}\sum_{i=1}^{m} \mathrm{E}\left(\ln 2\pi + \ln\sigma_i^2 + \beta_i^2 \frac{1}{\sigma_i^2}\right) \\
&= -\frac{1}{2}\sum_{i=1}^{m} \mathrm{E}\left(\beta_i^2 \frac{1}{\sigma_i^2}\right) + R \\
&= -\frac{1}{2}\sum_{i=1}^{m} \beta_i^2 \left(\frac{\beta_i^{2[t]} + v_\beta S_\beta}{v_\beta^*}\right)^{-1} + R,
\end{aligned} \tag{5.75}$$

where R includes terms that do not involve $\alpha, \beta, \sigma_e^2$. The expectation in the third line is obtained from (5.73) where β_i^2 is factored out as a constant, and this operation leads to the fourth line. Then the Q function ignoring an additive constant reduces to

$$
\begin{aligned}
Q &= \ln p\left(y|\alpha, \beta, \sigma_e^2\right) - \frac{1}{2}\sum_{i=1}^{m} \beta_i^2 \left(\frac{\beta_i^{2[t]} + \upsilon_\beta S_\beta}{\upsilon_\beta^*}\right)^{-1} + R \\
&= \ln\left\{\left(\sigma_e^2\right)^{-\frac{n}{2}} \exp\left[-\frac{1}{2\sigma_e^2}(y - Z\alpha - X\beta)'(y - Z\alpha - X\beta)\right]\right\} - \frac{1}{2}\beta' D^{[t]}\beta + R \\
&= -\frac{n}{2}\ln\sigma_e^2 - \frac{1}{2\sigma_e^2}(y - Z\alpha - X\beta)'(y - Z\alpha - X\beta) - \frac{1}{2}\beta' D^{[t]}\beta + R \qquad (5.76)
\end{aligned}
$$

where

$$
D^{[t]} = \text{diag}\left[\left(\frac{\beta_i^{2[t]} + \upsilon_\beta S_\beta}{\upsilon_\beta^*}\right)^{-1}\right], \quad i = 1, \ldots, m.
$$

M-Step

The M-step involves maximisation of (5.76) with respect to σ_e^2, α and β. Maximising with respect to σ_e^2, setting the derivative equal to zero, and solving for σ_e^2 yield

$$
\sigma_e^{2[t+1]} = \frac{1}{n}\left(y - Z\alpha^{[t+1]} - X\beta^{[t+1]}\right)'\left(y - Z\alpha^{[t+1]} - X\beta^{[t+1]}\right). \qquad (5.77)
$$

Similarly for α

$$
\begin{aligned}
\frac{\partial Q}{\partial \alpha} &= \frac{1}{\sigma_e^2} Z'(y - Z\alpha - X\beta) = 0, \\
Z'Z\alpha + Z'X\beta &= Z'y, \qquad (5.78)
\end{aligned}
$$

and for β

$$
\begin{aligned}
\frac{\partial Q}{\partial \beta} &= \frac{1}{\sigma_e^2} X'(y - Z\alpha - X\beta) - D^{[t]}\beta = 0, \\
X'Z\alpha + \left[X'X + \sigma_e^2 D^{[t]}\right]\beta &= X'y. \qquad (5.79)
\end{aligned}
$$

Equations (5.78) and (5.79) result in the system

$$
\begin{bmatrix} Z'Z & Z'X \\ X'Z & X'X + \sigma_e^{2[t]} D^{[t]} \end{bmatrix} \begin{bmatrix} \alpha^{[t+1]} \\ \beta^{[t+1]} \end{bmatrix} = \begin{bmatrix} Z'y \\ X'y \end{bmatrix}. \qquad (5.80)
$$

These equations have the same structure as those that originate from a mixed model where the random effect β has variance

$$\left(D^{[t]}\right)^{-1} = diag\left[\left(\frac{\beta_i^{2[t]} + \upsilon_\beta S_\beta}{\upsilon_\beta^*}\right)\right], \quad i = 1, \ldots, m. \tag{5.81}$$

An Alternative Parametrisation

The system of Eqs. (5.80) has dimension $(p+m) \times (p+m)$. When the number of individuals (n) is smaller than the number of markers (m) the following strategy is computationally more attractive. Define genomic values as $g = X\beta$ and assume that

$$g|X \sim N\left(0, X\left(D^{[t]}\right)^{-1} X'\right),$$

where $\left(D^{[t]}\right)^{-1}$ is defined in (5.81). Secondly, find the maximiser of $p\left(y|\alpha, g, \sigma_e^2\right) p\left(g|D^{[t]}\right)$. This requires

$$\frac{\partial p\left(y|\alpha, g, \sigma_e^2\right) p\left(g|D^{[t]}\right)}{\partial \alpha} = 0,$$

$$\frac{\partial p\left(y|\alpha, g, \sigma_e^2\right) p\left(g|D^{[t]}\right)}{\partial g} = 0.$$

The maximisation involves the following operations:

$$\frac{\partial\left[-\frac{1}{2\sigma_e^2}(y - X\alpha - g)'(y - X\alpha - g) - \frac{1}{2}g'G^{-1}g\right]}{\partial \alpha} = 0,$$

$$\frac{\partial\left[-\frac{1}{2\sigma_e^2}(y - X\alpha - g)'(y - X\alpha - g) - \frac{1}{2}g'G^{-1}g\right]}{\partial g} = 0,$$

where $G = X\left(D^{[t]}\right)^{-1} X'$. This yields

$$\begin{bmatrix} Z'Z \; Z' \\ I \quad I + \sigma_e^2 G^{-1} \end{bmatrix} \begin{bmatrix} \alpha^{[t+1]} \\ g^{[t+1]} \end{bmatrix} = \begin{bmatrix} Z'y \\ y \end{bmatrix}. \tag{5.82}$$

The system of equations is of dimension $(p+n) \times (p+n)$. To retrieve marker effects from (5.82), use (5.81), and write

$$\begin{bmatrix} \beta^{[t+1]} \\ g^{[t+1]} \end{bmatrix} \sim SN\left(\begin{bmatrix} 0 \\ 0 \end{bmatrix}, \begin{bmatrix} \left(D^{[t]}\right)^{-1} & \left(D^{[t]}\right)^{-1} X' \\ X\left(D^{[t]}\right)^{-1} & X\left(D^{[t]}\right)^{-1} X' \end{bmatrix}\right).$$

Then

$$E\left(\beta^{[t+1]}|g^{[t+1]}\right) = \left(D^{[t]}\right)^{-1} X' \left(X \left(D^{[t]}\right)^{-1} X'\right)^{-1} g^{[t+1]}. \tag{5.83}$$

5.7 Example: Bayesian Analysis of the Truncated Normal Model

The setup here is the same as on page 134 where a likelihood analysis was implemented with the EM algorithm. To recapitulate, the data available originates from an initial draw of N records from the normal distribution $\mathcal{N}(\mu, \sigma^2)$. In a second stage, only records larger than an observed threshold C are kept, and those smaller than C are discarded. After this procedure, one can confirm that there are n observable records and that m are missing. The complete data z consist of $N = m+n$ records, $z' = \left((y^*)', y'\right)$ where y^* are the m (unobserved) missing records and y are the n observed records.

A Bayesian McMC analysis can be easily implemented augmenting the parameter space with the missing data. An observed datum is conceptually generated from

$$y_i|\mu, \sigma^2 \sim \mathcal{N}\left(\mu, \sigma^2\right) I\left(z_i > C\right)$$

and a missing datum from

$$y_i^*|\mu, \sigma^2 \sim \mathcal{N}\left(\mu, \sigma^2\right) I\left(z_i \leq C\right)$$

where $I\left(x \in A\right)$ is the indicator function that takes the value 1 if $x \in A$ and zero otherwise. Assuming independence, the conditional density of the complete data is

$$p\left(z|\mu, \sigma^2\right) \propto \prod_{i=1}^{N} \mathcal{N}(z|\mu, \sigma^2)\left[I\left(z_i > C\right) + I\left(z_i \leq C\right)\right]. \tag{5.84}$$

The augmented posterior density is

$$\begin{aligned} p\left(\mu, \sigma^2, y^*|y\right) &\propto p\left(\mu, \sigma^2, y^*\right) p\left(y|\mu, \sigma^2, y^*\right) \\ &= p\left(\mu, \sigma^2\right) p\left(y, y^*|\mu, \sigma^2\right). \end{aligned} \tag{5.85}$$

Assuming independent uniform prior distributions for μ and σ^2, this augmented posterior is proportional to (5.84). The parameters of the augmented Bayesian model are $\left(\mu, \sigma^2, y^*\right)$, and a Gibbs sampling implementation requires drawing from $\left[y^*|D\right]$, $\left[\mu|D\right]$ and $\left[\sigma^2|D\right]$, where, as before, D is a vector containing the observations y and all the parameters of the model except the one to be updated.

The update of y^* requires choosing the terms that include y^* from the augmented posterior (5.84). The resulting fully conditional posterior distribution is

$$[y^*|D] \propto \prod_{i=1}^{m} p\left(y_i^*|\mu, \sigma^2\right) I\left(y_i^* \leq C\right)$$

that is in the form of a truncated normal distribution with mean μ and variance σ^2 with probability density function

$$p\left(y_i^*|\mu, \sigma^2, y_i^* \leq C\right) = \frac{p\left(y_i^*|\mu, \sigma^2\right)}{\Phi(c)}, \quad c = \frac{C-\mu}{\sigma},$$

where $\Phi(\cdot)$ is the cdf of the $\mathcal{N}(0, 1)$. An efficient algorithm to draw from truncated distributions was given in (5.27).

The update for σ^2 is again based on the construction of its fully conditional distribution $\left[\sigma^2|D\right]$. From (5.84), this takes the form

$$\left[\sigma^2|D\right] \propto \prod_{i=1}^{N} p\left(z_i|\mu, \sigma^2\right).$$

The density is proportional to

$$\left(\sigma^2\right)^{-\frac{N}{2}} \exp\left[-\frac{\sum_{i=1}^{N}(z_i-\mu)^2}{2\sigma^2}\right].$$

This is in the form of a scaled inverted chi-square distribution with scale parameter $S = \sum_{i=1}^{N}(z_i-\mu)^2$ and $N-2$ degrees of freedom. Therefore,

$$\left[\sigma^2|D\right] \sim \chi^{-2}(N-2, S).$$

To obtain a draw first sample from a chi-square distribution with $N-2$ degrees of freedom, invert this number, and multiply by S.

The final update requires drawing from $[\mu|D]$. This distribution is proportional to (5.84), now seen as a function of μ. Adding a subtracting $\overline{z} = \sum_{i=1}^{N} z_i \Big/ N$ in the squared term and dropping the terms that do not contain μ result in

$$\frac{\sum_{i=1}^{N}\left[(z_i-\overline{z})+(\overline{z}-\mu)\right]^2}{2\sigma^2} = \frac{N(\overline{z}-\mu)^2}{2\sigma^2} + k,$$

where k is an additive constant. Regarded as a function of μ, this expression is recognised as the logarithm of the kernel of $N\left(\overline{z}, \sigma^2/N\right)$. Therefore,

$$[\mu|D] \sim N\left(\overline{z}, \sigma^2\Big/N\right).$$

The R-code below executes a Gibbs sampler using the same data as used on page 134:

```
# CODE0503
# EM FOR TRUNCATED DATA; ESTIMATE MEAN OF UNTRUNCATED
# GENERATE Y ~ N(MEAN,VAR)
# TRUNCATE AT T SO THAT Z = Y > T ARE OBSERVED
# Y < T ARE MISSING (KNOWN INFORMATION)
rm(list=ls()) # CLEAR WORKSPACE
set.seed(12371)
nindiv<-50000
mean <- 10
var <- 3
T <- mean + 1.5*sqrt(var) # ASSUMED KNOWN
# CREATE COMPLETE DATA
y <- rnorm(nindiv,mean,sqrt(var))
# TRUNCATE: OBSERVED DATA
z <- y[y>T]
#length(z)
m <- length(y)-length(z)
#mean(y)
#mean(z)
#var(z)
#####################  McMC ######################
nrep <- 1000
resmc <- matrix(data=NA,nrow=nrep,ncol=2)
w <- rep(0,m)
# START VALUES FOR MEAN (mu) AND VARIANCE (sigmasq)
mu <- 0
sigmasq <- 2
sigma <- sqrt(sigmasq)
ptm <- proc.time()
for (j in 1:nrep){
#  print(j)
  T_star <- (T-mu)/sigma
  std <- sqrt(var)
# sample m missing records in one go (left from threshold T)
  w <- mu + std*qnorm(runif(m)*pnorm(T_star))
# sample the variance
  scale <- sum((w-mu)^2) + sum((z-mu)^2)
  sigmasq <- scale/rchisq(1,length(y)-2)
  sigma <- sqrt(sigmasq)
# sample the mean
  xbar <- (sum(w)+sum(z))/(length(w)+length(z))
  disp <- sigmasq/(length(w)+length(z))
  mu <- rnorm(1,xbar,sqrt(disp))
  resmc[j,] <- c(mu,sigmasq)
}
proc.time()-ptm
```

```
##    user  system elapsed
##    2.81    0.14    2.96
```

```
postmean <- mean(resmc[100:nrep,1])
postvar <- mean(resmc[100:nrep,2])
postmean
```

```
## [1] 10.03055
```

```
postvar
```

```
## [1] 3.005774
```

```
# 95% POSTERIOR INTERVAL FOR THE MEAN
pimean <- quantile(resmc[100:nrep,1],c(0.025,0.975))
# 95% POSTERIOR INTERVALFOR THE VARIANCE
pivar <- quantile(resmc[100:nrep,2],c(0.025,0.975))
pimean
```

```
##       2.5%     97.5%
## 10.00086 10.06202
```

```
pivar
```

```
##       2.5%     97.5%
## 2.953886 3.059747
```

The MC estimates of the posterior mean and variance are 10.031 and 3.006, in good agreement with the parameters of the untruncated distribution (and with the ML estimates introduced on page 134). The sampler generates the complete joint posterior distribution of $(\mu, \sigma^2, y^\star)$, and the margins correspond to the marginal posterior distribution of each variable in turn. MC estimates of posterior intervals are obtained from the `nrep` draws (here I exclude the first 100 draws as burn-in). The MC estimate of the 95% posterior interval for the posterior mean is (10.001, 10.062) and for the posterior variance (2.954, 3.06).

5.8 A Digression on Model Comparison

The general problem of model criticism and model choice occupies a vast literature including classical and Bayesian approaches. Here, the topic is visited briefly, and it is shown that within an McMC environment, a few extra calculations lead to a useful and versatile measure of model comparison: the *pseudo-log-marginal probability of the data*. The basic ideas are summarised and an example is provided.

The *pseudo-log-marginal probability of the data* is a standard measure of model comparison (Gelfand 1996) and is defined and computed as follows: Consider data vector $y' = (y_i, y'_{-i})$, $i = 1, \cdots, n$, where y_i is the ith datum and y_{-i} is the vector of data with the ith datum deleted. The *conditional predictive distribution* of y_i has density

$$p(y_i|y_{-i}) = \int p(y_i|\theta, y_{-i}) f(\theta|y_{-i}) d\theta, \tag{5.86}$$

where θ is the vector of parameters of the model. This density can be interpreted as the probability of each data point given the remainder of the data; a low value indicates that the datum is poorly fit by the model. The actual value of $p(y_i|y_{-i})$ is known as the *conditional predictive ordinate* (CPO) for the ith observation. A plot of the CPOs versus the index for the observations can serve as a useful tool for outlier detection. The *pseudo-log-marginal probability of the data* or *pseudo-marginal likelihood* is given by

$$\sum_i \ln p(y_i|y_{-i}). \tag{5.87}$$

The *pseudo-Bayes factor* for comparing two models M_1 and M_2 (Gelfand et al 1992; Gelfand 1996) is

$$PBF_{12} = \prod_{i=1}^{n} \frac{\Pr(Y_i = y_i|y_{-i}, M_1)}{\Pr(Y_i = y_i|y_{-i}, M_2)}. \tag{5.88}$$

A Monte Carlo approximation of the CPO (5.86) for observation i is given by (Gelfand 1996)

$$\widehat{p}(y_i|y_{-i}, M_k) = N\left[\sum_{j=1}^{N} \frac{1}{p\left(y_i|y_{-i}, \theta^{(j)}, M_k\right)}\right]^{-1}, \tag{5.89}$$

where N is the number of McMC draws, M_k is a label for model k and $\theta^{(j)}$ is the jth draw from the posterior of θ under model M_k. Often, the y_is are conditionally independent given θ; then the term $p\left(y_i|y_{-i}, \theta^{(j)}, M_k\right)$ in (5.89) simplifies to $p\left(y_i|\theta^{(j)}, M_k\right)$.

The so-called $LogCPO$s are based on

$$\sum_i \ln \hat{p}(y_i|y_{-i}, M_k). \tag{5.90}$$

Larger values indicate a relative better fit.

An appealing side of (5.89) is that only one analysis is required, rather than n (n is sample size), where one out of n observations is left out in each of the n analyses.

A very useful property is that $p(y_i|y_{-i})$ is always a proper density, provided the posterior density is proper. The models under comparison do not need to be nested, and since asymptotics are not involved, there is no problem with testing values of parameters that lie on the border of the parameter space.

The derivation of (5.89) is instructive and straightforward. The details are as follows (Gelfand 1996):

$$
\begin{aligned}
p\left(y_i|y_{-i}, M_r\right) &= \frac{p\left(y|M_r\right)}{p\left(y_{-i}|M_r\right)} \\
&= \frac{1}{\frac{p(y_{-i}|M_r)}{p(y|M_r)}} \\
&= \frac{1}{\int p\left(y_{-i}, \theta_r|M_r\right) \frac{1}{p(y|M_r)} d\theta_r} \\
&= \frac{1}{\int p\left(y_{-i}, \theta_r|M_r\right) \frac{p(\theta_r|y,M_r)}{p(\theta_r,y|M_r)} d\theta_r} \\
&= \frac{1}{\int \frac{p(\theta_r,y|M_r)}{p(y_i|y_{-i},\theta_r,M_r)} \frac{p(\theta_r|y,M_r)}{p(\theta_r,y|M_r)} d\theta_r} \\
&= \frac{1}{\int \frac{1}{p(y_i|y_{-i},\theta_r,M_r)} p\left(\theta_r|y, M_r\right) d\theta_r},
\end{aligned} \tag{5.91}
$$

where the fifth line is a consequence of the equality

$$
p\left(y_i|y_{-i}, \theta_r, M_r\right) = \frac{p\left(\theta_r, y|M_r\right)}{p\left(y_{-i}, \theta_r|M_r\right)}.
$$

A Monte Carlo estimator of (5.91) is (5.89).

Example

Data are available from the following "true" model

$$
\begin{aligned}
y_{ij}|\mu, f_i &\overset{iid}{\sim} N\left(\mu + f_i, \sigma_e^2\right), \\
f_i|\sigma_f^2 &\overset{iid}{\sim} N\left(0, \sigma_f^2\right), \\
i = 1, \ldots, n_f, &\quad j = 1, \ldots, n.
\end{aligned}
$$

There are n_f full-sib families, with n full-sibs per family and a total of $N = n_f \times n$ records. In matrix notation, the model is

$$
y = 1\mu + Zf + e,
$$

where 1 is a vector of ones of length N, Z is of dimension $N \times n_f$ and f and e contain family and residual effects, respectively.

The data set is analysed using the "true" model and using a "wrong" model that assumes $\sigma_f^2 = 0$. This model is

$$y_i|\mu \overset{iid}{\sim} N\left(\mu, \sigma_\varepsilon^2\right),$$

$$i = 1, \ldots, N.$$

The two Bayesian models will be implemented with the Gibbs sampler and the *pseudo-marginal likelihood* (5.90) will be calculated for each.

The values of n and n_f are set equal to 3 and 400, respectively, leading to $N = 1200$ records. Variance components are assumed to be $\sigma_f^2 = 10$ and $\sigma_e^2 = 50$.

A Gibbs Sampler for the "True" Model

The Bayesian model assumes improper, uniform prior distributions for $\left(\mu, \sigma_f^2, \sigma_e^2\right)$. The posterior density is

$$p\left(\mu, \sigma_f^2, \sigma_e^2|y\right) \propto p\left(f|\sigma_f^2\right) p\left(y|\mu, \sigma_f^2, \sigma_e^2\right)$$

$$\propto \left(\sigma_e^2\right)^{-\frac{N}{2}} \left(\sigma_f^2\right)^{-\frac{n_f}{2}} \exp\left(-\frac{1}{2\sigma_f^2} f'f\right) \exp\left(\frac{1}{2\sigma_e^2} (y-1\mu-Zf)' (y-1\mu-Zf)\right)$$

$$= \left(\sigma_e^2\right)^{-\frac{N}{2}} \left(\sigma_f^2\right)^{-\frac{n_f}{2}} \exp\left[-\frac{1}{2\sigma_e^2} \left\{(y-1\mu-Zf)' (y-1\mu-Zf) + kf'f\right\}\right],$$

where $k = \sigma_e^2/\sigma_f^2$. The Gibbs sampler is implemented drawing from the following fully conditional posterior distributions:

$$[\mu|D] \sim N\left(\widehat{\mu}, \left(1'1\right)^{-1} \sigma_e^2\right),$$

$$\widehat{\mu} = \left(1'1\right)^{-1} 1' (y - Zf),$$

$$[f|D] \sim N\left(\widehat{f}, \left(Ik + Z'Z\right)^{-1} \sigma_e^2\right),$$

$$\widehat{f} = \left(Ik + Z'Z\right)^{-1} Z' (y - 1\mu),$$

$$\left[\sigma_f^2|D\right] \sim \frac{f'f}{\chi^2 \left(n_f - 2\right)},$$

$$\left[\sigma_e^2|D\right] \sim \frac{(y - 1\mu - Zf)' (y - 1\mu - Zf)}{\chi^2 (N - 2)}.$$

The computation of (5.90) requires a few intermediate calculations. In each Gibbs round, one first calculates

$$p\left(y_i|\mu^{[j]}, f^{[j]}, \sigma_e^{2[j]}, M_{correct}\right) \quad i = 1, \ldots, N.$$

This can be accomplished for all the data (vector y) in one go using the `R` function:

```
dnorm(y,mean=mu+Z%*%f,sd=sqrt(ve))
```

Specifically, for each round, compute

```
pyinvt <- -1/(dnorm(y,mean=mu+Z%*%f,sd=sqrt(ve)))
sumpyinvt <- sumpyinvt+pyinvt
```

Once all the iterations have been executed, one computes (5.89) and finally (5.90). The code is

```
phatyt <- rep*(sumpyinvt)^(-1)}
logcpot <- sum(log(phatyt))
```

A Gibbs Sampler for the “Wrong” Model

Using improper uniform priors for $(\mu, \sigma_\varepsilon^2)$, the posterior density under the incorrect model is

$$\begin{aligned} p\left(\mu, \sigma_\varepsilon^2|y\right) &\propto p\left(y|\mu, \sigma_\varepsilon^2\right) \\ &\propto \left(\sigma_\varepsilon^2\right)^{-\frac{N}{2}} \exp\left(-\frac{1}{2\sigma_\varepsilon^2}(y - 1\mu)'(y - 1\mu)\right). \end{aligned}$$

It is straightforward to derive the fully conditional posterior distributions. These are

$$\begin{aligned} [\mu|D] &\sim N\left(\widehat{\mu}, \left(1'1\right)^{-1}\sigma_e^2\right), \\ \widehat{\mu} &= \left(1'1\right)^{-1}1'y, \\ \left[\sigma_e^2|D\right] &\sim \frac{(y - 1\mu)'(y - 1\mu)}{\chi^2(N - 2)}. \end{aligned}$$

The `R-code` below generates data under the true model and then runs a Gibbs sampler with the two models. The *pseudo-marginal likelihoods* (5.90) are calculated for each model. The code is spelled out line by line avoiding the use of more efficient and more compact programming that is to be preferred with more demanding computations:

```
# CODE0504
#CPO EXAMPLE
rm(list=ls()) # Clear the workspace
set.seed(123771)
ptm<-proc.time()
require(graphics)
# GENERATE CORRELATED (FULL-SIBS DATA
#install.packages("MCMCpack", .libPaths()[1])
#install.packages("mvtnorm", .libPaths()[1])
library(MCMCpack)
# INITIALISE PARAMETERS
mus<-10 # MEAN
vfs<-10 #VARIANCE BETWEEN FULL-SIBS
# RESIDUAL VARIANCE
ves<-50
nf<-400 # NUMBER OF FULL-SIB FAMILIES
n<-3 #FULL-SIB FAMILY SIZE
N<-nf*n
y<-matrix(data=0,nrow=nf*n,ncol=1)
# z IS COLUMN MATRIX WITH FAMILY ID (ID=1,.,nfs)
z<-matrix(data=0,nrow=nf*n,ncol=1)
# GENERATE nf FULL-SIB EFFECTS f
fs<-rnorm(nf,mean=0,sd=sqrt(vfs))
# GENERATE nf*n RESIDUAL EFFECTS f
es<-rnorm(nf*n,mean=0,sd=sqrt(ves))
# GENERATE FULL SIBS (CAN CHOOSE MORE TRANSPARENT LOOP ABOVE)
z <- rep(1:nf,each=n)
y <- mus+fs[z]+es
d<-data.frame(y,z)
# GENERATE INCIDENCE MATRIX Z
family<-z
family <- as.factor(family)
Z<-model.matrix(~0+family)
# WITH INDEPENDENT FAMILIES Z'Z IS DIAGONAL
ztz<-rep(n,nf)
#END OF GENERATION OF DATA Y
#CHOOSE LENGTH OF CHAIN
rep<-1000
resultt<-matrix(data=NA,nrow=rep,ncol=4)
resultw<-matrix(data=NA,nrow=rep,ncol=3)
#INITIALISE THE VECTOR OF FAMILY EFFECTS f
f<-rep(0,nf)
# INITIALISE BETWEEN FAMILY VARIANCE COMPONENT vf
vf<-5
# INITIALISE RESIDUAL VARIANCE
ve<-5
# INITIALISE k
k<-ve/vf
# INITIALISE THE MEAN
mu<-0
sumpyinvt<-0
#START GIBBS LOOP TRUE MODEL
for (i in 1:rep)
{
  # SAMPLE mu
  meanmu<-sum(y-Z%*%f)/(nf*n)
  mu<-rnorm(1,mean=meanmu,sd=sqrt(ve/(nf*n)))
  # SAMPLE FAMILY EFFECTS f
  varf<-(k+n)^(-1)
```

```
  fmean<- varf*(t(Z)%*%(y-mu))
  f<-rnorm(nf,mean=fmean, sd=sqrt(varf*ve))
  #SAMPLE vf
  #COMPUTE SCALE
  ftf<-sum(f*f)
  vfx<-ftf/rchisq(1,nf-2)
  vf<-as.numeric(vfx)
  # SAMPLE ve
  # COMPUTE SCALE
  e<-(y-mu-Z%*%f)
  ete<-t(e)%*%e
  vex<-ete/rchisq(1,N-2)
  ve<-as.numeric(vex)
  k<-ve/vf
  resultt[i,]<-c(i,mu,vf,ve)
#  print(resultt[i,])
  # COMPUTE CPOs FOR TRUE MODEL
  pyinvt<-1/(dnorm(y,mean=mu+Z%*%f,sd=sqrt(ve)))
  sumpyinvt<-sumpyinvt+pyinvt
}
phatyt<-rep*(sumpyinvt)^(-1)
logcpot<-sum(log(phatyt))
proc.time()-ptm
```

```
##    user  system elapsed
##   14.14    0.41    3.67
```

```
#START GIBBS LOOP WRONG MODEL
vew<-20
sumpyinvw<-0
for (i in 1:rep)
{
  # SAMPLE muw
  meanmuw<-sum(y)/N
  muw<- rnorm(1,mean=meanmuw,sd=sqrt(vew/(N)))
  # SAMPLE vew
  vew<-sum((y-muw)*(y-muw))/rchisq(1,N-2)
  resultw[i,]<-c(i,muw,vew)
  # COMPUTE CPOs FOR WRONG MODEL
  pyinvw<-1/(dnorm(y,mean=mu,sd=sqrt(vew)))
  sumpyinvw<-sumpyinvw+pyinvw
}
phatyw<- rep*(sumpyinvw)^(-1)
logcpow<-sum(log(phatyw))
# PRINT OUTPUT
# LOG CPO TRUE MODEL
logcpot
```

```
## [1] -4176.398
```

```
# 95% POSTERIOR INTERVALS FOR THE MEAN AND THE TWO VARIANCES
quantile(resultt[,2],c(0.025,0.975))
```

```
##       2.5%     97.5%
##  9.344941 10.410468
```

```
quantile(resultt[,3],c(0.025,0.975))
```

```
##       2.5%     97.5%
##  5.504737 13.107484
```

```
quantile(resultt[,4],c(0.025,0.975))
```

```
##     2.5%    97.5%
## 49.94121 60.39323
```

```
# LOG CPO FOR WRONG MODEL
logcpow
```

```
## [1] -4194.195
```

```
# 95% POSTERIOR INTERVALS FOR THE MEAN AND VARIANCE
quantile(resultw[,2],c(0.025,0.975))
```

```
##       2.5%     97.5%
##  9.422171 10.308375
```

```
quantile(resultw[,3],c(0.025,0.975))
```

```
##     2.5%    97.5%
## 58.87403 69.06082
```

McMC estimates of 95% posterior intervals for the parameters of both models are printed at the bottom of the code. The $\ln CPO$ based on (5.90) for the true model is −4176 and for the incorrect model is −4194. The difference in favour of the true model is very large. This may not be the case if σ_f^2 is small. For example, running the program with $\sigma_f^2 = 5$ results in a $\ln CPO$ for the true model equal to −4045 and for the wrong model, $LogCPO = -4048$. The difference is markedly smaller.

Part II
Prediction

Chapter 6
Fundamentals of Prediction

This chapter provides an overview of prediction with examples taken from quantitative genetics. The first part summarises *best prediction* and *best linear prediction* and offers a brief tour of the standard linear least squares regression. Many important ideas related to prediction can be introduced using the simple least squares setting. The more specific topics on prediction are introduced in Sect. 6.4, where the central question that dominates the remaining of the chapter is posed: suppose that observations in the form of $(y_1, x_1), (y_2, x_2), \ldots, (y_n, x_n)$ are available, where scalars y_i are outcomes or responses and vectors $x_i \in \mathbb{R}^p$ are signals, covariates or features that constitute the feature space. The goal is to construct a predictor that specifies the form of the relationship between the response and the covariate and generates a prediction for a future y given an observed x. The construction of the predictor involves estimation of parameters of the model that describes the relationship between y and x. Among the topics discussed is the accuracy with which future observations can be predicted, how this accuracy is measured and what are the factors affecting it. A distinction is made between the ability to predict (or to fit) those same observations that were used for estimation of parameters or for prediction of new, yet-to-be observed outcomes.

The body of the chapter deals with prediction from a classical/frequentist perspective. Bayesian prediction is illustrated in several examples in the coming chapters and particularly in Chap. 10.

6.1 Best Predictor and Best Linear Predictor

Consider a scalar random variable y that can represent data such as height measurements in humans and a scalar or vector random variable x that can represent explanatory variables or covariates such as genetic marker genotypes. One may wish to establish the form of the association between y and x either to make inferences

D. Sorensen, *Statistical Learning in Genetics*, Statistics for Biology and Health,
https://doi.org/10.1007/978-3-031-35851-7_6

of the parameters that describe this association or to make predictions of a yet-to-be observed y_0 using a predictor $f(x)$, a function of x. One way of deriving the predictor f is by minimising the expected value of the *mean squared error of prediction* over the joint distribution $[y_0, x]$:

$$\begin{aligned} \mathrm{E}_{y_0x}\left[\mathrm{MSE}\left(f\left(x\right)\right)\right] &= \mathrm{E}_{y_0x}\left[\left(y_0 - f\left(x\right)\right)^2\right] \\ &= \mathrm{E}_x\left[\mathrm{E}_{y_0|x}\left(y_0 - f\left(x\right)\right)^2 |x\right] \\ &= \mathrm{E}_x\left[\mathrm{Var}\left(y_0|x\right) + \left(\mathrm{E}(y_0|x) - f(x)\right)^2\right], \end{aligned} \tag{6.1}$$

where the notation E_{y_0x} means expectation over the joint distribution $[y_0, x]$. Adding and subtracting $\mathrm{E}\left(y_0|x\right)$ in $\left(y_0 - f\left(x\right)\right)$ in the second line, expanding the square and taking expectations lead to the third line. The first term in the right-hand side does not depend on f and therefore (6.1) is minimised when

$$f\left(x\right) = \mathrm{E}\left(y_0|x\right). \tag{6.2}$$

This is the *best predictor*; it minimises $\mathrm{E}_{y_0x}\left[\mathrm{MSE}\left(f\left(x\right)\right)\right]$ with respect to both, the joint distribution of (x, y_0) and the conditional distribution of y_0 given x. In Gaussian linear models, the conditional distribution of y_0 given x depends on x only through the conditional mean. In the case of other distributions, such as the Bernoulli where independent binary data are drawn with probability $p\left(x\right)$, both the conditional mean $\mathrm{E}\left(y_0|x\right) = p\left(x\right)$ and the conditional variance $\mathrm{Var}\left(y_0|x\right) = p\left(x\right)\left[1 - p\left(x\right)\right]$ depend on x.

The error of prediction is

$$e = y_0 - \mathrm{E}\left(y_0|x\right)$$

that motivates the model for y_0

$$y_0 = \mathrm{E}\left(y_0|x\right) + e.$$

Therefore, by construction,

$$\mathrm{E}\left(e|x\right) = \mathrm{E}\left[\left(y_0 - \mathrm{E}\left(y_0|x\right)|x\right)\right] = 0. \tag{6.3}$$

It does not follow that the distribution of e is independent of x unless it is imposed by assumption.

Using the law of iterated expectations reveals that the unconditional mean is also zero:

$$\mathrm{E}\left(e\right) = \mathrm{E}_x\left[\mathrm{E}\left(e|x\right)\right] = 0. \tag{6.4}$$

The conditional expectation is typically not available and neither is the form of the joint distribution of (x, y_0). To make progress requires choice of a model. If the choice falls on a linear relationship, whereby the scalar y_0 is predicted using $\alpha + x'\beta$ where α is an intercept, β is the $(p-1) \times 1$ column vector of population coefficients and x is the $(p-1) \times 1$ column vectors of covariates, then, minimising

$$E_{y_0,x}\left[\left(y_0 - \alpha - \beta' x\right)^2\right] \tag{6.5}$$

with respect to α and β yields the population parameters

$$\beta = \text{Var}(x)^{-1}\,\text{Cov}(y_0, x), \quad \alpha = E(y_0) - \beta' E(x). \tag{6.6}$$

The dimensions are as follows: Var (x) is $(p-1)\times(p-1)$, Cov (y_0, x) is $(p-1)\times 1$, β is $(p-1) \times 1$, and E (x) is $(p-1) \times 1$.

The resulting *best linear predictor* is

$$\hat{y}_0 = E(y_0) + \text{Cov}\left(y_0, x'\right)[\text{Var}(x)]^{-1}(x - E(x)) \tag{6.7}$$

that is in the best linear approximation of y_0 given x. Writing

$$y_0 = \hat{y}_0 + \epsilon, \tag{6.8}$$

then the following properties can be derived:

$$E(\epsilon) = E(y_0) - E(\hat{y}_0) = 0, \tag{6.9a}$$

$$\text{Cov}\left(\epsilon, \hat{y}_0\right) = \text{Cov}\left(y_0, \hat{y}_0\right) - \text{Var}\left(\hat{y}_0\right) = 0, \tag{6.9b}$$

$$\text{Var}(\epsilon) = \text{Var}(y_0) - \text{Var}\left(\hat{y}_0\right). \tag{6.9c}$$

In (6.9), expectations are taken over $[y_0, x]$. The first equality (6.9a) indicates that the marginal expectation of ϵ over $[y_0, x]$ is zero, but using (6.8) and writing $E(\epsilon|x) = E(y_0|x) - E(\hat{y}_0|x)$ reveals that in contrast to (6.3), the expectation of ϵ given x is not zero in general, unless $E(y_0|x)$ is linear.

The best linear predictor shares properties (6.9) with the best predictor as indicated below in (6.10).

A classical animal breeding setup where some of these concepts are illustrated is in Part III, on pages 563 and 651.

Properties of the Best Predictor

Let $\mathrm{E}(y_0|x) = \tilde{y}_0$. Properties of the best predictor are

$$\mathrm{E}_x[\tilde{y}_0] = \mathrm{E}_x[\mathrm{E}(y_0|x)] = \mathrm{E}(y_0). \tag{6.10a}$$

$$\mathrm{Cov}_{y_0x}\left[y_0, \tilde{y}_0\right] = \mathrm{Var}_x\left[\tilde{y}_0\right]. \tag{6.10b}$$

$$\begin{aligned}\mathrm{PEV}(\tilde{y}_0) &= \mathrm{Var}(\tilde{y}_0 - y_0)\\ &= \mathrm{E}_x\left\{\mathrm{Var}\left[(\tilde{y}_0 - y_0)|x\right]\right\} + \mathrm{Var}_x\left\{\mathrm{E}\left[(\tilde{y}_0 - y_0)|x\right]\right\}\\ &= \mathrm{E}_x\left[\mathrm{Var}(y_0|x)\right]\end{aligned} \tag{6.10c}$$

where PEV is the prediction error variance, the variance of the deviation between the predictand and the conditional mean. The corresponding term in (6.9c) is the variance of the deviation between the *predictand* and its best linear approximation.

The top line (6.10a) indicates that the expectation of the predictor is equal to the expectation of the predictand, the classical frequentist definition of unbiasedness for "prediction of random variables".

The proof of (6.10b) is based on the equality

$$\begin{aligned}\mathrm{Cov}_{y_0x}\left[y_0, \mathrm{E}(y_0|x)\right] &= \mathrm{E}_x\left[\mathrm{Cov}(y_0, \mathrm{E}(y_0|x)|x)\right] + \mathrm{Cov}_x\left[\mathrm{E}(y_0|x), \mathrm{E}(y_0|x)\right]\\ &= \mathrm{Var}_x\left[\tilde{y}_0\right].\end{aligned}$$

The squared correlation between predictand and predictor is

$$\begin{aligned}r^2_{y_0\tilde{y}_0} &= \frac{\left[\mathrm{Cov}_{y_0x}(y_0, \tilde{y}_0)\right]^2}{\mathrm{Var}(y_0)\,\mathrm{Var}(\tilde{y}_0)} = \frac{\mathrm{Var}_x(\tilde{y}_0)}{\mathrm{Var}(y_0)}\\ &= 1 - \frac{\mathrm{PEV}(\tilde{y}_0)}{\mathrm{Var}(y_0)}.\end{aligned} \tag{6.11}$$

For simplicity, take x to be a scalar. Then

$$\begin{aligned}\mathrm{E}_{y_0x}(y_0x) &= \mathrm{E}_x\left[\mathrm{E}_{y_0|x}(y_0x|x)\right]\\ &= \mathrm{E}_x\left[x\,\mathrm{E}_{y_0|x}(y_0|x)\right]\\ &= \mathrm{E}_x(\tilde{y}_0x).\end{aligned}$$

When the equality $\mathrm{E}_{y_0x}(y_0x) = \mathrm{E}_x(\tilde{y}_0x)$ is written as

$$\mathrm{E}_{y_0x}\left[x(y_0 - \tilde{y}_0)\right] = 0$$

it takes the form of an orthogonality condition indicating that the prediction errors $(y_0 - \tilde{y}_0)$ are uncorrelated with the covariate x.

Example: Additive Genetic Values as Best Linear Predictors

Consider a continuously distributed trait such as height in humans, $y_i, i = 1, \ldots, n$, measured on n individuals. Assume that the trait is genetically determined by q biallelic loci coded as z_{ij}, $j = 1, \ldots, q$, and that the genotypic codes are centred so that $\mathrm{E}\left(z_{ij}\right) = 0$ for all individuals and loci. The genetic value of individual i is the deviation between the expected phenotypic value y_i, given z_i and the population mean, where $\mathrm{E}\left(y_i|z_i\right) = \mu + G\left(z_i\right)$, μ is the population mean, $G(z_i)$ is the genetic value and $z_i' = \left(z_{i1}, \ldots, z_{iq}\right)$ is the row vector of genotype codes for individual i. The centred genotype codes are discrete random variables denoted allele contents. The genetic values of each of the q biallelic loci can take three modalities corresponding to the three possible genotypes.

The phenotype is assumed to be the result of the additive contribution of a mean, the genetic value and an environmental effect

$$y_i = \mu + G\left(z_i\right) + e_i \tag{6.12}$$

where the continuously distributed environmental effect has mean zero and variance σ^2; $e_i \sim \left(0, \sigma^2\right)$.

The conditional expectation $\mu + G(z_i)$ may not be linear on z_i due to the genetic mechanism operating within and across loci. However, one can always define a linear relationship of the form

$$\mu + \alpha' z_i. \tag{6.13}$$

Then μ and α are obtained minimising

$$\mathrm{E}\left[\left(y_i - \mu - \alpha' z_i\right)^2\right]$$

with respect to μ and α. The expected squared error is a minimum with

$$\begin{aligned}\alpha &= \left[\mathrm{Var}\left(z_i\right)\right]^{-1}\mathrm{Cov}\left(y_i, z_i\right)\\ &= \left[\mathrm{Var}\left(z_i\right)\right]^{-1}\mathrm{Cov}\left(z_i, G\left(z_i\right)\right)\end{aligned} \tag{6.14}$$

and

$$\mu = \mathrm{E}\left(y_i\right) - \alpha' \mathrm{E}\left(z_i\right). \tag{6.15}$$

The equality in the second line of (6.14) is the result of the following:

$$\begin{aligned}\mathrm{Cov}\left(y_i, z_i\right) &= \mathrm{E}\left(y_i z_i\right)\\ &= \mathrm{E}_{z_i}\left[\mathrm{E}\left(y_i z_i | z_i\right)\right]\end{aligned}$$

$$
\begin{aligned}
&= \mathrm{E}_{z_i}\left[z_i\,\mathrm{E}\left(y_i|z_i\right)\right] \\
&= \mathrm{E}_{z_i}\left[z_i\left(\mu + G\left(z_i\right)\right)\right] \\
&= \mathrm{Cov}\left(z_i, G\left(z_i\right)\right).
\end{aligned}
$$

Substituting (6.14) and (6.15) in (6.13) yields the best linear predictor of the phenotypic value

$$
\begin{aligned}
\hat{y}_i &= \mathrm{E}\left(y_i\right) + \mathrm{Cov}\left(z_i, G\left(z_i\right)\right)\left[\mathrm{Var}\left(z_i\right)\right]^{-1}\left(z_i - \mathrm{E}\left(z_i\right)\right) \\
&= \mu + \alpha' z_i
\end{aligned}
$$

because $E\left(z_i\right) = 0$. The column vector α has q elements representing the multiple regression coefficients of phenotype on allele content. They are the additive genetic effects of the causal loci or effects of gene substitutions for each of the q loci affecting the trait (Falconer and Mackay 1996), and $\alpha' z_i$ is the additive genetic value of individual i, also known as the breeding value, the best linear approximation describing the relationship between genetic value and allele content.

Using a Biased Predictor

Imagine that a biased predictor is used instead of the best predictor $\mathrm{E}\left(y_0|x\right) = \tilde{y}_0$. It is not unusual that a best predictor cannot be derived, and compromises must be sought. Assume that the biased predictor is equal to $b\tilde{y}_0$, $0 \le b \le 1$, proportional to the best predictor $\tilde{y}_0$. How does this affect prediction error variance and the squared correlation between predictand and predictor? The prediction error variance is

$$
\mathrm{Var}\left(y_0 - b\tilde{y}_0\right) = \mathrm{Var}\left(y_0\right) + b^2\,\mathrm{Var}\left(\tilde{y}_0\right) - 2b\,\mathrm{Cov}\left(y_0, \tilde{y}_0\right).
$$

Using $\mathrm{Cov}\left(y_0, \tilde{y}_0\right) = \mathrm{Var}\left(\tilde{y}_0\right)$ shows that the prediction error variance using the biased predictor increases to

$$
\mathrm{Var}\left(y_0 - b\tilde{y}_0\right) = \mathrm{PEV}\left(\tilde{y}_0\right) + \mathrm{Var}\left(\tilde{y}_0\right)(1-b)^2.
$$

Similar algebra yields

$$
r^2_{y_0, b\tilde{y}_0} = r^2_{y_0\tilde{y}_0}\left[1 - (1-b)^2\right],
$$

indicating that the squared correlation between predictor and predictand is reduced by a factor $\left[1 - (1-b)^2\right]$. The topic is elaborated in de los Campos et al (2013b).

6.2 Estimating the Regression Function in Practice: Least Squares

The population parameters of the best linear predictor (6.7) are typically estimated using a sample. Anticipating a terminology to be used later in the book, this sample used to estimate parameters is labelled *training data*. This may consist of independent draws $\{y_i, x_i\}_{i=1}^n$ from some distribution, where $x_i \in \mathbb{R}^p$, and in this section, they are assumed to have mean zero for all i. Collecting the responses onto the vector $y = \{y_i\}_{i=1}^n \in \mathbb{R}^n$ and the vector of predictors onto the rows of a full rank matrix x,

$$x = \begin{bmatrix} x_1' \\ x_2' \\ \vdots \\ x_n' \end{bmatrix} \in \mathbb{R}^{n \times p},$$

the linear regression model can be written as

$$y = 1\alpha + x\beta + e, \quad e \sim \left(0, \sigma^2 I\right), \tag{6.16}$$

where α is an intercept, e is independent of x (here an assumption) and 1 is an $n \times 1$ column vector of ones. In (6.16), x is a matrix of dimension $n \times (p-1)$. Including the intercept in β and taking the first column of x to be the vector of ones, the model is

$$y = x\beta + e \tag{6.17}$$

and now, including the vector of 1s, x has dimension $n \times p$ and β is a column vector with dimension $p \times 1$. The sum of squared errors is

$$\sum_{i=1}^{n} \left(y_i - x_i'\beta\right)^2. \tag{6.18}$$

The sample regression coefficient (or least squares estimator) is obtained by minimising (6.18):

$$\begin{aligned} \hat{\beta} &= \left(x'x\right)^{-1} x'y \\ &= \left(x'x\right)^{-1} x' \left(x\beta + e\right) \\ &= \beta + \left(x'x\right)^{-1} x'e \end{aligned} \tag{6.19}$$

provided the relationship (6.17) holds. From (6.3), e has conditional mean zero, and the conditional expectation is

$$\begin{aligned} \mathrm{E}\left(\hat{\beta}|x\right) &= \beta + \left(x'x\right)^{-1} x' \mathrm{E}\left(e|x\right) \\ &= \beta \end{aligned}$$

and therefore, unconditionally, $\mathrm{E}\left(\hat{\beta}\right) = \beta$.

A fitted value is defined as

$$\hat{y}_i = x_i' \hat{\beta}$$

and the estimated residual is

$$\hat{e}_i = y_i - \hat{y}_i = y_i - x_i' \hat{\beta}.$$

Each datum can also be expressed as

$$y_i = x_i' \hat{\beta} + \hat{e}_i.$$

The conditional variance-covariance of the least squares estimator is readily seen to be

$$\begin{aligned} \mathrm{Var}\left(\hat{\beta}|x\right) &= \sigma^2 \left(x'x\right)^{-1} \\ &= \frac{\sigma^2}{n} \left(\frac{1}{n} x'x\right)^{-1}. \end{aligned}$$

As the number of observations increases, the term inside the brackets converges to a finite nonsingular matrix. Then the variance of the estimates of the regression coefficients will (1) decrease as sample size n increases, (2) increase with larger σ^2 indicating a poor fit of the linear regression, and (3) decrease as the predictor variables x have larger sampling variances and are uncorrelated with each other. This latter point can be illustrated for the case of two covariates. Let

$$X'X = J = \begin{bmatrix} J_{11} & J_{12} \\ J_{12} & J_{22} \end{bmatrix}.$$

If the covariables are uncorrelated, $J_{12} = 0$, and the variance of the estimator of the first regression is proportional to $(J_{11})^{-1}$. When the covariates are correlated, $J_{12} \neq 0$, and the variance of the estimator of the first regression is proportional to J_{22}/Δ where

$$\left(X'X\right)^{-1} = J^{-1} = \frac{1}{\Delta} \begin{bmatrix} J_{22} & -J_{12} \\ -J_{12} & J_{11} \end{bmatrix}.$$

The determinant $\Delta = J_{11}J_{22} - (J_{12})^2 > 0$ because $X'X$ is positive definite. This implies that $J_{11}J_{22} > \Delta$. Dividing both sides by Δ and multiplying by $(J_{11})^{-1}$, we get

$$\frac{J_{22}}{\Delta} > (J_{11})^{-1}$$

indicating that when the covariables are correlated the uncertainty increases. When $J_{12} = 0$, $J_{22}/\Delta = (J_{11})^{-1}$.

If x is allowed to vary, then by the law of total variance

$$\begin{aligned}
\text{Var}\left(\hat{\beta}\right) &= \text{E}_x\left[\text{Var}\left(\hat{\beta}|x\right)\right] + \text{Var}_x\left[\text{E}\left(\hat{\beta}|x\right)\right] \\
&= \text{E}_x\left[\text{Var}\left(\hat{\beta}|x\right)\right] \\
&= \text{E}_x\left[\frac{\sigma^2}{n}\left(\frac{1}{n}x'x\right)^{-1}\right] \\
&= \frac{\sigma^2}{n}\text{E}_x\left[\left(n^{-1}x'x\right)^{-1}\right].
\end{aligned}$$

Omitting the intercept, as $n \to \infty$, $n^{-1}x'x \to \text{Var}(x)$ and (see note0602.pdf at https://github.com/SorensenD/SLGDS)

$$\text{Var}\left(\hat{\beta}\right) \to n^{-1}\sigma^2 (\text{Var}(x))^{-1}. \tag{6.20}$$

The conditional variance of y_i in the linear model specified in (6.16), given x, is

$$\begin{aligned}
\text{Var}(y_i|x_i) &= \text{E}\left[(y_i - \text{E}(y_i|x_i))^2\right] \\
&= \text{E}\left(e_i^2|x_i\right)
\end{aligned}$$

that by assumption, is independent of x and therefore $\text{Var}(y_i|x_i) = \text{E}\left(e_i^2|x_i\right) = \sigma^2$, for all i. This defines a homoscedastic error. In a heteroscedastic model, the conditional variance depends on the covariate x.

A more general approach to study factors affecting the conditional variance is to specify a model of the form

$$\begin{aligned}
y_i|x_i &\sim N\left(x_i'\beta, \sigma^2(x_i)\right), \\
\ln\left(\sigma^2(x_i)\right) &= x_i'\beta^*
\end{aligned}$$

where the logerror variance is modelled explicitly as in San Cristobal-Gaudy et al (1998). When $\ln\left(\sigma^2(x_i)\right) = \beta^*$, a parameter in $\mathbb{R}$, the model reduces to its homoscedastic form (6.16). This is a fruitful area of research in quantitative and evolutionary genetics that is briefly revisited in Chapter 10 on page 435.

The unconditional error variance is, by construction, independent of x. Indeed, the unconditional variance is

$$\mathrm{E}\left(e_i^2\right) = \mathrm{E}_{x_i}\left[\mathrm{E}\left(e_i^2|x_i\right)\right],$$

the average conditional variance.

Least Squares Linear Regression and the "Hat" Matrix

The fitted value of y (a column vector with n elements) is

$$\begin{aligned}\hat{y} &= x\hat{\beta} \\ &= x\left(x'x\right)^{-1}x'y \\ &= Hy \end{aligned} \tag{6.21}$$

where the $n \times n$ idempotent, symmetric, positive semidefinite matrix $H = x\left(x'x\right)^{-1}x'$ is known as the "hat" matrix (an idempotent matrix is a matrix which, when multiplied by itself, yields itself) and matrix x, whose first column is a vector of 1s to accommodate the constant term, is of dimension $n \times p$, where $p - 1$ is the number of covariates. Since H is positive semidefinite, its ith diagonal element $h_{ii} \geq 0$. Other properties of H are

1. $Hx = x$.
2. Partition x as $x = (x_1\, x_2)$. Then $Hx_1 = x_1$.
3. If matrix x contains a constant term, $H1 = 1$, where 1 is a column vector of 1s. It follows that $\sum_{j=1}^{n} h_{ij} = \sum_{i=1}^{n} h_{ij} = 1$.
4. If matrix x contains a constant term, $n^{-1} \leq h_{ii} \leq 1$, and in general, $0 \leq h_{ii} \leq 1$.

Writing

$$y = x\hat{\beta} + \hat{e} = \hat{y} + \hat{e}, \tag{6.22}$$

then

$$x'\hat{e} = x'\left(y - \hat{y}\right) = x'\left(y - Hy\right) = x'y - x'x\left(x'x\right)^{-1}x'y = 0$$

and therefore, the estimated residuals $\hat{e}$ are orthogonal to the covariable x.

The estimated residuals are also orthogonal to the predicted values. Indeed,

$$\hat{e}'\hat{y} = \left(\left(I - H\right)y\right)' Hy = y'Hy - y'HHy = 0,$$

because H is idempotent and therefore $HH = H$. The estimated residuals $\hat{e} = y - \hat{y}$ can be generated using

$$\begin{aligned}\hat{e} &= (I - H)\, y \\ &= (I - H)\,(x\beta + e) \\ &= (I - H)\, e,\end{aligned} \tag{6.23}$$

because $Hx = x$. For the homoscedastic model, the variance of the vector of estimated residuals is

$$\mathrm{Var}\left(\hat{e}|x\right) = \mathrm{Var}\left((I - H)\, e|x\right) = \sigma^2\,(I - H)$$

because $I - H$ is idempotent, indicating that estimated residuals are heteroscedastic and correlated despite homoscedasticity of the uncorrelated errors e. In particular for record i,

$$\mathrm{Var}\left(\hat{e}_i|x\right) = \sigma^2\,(1 - h_{ii}) \tag{6.24}$$

where h_{ii} is the ith diagonal element of H.

An estimator of the conditional variance given x is

$$\begin{aligned}\hat{\sigma}^2 &= \frac{1}{n}\hat{e}'\hat{e} \\ &= \frac{1}{n}y'\,(I - H)\,(I - H)\,y \\ &= \frac{1}{n}y'\,(I - H)\,y,\end{aligned}$$

because $(I - H)$ is idempotent. This is a method of moment estimator equal to the maximum likelihood estimator in the normal regression model. Using $Hx = x$ and the formula for the expectation of a quadratic form (see Note on page 134), one can show that the expected value of the estimator is

$$\mathrm{E}\left(\hat{\sigma}^2\right) = \frac{n-p}{n}\sigma^2,$$

indicating that there is a downward bias. The usual unbiased estimator is $\frac{1}{n-p}\hat{e}'\hat{e}$.

Using the hat matrix, it is easy to show that

$$\mathrm{E}\left(\hat{y}|x\right) = \mathrm{E}\left(y|x\right)$$

and that

$$\mathrm{Var}\left(\hat{y}|x\right) = \mathrm{Var}\left(Hy|x\right) = HVar\left(y|x\right)H = \sigma^2 H. \tag{6.25}$$

Also,

$$\mathrm{Cov}\left(y, \hat{y}'|x\right) = \mathrm{Cov}\left(y, y'H|x\right) = \sigma^2 H. \tag{6.26}$$

For the ith record,

$$\begin{aligned}\mathrm{Cov}\left(y_i, \hat{y}_i|x\right) &= \mathrm{Cov}\left(y_i, \sum\nolimits_{j=1}^{n} h_{ij} y_j|x\right) \\ &= \mathrm{Cov}\left(y_i, h_{ii} y_i|x\right) = h_{ii}\sigma^2,\end{aligned} \tag{6.27}$$

where h_{ij} is the element in row i and column j of the hat matrix H, known as the leverage. The equality in the second line follows because $\mathrm{Cov}\left(y_i, h_{ij} y_j|x\right) = 0$ for $i \neq j$. This covariance can be interpreted as the influence that a datum has on its own prediction. It plays an important role in understanding the relationship between overfitting and the model's prediction ability and is related to the concept of effective number of parameters of a model (Hastie et al 2009).

Also,

$$\mathrm{Var}\left(\hat{y}_i|x\right) = h_{ii}\sigma^2. \tag{6.28}$$

Then it follows that conditional on x,

$$r^2_{y_i,\hat{y}_i} = \frac{\left[\mathrm{Cov}\left(y_i, \hat{y}_i|x\right)\right]^2}{\mathrm{Var}\left(y_i|x\right)\mathrm{Var}\left(\hat{y}_i|x\right)} = \frac{\mathrm{Var}\left(\hat{y}_i|x\right)}{\mathrm{Var}\left(y_i|x\right)} = h_{ii}, \tag{6.29}$$

indicating the two related interpretations of $r^2_{y_i,\hat{y}_i}$: (i) as the proportion of the variance in y explained by the linear predictor and (ii) the squared correlation between predictor $\hat{y}_i$ and predictand y_i. The ratio of variances in the second equality emphasises that $r^2_{y_i,\hat{y}_i}$ is a measure of how much the full model (6.16) improves on the reduced model that only includes an intercept, and not on how good the full model is in an absolute sense. It also indicates that the full and reduced models are nested.

The correlation can be regarded as a parameter of the joint (bivariate) distribution of fitted values $\hat{y}_i$ and the data y_i used to obtain the fitted values (the training data). Indeed, assuming normality

$$\left(\begin{matrix}\hat{y}_i \\ y_i\end{matrix}|x\right) \sim N\left[\left(\begin{matrix}x_i'\beta \\ x_i'\beta\end{matrix}\right), \left(\begin{matrix}h_{ii}\sigma^2 & h_{ii}\sigma^2 \\ h_{ii}\sigma^2 & \sigma^2\end{matrix}\right)\right] \tag{6.30}$$

with correlation between $\hat{y}_i$ and y_i equal to $\sqrt{h_{ii}}$ and with regression of $\hat{y}_i$ on y_i equal to h_{ii}.

With n records, the average squared correlation in a model with intercept is

$$\frac{\sum_{i=1}^{n} h_{ii}}{n} = \frac{tr\,(H)}{n} = \frac{p}{n},$$

indicating that the regression and the squared correlation increase towards 1 as p increases towards n reflecting overfitting. The proportion of variance explained of the training records or squared correlation between $\hat{y}_i$ and y_i depends only on the structure of the covariates for a particular data set and is a poor reflection of the model's ability to predict records from an independent set of data.

The sum of squared deviations between the observations and their fitted values (the residual sum of squares) is

$$\begin{aligned}\sum_{i=1}^{n} (y_i - \hat{y}_i)^2 &= (y - \hat{y})' (y - \hat{y}) \\ &= y' (I - H) y \\ &= \sum_{i=1}^{n} y_i^2 (1 - h_{ii}) - \sum_{i \neq j} y_i y_j h_{ij}.\end{aligned}$$

As p increases towards n, h_{ii} increases towards 1, and h_{ij}, $(i \neq j)$ and the residual sum of squares decrease towards 0.

The estimator (6.19) was obtained by minimising the residual sum of squares (6.18), here with $f(x_i) = x_i'\beta$, a linear function of the parameters. In principle, any arbitrary function passing through the data points (x_i, y_i) may represent a solution to the minimisation problem. However, we shall see that such solutions yielding perfect or nearly perfect fits obtained using a given sample of data may be poor predictors of data points from a new sample of data.

Prediction of Out-of-Sample Values

Given a value of the covariate x_0, one may wish to obtain a prediction of a yet-to-be-observed scalar y_0 (the *validating data*). The model for y_0 is

$$y_0 = x_0'\beta + e_0,$$

where e_0 is homoscedastic with variance σ^2 and assumed to be independent of x_0. This new datum is assumed to be independent of the data used to estimate β, the *training data* and drawn from the same distribution. The model for the training data is $y = x\beta + e$. The classical approach is to use an estimate of the conditional mean as predictor

$$\hat{y}_0 = \mathrm{E}\left(y_0|x_0, \beta = \hat{\beta}\right) = x_0'\hat{\beta}$$

where $\hat{\beta}$ is estimated using the training data. This is a point prediction of the average value of yet-to-be-observed data evaluated at x_0. It does not account for any form for uncertainty, but can be applied in an initial stage to compare predictors derived from different models. Conditional on training and validating data, the predictor is a point mass at $x_0'\hat{\beta}$. The variance of the predictor over conceptual replications of training data is

$$\begin{aligned}\mathrm{Var}\left(\hat{y}_0|x_0\right) &= \mathrm{Var}_{\hat{\beta}}\left[\mathrm{E}\left(\hat{y}_0|x_0, \hat{\beta}\right)\right] \\ &= x_0'\left(x'x\right)^{-1}x_0\sigma^2.\end{aligned}$$

Averaging over conceptual replications of validating data as well, the variance of the error of prediction is

$$\begin{aligned}\mathrm{Var}\left(y_0 - \hat{y}_0|x_0\right) &= \mathrm{Var}\left(y_0|x_0\right) + \mathrm{Var}\left(\hat{y}_0|x_0\right) \\ &= \sigma^2\left[1 + x_0'\left(x'x\right)^{-1}x_0\right]. \qquad (6.31)\end{aligned}$$

The covariance term cancels due to the independence of y_0 and $\hat{y}_0$. This expression has an extra term compared to (6.10c) accounting for the use of an estimate of the conditional expectation.

A Justification for the Choice of a Linear Relationship

The conditional expectation was approximated using the linear function $x\beta$. While this may appear arbitrary, some justification can be found in the following argument. If f is a smooth function, a first-order Taylor expansion about a chosen point x^* results in

$$f(x) = f\left(x^*\right) + \sum_{i=1}^{p}\left(\left.\frac{\partial f}{\partial x_i}\right|_{x=x^*}\right)\left(x_i - x_i^*\right) + \text{higher-order terms in } \left(x_i - x_i^*\right),$$

where the partial derivatives of the function become the regression coefficients. If x^* is close to x, higher-order terms are small, and the linear approximation is (locally) acceptable.

Some Caveats of the Linear Regression Model

Several issues arise when the linear regression model is a poor description of the model that generated the data:

- The first issue is concerned with the dependence of the population regression coefficients (6.6) on the distribution of the input variables x. Specifically,

$$\begin{aligned} &\mathrm{Var}(x)^{-1}\,\mathrm{Cov}(x, y) \\ &= \mathrm{Var}(x)^{-1}\,\mathrm{Cov}\left(x, x'\beta + e\right) \\ &= \mathrm{Var}(x)^{-1}\,(\mathrm{Var}(x)\,\beta + \mathrm{Cov}(x, e)) \\ &= \beta, \end{aligned}$$

  if the true model is linear and the error terms are uncorrelated with x. If this is not the case, the coefficient β is sensitive to changes in the input variable x.
- Another potential problem arises when the regression model omits variables that are part of the true model. If one postulates,

$$y = x\beta + e \tag{6.32}$$

  where e has zero mean and is independent of x, when the true model is

$$y = x\beta + z\gamma + \varepsilon \tag{6.33}$$

  where ε has zero mean and is independent of x and z, then the residual of the postulated model is $e = z\gamma + \varepsilon$. This residual may depend on x if z and x are associated. A consequence is that the estimator (6.19) is biased with respect to β. Indeed,

$$\begin{aligned} \hat{\beta} &= \left(x'x\right)^{-1} x'y \\ &= \left(x'x\right)^{-1} x'\,(x\beta + z\gamma + \varepsilon) \end{aligned}$$

  and

$$\begin{aligned} \mathrm{E}\left(\hat{\beta}|x, z\right) &= \beta + \left(x'x\right)^{-1} x'z\gamma + \mathrm{E}\,(\varepsilon|x, z) \\ &= \beta + \left(x'x\right)^{-1} x'z\gamma \end{aligned}$$

  which is biased unless $x'z = 0$.
- On the other hand, if γ in (6.33) is a random variable with zero mean so that $\mathrm{E}(y) = x\beta$ and the fitted model is (6.32) (that ignores the random variable γ), then $\mathrm{E}(\hat{\beta}|x) = (x'x)^{-1}x'\,\mathrm{E}(y) = \beta$, an unbiased estimator. The variance of

this estimator is larger than the variance of the correct, generalised least squares estimator (the blue, best linear unbiased estimator).

- Adding superfluous variables to a linear regression model does not introduce bias but increases the variance of estimation. Specifically, consider now that the true model is defined in (6.32), but we fit

$$y = x\beta_1 + z\beta_2 + \varepsilon. \tag{6.34}$$

Above, z is redundant when $\beta_2 = 0$. I sketch the proof that including z in the model does not bias the estimator of β_1, which has the same form as the estimator of β based on the true model (6.32). The least squares equations based on (6.34) are

$$\begin{bmatrix} x'x & x'z \\ z'x & z'z \end{bmatrix} \begin{bmatrix} \hat{\beta}_1 \\ \hat{\beta}_2 \end{bmatrix} = \begin{bmatrix} x'y \\ z'y \end{bmatrix}. \tag{6.35}$$

Carrying out the matrix multiplication yields after simple algebra

$$\hat{\beta}_1 = \left(x'x\right)^{-1} x' \left(y - z\hat{\beta}_2\right). \tag{6.36}$$

A little more effort (involving substituting $\hat{\beta}_1$ in terms of $\hat{\beta}_2$) leads to

$$\hat{\beta}_2 = \left[z'\left(I - H\right)z\right]^{-1} z'\left(I - H\right)y \tag{6.37}$$

where the hat matrix defined in (6.21) is

$$H = x\left(x'x\right)^{-1} x'.$$

Taking expectation under the true model in (6.32) yields as the $\mathrm{E}\left(\hat{\beta}_2 | x\right)$,

$$\begin{aligned} \mathrm{E}\left\{\left[z'\left(I - H\right)z\right]^{-1} z'\left(I - H\right)y|x\right\} &= \left[z'\left(I - H\right)z\right]^{-1} z'\left(I - H\right)x\beta \\ &= \left[z'\left(I - H\right)z\right]^{-1} z'\left(x - x\left(x'x\right)^{-1} x'x\right)\beta = 0. \end{aligned}$$

Therefore, the expectation of $\hat{\beta}_1$ in (6.36) is

$$\mathrm{E}\left[\left(x'x\right)^{-1} x'\left(y - z\hat{\beta}_2\right)|x\right] = \left(x'x\right)^{-1} x'x\beta = \beta, \tag{6.38}$$

showing that $\hat{\beta}_1$ is an unbiased estimator.

The sampling variance of $\hat{\beta}_1$ under (6.34) is proportional to the upper left hand block of the inverse of the coefficient matrix constructed on the basis of (6.34). Using results from partitioned matrices, it is easy to show that this variance is larger than the variance under the true model (6.32) (which is proportional to $\left(x'x\right)^{-1}$), unless $x'z = 0$.

- Finally, consider fitting the operational model to data y (a vector with n elements),

$$y = z\gamma + \epsilon \tag{6.39}$$

where the regression parameter $\gamma = 0$ (a scalar), when the true model is

$$y = x\beta + e, \quad \mathrm{E}(y|x) = x\beta.$$

Assume that ε has mean 0 and is independent of z and assume e is independent of z.

The least squares estimator of γ is

$$\hat{\gamma} = \left(z'z\right)^{-1} z'y = \left(\frac{1}{n}z'z\right)^{-1} \frac{1}{n}z'y$$

As n increases, $n^{-1}z'z \to \sigma_z^2$, $n^{-1}z'y \to \beta\sigma_{xz}$, and the least squares estimator converges to

$$\hat{\gamma} \to \frac{\sigma_{xz}}{\sigma_z^2}\beta. \tag{6.40}$$

This situation mimics a GWAS study, where the effects of multiple marker loci are studied by fitting one marker at a time. In this case, the causal locus is x, and z is a marker genotype that is correlated with x ($\sigma_{xz} \neq 0$; the marker is in linkage disequilibrium with x and does not have a direct effect on the dependent variable y). The consequence of the single marker regression is that neutral markers that are correlated with causal loci may be assigned phantom significant $p-$values, a problem that is aggravated with sample size (large sample increases the power of detection). This issue is briefly revisited on page 365 in connection with the computation of false discovery rates.

Least Squares Prediction as an Approximation to Best Linear Prediction

Let (x, y, e) be a random vector, where y and e are scalars on the real line $\mathbb{R}$, and vector x, of order $((p+1) \times 1)$, takes values in $\mathbb{R}^{p+1}$. The first component of $x = \left(x_0, x_1, \ldots, x_p\right)'$ is the constant $x_0 = 1$. The conditional expectation of y given x, $\mathrm{E}(y|x)$, may not be linear, but as mentioned in connection with (6.7), one can find the best linear approximation of the form $x'b$ for some choice of

$b = (b_0, b_1, \ldots, b_p)'$. This can be achieved by minimising the expected squared error

$$\mathrm{E}\left[(y - x'b)^2\right]$$

with respect to b. Differentiating with respect to b shows that the solution b^* must satisfy

$$\mathrm{E}\left[x\left(y - x'b^*\right)\right] = 0$$

and

$$b^* = \mathrm{E}\left(xx'\right)^{-1} \mathrm{E}\left(xy\right)$$

provided xx' is invertible. The scalar $x'b^*$ is the best linear predictor.

Consider an iid sequence of random vectors $(x_1, y_1), \ldots, (x_n, y_n)$ drawn from the same distribution as the random variable (x, y). If one postulates a linear model

$$y = x'b + e,$$

then the least squares estimator of b, $\hat{b}$, is the minimiser of

$$\frac{1}{n}\sum_{i=1}^{n}\left(y_i - x_i'\hat{b}\right)^2.$$

Carrying out the minimisation yields

$$\hat{b} = \left(\frac{1}{n}\sum_{i=1}^{n} x_i x_i'\right)^{-1}\left(\frac{1}{n}\sum_{i=1}^{n} x_i y_i\right).$$

Results from asymptotic theory tell us that as $n \to \infty$ sample moments converge to their population expectations and $\hat{b} \to b^*$. The message is that regardless of the form of the relationship between y and x, the least squares predictor $x'\hat{b}$ approaches the best linear predictor $x'b^*$ when sample size n is large.

The Linear Regression as a Linear Smoother

Consider the case where x_i is a scalar and assume the regression through the origin model

$$y_i = \beta x_i + e_i.$$

Then

$$\hat{\beta} = \frac{\sum_{i=1}^{N} x_i y_i}{\sum_{i=1}^{N} x_i^2}$$

and

$$\begin{aligned}\hat{f}(x_i) &= x_i \hat{\beta} \\ &= x_i \frac{\sum_{j=1}^{N} x_j y_j}{\sum_{j=1}^{N} x_j^2} \\ &= \sum_{j=1}^{N} h_{ij} y_j,\end{aligned} \tag{6.41}$$

where h_{ij} is the element in the ith row and jth column of the hat matrix H. Expression (6.41) indicates that the prediction based on least squares, $\hat{f}(x_i)$, is a smoothing of the data whereby the prediction is a weighted average of the output variables y_j with weights proportional to x_j.

The influence that output y_j has on the smoother's fitted value $\hat{f}(x_i)$ is h_{ij}. In general for a linear smoother, the influence that each data point y_i has on its own predicted value $\hat{y}_i$ is equal to $h_{ii}\sigma^2$. For the case of this regression through the origin model $h_{ii} = x_i^2 / \sum_i x_i^2, i = 1, 2, \ldots, N$.

To summarise, the general form for a linear smoother or a linear fitting method is given by

$$\hat{f}(x) = Hy, \tag{6.42}$$

where the $n \times n$ hat matrix H depends on x but not on y.

6.3 Overview of Things to Come

In this chapter, models are regarded as prediction machines as opposed to descriptions of the real world. Having collected data and obtained a predictor, one may wish to know how accurate it is for predicting future observations. The estimation of parameters used to construct the predictor is accomplished using *training data* consisting of N iid observations $\{y_i, x_i\}_{i=1}^{N}$ drawn from some joint distribution. The conditional mean and variance of the distribution of a training datum $[y_i | x_i]$ are $f(x_i)$ and σ^2, respectively. Clearly, the predictor is a function of the training data (y, x); its quality can be studied in a second stage using the *validating* or *testing data* $\{y_{v,i}, x_i\}_{i=1}^{N}$. The validating data are an independent sample from the same distribution that generated the training data (here, training and validating data are of the same length N but need not be, and the subscript v is omitted in the covariate

x_i). Let $\hat{f}(x_i)$ be the predictor of the ith data point evaluated at x_i, $i = 1, 2, \ldots, N$, obtained using the training data y. How good is $\hat{f}(x_i)$ to predict the validating datum $y_{v,i}$? A common measure of prediction ability is the *expected validating squared error* of prediction, which evaluated at x_i is given by

$$E[\text{MSE}_v(x_i)] = E_{y_v y}\left(\left(y_{v,i} - \hat{f}(x_i)\right)|x_i\right)^2. \tag{6.43}$$

(Expectations are conditional on covariables x but this is not mandatory). It will be shown that when the expectation is taken over both training and validating data, this measure of prediction ability of a future record has three terms:

$$E[\text{MSE}_v(x_i)] = \text{Var}\left(y_{v,i}|x_i\right) + \text{Var}\left(\hat{f}(x_i)|x_i\right) + \text{bias}^2, \tag{6.44}$$

where the squared bias is defined as $\left(E_{y_v}\left(y_{v,i}|x_i\right) - E_y\left(\hat{f}(x_i)\right)\right)^2$. The first term, $\text{Var}\left(y_{v,i}|x_i\right)$, is the conditional variance of a future record, also labelled irreducible variance since it reflects sampling variation beyond the control of the experimenter. The second term $\text{Var}\left(\hat{f}(x_i)|x_i\right)$ is the variance of the predictor. It describes sampling uncertainty associated with estimation of $\hat{f}$ due to finite size of the training data. The bias term reflects average systematic discrepancy between the predictor and the predictand. Increasing the number of parameters of a model tends to reduce bias and to increase the variance of the predictor. The choice of a prediction model often involves a trade-off involving these two terms.

Expression (6.44) cannot be obtained directly because parameters are not known. In practice, the sample *mean squared error* is computed. For a particular realisation of validating data, the sample validating *mean squared error* is

$$\text{MSE}_v = \frac{1}{N}\sum_{i=1}^{N}\left(y_{v,i} - \hat{f}(x_i)\right)^2, \tag{6.45}$$

where $\hat{f}$ is the predictor calculated using the training data. The *expected mean squared error* in the validating data is

$$E_{y_v y}(\text{MSE}_v), \tag{6.46}$$

with a corresponding expression for the *expected mean squared error* in the training data

$$E_y(\text{MSE}_t) = E_y\left[\frac{1}{N}\sum_{i=1}^{N}\left(y_i - \hat{f}(x_i)|x_i\right)^2\right]. \tag{6.47}$$

The expectation in (6.46) is taken over conceptual replications of training and validating data and in (6.47) over the training data. Monte Carlo estimates of these expectations can be obtained using the bootstrap or simulating the data, given a parametric model. Monte Carlo methods generate estimates of the complete sampling distribution of MSE. An example is discussed on page 430.

An important theme of this chapter is the distinction between the predictive ability of $\hat{f}(x_i)$ quantified using either the training data used to estimate $\hat{f}(x_i)$ or the validating data. It will be shown that the former overestimates predictive performance in the validating data, on average, as reflected in the expectations (6.46) and (6.47). These are given by

$$E_{y_v y}(\text{MSE}_v) = \text{Var}\left(y_{v,i}|x_i\right) + \text{bias}^2 + \frac{1}{N}\text{Var}\left(\hat{f}(x_i)|x_i\right)$$

and

$$E_{y_v y}(\text{MSE}_t) = E_{y_v y}(\text{MSE}_v) - \frac{2}{N}\sum_i \text{Cov}\left(y_i, \hat{f}(x_i)\right).$$

The covariance term in the second equation was introduced in (6.27) and reflects the influence that observation i has on its own prediction. It is the amount by which, on average, the training mean squared error underestimates the validating mean squared error and is labelled *expected optimism* by Efron (1986). The following section elaborates on these ideas.

6.4 The Bias-Variance Trade-Off

Consider two random variables (Y, X) that have a joint distribution and suppose that one wishes to predict Y from X using a function evaluated at $X = x$, $f(x)$. This unknown function specifies the true association between Y and X and is estimated using N training observations giving rise to the predictor $\hat{f}$. The N training observations are realisations from the joint distribution $[Y, X]$. The model for datum i in the training data is

$$y_i = f(x_i) + e_i, \quad e_i \overset{iid}{\sim} (0, \sigma^2), \quad i = 1, 2, \ldots, N,$$

where residuals e have mean zero and variance σ^2 independent of x. The assumption of normality at this stage is not required. According to the model, $E(y_i|x_i) = f(x_i)$.

Consider another sample (y_0, x_0) drawn independently from the same joint distribution $[Y, X]$. This constitutes the validating data. How well does the function $\hat{f}$ constructed using the training data $(y_1, x_1), \ldots, (y_N, x_N)$ predict the new test point y_0 at the input point x_0? The predicted value is $\hat{y}_0 = \hat{f}(x_0)$. For a particular datum and its prediction, the *validating mean squared error* is

$$\mathrm{MSE}_v = \left(y_0 - \hat{y}_0\right)^2 = y_0^2 + \hat{y}_0^2 - 2y_0\hat{y}_0, \tag{6.48}$$

and the measure of prediction ability of $\hat{y}_0$ under enquiry is the *expected validating mean squared error* evaluated at $X = x_0$, denoted $\mathrm{E}\left[\mathrm{MSE}_v\left(x_0\right)\right]$.

The expected validating mean squared error is studied in three scenarios. These determine whether sampling uncertainty of the predictor and/or sampling uncertainty of new records contribute to the expectation of the validating mean squared error.

Training Data Treated as Fixed and Validating Data as Random

The first scenario accounts for variation of single records; expectations are taken over the distribution of validating data, conditional on the training data y. Therefore, the predictor $\hat{y}_0$ is regarded as a fixed constant. In this case,

$$\begin{aligned}
\mathrm{E}_{y_0}\left[\mathrm{MSE}_v\left(x_0\right)\right] &= \mathrm{E}\left(y_0^2\right) + \hat{y}_0^2 - 2\hat{y}_0\,\mathrm{E}\left(y_0\right) \\
&= \mathrm{Var}\left(y_0\right) + \left(\mathrm{E}\left(y_0\right)\right)^2 + \hat{y}_0^2 - 2\hat{y}_0\,\mathrm{E}\left(y_0\right) \\
&= \mathrm{Var}\left(y_0\right) + \left(\mathrm{E}\left(y_0\right) - \hat{y}_0\right)^2 .
\end{aligned} \tag{6.49}$$

The first term accounts for the conditional sampling variance (given covariates) of new validating data, and the second reflects the squared discrepancy between the predictor $\hat{y}_0$ and the mean of the distribution of validating data. The squared discrepancy term depends on the particular realisation of the training data used to obtain $\hat{y}_0$.

The focus is on the quality of a predictor, measured by $\mathrm{E}(\mathrm{MSE}_v)$, computed from a single sample of training data, measured over conceptual replications of validating data. Single observations are predicted accounting for their sampling uncertainty but ignoring sampling uncertainty of the parameters that index the predictor.

Training Data Treated as Random and Validating Data as Fixed

The second scenario accounts for variation of the predictor, and the mean validating squared error is averaged over conceptual replications of training data, keeping validating data fixed. Using similar algebra leads to

$$\mathrm{E}_y\left[\mathrm{MSE}_v\left(x_0\right)\right] = \mathrm{Var}\left(\hat{y}_0\right) + \left(\mathrm{E}\left(\hat{y}_0\right) - y_0\right)^2 \tag{6.50}$$

indicating that there is a component from the sampling variation of the predictor over conceptual replications of training data and a squared discrepancy between the validating data y_0 and the mean of the predictor (averaged over training data).

The focus here is on the quality of a predictor, computed repeatedly from conceptual replications of training data, measured in a single sample of validating data. Average values of new observations, all evaluated at $X = x_0$, are predicted, accounting for sampling variation of the parameters that index the predictor.

Training and Validating Data Are Treated as Random

The third scenario accounts for variation of single records and of the predictor, and the average performance of the MSE_v is studied over conceptual replications of training and validating data, given $X = x_0$, as in (6.44):

$$\begin{aligned} E_{yy_0}\left[\text{MSE}_v(x_0)\right] &= E_{y_0}\left(y_0^2\right) + E_y\left(\hat{y}_0^2\right) - 2\,E_{yy_0}\left(y_0\hat{y}_0\right) \\ &= \text{Var}(y_0) + (E(y_0))^2 + \text{Var}\left(\hat{y}_0\right) + \left(E\left(\hat{y}_0\right)\right)^2 - 2\,E(y_0)\,E\left(\hat{y}_0\right) \\ &= \text{Var}(y_0) + \text{Var}\left(\hat{y}_0\right) + \left(E(y_0) - E\left(\hat{y}_0\right)\right)^2 \end{aligned} \tag{6.51}$$

where the second line is the result of using $E_{yy_0}\left(y_0\hat{y}_0\right) = E(y_0)\,E\left(\hat{y}_0\right)$ due to the independence of training and validating data in this simple least squares setting.

The focus here is on the quality of a predictor, computed repeatedly from conceptual replications of training data and measured over conceptual replications of validating data. Single observations are predicted, accounting for their sampling uncertainty and also accounting for sampling uncertainty of the parameters that index the predictor.

Expression (6.51) indicates how the expected validating mean squared error $E_{yy_0}\left[\text{MSE}_v(x_0)\right]$ evaluated at x_0 is affected by the inherent variation of y_0, by the variance of the predictor $\text{Var}\left(\hat{y}_0\right)$ that describes how the values of $\hat{y}_0$ for the different training data sets vary around their average and the squared bias of the predictor. As the complexity of a model increases, the variance of the predictor tends to increase, and the bias term tends to decrease. In practice, most operational models are distortions of the true state of nature and hence are biased. In a prediction setup, one often chooses the degree of complexity of a predictive model to trade bias off with variance in such a way as to minimise $E\left[\text{MSE}_v(x_0)\right]$. Therefore, choosing a predictive model amounts to striking a balance between bias and variance. This is particularly relevant with the advent of new technologies that allows vast amounts of covariates to be collected. Even when the covariates are relevant, the variance incurred in estimating their effects may outweigh the reduction in bias. The “true model” may not lead to the best predictive tool.

For a given number of observations, positively correlated data due to underlying random effects, such as family effects, lead to a similar decomposition as (6.51) except that the terms $\text{Var}(y_0)$ (conditional on covariates but averaged over the distribution of random effects) and $\text{Var}\left(\hat{y}_0\right)$ are larger, and often validating and training data are correlated resulting in $E_{yy_0}\left(y_0\hat{y}_0\right) = \text{Cov}\left(y_0, \hat{y}_0\right) + E(y_0)\,E\left(\hat{y}_0\right)$.

The three scenarios described above, developed from a frequentist perspective, have a Bayesian counterpart where all expectations are conditional on training and validating data. The topic is discussed in Chapter 10, in the Exercises on Prediction on pages 562 and 651 and also on page 430. Bayesian expectations of mean squared error are derived on page 428.

Briefly, with y_0 representing validating data, y representing training data and θ the parameter of the model used to construct the predictor $\hat{y}_0$ of the validating data, the Bayesian scenario 1 corresponds to drawing the predictor $\hat{y}_0$ from $\left[y_0|\hat{\theta}, y, x_0\right]$, where $\hat{\theta}$ is some point estimate of θ. In this situation, account is taken of sampling variation of the predictor of single records $\hat{y}_0$, and this variation is one of the components of the Bayesian expected validating mean squared error. By replicating data from $\left[y_0|\hat{\theta}, y, x_0\right]$, we are studying the frequency properties of the Bayesian procedure.

The second scenario corresponds to constructing the predictor of an average of records using a function f, $\hat{y}_0 = f(\theta^\star, x_0)$, where $\theta^\star$ is a draw from the posterior distribution $[\theta|y]$. In this case, account is taken of posterior uncertainty of θ, and this uncertainty is a component of the Bayesian expected validating mean squared error.

The third scenario takes care of uncertainty of θ and of sampling variation of a new record $\hat{y}_0$. In this case, $\theta^\star$ is drawn from $[\theta|y]$, and given $\theta^\star$, $\hat{y}_0$ is drawn from $\left[y_0|\theta^\star, y, x_0\right]$. Both sources of uncertainty are incorporated in the Bayesian expected validating mean squared error.

An alternative description of the three scenarios is as follows: Each scenario gives rise to a predictor that is characterised by its distribution. The mean squared errors are transformations with their own distributions that here are summarised by their expected value. This general narrative holds for the frequentist and Bayesian approaches with the obvious caveat that what is considered random and uncertain and what is considered non-random is peculiar to each approach to inference.

An attraction of the Bayesian approach implemented in an McMC environment is the ease with which the marginal posterior distribution of MSE can be estimated. If two models are evaluated in terms of their predictive ability, the computation of the marginal posterior distribution of the difference in MSE between the models only requires a few extra lines of code. This can be particularly relevant when the merits of different prediction machines are to be judged on the basis of relatively small differences. However, a usual way of comparing models is to estimate parameters and to use these to construct the point predictor ignoring uncertainty.

6.5 Estimation of Validation MSE of Prediction in Practice

The measure of prediction ability often used in practice is the sample mean squared error. A possibility is to use the same training data $\{y_i, x_i\}_{i=1}^N$, from which the predictions $\hat{f}(x_i)$ were estimated, in order to compute the sample *training mean squared error* :

$$\text{MSE}_t = \frac{1}{N}\sum_{i=1}^{N}\left(y_i - \hat{f}(x_i)\right)^2. \tag{6.52}$$

This is the average squared discrepancy between observations and their predictions. The training mean squared error is not a good measure of prediction ability because it has the property of decreasing consistently with model complexity. A model with close to zero training squared error is overfit to the training data, as it captures not only true signal but also the noise peculiar to the training data and will predict new, independent observations poorly.

A better proposition for model validation or model selection would be to study the predictive ability of $\hat{f}$, computed using the training data, in an independent set of data: the testing or validating data $\{y_{v,i}, x_{v,i}\}_{i=1}^{N}$. This validating data is conceptually a new realisation from the same distribution that generated the training data. Using the validating data, one can compute

$$\text{MSE}_v = \frac{1}{N}\sum_{i=1}^{N}\left(y_{v,i} - \hat{f}\left(x_{v,i}\right)\right)^2, \tag{6.53}$$

the sample *testing or validating mean squared error*. However, with limited data or without access to real validating data, one must resort to alternative procedures. These consist of artificially splitting the data into training and validating sets or to use approximations derived from theory without splitting the available data.

One possibility is to use *leave-one-out cross-validation*. Each of the N observations (y_i, x_i) is held out in turn, and the function $\hat{f}^{(-i)}$ is estimated with the remaining $N-1$ records from the training data, with the ith pair (y_i, x_i) omitted. Then the squared discrepancy between y_i and $\hat{f}^{(-i)}(x_i)$ is computed, and this is repeated N times. The resulting estimate of the leave-one-out prediction mean squared error is

$$\text{MSE}_1 = \frac{1}{N}\sum_{i=1}^{N}\left(y_i - \hat{f}^{(-i)}(x_i)\right)^2. \tag{6.54}$$

A more common practice is to use *K–fold cross-validation* whereby the complete set of N records is divided into K groups or folds of approximately equal size. For the kth fold, the model is fitted to the other $K-1$ folds of the data (this yields $\hat{f}^{(-k)}$ the estimate of the predictive function based on the $K-1$ folds of the data with the kth fold excluded), and the prediction mean squared error of the fitted model is estimated using data from the kth fold. This is repeated for the K groups, and the K estimates of prediction mean squared error are combined to obtain the cross-validation estimate of prediction mean squared error. Specifically, *K–fold cross-validation* proceeds as follows:

1. Divide the N records into K folds of approximately equal size n_k; $N = \sum_{k=1}^{K} n_k$.
2. Fit the model to the remaining $K-1$ parts of the data (with fold k excluded), and obtain the estimate $\hat{f}^{(-k)}$.

3. Using the records from the kth fold only, estimate the mean squared error:

$$\mathrm{MSE}_k\left(\hat{f}^{(-k)}\right) = \frac{1}{n_k} \sum_{i \in k\text{th } fold} \left(y_i - \hat{f}^{(-k)}(x_i)\right)^2$$

4. Repeat using all the K folds to obtain $\mathrm{MSE}_1\left(\hat{f}^{(-1)}\right) \ldots \mathrm{MSE}_K\left(\hat{f}^{(-K)}\right)$.
5. Obtain the $K-$fold cross-validation combined (weighted according to fold size n_k) estimate of mean squared error:

$$\mathrm{MSE}_K = \frac{1}{N} \sum_{k=1}^{K} n_k \, \mathrm{MSE}_k\left(\hat{f}^{(-k)}\right). \tag{6.55}$$

If all groups are of equal size n, $N = Kn$, and this reduces to

$$\mathrm{MSE}_K = \frac{1}{K} \sum_{k=1}^{K} \mathrm{MSE}_k\left(\hat{f}^{(-k)}\right). \tag{6.56}$$

Common choices for K are $K = 5$ or $K = 10$. If $n = 1$, the procedure recovers *leave-one-out cross-validation*. The $K-$fold cross-validation may be repeated several times by reconstructing the folds at random to obtain estimates of the distribution of MSE_K.

Compared to $K-$fold cross-validation, leave-one-out cross-validation can suffer from highly correlated out-of-sample predictions. Indeed, with N observations, there are N forecasts, and any two training sets share $N-2$ points. Therefore, despite averaging N forecasts, these tend to be highly correlated, and therefore the leave-one-out estimator tends to have high variance (taken over conceptual replications of training data). With $K-$fold cross-validation, the folds share relatively fewer observations and are less correlated. This tends to reduce the variance of the estimator. However, a relatively small number of the observations in the training data are included to estimate parameters, and the $K-$fold estimator may be biased upwards with respect to the squared error obtained using the entire data set. This is not the case for leave-one-out cross-validation, since the method repeatedly trains the model on $N - 1$ observations, almost as many as in the entire training data.

Implementation Shortcut for Leave-One-Out Cross-Validation

For linear smoothers or linear fitting methods as defined in connection with (6.21), a computational advantage of leave-one-out cross-validation is that it requires running through the training data only once, rather than N times (Seber and Lee 2003). This relies on a little trick that works as follows: From (6.21), the ith predicted value at x_i is

$$\hat{y}_i = h_{i1} y_1 + \cdots + h_{ii} y_i + \cdots + h_{iN} y_N$$

with $\sum_{j=1}^{N} h_{ij} = 1$ when the model includes an intercept. Eliminating the ith datum and rescaling so that the $N-1$ weights sum to one yield

$$\hat{y}_{-i} = \frac{\hat{y}_i - h_{ii} y_i}{1 - h_{ii}}.$$

A more formal proof of this identity can be found in Seber and Lee (2003).

The leave-one-out residual or prediction error is

$$\begin{aligned} r_i &= y_i - \hat{y}_{-i} \\ &= y_i - \frac{\hat{y}_i - h_{ii} y_i}{1 - h_{ii}} \\ &= \frac{y_i (1 - h_{ii}) - (\hat{y}_i - h_{ii} y_i)}{1 - h_{ii}} \\ &= \frac{y_i - \hat{y}_i}{1 - h_{ii}} \end{aligned} \tag{6.57}$$

and the leave-one-out cross-validation prediction mean squared error (6.54) reduces to

$$MSE_1 = \frac{1}{N} \sum_{i=1}^{N} (y_i - \hat{y}_{-i})^2 = \frac{1}{N} \sum_{i=1}^{N} \left(\frac{y_i - \hat{y}_i}{1 - h_{ii}} \right)^2, \tag{6.58}$$

that also holds when the model does not have an intercept (Seber and Lee 2003). The calculation of (6.58) requires one to pass through the training data rather than the N implied in the left-hand side.

A computationally faster approximation, known as the *generalised cross-validation*, is

$$\frac{1}{N} \sum_{i=1}^{N} (y_i - \hat{y}_{-i})^2 = \frac{1}{N \left(1 - N^{-1} \operatorname{tr}(H)\right)^2} \sum_{i=1}^{N} (y_i - \hat{y}_i)^2 \tag{6.59}$$

whereby rather than weighting each term by 1 minus the ith diagonal element of the hat matrix H, all terms are given the same weight (1 minus the average of the diagonal elements of H). The generalised cross-validation can be a numerically more robust alternative to (6.58) in cases where a few records may have a strong influence due to the terms $1 - h_{ii}$ in the denominator.

Using the approximation

$$\frac{1}{(1-x)^2} \approx 1 + 2x$$

allows expressing (6.59) as

$$\frac{1}{N}\sum_{i=1}^{N}\left(y_i - \hat{y}_{-i}\right)^2 \approx \frac{\sum_{i=1}^{N}\left(y_i - \hat{y}_i\right)^2}{N}\left(1 + 2\frac{\mathrm{tr}\,(H)}{N}\right). \tag{6.60}$$

For the least squares linear regression with p parameters, $tr\,(H) = p$, and (6.60) reduces to

$$\frac{1}{N}\sum_{i=1}^{N}\left(y_i - \hat{y}_{-i}\right)^2 \approx \frac{\sum_{i=1}^{N}\left(y_i - \hat{y}_i\right)^2}{N}\left(1 + 2\frac{p}{N}\right) \tag{6.61}$$

which establishes a connection between leave-one-out cross-validation and Mallows' C_p (Mallows 1973; see (6.75) below).

This is an instance where prediction ability is evaluated using training data only, without using cross-validation.

6.6 On Average Training MSE Underestimates Validation MSE

The within sample prediction ability based on $\hat{f}$ will be overstated, in expectation, relative to the out-of-sample prediction, leading to $\mathrm{MSE}_t < \mathrm{MSE}_v$. The algebra underlying this result is simple and illuminating and is described in this section.

Independently Distributed Data

Let $\{y_i, x_i\}_{i=1}^{N}$ and $\left\{y_{v,i}, x_{v,i}\right\}_{i=1}^{N}$ denote the N observations in the training and validating data, respectively. Assume that under the true model, the observations have means $f(x_i)$ and variance σ^2; that is,

$$y_i = f(x_i) + e_i,\ e_i \sim \left(0, \sigma^2\right),$$

$$y_{v,i} = f\left(x_{v,i}\right) + e_{v,i},\ e_{v,i} \sim \left(0, \sigma^2\right).$$

In the derivation that follows, (y_i, x_i) and $(y_{v,i}, x_{v,i})$ are different realisations from the same probability model. The random quantities are $(y_i, y_{v,i})$, whereas the covariates x are treated as fixed and known. Importantly, records are assumed to be independent. The modification needed for correlated data is discussed below.

The true model is unknown, and a number of operational or instrumental models may be available to choose as prediction machines. The prediction ability will be studied specifically as follows:

1. Fit the operational model to the training data $\{y_i, x_i\}$, and obtain predictive values for the validating data $\hat{y}_{v,i} = \hat{f}(x_{v,i})$, $i = 1, 2, \ldots, N$.
2. Obtain predictions for the training records $\hat{y}_i = \hat{f}(x_i)$, $i = 1, 2, \ldots, N$, and compute the training mean squared error

$$\text{MSE}_t = \frac{1}{N}\sum_i (y_i - \hat{y}_i)^2 \tag{6.62}$$

3. Use the validating data $\{y_{v,i}, x_{v,i}\}$ to compute the testing or validating mean squared error:

$$\text{MSE}_v = \frac{1}{N}\sum_i (y_{v,i} - \hat{y}_{v,i})^2 \tag{6.63}$$

The expected value of MSE_v is derived first. Taking expectations of (6.63) with respect to the distribution of (y, y_v)

$$\text{E}_{y,y_v}\left[\frac{1}{N}\sum_i (y_{v,i} - \hat{y}_{v,i})^2\right] = \frac{1}{N}\sum_i \text{E}_{y_i y_{v,i}} (y_{v,i} - \hat{y}_{v,i})^2 \tag{6.64}$$

The squared term is expanded as in (6.48) and going through the same algebra used in the derivation of (6.51) leads to the expected validating mean squared error:

$$\begin{aligned}
\text{E}_{y,y_v}(\text{MSE}_v) &= \text{E}_{y,y_v}\left[\frac{1}{N}\sum_i (y_{v,i} - \hat{y}_{v,i})^2\right] \\
&= \frac{1}{N}\sum_i \sigma^2 + \frac{1}{N}\sum_i \text{bias}^2(i) + \frac{1}{N}\sum_i \text{Var}(\hat{y}_{v,i}) \\
&= \sigma^2 + \text{bias}^2 + \frac{1}{N}\sum_i \text{Var}(\hat{y}_{v,i}),
\end{aligned} \tag{6.65}$$

where $\text{bias}^2(i) = \left(\text{E}(y_{v,i}) - \text{E}(\hat{y}_{v,i})\right)^2$.

The expected value of the training mean squared error is derived in a similar manner. Taking expectations of (6.62) with y random and covariates x fixed yields

$$\mathrm{E}_y(\mathrm{MSE}_t) = \mathrm{E}_y\left[\frac{1}{N}\sum_i (y_i - \hat{y}_i)^2\right] = \frac{1}{N}\sum_i \mathrm{E}_{y_i}(y_i - \hat{y}_i)^2 .$$

Expanding the square results in

$$\mathrm{E}_{y_i}(y_i - \hat{y}_i)^2 = \mathrm{E}_{y_i}\left(y_i^2 + \hat{y}_i^2 - 2y_i\hat{y}_i\right)^2$$

$$= \mathrm{Var}(y_i) + (\mathrm{E}(y_i))^2 + \mathrm{Var}(\hat{y}_i) + (\mathrm{E}(\hat{y}_i))^2 - 2\,\mathrm{Cov}(y_i, \hat{y}_i) - 2\,\mathrm{E}(y_i)\,\mathrm{E}(\hat{y}_i)$$

$$= \mathrm{Var}(y_i) + \mathrm{Var}(\hat{y}_i) + (\mathrm{E}(y_i) - \mathrm{E}(\hat{y}_i))^2 - 2\,\mathrm{Cov}(y_i, \hat{y}_i) .$$

Therefore,

$$\mathrm{E}_y(\mathrm{MSE}_t) = \sigma^2 + \frac{1}{N}\sum_i \mathrm{Var}(\hat{y}_i) + \mathrm{bias}^2 - \frac{2}{N}\sum_i \mathrm{Cov}(y_i, \hat{y}_i) \tag{6.66}$$

and from (6.65)

$$\mathrm{E}_y(\mathrm{MSE}_t) = \mathrm{E}_{yy_v}(\mathrm{MSE}_v) - \frac{2}{N}\sum_i \mathrm{Cov}(y_i, \hat{y}_i) , \tag{6.67}$$

Efron (1986) indicating that on average, the training MSE_t underestimates validating MSE_v especially if y_i and $\hat{y}_i$ are highly correlated. This happens when the operational model has a large number of parameters that result in a very close fit, with little discrepancy between observations y_i and fitted values $\hat{y}_i$. The term $\mathrm{Cov}(y_i, \hat{y}_i)$ is not observable and unless an analytic form is available (as for linear smoothers) must be estimated using, for example, the bootstrap or Monte Carlo. This subject is revisited in the Examples Section, page 570.

The form of (6.67) encompasses two terms, reflecting model fit, $\mathrm{E}_y(\mathrm{MSE}_t)$ and model complexity, $\sum_i \mathrm{Cov}(y_i, \hat{y}_i)$. The second term on the right-hand side of (6.67) is labelled *expected optimism* by Efron (1986) :

$$\mathrm{E}_{yy_v}(\mathrm{MSE}_v) - \mathrm{E}_y(\mathrm{MSE}_t) = \frac{2}{N}\sum_i \mathrm{Cov}(y_i, \hat{y}_i) = \frac{2}{N}\,\mathrm{tr}\left(\mathrm{Cov}(y, \hat{y}')\right), \tag{6.68}$$

the difference between the expected validating mean squared error and the expected training mean squared error.

Invoking the law of large numbers, a natural estimate of the expected validating mean squared error obtained using the training data is

$$\hat{\mathrm{E}}\left(\mathrm{MSE}_v\right) = \frac{1}{N}\sum_{i=1}^{N}\left(y_i - \hat{y}_i\right)^2 + \frac{2}{N}\sum_{i=1}^{N}\widehat{\mathrm{Cov}}\left(y_i, \hat{y}_i\right). \tag{6.69}$$

The first term on the right-hand side is a sample estimate of $\mathrm{E}_{yy_v}\left(\mathrm{MSE}_t\right)$, and $\widehat{\mathrm{Cov}}\left(y_i, \hat{y}_i\right)$ is some estimate obtained using Monte Carlo simulation or the bootstrap. Expression (6.69) obviates estimation of σ^2 and is simple to compute if an analytic form is available for $\mathrm{Cov}\left(y_i, \hat{y}_i\right)$. Otherwise, one incurs the cost of a bootstrap or a Monte Carlo analysis. An important issue with estimator (6.69) is that the rate of convergence to its expectation depends on the ratio p/N.

An estimate of expected optimism provides an approximation to the evaluation of prediction ability without incurring the cost of cross-validation.

Correlated Data

Often, data are correlated such as in genetic studies where family structure is a classical source of correlation. The correlation structure can be patterned and simple as in independent full-sib families, where correlation arises between members of the same family, or can be highly complex with pedigrees spanning several overlapping generations.

With correlated data, the expected training mean squared error (6.66) is partitioned into the same four components, while the expectation of the validating mean squared error (6.65) has an extra term that describes the covariance between an observation in the validating set and its predictor. The degree of this association depends on the correlation structure in the data. With highly correlated data as found in animal breeding, a predicted value is constructed using information in the training data from many related individuals, and the validating datum correlates with these training data. The expected value of the validating mean squared error takes the form

$$\mathrm{E}_{y,y_v}\left(\mathrm{MSE}_v\right) = \sigma^2 + \mathrm{bias}^2 + \frac{1}{N}\sum_{i}\mathrm{Var}\left(\hat{y}_{v,i}\right) - \frac{2}{N}\sum_{i=1}^{N}\mathrm{Cov}\left(y_{v,i}, \hat{y}_{v,i}\right) \tag{6.70}$$

and the expected optimism is

$$\mathrm{E}_{y,y_v}\left(\mathrm{MSE}_v\right) - \mathrm{E}_y\left(\mathrm{MSE}_t\right) = \frac{2}{N}\left[\sum_{i=1}^{N} \mathrm{Cov}\left(y_i, \hat{y}_i\right) - \sum_{i=1}^{N} \mathrm{Cov}\left(y_{v,i}, \hat{y}_{v,i}\right)\right]. \tag{6.71}$$

Here, it is assumed that training and validating data are of size N.

Estimating Optimism of the Training Sample

Often, the covariance term $\mathrm{Cov}\left(y_i, \hat{y}_i\right)$ cannot be obtained analytically due to the nature of the fitting procedure but can be estimated either using the nonparametric or the (model dependent) parametric bootstrap (Efron and Hastie 2016). The parametric approach is easier to use when data are correlated. In the parametric bootstrap, repeated samples of the vector of data are drawn from the parametric model used to analyse the original observations, conditional on parameter estimates. The latter are obtained using these original observations.

If a nonparametric bootstrap is chosen, at least one of two strategies can be followed. Assume that the training data consist of independent samples $(x_1, y_1), (x_2, y_2), \ldots, (x_N, y_N)$. Then a **first** bootstrap strategy is to cycle through a loop from $b = 1$ to B ($B = 1000$, say)

1. Draw N bootstrap samples (i.e., sampling with replacement from the data)

$$\left(x_i^b, y_i^b\right), \quad i = 1, \ldots, N,$$

   calculate the regression function $\hat{f}^b$ using $\left(x_i^b, y_i^b\right), \quad i = 1, \ldots, N$, and obtain the fitted values $\hat{y}_i^b = \hat{f}^b\left(x_i^b\right), \quad i = 1, \ldots, N$.
2. Store $y_i^b, \quad i = 1, \ldots, N$ and $\hat{y}_i^b, \quad i = 1, \ldots, N$
3. After B cycles, calculate the empirical covariance between y_i and $\hat{y}_i$

$$\widehat{\mathrm{Cov}}\left(y_i, \hat{y}_i\right) = \frac{1}{B}\sum_{b=1}^{B}\left(y_i^b - \overline{y}_i^b\right)\left(\hat{y}_i^b - \overline{\hat{y}}_i^b\right),$$

   where

$$\overline{y}_i^b = \frac{1}{B}\sum_{b=1}^{B} y_i^b, \quad \overline{\hat{y}}_i^b = \frac{1}{B}\sum_{b=1}^{B} \hat{y}_i^b.$$

4. Finally, sum these over $i = 1, \ldots, N$ to obtain the bootstrap estimate

$$\sum_{i=1}^{N} \widehat{\text{Cov}}\left(y_i, \hat{y}_i\right).$$

For continuous data, a **second** bootstrap strategy is to use the *residual bootstrap* that proceeds as follows:

1. Using the training data $(x_1, y_1), (x_2, y_2), \ldots, (x_N, y_N)$, calculate $\hat{y}_i = \hat{f}(x_i), \quad i = 1, \ldots, N$, and compute the empirical residuals

$$\hat{e}_i = y_i - \hat{y}_i, \quad i = 1, \ldots, N.$$

Then cycle through the loop from $b = 1$ to B:
2. Draw N samples with replacement of the empirical residuals

$$\hat{e}_i^b, \quad i = 1, \ldots, N. \tag{6.72}$$

3. Using the N boostrap residuals (6.72), calculate

$$y_i^b = \hat{y}_i + \hat{e}_i^b, \quad i = 1, \ldots, N.$$

4. Using $\left(x_i, y_i^b\right), \quad i = 1, \ldots, N$, calculate the regression function $\hat{f}^b$, and obtain the fitted values $\hat{y}_i^b = \hat{f}^b\left(x_i^b\right), \quad i = 1, \ldots, N$. Notice that the x_i are not sampled; only the residuals are sampled to compute $y_i^b = \hat{y}_i + \hat{e}_i^b$.
5. Store $y_i^b, \quad i = 1, \ldots, N$ and $\hat{y}_i^b, \quad i = 1, \ldots, N$
6. After B cycles, calculate the empirical covariance between y_i^b and $\hat{y}_i^b$

$$\widehat{\text{Cov}}\left(y_i^b, \hat{y}_i^b\right) = \frac{1}{B} \sum_{b=1}^{B} \left(y_i^b - \overline{y}_i^b\right)\left(\hat{y}_i^b - \overline{\hat{y}}_i^b\right)$$

where

$$\overline{y}_i^b = \frac{1}{B} \sum_{b=1}^{B} y_i^b, \quad \overline{\hat{y}}_i^b = \frac{1}{B} \sum_{b=1}^{B} \hat{y}_i^b$$

7. Finally, sum these over $i = 1, \ldots, N$ to obtain the bootstrap estimate

$$\sum_{i=1}^{N} \widehat{\text{Cov}}\left(y_i, \hat{y}_i\right). \tag{6.73}$$

An alternative approach to use when data are correlated or have a complicated structure is the parametric bootstrap. In the Prediction Exercises, page 570, you are asked to derive an expression for the expected optimism for a hierarchical model, to obtain numerical results using simulated data and to write a code to obtain a Monte Carlo estimate of expected optimism.

The two nonparametric bootstrap strategies described in this section are drawn from lecture notes of the course Advanced Methods for Data Analysis (R. Tibshirani) and can be found at https://www.stat.cmu.edu/~ryantibs/advmethods/.

6.7 Least Squares Prediction

The form of covariance penalty $\text{Cov}\left(y_i, \hat{y}_i\right)$ depends on the operational model used for prediction. Suppose that the operational model takes the standard regression form (6.17) and that β (of dimension p) is estimated by least squares. In this case as indicated in (6.26)

$$\text{Cov}\left(y_i, \hat{y}_i\right) = h_{ii}\sigma^2$$

where h_{ii} is the ith diagonal element of the hat matrix H. Then the penalty term is

$$\begin{aligned}
\frac{2}{N}\sum_i \text{Cov}\left(y_i, \hat{y}_i\right) &= \frac{2}{N}\,\text{tr}\,(H)\,\sigma^2 \\
&= \frac{2}{N}\,\text{tr}\left(X\left(X'X\right)^{-1}X'\right)\sigma^2 \\
&= \frac{2}{N}\,\text{tr}\left(\left(X'X\right)^{-1}X'X\right)\sigma^2 \\
&= 2\frac{p}{N}\sigma^2
\end{aligned}$$

and (6.67) takes the form

$$\text{E}_{yy^v}\left(\text{MSE}_t\right) = \text{E}_{yy^v}\left(\text{MSE}_v\right) - 2\frac{p}{N}\sigma^2. \tag{6.74}$$

In view of this result and (6.69), a natural unbiased estimator of the validating mean squared error for the standard regression model, using the training mean squared error, could be

$$\begin{aligned}
\hat{\text{E}}\left(\text{MSE}_v\right) &= \frac{1}{N}\sum_{i=1}^{N}\left(y_i - \hat{y}_i\right)^2 + \frac{2}{N}\,\text{tr}\,(H)\,\sigma^2 \\
&= \frac{1}{N}\sum_{i=1}^{N}\left(y_i - \hat{y}_i\right)^2 + 2\frac{p}{N}\sigma^2
\end{aligned} \tag{6.75}$$

which is a version of Mallows' C_p (Mallows 1973). When σ^2 is not known, it must be replaced by an estimate. The unbiased property holds if an unbiased estimator of σ^2 is available.

Example: Prediction Ability of the Least Squares Estimator

Some of the ideas discussed so far are illustrated using simulated data. These consist of $N = 55$ observations from a training set and $N = 55$ from a testing or validating set. The structure of the simulated data (not its size!) is representative of a genomic data set. A matrix X of genetic marker genotypes of dimension $N \times p_{\text{max}}$ (centred and scaled) was generated representing genotype codes whose elements consist of *iid* (identically and independently distributed) draws from a binomial distribution with parameters $n = 2$ and probability 0.5. The number of columns of X in the example is set to $p_{\text{max}} = 50$, and the first $p_\ell = 10$ columns are assigned as quantitative trait loci (QTL) affecting the trait.

Since this is the first example where marker genotypes are simulated, a comment is in order. In a typical situation, matrix X consists of correlated observed marker genotypes that are not themselves causal. The idea is that the correlation structure induces a correlation between the marker loci and the unobserved causal loci. In this way, the marker genotypes may inform on the underlying genetics of a trait. Here, the simulated marker genotypes are uncorrelated, on average, but importantly, the "unobserved" causal loci are part of the marker panel. As mentioned in the appendix of Chap. 1, when the causal loci are included in the marker panel, the genetic markers capture the genetic variation of the trait. Including information from non-causal markers inflates the validating mean squared error, and the prediction ability of the model is compromised.

The true model for the training data is

$$y_t = X_\ell b_\ell + e, \tag{6.76}$$

where y_t is an $N \times 1$ vector of training observations, X_ℓ is the $N \times p_\ell$ matrix of QTL genotype codes and b_ℓ is an $p_\ell \times 1$ vector of allele substitution effects of causal loci with elements set equal to 0.5. Due to centring and scaling, the elements of X are random variables with mean 0 and variance 1. The resulting equilibrium additive genetic variance is equal to $\text{Var}(x'_{i,\ell} b_\ell | b_\ell) = 10 \times (1 \times 0.5^2) = 2.5$ squared units, where $x'_{i,\ell}$ is the ith row of matrix X_ℓ. The vector e consists of random residuals $e \sim N(0, I\,5)$ and is of dimension $N \times 1$. The proportion of variance of y explained by the true model (the R^2 or equilibrium heritability) is expected to be 33% $((10 \times 1 \times 0.5^2/7.5) \times 100)$. Training and validating data are generated using the same model and different samples of X and of e.

Training data y_t are analysed using operational models that regress y_t on marker genotypes X as covariates. Ten operational linear models are used to analyse y_t of

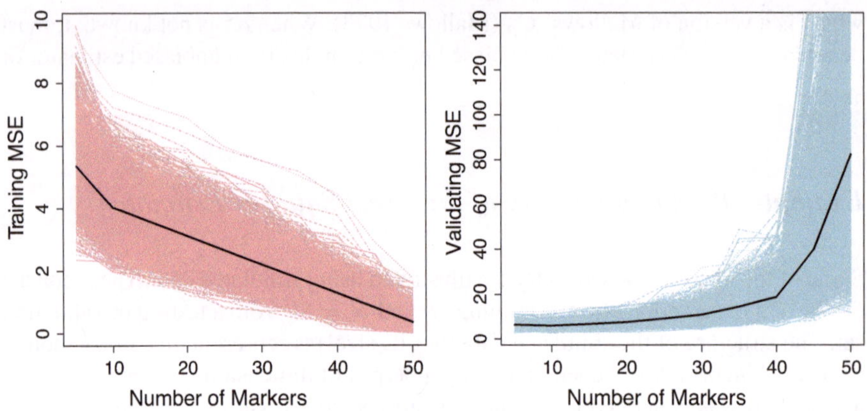

Fig. 6.1 Left panel, red: training mean squared error (6.52). Right panel, blue: validating mean squared error (6.53) of the least squares predictor using simulated data generated with the linear model. Results from 2000 training/validating replicates; black lines, average over replicates

the form

$$y_t = X_j b_j + \varepsilon_j, j = 1, 2, \ldots, 10 \tag{6.77}$$

that differ in the number of markers included. This is reflected in the number of columns of X_j and the number of elements of b_j, which range from 5 to 50 in increments of 5, and the subscript j indexes the model. When $j = 2$, the operational model is the same as the true model including ten covariates (the 10 QTL).

Figure 6.1 displays the training (left panel, red) and validating (right panel, blue) sample mean squared errors (6.52) and (6.53) using data analysed with the ten linear models (6.77) with increasing number of covariates. The figures show the behaviour of individual replicates (2000 replications of training/validating records, with genetic marker covariates X fixed at the initial simulated value) as well as the average MSE over replicate lines that is a Monte Carlo estimate of the expected MSE. The training mean squared errors (left panel) show a marked decline as the number of covariates increases from 5 to 10. At this point, the true model and the operational model are identical. As more covariates are included, the training squared error falls, albeit at a lower rate. The validating squared error decreases slightly from its initial value, reaches a minimum when the number of covariates is 10 and increases subsequently as additional covariates are included. The rate of increase is very pronounced when 40 or more covariates are included in the operational models.

The figures emphasise an important feature of the training-testing split as a tool to monitor prediction ability: the large variation among replications. A correct specification of this variation may be especially relevant in a situation where the goal is to learn how well a given statistical model will perform on independent data. On the other hand, one may be interested in comparing models, and therefore locating

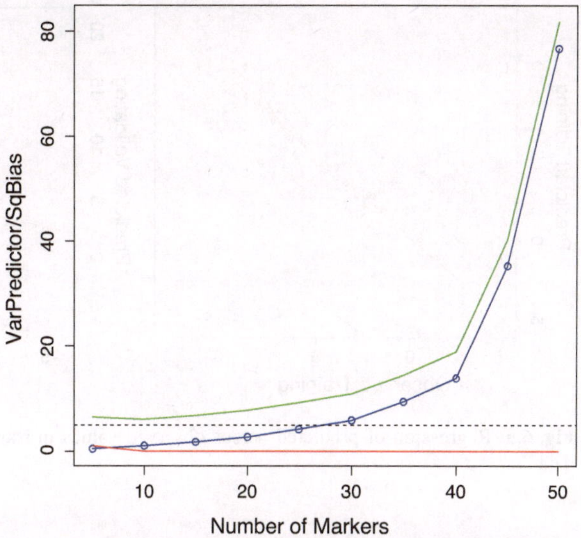

Fig. 6.2 The expected validating mean squared error (green) and its three components (6.51): expected squared bias (red), variance of predictor (blue) and residual variance (dashed lines). Data simulated with the linear model

the minimum of the MSE curve may be an adequate goal, rather than the actual value of the MSE. The curves of the different replicates in Fig. 6.1, right panel, seem to identify the correct model with ten covariates despite the fact that each replicate may deviate quite markedly from the average MSE.

The expected validating mean squared error and its three components (6.65) are shown in Fig. 6.2. The expected validating mean squared error agrees well with the simulated value displayed in the right panel of Fig. 6.1. The variance of the predictor increases with the number of covariates, particularly when this exceeds 20 and the increase becomes very pronounced with 40 or more covariates. The expected squared bias is positive when the number of covariates is equal to 5. When this number is equal to 10, the true model and the operational model are the same, and the bias falls to zero. As the number of covariates increases beyond 10, the expected squared bias remains zero because in the case of the present example, the least squares estimator using models with ten or more covariates is unbiased (see (6.38)). The dashed line represents the true residual variance of the observations.

Quantifying Optimism

The simulation example compares the expected optimism (6.68) with its estimate based on (6.75). When the number of records in training and validating data is equal to 55, the average difference between the validating (6.63) and training (6.62) mean squared errors (over 500 Monte Carlo replicates) is as follows: Using an operational model with 10, 15, 25 or 35 covariates, the observed average difference (O) and the predicted (P) based on $2(\hat{\sigma}^2/N)p$, where $\hat{\sigma}^2 = 4.99$ and $\sigma^2 = 5$, is (O; P): (3.1; 1.8), (3.5; 2.7) , (6.9; 4.5), (12.7; 6.6). When instead of 55 records,

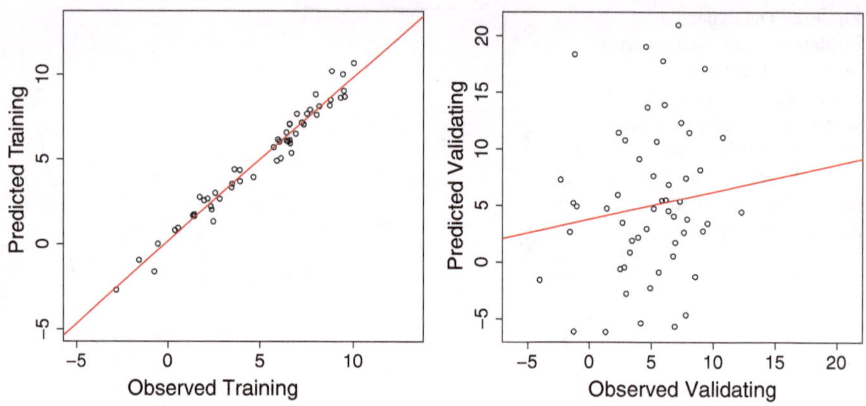

Fig. 6.3 Regression of predicted versus observed values in training (left) and validating (right) data

550 are used, with $\hat{\sigma}^2 = 5.00$, these figures are (O; P): (0.18, 0.18), (0.27; 0.28), (0.46; 0.45), (0.63; 0.64). This illustrates the influence of the ratio p/N on the convergence of estimator $\frac{2}{N}\sum_{i=1}^{N}\widehat{\text{Cov}}\left(y_i, \hat{y}_i\right)$ to its expectation.

Another way of visualising optimism is illustrated in Fig. 6.3 that shows the regression of predictions versus observations in the training and validating data (55 observations) for a randomly chosen simulation replicate, using 50 covariates. In the validating data, the regression is not significantly different from 0, and the linear model explains 1.5% of the variance of the predicted values. In contrast, in the training data, the regression is equal to 0.97 (not significantly different from 1), and the linear model explains 97% of the variance of the predicted values.

Example: What Measure of Prediction Error Does Leave-One-Out Estimate?

This example illustrates the ability of leave-one-out cross-validation to estimate either the conditional validating mean squared error (6.49) (averaging MSE_v over validating data, with training data fixed) or the unconditional validating squared error (6.51) (averaging MSE_v over training and validating data). The setup is the same as in the previous example, with 55 observations in training and testing data, ten covariates assigned as loci affecting the trait and ten operational models that differ in the number of covariates used for analysis.

The conditional validating squared error (6.49) has two terms: the irreducible error (equal to five in the simulation) and the squared term representing systematic difference between predictor and the expectation of the predictand. In least squares linear regression, this last quantity is zero on average over replications of training data, but it is not zero in any one realisation. For a given realisation of training data,

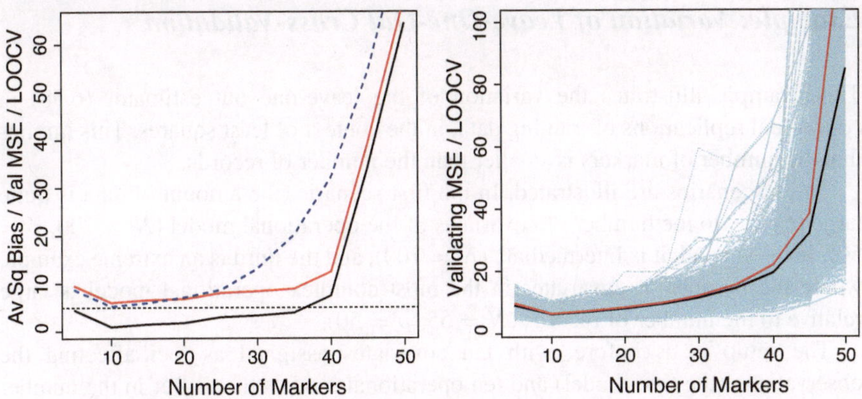

Fig. 6.4 Left: components of Monte Carlo estimates of E(MSE) obtained by averaging over replications of validating data, given a single sample of training data. Dashed horizontal black line: conditional variance of a datum based on the true model (five squared units). Solid black line: average (over 1000 replicated validating data sets; a single training data is simulated) squared discrepancies between predictor and the expectation of the predictand y_0, obtained using operational models with increasing number of covariates. Solid red line: average (over 1000 replicated validating data sets) conditional validating mean squared error, given training data. Dashed blue line: leave-one-out cross-validation estimate based on the single training data. Right, blue: validating mean squared errors for each of the 1000 replicated training and testing data sets; solid black line, average validating mean squared error over the 1000 replicated training and testing data sets; solid line red, average leave-one-out estimate of mean squared error over 1000 replicated training data sets

as the number of covariates in the model increases, the variance of the predictor becomes larger and so does the size of the second term in (6.49). This is illustrated in the left panel of Fig. 6.4. The setup is one where there is a single realisation of training data and 1000 simulation replicates of validating data. The average (over 1000 simulation replicates of validating data) conditional validating squared error, shown in red, is equal to the conditional variance of the record (5 squared units), plus the average (over the 55 observations and 1000 simulation replicates of validating data) of the squared discrepancies between predictor and average predictand (shown in black). The figure also shows that the leave-one-out estimate of squared error (dashed line, blue) based on the single realisation of training data overestimates the conditional validating mean squared error (6.49).

The right panel of Fig. 6.4 displays the validating squared error for each of the 1000 replications of training and testing data (blue); the average validating squared error over the 1000 replicates (thick black line); which is an estimate of (6.51); and the average leave-one-out estimate of squared error over the 1000 replications of training data (red line). The average leave-one-out estimator follows closely the average unconditional validating squared error (6.51), with a little overestimation particularly as the number of markers approaches the size of the training data.

Example: Variation of Leave-One-Out Cross-Validation

This example illustrates the variation of the leave-one out estimator (6.58) in conceptual replications of training data, in the context of least squares. This implies that the number of markers is smaller than the number of records.

Three scenarios are illustrated. In the first scenario, the amount of data is quite large relative to the number of covariates of the operational model ($N = 200$, $p = 50$). In the second, it is intermediate ($N = 100$), and the third is an extreme example where the number of covariates in the most complex operational model is large relative to the number of records ($N = 55$, $p = 50$).

The setup is as before, with ten covariates assigned as loci affecting the observations (the true model) and ten operational models that differ in the number of covariates used for analysis, ranging from 5 to 50 in steps of 5. The leave-one-out cross-validation estimate of MSE (6.58) for the 1000 simulation replicates of training data is shown in Fig. 6.5. With $N = 55$ records the degree of uncertainty is very large and it explodes as p approaches N. The variance of the leave-one-out cross-validation estimate of MSE decreases as the ratio number of makers (p) to size of training data (N) decreases (centre and right panels of Fig. 6.5).

Table 6.1 displays the average validating MSE over 2000 simulated training and validating data sets and the average leave-one-out estimates of MSE based on expression (6.58) over 2000 simulated training data for various combinations of number of records N and covariates p. The leave-one-out estimates use the N training records, whereas computation of the validating mean squared error requires the N training and N validating records.

In general, the leave-one-out estimates have a tendency to overpredict the average validating mean squared error when the ratio p/N approaches 1. This is not a restriction when leave-one-out cross-validation is used with linear smoothers that use shrinkage, the topic of the next chapter.

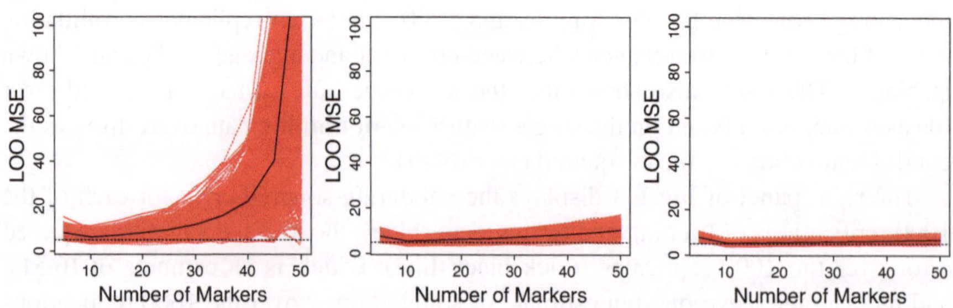

Fig. 6.5 Leave-one-out cross-validation estimates of mean squared error (6.58) for three sizes of training data (left panel, $N = 55$; centre panel, $N = 100$; right panel, $N = 200$) and operational models with increasing number of covariates, ranging from 5 to 50 in steps of 5. The average over 1000 simulation replicates of training and validating data is shown as solid black lines. The environmental variance, equal to 5 squared units, is shown as horizontal dashed lines

Table 6.1 Average (over 2000 simulated training data sets) leave-one-out estimate of MSE (X) and average over training and validating data of the validating MSE (Y), for three sizes of training and validating records N. Figures in the body of table correspond to X; Y

	Number of covariates (p)		
N	10	20	50
55	6.3; 6.3	8.3; 8.0	100.8; 59.2
100	5.6; 5.6	6.4; 6.2	10.4; 10.1
200	5.3; 5.3	5.6; 5.6	6.7; 6.7

Chapter 7
Shrinkage Methods

Expression (6.51) indicates how prediction ability is governed by bias and variance. As models become more complex, local noise can be captured, but coefficient estimates suffer from higher variance as more terms are included in the model. In the context of the traditional regression model $y = Xb + e$, $e \sim N\left(0, I\sigma^2\right)$, when the number of covariates (number of columns of X) p is large relative to the number of records/individuals n (number of rows of X), the columns of matrix X may become rank-deficient (X may not be or is close to not being of full column rank). In this case, even when $p < n$, it is difficult to separate the effect of individual covariates.

Shrinkage methods are typically used to confront this problem that emerges in highly dimensional models. Shrinkage estimators are obtained as the solution to an optimisation problem that balances bias and variance or model complexity. The general form of the optimisation problem is

$$\hat{b} \underset{\arg\min b}{=} \{L(y, b) + \lambda C(b)\}, \tag{7.1}$$

where $L(y, b)$ is a function that measures the lack of fit of the model to the data, $C(b)$ is a measure of model complexity and $\lambda \geq 0$ is a regularisation parameter controlling the trade-off between model fit and model complexity. This chapter starts by describing one of the simplest shrinkage estimators known as ridge regression. The balance between model fit and model complexity is achieved by shrinking all coefficients towards a common point; none is set to zero, and coefficients tend to resemble each other.

The second shrinkage estimator described in this chapter is the *lasso* (least absolute shrinkage and selection operator). Lasso solves a critically different optimisation problem (lasso and ridge regression use different expressions for the model complexity parameter) and generates solutions where some of the regression estimates are exactly zero. Efficient algorithms have been developed that allow application of the lasso in models with a vast number of variables (typically larger

D. Sorensen, *Statistical Learning in Genetics*, Statistics for Biology and Health,
https://doi.org/10.1007/978-3-031-35851-7_7

than the number of records). The lasso can be used for both prediction and model selection.

The chapter concludes with a description of a fully Bayesian-McMC spike and slab model. The model assigns a two-component prior mixture distribution to the covariate parameters. One of the components is a point mass at zero; the other is a normal distribution with mean zero and unknown variance. The model outputs the posterior probability that a covariate has no effect on the observations for each covariate. The spike and slab model can be used for prediction and as a tool to isolate a handful of promising covariates for further study, among thousands or hundred of thousands present. Examples with simulated data mimicking genomic models illustrate various features of these models.

7.1 Ridge Regression

Ridge regression (Hoerl and Kennard 1970b,a) is a particular case of (7.1) with $L(y, b) = (y - Xb)'(y - Xb)$ and $C(b) = \sum_{j \in S} b_j^2$, where S is the set of coefficients to be penalised. Typically, some of the regression coefficients such as the intercept are not penalised. The ridge coefficients are obtained as

$$\hat{b}_{rr} \underset{\text{arg min } b}{=} (y - Xb)'(y - Xb) + \lambda b' \Delta b. \tag{7.2}$$

Here, Δ is a diagonal matrix whose entries are 1 for $j \in S$ and zero elsewhere. Taking first derivatives of (7.2) with respect to b, setting the linear system equal to zero, and solving for b yields

$$\hat{b}_{rr} = \left(X'X + \lambda \Delta\right)^{-1} X'y, \tag{7.3}$$

a linear function of y. An equivalent way of writing (7.2) is

$$\hat{b}_{rr} \underset{\text{arg min } b}{=} (y - Xb)'(y - Xb)$$

subject to

$$b' \Delta b \leq t, \quad t \geq 0,$$

showing the size constraint on the parameters. There is a one-to-one correspondence between t and λ. The constrained minimisation uses Lagrange multipliers.

Relative to least squares, shrinkage adds a constant λ to the diagonal entries of $X'X$ which guarantees that the inverse in (7.3) exists even when $X'X$ is singular; there is always a unique solution to $\hat{b}_{rr}$. The form of the solution shows that as $\lambda \to \infty$, all the coefficients tend to zero $\hat{b}_{rr} \to 0$ and when $\lambda \to 0$, $\hat{b}_{rr}$ approaches the least squares estimator. The coefficients are shrunk towards zero at a rate depending of λ that controls the size of the coefficients.

The relationship between the ridge estimator and the least squares estimator $\hat{b}$ is

$$\hat{b}_{rr} = \left[I + \lambda(X'X)^{-1}\Delta\right]^{-1}\hat{b}$$

that can be verified by replacing above $\hat{b} = (X'X)^{-1}X'y$ and using $B^{-1}A^{-1} = (AB)^{-1}$. Here, $X'X$ is assumed to be non-singular.

The ridge solutions depend on the scaling of the inputs, and therefore, X is typically standardised (to have sample variance 1) and centred, so that $1'X = 0$, $\frac{1}{n}x_i'x_i = 1, i = 1, 2, \ldots, p$ and $X'X$ is in "correlation form". Also, in order to avoid fitting the intercept but leaving it unpenalised (Brown 1977), y is often also centred. When this is the case, matrix X is $n \times p$ instead of $n \times (p+1)$ when the intercept is included and $\Delta = I$. If X is orthonormal such that $X'X = I$ and in models without the intercept,

$$\hat{b}_{rr} = [I + \lambda I]^{-1}\hat{b} = [(1+\lambda)I]^{-1}\hat{b} = \frac{1}{1+\lambda}\hat{b}$$

showing that in this situation, the ridge estimator is a downweighted version of the least squares estimator $\hat{b}$.

Ridge estimation leads to biased estimators of b. With $\mathrm{E}(y|X) = Xb$ and $\Delta = I$, using (7.3),

$$\begin{aligned}
\mathrm{E}\left(\hat{b}_{rr}|X\right) &= \left(X'X + \lambda I\right)^{-1} X'Xb \\
&= \left(X'X + \lambda I\right)^{-1} \left(X'X + \lambda I - \lambda I\right) b \\
&= b - \lambda \left(X'X + \lambda I\right)^{-1} b
\end{aligned}$$

with bias

$$\mathrm{E}\left(\hat{b}_{rr}|X\right) - b = \lambda \left(X'X + \lambda I\right)^{-1} b,$$

proportional to λ. On the other hand, the ridge estimator has smaller variance than the least squares estimator. Let $X = UDV'$ be the singular value decomposition (SVD) of X of dimension $n \times p$. In the SVD, for $n > p$, U is $n \times p$, $U'U = I$, $D = \mathrm{diag}\left(d_1, d_2, \ldots, d_p\right)$, $d_i > 0, i = 1, \ldots, p$ is a diagonal matrix with positive eigenvalues (when X is of full column rank), and V is $p \times p$, $V'V = I$. The variance of the least squares estimator is

$$\begin{aligned}
\mathrm{Var}\left(\hat{b}|X\right) &= \sigma^2 \left(X'X\right)^{-1} \\
&= \sigma^2 \left(VDU'UDV'\right)^{-1}
\end{aligned}$$

$$= \sigma^2 \left(VD^2V'\right)^{-1}$$
$$= \sigma^2 VD^{-2}V',$$

where D is a diagonal matrix with elements $1/d_i^2$. To go from the third to the fourth line use $(ABC)^{-1} = C^{-1}B^{-1}A^{-1}$ and $V^{-1} = V'$. On the other hand, the variance of the ridge estimator is

$$\begin{aligned}
\text{Var}\left(\hat{b}_{rr}|X\right) &= \sigma^2 \left(X'X + \lambda I\right)^{-1} X'X \left(X'X + \lambda I\right)^{-1} \qquad (7.4) \\
&= \sigma^2 \left(VD^2V' + \lambda I\right)^{-1} VD^2V' \left(VD^2V' + \lambda I\right)^{-1} \\
&= \sigma^2 \left[V\left(D^2 + \lambda I\right)V'\right]^{-1} VD^2V' \left[V\left(D^2 + \lambda I\right)V'\right]^{-1} \\
&= \sigma^2 V \left(D^2 + \lambda I\right)^{-1} D^2 \left(D^2 + \lambda I\right)^{-1} V' \\
&= \sigma^2 VWV',
\end{aligned}$$

where $W = \left(D^2 + \lambda I\right)^{-1} D^2 \left(D^2 + \lambda I\right)^{-1}$ is a diagonal matrix with elements $d_i^2/(d_i^2 + \lambda)^2$ indicating that $\text{Var}\left(\hat{b}\right) \geq \text{Var}\left(\hat{b}_{rr}\right)$, with equality if $\lambda = 0$ in the absence of shrinkage. The variance decreases with λ and vanishes as $\lambda \to \infty$.

Ridge regression performs particularly well in terms of mean squared error (or prediction error variance) relative to standard regression when some of the regression parameters are small or even zero. This advantage is less marked when the true parameters take larger values.

A Toy Example

This example is elaborated from a version in unpublished notes by G. de los Campos, Michigan State University.

Consider a simple one way classification with two levels parametrised as a linear regression model

$$y_i = x_{1i}b_1 + x_{2i}b_2 + e_i,$$

where $x_{1i} = 1$ if record i belongs in treatment 1 and 0 otherwise and $x_{2i} = 1 - x_{1i}$. The treatment effects are b_1 and b_2, respectively. The least squares equations are

$$\begin{bmatrix} \sum_i x_{1i}^2 & \sum_i x_{1i}x_{2i} \\ \sum_i x_{1i}x_{2i} & \sum_i x_{2i}^2 \end{bmatrix} \begin{bmatrix} \hat{b}_1 \\ \hat{b}_2 \end{bmatrix} = \begin{bmatrix} \sum_i x_{1i}y_i \\ \sum_i x_{2i}y_i \end{bmatrix}.$$

Since $\sum_i x_{1i}^2$ and $\sum_i x_{2i}^2$ are the number of records in treatments 1 and 2 (denoted n_1 and n_2), $\sum_i x_{1i}x_{2i} = 0$, and $\sum_i x_{1i}y_i$ and $\sum_i x_{2i}y_i$ are the sum of records in treatments 1 and 2, the least squares estimators are

$$\begin{bmatrix} \hat{b}_1 \\ \hat{b}_2 \end{bmatrix} = \begin{bmatrix} n_1 & 0 \\ 0 & n_2 \end{bmatrix}^{-1} \begin{bmatrix} \sum x_{1i}y_i \\ \sum_i x_{2i}y_i \end{bmatrix}.$$

The least squares estimators of the treatment effects are the treatment means

$$\begin{bmatrix} \hat{b}_1 \\ \hat{b}_2 \end{bmatrix} = \begin{bmatrix} \frac{\sum x_{1i}y_i}{n_1} \\ \frac{\sum_i x_{2i}y_i}{n_2} \end{bmatrix}$$

with sampling variance

$$\text{Var}\left(\hat{b}_j | n_j\right) = \frac{\sigma^2}{n_j}, \quad j = 1, 2. \tag{7.5}$$

The ridge regression estimates are

$$\begin{bmatrix} \hat{b}_{rr1} \\ \hat{b}_{rr2} \end{bmatrix} = \begin{bmatrix} n_1 + \lambda & 0 \\ 0 & n_2 + \lambda \end{bmatrix}^{-1} \begin{bmatrix} \sum_i x_{1i}y_i \\ \sum_i x_{2i}y_i \end{bmatrix}$$

$$= \begin{bmatrix} \frac{\sum_i x_{1i}y_i}{n_1+\lambda} \\ \frac{\sum_i x_{2i}y_i}{n_2+\lambda} \end{bmatrix} = \begin{bmatrix} \frac{n_1}{n_1+\lambda}\hat{b}_1 \\ \frac{n_2}{n_2+\lambda}\hat{b}_2 \end{bmatrix}.$$

Adding λ on the diagonal shrinks the estimates towards zero. The extent of shrinkage depends on the value of λ relative to the sample size.

The ridge estimator is biased:

$$\text{E}\left(\hat{b}_{rrj} | n_j\right) = \frac{n_j}{n_j + \lambda} \text{E}\left(\hat{b}_j | n_j\right) = \frac{n_j}{n_j + \lambda} b_j \tag{7.6}$$

because $\text{E}\left(\hat{b}_j | n_j\right) = b_j$. The bias is

$$\text{E}\left(\hat{b}_{rrj} | n_j\right) - b_j = b_j \left(\frac{n_j}{n_j + \lambda} - 1\right). \tag{7.7}$$

The sampling variance is

$$\text{Var}\left(\hat{b}_{rrj} | n_j\right) = \frac{n_j}{\left(n_j + \lambda\right)^2}\sigma^2, \tag{7.8}$$

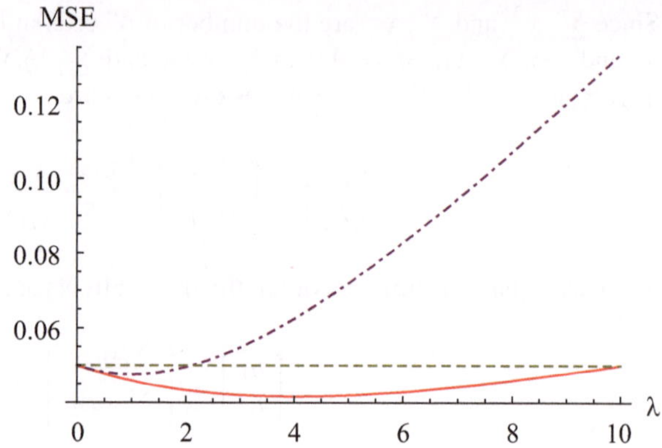

Fig. 7.1 Mean squared error of the ridge regression estimator versus the shrinkage parameter λ. The population size is $n = 20$ and the variance $\sigma^2 = 1$. Red, $b_j = 0.5$; dot dashed, $b_j = 1.0$; dashed, MSE of the least squares estimator ($\lambda = 0$), Eq. (7.5)

which reduces to the variance of the least squares estimator when $\lambda = 0$. For $\lambda > 0$, $\text{Var}\left(\hat{b}_{rri}\right) < \text{Var}\left(\hat{b}_j\right)$. The smoothing process will reduce mean squared error if it gets rid of more variance than it adds bias. Specifically, the mean squared error is given by adding the square of (7.7) and (7.8). This gives

$$\text{E}\left[\left(\hat{b}_{rrj} - b_j\right)^2 |n_j\right] = b_j^2\left(\frac{n_j}{n_j + \lambda} - 1\right)^2 + \frac{n_j}{\left(n_j + \lambda\right)^2}\sigma^2. \tag{7.9}$$

In the absence of shrinkage, $\lambda = 0$, the first term vanishes, and the second term is equal to (7.5). When λ is very large, the variance term approaches zero, and the bias achieves its maximum value at b_j^2 (when the ridge coefficients are strongly regressed towards zero).

Figure 7.1 illustrates the trade-off variance versus bias of the ridge regression using (7.9). For a given population size, the MSE (7.9) has a minimum at $\lambda = \sigma^2/b^2$. The figure shows for $n = 20$, $\sigma^2 = 1$, that MSE of the ridge estimator, for a true $b_j = 0.5$, outperforms the MSE of the least squares estimator up to $\lambda < 10$. For a true $b_j = 1.0$, the ridge estimator does better up to $\lambda \approx 2$, with a minimum MSE at $\lambda = 1$.

Choice of Shrinkage Parameter

A natural choice of λ could be based on minimising the mean squared error of the estimator. Glancing at (7.6) and (7.8) and recalling that mean squared error is a function of the bias and the variance indicate that the optimal choice of λ depends on the values of b and σ^2. In a prediction context, since these are typically unknown, λ can be chosen using cross-validation.

Bayesian View of Ridge Regression

In a Bayesian framework, assume the regression coefficients are assigned the prior distribution $b|\sigma_b^2 \sim N\left(0, I\sigma_b^2\right)$ and the sampling model for the data is the normal process

$$y|X, b, \sigma_e^2 \sim N\left(Xb, I\sigma_e^2\right).$$

A standard result (Lindley and Smith 1972) sketched below states that the conditional posterior distribution of b is

$$b|X, \sigma_b^2, \sigma_e^2, y \sim N\left(\hat{b}_{rr}, \left(X'X + \lambda I\right)^{-1} \sigma_e^2\right), \tag{7.10}$$

where $\hat{b}_{rr}$ is the ridge regression estimator

$$\hat{b}_{rr} = \left(X'X + \lambda I\right)^{-1} X'y \tag{7.11}$$

and $\lambda = \sigma_e^2/\sigma_b^2$. Therefore, the ridge regression estimator is equivalent to the posterior mode (or mean, given normality) of the regression coefficients in the following Bayesian model:

$$\text{Likelihood:} \quad y|X, b, \sigma_e^2 \sim N\left(Xb, I\sigma_e^2\right)$$

$$\text{Prior:} \quad b|\sigma_b^2 \sim N\left(0, I\sigma_b^2\right).$$

Numerically, the solution (7.11) is also the best linear unbiased predictor (BLUP) of b.

The ratio σ_e^2/σ_b^2 is equivalent to λ in the ridge regression context. In a fully Bayesian analysis, the variance components are assigned prior distributions and can be inferred jointly with b.

Note

The proof of (7.10) is based on combining two quadratic forms, used repeatedly in the book (see also page 145). Using Bayes theorem, the posterior distribution of b is

$$\begin{aligned} p\left(b|X, \sigma_b^2, \sigma_e^2, y\right) &\propto p\left(y|X, b, \sigma_e^2\right) p\left(b|\sigma_b^2\right) \\ &\propto \exp\left[-\frac{1}{2\sigma_e^2}(y - Xb)'(y - Xb)\right] \exp\left[-\frac{1}{2\sigma_b^2} b'b\right] \end{aligned}$$

$$= \exp\left[-\frac{1}{2\sigma_e^2}\left\{(y - Xb)'(y - Xb) + \lambda b'b\right\}\right]$$

$$\propto \exp\left[-\frac{1}{2\sigma_e^2}\left(-2b'X'y + b'X'Xb + \lambda b'b\right)\right]$$

$$= \exp\left[-\frac{1}{2\sigma_e^2}\left(-2b'X'y + b'\left(X'X + \lambda I\right)\right)b\right] \qquad (7.12)$$

where $\lambda = \sigma_e^2/\sigma_b^2$. Let

$$\left(X'X + \lambda I\right)\hat{b} = X'y, \qquad (7.13)$$

and replace $X'y$ in (7.12) by $\left(X'X + \lambda I\right)\hat{b}$. Add and subtract $\hat{b}'\left(X'X + \lambda I\right)\hat{b}$ and keep terms in b only. Then a little manipulation yields

$$p\left(b|X, \sigma_b^2, \sigma_e^2, y\right) \propto \exp\left[-\frac{1}{2\sigma_e^2}\left(b - \hat{b}\right)'\left(X'X + \lambda I\right)\left(b - \hat{b}\right)\right]$$

which is the kernel of a normal distribution with mean $\hat{b}$ and variance $\left(X'X + \lambda I\right)^{-1}\sigma_e^2$, as in (7.10).

There is a difference between the posterior variance of the Bayesian model and the variance of the ridge estimator (7.4). Using the singular value decomposition of X, it is easy to show that

$$\left(X'X + \lambda I\right)^{-1}\sigma_e^2 = V\widetilde{W}V'\sigma_e^2,$$

where $\widetilde{W}$ is a $p \times p$ diagonal matrix with elements $\left(d_i^2 + \lambda\right)^{-1}$. On the other hand, as indicated in (7.4), the ridge regression estimator has sampling variance $\sigma^2 VWV'$, where W is a $p \times p$ diagonal matrix with elements $d_i^2/(d_i^2 + \lambda)^2$.

Example: Prediction Ability of the Ridge Estimator

The prediction ability of the ridge estimator is illustrated with simulated data ($n =$ 55 observations) similar in structure to the one used in the example of page 291. In contrast to that example, the substitution effects b_ℓ of the $p_\ell = 10$ QTL are drawn from $N(0, \sigma_b^2 = 0.25)$ leading to a genomic variance (at the level of the operational model) equal to 2.5 squared units as explained below.

The data are analysed with two sets of operational models. In the first set, the QTL are part of the marker panel. In the second set, the QTL are not part of the marker panel. In the latter scenario, observations and covariates are uncorrelated. This creates an extreme case where the operational model has no prediction ability.

The *genomic value* of individual i is $g_i = x'_{i,\ell} b_\ell$, were $x'_{i,\ell}$ is the ith row of the $n \times p_\ell$ matrix X_ℓ of QTL genotypes. Due to centring and scaling, the elements of X are random variables with mean 0 and variance 1.

At the level of the operational models,

$$\mathrm{Var}\,(g_i\,|x_i) = x'_i x_i \sigma_b^2$$

and the genomic variance is

$$\begin{aligned}
\mathrm{Var}\,(g_i) = \sigma_g^2 &= \mathrm{E}_{x_i}\,[\mathrm{Var}\,(g_i\,|x_i)] + \mathrm{Var}_{x_i}\,[\mathrm{E}\,(g_i\,|x_i)] \\
&= \mathrm{E}_{x_i}\,[\mathrm{Var}\,(g_i\,|x_i)] \\
&= \sigma_b^2\,\mathrm{E}_{x_i}\,\left(x'_i x_i\right) \\
&= p\sigma_b^2,
\end{aligned} \tag{7.14}$$

where p is the number of markers of the operational model. When $p = p_\ell = 10$, $\sigma_g^2 = 10 \times 0.25 = 2.5$.

In view of (7.14), the shrinkage parameter in (7.13) is

$$\lambda = \frac{\sigma_e^2}{\sigma_b^2} = \frac{\sigma_e^2}{\sigma_g^2} p. \tag{7.15}$$

When σ_e^2 and σ_g^2 are assumed known and fixed with values 5 and 2.5, respectively, the shrinkage parameter is therefore set to increase linearly with the number of markers $p = 5, 10, \ldots, 45, 50$ included in the operational model (7.13) with slope σ_e^2/σ_g^2.

The prediction ability of the ridge estimator using data simulated with the linear model is displayed in Fig. 7.2. When the operational model includes the first 10 covariates, it coincides with the true model, and the validating mean squared error reaches its minimum. Relative to the least squares predictor, the increase in the validating mean squared error of the ridge predictor as a function of the number of covariates included in the operational model is attenuated by the shrinkage parameter that also increases as more covariates are included as indicated in (7.15). The training mean squared error shows a decline with increasing number of covariates. This decline is less pronounced than the one displayed in Fig. 6.1 for the least squares predictor.

Figure 7.3 displays the three components of the expected validating mean squared error. The variance of the ridge regression predictor of the validating record shows a steady increase as the number of covariates increases despite the increase of the penalty parameter that cannot compensate with increasing dimensionality of the predictive model. The bias of the predictor has a minimum when the operational model and the true model coincide and increases steadily when the operational model has more than *ten* covariates. As the penalty parameter increases, the ridge

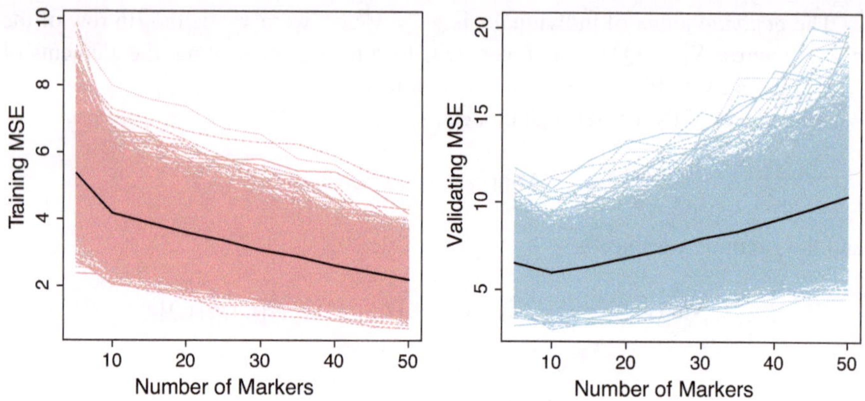

Fig. 7.2 Left (red): training mean squared error (6.52). Right (blue): validating mean squared error (6.53) of the ridge estimator using simulated data generated with a linear model with *ten* causal loci and analysed with operational linear models with increasing number of covariates that include the causal loci. Results from 2000 replicates; black lines, average over replicates

Fig. 7.3 The expected validating mean squared error (green) and its three components: expected squared bias (red), variance of predictor (blue) and residual variance (dashed lines). Data simulated using a linear model with *ten* causal loci and analysed with operational linear models with increasing number of covariates, which include the causal loci

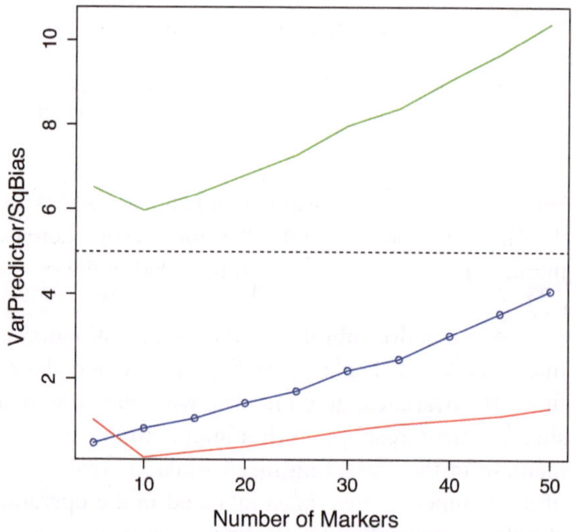

regression estimates are shrunk more heavily towards zero with a consequent increment in the bias. There is good agreement with the expected validating mean squared error and the mean validating mean squared error over the 2000 simulation replicated in Fig. 7.2.

The mean squared errors computed using validating and training data that result from fitting operational models with increasing number of covariates that do not include the causal loci are shown in Fig. 7.4. This is an example that mimics a trait whose expression is not affected by a genetic component: marker genotypes are uninformative about phenotype. Despite the lack of association between phenotypes

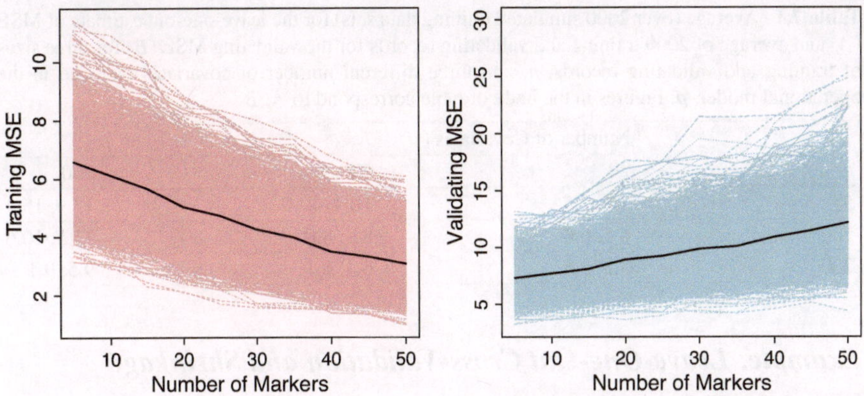

Fig. 7.4 Left (red), training mean squared error (6.52). Right (blue), validating mean squared error (6.53) of the ridge estimator using simulated data generated with a linear model with *ten* causal loci and analysed with operational linear models with increasing number of covariates that do not include the causal loci. Results from 2000 replicates; black lines, average over replicates

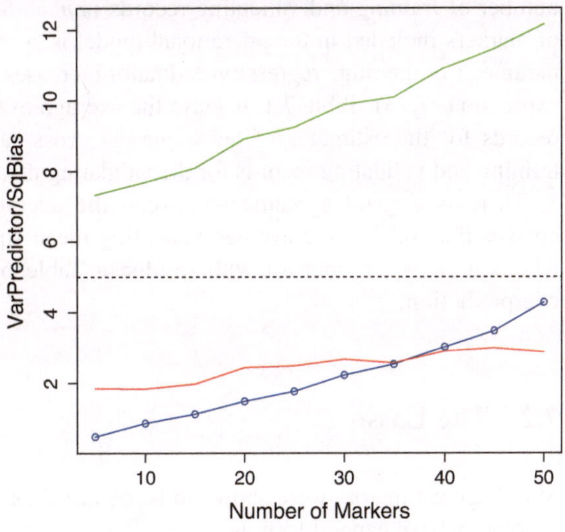

Fig. 7.5 The expected validating mean squared error (green) and its three components: expected squared bias (red), variance of predictor (blue) and residual variance (dashed lines). Data simulated with the linear model, *ten* causal loci and analysed with operational ridge regression models with increasing number of covariates that do not include the causal loci

and covariates the training mean squared errors show the expected decline due to overfitting. The validating mean squared errors increase steadily as more covariates are added to the operational model. There is a slight squared bias contribution to this increase, but most of it is driven by larger variances as the dimension of the operational model increases. Figure 7.5 illustrates this setup.

Table 7.1 Average (over 2000 simulated training data sets) for the leave-one-out estimate of MSE (A) and average of 2000 training and validating records for the validating MSE (B) for three sizes of training and validating records, n, and three different number of covariates included in the operational model, p. Figures in the body of table correspond to A; B

	Number of Covariates (p)		
n	20	40	200
55	7.8; 7.5	10.4; 10.0	15.9; 15.4
100	6.5; 6.4	8.1; 8.0	11.1; 10.6
200	5.5; 5.5	6.1; 6.1	9.5; 9.1

Example: Leave-One-Out Cross-Validation and Shrinkage

This example illustrates the behaviour of expression (6.58), the leave-one-out cross validation mean squared error, as an estimator of the validating mean squared error, averaged over training and validating data, when it is applied using ridge regression.

The setup is similar to the one used to construct Table 6.1 on page 297. The number of training and validating records is $n = 55, 100$ or 200, and the number of markers included in the operational model is $p = 20, 40$ or 200. The shrinkage parameter of the ridge regression estimator increases linearly with p, as indicated in expression (7.15). Table 7.1 displays the average over 2000 replications of training records for the estimates of leave-one-out cross-validation and the average over training and validating records for the validating mean squared errors.

There is a good agreement between the average estimates of leave-one-out cross-validation and the average validating mean squared error, even in situations where $p > n$ (in contrast with results in Table 6.1), with a slight tendency to overprediction.

7.2 The Lasso

Shrinkage estimators were shown to be obtained as the solution to an optimisation problem whose general form is

$$\hat{b} \underset{\arg\min b}{=} \{L(y, b) + \lambda C(b)\} \tag{7.16}$$

where $L(y, b)$ is a function that measures the lack of fit of the model to the data, $C(b)$ is a measure of model complexity and $\lambda \geq 0$ is a regularisation parameter controlling the trade-off between fitness and model complexity.

Let $b = (\beta_0, \beta) \in \mathbb{R}^{p+1}$. In ridge regression,

$$L(y, b) = \sum_{i=1}^{n} \left(y_i - \beta_0 - \sum_{j=1}^{p} x_{ij}\beta_j \right)^2, \quad i = 1, \ldots, n \tag{7.17}$$

and

$$\lambda C(b) = \lambda \sum_{j=1}^{p} \beta_j^2 \tag{7.18}$$

so that the ridge regression estimator is the solution (linear in y) to

$$\hat{b}_{rr} \underset{\arg\min \beta}{=} \left\{ \sum_{i=1}^{n} \left(y_i - \beta_0 - \sum_{j=1}^{p} x_{ij}\beta_j \right)^2 + \lambda \sum_{j=1}^{p} \beta_j^2 \right\}. \tag{7.19}$$

The term $\sum_{j=1}^{p} \beta_j^2$ (which does not include β_0) is known as the L_2 penalty. Relative to the classical least squares estimator, ridge regression reduces variability but can potentially cloud interpretation, as shrinkage is applied to all coefficients simultaneously. In ridge regression, all coefficients "resemble each other" (shrinkage is to the same point or distribution, when interpreted from a Bayesian perspective), and none of the coefficients is set to 0.

The lasso (Tibshirani 1996, "least absolute shrinkage and selection operator") has (7.17) in common with ridge regression but the L_2 penalty is replaced by the L_1 penalty, given by $\sum_{j=1}^{p} |\beta_j|$, and the lasso coefficients are the solutions (nonlinear in y) to

$$\hat{\beta}_{lasso} \underset{\arg\min \beta}{=} \left\{ \sum_{i=1}^{n} \left(y_i - \beta_0 - \sum_{j=1}^{p} x_{ij}\beta_j \right)^2 + \lambda \sum_{j=1}^{p} |\beta_j| \right\}. \tag{7.20}$$

In contrast with ridge regression, there is no closed form expression for the lasso.

The tuning parameter λ in (7.20) also controls the extent of the penalty. When $\lambda = 0$, the least squares solutions are obtained, and when $\lambda = \infty$, $\beta_j = 0$ for all j. In between these two extremes, coefficients are shrunk, but the effect of the L_1 penalty is to set some coefficients exactly to zero. As λ increases, more coefficients are set to zero (less variables are selected), and among the non-zero coefficients, shrinkage is stronger. The lasso is an attempt at combining features of subset selection and ridge regression simultaneously.

The lasso coefficients can also be obtained solving

$$\hat{\beta}_{lasso} \underset{\arg\min \beta}{=} \sum_{i=1}^{n} \left(y_i - \beta_0 - \sum_{j=1}^{p} x_{ij}\beta_j \right)^2$$

subject to

$$\sum_{j=1}^{p} \left|\beta_j\right| < t,$$

where $t \geq 0$ acts now as the tuning parameter. If t is chosen larger than $\sum_{j=1}^{p} \left|\hat{\beta}_j\right|$, where $\hat{\beta}_j$ is the least squares estimate, then the lasso estimates are the $\hat{\beta}_j^{\prime}s$. If t is smaller than $\sum_{j=1}^{p} \left|\hat{\beta}_j\right|$, shrinkage takes place, and some $\beta's$ are set to zero.

In the context of prediction where the lasso is typically deployed, the shrinkage parameter λ is chosen using cross-validation.

If the columns of x are centred, the estimator of the intercept is $\hat{\beta}_0 = \overline{y}$. When y is also centred, the model does not include an intercept term. In (7.20), the constraint treats all the coefficients equally, and therefore it makes sense to scale x, as with ridge regression. For centred and scaled x, $\sum_i x_{ij} = 0$, $\sum_i x_{ij}^2 = n$.

The Subdifferential

A number of algorithms are available to obtain the solution to (7.20) that is a quadratic programming problem with linear inequality constraints. Before sketching one of the algorithms, the notion of the subderivative or subdifferential is briefly introduced. The subdifferential generalises the derivative to convex functions that may not be differentiable. This is necessary because the (convex) function defined by (7.20) does not have a derivative at the point $\left|\beta_j\right|$ for $\beta_j = 0$.

To illustrate, consider the (convex, i.e., "holds water") absolute value function $f(x) = |x|$, $x \in \mathbb{R}$, where

$$|x| = \begin{cases} x, & x \geq 0, \\ -x, & x < 0. \end{cases} \tag{7.21}$$

For $x \neq 0$, the derivative of f with respect to x is $+1$ if $x > 0$ and -1 if $x < 0$. However, for $x = 0$ (that defines the point where f achieves its absolute minimum), the derivative does not exist. However, at $x = 0$, one can draw (many) lines that touch the function at $(0, f(0))$ that are everywhere below f except at the point

$x = 0$. This collection of slopes tangent to the function at the point $x = 0$ is the subdifferential of f at the point $x = 0$.

One can proceed a little more formally as follows: A scalar c is a subgradient of a function f at the point x_0 if for all points x the following holds:

$$f(x) \geq f(x_0) + c(x - x_0)$$

for $x \in \mathbb{R}$ (the real line) and belonging to the domain of f. Then

$$c(x - x_0) \leq f(x) - f(x_0), \tag{7.22}$$

and the collection of all slopes c of subtangent lines is called the subdifferential of f at the point $x = x_0$. The set of subdifferentials c at x_0 is a nonempty closed interval $[a, b]$ where a and b are the one-sided limits approached from the left and right of x, respectively, defined as

$$a = \lim_{x \to x_0^-} \frac{f(x) - f(x_0)}{x - x_0}, \tag{7.23a}$$

$$b = \lim_{x \to x_0^+} \frac{f(x) - f(x_0)}{x - x_0}, \tag{7.23b}$$

that satisfy $a \leq b$. Any c belonging in $[a, b]$ satisfies (7.22) and is a subdifferential of f at the point $x = x_0$, denoted $\partial f(x_0)$.

I return to the absolute value function $f(x) = |x|$ and find the set of subdifferentials at $x_0 = 0$. One way of approaching this is to use definition (7.22). Here, $f(x) = |x|$, $f(x_0) = |0| = 0$ and using (7.22) with $x_0 = 0$

$$cx \leq |x|, \quad \text{for all } x \in \mathbb{R}.$$

Since the inequality holds for all $x \in \mathbb{R}$, it also holds for $x = \pm 1$. Then for $x = +1$

$$\begin{aligned} c(1) &\leq |1| = 1, \\ c &\leq 1 \end{aligned}$$

and for $x = -1$

$$\begin{aligned} c(-1) &\leq |-1| = 1, \\ -c &\leq 1, \\ c &> -1. \end{aligned}$$

Therefore, c is any number defined in the closed interval

$$c \in [-1, 1]$$

and more generally, the subdifferential of f at $x = 0$ is written

$$\partial f(0) \subset [-1, 1].$$

For any x

$$\partial_x |x| = \begin{cases} -1, \text{ if } x < 0 \\ [-1, 1], \text{ if } x = 0 \\ 1, \text{ if } x > 0. \end{cases}$$

The limits of the closed set can also be obtained using (7.23). For the lower limit as x approaches 0 from the left,

$$\begin{aligned} a &= \lim_{x \to 0^-} \frac{f(x) - f(x_0)}{x - x_0} = \lim_{x \to 0^-} \frac{|x| - |0|}{x - 0} \\ &= \lim_{x \to 0^-} \frac{|x|}{x} = \lim_{x \to 0^-} \frac{x}{-x} = \lim_{x \to 0^-} -1 = -1 \end{aligned}$$

and for the upper limit, as x approaches 0 from the right,

$$\begin{aligned} b &= \lim_{x \to 0^+} \frac{f(x) - f(x_0)}{x - x_0} = \lim_{x \to 0^+} \frac{|x|}{x} \\ &= \lim_{x \to 0^+} \frac{x}{x} = \lim_{x \to 0^+} 1 = 1, \end{aligned}$$

where both cases make use of the fact that the limit of a constant (here the number -1 or 1) is the constant itself.

An Example with a Single Regression Parameter

Drawing from Friedman et al (2007), to get a little intuition for the behaviour of the lasso, consider a linear model with p covariates, where $(p - 1)$ are fixed at some value and β_j (a scalar) is the only parameter to be estimated. The vector of covariates x is scaled and centred, so that $x_j' x_j = n$, and the observations y are expressed as deviations from the mean, so that $\overline{y} = 0$. Therefore, the model does not include an intercept. Expression (7.20) now takes the form

$$f(\beta_j) = \frac{1}{n}\sum_{i=1}^{n}(r_i - x_{ij}\beta_j)^2 + \lambda\sum_{k\neq j}|\beta_k| + \lambda|\beta_j|$$

$$= \frac{1}{n}(r - x_j\beta_j)'(r - x_j\beta_j) + \lambda\sum_{k\neq j}|\beta_k| + \lambda|\beta_j| \tag{7.24}$$

where $r_i = y_i - \sum_{k\neq j} x_{ik}\beta_k$ and the $\beta_k' s$, $k \neq j$, are fixed at some value. The lasso estimate of β_j is the value of β_j that minimises (7.24). This gives

$$\partial f(\beta_j) = \begin{cases} \beta_j - x_j' r/n + \lambda, & \beta_j > 0, \\ \beta_j - x_j' r/n - \lambda, & \beta_j < 0, \\ [-\lambda, \lambda] - x_j' r/n, & \beta_j = 0. \end{cases}$$

Using $\hat{\beta}_j = x_j' r/n$ as the least squares estimate treating the residuals r as "data" and setting these equations equal to zero lead to the lasso estimate of β_j

$$\beta_j^L = \begin{cases} \hat{\beta}_j - \lambda, & \hat{\beta}_j > \lambda, \\ \hat{\beta}_j + \lambda, & \hat{\beta}_j < -\lambda, \\ 0, & -\lambda \leq \hat{\beta}_j \leq \lambda, \end{cases}$$

or more compactly

$$\beta_j^L = \begin{cases} 0, & \text{if } \left|\hat{\beta}_j\right| \leq \lambda, \\ sign\left(\hat{\beta}_j\right)\left(\left|\hat{\beta}_j\right| - \lambda\right), & \text{if } \left|\hat{\beta}_j\right| \geq \lambda. \end{cases} \tag{7.25}$$

The system (7.25) indicates that the smallest value for λ, such that all regression solutions $\beta^L = 0$, is $\max_j \left|\hat{\beta}_j\right|$, $j = 1, \ldots, p$, the largest least squares estimate. Clearly, the choice of the tuning parameter λ has a strong influence on the behaviour of the lasso. The choice can be guided using cross-validation.

A General Algorithm to Obtain Lasso Solutions

The following general algorithm can be implemented to obtain lasso solutions in models with p correlated covariates. The algorithm is known as the path-wise coordinate descent algorithm (Friedman et al 2007, 2010; Efron and Hastie 2016) and is based on updating regression coefficients one at a time as in the previous example. This is performed in an iterative fashion, and the process is repeated until it converges to the lasso solution.

The algorithm has an external loop that runs over the number of iterates and an internal loop that runs over the number of covariates, updating one covariate at a time. This requires updating the residuals. A computationally efficient manner of updating residuals is as follows (Friedman et al 2010): Consider a linear model with p covariates and write $Xb = (X_i, X_{-i})(b_i, b_{-i})'$. The residual including these p covariates is $r_1 = (y - X_i b_i - X_{-i} b_{-i})$ and the residual excluding the ith covariate is $r_0 = (y - X_{-i} b_{-i})$. Therefore,

$$r_0 = r_1 + X_i b_i,$$
$$r_1 = r_0 - X_i b_i.$$

The kernel of the internal loop is simply

INITIALISE $r_1 = y - \overline{y}, \quad b = 0$
LOOP OVER NUMBER COVARIATES $\quad i = 1, \ldots, p$
  $r_0 = r_1 + X_i b_i$ this updates $(y - 1\mu - X_{-i} b_{-i})$
  UPDATE b_i involves (7.25)
  $r_1 = r_0 - X_i b_i$ this updates $(y - 1\mu - Xb)$
END LOOP

The following R-code illustrates the implementation of the algorithm with a toy example. The bottom of the code includes a call to the package GLMNET (Hastie and Qian 2016) as a test on the output. Details regarding this package are deferred to the Example on page 319:

```
# CODE0701
# AN EXAMPLE USING SIMULATED CORRELATED X
rm(list=ls()) # CLEAR WORKSPACE
set.seed(3711)
#install.packages("glmnet", .libPaths()[1])
library(glmnet)
n <- 100
p <- 100
X <- matrix(rnorm(p*n),ncol=p)
X <- X*0.8 + X[,1]*0.3 # GENERATE CORRELATED COVARIATES
X <- scale(X)*sqrt(n)/sqrt(n-1)
beta <- rep(0,p)
betac <- rep(0,p)
beta <- sample(0:1,p,replace=TRUE,prob=c(2,1))
length(which(beta!=0))
```

```
## [1] 42
```

```
y <- X%*%beta + rnorm(n,sd=0.4)
y <- y - mean(y)
for(i in 1:p){ betac[i]=coef(lm(y~X[,i]))[2]}
lambda=max(abs(betac))*.1
lambda
```

```
## [1] 1.457124
```

```
niter <- 100
bL=matrix(nrow=niter,ncol=p)
bL[1,]=0 # initial lasso estimates set to zero
r1 <- y-mean(y)
for (i in 2:niter) {
  for(j in 1:p){
    r0 <- r1+X[,j]*bL[i-1,j]
    bLS <- crossprod(X[,j],r0)/n # LEAST SQUARES ESTIMATE
    if (abs(bLS) >= lambda){bL[i,j]<-sign(bLS)*(abs(bLS)-lambda)
    } else{
      bL[i,j] <- 0
    }
    r1 <- r0-X[,j]*bL[i,j]
  }
}
fm=glmnet(y=y,x=X,alpha=1,lambda=lambda)
# alpha=1: LASSO; alpha=0: RIDGE; 0<alpha<1: ELASTIC NET
# Number covariates included with GLMNET:
length(which(fm$beta!=0))
```

```
## [1] 13
```

```
# Number included with present code
length(which(bL[niter,]!=0))
```

```
## [1] 13
```

```
# PRINT A FEW ESTIMATES OBTAINED WITH GLMNET
round(fm$beta[which(fm$beta!=0)][1:7],3)
```

```
## [1] 11.162 0.311 1.345 0.028 0.355 0.620 1.256
```

```
# AND THE SAME WITH PRESENT CODE
round(bL[niter,which(bL[niter,]!=0)][1:7],3)
```

```
## [1] 11.16  0.312 1.344 0.028 0.355 0.620 1.256
```

Out of the $p = 100$ covariates in the full model, lasso retains 13.

A Bayesian Interpretation of the Lasso

Tibshirani (1996) indicated that Lasso estimates of the elements of β can be interpreted as posterior mode estimates, when these regression parameters have independent and identical Laplace (i.e., double exponential) priors of the form

$$p\left(\beta_j\right) = \frac{\tau}{2} \exp\left(-\tau \left|\beta_j\right|\right) \tag{7.26}$$

and observations are realisations from a normal process. More specifically, assume that centred data (no intercept required in the model) arise from

$$y|\beta, \sigma^2 \sim N\left(X\beta, I\sigma^2\right) \tag{7.27}$$

and

$$p\left(\beta|\tau\right) = \left(\frac{\tau}{2}\right)^p \exp\left(-\tau \sum\nolimits_{j=1}^{p} \left|\beta_j\right|\right). \tag{7.28}$$

In (7.27), y is a column vector with n elements, X is $n \times p$ matrix, and β is a column vector with p elements. Then the posterior density of β takes the form

$$\begin{aligned} p\left(\beta|\sigma^2, \tau, y\right) = \left(2\pi\sigma^2\right)^{-\frac{n}{2}} \exp\left[-\frac{1}{2\sigma^2}\left(y - X\beta\right)'\left(y - X\beta\right)\right]\left(\frac{\tau}{2}\right)^p \\ \exp\left(-\tau \sum\nolimits_{j=1}^{p} \left|\beta_j\right|\right) \end{aligned}$$

and the logposterior density, including only those terms that contain β, is given by

$$\ln\left(p\left(\beta|\sigma^2, \tau, y\right)\right) = -\frac{1}{2\sigma^2}\left(y - X\beta\right)'\left(y - X\beta\right) - \tau \sum\nolimits_{j=1}^{p} \left|\beta_j\right|. \tag{7.29a}$$

Multiplying out by $2\sigma^2$ the posterior mode results in the lasso solution

$$\begin{aligned} \hat{\beta}_L &\underset{\arg\max b}{=} -\left(y - X\beta\right)'\left(y - X\beta\right) - \lambda \sum\nolimits_{j=1}^{p} \left|\beta_j\right| \\ &\underset{\arg\min b}{=} \left(y - X\beta\right)'\left(y - X\beta\right) + \lambda \sum\nolimits_{j=1}^{p} \left|\beta_j\right| \end{aligned}$$

where $\lambda = 2\sigma^2\tau$. It is important to stress that the equivalence between the Bayesian lasso and the standard lasso occurs at the level of the modal value of the posterior distribution. The posterior mean is not equal to the posterior mode and does not produce the same lasso shrinkage of coefficients to zero. A fully Bayesian implementation of the lasso can be found in Park and Casella (2008).

7.3 An Extension of the Lasso: The Elastic Net

Some of the properties of the lasso are seen as shortcomings. For example, in the $p > n$ situation (n is the number of records and p is the number of covariates), lasso selects at most n covariates. This is not desirable when the analysis involves few observations and many predictors contribute to the response. Also, in models with many highly correlated predictors, lasso tends to pick one arbitrarily and ignores the rest. On the other hand, ridge regression tends to shrink coefficients towards each other, but all will be included in the model; none is set to zero. In the context of genomic models, there may be many highly correlated SNPs that cloud around unobserved causal variants. In such cases, lasso tends to choose one among the correlated set and eliminate the remaining predictors. Arguably, more of the signal can be captured by a linear combination of correlated predictors than by inclusion of only one. In contrast, ridge regression includes all the coefficients in the model despite the shrinkage, including the superfluous ones. The elastic net proposed by Zou and Hastie (2005) is a compromise between ridge regression and lasso and circumvents some of these limitations. The elastic net solves the following problem (Zou and Hastie 2005):

$$\min_{(\beta_0,\beta)\in\mathbb{R}^{p+1}} R_\lambda(\beta_0,\beta) = \min_{(\beta_0,\beta)\in\mathbb{R}^{p+1}} \left[\frac{1}{2N}\sum_{i=1}^{N}\left(y_i - \beta_0 - x_i'\beta\right)^2 + \lambda P_\alpha(\beta) \right] \tag{7.30}$$

where

$$P_\alpha(\beta) = \sum_{i=1}^{p} \left[\frac{1}{2}(1-\alpha)\beta_j^2 + \alpha\left|\beta_j\right| \right].$$

$P_\alpha(\beta)$ is the elastic net penalty and is a compromise between ridge regression penalty ($\alpha = 0$) and the lasso penalty ($\alpha = 1$). The quadratic part of the penalty removes the limitation on the number of variables selected and encourages grouping effects. The L_1 penalty associated with the lasso generates a sparse model. The public package **glmnet** (Friedman et al 2009) implemented in `R` allows the user to choose among ridge regression, lasso or elastic net and is illustrated as part of the example below.

7.4 Example: Prediction Using Ridge Regression and Lasso

This simulation example illustrates the prediction ability of ridge regression and lasso using two data sets that mimic two different genetic scenarios. The data in both scenarios consist of 1500 individuals and 1500 correlated marker genotypes at intermediate gene frequencies. The 1500 marker genotypes are in 75 independent blocks with 20 marker genotypes in each block. Marker genotypes are correlated within blocks and uncorrelated among blocks. The correlation between adjacent markers is approximately 0.60 and decays with marker distance.

The two data sets differ in the number of QTL. In the first set, 25 out of the 1500 markers are randomly chosen as QTL, and in the second set, this number is 250. The substitution effects are the same for all QTL within sets of data and are chosen such that the equilibrium additive genetic variance (defined as $2 \sum_{QTL} p_i (1 - p_i) \beta_i^2$, where β_i is the allele substitution effect of the ith QTL, the same for all i) is equal to 10 squared units in both sets of data. The environmental variance is equal to 30 squared units, and therefore the trait heritability is equal to 0.25. In the first data set, each of the 25 QTL explains 1% of the phenotypic variance; in the second, each of the 250 QTL explains 0.1% of the phenotypic variance.

The 1500 individuals were divided into a training set of size 750, and a validating or testing set of size 750 and a ridge regression or lasso was fitted using the package **glmnet**. The performance of the predictive models was measured by the sample validating mean squared error (MSE) and the correlation between the predicted phenotypes (using estimates of substitution effects obtained using the training data) and observed phenotypes (from the validating data). In the data set with 25 QTL, the lasso is expected to outperform ridge regression as the latter includes all 1500 markers in the model, including the redundant ones. In the data with 250 QTL, the lasso limits the choice of the number of markers to be included in the predictive model within the block of correlated loci, failing in this way to capture an important part of the QTL (causal) additive genetic variance. This will lead to a relatively larger bias in the predictions. On the other hand, ridge regression includes all markers in the predictive model and captures a larger proportion of the QTL additive genetic variance. The possible increase in variance of the predictions is attenuated by a diminution of the squared bias, leading to a better predictive performance as measured by the MSE.

These expectations are borne out by the results, albeit drawn from one replication. With 25 QTL, the MSE (correlation) of the lasso is 34.6 (0.30) and of the ridge regression 36.4 (0.20). The lasso includes 112 QTL in the model. With 250 QTL, the MSE (correlation) of the lasso is 43.6 (0.27) and of the ridge regression 40.9 (0.35). In this case, the lasso includes 145 QTL.

These numbers can be put in perspective using the true value of the QTL substitution effects β_{true}, as predictions of the phenotypes in the validating (or test) data. The predictions are $X_{test}\beta_{true}$, where X_{test} is the matrix of QTL genotypes in the test or validating data. The MSE (correlation) is 28.5 (0.48). In this case, the MSE and the correlation measure, respectively, the $\text{Var}(y|X)$ and the heritability of the trait realised in the training sample of 750 individuals. In the population, these parameters are 30 square units and $\sqrt{0.25} = 0.5$, respectively. This is as good a prediction as it gets!

An extract of the `R`-code (adapted from James et al 2017, pages 251–255) used in the implementation of **glmnet** is shown below. Documentation about **glmnet** can be found in Hastie and Qian (2016) and in Friedman et al (2010). In this example, the function **cv.glmnet** is executed in a first step on the training data in order to find the value of the tuning parameter (best value of λ) of the lasso (using $\alpha = 1$ as an argument) or of the ridge regression (using $\alpha = 0$ as an argument) that optimises prediction ability measured by MSE. Execution of **predict** in a second step yields predictions of phenotypes in the testing data and the associated validation MSE (mean((`fm.pred-y.test`)^2)):

```
# Lasso solutions using package glmnet
#install.packages("glmnet", .libPaths()[1])
library(glmnet)
# UPLOAD MATRIX OF MARKER GENOTYPES X AND RECORDS y
n<-nrow(X)
Xlasso<- X
train=sample(1:nrow(Xlasso),nrow(Xlasso)/2)
test=(-train)
y.test=y[test] # y is the phenotypic data generated separately

# **********  FOR PREDICTION   *****************
# STEP 1

cv.out=cv.glmnet(Xlasso[train,],y[train],alpha=1)

plot(cv.out)
bestlam=cv.out$lambda.min
bestlam

# TO OBTAIN THE NUMBER OF NON-ZERO COEFFICIENTS CAN EXECUTE:

length(which(as.vector(coef(cv.out, s=bestlam))!=0))

# STEP 2

# Having obtained the best lambda fit the model
# on the testing data using this lambda to obtain predictions

fm.pred=predict(cv.out,s=bestlam,newx=Xlasso[test,])
mean((fm.pred-y.test)^2) # VALIDATION MSE
correl<-cov(fm.pred,y.test)/sqrt((var(fm.pred)*var(y.test)))
# AS GOOD AS IT GETS:
predtrue<-Xlasso[test,]%*%be # PREDICTION BASED ON TRUE MODEL
mean((predtrue-y.test)^2) # MSE USING TRUE MODEL
denominator <- sqrt(var(predtrue)*var(y[test]))
correltrue<-cov(predtrue,y[test])/denominator)
```

7.5 A Bayesian Spike and Slab Model

The example on page 127, *A two-component mixture model*, assumes that records are realisations from either of two unobserved mixture components and the objective is to assign each *observed* data point to a specific component. The mixture components are both Gaussian with a common variance that differ in mean value.

In the hierarchical setup discussed here, conditional on marker effects, records are realisations from a Gaussian distribution, but at the second level of the hierarchy, marker effects arise from either of two unobserved mixture components. In contrast with the example of page 127, here, the objective is to allocate *unobserved* genetic marker effects to each mixture component. Genetic marker effects drawn from one of the components have an influence on the trait of interest, while those drawn from the alternative mixture component have no detectable influence. The allocation is not unambiguous: it uses the laws of probability.

There are a variety of alternatives to model mixture components. One option is to assume that the mixtures are two Gaussians that differ in their means and/or variances. In this case, the marker effects do not take the value zero, exactly, but their posterior means can be very small depending on the parameters of the Gaussian components. This makes it possible to distinguish the two mixture components. Another option takes the two-component mixture to be made up of a normal distribution and a degenerate distribution with a point mass at zero. This modelling option sets a number of marker effects exactly equal to zero, as in the lasso. An example of the first option implemented using Gibbs sampling is presented in George and McCulloch (1993), and an McMC implementation of the second can be found in Habier et al (2011).

In the context of prediction, mixtures involving Gaussian distributions are likely to outperform the alternative that sets marker effects exactly equal to zero. This is so because a larger proportion of small effects variants that contribute to the phenotype are bound to be included in the final prediction model. On the other hand, the model that includes a point mass at zero constitutes arguably a better detection tool. However, this statement, particularly in a prediction setup, must be interpreted in the light of the variance-bias trade-off, where the size of the training sample, statistical properties of the prediction model, covariance structure of the marker/causal genotypes and genetic architecture of the trait play important roles in defining the best choice (Hayes et al 2010).

In the so-called spike and slab model that is considered here, the SNP effects are a priori mutually independent and drawn from a two-component mixture distribution made up of a normal distribution with mean zero and variance σ_b^2 (the slab), $N\left(0, \sigma_b^2\right)$ and a degenerate distribution with a point mass at zero, Δ_0 (the spike). Specifically, if b_i is the ith SNP effect and π denotes the a priori probability that this effect is non-zero, then

$$p\left(b_i|\sigma_b^2, \pi\right) = N\left(0, \sigma_b^2\right)\pi + \Delta_0\left(1 - \pi\right), \tag{7.31}$$

where Δ_0 represents a point probability mass at zero.

This section provides a detailed development of a Bayesian-McMC implementation of the spike and slab model and concludes with an example in the context of prediction. The problem of detection of genetic markers that have an effect on the trait is deferred to an example on page 361.

The Mixture Model

A description of the model involves first the assumption that observed records y (vector of dimension $n \times 1$) are a realisation from the Gaussian process

$$y|\mu, b, \sigma^2 \sim N\left(1\mu + Xb, I\sigma^2\right), \tag{7.32}$$

where 1 is a vector of ones of dimension $n \times 1$, μ is a scalar, X is an observed matrix of dimension $n \times m$ representing genotypic codes of the SNPs (originally scored as 0, 1 or 2) and σ^2 is the conditional variance of y, given the location parameters of the model μ and b.

As a reminder, in the standard model ignoring the mixture, the m SNP effects are elements of the vector b (dimension $m \times 1$) that are iid realisations from a common distribution, assumed to be Gaussian with mean zero and variance σ_b^2. Therefore,

$$b|\sigma_b^2 \sim N\left(0, I\sigma_b^2\right). \tag{7.33}$$

The parameter σ_b^2 represents prior uncertainty about the SNP effect, the same for each SNP. The columns of matrix X are centred but not standardised. Then for each element of $X = \{x_{ij}\}$, we have

$$x_{ij}|p_j \sim \left(0, \text{Var}\left(x_{ij}|p_j\right)\right), \quad \text{Var}\left(x_{ij}|p_i\right) = 2p_j\left(1 - p_j\right).$$

A genomic variance is often associated with (7.32) and (7.33) and in the absence of a mixture is defined as (conditional on gene frequencies $p_j, \quad j = 1, \ldots, m$)

$$\begin{aligned}
\sigma_g^2 &= \text{Var}\left(X_i'b\right) \\
&= \text{E}[\text{Var}\left(X_i'b|X_i\right) + \text{Var}\left[\text{E}\left(X_i'b|X_i\right)\right] \\
&= \text{E}\left[X_i'X_i\right]\sigma_b^2 \\
&= \sigma_b^2 \sum_{j=1} \text{E}\left(x_{ij}^2\right), \quad \text{with } \text{E}\left(x_{ij}^2\right) = \text{Var}\left(x_{ij}\right) = 2p_j\left(1 - p_j\right).
\end{aligned} \tag{7.34}$$

In these expressions, X_i' is the ith row of matrix X, corresponding to the ith genomic value.

From (7.34),

$$\sigma_b^2 = \frac{\sigma_g^2}{\sum_{j=1} 2p_j\left(1-p_j\right)}. \tag{7.35}$$

The term $2p_j\left(1-p_j\right)$ can be approximated by computing the sampling variance among the elements of the jth column of X.

In order to introduce the mixture, it is convenient to define the ith SNP effect as

$$b_i = \alpha_i \delta_i, \quad i = 1, \ldots, m, \tag{7.36}$$

where the random variable α_i is distributed as $\alpha_i|\sigma_b^2 \overset{iid}{\sim} N\left(0, \sigma_b^2\right)$, independent of the binary (0, 1) indicator variable δ_i that is a priori Bernoulli distributed with probability π:

$$\begin{aligned} &\delta_i|\pi \overset{iid}{\sim} Br(\pi), \quad , i = 1, \ldots, m, \\ &\Pr(\delta_i = 1|\pi) = \pi, \quad \Pr(\delta_i = 0|\pi) = 1 - \pi. \end{aligned} \tag{7.37}$$

This binary indicator variable with its associated distribution $\Pr(\delta_i = 1|\pi)$ specifies the a priori probability that a marker effect b_i is non-zero; this prior probability is the same for all markers. As indicated below, this is in contrast with the posterior probability that marker effect b_i is non-zero, $\Pr(\delta_i = 1|y)$, distinct for each marker. One Monte Carlo estimator of this probability is given by the sampling average of the draws of δ_i from their posterior distribution.

The marginal prior density of b_i (marginalised over $[\delta_i|\pi]$) is

$$\begin{aligned} p\left(b_i|\sigma_b^2, \pi\right) &= \sum_{\delta} p\left(b_i, \delta_i|\sigma_b^2, \pi\right) \\ &= p\left(b_i|\delta_i = 1, \sigma_b^2, \pi\right) \Pr(\delta_i = 1|\pi) \\ &\quad + p\left(b_i|\delta_i = 0, \sigma_b^2, \pi\right) \Pr(\delta_i = 0|\pi) \\ &= N\left(0, \sigma_b^2\right)\pi + \Delta_0 (1 - \pi). \end{aligned} \tag{7.38}$$

retrieving (7.31). The mixture model specifies that when the unobserved indicator variable δ_i takes the value of 1, b_i is a realisation from a normal prior with mean zero and variance σ_b^2; when $\delta_i = 0$, $b_i = 0$ with probability 1. The probability π can be interpreted as the a priori proportion of marker loci with non-null effects.

The genomic variance under the mixture model is

$$\begin{aligned}\sigma_g^2 &= \mathrm{Var}\left(\sum_{j=1}^m x_{ij}\alpha_j\delta_j\right)\\ &= \mathrm{E}_{x\delta}\left[\mathrm{Var}\left(\sum_{j=1}^m x_{ij}\alpha_j\delta_j|x_{ij},\delta_j\right)\right]+\mathrm{Var}_{x\delta}\left[\mathrm{E}\left(\sum_{j=1}^m x_{ij}\alpha_j\delta_j|x_{ij},\delta_j\right)\right]\\ &= \mathrm{E}_{x\delta}\left[\sigma_b^2\sum_{j=1}^m x_{ij}^2\delta_j^2\right]\\ &= \left[\sigma_b^2\sum_{j=1}^m \mathrm{E}\left(x_{ij}^2\right)\mathrm{E}\left(\delta_j^2\right)\right]\\ &= \sigma_b^2\pi\sum_{j=1}^m 2p_j\left(1-p_j\right). \end{aligned} \tag{7.39}$$

Therefore, along the same lines as in (7.35),

$$\sigma_b^2 = \frac{\sigma_g^2}{\pi\sum_{j=1}2p_j\left(1-p_j\right)}. \tag{7.40}$$

The conditional distribution of the data takes the form

$$y|\mu,\alpha,\delta,\sigma^2 \sim N\left(1\mu+\sum_{i=1}^m X_i\left(\alpha_i\delta_i\right), I\sigma^2\right), \quad ,i=1,\ldots,m \tag{7.41}$$

where X_i is the ith column of X.

The probability π can be assumed to follow a beta distribution, a priori, with user-tuned hyperparameters α and β,

$$\pi|\alpha,\beta \sim Be\left(\alpha,\beta\right). \tag{7.42}$$

The hyperparameters can be chosen in order to assign a relatively higher probability mass to low values of π, reflecting prior information about the proportion of genetic markers likely to be associated with the trait.

The scalar μ can be assumed to follow an improper uniform distribution, a priori.

The two variance parameters can be assigned, a priori, scaled inverted chi-square distributions:

$$\begin{aligned}\sigma_b^2|S_b,v_b &\sim S_b\chi^{-2}\left(S_b,v_b\right),\\ \sigma^2|S,v &\sim S\chi^{-2}\left(S,v\right),\end{aligned}$$

where the $S's$ and the $v's$ are user-tuned hyperparameters.

With the above specification, the prior density factorises as follows:

$$p\left(\mu,\alpha,\delta,\pi,\sigma_b^2,\sigma^2\right) \propto p\left(\alpha|\sigma_b^2\right)p\left(\delta|\pi\right)p\left(\pi\right)p\left(\sigma_b^2\right)p\left(\sigma^2\right) \tag{7.43}$$

and the posterior density takes the form

$$p\left(\mu,\alpha,\delta,\pi,\sigma_b^2,\sigma^2|y\right) \propto p\left(y|\mu,\alpha,\delta,\sigma^2\right) p\left(\alpha|\sigma_b^2\right) p\left(\delta|\pi\right) p\left(\pi\right) p\left(\sigma_b^2\right) p\left(\sigma^2\right). \tag{7.44}$$

McMC Implementation

The fully conditional posterior distributions (fcpd) are of standard form, and the model can be implemented using Gibbs sampling. The required fcpd are

$$[\mu|D], [\alpha|D], [\delta|D], [\pi|D], \left[\sigma_b^2|D\right], \left[\sigma^2|D\right],$$

where, as before (see page 222), D is a vector containing the observations y and all the parameters of the model except the one to be updated.

Updating $[\mu|D]$

From (7.44), the fcpd of the scalar μ is proportional to

$$\begin{aligned} p(\mu|D) &\propto p\left(y|\mu,\alpha,\delta,\sigma^2\right) \\ &\propto \exp\left[-\frac{1}{2\sigma^2}(y-1\mu-Xb)'(y-1\mu-Xb)\right] \end{aligned}$$

The quadratic form in this expression, as a function of μ, can be manipulated into

$$(y-1\mu-Xb)'(y-1\mu-Xb) \propto \mu 1'1\mu - 2\mu 1'(y-Xb). \tag{7.45}$$

Defining

$$(1'1)\,\hat{\mu} = 1'(y-Xb) \tag{7.46}$$

and replacing $1'(y-Xb)$ in (7.45) by $(1'1)\,\hat{\mu}$,

$$(y-1\mu-Xb)'(y-1\mu-Xb) \propto \mu 1'1\mu - 2\mu\,(1'1)\,\hat{\mu}.$$

Adding and subtracting $\hat{\mu}\,(1'1)\,\hat{\mu}$ and keeping terms containing μ only,

$$(y-1\mu-Xb)'(y-1\mu-Xb) \propto (\mu-\hat{\mu})\,(1'1)\,(\mu-\hat{\mu})$$

and therefore,

$$p(\mu|D) \propto \exp\left[-\frac{1}{2\sigma^2}(\mu-\hat{\mu})(1'1)(\mu-\hat{\mu})\right],$$

leading to the final result

$$[\mu|D] \sim N\left(\hat{\mu}, \sigma^2/n\right). \tag{7.47}$$

Updating $[\alpha_i|D]$

From (7.44), the fcpd is proportional to $p\left(y|\mu,\alpha,\delta,\sigma^2\right)p\left(\alpha|\sigma_b^2\right)$ where the first term is given by (7.41). There are two cases to consider. When $\delta_i = 0$, the product $\alpha_i\delta_i$ in (7.41) is equal to zero, the conditional distribution of the data is not a function of α_i, and α_i is updated from

$$p(\alpha_i|D) \propto p\left(\alpha_i|\sigma_b^2\right)$$

the density of the normal distribution

$$[\alpha_i|D] \sim N\left(0, \sigma_b^2\right). \tag{7.48}$$

In this case, α_i is updated from its prior distribution. When $\delta_i = 1$,

$$p(\alpha_i|D) \propto p\left(y|\mu,\alpha,\delta,\sigma^2\right)p\left(\alpha|\sigma_b^2\right)$$

which is the product of two normal densities: (7.41) and $N\left(0, \sigma_b^2\right)$. This fcpd has the same form as in (5.34), Example 5.3, *A regression model for correlated binary data* on page 220. Then, following the steps in that example involving the combination of two quadratic forms leads to

$$[\alpha_i|D] \sim N\left(\hat{\alpha}_i, \left(X_i'X_i + k\right)^{-1}\sigma^2\right), \tag{7.49}$$

where $k = \sigma^2/\sigma_b^2$, X_i is the ith column of X and $\hat{\alpha}_i$ satisfies

$$\left(X_i'X_i + k\right)\hat{\alpha}_i = X_i'\left(y - 1\mu - X_{-i}b_{-i}\right). \tag{7.50}$$

In (7.50), X_{-i} is matrix X with the ith column deleted, and b_{-i} is vector b (whose elements are $\alpha_i\delta_i$) with the ith element deleted.

Updating $[\delta_i | D]$

From (7.44), the fcpd of δ_i is proportional to $p\left(y|\mu,\alpha,\delta,\sigma^2\right)p\left(\delta|\pi\right)$. when $\delta_i = 0$, $\alpha_i\delta_i = b_i = 0$, $p\left(\delta|\pi\right) = \Pr\left(\delta_i = 0|\pi\right) = (1-\pi)$ and

$$\Pr\left(\delta_i = 0|D\right) \propto \exp\left[-\frac{1}{2\sigma^2}\left(y - 1\mu - X_{-i}b_{-i}\right)'\left(y - 1\mu - X_{-i}b_{-i}\right)\right](1-\pi). \tag{7.51}$$

When $\delta_i = 1$, $\alpha_i\delta_i = \alpha_i$, $p\left(\delta|\pi\right) = \Pr\left(\delta_i = 1|\pi\right) = \pi$ and

$$\Pr\left(\delta_i = 1|D\right) \propto \exp\left[-\frac{1}{2\sigma^2}\left(y - 1\mu - Xb\right)'\left(y - 1\mu - Xb\right)\right]\pi. \tag{7.52}$$

Let $\left(y - 1\mu - X_{-i}b_{-i}\right)'\left(y - 1\mu - X_{-i}b_{-i}\right) = RSS0$, the residual sum of squares of the model without the ith SNP, and let $\left(y - 1\mu - Xb\right)'\left(y - 1\mu - Xb\right) = RSS1$ denote the residual sum of squares of the model with all the m SNPs. Then the fcpd of the probability that $\delta_i = 0$ is the Bernoulli process with probabilities

$$\Pr\left(\delta_i = 0|D\right) = \frac{\exp\left[-\frac{1}{2\sigma^2}RSS0\right](1-\pi)}{\exp\left[-\frac{1}{2\sigma^2}RSS0\right](1-\pi) + \exp\left[-\frac{1}{2\sigma^2}RSS1\right]\pi} \tag{7.53}$$

and the complement

$$\Pr\left(\delta_i = 1|D\right) = \frac{\exp\left[-\frac{1}{2\sigma^2}RSS1\right]\pi}{\exp\left[-\frac{1}{2\sigma^2}RSS0\right](1-\pi) + \exp\left[-\frac{1}{2\sigma^2}RSS1\right]\pi}. \tag{7.54}$$

A Monte Carlo estimator of the (marginal) posterior probability that marker i is non-zero is given by

$$\hat{\varphi}_i = \hat{\Pr}\left(\delta_i = 1|y\right) = \frac{1}{l}\sum_{j=1}^{l}\delta_{ij}, \tag{7.55}$$

where δ_{ij} is the sampled value of δ for marker i at iterate j of the Gibbs sampler and l is the length of the Gibbs chain. To draw $\delta_{ij}^{[t]}$ at round t of the McMC sampler, compute (7.53), and draw a random variable u from a uniform distribution in (0, 1). If u is less than or equal to (7.53), set $\delta_{ij}^{[t]} = 1$; otherwise, set $\delta_{ij}^{[t]} = 0$.

Remarks

1. Residuals $RSS0$ and $RSS1$ are efficiently updated as indicated on page 316.
2. In a computing environment, (7.53) and (7.54) can be obtained as follows: Let $\theta_i = \Pr(\delta_i = 1|D)$ denote the fully conditional posterior distribution that marker i is non-zero. Then

$$\ln\left(\frac{\theta_i}{1-\theta_i}\right) = \frac{1}{2\sigma^2}(RSS0 - RSS1) - (\ln(1-\pi) - \ln(\pi)) = K_i \quad (7.56)$$

and

$$\theta_i = \frac{\exp(K_i)}{1+\exp(K_i)}. \quad (7.57)$$

When the McMC algorithm converges, the draw at iteration j, $\theta_i^{[j]}$, is an extraction from the marginal posterior distribution $\Pr(\delta_i = 1|y)$. Therefore, the algorithm provides two characterisations of $\Pr(\delta_i = 1|y)$: one in terms of the mean of the Monte Carlo samples δ_{ij}, $\hat{\varphi}_i$ defined in (7.55) that yields a Monte Carlo point estimator of the mean of the posterior distribution $\Pr(\delta_i = 1|y)$ and the other in terms of a Monte Carlo description of its complete posterior distribution through the samples $\theta_i^{[j]}$ via (7.57). Estimator (7.57) can be used to construct a Monte Carlo estimator of the marginal posterior distribution of the *false discovery rate*, a topic discussed in the next chapter on page 351.

Updating $[\pi|D]$

From (7.44), the density of the fcpd of π is proportional to $p(\delta|\pi)\,p(\pi)$ given by

$$\begin{aligned} p(\pi|D) &\propto \pi^{\sum_{i=1}^m \delta_i}(1-\pi)^{m-\sum_{i=1}^m \delta_i}\pi^{\eta-1}(1-\pi)^{\beta-1} \\ &= \pi^{\sum_{i=1}^m \delta_i+\eta-1}(1-\pi)^{m-\sum_{i=1}^m \delta_i+\beta-1}, \end{aligned}$$

which is the kernel of the density of a beta distribution with shape parameters $\sum_{i=1}^m \delta_i + \eta$ and $m - \sum_{i=1}^m \delta_i + \beta$. Specifically,

$$[\pi|D] \sim Be\left(\sum_{i=1}^m \delta_i + \eta, m - \sum_{i=1}^m \delta_i + \beta\right). \quad (7.58)$$

Updating $\left[\sigma_b^2|D\right]$

From (7.44), density of the fcpd of σ_b^2 is proportional to $p\left(\alpha|\sigma_b^2\right)p\left(\sigma_b^2\right)$,

$$
\begin{aligned}
p\left(\sigma_b^2|D\right) &\propto \left(\sigma_b^2\right)^{-\frac{m}{2}} \exp\left[-\frac{\alpha'\alpha}{2\sigma_b^2}\right]\left(\sigma_b^2\right)^{-\left(1+\frac{\upsilon_b}{2}\right)} \exp\left[-\frac{\upsilon_b S_b}{2\sigma_b^2}\right] \\
&= \left(\sigma_b^2\right)^{-\left(\frac{\upsilon_b+m}{2}+1\right)} \exp\left[-\frac{\alpha'\alpha+\upsilon_b S_b}{2\sigma_b^2}\right] \\
&= \left(\sigma_b^2\right)^{-\left(\frac{\widetilde{\upsilon}_b}{2}+1\right)} \exp\left[-\frac{\widetilde{\upsilon}_b\widetilde{S}_b}{2\sigma_b^2}\right],
\end{aligned} \tag{7.59}
$$

where $\widetilde{\upsilon}_b = \upsilon_b + m$ and $\widetilde{S}_b = \left(\alpha'\alpha + \upsilon_b S_b\right)/\widetilde{\upsilon}_b$. This is the kernel of a scaled inverse chi-square distribution with hyperparameters $\widetilde{\upsilon}_b$ and $\widetilde{S}_b$

$$
\left[\sigma_b^2|D\right] \sim \widetilde{\upsilon}_b\widetilde{S}_b\chi^{-2}\left(\widetilde{\upsilon}_b\right). \tag{7.60}
$$

To obtain a sample from (7.60), draw from a chi-square distribution with $\widetilde{\upsilon}_b$ degrees of freedom and the reciprocal of this number is multiplied by $\widetilde{\upsilon}_b\widetilde{S}_b = \alpha'\alpha + \upsilon_b S_b$.

Updating $\left[\sigma^2|D\right]$

From (7.44), the density of the fcpd of σ^2 is proportional to $p\left(y|\mu,\alpha,\delta,\sigma^2\right)p\left(\sigma^2\right)$,

$$
\begin{aligned}
p\left(\sigma^2|D\right) &\propto \left(\sigma^2\right)^{-\frac{n}{2}} \exp\left[-\frac{(y-1\mu-Xb)'(y-1\mu-Xb)}{2\sigma^2}\right]\left(\sigma^2\right)^{-\left(1+\frac{\upsilon}{2}\right)} \\
&\quad \exp\left[-\frac{\upsilon S}{2\sigma^2}\right] \\
&= \left(\sigma^2\right)^{-\left(\frac{\upsilon+n}{2}+1\right)} \exp\left[-\frac{(y-1\mu-Xb)'(y-1\mu-Xb)+\upsilon S}{2\sigma^2}\right] \\
&= \left(\sigma^2\right)^{-\left(\frac{\widetilde{\upsilon}}{2}+1\right)} \exp\left[-\frac{\widetilde{\upsilon}\widetilde{S}}{2\sigma^2}\right],
\end{aligned}
$$

where $\widetilde{\upsilon} = \upsilon + n$ and $\widetilde{S} = \left((y-1\mu-Xb)'(y-1\mu-Xb)+\upsilon S\right)/\widetilde{\upsilon}$. This is the kernel of a scaled inverse chi-square distribution with hyperparameters $\widetilde{\upsilon}$ and $\widetilde{S}$,

$$
\left[\sigma^2|D\right] \sim \widetilde{\upsilon}\widetilde{S}\chi^{-2}\left(\widetilde{\upsilon}\right). \tag{7.61}
$$

To obtain a sample from (7.61), draw from a chi-square distribution with $\widetilde{v}$ degrees of freedom and the reciprocal of this number is multiplied by $\widetilde{v}\widetilde{S} = (y - 1\mu - Xb)'(y - 1\mu - Xb) + vS$.

Example: Spike and Slab Model

Using the example described on page 319, the spike and slab model is fitted to the same two data sets. As before, each data set is divided into 750 training (t) records and 750 validating (v) records. Let

$$W = (1\,X),$$
$$\theta' = (\mu, b').$$

Using obvious notation, a predictor of the validating records based on the spike and slab model is

$$\hat{y}_v = W_v \hat{\theta}, \tag{7.62}$$

where $\hat{\theta} = \hat{\mathrm{E}}(\theta|X_v, y_t)$, the Monte Carlo estimate of the marginal posterior mean using the training data y_t. In common with the examples used for the ridge regression and the lasso, this is also a point prediction that does not account for uncertainty but is useful for comparing the prediction ability of various models. It addresses the general question: Given the model, what is the predicted mean value of future data given $\hat{\theta}$, at the value of the covariates X_v?

In the first data set with 25 QTL, the MSE (correlation) obtained fitting the spike and slab model is 34.1 (0.29). In the second data set with 250 QTL, the MSE (correlation) is 41.3 (0.34). The predictive performance of the spike and slab model is very similar to the lasso in the first data set and slightly better than the lasso and similar to the ridge regression in the second data set. The spike and slab model involves inferring the $four$ parameters (over and above the SNP effects): the mean μ, the probability π and the two variance components σ_b^2 and σ^2. These were estimated using the training data and not using cross-validation, as was the case with lasso and ridge regression. Arguably, a better predictive performance can be expected if these parameters were chosen via cross-validation.

One can exploit the flexibility afforded by the McMC environment to compute the marginal posterior distribution of functions of the parameters of the model. Two such functions are mean squared errors that account for particular sources of uncertainty. An example is displayed in Fig. 7.6. The left panel is a histogram of the marginal posterior distribution of the validating mean squared error reflecting posterior uncertainty in the parameters of the Bayesian model. It is obtained by computing $\hat{y}_v^{[t]} = W_v \theta^{[t]}$ for each McMC cycle t and the associated validation MSE.

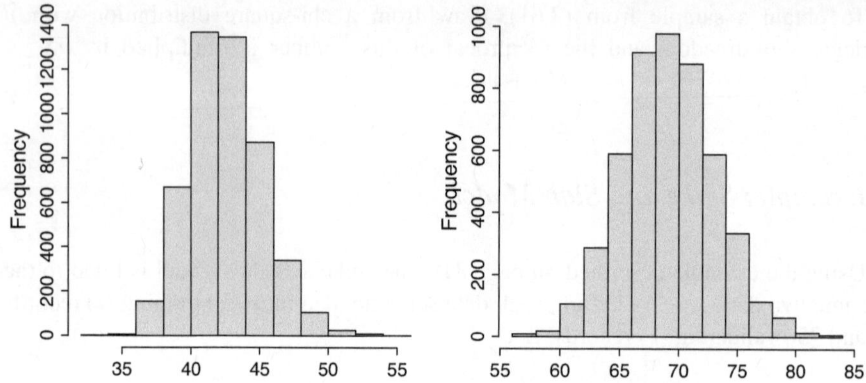

Fig. 7.6 Left: histogram of the Monte Carlo estimate of the marginal posterior distribution of the validating mean squared error accounting for posterior uncertainty of the parameters of the spike and slab model. Right: histogram of the Monte Carlo estimate of the marginal posterior distribution of the validating mean squared error accounting for posterior uncertainty of the parameters of the spike and slab model and for sampling variation of new validating data

The right panel of Fig. 7.6 shows the Monte Carlo estimate of the validating mean squared error calculated in yet another way. Here, predictions of single validating records are drawn from

$$\hat{y}_v^{[t]} \sim N\left(W_v\theta^{[t]}, I\sigma^{2[t]}\right)$$

and in this manner, account is taken of the uncertainty in (θ, σ^2) and of the sampling variation of new observations. The issue of prediction uncertainty is revisited in Chap. 10.

An attractive property of the spike and slab model is that it may be useful for detection of QTL. Before illustrating this property, I take a detour from prediction and provide an overview of the concept of false discovery rate in the next chapter. In general terms, the problem at hand is how to go about finding as many promising genetic markers as possible among thousands or millions observed while incurring a relatively low proportion of false positives.

Chapter 8
Digression on Multiple Testing: False Discovery Rates

A classical single hypothesis test proceeds by specifying α, the probability of a significant result, given the null hypothesis ($H = 0$) is true. This is also known as the probability of a false discovery and more commonly of a type I error. If m independent hypotheses H_i are tested, the so-called *family wise error rate* (FWER) is the probability of making one or more type I errors among the family of hypothesis tests is

$$\Pr(\text{at least 1 false positive result in } m \text{ tests}|H_1 = 0, \ldots, H_m = 0) =$$
$$1 - \Pr(\text{no false positive results in } m \text{ tests}|H_1 = 0, \ldots, H_m = 0)$$
$$= 1 - (1 - \alpha)^m. \tag{8.1}$$

With $\alpha = 0.05$ and $m = 50$, this yields

$$\Pr(\text{at least 1 significant result in } m \text{ tests}|H_1 = 0, \ldots, H_m = 0) = 0.92$$

by chance alone. In terms of p-values, a p-value threshold of 0.05 guarantees that the expected number of false positives over the whole family of m tests is less than or equal to $0.05\,m$, far too large for modern genomewide studies. Traditional methods for dealing with multiple testing call for adjusting α in some manner so that the FWER remains below a desired level. A popular method is the Bonferroni correction: for m hypothesis tests, each test is controlled so that the probability of a false positive is less than or equal to α/m. Then for the m tests, the overall FWER is less than or equal to α. This is a global test that addresses the question: Is there any null hypothesis that is rejected?

The Bonferroni result is readily derived using Boole's inequality and makes no assumptions about the degree of dependence among the tests (see Note 1). Therefore, in the example above with $m = 50$ tests, the (global) null hypothesis is rejected if the p-value for any particular hypothesis is less than $(0.05/50) =$

D. Sorensen, *Statistical Learning in Genetics*, Statistics for Biology and Health,
https://doi.org/10.1007/978-3-031-35851-7_8

0.001. Controlling FWER is useful when the number of hypotheses tested is small. However, in many modern genomewide studies where many hypotheses are tested, the Bonferroni correction can lead to a high rate of type II errors or false negatives (missing good candidates). The method tries too hard to make it unlikely that even one false rejection of the null is made.

An alternative that is particularly attractive for large-scale multiple testing is to identify as many significant features as possible while keeping the proportion of false discoveries as low as possible *among these significant features*. This is the goal of the *false discovery rate* (FDR) that was first proposed from a frequentist perspective in a much celebrated paper by Benjamini and Hochberg (1995). Since its publication, the method has played a very important role in multiple testing and has undergone a number of refinements.

This chapter provides an overview of FDR starting with the classical approach proposed by Benjamini and Hochberg (1995). The method can also be anchored in a Bayesian framework and two approaches are presented. The first is an empirical Bayes approach (Efron et al 2001) that uses as input summary statistics (such as p-values or z-values) embedded in a two-component mixture model. Secondly, a fully Bayesian hierarchical model is described and implemented with McMC. One output of this Bayesian model is a Monte Carlo estimate of the marginal posterior distribution of FDR.

The application of these methods is illustrated with two examples: one is based on the Gaussian, two-component mixture model introduced on page 127; the other is based on the spike and slab model introduced on page 321.

Note 1

The Bonferroni correction can be derived using Boole's inequality (Casella and Berger 1990). Assume that $H_1, H_2, \ldots, H_m$ hypotheses are to be tested, and let $p_1, p_2, \ldots, p_m$ be their p-values. Assume that m_0 are unknown true null hypotheses. The Bonferroni correction rejects the ith null hypothesis when $p_i \leq \alpha/m$ and, in so doing, controls the FWER at level $\leq \alpha$. The proof is as follows (recall that $\mathbb{P}(A \cup B) = \mathbb{P}(A) + \mathbb{P}(B) - \mathbb{P}(A \cap B)$, where $\cup$ is the union symbol and $\cap$ is the intersection symbol):

$$\text{FWER} = \Pr\left[\cup_{i=1}^{m_0}\left(p_i \leq \frac{\alpha}{m}\right)\right] \leq \sum_{i=1}^{m_0} \Pr\left(p_i \leq \frac{\alpha}{m}\right) = m_0\frac{\alpha}{m} \leq m\frac{\alpha}{m} = \alpha.$$

This result does not require knowledge of the number of true null hypotheses, nor does it assume independence of the tests.

An interesting alternative proof is as follows (Benjamini 2013, *Selective Inference and False Discovery Rate*, Public Lecture at UC Berkeley): The starting point is as above, where m hypotheses are to be tested, and among these, m_0 (unknown) are true null hypotheses. Let $R_i, i = 1, 2, \ldots, m$ denote the binary random variable equal to 1 if H_i is rejected (a discovery is made) or 0 if it is not rejected. Let $V_i = 1$ if $R_i = 1$ but H_i is true (a false discovery is made or a type I error) and 0 otherwise. Then $R = \sum R_i$ is the number of hypotheses rejected and $V = \sum V_i$

is the number of null hypotheses rejected in error (number of false discoveries). Consider a procedure that tests each of the m hypotheses separately at some level α_B. Then $\Pr(V_i = 1) = \mathrm{E}(V_i) \leq \alpha_B$, and the expected number of false discoveries over the m tests V, also known as the per family error rate (PFER), is

$$\mathrm{E}(V) = \mathrm{E}\Big(\sum_{i=1}^{m} V_i\Big) = \sum_{i=1}^{m} \mathrm{E}(V_i) \leq m_0\alpha_B \leq m\alpha_B.$$

Therefore, to achieve a level of the expected number of false discoveries for the family of tests, $\mathrm{E}(V)$, smaller than or equal to α, it is sufficient to choose $\alpha_B = \alpha/m$. Interestingly, the Bonferroni correction ensures that achieving $\mathrm{E}(V) \leq \alpha$ also guarantees that the FWER $= \Pr(V \geq 1) \leq \alpha$. Indeed,

$$\begin{aligned}
\mathrm{E}(V) &= 0\Pr(V=0) + 1\Pr(V=1) + 2\Pr(V=2) + \cdots + m\Pr(V=m) \\
&\geq 0 + 1\Pr(V=1) + 1\Pr(V=2) + \cdots + 1\Pr(V=m) \\
&= 0 + \Pr(V \geq 1) \leq \alpha.
\end{aligned}$$

Therefore, when using $\alpha_B = \alpha/m$ for individual tests

$$FWER = \Pr(V \geq 1) \leq \mathrm{E}(V) \leq \alpha.$$

8.1 Introduction

Consider an experiment where m hypotheses are tested. This could be a GWAS experiment designed to isolate a few SNPs with an effect on a trait of interest out of a very large number m tested. Let δ represent the unobserved binary random variable that takes the value $\delta_i = 0$ if the ith hypothesis is a true null hypothesis ($H_i = 0$) (e.g., the ith SNP has no effect on the trait), or $\delta_i = 1$ if the alternative hypothesis is true ($H_i = 1$) (the ith SNP has an effect on the trait) $i = 1, \ldots, m$. Let the observed binary random variable $r_i = 1$ if the ith hypothesis is rejected and 0 otherwise according to a particular rule. Given observation y_i, the decision to reject the null could be based on the rule $r_i = I(y_i \geq y_t)$ for some threshold value y_t or more commonly using "z-values". More generally, the rule can be to reject the null if a test statistic z falls in a rejection region A, $r_i = I(z_i \in A)$. The total number of rejections (an observed random variable) depends on the rule chosen and is $R = \sum_{i=1}^{m} r_i$ (equal to the number of discoveries or number of significant results and constitutes the discovery set). The number of true discoveries is $\sum_{i=1}^{m} r_i\delta_i$. The number of false discoveries or number of rejections in error is

$$V = \sum_{i=1}^{m} I(\delta_i = 0) \times I(r_i = 1) = \sum_{i=1}^{m} (1-\delta_i)r_i. \tag{8.2}$$

The FWER is equal to $\Pr(V \geq 1)$ and the *false discovery proportion*, the proportion of false discoveries among the discoveries, is

$$Q = \frac{V}{R}, \quad R > 0, \tag{8.3a}$$

$$= 0, \quad R = 0. \tag{8.3b}$$

Q is unobservable because V is unknown. It involves the unobserved parameters δ_i, the decisions taken on the basis of rule r_i and the data indirectly through the decisions.

A variety of approaches have been suggested to infer Q and three of these are outlined below. The first uses a frequentist paradigm and is the original false discovery rate of Benjamini and Hochberg (1995). It is defined as the expectation of (8.3). It treats hypotheses as not random. The second and third are based on the Bayesian paradigm where hypotheses are treated as unobserved random quantities. This chapter reviews an empirical Bayes approach (Efron et al 2001) and presents a fully Bayesian model driven with McMC.

8.2 Preliminaries

Before embarking on the subject, two standard results are described; these will be used in the coming sections. The first concerns the distribution of p-values and the second the use of the inverse transform method to generate a random variable with a desired distribution:

1. Consider a random variable $X \in \mathbb{R}$ with probability density function $f_X(x) = \frac{d}{dx}F(x)$, and let $y = F(x)$, such that $x = F^{-1}(y)$, where the continuous function F (a cumulative distribution function, invertible and differentiable) maps x from the real line $\mathbb{R}$ to the $(0, 1)$ interval; $F : \mathbb{R} \rightarrow (0, 1)$, and F^{-1} is the inverse transformation, $F^{-1} : (0, 1) \rightarrow \mathbb{R}$. The probability density function of Y, $f_Y(y)$, is obtained using the theory of transformations. This gives

$$f_Y(y) = f_X\left(F^{-1}(y)\right)\left|\frac{dF^{-1}(y)}{dy}\right|, \quad Y \in (0, 1). \tag{8.4}$$

   Noting that $\frac{dF^{-1}(y)}{dy}$ in (8.4) can be written

$$\frac{dx}{dF(x)} = \left(\frac{dF(x)}{dx}\right)^{-1}$$

$$= \frac{1}{f_X(x)}, \quad x = F^{-1}(y),$$

and substituting in (8.4):

$$f_Y(y) = f_X\Big(F^{-1}(y)\Big)\frac{1}{f_X\big(F^{-1}(y)\big)} = 1, \quad Y \in (0, 1) \tag{8.5}$$

which is the uniform distribution Un(0, 1).

The classical application of this result concerns the distribution of p-values under a null hypothesis. When the alternative hypothesis corresponds to small values of the test statistic X, the p-value is $F(x)$, where x is the observed value of the test statistic. The p-values are then uniformly distributed in the interval (0, 1), whatever the (continuous) distribution of X might be. A similar argument holds when the alternative hypothesis corresponds to large values of the test statistic X.

2. The following algorithm generates a random variable with desired distribution F:

- Generate $U \sim \text{Un}(0, 1)$
- Set $X = F_X^{-1}(U)$

and X has the desired distribution $F_X(\cdot)$. To see this, note that

$$\begin{aligned} \Pr(X \le x) &= \Pr\Big(F_X^{-1}(U) \le x\Big) \\ &= \Pr(U \le F_X(x)) \\ &= \int_0^{F_X(x)} 1\, du \\ &= F_X(x). \end{aligned} \tag{8.6}$$

This method of generating the random variable is known as the inverse transform.

An illustration is the computation of z-values from uniformly distributed p-values. Assume two samples drawn from independent normally distributed populations with unknown means μ_i, $i = 1, 2$ and a common, unknown variance. Let $\overline{x}_i$ and $\hat{Var}(\overline{x}_i)$ denote the sample mean and an estimate of the variance of the sample mean, respectively, calculated using n_i records, $i = 1, 2$. Then the observed so-called t-value is

$$t = \frac{\overline{x}_1 - \overline{x}_2}{\sqrt{\widehat{\text{Var}}(\overline{x}_1 - \overline{x}_2)}}.$$

Given $H_0 : \mu_1 = \mu_2$, the random variable T whose realised value is t has a student-t distribution with n_1+n_2-2 degrees of freedom. For a level $\alpha = \Pr(\text{reject } H_0|H_0)$ test, the null hypothesis is rejected if the p-value

-

$$p = \Pr(T \ge t|H_0) = 1 - F_0(t) \le \alpha \tag{8.7}$$

in favour of $H_1 : \mu_1 > \mu_2$,

- if

$$p = \Pr(T \leq t|H_0) = F_0(t) \leq \alpha \tag{8.8}$$

in favour of $H_1 : \mu_1 < \mu_2$ and

- if

$$p = 2(1 - \Pr(T \leq |t|\,|H_0)) = 2(1 - F_0(|t|)) \leq \alpha \tag{8.9}$$

in favour of $H_1 : \mu_1 \neq \mu_2$, where F_0 is the cumulative distribution function of T under H_0. In the last case, $\Pr(T \leq -t|H_0) = \Pr(T \geq t|H_0) = \frac{1}{2}\alpha$. Using (8.5), given H_0, the random variable p is uniformly distributed $\text{Un}(0, 1)$. Now, compute the z-value

$$z = \Phi^{-1}(p) \tag{8.10}$$

where Φ is the cumulative distribution function of the standard normal. It follows from (8.6) that

$$z \sim N(0, 1). \tag{8.11}$$

8.3 The Benjamini-Hochberg False Discovery Rate

Benjamini and Hochberg (1995) define the false discovery rate (FDR-BH) as the expectation of the false discovery proportion. From (8.3),

$$\begin{aligned}\text{FDR} = \text{E}(Q) &= \text{E}\left[\frac{V}{R}|R > 0\right]\Pr(R > 0) + 0\Pr(R = 0)\\ &= \text{E}\left[\frac{V}{R}|R > 0\right]\Pr(R > 0),\end{aligned} \tag{8.12}$$

where the expectation is taken over replications of the experiment. A decision rule r_q controls FDR at level q, with q a pre-chosen value between 0 and 1 if

$$\text{FDR}(r_q) \leq q. \tag{8.13}$$

In other words, the expected proportion of false discoveries *among the hypotheses rejected* is less than or equal to q. This is in contrast with multiple testing based on the traditional p-values, which leads to an expected proportion of false discoveries *among all the tests performed* less than or equal to q.

Benjamini and Hochberg (1995) show that the decision rule is as follows: Let p_i be the ith element ($i = 1, 2, \ldots, m$) among the ordered p-values $p_1 \leq p_2 \leq \ldots \leq p_m$ corresponding to the m hypotheses H_i. Let

$$r_q = \max\{i : p_i \leq (i/m)q\}, \quad \text{then,} \tag{8.14a}$$

$$\text{Reject } H_1, H_2, \ldots, H_{r_q}. \tag{8.14b}$$

The realised V and R depend on the random outcome of the p-values in repeated realisations of the experiment. The rule (8.14) can be implemented as follows: Choose q. Start from the largest p-value with $i = m$ and check whether inequality (8.14a) holds. Move towards decreasing values of i one step at a time, checking for the inequality at each i. Stop at the ith value the first time $p_i \leq (i/m)q$ and set $r_q = i$. This is the largest i for which the inequality holds. Any i smaller than r_q for which the inequality holds is ignored. Then reject $H_1, H_2, \ldots, H_{r_q}$. If no p-value satisfies inequality (8.14a), then no hypothesis test is declared significant. Common choices for q are 0.05 and 0.10.

Benjamini and Hochberg (1995) prove that if the p-values are independent of each other, then the rule r_q controls the expected false discovery proportion at level

$$\text{FDR}(r_q) = \pi_0 q \leq q, \tag{8.15}$$

where $\pi_0 = m_0/m$ is the proportion of null hypotheses (the true proportion of SNP that do not have an effect on the trait). Therefore, the FDR-BH controls the expected false discovery rate exactly at level $\pi_0 q$ and conservatively at level q. Since the number of null hypotheses m_0 is not known, it is usual to specify the inequality in terms of expression (8.13). In most modern applications, the ratio π_0 is close to 1, but it can be estimated as in Storey (2002) and Benjamini et al (2006) or adopting a likelihood or fully Bayesian approach. Benjamini and Yekutieli (2001) proved that the above theorem holds under certain type of stochastic dependence among the p-values, and Storey and Tibshirani (2007) argued that the theorem holds asymptotically as m becomes large, under any form of dependence.

For any given set of data, Q may or may not be less than or equal to q. Rather, it is the expectation of Q that is smaller than or equal to q over conceptual replications of the experiment.

Expression (8.14a) can be written as

$$r_q = \max\left\{i : \frac{mp_i}{i} \leq q\right\}, \tag{8.16}$$

where mp_i is the expected number of false discoveries and i is the number rejected. The ratio mp_i/i is an intuitive expression for the false discovery proportion. However, the ratio is not monotonically related to the p-values, and moving to a lower p-value may result in a higher mp_i/i. Therefore, the discovery sets for a given FDR obtained using the rule (8.14) and (8.16) may differ. In order to align both and enforce monotonicity, Yekutieli and Benjamini (1999) suggest the following procedure: Let $k = \max\{i : p_i \leq (i/m)q\}$. If such k exists, reject $H_1, \ldots, H_k$, as in (8.14a). The so-called q^*, the FDR-BH p-value adjustment, is

$$q_i^* = q(p_i) = \min_{i \leq k}\left(p_k \frac{m}{k}\right), \quad \text{reject } H_i \text{ if } q_i^* \leq q, \tag{8.17}$$

which leads to the same decision as that based on (8.14a). In order to use the p-value adjustment, one can proceed as follows: First, order all p-values from small to large, multiply each p-value by the total number of tests m, and divide by the rank order i, $i = 1, \ldots, m$. Start from the largest $p_i m/i$ at the bottom of the sequence, and move upwards towards smaller values. Second, make sure that the resulting sequence is non-increasing: if it increases, set the increased value equal to the preceding value (repeatedly, until the whole sequence becomes non-increasing). Label the ith element of the sequence q_i^*. If q_i^* is smaller than or equal to q, reject $H_1, \ldots, H_i$. The kernel of the R−code to implement the algorithm is shown below:

```
# pv are the sorted p-values used as input
# smallest at the top
q=0.15 # SET FDR
qstar[m] <- pv[m] # Initialise qstar = largest p-value
minqstar <- qstar[m] # Initialise the minimum qstar
startloop <- m-1
for(i in startloop:1){ # loop towards smaller p-values
  qstar[i] <- pv[i]*(m/i)
  if(qstar[i]>=minqstar){
   qstar[i] <- minqstar # If new qstar (moving upwards towards
                        # smaller p-values) is larger than the
                        # previous qstar, set new qstar = old qstar
   }
  minqstar <- qstar[i] # update the minimum qstar
  if(minqstar <= q){break}
}
discovset <- i # SIZE OF DISCOVERY SET
```

As a simple illustration, consider the computation of q^* and of the traditional FDR-BH rule (8.14) using as input a list of sorted p- values. A small subset of the relevant records is shown in Table 8.1.

A total of $m = 100$ hypotheses are tested, and the FDR-BH rule is applied for a value of $q = 0.15$ in (8.14). Starting from the largest p-value at $i = m = 100$, check rule (8.14) for $H_{100}, H_{99}, H_{98} \ldots$. At $i = 19$, p_i is smaller than $(i/m)q$ (coloured row in the table). Then reject $H_1, \ldots, H_{19}$. This rule controls FDR at level $q = 0.15$. Exactly the same discovery set is obtained if one chooses to use q^*. For $i = 19$, $q_{19}^* \leq q = 0.15$ which leads to the same rejection of $H_1, \ldots, H_{19}$. These constitute the discovery set that is expected to include $19 * 0.15 \approx 3$ false discoveries.

The use of (8.16) in column 4 of the table generates a value of $p_i(m/i)$ for $i = 17$ that is larger than for $i = 18$. This is remedied in the computation of q_i^* shown in the fifth column of the table where q_{17}^* is set equal to q_{18}^*.

Table 8.1 A subset of $m = 100$ hypotheses indicating control of FDR at level $q = 0.15$. The columns are the index for the hypothesis $(i, i = 1, \ldots, m)$, the sorted p-values (p_i), rule (8.14) $((i/m)q)$, the values of (8.17), $(p_i(m/i))$ and the computation of q^* in column 5, (q_i^*)

i	p_i	$(i/m)q$	$p_i(m/i)$	q_i^*
15	0.01246737	0.0225	0.08311579	0.08311579
16	0.01860016	0.0240	0.11625099	0.11625099
17	0.02470245	0.0255	0.14530855	0.14071686
18	0.02532904	0.0270	0.14071686	0.14071686
19	0.02844954	0.0285	0.14973443	0.14973443
20	0.03269821	0.0300	0.16349105	0.16349105
21	0.04587966	0.0315	0.21847455	0.21847455
22	0.06910327	0.0330	0.31410579	0.30235174

8.4 A Bayesian Approach for a Simple Two-Group Mixture Model

Much of the material in this and the coming section has been adapted from Efron and Hastie (2016).

When the multiple testing problem is embedded in a Bayesian framework, the false discovery rate is expressed in terms of a posterior probability, rather than as the expectation (8.12) over repeated sampling of the data.

This section starts by considering a simple two-group mixture model involving m identical hypotheses, of which m_0 are null $(H = 0)$ with probability $\Pr(H = 0) = \pi_0 = m_0/m$ and $m - m_0$ are non-null $(H = 1)$ with probability $\pi_1 = 1 - \pi_0$. The probability π_0 is the a priori probability of null for each of the m hypotheses. The quantity m is observed, but m_0 is unknown. In this simple setup, H is the only unknown random variable and the focus of enquiry. In modern genomewide studies, involving million of genetic markers π_0 is much bigger than π_1, say, $\pi_0 > 0.95$.

The random binary quantity H is Bernoulli distributed,

$$H \sim \text{Bernoulli}(\pi_1).$$

This is in contrast to the frequentist approach where H is considered non-random. In the simple two-group mixture model considered here, the data are m test statistics such as "z values" associated with the m hypotheses. The density functions under the null and non-null hypotheses are

$$p(z|H = 0) = p_0(z),$$
$$p(z|H = 1) = p_1(z).$$

Marginally, Z has the mixture distribution

$$p(z) = \pi_0 p_0(z) + \pi_1 p_1(z).$$

The mixture model is equivalent to the hierarchical model

$$p_j(z) = p(z|\delta = j), \quad \Pr(\delta = j) = \pi_j, \quad j = 0, 1,$$

where δ is a binary random variable representing the true state of nature.

Consider a fixed rejection region A. Let

$$F_0(A) = \Pr(Z \in A|H = 0) = \int_A p_0(z)dz,$$
$$F(A) = \Pr(Z \in A) = \int_A p(z)dz$$

denote probabilities of a set A under the null or marginal distributions.

Often, a critical value z_c is chosen, and A takes one of the following forms:

$$[z_c, \infty), \quad (-\infty, z_c] \cup [z_c, \infty), \quad (-\infty, z_c].$$

For example, for $A = [z_c, \infty)$, $F_0(A) = \Pr(Z \geq z_c|H_0) = 1 - F_0(z_c)$ where $F_0(z_c)$ is the cumulative distribution function of Z.

When $Z = z$ is observed to belong in this region ($z \in A$), the hypothesis is declared as non-null, flagging a discovery. Then the probability that a false discovery has been made is

$$\text{BFDR}(A) = \Pr(H = 0|z \in A). \tag{8.18}$$

BFDR(A) is the Bayes false discovery rate for A. The subscript i denoting the ith hypothesis is not used because under the setup considered here $\Pr(H_i = 0|z_i \in A)$ is the same for all $i = 1, \ldots, m$. A straightforward application of Bayes theorem yields

$$\begin{aligned}\Pr(H = 0|z \in A) &= \frac{\Pr(Z \in A|H = 0)\Pr(H = 0)}{\Pr(Z \in A)} \\ &= \frac{F_0(A)\pi_0}{F(A)},\end{aligned} \tag{8.19}$$

where

$$\begin{aligned}F(A) = \Pr(Z \in A) &= \Pr(Z \in A|H = 0)\Pr(H = 0) \\ &+ \Pr(Z \in A|H = 1)\Pr(H = 1).\end{aligned} \tag{8.20}$$

Expression (8.19) can be written as

$$\Pr(H = 0|z \in A) = \frac{\Pr(\text{Type I error of } A)\pi_0}{\Pr(\text{Type I error of } A)\pi_0 + (\text{Power of } A)\pi_1}, \tag{8.21}$$

indicating that BFDR(A) increases with increasing type I errors and decreases with power (Storey 2003).

Storey (2003) shows that under the assumptions outlined in this section (m identical hypotheses H_i and test statistics Z_i arising from a random mixture of null and alternative distributions, with (H_i, Z_i) *i.i.d.* random variables, a fixed significant region A), the BFDR(A) (8.18) can be written as

$$pBFDR(A) = \mathrm{E}\left[\frac{V(A)}{R(A)}|R(A) > 0\right] \tag{8.22}$$

that he calls the *positive false discovery rate* and is a departure from (8.12) in that the expectation is conditional on the fact that at least one discovery has been made. One proof is simple and instructive, and it goes as follows (Storey 2003):

$$\begin{aligned}\mathrm{E}\left[\frac{V(A)}{R(A)}|R(A) > 0\right] &= \sum_{k=1}^{m} \mathrm{E}\left[\frac{V(A)}{R(A)}|R(A) = k\right] \Pr(R(A) = k|R(A) > 0) \\ &= \sum_{k=1}^{m} \mathrm{E}\left[\frac{V(A)}{k}|R(A) = k\right] \Pr(R(A) = k|R(A) > 0),\end{aligned}$$

where k depends on the particular outcome of the p-values in any one replication of the experiment. With the independence assumption, the number of false discoveries $V(A)$, given k hypotheses have been rejected, $R(A) = k$, is binomially distributed $Bi(k, \Pr(H = 0|z \in A))$ with expectation $k \Pr(H = 0|z \in A)$. Therefore, the last line is

$$\begin{aligned}\mathrm{E}\left[\frac{V(A)}{R(A)}|R(A) > 0\right] &= \Pr(H = 0|z \in A) \sum_{k=1}^{m} \Pr(R(A) = k|R(A) > 0) \\ &= \Pr(H = 0|z \in A)\end{aligned} \tag{8.23}$$

since the last term sums to 1. The positive false discovery rate works by first fixing the rejection region A and then estimating the FDR, which is the opposite of the Benjamini and Hochberg's FDR-BH.

The BFDR (8.19) can also be interpreted as the expected proportion of false discoveries among all features declared significant. Indeed, the number of hypotheses rejected in error V is binomial $Bi(R, \Pr(H = 0|z \in A))$. Therefore, the expected number rejected in error is $R \Pr(H = 0|z \in A)$, and the number rejected is R. The ratio of the expected number rejected in error to the number rejected is the expected proportion of false discoveries among all features declared significant, equal to $\Pr(H = 0|z \in A)$. As such, the BFDR is a global measure and does not provide information about each hypothesis. This is taken care of by the *local false discovery rate* (Efron and Tibshirani 2002) discussed below.

8.5 Empirical Bayes Estimation

Empirical Bayes estimation of false discovery rate is introduced in the rather idealised framework of the two-group mixture model presented above, with m genes that either are null (an unknown number m_0 among these m have no effect on a response variable) with a priori probability $\pi_0 = m_0/m$ or non-null with a priori probability π_1. In the setting considered here, m is very large, m_0 is large, and the objective is to reduce this vast set to a handful of scientifically interesting genes for further study.

A glance at (8.19) and (8.20) reveals that computation of BFDR requires knowledge of π_0, F_0 and F. With very large datasets, these quantities can be estimated from the data, and then the estimates can be used to approximate (8.19). This is the essence of the *empirical Bayes* approach, that in the context of false discovery was proposed by Efron et al (2001) and summarised in Efron (2010). Efron and Tibshirani (2002) show the connection with the Benjamini and Hochberg FDR.

The method proceeds as follows: The starting point is the availability of simple t-tests for the m genes. These t-values represent the "data". One assumes

- F_0 is known and assumed to be the normal distribution $N(0, 1)$ that can be justified by transforming p-values derived from t-tests for each of the m hypotheses to z-values, as indicated in (8.11). Using the z-values as transformed "data",

$$F_0(A) = \Pr(Z \in A|H = 0) = \int_A N(0, 1). \tag{8.24}$$

- π_0 is "almost known" and in practice can be set equal to 1.
- F is unknown but can be consistently estimated as follows:

$$\begin{aligned} F(A) = \Pr(Z \in A) &= \int_A p(z)dz \\ &= \int_{-\infty}^{\infty} I(z \in A)p(z)dz \\ &= \mathrm{E}[I(z \in A)], \end{aligned} \tag{8.25}$$

and then the empirical estimator of $E[I(z \in A)]$ is

$$\widehat{F}(A) = \widehat{\Pr}(Z \in A) = \frac{1}{m}\sum_{i=1}^{m} I(z_i \in A) \tag{8.26}$$

(does not depend on i and is a consistent estimator also when the $Z's$ are correlated).

The empirical Bayes estimator is

$$\widehat{\text{BFDR}}(A) = \frac{\hat{\pi}_0 F_0(A)}{\widehat{F}(A)} \tag{8.27}$$

where $\hat{\pi}_0$ is some estimate of $\pi_0 = m_0/m$ (e.g., using maximum likelihood, unless it is set equal to 1 as a first approximation). This is an estimator of the posterior probability of false discoveries among the features whose z-values fall in the rejection region A; these constitute the discovery set.

One can also use p-values as "data". When a fixed threshold p_t, $0 < p_t \leq 1$ is chosen, the features whose p-values are less than or equal to the fixed p_t are called significant and constitute the discovery set. Then following the same reasoning as above, with (8.24) now replaced by the uniform $\mathcal{U}(0, 1)$ distribution

$$F_0(p_t) = \Pr(P_i \leq p_t | H_i = 0) = \int_0^{p_t} \mathcal{U}(0, 1) = p_t$$

with

$$F(p_t) = \Pr(P \leq p_t)$$

results in the Bayes FDR

$$\text{BFDR}(p_t) = \Pr(H = 0 | p \leq p_t) = \frac{\pi_0 F_0(p_t)}{F(p_t)}. \tag{8.28}$$

An empirical Bayes estimator is obtained using

$$\widehat{F}(p_t) = \widehat{\Pr}(P \leq p_t) = \frac{1}{m} \sum\nolimits_{i=1}^{m} I(p_i \leq p_t).$$

Then the empirical Byes estimator of the posterior probability of a false discovery is

$$\widehat{\text{BFDR}}(p_t) = \widehat{\Pr}(H = 0 | P \leq p_t) = \frac{\hat{\pi}_0 F_0(p_t)}{\widehat{F}(p_t)} \tag{8.29}$$

as in (8.27). A quantity closely related to (8.28), where the fixed threshold p_t is replaced by p_i corresponding to hypothesis i, is

$$\text{BFDR}(p_i) = \Pr(H = 0 | P \leq p_i) = \frac{\pi_0 F_0(p_i)}{F(p_i)} \tag{8.30}$$

that corresponds to Storey's q-value (Storey 2003) discussed below.

In contrast to the frequentist counterpart of Benjamini and Hochberg (1995), the derivation of the BFDR does not require independence of the hypotheses (Efron 2010). However, correlation among test statistics increases the (sampling) variance of $\widehat{\text{BFDR}}(p_t)$ (Efron 2010).

Connection with Benjamini-Hochberg False Discovery Rate

In order to establish a connection between (8.27) and FDR-BH, following Storey (2002) and Efron and Tibshirani (2002), assume that rather than a fixed rejection region A, ordered p-values $p_1 < p_2 < \cdots < p_m$ are available. This may require first mapping the z-values to p-values using (8.7) with z instead of t. The numerator of (8.27) is

$$\begin{aligned} p_i &= 1 - F_0(z_i) = \Pr(Z \geq z_i | H = 0) \\ &= \int_{z_i}^{\infty} N(0, 1). \end{aligned}$$

Using the ordered p-values from smallest to largest, the denominator of (8.27) is

$$\widehat{\Pr}(P_i \leq p_i) = \frac{1}{m} \sum_{k=1}^{m} I(p_k \leq p_i) = \frac{i}{m}$$

and

$$\widehat{\text{BFDR}}(p_i) = \frac{p_i}{\widehat{\Pr}(P_i \leq p_t)} \widehat{\pi}_0 = \left(\frac{mp_i}{i}\right) \widehat{\pi}_0. \tag{8.31}$$

The threshold condition (8.16) of the FDR-BH can be written in terms of the empirical Bayes estimate

$$\frac{\widehat{\text{BFDR}}(p_i)}{\widehat{\pi}_0} = \frac{mp_i}{i} \leq q$$

or

$$\widehat{\text{BFDR}}(p_i) = \widehat{\pi}_0 q \leq q, \tag{8.32}$$

as in (8.15). Expression (8.32) indicates that rule (8.16) generates a discovery set that includes those cases where the empirical Bayes posterior probability that the hypothesis is a true null is smaller than q.

Under the assumption of a fixed significant threshold p_t, there is another way of connecting $BFDR$ with (8.12). The marginal probability of a rejection is

$$\begin{aligned} \Pr(r_i = 1) &= \Pr(r_i = 1 | \delta_i = 0) \Pr(\delta_i = 0) + \Pr(r_i = 1 | \delta_i = 1) \Pr(\delta_i = 1) \\ &= \Pr(P_i \geq p_t | \delta_i = 0) \Pr(\delta_i = 0) + \Pr(P_i \geq p_t | \delta_i = 1) \Pr(\delta_i = 1) \\ &= \Pr(P_i \geq p_t). \end{aligned}$$

The first term in the right-hand side of the first line represents the probability of a false discovery (rejection in error), and the second term is the probability of a true discovery. The total expected number of hypotheses rejected is then

$$\begin{aligned} m\Pr(P_i \geq p_t) &= m\Pr(P_i \geq p_t|\delta_i = 0)\Pr(\delta_i = 0) \\ &+m\Pr(P_i \geq p_t|\delta_i = 1)\Pr(\delta_i = 1) \end{aligned}$$

where the expected number of hypotheses rejected in error is

$$m\Pr(P_i \geq p_t|\delta_i = 0)\Pr(\delta_i = 0) = m\Pr(P_i \geq p_t|\delta_i = 0)\pi_0. \tag{8.33}$$

In any one experiment, the false discovery proportion is

$$Q = \frac{V}{R}.$$

Replacing V and R by their expectation yields

$$\frac{\mathrm{E}(V)}{\mathrm{E}(R)} = \frac{\Pr(P_i \geq p_t|\delta_i = 0)\pi_0}{\Pr(P_i \geq p_t)} = \mathrm{BFDR}(p_t),$$

indicating the close connection between the empirical Bayes false discovery rate and (8.12). Indeed for large m,

$$\frac{\mathrm{E}[V]}{\mathrm{E}[R]} \approx \mathrm{E}\left[\frac{V}{R}\right], \tag{8.34}$$

(Storey and Tibshirani 2007). Efron (2010) shows that $\mathrm{BFDR}(p_t)$ is conservative, in the sense that, on average, it is upwardly biased for estimating $\mathrm{E}[V/R]$.

8.6 Local False Discovery Rates

The Benjamini and Hochberg FDR-BH and the Bayesian FDR (8.19) are concerned with identifying a discovery set. Efron and Tibshirani (2002) define the *local false discovery rate*

$$\mathrm{fdr}(y_j) = \Pr(H_j = 0|y_j) \tag{8.35}$$

that gives a probabilistic assessment of a false positive for the specific feature j given that y_j was observed, instead of providing a measure of the expected proportion of nulls among all rejections. The specificity issue is important here: the significance of the feature depends on the value y_j rather than on its inclusion in the significant region A as in (8.18). Although data y are used here, the common setup is to consider test statistics for each feature. In the context of the two-group mixture model with densities p_0 and p_1 for the null and non-null densities of the test statistics, using Bayes theorem

$$\Pr(H_j = 0|y_j) = \frac{p_0(y_j|H_j = 0)\pi_0}{p(y_j)} \tag{8.36}$$

where

$$p(y_j) = p_0(y_j|H_j = 0)\pi_0 + p_1(y_j|H_j = 1)(1 - \pi_0). \tag{8.37}$$

While the BFDR involves ratios of distribution functions as shown in (8.19), the local false discovery rate (8.36) involves ratios of densities. Therefore, the local false discovery rate can be interpreted as the BFDR for an infinitesimal rejection region around y_j. There is a relationship between BFDR and the local false discovery rate given by

$$BFDR(y_t) = \mathrm{E}[\mathrm{fdr}(y)|y \geq y_t] \tag{8.38a}$$

$$= \pi_0 \frac{\int_{y_t}^{\infty} p_0(y_j|H_j = 0)dy_j}{\int_{y_t}^{\infty} p(y_j)dy_j} \tag{8.38b}$$

$$= \frac{F_0(A)\pi_0}{F(A)}, \tag{8.38c}$$

as in (8.27), with $A = [y_t, \infty)$ and where y_t is some chosen threshold value, so BFDR(y_t) is the mean value of fdr(y) for $y \geq y_t$. As before, there is an implicit conditioning on p_0, p_1 and π_0, whereby these parameters can be estimated parametrically or nonparametrically as in Efron et al (2001).

Empirical Bayes approach requires estimation of the terms in the right-hand side of (8.36). Under the simple two-group mixture model, when the $y's$ are interpreted as z-values, p_0 is simply the value of the standard normal density at y_j, and π_0 can be set equal to 1 as an approximation. Getting a hand on the denominator is a little more delicate and requires density estimation. The topic is discussed in Efron (2010). In the example below, I illustrate estimation by maximum likelihood and using a more general approach based on McMC-Bayes.

The information supplied by FDR-BH can be combined with fdr, with the former identifying a discovery set and the latter providing specific probabilities of false discovery for each feature in the discovery set.

8.7 Storey's *q*-*Values*

Consider an experiment where the estimator of the Bayes false discovery rate (8.18) together with a rejection region give rise to a discovery set. A particular feature (e.g., a marker genotype) is included in the discovery set if the observed test statistic $T = t$ associated with this feature falls in the rejection region $A = (-\infty, c] \cup [c, \infty)$, $-\infty \leq c \leq \infty$, for a threshold c. The estimator of the Bayes false discovery rate provides a global measure informing on a whole set of significant features. For example, for a threshold $c = 3$, say, the pBFDR involves tests based on an entire rejection region $|t_i| \geq 3$. However, within the discovery set, a feature with $t_i = 5$ will be more significant than a feature with $t_i = 3.5$. One may be interested in a measure of significance attached to each feature in the discovery set that accounts for the fact that, within the discovery set, some features are more significant than others.

The *q-value* is designed to give each test statistic a measure of its significance in terms of the positive false discovery rate pBFDR. For a fixed rejection region A and an observed test statistic $T = t$, the q-value is formally defined as

$$q(t) = \min_{\{A: t \in A\}} \left[\text{pBFDR}(A)\right], \tag{8.39}$$

(Storey 2002, 2003). The q-value is the smallest pBFDR over all regions that reject H_i. It can be interpreted as the expected proportion of false positives among all features as or more extreme than the one observed.

The computations of $q^{*'}s$ defined in (8.17) and of Storey's *q-values* are closely related, as shown in (8.32). The empirical Bayes estimates of the q-values can be obtained using the same calculations as those for q^* except for the term π_0. If an estimate $\hat{\pi}_0$ is available, q_i-value $= q_i^* \hat{\pi}_0$, and this aligns it with the BFDR (8.29) (the same code as shown below (8.17) can be used, with the modification immediately below the do-loop:

```
qstar[i] <- pv[i] * (m/i) * pi_0hat
```

where `pi_0hat` is an estimate of π_0).

In the example of Table 8.1, where the test statistics are sorted p-values starting from the largest at the bottom of the list, the q_i^* is numerically equal to q_i-*value*$(1/\widehat{\pi}_0)$. The regions that reject H_{19} (p-values ≥ 0.02844954) involve all the q_i-*values* $\left(1/\widehat{\pi}_0\right)$, $i = 19, \ldots, 100$, and q_{19}-*value*$(1/\widehat{\pi}_0) = 0.14973443$ is the smallest such BFDR. Alternatively, $\widehat{\pi}_0$ can be set equal to 1, and then q_i^* and q_i-value are numerically identical quantities. In this case, if the null hypotheses is rejected for features $i = 15, 14, \ldots, 1$, then the figures in Table 8.1 indicate that $q_{15}^* = 0.083$. This is the estimate of BFDR for $i = 15$ and for the set comprising all the features that are this significant or more.

Incorporation of $\widehat{\pi}_0$ in the calculations leads to a larger discovery set for the same expected proportion of false positives.

The q-value is a pBFDR analogue of the p-value. The latter is a measure of significance in terms of the false positive rate, whereas the former is a measure of significance in terms of the pBFDR. This conceptual difference is readily seen by writing

$$\text{p-value} = \Pr(\text{Test Positive} \mid H_0) = \text{ False positive rate}$$

$$\text{q-value} = \Pr(H_0 \mid \text{Test Positive}) = \text{ False discovery rate.}$$

A $q = 0.05$, say, means that an estimated 5% of the significant results **in the discovery set** are false discoveries. In contrast, in classical hypothesis testing, a rejection of H_0 at $\alpha = 0.05$ in a single test does not mean that there is a 95% probability that a discovery has been made. It rather means that an estimated 5% of **all the tests performed** are false discoveries. Further and also in contrast with the p-value, the q-value provides a measure of each markers' significance automatically accounting for multiple testing.

There is a **qvalue** package in R that performs FDR estimation from a collection of p-values or test statistics. The package outputs estimates of q-values, proportion of true null hypotheses π_0 and local false discovery rates (Storey and Bass 2021).

The false discovery rate, the local false discovery rate and the q-values can be obtained using a fully parametrised Bayesian model. This is the subject of the next section.

8.8 Fully Bayesian McMC False Discovery Rate

The Bayesian FDR (8.18) was derived within the stylised example of the two-group mixture model and using z-values as data. The only unknown in this setting is the random quantity H and its associated "parameters". Here, an alternative development is followed that consists of inferring the parameters of (8.3) from their posterior distribution using the McMC implementation of the spike and slab model of Sect. 7.5. This allows more flexibility in modelling at the cost of heavier computations and parametrisation. Flexibility is important, particularly with sets of data involving a large number of highly correlated covariates and other explanatory variables found in modern genomic studies.

The spike and slab model is used for focusing the presentation, but the approach remains valid for many massive multiple comparison problems. The indicator δ is an ingredient of one estimator of the Bayesian FDR; it is inferred from its marginal posterior distribution $\Pr(\delta_i = 1|y)$, with the remaining parameters integrated out from the joint posterior distribution of the Bayesian model.

From (8.2) and (8.3),

$$\frac{V}{R} = \frac{\sum_{i=1}^{m}(1-\delta_i)r_i}{\sum_{i=1}^{m} r_i}$$
$$= 1 - \frac{\sum_{i=1}^{m}\delta_i r_i}{R}. \tag{8.40}$$

Conditionally on the data (indirectly, via the rule r), the only unknown quantity in (8.40) is δ that appears in the numerator, in contrast with the frequentist counterpart (8.12) that involves the expectation of a ratio of random variables. On taking expectations over the Bernoulli random variable $[\delta_i|y]$, given acceptance or rejection based on r, the fully Bayesian FDR is immediately obtained as

$$\frac{E(V|r)}{R} = 1 - \frac{\sum_{i=1}^{m}\Pr(\delta_i = 1|y)r_i}{R} \tag{8.41a}$$
$$= \frac{\sum_{i=1}^{m}\Pr(\delta_i = 0|y)r_i}{R} \tag{8.41b}$$

(Müller et al 2007; Whittemore 2007). The ith term in the numerator $\Pr(\delta_i = 0|y)r_i$ is the (fully) Bayesian local false discovery rate for the ith marker in the discovery set, so the Bayesian FDR is seen to be an average of the local false discovery rates in the discovery set, as in (8.38). It can also be interpreted as the expected proportion of false discoveries among the features in the discovery set which aligns it with (8.19).

In these expressions and the ones that follow, y represents the complete data. When the conditional distribution of δ_i given y_i is not independent of the remaining $y's$, the fully Bayesian local false discovery rate $\Pr(\delta_i = 0|y)r_i$ can differ from $\Pr(\delta_i = 0|y_i)r_i$ as defined in (8.35).

The marginalisations required to obtain terms $\Pr(\delta_i = 0|y)$ can be approximated using McMC. An McMC-based estimator of the Bayesian FDR is obtained using the McMC output as follows: First, calculate

$$\widehat{\varphi}_i = \widehat{\Pr}(\delta_i = 1|y) = \frac{1}{l}\sum_{j=1}^{l}\delta_{ij}, \tag{8.42}$$

where δ_{ij} is the sampled value of δ for marker i at iterate j of the Gibbs sampler and l is the length of the Gibbs chain. This yields a Monte Carlo estimator of the mean of the posterior distribution $\widehat{\Pr}(\delta_i = 1|y)$. Then the Bayesian-McMC estimator of FDR is

$$\frac{\widehat{E}\left(V|r, \widehat{\Pr}(\delta_i = 0|y)\right)}{R} = \frac{\sum_{i=1}^{m}\widehat{\Pr}(\delta_i = 0|y)r_i}{R}, \tag{8.43}$$

where $\widehat{\Pr}(\delta_i = 0|y) = 1 - \widehat{\Pr}(\delta_i = 1|y)$. Expression (8.43) is a point estimator of the Bayesian FDR.

Posterior Distribution of False Discovery Rate

An alternative approach provides a Monte Carlo estimator of the complete posterior distribution of the fully Bayesian FDR. This distribution reflects posterior uncertainty of the parameters of the Bayesian model that propagates on to the FDR. The protocol is as follows: First, execute the McMC sampler, and store the draws from (7.57) at each cycle. On completion, choose a discovery set; an informal approach could be based on selecting the marker genotypes whose Monte Carlo estimates of the posterior probability of non-zero effects are larger than a chosen value $\theta^{\star}$. These posterior probabilities could be obtained from φ in (8.42) or from the average of the draws from (7.57). The resulting discovery set includes R marker genotypes. Then, from the stored draws $\theta_i^{[j]}$ obtained from (7.57) *belonging to the discovery set*, where i refers to the marker genotype and j to the McMC cycle, compute at each McMC cycle the average local false discovery rate for all the marker genotypes in this discovery set. For example, at round j, compute

$$\frac{\sum_{i=1}^{m}\left(1-\theta_i^{[j]}\right)r_i}{R} \tag{8.44}$$

where m is the number of markers. The quantity (8.44) is the jth McMC draw from the posterior distribution of the Bayesian FDR for the discovery set chosen, here based on $r_i = I(\theta_i \geq \theta^*)$. Importantly, the discovery set is chosen only once and used repeatedly across all the draws of the McMC chain.

In the case of the two-component normal mixture model, rules of the type $r_i = I(y_i \geq y_t)$ or $r_i = I(\theta_i \geq \theta_t)$ are illustrated in the examples below, where θ_i is the marginal probability that δ_i is equal to 1, as in (7.57).

A recent McMC implementation of the Bayesian FDR has been described by de los Campos et al (2022).

8.9 Example: A Two-Component Gaussian Mixture

I return to the two-component mixture model of the example on page 127. Data are realisations from two unobserved normal mixture components, and the purpose of the analysis is to infer parameters and to allocate each data point to a specific component. The data could represent levels of a physiological marker in blood samples, and one wishes to isolate those individuals that were exposed to a particular condition from those that were not. This condition is known to generate a higher mean level of the physiological marker. Data are simulated from a two-component Gaussian mixture, where the first component has a mean of zero (9000 observations), the second (1000 observations) a mean of 2.5 units and both a standard deviation of 1. Figure 8.1 shows a histogram of the 10,000 records (left)

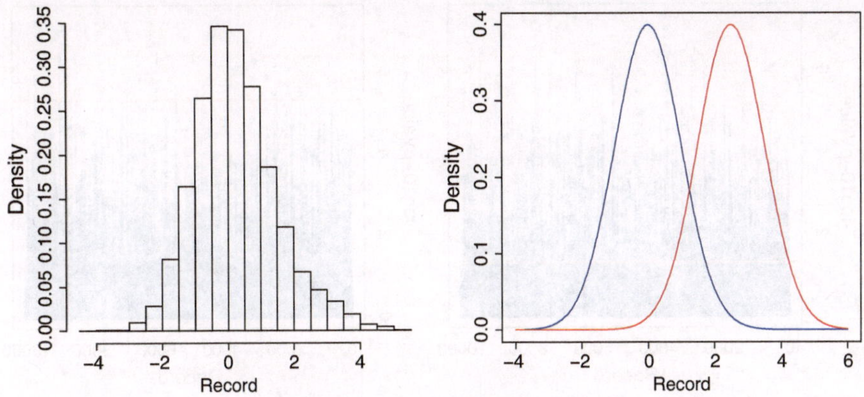

Fig. 8.1 Left: histogram of data. Right: plots of $N(0, 1)$, $N(2.5, 1)$

and plots the mixture distribution involving $N(0, 1)$ and $N(2.5, 1)$. The histogram reveals a certain amount of skewness due to the underlying mixture.

Information will be extracted from these data using five approaches:

1. A simple unadjusted test of the null hypothesis that each record is not different from zero
2. A test of the null hypothesis that each record is not different from zero using a Bonferroni correction
3. The classical Benjamini and Hochberg false discovery rate (FDR-BH)
4. An analysis of the two-component mixture model using maximum likelihood via de EM algorithm described on page 127
5. A Bayesian McMC analysis of the two-component mixture model described on page 234

The last two analyses use fully parametrised models, while the first three use p- or z-values as input.

The results of the simple test and of the Bonferroni test are shown in Fig. 8.2 for a nominal probability of type I error equal to $\alpha = 0.05$. The left plot displays the p-values against the 10,000 records, and the threshold for rejecting the null hypothesis at $\alpha = 0.05$ is indicated by the horizontal line in red. This simple analysis, uncorrected for multiple testing, rejects H_0 for 1278 records out of which 470 are false discoveries. For $\alpha = 0.10$, H_0 is rejected for 1829 records, of which 928 are false discoveries. The plot on the right panel shows the results based on the Bonferroni correction using a per datum significant level equal to 0.05/10,000. Only 35 records are declared significant with zero false discoveries among them. When the per datum significant level is 0.10/10,000, the size of the discovery set increases to 41 with zero false discoveries.

It is interesting to look more closely at the type I and type II errors in these cases. A total of 9000 records were drawn from H_0 and 1000 from H_1. The simple uncorrected test using $\alpha = 0.05$ results in 470 rejections of H_0, given that H_0 is

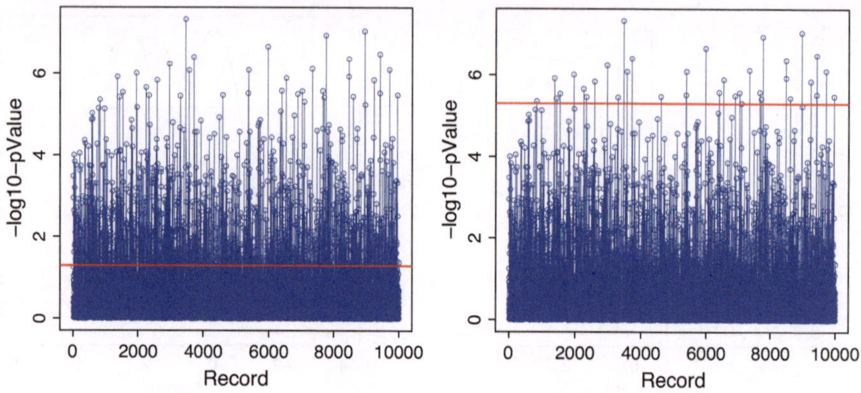

Fig. 8.2 Left, $-Log_{10}p$-values corresponding to 10,000 records for testing the null hypothesis that each record is a draw from a single normal distribution with null mean. The horizontal red line is the $-Log_{10}p$-value threshold corresponding to $p = 0.05$. Right: similar output where the horizontal red line is the Bonferroni threshold corresponding to $p = 0.05/10{,}000$

true. This yields a proportion of type I error equal to $470/9000 = 0.052$, very close to the nominal value of 0.05. The number of type II errors (false negatives) is 192 out of 1000 records drawn from $N(2.5, 1)$ or a power of the test of 81%.

For the Bonferroni test using a per datum significant level equal to 0.05/10,000, the number of type I errors is now 0 out of 9000, but the number of observed false negatives (type II errors) is 965 out of 1000 and a power of 3.5%! (not too far off from the expected power of *pnorm*(*4.42*, *2.5*, *1*, *lower.tail* = *F*) = 0.028).

Maximum Likelihood via EM

The likelihood analysis of this problem implemented via the EM algorithm is based on the iterative system described in connection with the mixture model on page 127:

$$\widehat{p}_{ij}^{[t+1]} = \Pr\left(Z_i = j|\pi^{[t]}, \theta_j^{[t]}, y_i\right) = \frac{p_j\left(y_i|\theta_j^{[t]}\right)\pi_j^{[t]}}{\sum_j p_j\left(y_i|\theta_j^{[t]}\right)\pi_j^{[t]}}, \tag{8.45a}$$

$$\pi_j^{[t+1]} = \frac{1}{n}\sum_{i=1}^{n}\widehat{p}_{ij}, \tag{8.45b}$$

$$\theta_j^{[t+1]} = \frac{\sum_{i=1}^{n}\widehat{p}_{ij}\, y_i}{\sum_{i=1}^{n}\widehat{p}_{ij}}, \tag{8.45c}$$

$$\sigma^{2[t+1]} = \frac{1}{n}\sum_{i=1}^{n}\sum_{j=0}^{j=1}\widehat{p}_{ij}\left(y_i - \theta_j\right)^2. \tag{8.45d}$$

The loop is over the number of iterates (t), and within each iterate, there is a loop over the number of records (i).

In this example, the true value of the parameters of the mixture model is $\theta_0 = 0$, $\theta_1 = 2.5$, $\sigma^2 = 1$ and $\pi = 0.1$. When the system converges after approximately 70 iterations, the ML estimates are $\widehat{\theta}_0 = 0.01$, $\widehat{\theta}_1 = 2.5$, $\widehat{\sigma}^2 = 1.02$ and $\widehat{\pi} = 0.10$. A useful off-shot of the EM implementation are the terms $\widehat{p}_{ij}$ that assign the probability that each datum i belongs to mixture component j; this allows the calculation of a local false discovery rate along the lines in (8.36).

There is a connection between (8.45a) and the Bayesian counterpart. In a likelihood setting, (8.45a) is a conditional posterior probability, given ML estimates $\widehat{\theta}$ and $\widehat{\pi}$. This can be regarded as an approximation to the Bayesian marginal posterior probability $\Pr(Z_i = j|y)$ (see page 155 for an explanation).

The kernel of the R-code to implement the EM iterates is shown below:

```
CREATE DATA SET AND INITIALISE PARAMETERS
prob<-vector()
# INITIALISE THETA1 AND THETA0
y0<-which(y < mean(y))
theta0<-mean(y[y0])
y1<-which(y > mean(y))
theta1<-mean(y[y1])
iter<-100
resultML<-matrix(data=NA,nrow=iter,ncol=n+4)
for (i in 1:iter){
  for (j in 1:n){
    denom<-(dnorm(y[j],theta1,sqrt(v))*pi)
    +(dnorm(y[j],theta0,sqrt(v))*(1-pi))
    prob[j]<-(dnorm(y[j],theta1,sqrt(v))*pi)/denom
  }
  pi<-(sum(prob))/n
  theta1<-(sum(prob*y))/sum(prob)
  theta0<-(sum((1-prob)*y))/sum(1-prob)
  v<-(sum(prob*(y-theta1)^2))/n+(sum((1-prob)*(y-theta0)^2))/n
  resultML[i,]<-c(v,pi,theta1,theta0,prob)
}

#             COMPUTE FDR FOR ML ANALYSIS
cumavr<-rep(0,n)
fdr<-rep(0,n)
sortprob<--sort(-prob)
true<-which(dataset$label==T)
fals<-which(dataset$label==F)
for (i in 1:n){
  cumavr[i]<-mean(sortprob[1:i])
  fdr[i]<-1-cumavr[i] # THIS COMPUTES THE (CUMULATIVE) EXPECTED
#  PROPORTION OF BFDR FOR EACH DISCOVERY SET OF SIZE 1 TO n
}
```

In the example, it is assumed that all the non-null observations arise from a common distribution. This may not be the case, particularly if the "observations" are expression of genes or gene effects. Accounting for this requires adding extra distributions for the alternative hypotheses.

Bayesian McMC Implementation

The Bayesian computation of (8.41) is based on the McMC algorithm detailed in Sect. 5.5 on page 234. The true value of the parameters of the mixture model are $\theta_0 = 0, \theta_1 = 2.5, \sigma^2 = 1$ and $\pi_1 = 0.1$. The Monte Carlo estimates of the posterior means are $\widehat{\theta_0} = 0.01$, $\widehat{\theta_1} = 2.49$, $\widehat{\sigma}^2 = 1.03$, $\widehat{\pi}_1 = 0.10$, in very close agreement with the true values and with the likelihood estimates.

The kernel of the R-code to implement the McMC-based Bayesian mixture model is shown below:

```
CREATE DATA SET AND INITIALISE PARAMETERS
# READ CHAIN LENGTH
rep <- 1000
resmcmc <-matrix(data=NA,nrow=rep,ncol=5)
acprob <- rep(0,rep)
avr<-rep(0,n)
fd<-rep(0,n)
# READ HYPERPARAMETERS OF SCALE INVERSE CHI-SQUARE PRIOR FOR v
nu_v <- 4.5
Sv <- 1
# READ HYPERPARAMETERS OF THE BETA PRIOR FOR pi
a1 <- 1.5
a2 <- 10
####################  GIBBS LOOP ####################
for (i in 1:rep) {
  #  print(i)
  # UPDATE THETA j
  theta1hat <- mean(y[z == 1])
  theta0hat <- mean(y[z == 0])
  v1 <- v / sum(z == 1)
  v0 <- v / sum(z == 0)
  theta1 <- rnorm(1, theta1hat, sqrt(v1))
  theta0 <- rnorm(1, theta0hat, sqrt(v0))
  # UPDATE THE n z's
  K <-
    (((y-theta0)^2-(y-theta1)^2)/(2*v))-(log(1-pi)-log(pi))
  un <- runif(n, 0, 1)
  z <- ifelse(log(un / (1 - un)) <= K, 1, 0)
  prob <- exp(K) / (1 + exp(K))
  acprob <- acprob + prob
  # UPDATE v
  #  ft <- sum(((y - theta1) ^ 2)[which(z == 1)])
  ft <- sum(((y - theta1) ^ 2)[z == 1])
  st <- sum(((y - theta0) ^ 2)[z == 0])
  Sc <- ft + st + (nu_v * Sv)
  df <- nu_v + n
  v <- Sc / rchisq(1, df)
  # UPDATE pi
  a1ny <- a1 + sum(z)
  a2ny <- n - sum(z) + a2
  pi <- rbeta(1, a1ny, a2ny)
  resprob[i,] <- prob
  resmcmc[i,] <- c(i, v, pi, theta1, theta0)
}
```

```
################################################################
postprob <- acprob/rep
################################################################
# COMPUTE FALSE DISCOVERY RATE USING McMC OUTPUT

ordpp<-order(-postprob)
sortpp<-postprob[ordpp]
localfdr <- 1-sortpp

for (i in 1:n){
  avr[i]<-mean(sortpp[1:i])
  fd[i]<-1-avr[i] # THIS IS THE FDR
###  ALTERNATIVELY MORE COMPACTLY: MEAN OF LOCAL FDR
#        IN DISCOVERY SET OF SIZE i
  fdloc[i] <- mean(localfdr[1:i])
# fd[] = fdloc[]
}
```

A summary of the main results using the five approaches is displayed in Table 8.2. The number and proportion of false discoveries reported in the table are true values based on the particular realisation of the simulated data. ML and Bayes-McMC give very similar results for all the features in the table. Increasing the nominal false discovery q from 0.05 to 0.20 leads to a larger discovery set (from 356 and 353 to 799 for FDR-ML and FDR-Bayes) and of course to a larger FDR. Notice that this increase in FDR is a consequence of a large increase in type I errors (almost a tenfold increase) and a relatively smaller increase in power (almost twofold from 34% to 64%, approximately), as revealed by inspection of expression (8.21).

For the given nominal FDR, FDR-BH is expected to lead to a little smaller size of discovery set for the same FDR than the other two implementations because it uses a more conservative approach (π_0 is set equal to one in expression (8.15), whereas it is included in the computations of the FDR in the likelihood and Bayesian approaches). This is only vaguely noticeable for a nominal $q = 0.20$, but the

Table 8.2 Proportion of type I and type II errors, size of discovery set/true number of false discoveries (SizeDiscov/FD) and true proportion of false discoveries (FDR) for (1) the uncorrected p-values (uncorrect) and (2) the Bonferroni correction (Bonferroni). In (1), the figures in brackets correspond to nominal probability of type I errors ($\alpha = 0.05$; $\alpha = 0.10$); in (2), these are ($\alpha = 0.05/10{,}000$; $\alpha = 0.10/10{,}000$), (3) the Benjamini-Hochberg FDR (FDR-BH), (4) the EM-likelihood analysis of the two-component mixture model (FDR-ML) and (5) the McMC Bayesian analysis of the two-component mixture model (FDR-Bayes). In (3), (4) and (5), the figures in brackets correspond to nominal FDR $q = 0.05$; $q = 0.20$). Simulated data: 1000 records from $N(0, 1)$ and 9000 records from $N(2.5, 1)$

	Type I	Type II	SizeDiscov / FD	FDR
Uncorrect	(0.052; 0.10)	(0.19; 0.10)	(1278/470); (1829/928)	(0.37; 0.51)
Bonferroni	(0; 0)	(0.97; 0.96)	(35/0); (41/0)	(0; 0)
FDR-BH	(0.002; 0.016)	(0.65; 0.36)	(364/14); (781/144)	(0.04; 0.18)
FDR-ML	(0.002; 0.017)	(0.66; 0.35)	(356/14); (799/154)	(0.04; 0.19)
FDR-Bayes	(0.002; 0.017)	(0.66; 0.35)	(353/14); (799/155)	(0.04; 0.19)

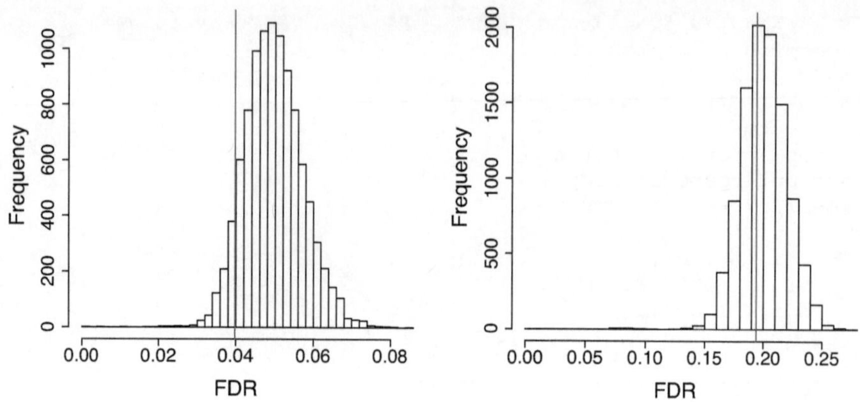

Fig. 8.3 Histograms of Monte Carlo estimates of posterior distributions of FDR. Left: nominal FDR of 0.05. Right: nominal FDR of 0.20. The vertical red lines are the true FDR (based on the simulation) for the particular realisation of the data

reverse is the case for $q = 0.05$. The explanation lies in the small overstatement of the true FDR by FDR-ML and FDR-Bayes observed at low FDR values. This is illustrated in Fig. 8.4, centre and rightmost panels. Moreover, the FDR estimated using likelihood or Bayesian methods are subject to the uncertainty (frequentist for ML and posterior for the Bayesian methods) of the estimates of parameters that feature in the computation of FDR.

The Bayesian McMC implementation provides a measure of uncertainty for the FDR (conditional on the realisation of the data at hand and on the discovery set) using the draws from the Monte Carlo estimate of the posterior distribution of FDR (8.44). The Monte Carlo estimates of the posterior distributions of the FDR-Bayes for $q = 0.05$ and $q = 0.20$ are shown in Fig. 8.3 in the form of histograms. These lead to discovery sets of size 353 and 799, respectively. For the particular realisation of the data and discovery sets, the red vertical lines show the true (realised in the simulation) FDR. The Monte Carlo estimates of posterior means and of 95% posterior intervals of the FDR-Bayes estimates for $q = 0.05$ are 0.050(0.036; 0.065) and for $q = 0.20$ 0.200(0.163; 0.239).

In the case of FDR-BH, for the particular realisation of the data, the true FDR is 0.04 and 0.18 for $q = 0.05$ and $q = 0.20$, respectively (Table 8.2), column 5).

Over repeated sampling of the data, FDR-BH controls FDR at level smaller than or equal to q as indicated by expression (8.15).

Results of a little closer comparison between the Benjamini-Hochberg FDR-BH and the fully parametric likelihood or Bayesian analyses are shown in Table 8.3. The data are analysed with the FDR-BH approach using as input the 10,000 sorted p-values. In the top half of the table, that corresponds to an expected FDR equal to $q = 0.05$, at $i = 364$ condition (8.16) is satisfied (equivalent to condition (8.17)) and hypotheses $H_1, \ldots, H_{364}$ are rejected leading to a discovery set of size 364. In the bottom half of the table that corresponds to an expected FDR equal to $q = 0.20$,

Table 8.3 A subset of results obtained by applying the Benjamini-Hochberg FDR-BH approach testing $m = 10{,}000$ hypotheses H_i with a false discovery rate set at level $q = 0.05$ (top five rows) or $q = 0.20$ (bottom five rows). The top coloured row corresponds to H_{364} (a discovery set of size $i = 364$) where condition (8.14) is satisfied and where $q^{\star}_{i=364} \leq q = 0.05$. The bottom coloured row corresponds to H_{781} (a discovery set of size 781) where condition (8.14) is satisfied and where $q^{\star}_{i=781} \leq q = 0.20$. The column headings are i, the index for the sorted p-values; Label=TRUE if the record is a draw from $N(2.5, 1)$ or FALSE from $N(0, 1)$; $Pr(Z = 1|y)$ the ML estimate of the conditional probability that the record is a draw from $N(2.5, 1)$; FDR-ML, the ML estimate of the false discovery rate for the particular discovery set i and the remaining three columns as in Table 8.1

i	Label	Record	$\Pr(Z=1\|y)$	FDR-ML	p−value	$(i/m)q$	q_i^*
362	TRUE	2.913	0.8612	0.05130	0.001789	0.0018	0.04936
363	TRUE	2.912	0.8609	0.05154	0.001794	0.0018	0.04936
364	TRUE	2.911	0.8608	0.05178	0.001796	0.0018	0.04936
365	TRUE	2.903	0.8583	0.05202	0.001846	0.0018	0.05058
366	TRUE	2.901	0.8579	0.05227	0.001855	0.0018	0.05068
779	TRUE	2.157	0.4978	0.1918	0.015491	0.016	0.1988
780	FALSE	2.156	0.4971	0.1922	0.01553	0.016	0.1991
781	TRUE	2.154	0.4960	0.1926	0.01560	0.016	0.1998
782	TRUE	2.151	0.4942	0.1930	0.01571	0.016	0.2007
783	FALSE	2.151	0.4942	0.1934	0.01572	0.016	0.2007

the Benjamini-Hochberg FDR-BH rejects $H_1, \ldots, H_{781}$. These rejection thresholds are coloured in green. The two columns with headings $\Pr(Z = 1|y)$ and FDR-ML are derived from the maximum likelihood analysis (the corresponding figures from the Bayesian analysis are very similar). Column $\Pr(Z = 1|y)$ displays the estimated probability that record i is a draw from $N(2.5, 1)$ and $1 - \Pr(Z = 1|y)$ is the ML estimate of the local false discovery rate for record i. For example, for record $y_{i=364} = 2.911798$, the local false discovery rate is $1 - 0.8608679 \approx 0.14$, but the FDR (the average local false discovery rate in the discovery set of size 364 comprising $y_1, \ldots, y_{364}$) is 0.052. The corresponding figures for $q = 0.20$ in the bottom half of the table are $1 - 0.4960449 \approx 0.50$ for the local FDR for record $i = 781$ and 0.19 for the FDR in the discovery set $y_1, \ldots, y_{781}$. The last column of the table, $q_i^{\star}$, is the Benjamini-Hochberg adjusted p-value (8.17) that in view of equivalence (8.31) is numerically identical to the empirical Bayes estimate of the FDR when π_0 is set equal to 1.

Knowledge of the true mixture proportions allows computation of the Bayesian FDR and the Bayesian local FDR. The following R−code can be used with the values in the table for $y_{364} = 2.911798$ and $y_{781} = 2.154387$:

```
# For q = 0.05:
FDR_THEOR05 <- (pnorm(2.911798,lower.tail=F)*0.9)/
  (pnorm(2.911798,lower.tail=F)*0.9 +
    pnorm(2.911798,2.5,lower.tail=F)*0.1)
LOCAL_THEOR05 <- (dnorm(2.911798)*0.9)/(dnorm(2.911798)*0.9 +
```

```
dnorm(2.911798,mean=2.5,sd=1)*0.1)
FDR_THEOR05
```

```
## [1] 0.04537123
```

```
LOCAL_THEOR05
```

```
## [1] 0.1237628
```

```
# For q = 0.20:
FDR_THEOR20 <- (pnorm(2.154387,lower.tail=F)*0.9)/
  (pnorm(2.154387,lower.tail=F)*0.9 +
     pnorm(2.154387,2.5,lower.tail=F)*0.1)
LOCAL_THEOR20 <- (dnorm(2.154387)*0.9)/(dnorm(2.154387)*0.9 +
dnorm(2.154387,mean=2.5,sd=1)*0.1)
FDR_THEOR20
```

```
## [1] 0.1810718
```

```
LOCAL_THEOR20
```

```
## [1] 0.4840697
```

This compares well with the ML estimates of the local FDR≈ 0.14 and BFDR=0.052 for $q = 0.05$ and local FDR≈ 0.50 and BFDR=0.19 for $q = 0.20$ reported in Table 8.3.

Examples of other output are shown in Fig. 8.4 for the likelihood analysis. The corresponding output from the Bayesian implementation is almost identical and is not shown. The panel on the left shows ML estimates of $\Pr\left(Z_i = 1|\widehat{\pi}, \widehat{\theta}_j, \widehat{\sigma}^2, y_i\right)$ for each of the 10,000 data points. Values of the data below 1 and larger than 3 are easily allocated to the correct mixture component since estimated probabilities are

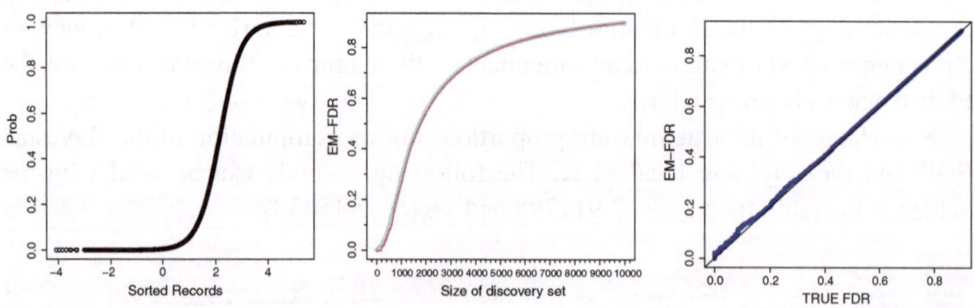

Fig. 8.4 Left: ML estimates of $\widehat{p}_{i1} = \Pr\left(Z_i = 1|\widehat{\pi}, \widehat{\theta}_j, \hat{\sigma}^2, y_i\right)$ versus the sorted records. Centre: ML estimates of FDR (light blue) and true FDR (red) against size of discovery set. Right: plot of ML estimates of false discovery rates versus true false discovery rates. The 45° line is shown in black

extreme, but there is ambiguity where the two components overlap. The panel on the centre shows FDR-ML (in light blue) and true FDR (in black) versus size of discovery set. The panel reveals a slight overestimation for low FDR values. This is also confirmed in the rightmost panel that displays a plot of FDR-ML versus true FDR.

In the example, the level q of the FDR-BH is set to either 0.05 or 0.20. This gives rise to a particular size of the discovery set. A similar strategy can be applied for a likelihood or Bayesian-McMC implementation. An alternative is to choose various discovery sets based on Monte Carlo estimates of $\Pr(Z_i = j|y)$, obtain the associated FDR for each and decide which discovery set to keep based on this information.

The conclusion from this example based on independent data is that all three methods show similar performance. One attraction of the McMC Bayesian approach is the possibility to fit complex hierarchical models capable of accounting for various sources of variation and to generate marginal posterior distributions of the parameters of interest, or of functions of these such as FDR, in a single coherent analysis. This comes at the cost of more elaborate computations.

The following example introduces a new problem. How do computations of false discovery rate are affected when input data are correlated? A classical example is the problem of distinguishing signals from noise in genome studies involving multiple testing of genetic markers that are in linkage disequilibrium with putative causal loci. The first part of the example deals with uncorrelated marker genotypes.

8.10 Example: The Spike and Slab Model with Genetic Markers

The problem discussed in this example concerns the detection of unobserved causal loci that have an effect on an observed continuous trait. This is studied in a variety of scenarios where in all cases the data consist of 1500 individuals and 1500 markers, and among these, 25 are randomly assigned as causal QTL. The phenotypic value for a datum is simulated by adding the contributions of a common mean, the effects of the 25 QTL and a normally distributed residual term. The operational models used for analysing the data included the 1500 marker genotypes.

The protocol is as follows:

1. The 1500 loci are independent (in linkage equilibrium):
   a. Heritability (h^2) is 0.10.
   b. Heritability is 0.25.
   c. Heritability is 0.50.
2. The 1500 loci are correlated (in linkage disequilibrium):
   a. Heritability is 0.10.

b. Heritability is 0.25.
c. Heritability is 0.50.

The 25 loci identified as QTL differ between the independent and the correlated groups but are the same for the three heritability levels within each group.

The three levels of h^2 result in different levels of power to detect marker effects. This is achieved by setting the additive genetic variance of the trait equal to 4, 10 and 20 squared units and keeping the phenotypic variance (variance of the records, y) constant at 40 squared units. This results in effects for each of the 25 QTL (in standard deviation units of y) equal to 0.09, 0.14 and 0.20, for the three heritability levels.

The 1500 independent marker genotypes were generated by drawing each marker genotype from a binomial distribution $Bi(2, 0.5)$.

The 1500 correlated marker genotypes are in 75 independent blocks of size 20 each, where markers are correlated within blocks and uncorrelated among blocks. The correlation between adjacent markers is approximately 0.60 and decays with marker distance.

For all scenarios, the data are first analysed using a GWAS approach (one marker at a time) using a Bonferroni correction for the probability of a false discovery per test set equal to $0.05/1500$.

The p-values from the GWAS were used to compute false discovery rates (FDR) based on the standard Benjamini-Hochberg algorithm (FDR-BH). The levels of FDR were set equal to $q = 0.10$, $q = 0.05$, $q = 0.01$ for $h^2 = 0.10$, $h^2 = 0.25$ and $h^2 = 0.50$, respectively, that give rise to discovery sets of a given size.

Finally, the data were analysed with a fully Bayesian spike and slab model that outputs the marginal posterior distribution of FDR. In the Bayesian approach, the discovery sets were constructed either by including those markers for which $\Pr(Z_i = 1|y) > 0.5$ or by choosing a particular size of discovery set. In the latter case, the size chosen was equal to that found in the FDR-BH approach to allow comparison between both methods of computation. In all cases, the true FDR realised in the simulated sample was compared to the result obtained from each of the three approaches.

Independent Marker Genotypes

The results of the GWAS using the Bonferroni correction are displayed in Fig. 8.5 for the three heritability levels and in the left block of Table 8.4. The number of loci correctly classified in the simulated sample is 1, 11 and 23 out of 25 QTL for h^2 equal to 0.10, 0.25 and 0.50, respectively, with zero false positive results.

The FDR-BH approach identifies a larger number of QTL than the Bonferroni approach at low heritability values (8 and 20, for h^2 equal to 0.10 and 0.25, respectively), as expected due to higher power, albeit at the expense of incurring on three false positive results. At $h^2 = 0.5$, in the simulated sample, both methods

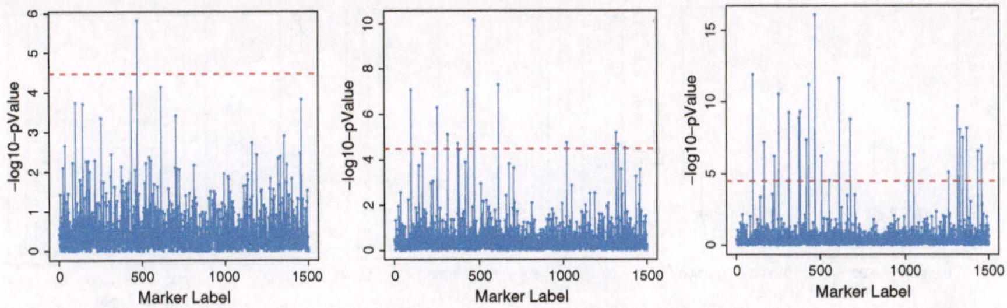

Fig. 8.5 GWAS analysis of 1500 individuals, 1500 **uncorrelated** marker genotypes with 25 causal loci and $h^2 = 0.10$ (left), $h^2 = 0.25$ (centre), $h^2 = 0.50$ (right). The Y-axes show $-log_{10}$ p-values and the X-axes the marker labels. The markers that reach $-log_{10}$ p-values beyond the Bonferroni threshold—horizontal line in red, set at $-log_{10}(0.05/1500)$—are declared as discoveries (1, 11 and 23 out of 1500 markers in the left, centre and right panels, respectively)

Table 8.4 *Size of discovery set/true (observed) number of false discoveries*, in one simulated sample, for uncorrelated and correlated marker genotypes, using (1) the Bonferroni correction (Bonferroni) with nominal probability of type I error per marker ($\alpha = 0.05/1500$, (2) the Benjamini-Hochberg FDR (FDR-BH) and (3) the McMC Bayesian analysis (FDR-Bayes). The nominal FDR for FDR-BH is $q = 0.10$ for $h^2 = 0.10$, $q = 0.05$ for $h^2 = 0.25$ and $q = 0.01$ for $h^2 = 0.50$

	Uncorrelated			Correlated		
h^2	0.10	0.25	0.50	0.10	0.25	0.50
Bonferroni	1/0	11/0	23/0	4/0	27/12	50/25
FDR-BH	8/3	20/3	23/0	27/14	50/27	70/45
FDR-Bayes[a]	8/3	20/2	23/0	27/14	50/25	70/45
FDR-Bayes[b]	10/5	25/5	28/3	13/4	20/1	27/2

[a] FDR-Bayes: The discovery set for FDR-Bayes is set equal to that of FDR-BH
[b] FDR-Bayes: the discovery set is obtained by choosing the markers with $\hat{\Pr}(Z_i = 1|y) > 0.5$

perform equally well, with 23 loci detected out of the 25 QTL and zero false positives (Table 8.4).

Figure 8.6 shows output from the Bayesian implementation for the three heritability levels, where MC estimates of $\Pr(Z_i = 1|y)$ for each marker are plotted against marker labels (see also Table 8.4). The discovery set is here arbitrarily defined by the group of markers that satisfy $\Pr(Z_i = 1|y) > 0.5$. The choice of threshold would typically not be fixed but would rather depend on an eyeball judgement of the particular plot, and several thresholds would be explored. Given the chosen threshold of 0.5, the size of discovery sets for h^2 equal to 0.10, 0.25 and 0.50 are 10, 25 and 28 markers with, respectively, 5, 5 and 3 false discoveries. The rightmost figure indicates that many markers reach $\Pr(Z_i = 1|y) \approx 1$. If for this heritability level the threshold is set equal to $\Pr(Z_i = 1|y) > 0.9$, then 25 markers are chosen with 0 false positives. Similarly, for the figure in the centre panel corresponding to $h^2 = 025$, if the threshold is $\Pr(Z_i = 1|y) > 0.7$, then 23 markers are chosen as part of the discovery set with three false discoveries.

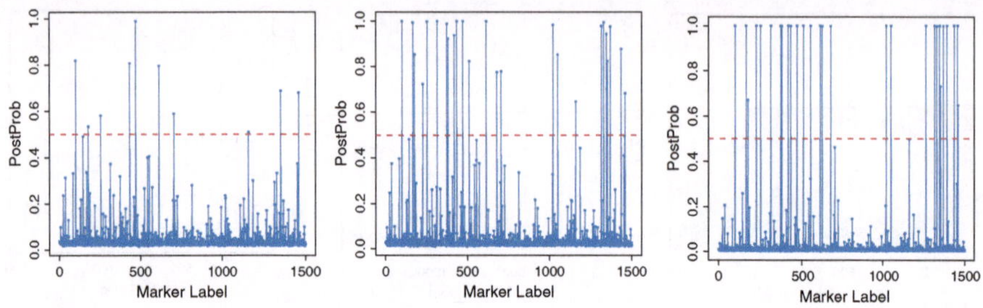

Fig. 8.6 Bayes-McMC implementation of a model with 1500 individuals, 1500 **uncorrelated** genetic markers of which 25 are chosen as QTL. The heritability is 10% (left), 25% (centre) and 50% (right). The figures display the MC estimates of $\Pr(Z_i = 1|y_i)$ against marker labels

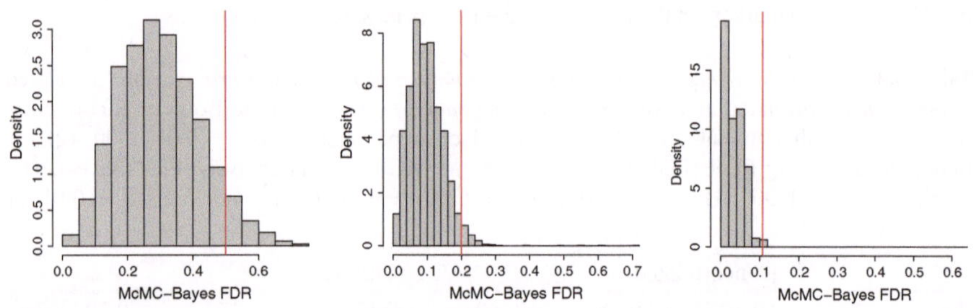

Fig. 8.7 Uncorrelated marker genotypes. The histograms are MC estimates of the marginal posterior distributions of false discovery rates for the three heritability levels. The vertical lines indicate the true number of false discoveries realised in the simulated sample. The heritability is 10% (left), 25% (centre) and 50% (right). Discovery set defined by those markers that satisfy $\Pr(Z_i = 1|y) > 0.5$

When the size of the discovery set of the Bayesian approach is set equal to that generated by FDR-BH, both methods yield almost identical results (Table 8.4).

The figures for FDR in Table 8.4 are based on the true quantities obtained for the sample of simulated data at hand. With real data, this information is not available. In such a situation, for the FDR-BH approach, one must draw conclusions based on the nominal values of q set by the user. In a discovery set of size N say, on average, one expects qN false discoveries. For the three heritability levels, the expected number of FD for the FDR-BH are $8 \times 0.1 \approx 1$, $20 \times 0.05 = 1$ and $23 \times 0.01 \approx 0$, not too far from the true realisations.

The Bayesian implementation yields the marginal posterior distribution of the FDR. Figure 8.7 displays histograms of MC estimates of these distributions for the three heritability levels, using the threshold corresponding to $\Pr(Z_i = 1|y) > 0.5$. The vertical lines indicate the true number of false discoveries realised in the simulated sample.

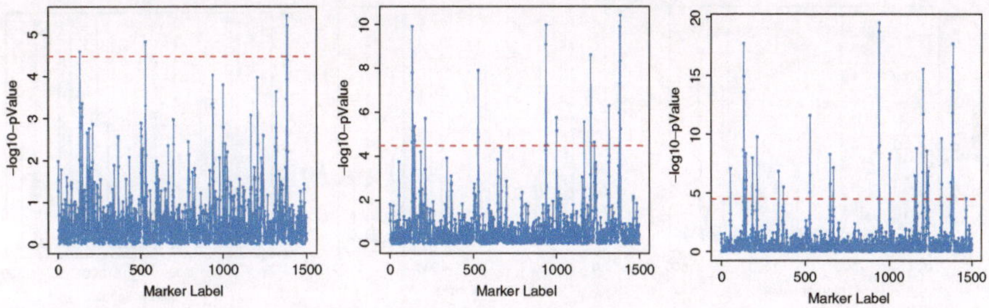

Fig. 8.8 GWAS analysis of 1500 individuals and 1500 **correlated** marker genotypes with 25 loci and $h^2 = 0.10$ (left), $h^2 = 0.25$ (centre), $h^2 = 0.50$ (right). The Y-axes shows $-log_{10}$ p-values and the X-axes the marker labels. The markers that reach $-log_{10}$ p-values beyond the Bonferroni threshold—horizontal line in red, set at $-log_{10}(0.05/1500)$—are declared as discoveries (4, 27 and 50 out of 1500 markers in the left, centre and right panels, respectively)

Correlated Marker Genotypes

In this case, account is taken of the more realistic situation where the 1500 genetic markers are in linkage disequilibrium (marker genotypes are correlated). This has important consequences for detection of QTL and on the behaviour of single marker regressions (GWAS).

The results of GWAS using the Bonferroni correction are displayed in Fig. 8.8 and in the right block of Table 8.4 for the three heritability levels. When $h^2 = 0.10$, $four$ markers reach the significant threshold. When $h^2 = 0.25$, 27 markers are declared significant, and 12 of these are false positive calls. At $h^2 = 0.5$, 50 markers are declared significant and 25 are false positive results. The FDR-BH approach declares 27, 50 and 70 markers as significant for $h^2 = 0.10$, $h^2 = 0.25$ and $h^2 = 0.50$, respectively, with 14, 27 and 45 false positives. These results are far from the expectations. The Bonferroni test should lead to 0 false positive results on average and FDR-BH, given the size of the discovery sets, to $27 \times 0.1 \approx 3$, $50 \times 0.05 \approx 3$ and $70 \times 0.01 \approx 1$ false positive results, for the three heritability levels, the three values of q (equal to $0.1, 0.05, 0.01$) and for the three heritability levels.

In contrast, the results based on the FDR-Bayes are more in line with those of the uncorrelated markers, as reflected in the bottom row of Table 8.4 and in Fig. 8.9.

The different behaviour of FDR-Bayes and FDR-BH is due to the consequence of fitting one marker at a time in situations where loci are correlated. As indicated on page 273, expression (6.40), fitting a single marker that has no direct effect on the dependent variable and that is correlated with a causal marker not included in the model, generates a bias that can result in a phantom significant p-value. When thousands of tests are performed in this way, the phantom p-values that are fed into the FDR-BH lead to incorrect inferences. This is not an inherent problem of FDR-BH; it is rather the single marker regression that cannot be used in this manner in conjunction with FDR-BH. The problem is well understood by practitioners of

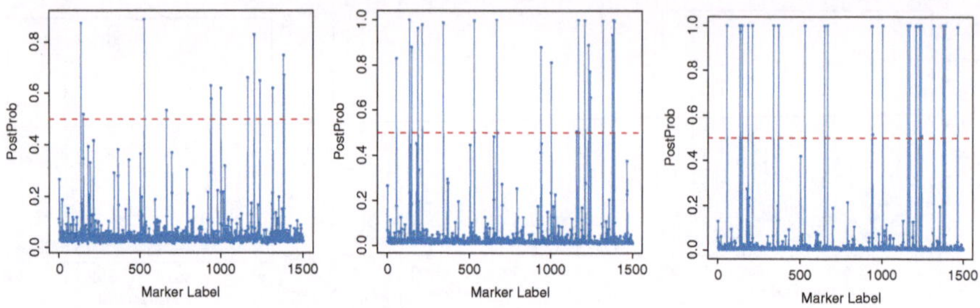

Fig. 8.9 Bayes-McMC implementation of a model with 1500 individuals, 1500 **correlated** genetic markers of which 25 are chosen as QTL. The heritability is 10% (left), 25% (centre) and 50% (right). The figures display the MC estimates of $\Pr(Z_i = 1|y_i)$ against marker labels

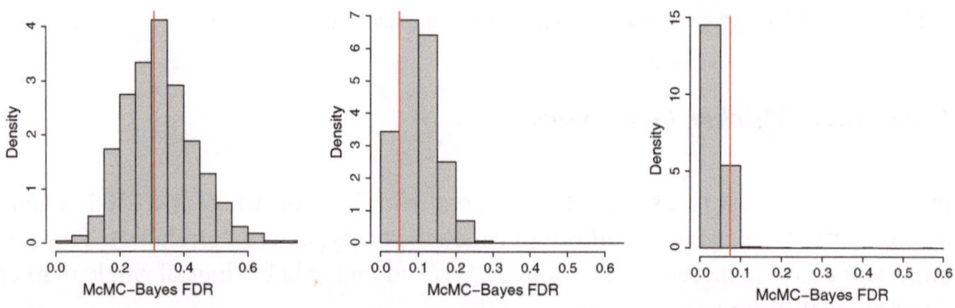

Fig. 8.10 Correlated marker genotypes. The histograms are MC estimates of the marginal posterior distributions of false discovery rates for the three heritability levels. The vertical lines indicate the true number of false discoveries realised in the simulated sample. The heritability is 10% (LEFT), 25% (CENTRE) and 50% (RIGHT). Discovery set defined by those markers that satisfy $\Pr(Z_i = 1|y) > 0.5$

genome studies, and several approximate solutions have been proposed (Yang et al 2012, Brzyski et al 2017).

On the other hand, a method that fits all the markers simultaneously such as the spike and slab model should retrieve more appropriate inferences and provides a unified approach for estimation, prediction and detection of promising covariates. In practice, the computational burden of the implementation of such a model using large data sets with millions of correlated covariates poses serious challenges. Many of these challenges are being met (Patxot et al 2021, de los Campos et al 2022), and very likely these are going to become the methods of choice in a not too distant future.

The MC estimates of the marginal posterior distributions of FDR obtained using FDR-Bayes are displayed in Fig. 8.10.

The R-code below shows the computation of the Bayesian FDR based on (8.44), the construction of the histograms in Figs. 8.7 and 8.10, and Monte Carlo estimates of posterior means and posterior intervals :

```
# FUNCTION TO COMPUTE BAYESIAN FDR AND TO DRAW HISTOGRAM OF FDR
# INPUT:
# 1. resultprobtheta: A FILE WITH DRAWS FROM (7.57)
# 2. truefd: THE TRUE PROPORTION OF FALSE DISCOVERIES OBSERVED
#    IN THE SAMPLE (USED TO DRAW VERTICAL LINE IN HISTOGRAM)
# 3. prob: CHOSEN THRESHOLD THAT DEFINES THE DISCOVERY SET
# 4. postprob: VECTOR OF POSTERIOR PROBABILITIES
#     FOR EACH GENETIC MARKER
fdrhist <- function(resultprobtheta,truefd,prob){
  discset <- which(postprob > prob)
  fdisc <- apply(1-(resultprobtheta[,discset]),1,mean)
# BAYESIAN FDR POSTERIOR MEAN AND POSTERIOR INTERVAL:
  avfdis <- mean(fdisc)
  quantilefdis <- quantile(fdisc,c(0.025,0.975))
# HISTOGRAM OF BAYESIAN FDR:
  hist(fdisc,breaks=30,xlab='McMC-Bayes FDR',
       main=NULL,freq=FALSE)
  abline(v=truefd[length(discset)],col="red",lwd=2)
# RETURNS: MC average of Bayesian FDR; posterior interval,
#  size of discovery set, number of false discoveries
  return<-c(avfdis,quantilefdis,length(discset),
            truefd[length(discset)]*length(discset))
}
out <- fdrhist(resultprobtheta,truefd,0.5)
out
```

Chapter 9
Binary Data

Many of the results derived under the assumption that observations are continuously distributed extend to dichotomous and categorical responses. There are some technical details that must be observed that are specific to discontinuous data. The chapter starts by illustrating the behaviour of training and validating mean squared error applied to binary records using operational logistic regression models with increasing number of covariates.

As the number of covariates in a prediction model increases relative to the number of records, the bias-variance trade-off calls for the introduction of shrinkage. Three modelling scenarios that incorporate shrinkage are described: penalised logistic regression, logistic lasso and the Bayesian-McMC driven spike and slab model introduced on page 321, extended here to deal with binary records.

The performance of a classifier can be gauged by studying the *true positive rate* and *false positive rate*. The *receiver operating characteristic curve* is often used to compare the performance of different binary classifiers and is the subject of Sect. 9.7.

A topic that is peculiar to data falling into discrete and mutually exclusive categories is the computation of the probability that a future observation falls in a given category, given some previous information. An application to binary occurrences discussed in Sect. 9.8 is the prediction of disease status of a genetic disease for an individual, given information on the disease status of its relatives.

The chapter ends with an appendix describing an approximate analysis of binary traits that can be useful as an initial investigating tool, before developing the full machinery needed for a more sophisticated approach.

A general framework for the analysis of binary data is introduced on page 84.

D. Sorensen, *Statistical Learning in Genetics*, Statistics for Biology and Health,
https://doi.org/10.1007/978-3-031-35851-7_9

9.1 Prediction for Binary Observations

Consider data $(Y_1, x_1), \ldots, (Y_n, x_n)$ where Y_i's are binary random variables and $x_i \in \mathbb{R}^p$ covariates. A binary classifier is a function $s(x_i)$ that transforms inputs x_i into a value between 0 and 1 interpreted as a probability. This can be expressed as a prediction $\hat{Y}_i$ that takes the values either 0 or 1, according to

$$\widehat{Y}_i = \begin{cases} 1 \text{ if } s(x_i) > t \\ 0 \text{ if } s(x_i) \leq t, \end{cases} \tag{9.1}$$

where $t \in [0, 1]$ is some threshold. When $s(x_i) = E(Y_i|x_i)$ and $t = 0.5$, this is known as Bayes rule (poor choice of terminology, commonly used, not to be confused with Bayes theorem) that minimises the overall probability of a misclassification $\Pr\left(\widehat{Y}_i \neq Y_i\right)$. For Bernoulli data we have

$$s(x_i) = \mathrm{E}(Y_i|x_i) = \Pr(Y_i = 1|x_i) \tag{9.2}$$

and given data and the threshold t, one predicts $\widehat{Y}_i$ according to the rule

$$\widehat{Y}_i = \begin{cases} 1 \text{ if } \widehat{s}(x_i) = \widehat{\Pr}(Y_i = 1|x_i) > t \\ 0 \text{ if } \widehat{s}(x_i) = \widehat{\Pr}(Y_i = 1|x_i) \leq t, \end{cases} \tag{9.3}$$

where $\widehat{s}(x_i) = \widehat{\Pr}(Y_i = 1|x_i)$ is an estimate of $s(x_i)$.

In the logistic model, the object of the modelling is the probability that Y is equal to 1, parametrised as

$$s(x_i) = \Pr(Y_i = 1|x_i) = \frac{\exp\left(x_i'\beta\right)}{1 + \exp\left(x_i'\beta\right)}, \tag{9.4}$$

or alternatively, as the logit or log odds

$$\ln\left[\frac{\Pr(Y_i = 1|x_i)}{\Pr(Y_i = 0|x_i)}\right] = x_i'\beta.$$

In these expressions β includes an intercept, and the first element of the $(p+1)$ row vector x_i' is a 1.

Imagine that estimates $\widehat{\beta}$ of the logistic coefficients are available. These yield estimates of the probability

$$\widehat{\Pr}(Y_i = 1|x_i) = \frac{\exp\left(x_i'\widehat{\beta}\right)}{1 + \exp\left(x_i'\widehat{\beta}\right)}.$$

If the choice falls on $t = 0.5$, we would predict $\widehat{Y}_i = 1$ if $\widehat{\Pr}(Y_i = 1|x_i) > \widehat{\Pr}(Y_i = 0|x_i)$, that is, if $\widehat{\Pr}(Y_i = 1|x_i) > 0.5$ or if $x_i'\hat{\beta} > 0$. The latter is readily derived as follows. Choose $\widehat{Y}_i = 1$ if

$$\frac{\widehat{\Pr}(Y_i = 1|x_i)}{\widehat{\Pr}(Y_i = 0|x_i)} > 1.$$

Taking logarithms on both sides,

$$\ln\left[\widehat{\Pr}(Y_i = 1|x_i)\right] - \ln\left[\widehat{\Pr}(Y_i = 0|x_i)\right] > 0.$$

This gives

$$\ln\left[\frac{\exp\left(x_i'\hat{\beta}\right)}{1+\exp\left(x_i'\hat{\beta}\right)}\right] - \ln\left[\frac{1}{1+\exp\left(x_i'\hat{\beta}\right)}\right] > 0,$$

$$x_i'\hat{\beta} > 0.$$

Prediction of binary outcomes is often known as *classification*.

9.2 Mean Squared Error

As indicated in (6.69), the estimate of the expected validating mean squared error using independent training data is

$$\hat{\mathrm{E}}\left(\mathrm{MSE}_v\right) = \frac{1}{N}\sum_{i=1}^{N}(y_i - \widehat{y}_i)^2 + \frac{2}{N}\sum_{i=1}^{N}\widehat{\mathrm{Cov}}\left(y_i, \widehat{y}_i\right), \tag{9.5}$$

where the second term is an estimate of the expected optimism. With binary observations

$$(y_i - \widehat{y}_i)^2 = \begin{cases} 1 \text{ if } y_i \neq \widehat{y}_i, \\ 0 \text{ if } y_i = \widehat{y}_i. \end{cases} \tag{9.6}$$

Therefore, the first term in the right hand side of (9.5) is the observed proportion of misclassifications (or error rate) in the training data. To compute the second term, proceed as follows:

$$\mathrm{Cov}\left(y_i, \widehat{y}_i\right) = \mathrm{E}\left(y_i\widehat{y}_i\right) - \mathrm{E}\left(y_i\right)\mathrm{E}\left(\widehat{y}_i\right), \tag{9.7}$$

where $\mathrm{E}(y_i) = \Pr(y_i = 1)$ and all expectations are conditional on x. In addition, we have

$$\begin{aligned}\mathrm{E}(\widehat{y_i}) &= \mathrm{E}_{y_i}\left[\mathrm{E}(\widehat{y_i}|y_i)\right] \\ &= \Pr(\widehat{y_i} = 1|y_i = 0)\Pr(y_i = 0) + \Pr(\widehat{y_i} = 1|y_i = 1)\Pr(y_i = 1),\end{aligned}$$

and

$$\begin{aligned}\mathrm{E}(y_i\widehat{y_i}) &= \mathrm{E}_{y_i}\left[\mathrm{E}(y_i\widehat{y_i}|y_i)\right] \\ &= \mathrm{E}_{y_i}\left[y_i\,\mathrm{E}(\widehat{y_i}|y_i)\right] \\ &= \Pr(y_i = 1)\Pr(\widehat{y_i} = 1|y_i = 1).\end{aligned}$$

Substituting these expressions in (9.7) gives

$$\begin{aligned}&\mathrm{Cov}(y_i, \widehat{y_i}) = \Pr(y_i = 1)(1 - \Pr(y_i = 1)) \\ &[\Pr(\widehat{y_i} = 1|y_i = 1) - \Pr(\widehat{y_i} = 1|y_i = 0)].\end{aligned}$$

The estimator of the expected validating mean squared error (9.5) for binary data takes the form

$$\begin{aligned}\hat{\mathrm{E}}(\mathrm{MSE}_v) &= \frac{\#(y_i \neq \widehat{y_i})}{N} + \frac{2}{N}\sum_{i=1}^{N}\widehat{\Pr}(y_i = 1)\left(1 - \widehat{\Pr}(y_i = 1)\right) \\ &\left[\widehat{\Pr}(\widehat{y_i} = 1|y_i = 1) - \widehat{\Pr}(\widehat{y_i} = 1|y_i = 0)\right]\end{aligned} \tag{9.8}$$

where $\#(y_i \neq \widehat{y_i})$ is the total number of misclassifications in the training data and the "hats" are estimates of the various probabilities. Expression (9.8) is compared to MC estimates of the expected validating mean squared error in the simulation example that follows. The second term in (9.8) is an estimator of expected optimism.

Much of the material in this section is taken from Efron and Hastie (2016) where more details on the subject can be found.

Example: Training and Validation MSE with Binary Data

This example illustrates in the context of binary data, how training and validating mean squared errors are affected as the number of covariates of the operational models increases. The example is in the same spirit as that on page 291.

Binary data are simulated using the logistic model (9.4) and consist of $N = 400$ records divided in equal numbers into training and validating data. The matrix X of genetic marker codes is generated as in the Example on page 291, here with $p_\ell = 10$ loci and includes a column vector of 1s to accommodate an intercept.

Training data are analysed with 5 operational models with numbers of covariates equal to 5, 10, 15, 20 or 25. Model 2 with 10 covariates is the true model. The regression coefficients of the true effects of the 10 loci were sampled from $N(0, 0.1)$. The intercept μ was chosen to generate either $\Pr(Y_i = 1|\mu) = 0.5$ or $\Pr(Y_i = 1|\mu) = 0.2$. In all, 400 replicates are simulated of both training and validating data.

Figure 9.1 displays the mean squared errors for the validating and the training data for each replicate, as well as the average over replications. The top panel

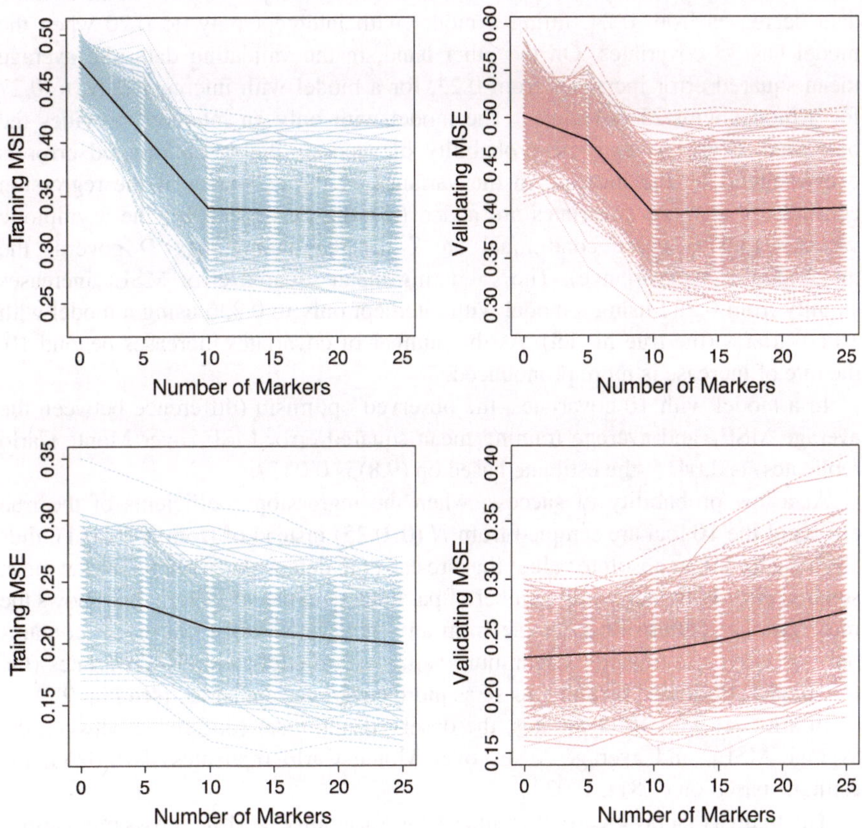

Fig. 9.1 Mean squared error (MSE) per replicate (400 Monte Carlo simulations) and average MSE shown as bold black lines. True marker effects drawn from $N(0, 0.1)$. Top panel, $\Pr(Y_i = 1|\mu) = 0.5$; bottom panel, $\Pr(Y_i = 1|\mu) = 0.2$. The left panels correspond to the training data and the right to the validating data

corresponds to $\Pr(Y_i = 1|\mu) = 0.5$ and the bottom panel to $\Pr(Y_i = 1|\mu) = 0.2$. In the training data, the mean squared error decreases with increasing number of parameters. For $\Pr(Y_i = 1|\mu) = 0.5$ the rate of decline is largest up to 10 markers (the true model) and thereafter becomes markedly smaller.

In the validating data, the pattern of the evolution of the mean squared error with increasing number of parameters depends on the true distribution of the data. At intermediate probability of success, the average mean squared error in the validating data decreases from 0.50 using a model with intercept only to 0.396 using a model with 10 covariates (the true model). Beyond 10 covariates there is a slight increase up to a value of 0.403 with 25 covariates. The observed optimism (difference between the average validating MSE_v and average training MSE_t over Monte Carlo replicates) is 0.0623 in a model with 10 covariates and the estimate based on (9.8) is 0.0617.

At the low probability of success, the average mean squared error in the training data decreases from 0.24, fitting a model with intercept only, to 0.20 when the model has 25 covariates. On the other hand, in the validating data, the average mean squared error increases from 0.23, for a model with intercept only, to 0.27 for a model with 25 covariates. The model with only an intercept provides the best predictions. At extreme probability of success, the mean squared error is overwhelmed by the increase in the variance of the estimates of the regression coefficients as more covariates are added (see the expression for the asymptotic variance (3.37), where conditional on X the diagonal matrix D governs the magnitude of the variance). The validating mean squared error MSE_v increases slightly from 0.232 using a model with intercept only to 0.236 using a model with 10 covariates (the true model). As the number of covariates increases beyond 10, the rate of increase is more pronounced.

In a model with 10 covariates, the observed optimism (difference between the average MSE_v and average training mean squared error MSE_t over Monte Carlo replicates) is 0.0415; the estimate based on (9.8) is 0.0477.

At a low probability of success, when the regression coefficients of the true effects of the 10 loci are sampled from $N(0, 0.25)$ instead of from $N(0, 0.1)$, thus allowing for larger absolute values that are easier to detect, the evolution of the mean squared error with increasing number of parameters in the validating data shows the more familiar pattern of a decline from an initial value when the model contains only an intercept, towards a minimum value with models with 10 covariates (the true model), followed by an increase as more covariates are added (see Fig. 9.2).

In a model with 10 covariates, the observed optimism (difference between the average MSE_v and average MSE_t over Monte Carlo replicates) is 0.0253; the estimate based on (9.8) is 0.0273.

The figures display clearly the rather large variability in MSE across replications of training and validating data.

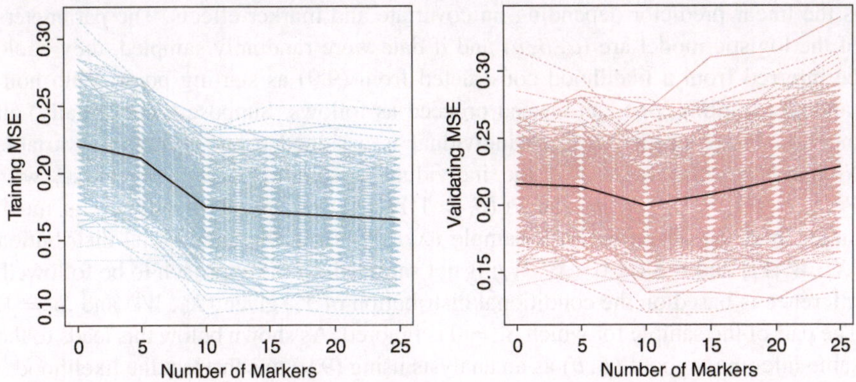

Fig. 9.2 Mean squared error (MSE) per replicate (400 Monte Carlo simulations) and average MSE in black, bold. Probability of success is $\Pr(Y_i = 1|\mu) = 0.2$. True marker effects drawn from $N(0, 0.25)$. The left panel corresponds to the training data and the right to the validating data

9.3 Logistic Regression with Non-random Sampling

A major concern in epidemiological studies is to determine factors that are linked to a condition or disease in a population. A variety of designs can be considered. When the condition is at very low frequency in the population, a common study design is to over-sample individuals with the condition (known as successes or cases) and then sample a control group (failures or controls) from a similar segment of the population. This type of design is known as a case-control study. A typical example might involve recording relevant covariates from people who died from a certain condition and from other patients who act as controls. If sampling is at random, the relationship between the variables in the dataset would be representative of the same relationships in the population, but this may not be the case with non-random samples of a case-control design. Correct inferences require that the statistical analysis accounts for the non-random sampling mechanism (Prentice and Pyke 1979).

Consider a population of individuals $i = 1, \ldots, n$ with disease status Y_i, covariate vector X_i and marker genotypes W_i. Assume that the conditional distribution of Y_i given $(X_i, W_i) = (x_i, w_i)$ is the logistic regression

$$\Pr(Y_i = 1|x_i, w_i) = \frac{\exp(\eta_i)}{1 + \exp(\eta_i)} \tag{9.9}$$

where

$$\eta_i = \alpha + x_i\beta + w_i b$$

is the linear predictor depending on covariate and marker effects. The parameters of the logistic model are (α, β, b) and if data were randomly sampled, they could be inferred from a likelihood constructed from (9.9) as starting point. With non-random sampling, one can instead proceed as follows. Suppose (conceptually) all individuals are observed. Each individual is assigned a binary random variable S_i that takes the value 1 if the individual is sampled and 0 otherwise, with $\Pr(S_i = 1|Y_i = 0) = p_0$ and $\Pr(S_i = 1|Y_i = 1) = p_1$. Typically, p_1 is much larger than p_0. Given $S_i = 1$, sample (X_i, W_i) from the conditional distribution $[X_i, W_i|Y_i]$ and if $S_i = 0$, (X_i, W_i) is not sampled. In the approach to be followed, inference is based on the conditional distribution of Y_i, given (X_i, W_i) and $S_i = 1$. The part of the sample for which $S_i = 0$ is ignored. As shown below this leads to the same inferences about (β, b) as an analysis using (9.9) to construct the likelihood.

The conditional distribution of Y_i given (X_i, W_i) and $S_i = 1$ is

$$\Pr(Y_i = 1|x_i, w_i, S_i = 1) =$$
$$\frac{\Pr(S_i = 1|Y_i = 1, x_i, w_i)\Pr(Y_i = 1|x_i, w_i)}{\Pr(S_i = 1|x_i, w_i)}.$$

Now make the important assumption that $\Pr(S_i = 1|Y_i = 1, x_i, w_i) = \Pr(S_i = 1|Y_i = 1)$, that is, S_i and (X_i, W_i) are conditionally independent given Y_i. In a missing data framework, (X_i, W_i) are missing at random, in the sense defined by Rubin (1976). Given this assumption the above can be written

$$\Pr(Y_i = 1|x_i, w_i, S_i = 1) = \frac{p_1 \exp(\eta_i)}{p_1 \exp(\eta_i) + p_0}$$
$$= \frac{(p_1/p_0)\exp(\eta_i)}{(p_1/p_0)\exp(\eta_i) + 1}. \tag{9.10}$$

Since

$$(p_1/p_0)\exp(\eta_i) = \exp(\alpha^* + x_i\beta + w_i b)$$

where $\alpha^* = \alpha + \ln(p_1/p_0)$, we have

$$\Pr(Y_i = 1|x_i, w_i, S_i = 1) = \frac{\exp(\alpha^* + x_i\beta + w_i b)}{1 + \exp(\alpha^* + x_i\beta + w_i b)}$$

as in (9.9) but with a new intercept. Therefore, the fact that the data were not sampled at random (they were collected retrospectively) can be ignored in logistic regression analyses, provided that the linear predictor contains an intercept and the sampling mechanism is independent of the explanatory variables. This is a property peculiar to the logistic link and does not carry over to other link functions.

9.4 Penalised Logistic Regression

Logistic regression can be modified in the same spirit as ridge regression to deal with the $p > n$ case (number of covariates p larger than the number of records n), by maximisation of a concave penalised loglikelihood (or, equivalently, by minimisation of a convex penalised cost function). Here details are provided for a Newton-Raphson implementation; a full discussion can be found in a study by Park and Hastie (2008).

The logistic regression model for the n binary responses y is parametrised as

$$\Pr(y_i = 1|x_i, \mu, \beta) = \frac{\exp(\mu + x_i'\beta)}{1 + \exp(\mu + x_i'\beta)}, \quad i = 1, \ldots, n. \tag{9.11}$$

In this expression, μ is a scalar intercept, x_i' is a row vector of p covariates and β is a column vector with p regression parameters. The convex function is the negative of the loglikelihood subject to a size constraint on the L_2 norm of β; the intercept μ is not penalised. Then μ and β are obtained as the minimisation of the cost function

$$J(\mu, \beta) = -\ell(\mu, \beta|y, x) + \frac{\lambda}{2}\beta'\beta,$$

where λ is the regularisation parameter and

$$-\ell(\mu, \beta|y, x) = -\sum_{i=1}^{n}\left\{y_i(\mu + x_i'\beta) - \ln\left[1 + \exp(\mu + x_i'\beta)\right]\right\} \tag{9.12}$$

is the negative of the loglikelihood.

I shall use

$$f(x_i) = \mu + x_i'\beta$$

and

$$\pi_i = \Pr(y_i = 1|x_i, \mu, \beta) = \frac{\exp(f(x_i))}{1 + \exp(f(x_i))}.$$

The first derivatives are

$$\frac{\partial J}{\partial \mu} = -\sum_{i=1}^{n}(y_i - \pi_i) = -1'(y - \pi), \tag{9.13a}$$

$$\frac{\partial J}{\partial \beta} = -\sum_{i=1}^{n}(y_i - \pi_i)x_i + \lambda\beta = -X'(y - \pi) + \lambda\beta, \tag{9.13b}$$

where $1'$ is a $(1 \times n)$ row vector of ones, y and π are $n \times 1$ column vectors with elements y_i and π_i, respectively, and X is an $n \times p$ matrix of covariates with row vectors x_i'.

The computation of second derivatives uses

$$\frac{\partial \pi_i}{\partial \mu} = \frac{\partial \pi_i}{\partial f(x_i)} \frac{\partial f(x_i)}{\partial \mu},$$
$$\frac{\partial \pi_i}{\partial \beta} = \frac{\partial \pi_i}{\partial f(x_i)} \frac{\partial f(x_i)}{\partial \beta}.$$

Application of the chain rule yields

$$\frac{\partial^2 J}{(\partial \mu)^2} = \sum_{i=1}^{n} \pi_i (1 - \pi_i) = 1'W1,$$
$$\frac{\partial^2 J}{\partial \mu \partial \beta'} = \sum_{i=1}^{n} \pi_i (1 - \pi_i) x_i = 1'WX,$$
$$\frac{\partial^2 J}{\partial \beta \partial \beta'} = \sum_{i=1}^{n} x_i x_i' \pi_i (1 - \pi_i) + I\lambda = X'WX + I\lambda,$$

where I is the $p \times p$ identity matrix and

$$W = \text{diag}\left\{\pi_i (1 - \pi_i)\right\}, \quad i = 1, \ldots, n,$$

a diagonal matrix of dimension $n \times n$. The matrix of second derivatives is

$$H = \begin{bmatrix} \frac{\partial^2 J}{(\partial \mu)^2} & \frac{\partial^2 J}{\partial \mu \partial \beta'} \\ \frac{\partial^2 J}{\partial \beta \partial \mu} & \frac{\partial^2 J}{\partial \beta \partial \beta'} \end{bmatrix}_{(p+1)\times(p+1)} = \begin{bmatrix} 1'W1 & 1'WX \\ X'W1 & X'WX + I\lambda \end{bmatrix} = Z'WZ + \Lambda \tag{9.14}$$

of order $(p + 1) \times (p + 1)$, where

$$Z = [1, X].$$

This is an $n \times (p + 1)$ matrix whose first column is a vector of ones and

$$\Lambda = diag\left\{0, \lambda, \lambda, \ldots, \lambda\right\},$$

is a $(p + 1) \times (p + 1)$ diagonal matrix. Let $\theta = \left(\mu, \beta'\right)'$. The Newton-Raphson algorithm is

$$\theta_{t+1} = \theta_t - (H)^{-1} S(\theta_t) \tag{9.15}$$

where, from (9.13),

$$S(\theta_t) = -Z'(y - \pi) + \Lambda\theta_t. \tag{9.16}$$

Using (9.14) and (9.16) in (9.15) yields

$$\begin{aligned}
\theta_{t+1} &= \theta_t - \left(Z'WZ + \Lambda\right)^{-1}\left(-Z'(y - \pi) + \Lambda\theta_t\right) \\
&= H^{-1}\left(Z'WZ + \Lambda\right)\theta_t + H^{-1}\left(Z'WW^{-1}(y - \pi) - \Lambda\theta_t\right) \\
&= H^{-1}Z'W(Z\theta_t) + H^{-1}\Lambda\theta_t + H^{-1}Z'W\left(W^{-1}(y - \pi)\right) - H^{-1}\Lambda\theta_t \\
&= H^{-1}Z'W\left(Z\theta_t + W^{-1}(y - \pi)\right) \\
&= H^{-1}Z'Wr,
\end{aligned} \tag{9.17}$$

where $r = Z\theta_t + W^{-1}(y - \pi)$ and $H^{-1} = \left(Z'WZ + \Lambda\right)^{-1}$. This implementation of the Newton-Raphson algorithm is the *iteratively reweighted ridge regression* algorithm and requires inversion of the Hessian, an $(p + 1) \times (p + 1)$ matrix. In highly dimensional settings with $p \gg n$, alternative strategies must be sought. The aim could be either to avoid matrix inversion or to reduce the dimensionality of the system from order p to order n. An example of the first type is the use of gradient descent that uses first derivatives of the cost function. An approach that works in an n-dimensional space involves kernel methods discussed on page 482. It is shown that use of a particular kernel known as the *linear kernel* allows a reparametrisation of the penalised logistic regression model that leads to an iterative system along the lines of (9.17), while operating in an $n + 1$-dimensional space.

9.5 The Lasso with Binary Records

The use of the lasso is illustrated for the case of binary responses parametrised as in (9.4). In this formulation the loglikelihood takes the form

$$\ell(\beta_0, \beta | x, y) = \sum\nolimits_{i=1}^{n} y_i\left(\beta_0 + x_i'\beta\right) - \ln\left[1 + \exp\left(\beta_0 + x_i'\beta\right)\right]$$

where here, contrary to (9.4), x_i' is $(1 \times p)$ and β is $(p \times 1)$. The lasso coefficients are the solutions to

$$\max_{\beta_0,\beta}\left[\sum_{i=1}^{n} y_i\left(\beta_0 + x_i'\beta\right) - \ln\left[1 + \exp\left(\beta_0 + x_i'\beta\right)\right]\right] - \lambda\sum_{j=1}^{p}\left|\beta_j\right|. \tag{9.18}$$

As before in (7.19), the intercept term is not penalised and the predictors x are standardised. An efficient computational algorithm is the path-wise coordinate descent described in connection with the linear model, with modifications to accommodate the lack of piece-wise linearity of the coefficients of the logistic likelihood (Efron and Hastie 2016).

9.6 A Bayesian Spike and Slab Model for Binary Records

An alternative to the penalised logistic regression and the lasso is the spike and slab model that postulates a two-component mixture distribution for the SNP effects consisting of a normal distribution and a point mass at zero. Section 7.5 provides a full description of the model and an McMC implementation for continuous data. Here, I indicate the modification of the McMC algorithm required for the analysis of binary responses.

The binary records are described with the probit model that assumes an underlying unobserved liability u. For the ith record, the liability is normally distributed $u_i|\mu, x_i, b \sim N\left(\mu + x_i'b, \sigma^2 = 1\right)$. The linear model for the liability of record i can be written

$$u_i = \mu + x_i'b + e_i.$$

Above, μ is an intercept, $x_i \in \mathbb{R}^p$ is a column vector of p observed marker genotypes, b is a $p \times 1$ column vector of unobserved marker effects and residuals are $e_i \overset{iid}{\sim} N(0, 1)$. For the ith record $(i = 1, 2, \dots, n)$

$$\begin{aligned} \Pr(y_i = 1|\mu, x_i, b) &= \Pr(u_i > 0|\mu, x_i, b) \\ &= \Pr\left(e_i < \mu + x_i'b\right) \\ &= \Phi\left(\mu + x_i'b\right), \end{aligned} \tag{9.19}$$

where $\Phi(t)$ is the distribution function of the standard normal distribution and I used the symmetry of the normal distribution to go from line 1 to line 2. The conditional likelihood function for record i, given μ, x_i, b, is proportional to (9.19). The prior distributions for the binary case are assumed to be identical to those in (7.43) on page 321, except for σ^2, that here takes the fixed value $\sigma^2 = 1$. As in the continuous case, the SNP effects are expressed as

$$b_i = \alpha_i \delta_i, \quad i = 1, \dots, p, \tag{9.20}$$

where $\alpha_i|\sigma_b^2 \overset{iid}{\sim} N\left(0, \sigma_b^2\right)$ is independent of the binary (0, 1) indicator variable δ_i that has an a priori Bernoulli distribution with probability π:

$$\delta_i|\pi \overset{iid}{\sim} Br(\pi), \quad , i = 1, \ldots, m,$$

$$\Pr(\delta_i = 1|\pi) = \pi, \quad \Pr(\delta_i = 0|\pi) = 1 - \pi.$$

This binary indicator variable with its associated distribution $\Pr(\delta_i = 1|\pi)$ specifies the a priori probability that a marker effect b_i is non-zero, and this prior probability is the same for all markers. The variance component σ_b^2, that describes the prior uncertainty of the SNP effect drawn from the slab component is assumed to have the scaled inverse chi-square prior distribution

$$\sigma_b^2|S_b, v_b \sim S_b\chi^{-2}(S_b, v_b),$$

where the S_b and the v_b are user-tuned hyperparameters.

It is computationally convenient to augment the vector of parameters with the unobserved liabilities u (see page 215). The a priori distribution of the random vector u is

$$u|\mu, b \sim N(1\mu + Xb, 1)$$

and the conditional pmf of the data, given u, takes the degenerate form (see page 216)

$$\Pr(Y_i = y_i|u_i) = I(u_i > 0)^{y_i} + I(u_i < 0)^{1-y_i}.$$

The posterior density for the binary case is

$$p\left(u, \mu, \alpha, \delta, \pi, \sigma_b^2|y\right) \propto p(y|u)\, p(u|\mu, b)\, p\left(\alpha|\sigma_b^2\right) p(\delta|\pi)\, p(\pi)\, p\left(\sigma_b^2\right) \tag{9.21}$$

with the b's defined in (9.20).

The fully conditional posterior distributions for the Gibbs sampling implementation are derived from (9.21) and with the exception of the fully conditional of u, $[u|D]$ (see page 222 for definition of D) and of the fact that the conditional variance of the liability is fixed to $\sigma^2 = 1$, are identical to those for the continuous data and will not be repeated here.

The fully conditional posterior density of the liability is obtained by extracting those terms in the posterior density (9.21) that contain u. That is,

$$\begin{aligned} p(u_i|D) &\propto p(u_i|\mu, b)\, p(y_i|u_i) \\ &= N(u_i|1\mu + Xb, 1)\left[I(u_i > 0)^{y_i} + I(u_i < 0)^{1-y_i}\right]. \end{aligned} \tag{9.22}$$

This expression indicates that when $y_i = 1$, $u_i > 0$ and the fully conditional posterior density of u_i is a truncated normal distribution, with mean $x_i'\beta$, variance 1 and support $(0, \infty)$. When $y_i = 0$, then $u_i < 0$ and the fully conditional posterior density of u_i is a truncated normal distribution, with mean $x_i'\beta$, variance 1 and support $(-\infty, 0)$.

A general method to sample the variable u from any univariate distribution truncated in the interval $[a, b]$ can be found in Devroye (1986), page 38, Example 10:

$$u = F^{-1}\left\{F(a) + U\left[F(b) - F(a)\right]\right\}, \tag{9.23}$$

where F is the distribution function of the untruncated variable and U is a draw from a uniform distribution in the interval [0, 1].

An R-code to draw all the U's in one pass is as follows. Assume that the mean of the vector of untruncated liability is given by mu+Xb, the variance is the identity matrix I and y represents the complete binary data vector. Then the code is

```
mean <- mu+Xb
sd <- 1
intermediate <- y*pnorm(0,mean=mean,sd=sd)+
    runif(length(y))*(pnorm(0,mean=mean,sd=sd)*(1-y) +
    (1-pnorm(0,mean=mean,sd=sd))*y)
u <- qnorm(intermediate,mean=mean,sd=sd)
```

The general algorithm to execute the Gibbs sampler for binary records is identical to that described on page 321, except that the liability u has to be generated with each new cycle of the Gibbs sampler. This is achieved by executing the code above. Further, the residuals have to be defined appropriately at the level of the liability, and σ^2 is fixed at the value 1.

Example: Prediction and QTL Detection Using Genetic Marker Information

This example investigates the predictive ability of the penalised logistic regression, the logistic lasso and the binary spike and slab model, using 2000 simulated binary records representing nominally unrelated individuals. The package **glmnet** described in Example 7.4 on page 319 was used to perform the computations for the logistic lasso.

The data are generated using a probit model where the unobserved liability u_i for the ith record has the linear structure

$$u_i = \mu + z_i'b + e_i, \quad i = 1, 2, \ldots, 2000. \tag{9.24}$$

Above, $\mu = -0.67$ is chosen to generate a proportion of $1's$ of approximately 0.25, z_i is the column vector for individual i of 50 centred QTL genotype codes, b is the column vector of the 50 substitution effects and the $e_i's$ are $N(0, 1)$ independent residual effects. The additive genetic variance in the underlying scale is $\sigma_a^2 = Var(z_i'b|b)$. This variance was set equal to 1 and the substitution effects b_i were chosen accordingly, the same value for the 50 loci (≈ 0.23 units).

The binary records are analysed with an operational probit model involving 1500 centred genotypic marker codes. The 50 loci of the true model in (9.24) were randomly sampled from these 1500 and were assigned as QTL. For individual i, the operational model of the underlying liability is

$$u_i = m + x_i'\beta + \varepsilon_i, \quad i = 1, 2, \ldots, 2000,$$

where m is an intercept, x_i is the (observed) column vector for individual i of the 1500 centred marker genotypic codes, β is the (unobserved) vector with the 1500 unknown marker effects and the $\varepsilon_i's$ are independently distributed standard normal random variables.

QTL Detection

Before studying prediction ability, we look into the performance of the spike and slab model as a QTL detection tool and compare it with a simple GWAS approach based on a probit likelihood, one marker at a time. Having generated the binary data vector y of length 2000 and the matrix of centred marker genotypes Xc, the kernel of the R-code for the GWAS analysis is shown below:

```
# PERFORM A GWAS ON THE DATA USING A PROBIT REGRESSION
# Xc is the 2,000 by 1,5000 matrix of centred marker
# genotype codes
GWAS=matrix(nrow=ncol(Xc),ncol=4)
colnames(GWAS)=c('estimate','SE','t-value','p-value')
for(i in 1:ncol(Xc)){
fm=glm(y~Xc[,i],family = binomial(link = "probit"))
GWAS[i,]= summary(fm)$coef[2,]
}
plot(-log10(GWAS[,4]),type='o',ylab='-log10-pValue',
     cex=.5,col=4)
abline(h=-log10(0.05/nmark),lty=2,col=2)
# log Bonferroni bound:
cat('Bonferroni',-log10(0.05/nmark),'\n')
GWASdetct<-which(-log10(GWAS[,4]) > -log10(0.05/nmark))
length(GWASdetct) # SIZE OF DISCOVERY SET
```

The computation of the p-values is obtained appealing to the asymptotic properties of the likelihood estimator of marker effects; this leads to approximate t-distributed test statistics and uniformly distributed p-values $Un(0, 1)$ under H_0 (see page 337).

The result for the GWAS based on a Bonferroni bound for an individual test equal to $0.05/1500$ is shown in the left panel of Fig. 9.3. Only 13 of the 50 QTL are declared significant, and in this discovery set, there are no false discoveries. The classical test without the Bonferroni correction declares 133 markers as significant, and among these, 86 are false discoveries.

The size of the discovery set based on the Benjamini and Hochberg FDR rule (8.14b), setting the level of the FDR $q = 0.15$, is equal to 50, and among these, the realised number of false discoveries is 11, leading to a realised false discovery proportion equal to $11/50 = 0.22$. The R-code that implements this computation is shown on page 340.

The right panel of Fig. 9.3 shows McMC estimates of the posterior probabilities that each marker effect is not equal to zero, generated from the Bayesian McMC spike and slab model using the 2000 binary records. A discovery set was obtained by arbitrarily choosing the markers with non-zero posterior probabilities larger than 0.8. The ensuing discovery set included 44 genetic markers, and among these, 2 were true false positives, leading to a true FDR equal to $2/44 = 0.045$. The McMC estimate of the mean of the posterior distribution of the Bayes-FDR for this discovery set of size 44 is 0.033, based on (8.43) or on (8.44) on page 351. The 95% posterior interval obtained from the McMC draws from (8.44) is $(4.56 * 10^{-5}; 9.11 * 10^{-2})$.

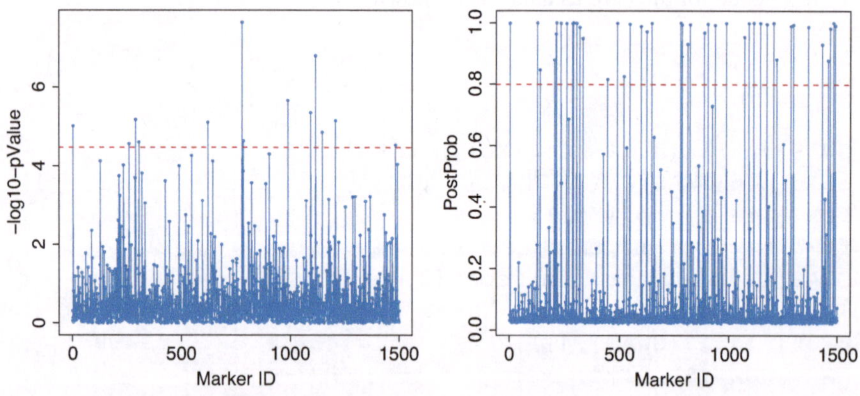

Fig. 9.3 Analysis of 2000 binary records. Left: $-\log_{10}$ p-values for testing the null hypothesis that each of 1500 marker effects is equal to zero. The horizontal red line is the Bonferroni threshold corresponding to $-\log_{10}(0.05/1500)$. Right: posterior probabilities that marker effects are not zero, for each of the 1500 markers, based on the Bayesian McMC spike and slab model. The horizontal line is the threshold corresponding to a posterior probability equal to 0.8

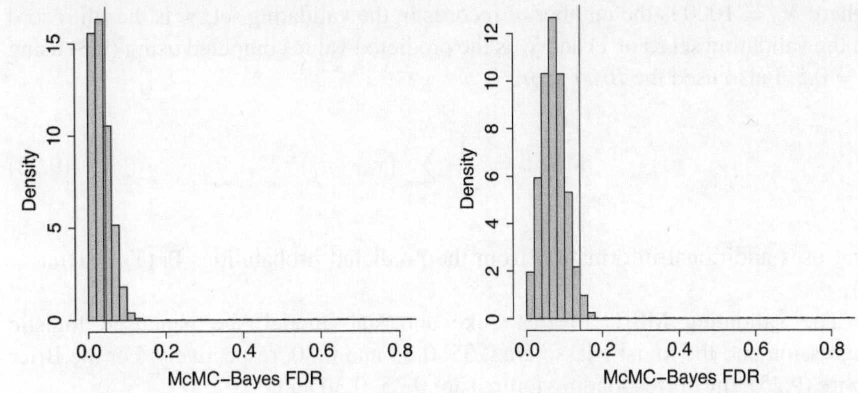

Fig. 9.4 Histograms of the McMC Bayes posterior distribution of the Bayes-FDR choosing as discovery set the set of markers whose marginal posterior probability of non-zero effects is larger than 0.80 (LEFT) or larger than 0.56 (RIGHT). The vertical lines show the true FDR realised in the simulated data

For a discovery set of size 50, obtained by choosing the markers with non-zero posterior probabilities larger than 0.56, the Bayes-FDR estimate is 0.072 with a 95% posterior interval equal to (0.013; 0.142). The true number of false discoveries is 7, resulting in a true FDR realised in the sample equal to 0.14. Figure 9.4 displays the histograms of the McMC posterior distribution of these Bayes-FDR. The vertical lines show the realised (true) FDR obtained in the particular simulation for each discovery set.

The R-code to produce the histograms and Monte Carlo estimates of posterior means and posterior intervals of the false discovery rates is shown on page 367.

Several other pieces of information can be extracted from the Bayesian model. For each marker one can extract the probability of a non-zero effect conditional on observed data, and the draws from (7.57) retrieve a Monte Carlo estimate of the posterior probability distribution of this probability of non-zero effect. For instance, the Monte Carlo estimate of the mean of the posterior probability that marker 141 is non-zero is 0.83, and the Monte Carlo estimate that this probability is larger than 0.6, say, is 0.96. A similar picture can be obtained for the local false discovery rate for each marker genotype in the discovery set.

Prediction

The 2000 binary records are divided in equal proportions into training and validating sets. The criterion to evaluate predictive ability is the proportion of mistaken predictions in the validating data or error rate, given by

$$\text{MSE}_v = \frac{1}{N_v} \sum_{i=1}^{N_v} (y_i - \widehat{y}_i)^2$$

where $N_v = 1000$ is the number of records in the validating set, y_i is the ith record in the validating set (0 or 1) and $\widehat{y}_i$ is the predicted value computed using (9.3) using $t = 0.5$. I also used the *Brier score*

$$\text{MSE}_v^B = \frac{1}{N_v} \sum_{i=1}^{N_v} (y_i - \hat{\pi}_i)^2 \tag{9.25}$$

that uses additional information from the predicted probabilities $\widehat{\Pr}(Y_i = 1|\mu) = \hat{\pi}_i$.

The validating MSE_v of the spike and slab model, the penalised logistic regression and the logistic lasso are 0.35, 0.35 and 0.30, respectively. For the Brier score (9.25), the corresponding figures are 0.25, 0.30 and 0.20.

The Bayesian McMC implementation produces automatically a Monte Carlo estimate of the marginal posterior distribution of the validating MSE_v. The McMC estimate of the 95% posterior interval of the validating MSE_v is (0.321; 0.386) and for the Brier score (0.227; 0.275). This quantifies the uncertainty in the measures of MSE due to uncertainty in the predictors, conditional on the data.

In order to have a point of reference, these error rates can be compared to the error rate of the null model, defined as

$$u_i = \mu + \varepsilon_i, \quad i = 1, 2, \ldots, 2000,$$

that results in a loglikelihood equal to

$$\ell(\mu|y) = \sum_{i=1} y_i \mu - n \ln\left[1 + \exp(\mu)\right],$$

where n is the number of records in the dataset. A differentiation yields

$$\begin{aligned}\frac{d\,\ell(\mu|y)}{d\mu} &= \sum_{i=1} y_i - n\frac{\exp(\mu)}{1+\exp(\mu)} \\ &= \sum_{i=1} y_i - n \Pr(Y_i = 1|\mu).\end{aligned}$$

On setting the derivative equal to zero,

$$\widehat{\Pr}(Y_i = 1|\mu) = \sum_{j=1} y_j/n = \hat{\pi}, \text{ for all } i,$$

equal to the proportion of $1'$s in the data. The predicted value is

$$\widehat{y}_i = \begin{cases} 1 & \text{if } \hat{\pi} > 0.5 \\ 0 & \text{if } \hat{\pi} \le 0.5 \end{cases} \quad \text{for all } i.$$

Depending on the value of $\hat{\pi}$, this amounts to setting all the predictions to either 0 or 1. Therefore, the error rate of the null model is $\hat{\pi}$ or $1 - \hat{\pi}$, whichever is smaller. In this example, the null model leads to an error rate equal to 0.35, identical to that obtained for the spike and slab model and a little higher than the value obtained using the logistic lasso. The Brier score for the null model is $\frac{1}{n_v} \sum_i (y_i - 0.35)^2 = 0.23$.

When the data are more informative, things look a little better. For instance, increasing the total number of records to 5000 and keeping 1500 markers, MSE_v for the spike and slab model, the penalised logistic regression and the logistic lasso are 0.29, 0.38 and 0.27, respectively. Using the Brier score, the values for MSE_v^B are 0.19, 0.35 and 0.18.

The kernel of the R-code for running the logistic lasso using the package **glmnet** is shown below.

```
# THE TRAINING DATA y AND MATRIX OF COVARIATES X
# BOTH CREATED IN A SECTION OF THE CODE NOT SHOWN
library(glmnet)
train=sample(1:nrow(X),nrow(X)/2)
test=(-train)
y.test=y[test]

# STEP 1

cv.out=cv.glmnet(X[train,],y[train],alpha=1,intercept=TRUE,
family="binomial",type = "class")
plot(cv.out)
bestlam=cv.out$lambda.min
bestlam
#NUMBER NON-ZERO COVARIATES:
length(which(as.vector(coef(cv.out,s=bestlam))!=0))

# STEP 2

# OBTAIN PREDICTIONS BASED ON CLASS LABELS
fm.predclass=predict(cv.out,s=bestlam,newx=X[test,],
family="binomial",type="class")

# OBTAIN PREDCTIONS BASED ON PROBABILITIES (Brier score)
fm.predresp=predict(cv.out,s=bestlam,newx=X[test,],
family="binomial",type="response")
 #ERROR RATE (CLASS LABELS):
mean((as.numeric(fm.predclass)-y.test)^2
 #ERROR RATE (FITTED PROBS):
mean((as.numeric(fm.predresp)-y.test)^2)

# ERROR RATE OF NULL MODEL: liability = mu + e
mean(y.test)
```

The kernel of the R-code for executing the penalised logistic regression using gradient descent is shown below.

```
nit <- 1500 # Number of gradient descent iterations

nrep <- 5 # Number of replications
# (To compute variability of MSE across replicates)

#lambda <- 0.0 # ZERO PENALTY !!!!!!!!
lambda <- 0.4

gama <- 0.008 # Learning rate
c <- rep(0,nit)
newcost <- rep(0,nit)
newcostv <- rep(0,nrep)

result <- matrix(data=NA, nrow=nit,ncol=8)
resulttv <- matrix(data=NA, nrow=nit,ncol=8)
resulttvnr <- matrix(data=NA, nrow=nitnr,ncol=8)

res <- matrix(data=NA, nrow=nrep,ncol=9)
resnr <- matrix(data=NA, nrow=nrep,ncol=8)

msev <- rep(NA,nrep)
msevnr <- rep(NA,nrep)

##################################################
# FUNCTION TO COMPUTE Pr[Y = 1]
prob1 <- function(miu,beta,X){
  pr <- exp(miu+X%*%beta)/(1+exp(miu+X%*%beta))
}
# FUNCTION TO COMPUTE THE LOSS FUNCTION
cost <- function(miu,beta,X,y)
      {-sum(y*(miu+X%*%beta) - log(1 + exp(miu+X%*%beta)))
      + crossprod(beta)*(lambda/2)}
c[1]<- cost(miu,beta,X,y)
ptm <- proc.time()

#####################################################
#########         GRADIENT DESCENT        #############
### FIT MODEL TO TRAINING DATA; TEST IN VALIDATING DATA

set.seed(771311)
numone <- sum(y)
numzero <- length(y) - numone
nindiv <- length(y)
ptm <- proc.time()
for (i in 1:nrep){
  cat(i, "\n",sep="")
  train=sample(1:nrow(X),floor(0.5*nrow(X)))
  Xtrain <- X[train,]
  Xval <- X[-train,]
  ytrain <- y[train]
  yval <- y[-train]
  miu <- 0.0
  beta <-  rep(0.0,ncol(Xtrain))
  for(j in 1:nit){
    fdmiu <- -sum(ytrain - prob1(miu, beta,Xtrain))
    fdbeta <- -t(Xtrain) %*%
       (ytrain - prob1(miu, beta,Xtrain)) + lambda * beta
    fd <- matrix(c(fdmiu,fdbeta),nrow=length(beta)+1,ncol=1)
    sol0 <- matrix(c(miu,beta),nrow=length(beta)+1,ncol = 1)
```

```
    sol1 <- sol0 - gama * fd
    miu <- sol1[1,1]
    beta <- sol1[-1,1]
    newcostv[j] <- cost(miu, beta,Xtrain,ytrain)
    resulttv[j,] <- c(j,newcostv[j],miu,beta[1:5])
  }
probval <- probl(miu,beta,Xval)
y_predval <- as.numeric(ifelse(probval > 0.5,1,0))
msev[i] <- mean((y_predval-yval)^2)
res[i,] <- c(i,j,newcostv[j],miu,beta[1:5])
}
proc.time()-ptm
tail(resulttv)
tail(res)
summary(msev) # SUMMARY MISCLASSIFICATION ACROSS REPLICATES
```

The complete R-code that performs the computations of this example is on the GitHub site in the folder named "codes". The web address is https://github.com/SorensenD/SLGDS/codes. The specific R-code is labelled code0901.R.

9.7 Area Under the Curve

Consider binary data and a classifier such as logistic regression that generates predictions or classifications $\hat{Y}$ taking values 0 or 1 according to the rule (9.3), given covariates x. The result of applying the classifier depends critically on the threshold t used. In the case of (9.3), setting $t = 0.5$ leads to Bayes' rule that minimises $\Pr\left(\hat{Y} \neq Y\right)$, the *overall error rate*.

The outcome of the binary classifier can be displayed in the so-called confusion matrix shown in Table 9.1. For a particular experiment, the proportions that fall in the body of the table are empirical realisations of the 4 joint probabilities that provide 3 independent pieces of information. Different ways of summarising these joint probabilities give rise to different methods of classification.

Table 9.1 A typical *confusion matrix* for a binary classifier with the 4 joint probabilities in the body of the table

		True status		
		$Y = 1$	$Y = 0$	Marginal
Predicted	$\hat{Y} = 1$	$\Pr\left(\hat{Y} = 1, Y = 1\right)$	$\Pr\left(\hat{Y} = 1, Y = 0\right)$	$\Pr\left(\hat{Y} = 1\right)$
Status	$\hat{Y} = 0$	$\Pr\left(\hat{Y} = 0, Y = 1\right)$	$\Pr\left(\hat{Y} = 0, Y = 0\right)$	$\Pr\left(\hat{Y} = 0\right)$
	Marginal	$\Pr(Y = 1)$	$\Pr(Y = 0)$	
	$\Pr\left(\hat{Y} = k \mid Y = k\right)$	$k = 1$: Sensitivity	$k = 0$: Specificity	

One way of summarising the 4 joint probabilities in the body of the table is by the following 2 conditional probabilities and 1 marginal probability:

1. $\Pr\left(\hat{Y}=1|Y=0\right)$ known as the *false positive rate* (FPR) (type I error) or false discovery rate. In epidemiology the term $1-\Pr\left(\hat{Y}=1|Y=0\right)=\Pr\left(\hat{Y}=0|Y=0\right)$ (true negative) is referred to as *specificity*
2. $\Pr\left(\hat{Y}=1|Y=1\right)$ known as the *true positive rate* (TPR) (power) or *sensitivity* in epidemiology. The term $1-\Pr\left(\hat{Y}=1|Y=1\right)=\Pr\left(\hat{Y}=0|Y=1\right)$ is known as the *false negative rate* (FNR). These conditional probabilities depend on the threshold t.
3. $\Pr(Y=1)$ is sometimes known as *prevalence* or *incidence* if 1 means having the condition.

The overall probability of misclassification or *overall error rate* is

$$\begin{aligned}\Pr\left(\widehat{Y}\neq Y\right)&=\Pr\left(\widehat{Y}=1|Y=0\right)\Pr(Y=0)+\Pr\left(\widehat{Y}=0|Y=1\right)\Pr(Y=1)\\&=\text{FPR}\,(1-\Pr(Y=1))+\text{FNR}\Pr(Y=1),\end{aligned}\tag{9.26}$$

equal to the sum of the off-diagonal cells in the body of Table 9.1. Given the incidence in the population, the two components of the overall error rate, FPR and FNR, are critically dependent on the threshold t and they are complementary. As the threshold t is reduced, FPR increases and FNR decreases. A classifier that uses $t=0.5$ is known to have the lowest overall error rate, but depending on the case at hand, one may choose a threshold to reduce FNR. A typical example is a rule used to classify individuals as diseased or not diseased, where one is more concerned about the error of classifying a diseased individual as not diseased than about the error of classifying a not diseased individual as diseased. The optimal threshold must be chosen on the basis of the "cost" associated with the disease.

The receiver operating characteristic curve (ROC) for data Y is a graphic often used in engineering, psychology and medicine to assess the performance of diagnostic systems for simultaneously displaying the two types of errors for all possible thresholds. A ROC curve is constructed by plotting true positive rate (or sensitivity), $\text{TPR}(t)=\Pr(s(x)>t|Y=1)$ (equal to $1-\text{FNR}$) versus the false positive rate (or 1-specificity), $\text{FPR}(t)=\Pr(s(x)>t|Y=0)$, for all possible values of the threshold t that defines disease status, where the scoring rule $s(x)=\Pr(Y=1|x)$ is defined in (9.2). The plot is displayed on the unit square, and the area under the curve (AUC) represents one measure of the performance of the classifier. Values of AUC close to 1 denote a good classifier, and one that performs not better than chance has an AUC of 0.5. When a ROC curve has an AUC close to 1, there is a high probability that a diseased individual will be correctly classified and a small probability that a healthy individual will be misdiagnosed as diseased. ROC curves for two classifiers display their ability to detect disease status.

An important detail is that TPR is computed using data from the diseased fraction only, and FPR is computed using data from the non-diseased fraction only. These quantities are conditional probabilities. Therefore, the ROC plot is independent of the prevalence of the disease in the population. This is in marked contrast to the so-called predicted values that involve the conditional probabilities $\Pr\left(Y = y|\hat{Y} = \hat{y}\right)$, reflecting the clinical value of the test. The predicted values depend on the prevalence of the disease in the population, $\Pr(Y = y)$, as illustrated below.

Predicted Disease Status

Sensitivity and specificity are often used to describe test performance. From a diagnostic point of view, one is interested in knowing how well a test result predicts disease status. The positive predicted value is equal to the conditional probability that an individual has the disease, given a positive test: $\Pr\left(Y = 1|\hat{Y} = 1\right)$. An important quantity is $\Pr\left(Y = 1|\hat{Y} = 0\right)$, the conditional probability that the disease is present given that the test is negative. These conditional probabilities are obtained using Bayes theorem

$$\Pr\left(Y = 1|\hat{Y} = 1\right) = \frac{\Pr\left(\hat{Y} = 1|Y = 1\right)\Pr(Y = 1)}{\Pr\left(\hat{Y} = 1\right)}, \tag{9.27}$$

where

$$\Pr\left(\hat{Y} = 1\right) = \Pr\left(\hat{Y} = 1|Y = 1\right)\Pr(Y = 1) + \Pr\left(\hat{Y} = 1|Y = 0\right)\Pr(Y = 0).$$

Similarly,

$$\Pr\left(Y = 0|\hat{Y} = 0\right) = \frac{\Pr\left(\hat{Y} = 0|Y = 0\right)\Pr(Y = 0)}{\Pr\left(\hat{Y} = 0\right)}, \tag{9.28}$$

where

$$\Pr\left(\hat{Y} = 0\right) = \Pr\left(\hat{Y} = 0|Y = 0\right)\Pr(Y = 0) + \Pr\left(\hat{Y} = 0|Y = 1\right)\Pr(Y = 1).$$

As a specific application, consider a test that has a specificity equal to $\Pr\left(\hat{Y}=0|Y=0\right) = 1$ and a sensitivity equal to $\Pr\left(\hat{Y}=1|Y=1\right) = 0.95$. Consider a person who feels who has been exposed to an infected relative and who is evaluated as having a pre-test probability of contracting the disease (prior probability) equal to $\Pr(Y=1) = 0.2$. Using the complement of (9.28), the post-test probability (posterior probability) of infection, given a negative test result, is $\Pr\left(Y=1|\hat{Y}=0\right) = 0.012$. With a prior probability equal to $\Pr(Y=1) = 0.5$, this changes to $\Pr\left(Y=1|\hat{Y}=0\right) = 0.048$.

On the other hand, if sensitivity of the test is $\Pr\left(\hat{Y}=1|Y=1\right) = 0.70$, with $\Pr(Y_i=1) = 0.2$, the posterior probability is $\Pr\left(Y=1|\hat{Y}=0\right) = 0.07$ and with a prior equal to $\Pr(Y=1) = 0.5$, the posterior probability is $\Pr\left(Y=1|\hat{Y}=0\right) = 0.23$.

Infection Prevalence

The prevalence of a disease is the proportion of individuals in the population that have the disease, $\Pr(Y=1)$. Imagine that in a random sample, n individuals are tested, that predictions $\hat{Y}$ are generated using some rule $s(x)$ and that T are declared positive. Then the ratio

$$\hat{\Pr}\left(\hat{Y}=1\right) = \frac{T}{n} \tag{9.29}$$

is taken as an estimator of prevalence $\Pr(Y=1)$. However, this estimate is biased unless the test is perfect with specificity and sensitivity equal to 1. To see this, let $\Pr\left(\widehat{Y}=1\right)$ be the expected value of $\hat{\Pr}\left(\widehat{Y}=1\right)$. The relationship between $\Pr\left(\hat{Y}=1\right)$ and $\Pr(Y=1)$ is given by

$$\begin{aligned}
\Pr\left(\hat{Y}=1\right) &= \Pr\left(\hat{Y}=1|Y=1\right)\Pr(Y=1) + \Pr\left(\hat{Y}=1|Y=0\right)\Pr(Y=0) \\
&= sen\Pr(Y=1) + (1-spe)(1-\Pr(Y=1)) \\
&= (1-spe) + (sen+spe-1)\Pr(Y=1),
\end{aligned}$$

(Diggle 2011) where $sen = \Pr\left(\hat{Y}=1|Y=1\right)$ and $1-spe = \Pr\left(\hat{Y}=1|Y=0\right)$. If the test is perfect, $spe = sen = 1$, and substituting above, yields $\Pr\left(\hat{Y}=1\right) = \Pr(Y=1)$. Otherwise, with an imperfect test, $\Pr\left(\hat{Y}=1\right)$ is a linear, increasing function of $\Pr(Y=1)$ (if $sen+spe > 1$, implying that the test is superior to the toss of a coin). An unbiased estimator of incidence (given n, sen, spe and bounded between 0 and 1) is

$$\frac{\Pr\left(\widehat{Y}=1\right)-(1-spe)}{sen+spe-1}=\frac{T-n\,(1-spe)}{n\,(sen+spe-1)}. \tag{9.30}$$

Example: Estimation of Prevalence Using an Imperfect Test

As an illustration assume that an imperfect test is implemented on $n = 10{,}000$ individuals and that $T = 900$ show a positive result. If this test has a sensitivity equal to 0.85 and a specificity equal to 0.95, then the estimate of prevalence based on (9.30) is $\widehat{\Pr}\,(Y = 1) = 0.05$.

Given n, spe and sen, a confidence interval for this estimator of prevalence is readily obtained from the fact that T is binomially distributed. Using R, the lower and the upper bounds for a 95% confidence interval for T are

```
lb <- qbinom(0.025,n,(T/n))
up <- qbinom(0.975,n,(T/n))
```

and the corresponding 95% (frequentist) confidence interval for $\Pr\,(Y = 1)$ is

$$\Pr\left[\frac{lb-n\,(1-spe)}{n\,(sen+spe-1)}<\Pr\,(Y=1)<\frac{ub-n\,(1-spe)}{n\,(sen+spe-1)}\right]=0.95.$$

For instance, for $T = 900$ tested positive out of $n = 10{,}000$ with $spe = 0.95$ and $sen = 0.85$, the above gives

$$\Pr\,(0.043<\Pr\,(Y=1)<0.057)=0.95. \tag{9.31}$$

This calculation assumes that spe and sen are known without error. Incorporation of uncertainty in these parameters can be accommodated in a straightforward manner adopting a Bayesian approach and is expected to result in a confidence interval wider than that given by (9.31).

As shown below it is quite simple to implement a Monte Carlo-based Bayesian approach. Let $\Pr\,(Y = 1) = \theta$ and $\Pr\left(\widehat{Y} = 1\right) = \phi$. The strategy adopted is to parametrise the model in terms of (ϕ, sen, spe), to draw posterior samples from $[\phi, sen, spe|t, n]$ and to construct θ from these draws using (9.30) at each round. These are draws from the desired posterior distribution $[\theta|t, n]$. This Monte Carlo method is known as *composition* (see page 153). The Bayesian model is

$$p\,(\phi, sen, spe|t, n) \propto p\,(\phi, sen, spe)\, p\,(t|n, \phi)\,. \tag{9.32}$$

Assume that ϕ, sen and spe are a priori independent, with $\phi \sim Un\,(0, 1)$, $sen \sim Un\,(0.8, 1)$ and $spe \sim Un\,(0.925, 1)$. Thus, sen is constrained in the interval [0.8, 1] and spe in the interval [0.925, 1]. With these assumptions,

$$p\,(\phi, sen, spe|t, n) \propto p\,(sen)\, p\,(spe)\, p\,(t|n, \phi) \tag{9.33}$$

where the binomial likelihood is of the form

$$p\,(t|n, \phi) \propto \phi^{t}\,(1-\phi)^{n-t}\,. \tag{9.34}$$

Viewed as a function of ϕ, this is the kernel of a beta distribution $Be\,(t+1, n-t+1)$. For $n = 10{,}000$ and $t = 900$, this is a very peaked (informative) likelihood. The algorithm proceeds as follows:

1. Draw sen from $Un\,(0.8, 1)$
2. Draw spe from $Un\,(0.925, 1)$
3. Draw ϕ from $Be\,(t+1, n-t+1)$
4. Calculate θ from (9.30) and repeat from 1 until the desired chain length is achieved.

Steps 1 and 2 could be replaced with scaled beta distributions $Be\,(a, b)$ where a and b are chosen to generate the appropriate mode (e.g. 0.85 for sen and 0.95 for spe). The scaling is chosen to guarantee that the draws are within specific intervals (e.g. [0.8, 1] for sen and [0.925, 1] for spe and the lower limit 0.925 guarantees that the estimator of θ is positive). The code uses an algorithm due to Devroye (1986) to sample from truncated distributions (see page 218).

It is clear from (9.33) and (9.34) that the data contain information neither about sen nor about spe. The choice of the range of values from which these parameters are sampled propagates into the posterior uncertainty of $[\theta|t, n]$.

The R-code below implements the algorithm.

```
# CODE0902
rm(list=ls()) # CLEAR WORKSPACE
set.seed(3033)
rep<-10000
se<-0.85
sp<-0.95
a_se<-4
b_se<-1.5
a_sp<-10
b_sp<-1.5

# THEORETICAL MODE OF BETA PRIORS
theoretmodese<-(a_se-1)/(a_se+b_se-2)
theoretmodesp<-(a_sp-1)/(a_sp+b_sp-2)
resultg<-matrix(data=NA,nrow=rep,ncol=4)

# NUMBER OF TESTS: n
n<-10000
# NUMBER OF POSITIVE TESTS: t
t<-900
```

```
for (i in 1:rep){
psi<-rbeta(1,(t+1),(n-t+1))
a<-1-sp
b<-se+sp-1
if((psi-a) > 0)
  {
  theta <- (psi-a)/b
  }
  else {theta <- 0}

# DRAW SE SP FROM TRUNCATED BETA'S USING DEVROYE'S ALGORITHM
se<-qbeta(runif(1,pbeta(0.8,a_se,b_se),pbeta(1,a_se,b_se)),
 a_se,b_se)
sp<-qbeta(runif(1,pbeta(0.925,a_sp,b_sp),pbeta(1,a_sp,b_sp)),
 a_sp,b_sp)
#se <- 0.85
#sp <- 0.95
# OR
# DRAW SE AND SP FROM APPROPRIATE UNIFORM DISTRIBUTIONS
#se<-runif(1,0.8,1)
#sp<-runif(1,0.925,1)
resultg[i,]<-c(i,theta,se,sp)
}
# BAYESIAN POSTERIOR MEAN OF THETA AND POSTERIOR INTERVAL
postmean <- mean(resultg[,2])
postinterval <- quantile(resultg[,2],c(0.025,0.975))
postmean
```

```
## [1] 0.05754302
```

```
postinterval
```

```
##        2.5%      97.5%
## 0.01928466 0.09916501
```

Execution of the code choosing the scaled beta distributions for sen and spe leads to a Monte Carlo estimate of the posterior mean of θ equal to 0.058, with a 95% Bayesian posterior interval (0.019, 0.099). This is wider than the classical interval generated by (9.31) and reflects the prior uncertainty in spe and sen. When sen and spe are fixed at 0.85 and 0.95, respectively, the Monte Carlo output yields an estimate of the posterior mean of θ equal to 0.050 and a 95% Bayesian posterior interval (0.043, 0.057), in agreement with the classical interval (9.31).

Probabilistic Interpretation of AUC

Given an observation $Y_i = 1$ (class 1) and an observation $Y_i' = 0$ (class 0), the AUC can be interpreted as the probability that the classification $s = s(x_i)$ will assign a higher score to Y_i than to Y_i' (assuming that 1's rank higher than 0's, so that if the score $s(x_i)$ is larger than a threshold t, a prediction based on x_i will be classified as a member of class 1. The classification score s can be defined at the level of the liability as in (3.74) on page 109).

It will be convenient to define the following quantities:

- TPR: $\Pr(s > t|Y = 1) = 1 - \Pr(s \leq t|Y = 1) = 1 - F_1(t)$
- FPR: $\Pr(s > t|Y = 0) = 1 - \Pr(s \leq t|Y = 0) = 1 - F_0(t)$
- $\varepsilon = 1 - F_0(t) \Longrightarrow F_0(t) = 1 - \epsilon \Longrightarrow t = F_0^{-1}(1 - \epsilon)$, where F_0^{-1} is the inverse cumulative distribution function that maps a FPR to a given threshold
- $f_0(t) = \frac{dF_0(t)}{dt} = p(t|Y = 0)$, the probability density function of s for $Y = 0$ (class 0)
- $\frac{d\epsilon}{dt} = -f_0(t) \Longrightarrow d\epsilon = -f_0(t)\,dt$
- $f_1(t) = \frac{dF_1(t)}{dt} = p(t|Y = 1)$, the probability density function of s for $Y = 1$ (class 1)

Using this notation the ROC curve can be written as a function of $\varepsilon = 1 - F_0(t)$ as follows:

$$ROC(\varepsilon) = 1 - F_1\left(F_0^{-1}(1 - \varepsilon)\right). \tag{9.35}$$

The area under the ROC curve (AUC) is defined as $\int_0^1 ROC(\varepsilon)\,d\epsilon$. Substituting (9.35) yields

$$\begin{aligned} AUC &= \int_0^1 \left(1 - F_1\left(F_0^{-1}(1 - \epsilon)\right)\right) d\epsilon \\ &= \int_{\infty}^{-\infty} \left[1 - F_1\left[F_0^{-1}(F_0(t))\right]\right] (-f_0(t))\,dt \\ &= \int_{-\infty}^{\infty} [1 - F_1(t)]\, f_0(t)\,dt. \end{aligned} \tag{9.36}$$

In the second line, the integration is from ∞ to $-\infty$ because $F_0^{-1}(1) = \infty$ and $F_0^{-1}(0) = -\infty$. The third line changes the sign of f_0 and the limits of integration are reversed. Substituting

$$1 - F_1(t) = \int_t^{\infty} f_1(u)\,du,$$

results in

$$\begin{aligned} AUC &= \int_{-\infty}^{\infty} \int_t^{\infty} f_1(u)\; f_0(t)\; du\,dt \\ &= \int_{-\infty}^{\infty} \int_{-\infty}^{\infty} I(u > t)\; f_1(u)\; f_0(t)\; du\,dt \end{aligned} \tag{9.37}$$

that can be read as the probability that the score assigned to a draw Y from class 1 is larger than the score assigned to a draw Y' from class 0.

Example: ROC Curves

ROC curves and AUC are illustrated with two simulated binary datasets that differ in the level of heritability on the underlying scale. These are $h^2 = 0.5$ and $h^2 = 0.23$. In both sets of data, there are 2000 phenotypes and 500 genetic markers, out of which 20 are causal (have an effect on the value of liability). The proportion of 1's is approximately 0.30, and this is controlled by setting the mean μ at the level of liability equal to -0.67. The effects of causal loci are obtained to ensure that the additive variance on the underlying scale is either 0.3 (for $h^2 = 0.23$) or 1 (for $h^2 = 0.5$). The residual effects at the level of liability are iid $N(0, 1)$ and phenotypes are simulated using a probit model. The R−code below generates the data for $h^2 = 0.23$.

```
# *******   GENERATE BINARY DATA - PROBIT MODEL *******
# CODE0903
rm(list=ls()) # CLEAR WORKSPACE
library(glmnet)
set.seed(30337)
va<-0.3
#va <- 1
p<-0.25
#p <- 0.5
mu <- qnorm(p)
ve <- 1
nindiv<-2000
nloci<-20
nmarker<-500
be<-matrix(data=0.0,nrow=nmarker,ncol=1) # parameter true model
IDq<-sample(1:nmarker,nloci,replace=F) # choose nloci as QTL
WT<-matrix(nrow=nindiv,ncol= nmarker,
           rbinom(n=nindiv*nmarker,size=2,p=.5))
XT<-matrix(data=NA,nrow=nindiv,ncol=nmarker) # NO INTERCEPT
XTi<-matrix(data=NA,nrow=nindiv,ncol=nmarker+1) # INTERCEPT
# CENTER MARKER MATRIX
cm<-colMeans(WT)
for (i in 1:nmarker)
{
  XT[,i]<-(WT[,i]-cm[i])
}
meanXT <- apply(XT,2,mean)
varXT<-apply(XT,2,var)
# Compute 2*Sum(p(1-p)); Sum over nqtl QTL:
sumvar<-sum(varXT[IDq])
QTLeff<-sqrt(va/sumvar) # QTL effect
#     so that genetic variance is VA
be[IDq] <- QTLeff # QTL EFFECT
xb<-XT%*%be
p1 <- pnorm(mu+xb) # PROBIT MODEL
```

```
be[IDq]<-QTLeff # TRUE MARKER EFFECT = QTLeff; REST ARE ZERO
var(XT%*%be)
```

```
##             [,1]
## [1,] 0.3103757
```

```
y<-rep(0,nindiv)
y <- rbinom(nindiv,1,p1)
mean(y)
```

```
## [1] 0.288
```

Two models are fitted to the binary data. The first is a logistic lasso implementation and the second is a standard logistic likelihood that includes the full set of 500 genetic markers (referred to as the *full model*). Both models are fitted using the package GLMNET. The full model is retrieved including $s = 0$ as an argument to the call of the function GLMNET. The code to run GLMNET, for $h^2 = 0.23$, is as follows:

```
#CODE0903(cont)
####### FITTING FULL MODEL AND LASSO MODEL WITH GLMNET ########
set.seed(3337)
#library(glmnet)

train=sample(1:nrow(XTi),nrow(XTi)/2)
test=(-train)
y.test=y[test]
y.train<-y[train]
# ********** FIT GLMNET TO FIND THE BEST LAMBDA ***************
# STEP 1
cv.out=cv.glmnet(
XT[train,],y[train],alpha=1,standardize=TRUE,family="binomial")
#plot(cv.out)
bestlam=cv.out$lambda.min
length(which(as.vector(coef(cv.out,s=bestlam))!=0))
```

```
## [1] 56
```

```
# STEP 2

fm.pred0=predict(
  cv.out,s=0,newx=XT[test,],family="binomial",type="class")
fm.pred=predict(
  cv.out,s=bestlam,newx=XT[test,],family="binomial",
  type="class")


mean((as.numeric(fm.pred)-y.test)^2) # VALIDATION MSE: LASSO
```

```
## [1] 0.273
```

```
# VALIDATION MSE: FULL MODEL:

mean((as.numeric(fm.pred0)-y.test)^2)
```

```
## [1] 0.374
```

```
# ERROR RATE OF NULL MODEL: y = mu + e
pnull<-mean(y.test)
pnull
```

```
## [1] 0.296
```

```
ynull<-rep(0,length(y.test))
if(pnull > 0.5){ynull<-1}
mean((ynull-y.test)^2) # VALIDATION MSE BASED ON NULL MODEL
```

```
## [1] 0.296
```

ROC curves can be generated using the package `pROC`. This package can use the output from `GLMNET` (step 2, `fm` or `fm0`). Various pieces of information can be extracted and one of these is displayed in Fig. 9.5. The curves illustrate that the classifiers' performance is higher at heritability $h^2 = 0.5$ than at $h^2 = 0.23$ due to an increase in power. The lasso classifier is consistently superior to the full model classifier (which includes the 500 marker covariates) over the full range of the values of the threshold t.

The `R`-code below generates Fig. 9.5 using the package `pROC`.

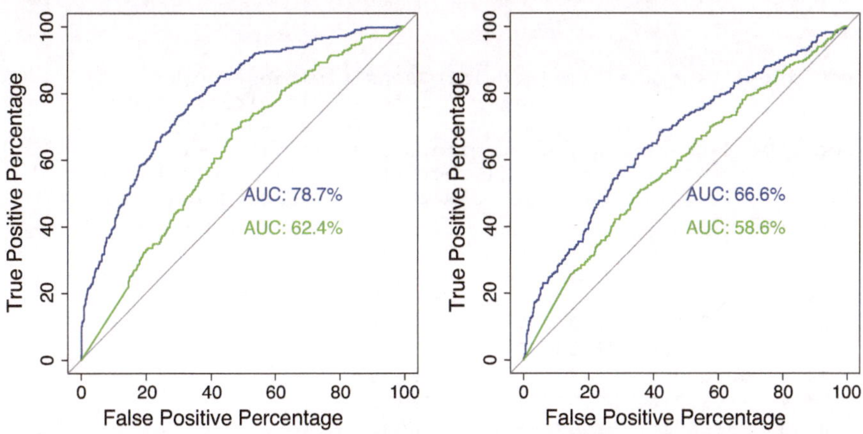

Fig. 9.5 ROC curves using simulated data. Left: $h^2 = 0.50$. Right: $h^2 = 0.23$. Blue line, lasso regression; green line, full likelihood model with all 500 covariates

```
# CODE0903(cont)
#install.packages("pROC", .libPaths()[1])
library(pROC)
set.seed(420)
par(pty="s") # GENERATES "PRETTY" FIGURES

fm.pred=predict(
cv.out,s=bestlam,newx=XT[test,],family="binomial",
type="response")
fm.pred0=predict(
 cv.out,s=0,newx=XT[test,],family="binomial",type="response")
pred<-as.numeric(fm.pred)
pred0<-as.numeric(fm.pred0)
roc.info<-roc(y.test, pred,plot=FALSE,legacy.axes=TRUE) # LASSO
```

```
## Setting levels: control = 0, case = 1

## Setting direction: controls < cases
```

```
# FULL MODEL:
roc.info0<-roc(y.test, pred0,plot=FALSE,legacy.axes=TRUE)
```

```
## Setting levels: control = 0, case = 1
## Setting direction: controls < cases
```

```
s <-roc(y.test, pred,plot=FALSE,legacy.axes=TRUE,percent=TRUE,
xlab="False Positive Percentage",
ylab="True Positive Percentage",
col="blue",print.auc=TRUE,cex.lab=1.3)
```

```
## Setting levels: control = 0, case = 1
## Setting direction: controls < cases
```

```
# CREATE DATA FRAME TO BE USED IN THE NEXT CODE
# THAT CONTAINS TPOS, FPOS FOR THE COMPLETE RANGE
# OF THRESHOLDS t
roc.df<-data.frame(tpos=roc.info$sensitivities*100,
                   fpos=(1-roc.info$specificities)*100,
                   thresholds=roc.info$thresholds)
roc.df0<-data.frame(tpos0=roc.info0$sensitivities*100,
                    fpos0=(1-roc.info0$specificities)*100,
                    thresholds0=roc.info0$thresholds)
head(roc.df)
```

```
##        tpos      fpos thresholds
## 1 100.00000 100.00000       -Inf
## 2 100.00000  99.85795 0.04341405
## 3 100.00000  99.71591 0.06056188
## 4  99.66216  99.71591 0.06751315
## 5  99.66216  99.57386 0.06976092
## 6  99.66216  99.43182 0.07345414
```

```
f50<-roc.df[min(which(roc.df$thresholds > 0.49995)),]
f50_0<-roc.df0[min(which(roc.df0$thresholds > 0.49995)),]
par(pty="m")
f50
```

```
##        tpos     fpos thresholds
## 950 12.5 1.988636 0.5000075
```

```
mean(y.test)
```

```
## [1] 0.296
```

The estimate of AUC can be retrieved typing `s$auc`. This yields 66.597 in agreement with the value displayed in the right panel of Fig. 9.5 corresponding to the lasso regression.

Estimates of AUC can also be obtained by numerical integration of the AUC curve. The following code performs this operation using the trapezoid rule:

```
# CODE0903(cont)
# **********  COMPUTE AUC BY NUMERICAL INTEGRATION ******
tpos <- roc.df$tpos/100
fpos <- roc.df$fpos/100
n<-length(tpos)
sum((tpos[-1]+tpos[-n])/2*(fpos[-n]-fpos[-1]))
```

```
## [1] 0.6659724
```

```
# OR AS A FUNCTION
auc <- function(fpos,tpos){
  n <- length(tpos)
  abs(sum((tpos[-1]+tpos[-n])/2*(fpos[-n]-fpos[-1])))
}
with(roc.df,auc(fpos/100,tpos/100)) # AUC FOR LASSO
```

```
## [1] 0.6659724
```

```
with(roc.df0,auc(
  fpos0/100,tpos0/100)) # AUC FULL MODEL: glmnet: LAMBDA=0
```

```
## [1] 0.5932389
```

A threshold t equal to 0.5 minimises the overall (validation) error rate (9.26). This can be confirmed by extracting the necessary information from data frames generated by the package **pROC**, executed by the section of `R`-code CODE0903 starting on page 399. The last two lines of this code output the proportion of *true positives*, *false positives* for the complete range of values of the *thresholds*, t, for the LASSO (`f50`) and for the full model (`f50_0`). For example, using the output linked to `f50`, the estimate of the validating mean squared error or error rate is

$$(100 - 12.5) \times 0.296 + 1.988636 \times (1 - 0.296) = 27.3$$

where the proportion of $1's$ in the validating data is 0.296. The result is in agreement with `mean((as.numeric(fm.pred)-y.test)^2)`=0.273 in the `R`-code starting on page 398.

9.8 Prediction of Disease Status of Individual Given Disease Status of relatives

The question addressed here is how to compute the probability that an individual will show a genetic disease, given sources of information that can be incorporated in a probability model. This could include family history and known susceptibility loci (So et al 2011). In the development that follows, information is restricted to the disease status of the individual's father and mother.

The Model

A simple binary trait (disease/not-disease) and a threshold probit model are assumed. In the parametrisation used (see page 109 for the alternative parametrisation), if u_i exceeds the threshold t, then the individual shows the disease and $Y_i = 1$; otherwise, if u_i is smaller than t, $Y_i = 0$. Assume the linear model for the liability

$$u_i = \mu + g_i + e_i, \tag{9.38}$$

where $g_i \sim N\left(0, \sigma_g^2\right)$ is the additive genetic value of individual i, σ_g^2 is the additive genetic variance, $e_i \sim N(0, 1)$ is the environmental deviation and μ is the population mean. The phenotypic variance of the marginal distribution of the liability is $\sigma^2 = \sigma_g^2 + 1$, and the heritability on the underlying scale is $h^2 = \sigma_g^2 \big/ \sigma^2$.

Calculation of Disease Status

The general result is motivated with a specific case. Consider the joint distribution of the liabilities of an offspring (u_o) and of its unrelated parents (u_f, u_m)

$$\left(u_o, u_f, u_m\right) \sim N\left(\begin{bmatrix} \mu \\ \mu \\ \mu \end{bmatrix}, \begin{bmatrix} 1 & 0.5h^2 & 0.5h^2 \\ 0.5h^2 & 1 & 0 \\ 0.5h^2 & 0 & 1 \end{bmatrix} \sigma^2\right). \tag{9.39}$$

An additive genetic model is assumed, where the covariance of the offspring with each parent is $\sigma_g^2/2 = h^2\sigma^2/2$.

Given this model, what is the probability that the offspring will show the disease, given the disease status of father and mother? This conditional probability model, given information on parents, will be referred to as the *selection model*, introduced by the set of Eqs. (2.53) on page 71. Two approaches are described. One uses standard probability theory where computations rely on multivariate normal numerical integration. The other is an approximation that avoids the numerical computation of multiple integrals; it is based on results of truncation selection of normal variates dating back to Pearson (1903) and Aitken (1934). The essentials of the latter approach are summarised in NOTE 2 on page 411.

Consider the case where the father has the disease and the mother does not. The conditional probability that the offspring is affected by the disease, given this information, is

$$\begin{aligned}
\Pr\left(Y_o = 1|Y_f = 1, Y_m = 0\right) &= \frac{\Pr\left(Y_o = 1, Y_f = 1, Y_m = 0\right)}{\Pr\left(Y_f = 1, Y_m = 0\right)} \\
&= \Pr\left(u_o > t|u_f > t, u_m < t\right) \\
&= \frac{\int_t^\infty \int_t^\infty \int_{-\infty}^t p\left(u_o, u_f, u_m\right) du_m du_f du_o}{\int_{-\infty}^\infty \int_t^\infty \int_{-\infty}^t p\left(u_o, u_f, u_m\right) du_m du_f du_o} \\
&= \frac{\int_t^\infty \int_t^\infty \int_{-\infty}^t p\left(u_o, u_f, u_m\right) du_m du_f du_o}{\int_t^\infty \int_{-\infty}^t p\left(u_f, u_m\right) du_m du_f}. \qquad (9.40)
\end{aligned}$$

In these expressions, $p\left(u_o, u_f, u_m\right)$ is the joint probability density function of the distribution defined in (9.39), and $p\left(u_f, u_m\right) = \int_{-\infty}^\infty p\left(u_o, u_f, u_m\right) du_0$. The evaluation of (9.40) requires multivariate normal numerical integration. Several software routines are available to compute the numerical integration; here I use the `R` function `pmvnorm` that is part of the library `mvtnorm`.

A little insight into expression (9.40) can be gained as follows. Under the selection model, the joint probability density function of the trivariate normal distribution is

$$p^s\left(u_o, u_f, u_m\right) = \frac{p\left(u_o, u_f, u_m\right)}{\int_{-\infty}^\infty \int_t^\infty \int_{-\infty}^t p\left(u_o, u_f, u_m\right) du_m du_f du_o}$$

and the marginal density of u_o under the selection model is

$$p^s\left(u_o\right) = \int_t^\infty \int_{-\infty}^t p^s\left(u_o, u_f, u_m\right) du_m du_f.$$

Integration of $p^s\left(u_o\right)$ over the interval $(t < u_o < \infty)$ yields (9.40).

To compute the marginal probability (under the selection model) that $Y_o = 1$ using the Pearson-Aitken formula (see NOTE 2 on page 411), proceed as follows. In terms of expression (9.58), let $x = \left(u_f, u_m\right)$ and $y = u_o$. Then using (9.54) and (9.56)

$$\mu_x^* = \mathrm{E}\left[\left(u_f, u_m\right) | u_f > t, u_m < t\right] = \begin{bmatrix} \mu + \sigma \frac{\phi(\alpha)}{1-\Phi(\alpha)} \\ \mu - \sigma \frac{\phi(\beta)}{\Phi(\beta)} \end{bmatrix} \tag{9.41}$$

and using (9.55) and (9.57),

$$V_x^* = \mathrm{Var}\left[\left(u_f, u_m\right) | u_f > t, u_m < t\right] =$$

$$= \begin{bmatrix} \sigma^2 \left[1 - \frac{\phi(\alpha)}{1-\Phi(\alpha)} \left(\frac{\phi(\alpha)}{1-\Phi(\alpha)} - \alpha\right)\right] & 0 \\ 0 & \sigma^2 \left[1 - \frac{\phi(\beta)}{\Phi(\beta)} \left(\frac{\phi(\beta)}{\Phi(\beta)} + \beta\right)\right] \end{bmatrix}. \tag{9.42}$$

(see NOTE 1 on page 407 for the definition of α and β). Then using (9.59) the mean of the offspring in the selection model is

$$\mathrm{E}\left(u_0 | u_f > t, u_m < t\right) = \mu_y^* = \mu_y + C_{yx} V_x^{-1} \left(\mu_x^* - \mu_x\right), \tag{9.43}$$

and using the block in the second row and second column of (9.60), the variance of the offspring in the selection model is

$$\mathrm{Var}\left(u_0 | u_f > t, u_m < t\right) = V_y - C_{yx} \left(V_x^{-1} - V_x^{-1} V_x^* V_x^{-1}\right) C_{xy}. \tag{9.44}$$

Finally, the marginal probability that the offspring is affected (in the selection model) is obtained by numerically integrating the (univariate) normal distribution with mean given by (9.43) and variance given by (9.44) over the range: lower limit $= t$, upper limit $=$ infinity.

It is important to notice that results (9.41) and (9.42) are exact. However, due to the correlation of u_0 with the truncated variables u_f and u_m, the marginal distribution of u_o after truncation is not normal. The Pearson-Aitken formula is used to compute the mean and variance of this distribution and then integrate over the normal distribution with this mean and variance. The assumption of normality in this last step is an approximation that works remarkably well in the example below, despite the rather extreme incidence for one of the cases considered.

Example: Prediction of a Genetic Disease

The computations are illustrated with a binary trait whose heritability on the underlying scale is $h^2 = 1/3$ and the phenotypic variance is $\sigma^2 = 3/2$. The objective is to predict the disease status of an offspring, given knowledge of the disease status of the father and the mother. The incidence in the population is either 2% or 0.2%.

The R-code to perform the "exact" computations and the Pearson-Aitken formula is shown below. The code is divided in two parts; the first part performs numerical integration and outputs the conditional probability that an offspring shows the disease given

1. CASE 1: father affected, mother affected
2. CASE 2: father affected, mother not affected
3. CASE 3: father unaffected, mother unaffected

```
# CODE0904
# NUMERICAL INTEGRALS
rm(list=ls()) # CLEAR WORKSPACE
library(mvtnorm)
# EXAMPLE: FATHER-MOTHER-CHILD
p<-0.02 # INCIDENCE IN THE POPULATION
# p <- 0.002
mean<-c(0,0,0) # MEAN OF THE THREE LIABILITIES
her <- 1/3 # heritability
var <- 1.5 # variance of liability
cov <- 0.5*her*var # covariance single parent-child
t<-qnorm((1-p),mean=0,sd=sqrt(1.5)) # THRESHOLD
# # VAR-COV MATRIX OF LIABILITY:
sigma <- matrix(c(var,cov,cov,cov,var,0,cov,0,var),3,3)
sigma
```

```
##      [,1] [,2] [,3]
## [1,] 1.50 0.25 0.25
## [2,] 0.25 1.50 0.00
## [3,] 0.25 0.00 1.50
```

```
# CASE 1: FATHER AFFECTED, MOTHER AFFECTED
den<-pmvnorm(lower=c(-Inf,t,t),upper=c(Inf,Inf,Inf),
             mean=mean,sigma=sigma)
num<-pmvnorm(lower=c(t,t,t),upper=c(Inf,Inf,Inf),
             mean=mean,sigma=sigma)
proboffsprcase1 <- num/den
#proboffsprcase1
# CASE 2: FATHER AFFECTED, MOTHER UNAFFECTED
den<-pmvnorm(lower=c(-Inf,t,-Inf),upper=c(Inf,Inf,t),
             mean=mean,sigma=sigma)
num<-pmvnorm(lower=c(t,t,-Inf),upper=c(Inf,Inf,t),
             mean=mean,sigma=sigma)
proboffsprcase2 <- num/den
```

```
#proboffsprcase2
# CASE 3: FATHER UNAFFECTED, MOTHER UNAFFECTED
den<-pmvnorm(lower=c(-Inf,-Inf,-Inf),upper=c(Inf,t,t),
             mean=mean,sigma=sigma)
num<-pmvnorm(lower=c(t,-Inf,-Inf),upper=c(Inf,t,t),
             mean=mean,sigma=sigma)
proboffsprcase3 <- num/den
#proboffsprcase3
```

The three conditional probabilities are as follows: CASE 1, 0.101; CASE 2, 0.046; and CASE 3, 0.019.

The second part computes the same conditional probabilities using the Pearson-Aitken formula.

```
# CODE0905
#              Pearson-Aitken formula
# Input
mean<-0 # marginal mean of liability
var<-1.5 # marginal variance of liability
varg<-0.5 # additive genetic variance
p<-0.02 # Incidence in population
t<-qnorm((1-p),mean=0,sd=sqrt(1.5)) # THRESHOLD
varfm<-var*diag(2) # VAR-COV PARENTS
covop<-matrix(0.5*varg,nrow=1,ncol=2) # COV OFFS-PARENTS
# CASE 1: FATHER AFFECTED, MOTHER AFFECTED
# FATHER:
alfa<-(t-mean)/sqrt(var) # LOWER TRUNCATION
p_alfa<-dnorm(alfa) # PDF AT LOWER TRUNCATION POINT
cum_alfa<-pnorm(alfa) # CUM. DIST. FUNCTION AT LOWER TRUNCATION
intsel<-p_alfa/(1-cum_alfa) # "INTENSITY OF SELECTION"
# MEAN AND VARIANCE OF (SELECTED) FATHER (MOTHER BELOW)
minfather<-mean+sqrt(var)*intsel
varfather<-var*(1-intsel*(intsel-alfa))
# MOTHER
minmother<-minfather
varmother<-varfather
# MEAN AND VARIANCE OF SELECTED FATHER AND MOTHER
expcase1 <- matrix(c(minfather,minmother),2,1)
varcase1 <- matrix(c(varfather,0,0,varmother),2,2)
# CONDITIONAL MEAN AND VARIANCE OF LIABILITY OF OFFSPRING
condmin<-mean+covop%*%solve(varfm)%*%(expcase1)
int1<-solve(varfm)-(solve(varfm)%*%varcase1%*%solve(varfm))
condvar<-var-(covop%*%(int1)%*%t(covop))
proboffPAcase1<-1-pnorm(t,mean=condmin,sd=sqrt(condvar))
# ***********************************************************
# CASE 2: FATHER AFFECTED, MOTHER UNAFFECTED
# FATHER: AS IN CASE 1
beta<-(t-mean)/sqrt(var) # upper TRUNCATION
p_beta<-dnorm(beta)
cum_beta<-pnorm(beta)
# MOTHER MEAN AND VARIANCE
minmother<-mean-sqrt(var)*(p_beta/cum_beta)
varmother<-var*(1-(p_beta/cum_beta)*((p_beta/cum_beta)+beta))
expcase2 <- matrix(c(minfather,minmother),2,1)
varcase2 <- matrix(c(varfather,0,0,varmother),2,2)
```

```
# CONDITIONAL MEAN AND VARIANCE OF LIABILITY OF OFFSPRING
condmin<-mean+covop%*%solve(varfm)%*%(expcase2)
int2<-solve(varfm)-(solve(varfm)%*%varcase2%*%solve(varfm))
condvar<-var-(covop%*%(int2)%*%t(covop))
proboffPAcase2<-1-pnorm(t,mean=condmin,sd=sqrt(condvar))
# ***************************************************
# CASE 3: FATHER UNAFFECTED, MOTHER UNAFFECTED
# FATHER: AS MOTHER IN CASE 2
expcase3 <- matrix(c(minmother,minmother),2,1)
varcase3 <- matrix(c(varmother,0,0,varmother),2,2)

# CONDITIONAL MEAN AND VARIANCE OF LIABILITY OF OFFSPRING
condmin<-mean+covop%*%solve(varfm)%*%(expcase3)
int3<-solve(varfm)-(solve(varfm)%*%varcase3%*%solve(varfm))
condvar<-var-(covop%*%(int3)%*%t(covop))
proboffPAcase3<-1-pnorm(t,mean=condmin,sd=sqrt(condvar))
```

The three conditional probabilities are as follows: CASE 1, 0.1; CASE 2, 0.046; and CASE 3, 0.019, in good agreement with the results based on numerical integration.

The results are summarised in Table 9.2. The results show that, for this example, both ways of computing the probabilities lead to identical results, to 3 decimal digits. Moving through the rows from right to left indicates that the probability of disease status of the offspring increases from a value that corresponds to the population incidence, when neither parent is affected, to a value that is 5 times larger or 15 times larger, for incidences in the population of 2% or 0.2%, respectively, when both parents are affected.

Note 1: Mean and Variance of the Truncated Normal Distribution

The pdf of the normal distribution with mean μ and variance σ^2 truncated between a and b, $a < b$, is

$$p(y|a < y < b) = \frac{\left(2\pi\sigma^2\right)^{-\frac{1}{2}} \exp\left(-\frac{(y-\mu)^2}{2\sigma^2}\right)}{\Pr(a < y < b)} \tag{9.45}$$

Table 9.2 Probability of disease of an offspring, given the disease status of its parents

$Pr(Y = 1)$		$(Y_f = 1; Y_m = 1)$	$(Y_f = 1; Y_m = 0)^a$	$(Y_f = 0; Y_m = 0)$
0.02	PA	0.100	0.046	0.019
	N	0.100	0.046	0.019
0.002	PA	0.031	0.0086	0.002
	N	0.031	0.0086	0.002

$\Pr(Y = 1)$ incidence in population, PA computation with the Pearson-Aitken formula, N numerical integration

[a] Disease status of parents: for example, $(Y_f = 1; Y_m = 0)$ symbolises the case where the father shows the disease $(Y_f = 1)$ and the mother does not $(Y_m = 0)$

where

$$\begin{aligned}\Pr(a < y < b) &= \Pr\left(\frac{a-\mu}{\sigma} < \frac{y-\mu}{\sigma} < \frac{b-\mu}{\sigma}\right)\\ &= \Pr(\alpha < z < \beta), \quad \alpha = \frac{a-\mu}{\sigma}; \beta = \frac{b-\mu}{\sigma}; z = \frac{y-\mu}{\sigma}\\ &= \Phi(\beta) - \Phi(\alpha).\end{aligned}$$

The expected value is

$$\mathrm{E}(y|a < y < b) = \frac{\left(2\pi\sigma^2\right)^{-\frac{1}{2}}}{\Phi(\beta) - \Phi(\alpha)} \int_a^b y \exp\left(-\frac{(y-\mu)^2}{2\sigma^2}\right) dy. \tag{9.46}$$

Since $y = \mu + \sigma z$ and $dy = \sigma dz$, this can be written as

$$\begin{aligned}\mathrm{E}(y|a < y < b) &= \frac{\left(2\pi\sigma^2\right)^{-\frac{1}{2}}}{\Phi(\beta) - \Phi(\alpha)} \int_\alpha^\beta (\mu + \sigma z) \exp\left(-\frac{z^2}{2}\right) \sigma dz\\ &= \frac{(2\pi)^{-\frac{1}{2}}}{\Phi(\beta) - \Phi(\alpha)} \mu \int_\alpha^\beta \exp\left(-\frac{z^2}{2}\right) dz\\ &\quad + \frac{(2\pi)^{-\frac{1}{2}}}{\Phi(\beta) - \Phi(\alpha)} \sigma \int_\alpha^\beta z \exp\left(-\frac{z^2}{2}\right) dz\\ &= \mu + \frac{(2\pi)^{-\frac{1}{2}}}{\Phi(\beta) - \Phi(\alpha)} \sigma \int_\alpha^\beta z \exp\left(-\frac{z^2}{2}\right) dz.\end{aligned} \tag{9.47}$$

Let $f(z) = \exp\left(-\frac{z^2}{2}\right)$; then $f'(z) = -z\exp\left(-\frac{z^2}{2}\right)$ and $\int_\alpha^\beta z \exp\left(-\frac{z^2}{2}\right) dz = -\int_\alpha^\beta f'(z)\, dz = -(f(\beta) - f(\alpha))$. Expression (9.47) can be written as

$$\begin{aligned}\mathrm{E}(y|a < y < b) &= \mu + \frac{(2\pi)^{-\frac{1}{2}}}{\Phi(\beta) - \Phi(\alpha)} \sigma \left[\exp\left(-\frac{\alpha^2}{2}\right) - \exp\left(-\frac{\beta^2}{2}\right)\right]\\ &= \mu + \sigma \frac{\phi(\alpha) - \phi(\beta)}{\Phi(\beta) - \Phi(\alpha)},\end{aligned} \tag{9.48}$$

where $\phi(x)$ is the pdf of the standard normal distribution evaluated at x.

The conditional variance is

$$\begin{aligned}\mathrm{Var}(y|a < y < b) &= \mathrm{Var}(\mu + \sigma z|\alpha < z < \beta)\\ &= \sigma^2 Var(z|\alpha < z < \beta).\end{aligned} \tag{9.49}$$

Then

$$\mathrm{Var}\,(z|\alpha < z < \beta) = \mathrm{E}\left[z^2|\alpha < z < \beta\right] - [\mathrm{E}\,(z|\alpha < z < \beta)]^2\,,$$

where $\mathrm{E}\,(z|\alpha < z < \beta) = \frac{\phi(\alpha)-\phi(\beta)}{\Phi(\beta)-\Phi(\alpha)}$. The first term is

$$\mathrm{E}\left[z^2|\alpha < z < \beta\right] = \frac{(2\pi)^{-\frac{1}{2}}}{\Phi\,(\beta) - \Phi\,(\alpha)} \int_{\alpha}^{\beta} z^2 \exp\left(-\frac{z^2}{2}\right) dz. \tag{9.50}$$

A simple way of computing the integration is to note that

$$(2\pi)^{-\frac{1}{2}}\, z^2 \exp\left(-\frac{z^2}{2}\right) = \phi''\,(z) + \phi\,(z)\,,$$

where $\phi''\,(z)$ is the second derivative of the pdf of the standard normal random variable z with respect to z. Then (9.50) can be written as

$$\begin{aligned}\mathrm{E}\left[z^2|\alpha < z < \beta\right] &= \frac{1}{\Phi\,(\beta) - \Phi\,(\alpha)} \int_{\alpha}^{\beta} \left[\phi''\,(z) + \phi\,(z)\right] dz \\ &= \frac{1}{\Phi\,(\beta) - \Phi\,(\alpha)} \left[\phi'\,(z)\big|_{\alpha}^{\beta} + (\Phi\,(\beta) - \Phi\,(\alpha))\right].\end{aligned} \tag{9.51}$$

Using

$$\begin{aligned}\phi'\,(z)\big|_{\alpha}^{\beta} &= (2\pi)^{-\frac{1}{2}}\,\alpha \exp\left(-\frac{\alpha^2}{2}\right) - (2\pi)^{-\frac{1}{2}}\,\beta \exp\left(-\frac{\beta^2}{2}\right) \\ &= \alpha\phi\,(\alpha) - \beta\phi\,(\beta)\,,\end{aligned}$$

and substituting in (9.51), yields

$$\mathrm{E}\left[z^2|\alpha < z < \beta\right] = \frac{\alpha\phi\,(\alpha) - \beta\phi\,(\beta)}{\Phi\,(\beta) - \Phi\,(\alpha)} + 1.$$

Then

$$\mathrm{Var}\,(z|\alpha < z < \beta) = 1 + \frac{\alpha\phi\,(\alpha) - \beta\phi\,(\beta)}{\Phi\,(\beta) - \Phi\,(\alpha)} - \left(\frac{\phi\,(\alpha) - \phi\,(\beta)}{\Phi\,(\beta) - \Phi\,(\alpha)}\right)^2$$

and

$$\operatorname{Var}(y|a < y < b) = \sigma^2 \left[1 + \frac{\alpha\phi(\alpha) - \beta\phi(\beta)}{\Phi(\beta) - \Phi(\alpha)} - \left(\frac{\phi(\alpha) - \phi(\beta)}{\Phi(\beta) - \Phi(\alpha)} \right)^2 \right]. \tag{9.52}$$

As a special case of (9.48) and (9.52), consider first $y > t$. Then $a = t$ and $b = +\infty$. Use the following result:

$$\lim_{z -> \infty} z \exp\left(-\frac{z^2}{2}\right) = 0. \tag{9.53}$$

Then $\phi(\beta) = 0$, $\Phi(\beta) = 1$ and using (9.53), $\beta\phi(\beta) = 0$. The conditional expectation is

$$\operatorname{E}(y|y > t) = \mu + \sigma \frac{\phi(\alpha)}{1 - \Phi(\alpha)}, \tag{9.54}$$

where $1 - \Phi(\alpha)$ is the proportion larger than t (the proportion selected, in quantitative genetic parlance) and $\phi(\alpha) / (1 - \Phi(\alpha))$ is known as the intensity of selection. The conditional variance is

$$\begin{aligned}\operatorname{Var}(y|y > t) &= \sigma^2 \left[1 + \frac{\alpha\phi(\alpha)}{1 - \Phi(\alpha)} - \left(\frac{\phi(\alpha)}{1 - \Phi(\alpha)} \right)^2 \right] \\ &= \sigma^2 \left[1 - \frac{\phi(\alpha)}{1 - \Phi(\alpha)} \left(\frac{\phi(\alpha)}{1 - \Phi(\alpha)} - \alpha \right) \right].\end{aligned} \tag{9.55}$$

As a second case, assume $y < t$. Then $a = -\infty$, $b = t$, $\phi(\alpha) = 0$, $\Phi(\alpha) = 0$, $\alpha\phi(\alpha) = 0$. The conditional expectation is

$$\begin{aligned}\operatorname{E}(y|y < t) &= \mu + \sigma \frac{-\phi(\beta)}{\Phi(\beta)} \\ &= \mu - \sigma \frac{\phi(\beta)}{\Phi(\beta)}.\end{aligned} \tag{9.56}$$

The conditional variance is

$$\begin{aligned}\operatorname{Var}(y|y < t) &= \sigma^2 \left[1 - \frac{\beta\phi(\beta)}{\Phi(\beta)} - \left(\frac{\phi(\beta)}{\Phi(\beta)} \right)^2 \right] \\ &= \sigma^2 \left[1 - \frac{\phi(\beta)}{\Phi(\beta)} \left(\frac{\phi(\beta)}{\Phi(\beta)} + \beta \right) \right].\end{aligned} \tag{9.57}$$

Note 2: The Pearson-Aitken Formula

The Pearson-Aitken formula describes how a mean vector and a covariance matrix of a set of variables are affected by selection on a subset of variables. It has been much used in animal breeding, notably by Henderson (1975). The normally distributed random variables are the vectors x and y with joint distribution

$$(x, y) \sim N\left[\begin{pmatrix} \mu_x \\ \mu_y \end{pmatrix}, \begin{pmatrix} V_x & C_{xy} \\ C_{yx} & V_y \end{pmatrix}\right]. \tag{9.58}$$

If selection operates on x and its expected mean changes to μ_x^*, then the expected mean of vector y changes to

$$\mu_y^* = \mu_y + C_{yx} V_x^{-1} \left(\mu_x^* - \mu_x\right). \tag{9.59}$$

If selection changes the variance of x to V_x^*, then the covariance matrix of (x, y) is changed to

$$\begin{pmatrix} V_x^* & V_x^* V_x^{-1} C_{xy} \\ C_{yx} V_x^{-1} V_x^* & V_y - C_{yx}\left(V_x^{-1} - V_x^{-1} V_x^* V_x^{-1}\right) C_{xy} \end{pmatrix}. \tag{9.60}$$

In the absence of selection (or with random selection), $V_x^* = V_x$, and the covariance (9.60) reduces to the covariance structure of (9.58).

9.9 Appendix: Approximate Analysis of Binary Traits

Binary traits can be analysed ignoring their discrete nature using a linear model. This can be a useful approximation for an initial analysis that may work adequately in large datasets, where the proportion of $1'$s is not extreme. Here, I provide the rationale for such an approximation and establish the relationship between the parameters of the Bayesian linear (approximate) model and the Bayesian (true) probit model, where parameters are defined on the unobserved underlying scale.

As before the binary datum is classified as "survival" versus "death", say. Let u represent the unobserved liability and t the threshold. If $u > t$ then the individual survives and the binary variable (the observed datum) takes the value $y = 1$. If $u \leq t$ the individual dies and the observed datum is $y = 0$. The liability associated with datum ij is u_{ij}; expressed in terms of the linear model, it takes the form

$$\begin{aligned} u_{ij} &= \mu + f_i + e_{ij}, \quad i = 1, 2, ..., n_f, \quad j = 1, \ldots, n, \\ f_i|\sigma_f^2 &\overset{iid}{\sim} N\left(0, \sigma_f^2\right), \quad e_{ij} \overset{iid}{\sim} N(0, 1), \end{aligned} \tag{9.61}$$

where μ is the mean, f_i is a random full-sib family effect and e_{ij} is a random residual $N(0, 1)$. There are n_f full-sib families and n full-sibs per family. The probability of survival of individual ij (which is the pmf of the random variable Y_{ij}) is

$$\Pr\left(y_{ij} = 1|\mu, f_i\right) = \Pr\left(u_{ij} > t|\mu, f_i\right) = \Pr\left(u_{ij} - \mu - f_i > t - \mu - f_i|\mu, f_i\right)$$

$$= \Pr\left(e_i > t - \mu - f_i|\mu, f_i\right) = \int_{t-\mu-f_i}^{\infty} p\left(e_{ij}\right) de_{ij}$$

$$= \int_{-\infty}^{\mu+f_i-t} p\left(e_{ij}\right) de_{ij} = \Phi\left(\mu + f_i - t\right). \quad (9.62)$$

The liabilities cannot be observed and a convenient origin is to set the value of the threshold to $t = 0$. Hence,

$$\Pr\left(y_{ij} = 1|\mu, f_i\right) = \Phi\left(\mu + f_i\right). \quad (9.63)$$

This constraint makes the likelihood model identifiable and the Hessian becomes negative definite. For the ijth datum, the conditional pmf is

$$\left[\Phi\left(\mu + f_i\right)\right]^{y_{ij}} \left[1 - \Phi\left(\mu + f_i\right)\right]^{1-y_{ij}}.$$

To derive the approximation, the following is needed:

$$\Phi\left(\mu + f_i\right)|_{f_i=0} = \Phi(\mu),$$

$$\left.\frac{\partial\Phi\left(\mu + f_i\right)}{\partial f_i}\right|_{f_i=0} = (2\pi)^{-\frac{1}{2}} \exp\left(-\frac{\mu^2}{2}\right) = \phi(\mu),$$

where Φ and ϕ are the cumulative distribution function and the density function of the standard normal distribution, respectively. Expanding the conditional likelihood (9.63) using a first-order Taylor series about $f_i = 0$, one obtains

$$\Pr\left(y_{ij} = 1|\mu, f_i\right) \approx \Phi(\mu) + f_i \left.\frac{\partial\Phi\left(\mu + f_i\right)}{\partial f_i}\right|_{f_i=0}$$

$$= \Phi(\mu) + \phi(\mu) f_i. \quad (9.64)$$

Since the data stem from a binary process,

$$\mathrm{E}\left(Y_{ij}|\mu, f_i\right) = \Pr\left(y_{ij} = 1|\mu, f_i\right) \approx \Phi(\mu) + \phi(\mu) f_i,$$

$$\mathrm{Var}\left(Y_{ij}|\mu, f_i\right) = \Pr\left(y_{ij} = 1|\mu, f_i\right)\left(1 - \Pr\left(y_{ij} = 1|\mu, f_i\right)\right)$$

$$\approx \left[\Phi(\mu) + \phi(\mu) f_i\right]\left[(1 - \Phi(\mu)) - \phi(\mu) f_i\right]$$

$$= \Phi(\mu)(1 - \Phi(\mu)) + \phi(\mu) f_i - \phi(\mu)^2 f_i^2 - 2\Phi(\mu)\phi(\mu) f_i. \quad (9.65)$$

Also,

$$\operatorname{Cov}\left(Y_{ij}, Y_{ik}|\mu, f_i\right) = 0.$$

Marginally with respect to f_i,

$$\mathrm{E}\left(Y_{ij}|\mu\right) = \mathrm{E}_{f_i}\left[E\left(Y_{ij}|\mu, f_i\right)\right] \approx \Phi\left(\mu\right). \tag{9.66}$$

$$\operatorname{Var}\left(Y_{ij}|\mu\right) = \operatorname{Var}_{f_i}\left(\mathrm{E}\left(Y_{ij}|\mu, f_i\right)\right) + \mathrm{E}_{f_i}\left(\operatorname{Var}\left(Y_{ij}|\mu, f_i\right)\right)$$

The first term in the right-hand side is

$$\operatorname{Var}_{f_i}\left(E\left(Y_{ij}|\mu, f_i\right)\right) \approx \operatorname{Var}_{f_i}\left[\Phi\left(\mu\right) + \phi\left(\mu\right) f_i\right] = \phi\left(\mu\right)^2 \sigma_f^2.$$

The second term in the right-hand side is

$$\begin{aligned}
\mathrm{E}_{f_i}\left(\operatorname{Var}\left(Y_{ij}|\mu, f_i\right)\right) &\approx E_{f_i}\left[\Phi\left(\mu\right)\left(1-\Phi\left(\mu\right)\right) - \phi\left(\mu\right) f_i - \phi\left(\mu\right)^2 f_i^2\right] \\
&= \Phi\left(\mu\right)\left(1-\Phi\left(\mu\right)\right) - \phi\left(\mu\right)^2 \sigma_f^2.
\end{aligned}$$

Therefore,

$$\begin{aligned}
\operatorname{Var}\left(Y_{ij}|\mu\right) &= \phi\left(\mu\right)^2 \sigma_f^2 + \Phi\left(\mu\right)\left(1-\Phi\left(\mu\right)\right) - \phi\left(\mu\right)^2 \sigma_f^2 \\
&= \Phi\left(\mu\right)\left(1-\Phi\left(\mu\right)\right).
\end{aligned} \tag{9.67}$$

Similarly,

$$\begin{aligned}
\operatorname{Cov}\left(Y_{ij}, Y_{ik}|\mu\right) &= \operatorname{Cov}\left[\mathrm{E}\left(Y_{ij}|\mu, f_i\right), \mathrm{E}\left(Y_{ik}|\mu, f_i\right)\right] + \mathrm{E}\left[\operatorname{Cov}\left(Y_{ij}, Y_{ik}|\mu, f_i\right)\right] \\
&= \phi\left(\mu\right)^2 \sigma_f^2.
\end{aligned} \tag{9.68}$$

Invoking the normal approximation to the likelihood, the hierarchical model becomes

$$\begin{aligned}
Y_{ij}|\mu, f_i &\sim N\left(\left[\Phi\left(\mu\right) + \phi\left(\mu\right) f_i\right], \left[\Phi\left(\mu\right)\left(1-\Phi\left(\mu\right)\right) - \phi\left(\mu\right)^2 \sigma_f^2\right]\right), \\
f_i|\sigma_f^2 &\sim N\left(0, \sigma_f^2\right),
\end{aligned}$$

that can be expressed as

$$\begin{aligned}
Y_{ij}|\mu, f_i &\sim N\left(\mu^* + f_i^*, \sigma_e^{2*}\right), \\
f_i^*|\sigma_f^2 &\sim N\left(0, \sigma_f^{2*}\right),
\end{aligned}$$

where

$$\begin{aligned}
\mu^* &= \Phi(\mu), \\
f_i^* &= \phi(\mu) f_i, \\
\sigma_e^{2*} &= \Phi(\mu)(1-\Phi(\mu)) - \sigma_f^{2*}, \\
\sigma_f^{2*} &= \phi(\mu)^2 \sigma_f^2, \\
h_o^2 &= \frac{2\sigma_f^{2*}}{\sigma_f^{2*}+\sigma_e^{2*}} = \frac{2\phi(\mu)^2\sigma_f^2}{\Phi(\mu)(1-\Phi(\mu))}.
\end{aligned} \tag{9.69}$$

These equations establish the association between the parameters of the linear approximation and those of the probit model on the underlying scale.

Assuming improper uniform prior distributions for $\left(\mu, \sigma_f^2\right)$, the approximate posterior density is

$$\begin{aligned}
p\left(\mu, \sigma_f^2\right) &\propto p\left(y|\mu, \sigma_f^2\right) p\left(f|\sigma_f^2\right) \\
&\propto \left(\sigma_e^{2*}\right)^{-\frac{N}{2}} \exp\left[-\frac{1}{2\sigma_e^{2*}}\left(y - 1\mu^* - Zf^*\right)'\left(y - 1\mu^* - Zf^*\right)\right] \\
&\quad \left(\sigma_f^{2*}\right)^{-\frac{n_f}{2}} \exp\left(-\frac{1}{2\sigma_f^{2*}} f^{*'} f^*\right), \quad N = n_f n,
\end{aligned} \tag{9.70}$$

where f^* is the column vector with n_f elements f_i^*.

Example

The approximation is implemented to analyse simulated binary data with a full-sib family structure. There are 1000 full-sib families, 3 offspring per family. On the underlying scale of the true model, the intraclass correlation between full-sibs is 0.15 leading to a heritability on the underlying scale equal to 0.30. The proportion of 1′s among the 3000 binary records is approximately 0.16. The true model is executed using the `R`-code on page 642. Based on a single chain of length 2000, the Gibbs sampler generated the following estimates of posterior means:

$$\begin{aligned}
\widehat{\mathrm{E}}(\mu|y) &= -1.07, \\
\widehat{\mathrm{E}}\left(\sigma_f^2|y\right) &= 0.173, \\
\widehat{\mathrm{E}}\left(h^2|y\right) &= 0.293.
\end{aligned}$$

The analysis based on the normal approximation generates the following estimates of posterior means of parameters on the observed scale:

$$\widehat{\mathrm{E}}\,(\mu|y) = 0.160,$$
$$\widehat{\mathrm{E}}\left(\sigma_f^2|y\right) = 0.0102,$$
$$\widehat{\mathrm{E}}\left(\sigma_e^2|y\right) = 0.126,$$
$$\widehat{\mathrm{E}}\left(h^2|y\right) = 0.149.$$

These can be transformed to the estimates on the underlying scale using (9.69):

$$\widehat{\mathrm{E}}\,(\mu|y) = -0.995,$$
$$\widehat{\mathrm{E}}\left(\sigma_f^2|y\right) = 0.174,$$
$$\widehat{\mathrm{E}}\left(h^2|y\right) = 0.291.$$

in good agreement with the results of the exact analysis. However, uncertainty is a little underestimated with the normal model. For instance, for $[\sigma_f^2|y]$, the 95% posterior interval obtained with the exact analysis is (0.090; 0.276) and with the approximation (0.106; 0.268). This is probably because the approximation is based on a first-order Taylor expansion and the variance is a second-order term. In general, the approximation deteriorates when the proportion of $1'$s is more extreme and datasets are smaller.

The R-code below fits the normal approximation:

```
# CODE0906
# GAUSSIAN FULL SIB-FAMILY MODEL; SINGLE-SITE GIBBS SAMPLING
# AS AN APPROXIMATION TO THE TRUE PROBIT MODEL
rm(list=ls()) # Clear the workspace
set.seed(12345)
require(graphics)
# GENERATE CORRELATED (FULL-SIBS DATA
#install.packages("MCMCpack", .libPaths()[1])
#install.packages("mvtnorm", .libPaths()[1])
library(MCMCpack)
# INITIALISE PARAMETERS
p0 <- 0.15
mu <- qnorm(p0)
iccfs<-0.15 #INTRACLASS CORRELATION FS
# VARIANCE BETWEEN FAMILIES: iccfs /(1- iccfs)
vfs <- iccfs/(1-iccfs)
nfs<-1000 # NUMBER OF FULL-SIB FAMILIES

fs<-3 #FULL-SIB FAMILY SIZE
N<-nfs*fs

c<-0
##########################################################
```

```
####  GENERATE BINARY RECORDS Y
f<-rnorm(nfs,mean=0,sd=sqrt(vfs))
p <- pnorm(mu+f)
y <- rbinom(N,1,rep(p,each=fs))
w <- rep(1:nfs,each=fs)
d<-data.frame(w,y)
family <- as.factor(w)
Z<-model.matrix(~0+family)
#########################################################
ztz<-t(Z)%*%Z
rep<-2000
resultap<-matrix(data=NA,nrow=rep,ncol=5)
transf<-matrix(data=NA,nrow=rep,ncol=4)

#INITIALISE THE VECTOR OF FAMILIY EFFECTS fe
# (Not to be confused with the TRUE family effects f)
fe<-rep(0,nfs)
# INITIALISE BETWEEN FAMILY VARIANCE COMPONENT vf
vf<-1
# INITILISE RESIDUAL VARIANCE
ve<-1
# INITIALISE k
k<-ve/vf
# INITIALISE THE MEAN
mu<-0
sumpyinvt<-0
#START GIBBS LOOP NORMAL MODEL
ptm <- proc.time()
for (i in 1:rep)
{
  print(i)
  # SAMPLE mu
  meanmu<-sum(y-Z%*%fe)/(nfs*fs)
  mu<-rnorm(1,mean=meanmu,sd=sqrt(ve/(nfs*fs)))
  # SAMPLE FAMILY EFFECTS f
  varf<-(k+fs)^(-1)
  fmean<- varf*(t(Z)%*%(y-mu))
  fe<-rnorm(nfs,mean=fmean, sd=sqrt(varf*ve))
  #SAMPLE vf
  #COMPUTE SCALE
  ftf<-sum(fe*fe)
  vfx<-ftf/rchisq(1,nfs-2)
  vf<-as.numeric(vfx)
  # SAMPLE ve
  # COMPUTE SCALE
  e<-(y-mu-Z%*%fe)
  ete<-t(e)%*%e
  vex<-ete/rchisq(1,N-2)
  ve<-as.numeric(vex)
  k<-ve/vf
  her <- (2*vf)/(vf+ve)
  resultap[i,]<-c(i,mu,vf,ve,her)
  # TRANSFORM TO PARAMETERS IN UNDERLYING SCALE
  mut <- qnorm(mu)
  vft <- vf/(dnorm(mut)**2)
  hert <- (2*vft)/(vft+1)
  transf[i,] <- c(i,mut,vft,hert)
}
proc.time()-ptm
```

Chapter 10
Bayesian Prediction and Model Checking

Aspects of Bayesian prediction have been addressed in previous chapters. In particular, Chaps. 7 and 9 show a Bayesian implementation of the spike and slab model for continuous and binary records, respectively, and illustrate how the marginal posterior distribution of validating mean squared errors can easily be computed in an McMC environment (pages 331 and 386).

This chapter brings several elements of Bayesian prediction into focus. After a brief discussion of the type of uncertainty accounted for by different predictors, the chapter introduces the prior and the posterior predictive distributions that are key ingredients of Bayesian prediction. Two examples, one using count data and the other continuous data, illustrate the kind of predictive inferences that are possible when posterior distributions are known. One example uses an analytical approach, whereas the other extracts Monte Carlo samples from known distributions using the method of composition. The Monte Carlo samples drawn from either the marginal posterior distributions of the parameters used to construct the predictor or from the posterior predictive distribution of the predictor are used to generate Monte Carlo estimates of the complete marginal posterior distributions of validating mean squared errors. This makes use of the important property of ergodic averages mentioned on page 186.

Section 10.3 offers a breakdown of the Bayesian expectation of training and validating mean squared errors of prediction. The concept of expected optimism, previously introduced in Chap. 6 in the frequentist framework, is given a Bayesian interpretation.

This is followed by an example of a logistic model using Bayesian-McMC and maximum likelihood implementations. The predictive ability of the model quantified using validating mean squared errors is obtained with both approaches. The uncertainty associated with the mean squared errors is also illustrated with both methods of inference. The chapter concludes with the topic of model checking using posterior predictive simulations with an application in an McMC environment.

D. Sorensen, *Statistical Learning in Genetics*, Statistics for Biology and Health,
https://doi.org/10.1007/978-3-031-35851-7_10

10.1 Levels of Uncertainty

One of the attractions of Bayesian methods is the natural manner of accounting for uncertainty and the possibility of doing so in a single, coherent analysis. Consider the classical regression model with the $n \times 1$ training data vector y modelled as

$$y|b, \sigma^2 \sim N\left(xb, I\sigma^2\right), \tag{10.1}$$

where b is a $p \times 1$ vector of unknown regression coefficients, x is an observed $n \times p$ matrix of covariates, I is the identity matrix of dimension $n \times n$ and the scalar σ^2 is the unknown conditional variance component of a datum, the same for all data.

Data are divided into training and validating sets, fitting the model using the former and evaluating its predictive performance with the latter.

Let $\hat{b}$ represent an estimator of b using training data y and covariates x. One possible predictor (given the sampling model (10.1)) evaluated at $x = x_0$ (here a $p \times 1$ vector) that has been explored in previous chapters is

$$\widehat{y_0} = x_0'\widehat{b}. \tag{10.2}$$

This predictor represents a point estimate of the average value of y_0 (evaluated at $x = x_0$) and ignores uncertainty.

A second possibility is to construct a predictor that accounts for the uncertainty associated with the unknown b. A Bayesian could do this in an McMC environment by replicating the following two steps:

1. Draw b^* from $[b|y]$
2. Construct the predictor $\hat{y}_0 = x_0'b^*$

Rather than a point prediction as in the previous case, the draws from $[b|y]$ and the derived $\hat{y}_0$'s generate a distribution of $\hat{y}_0$. This distribution represents the propagated posterior uncertainty of b on to $\hat{y}_0$, but as (10.2), $\hat{y}_0 = x_0'b^*$ predicts an average value y_0, given covariate x_0 and b^*.

The predictor can be used to compute posterior distributions of any function, such as validating mean squared errors. The inferential uncertainty in b is a component of the expected validating mean squared error.

A third possibility is to construct a predictor of new validating records y_0 that accounts for uncertainty about b and now also for sampling uncertainty of the new records. Again in an McMC environment, this is achieved by repeating the following:

1. Draw b^* from $[b|y]$
2. Draw the predictor $\hat{y}_0^*$ from $\left[y_0|b^*, x_0, y\right]$

This generates draws from the joint posterior distribution $[y_0, b|x_0, y]$. In this case, both sources of uncertainty are components of the expected validating mean squared error. Generation of new data from $\left[y_0|b^*, x_0, y\right]$ is accounting for sampling variation, and therefore, frequency properties of the Bayesian procedure are implicitly incorporated.

The Bayesian approach leads to a partitioning of the expected validating mean squared error in line with the frequentist counterpart. However, the interpretation is different. For instance, in the third case in the Bayesian context, the expectation is taken over the posterior predictive distribution of predicted data, with training and validating data fixed. This results in three terms contributing to the expected validating mean squared error. In common with the frequentist approach shown in (6.51) on page 279, the first term represents sampling uncertainty of replicated (validating) data. The second represents posterior uncertainty of the predictor (see expression (10.4b) below) and the third is the average (over the posterior predictive distribution) squared discrepancy between the predictor and the observed validating data that is held fixed.

In a frequentist setting, the sampling uncertainty of the estimate $\widehat{b}$ can be quantified by replicating simulation of data, estimating b, and constructing a prediction $\widehat{y}_0$ for each replicate. Using the $\widehat{y}_0$'s in the construction of the mean squared error accounts for the sampling uncertainty of $\widehat{b}$. An alternative to simulating new data is to resample the observed data. When a simulation strategy is chosen, parameters are replaced by their estimated values obtained from the original training data. Under both strategies, the sampling uncertainty of a new datum can be accounted for by adding an extra step consisting of drawing the predictors $\widehat{y}_0^*$ from $\left[y_0|\widehat{b}, x_0, y\right]$.

The concept of uncertainty and its propagation in the computation of validating mean squared errors, from a Bayesian and from a frequentist perspective, is illustrated with the logistic model example on page 430. Details of the algebra of the Bayesian expectation can be found on page 428 and in the prediction problems on pages 562, 651 and 654.

10.2 Prior and Posterior Predictive Distributions

In the standard setup, we observe an n−dimensional response vector y and the associated $(n \times p)$ matrix of covariates x. The conditional mean of y is $f(x)$ for some function f and the objective could be to obtain a prediction for a yet-to-be-observed response y_0, a scalar drawn from the same distribution as y, using the estimated function evaluated at a value of the covariate equal to x_0. The classical solution to this problem has been to use the training data $z_i = (x_i, y_i)$, $i = 1, 2, \ldots, n$ and to construct the predictor $\widehat{y}_0 = \widehat{f}(z, x_0)$, where $\widehat{f}$ is some estimate of f. For instance, in standard least squares linear regression, $f(x_i) = x_i'b$, $\widehat{f}(z, x_0) = x_0'\widehat{b}$, where $\widehat{b} = \left(x'x\right)^{-1} x'y$ and x_i' is the ith row of the $(n \times p)$ matrix x. In other words, the frequentist solution is to use $\widehat{y}_0 = E\left(y_0|x_0, \widehat{b}\right)$ as the predictor, where the parameter b that indexes the conditional distribution is replaced by its least squares estimator $\widehat{b}$ in this case. This prediction can be regarded as a point estimate of the average response in the population, conditional on x_0 and $\widehat{b}$.

In a more general formulation, given a vector of unknown parameters θ, the Bayesian solution to this prediction problem is to construct the *posterior predictive density* of y_0

$$\begin{aligned} p(y_0|y, x_0) &= \int p(y_0, \theta|y, x_0)\, d\theta \\ &= \int p(y_0|\theta, y, x_0)\, p(\theta|y)\, d\theta \\ &= \int p(y_0|\theta, x_0)\, p(\theta|y)\, d\theta, \end{aligned} \tag{10.3}$$

wherein, going from line 2 to line 3, it is assumed that given θ, y_0 is conditionally independent of y and that θ and x_0 are independent. Notice that even though y_0 is conditionally independent of y, they are not marginally independent. If in the last line, $p(\theta|y)$ is replaced by the prior density $p(\theta)$, (10.3) is known as the *prior predictive distribution* of y_0.

Having obtained (10.3), one may wish to use the mean or the mode as the point predictor $\widehat{y}_0$. Consider the linear regression example, where $y_0|b, x_0, \sigma^2 \sim N(x_0'b, \sigma^2)$. Assume that the variance σ^2 is known and that there is a prior distribution for b which is left unspecified. The mean and variance of the posterior predictive distribution are

$$\mathrm{E}\left(y_0|y, x_0, \sigma^2\right) = \mathrm{E}_{b|y,\sigma^2}\left[E(y_0|b, x_0)\right] = \mathrm{E}_{b|y,\sigma^2}\left(x_0'b\right), \tag{10.4a}$$

$$\begin{aligned} \mathrm{Var}\left(y_0|y, x_0, \sigma^2\right) &= \mathrm{E}_{b|y,\sigma^2}\left[\mathrm{Var}\left(y_0|b, x_0, \sigma^2\right)\right] + \mathrm{Var}_{b|y,\sigma^2}\left[\mathrm{E}\left(y_0|b, x_0\right)\right] \\ &= \sigma^2 + \mathrm{Var}_{b|y,\sigma^2}\left(x_0'b\right). \end{aligned} \tag{10.4b}$$

The mean can be seen to be an average, over the posterior distribution of b (given σ^2 since this variance is assumed known), of the conditional mean of y_0 given b. The variance contains two terms: one representing the sampling uncertainty of the new record and the other the posterior uncertainty of $x_0'b$. As the dimension of y increases and that of b remains constant, the second term in the last line (the variance of the posterior distribution of $x_0'b$) vanishes but the first term does not.

Posterior predictive distributions account for sampling uncertainty of records that could have been observed, given a model. In this sense, frequency properties of the Bayesian procedure are implicitly incorporated. This applies to any function of replicated data, such as mean squared errors.

The classical frequentist approach to prediction can be regarded as an approximation to a Bayesian prediction based on the posterior predictive density. Details are found in the Note on page 155 where it is shown that in large samples,

$$p(y_0|y, x_0) \approx p\left(y_0|\widehat{\theta}, x_0\right)$$

and therefore

$$\begin{aligned}
\mathrm{E}\,(y_0|y, x_0) &= \int y_0 p\,(y_0|y, x_0)\, dy_0 \\
&\approx \int y_0 p\left(y_0|\widehat{\theta}, x_0\right) dy_0 \\
&= \mathrm{E}\left[y_0|\widehat{\theta}, x_0\right] \\
&= x_0'\widehat{\theta},
\end{aligned}$$

where $\widehat{\theta}$ is the maximum likelihood estimator of θ (or the least squares estimator in this linear regression setting).

Example: Binary Data

The following stylised example based on binary records ($y \in 0, 1$) illustrates properties of the prior and posterior predictive distributions.

Before data Y are observed, the distribution of a future binary observation, given its prior distribution, takes the form $\Pr(Y = y|\theta) = \theta^y (1-\theta)^{1-y}$, where $E\,(Y|\theta) = \Pr(Y = 1|\theta) = \theta$ is the probability that the outcome is equal to 1. Then the pmf of the *prior predictive distribution* of the future record is

$$\begin{aligned}
\Pr(Y = y) &= \int_0^1 p\,(y, \theta)\, d\theta \\
&= \int_0^1 p\,(y|\theta)\, p\,(\theta)\, d\theta \\
&= \int_0^1 \theta^y (1-\theta)^{1-y}\, p\,(\theta)\, d\theta. \qquad (10.5)
\end{aligned}$$

I will assign a beta distribution with hyperparameters a and b as the prior for θ, $Be\,(\theta|a, b)$. These hyperparameters define how the probability mass is allocated through the support of the distribution and define its moments. The pdf of $Be\,(\theta|a, b)$ is

$$p\,(\theta|a, b) = \frac{\Gamma\,(a+b)}{\Gamma\,(a)\,\Gamma\,(b)}\theta^{a-1}(1-\theta)^{b-1} \propto \theta^{a-1}(1-\theta)^{b-1}, \quad \theta \in [0, 1], \qquad (10.6)$$

where $\Gamma(x+1) = x!$ is the gamma function that generalises the factorial function to all real numbers $x > 0$. The mean and variance are

$$\mathrm{E}(\theta|a,b) = \frac{a}{a+b},$$

$$\mathrm{Var}(\theta|a,b) = \frac{ab}{(a+b)^2(a+b+1)}.$$

When $a = 1, b = 1$, the beta distribution becomes a proper uniform distribution in $[0, 1]$.

Using the beta distribution as prior, (10.5) becomes

$$\begin{aligned}
\Pr(Y=y) &= \frac{\Gamma(a+b)}{\Gamma(a)\Gamma(b)} \int_0^1 \theta^{y+a-1}(1-\theta)^{b-y}\, d\theta \\
&= \frac{\Gamma(a+b)}{\Gamma(a)\Gamma(b)} \int_0^1 \theta^{a^*-1}(1-\theta)^{b^*-1}\, d\theta, \quad a^* = a+y, b^* = 1+b-y, \\
&= \frac{\Gamma(a+b)}{\Gamma(a)\Gamma(b)} \frac{\Gamma(a^*)\Gamma(b^*)}{\Gamma(a^*+b^*)} \int_0^1 \frac{\Gamma(a^*+b^*)}{\Gamma(a^*)\Gamma(b^*)} \theta^{a^*-1}(1-\theta)^{b^*-1}\, d\theta \\
&= \frac{\Gamma(a+b)}{\Gamma(a)\Gamma(b)} \frac{\Gamma(a^*)\Gamma(b^*)}{\Gamma(a^*+b^*)} = \frac{\Gamma(a+b)}{\Gamma(a)\Gamma(b)} \frac{\Gamma(a+y)\Gamma(b+1-y)}{\Gamma(a+b+1)}.
\end{aligned} \tag{10.7}$$

wherein going from the second to the third line, I multiplied and divided by the constant of integration and the fourth line follows because the integral is over a proper pdf and equals 1. Expression (10.7) is a special case of the beta-binomial distribution $Bb(x|a,b,n)$ that is generated by the mixture

$$Bb(x|a,b,n) = \int_0^1 Bi(x|n,\theta)\, Be(\theta|a,b)\, d\theta, \tag{10.8}$$

where $n = 1$ for the case of the Bernoulli trial.

The probability that $Y = 1$ derived from the prior predictive distribution is obtained from (10.7), setting $y = 1$ and using $\Gamma(x+1)/\Gamma(x) = x\Gamma(x)$,

$$\begin{aligned}
\Pr(Y=1) &= \frac{\Gamma(a+b)}{\Gamma(a)\Gamma(b)} \frac{\Gamma(a+1)\Gamma(b)}{\Gamma(a+b+1)} \\
&= \frac{a}{a+b},
\end{aligned}$$

which is equal to the prior mean, as expected, since the prior is the only source of information. This result can also be arrived at more expediently using the mixture formulation (10.8),

$$\Pr(Y=1) = \mathrm{E}(y) = \mathrm{E}_\theta\left[\mathrm{E}(y|\theta)\right] = \mathrm{E}_\theta(n\theta) = \frac{na}{a+b},$$

where $n = 1$ in the case of the Bernoulli random variable. Since the Bernoulli distribution is completely characterised by knowledge of $\Pr(Y=1)$ (from which $\Pr(Y=0) = 1 - \Pr(Y=1)$ is obtained), one can also write

$$\Pr(Y=1) = \int_0^1 \Pr(Y=1|\theta)\, p(\theta)\, d\theta = \int_0^1 \theta p(\theta)\, d\theta = \mathrm{E}(\theta), \tag{10.9}$$

again indicating the dependence of the prior prediction on the prior distribution.

Imagine now that m binary responses $y = (y_1, y_2, \ldots, y_m)'$ have been observed in a first sample. What is the *posterior predictive distribution* for a scalar y_{m+1} in a second sampling? With y representing the $m \times 1$ data vector from the first sample, the posterior predictive pmf for the Bernoulli example is characterised by computing

$$\Pr(Y_{m+1}=1|y) = \int_0^1 \Pr(Y_{m+1}=1|\theta)\, p(\theta|y)\, d\theta = \int_0^1 \theta p(\theta|y)\, d\theta = \mathrm{E}(\theta|y), \tag{10.10}$$

equal to the posterior mean. In addition, using (10.8)

$$\mathrm{E}(Y_{m+1}|y) = \mathrm{E}_{\theta|y}\left[\mathrm{E}(Y_{m+1}|\theta)\right] = \mathrm{E}(\theta|y).$$

Applying Bayes theorem, the posterior density is

$$\begin{aligned} p(\theta|y) &= \frac{p(y|\theta)\, p(\theta)}{p(y)} \\ &= \frac{1}{p(y)} \prod_{i=1}^m \theta^{y_i} (1-\theta)^{1-y_i} \frac{\Gamma(a+b)}{\Gamma(a)\Gamma(b)} \theta^{a-1} (1-\theta)^{b-1} \\ &= K\, \theta^{\sum_{i=1}^m y_i} (1-\theta)^{m-\sum_{i=1}^m y_i} \theta^{a-1} (1-\theta)^{b-1} \\ &\propto \theta^{a+\sum_{i=1}^m y_i - 1} (1-\theta)^{b+m-\sum_{i=1}^m y_i - 1}, \end{aligned}$$

$$K = \frac{\Gamma(a+b)}{p(y)\,\Gamma(a)\,\Gamma(b)}, \tag{10.11}$$

which is the kernel of a $Be\left(\widetilde{a}, \widetilde{b}\right)$ distribution, where

$$\widetilde{a} = a + \sum_{i=1}^{m} y_i,$$

$$\widetilde{b} = b + m - \sum_{i=1}^{m} y_i.$$

Then (10.10) is

$$\begin{aligned}\Pr(Y = 1) &= \mathrm{E}(\theta|y) \\ &= \frac{\widetilde{a}}{\widetilde{a} + \widetilde{b}} = \frac{a + \sum_{i=1}^{m} y_i}{a + b + m}.\end{aligned} \tag{10.12}$$

Given the model, the predictive posterior distribution is affected by the prior input through a and b, but the influence dissipates as sample size m increases, since a and b are usually small numbers relative to m and $\sum_{i=1}^{m} y_i$.

Example: Continuous Data

This example considers the linear regression model

$$y|b, \sigma^2 \sim N\left(xb, I\sigma^2\right) \tag{10.13}$$

with independent improper uniform prior distributions associated with the parameters $\left(b, \sigma^2\right)$.

The objective of this example is to illustrate Monte Carlo estimation of the distribution of 3 different validating mean squared errors that account for different sources of uncertainty of the predictor. The example is essentially an illustration of the construction of the distribution of one random variable (the mean squared error) using the known distribution of another random variable (the predicted values) and how easily this is accomplished using Monte Carlo methods.

The data (a vector with 2000 records) are simulated as follows:

- the vector b of regression coefficients, of dimension 150×1, is drawn from a normal distribution $b \sim N(0, 4 \times I)$
- the elements of the full rank matrix x of covariates, of dimension 2000×151, are drawn from a binomial distribution $x \sim Bin(2, 0.5)$, with an extra column vector of $1'$s added to account for an intercept which is set equal to $\mu = 10$
- vector of records y, of dimension the 2000×1, is drawn from a normal distribution $y \sim N(xb, 20 \times I)$

The 2000 records are divided into a training (t) and a validating (v) set of equal size, so that $y = \left(y_t', y_v'\right)'$.

The parameters of the model are drawn from the appropriate posterior distribution, given training data y_t, using the method of composition (see page 153) as follows:

- Draw σ^{2*} from the marginal posterior distribution

$$\left[\sigma^2 | y_t\right] \sim \chi^{-2}(v, S),$$

a scale inverted chi-square distribution with $n - (p+1) - 2$ degrees of freedom and scale parameter

$$\frac{1}{v}\left(y - x\widehat{b}\right)'\left(y - x\widehat{b}\right),$$

where $\widehat{b} = \left(x'x\right)^{-1} x'y$, as in (4.46) on page 161 and $p = 150$
- Draw b^* from the conditional posterior distribution

$$\left[b|\sigma^{2*}, y_t\right] \sim N\left(\widehat{b}, \left(x'x\right)^{-1} \sigma^{2*}\right),$$

a normal distribution (see page 156 for a derivation)
- Draw $y^*_{v,pred}$, the 1000×1 vector of predicted validating records from the conditional posterior distribution

$$\left[y_{v,pred}|b^*, \sigma^{2*}\right] \sim N\left(x_v b^*, I\sigma^{2*}\right),$$

a normal distribution (given $\left(b^*, \sigma^{2*}\right)$, $y_{v,pred}$ is independent of y_t).

These steps are repeated 1000 times to obtain the required sample. In each cycle, $\left(\sigma^{2*}, b^*, y_v^*\right)$ is a sample from the joint posterior distribution $\left[\sigma^2, b, y_v|y_t\right]$, while y_v^* is a sample from the posterior predictive distribution $\left[y_{v,pred}|y_t\right]$.

Using the draws from the joint posterior distribution, the following functions are calculated. First,

$$\text{MSE}_{v,1} = \frac{1}{750}\left(y_v - x_v\overline{b}^*\right)'\left(y_v - x_v\overline{b}^*\right), \tag{10.14}$$

a point estimator of the validating mean squared error of average predictions, where $\overline{b}^* = \widehat{E}\left(b|y\right)$, the Monte Carlo estimate of the posterior mean of b obtained using the Monte Carlo draws b^* from the marginal posterior distribution $[b|y]$.

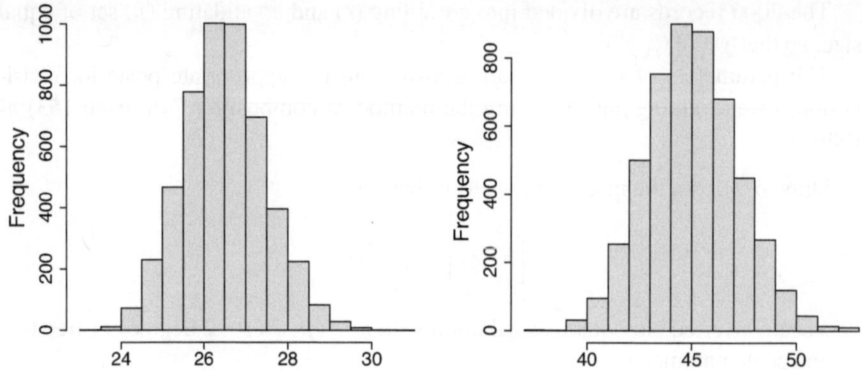

Fig. 10.1 Left: histogram of the marginal posterior distribution of $\text{MSE}_{v,2}$; the validating mean squared error accounting for posterior uncertainty of the parameters of the Bayesian model. Right: histogram of the marginal posterior distribution of $\text{MSE}_{v,3}$; the validating mean squared error accounting for posterior uncertainty of the parameters of the Bayesian model and for sampling uncertainty of new validating records

Second, for each of the 1000 cycles using the draws b^*, compute

$$\text{MSE}_{v,2} = \frac{1}{750}\left(y_v - x_v b^*\right)'\left(y_v - x_v b^*\right). \tag{10.15}$$

The elements of (10.15) constructed in this way constitute extractions from the Monte Carlo estimate of the marginal posterior distribution of $\text{MSE}_{v,2}$. This posterior distribution reflects the propagated posterior uncertainty of b.

Third, for each of the 1000 cycles, compute

$$\text{MSE}_{v,3} = \frac{1}{750}\left(y_v - y^*_{v,pred}\right)'\left(y_v - y^*_{v,pred}\right), \tag{10.16}$$

This estimator of validating mean squared error of predictions of individual data points accounts for posterior uncertainty of the parameters of the Bayesian model and for sampling uncertainty of the new data. The draws (10.16) constitute extractions from the Monte Carlo estimate of the marginal posterior distribution of $\text{MSE}_{v,3}$. The Monte Carlo estimates of the marginal posterior distributions of $\text{MSE}_{v,2}$ and $\text{MSE}_{v,3}$ are displayed as histograms in Fig. 10.1.

The `R`-code below simulates the data, executes the method of composition and computes the validating mean squared errors.

```
# CODE1001
rm(list=ls()) # CLEAR WORKSPACE
set.seed(123)

nindiv <- 2000
nmark <- 150
```

```
nsamples <- nindiv*nmark
# GENERATE COVARIATE MATRIX FROM BINOMIAL DISTRIBUTION
X<-matrix(nrow=nindiv,ncol=nmark,
   rbinom(n=nsamples,size=2,p=.5))
###########################################################
# CHOOSE VALUE FOR MEAN mu
mu <- 10
# CHOOSE VALUE FOR ENVIRONMENTAL VARIANCE ves
ves<-20
b<-matrix(data=0.0,nrow=nmark,ncol=1) # b from operational model

et<- rnorm(nindiv,mean=0,sd=sqrt(ves))
b <- rnorm(nmark,mean=0,sd=2)
y <- mu + X %*%b + et
train <- sample(1:nrow(X),floor(0.5*nrow(X)))
Xt <- X[train,]
yt <- y[train]
Xv <- X[-train,]
yv <- y[-train]
Zt <- cbind(1,Xt)
Zv <- cbind(1,Xv)
###################################################
#####   coefficient matrix LHSt, rhs & solution solt
RHSt <- crossprod(Zt,yt)
LHSt <- crossprod(Zt)
solt <- solve(LHSt,RHSt)
e <- yt-Zt%*%solt
###################################################
rep <- 5000 # NUMBER OF DRAWS USING COMPOSITION
ystartrain <- matrix(data=NA,nrow=length(yt),ncol=1)
ystarval <- matrix(data=NA,nrow=length(yt),ncol=1)

resMSE <- matrix(data=NA,nrow=rep,ncol=3)
res <- matrix(data=NA,nrow=rep,ncol=(nmark+2))

scale <- sum(e^2)
Cinv <- solve(LHSt)
ch <- chol(Cinv)
ptm<-proc.time()

for (i in 1:rep){
#  print(i)
  df <- length(yt)-(nmark+1)-2
# DRAW RESIDUAL VARIANCE
  varstar <- scale/rchisq(1,df)
  resid <- rnorm(length(solt),0,1)
# DRAW LOCATION PARAMETERS
  bstar <- solt + t(ch)%*% resid*sqrt(varstar)
  ystartrain <- Zt%*%bstar +
    rnorm(length(ystartrain),0,sd=sqrt(varstar))
# # DRAW VALIDATING DATA
  ystarval <- Zv%*%bstar +
    rnorm(length(ystarval),0,sd=sqrt(varstar))
  msevalystar <- mean((ystarval-yv)^2) # MSE_3
  msetrainystar <- mean((ystartrain-yt)^2)
  msevalbstar <- mean((Zv%*%bstar-yv)^2) # MSE_2
  resMSE[i,] <- c(msetrainystar,msevalystar,msevalbstar)
  res[i,] <- c(varstar,bstar)
}
proc.time()-ptm
```

```
##     user  system elapsed
##    17.62    0.02    4.23
```

```
av <- apply(res,2,mean)
bstarhat <- av[2:(nmark+2)]
ystarhatval <- Zv%*%bstarhat
mse_1 <- mean((ystarhatval-yv)^2) # POINT PREDICTOR
mserrors <- apply(resMSE,2,mean)
mse_2 <- mserrors[3]
mse_3 <- mserrors[2]
ci_2 <- quantile(resMSE[,3],c(0.025,0.975))
ci_3 <- quantile(resMSE[,2],c(0.025,0.975))
```

The Monte Carlo estimate of the point estimator $\text{MSE}_{v,1}$ is equal to 23.2. The mean of the marginal posterior distributions of $\text{MSE}_{v,2}$ and the 95% posterior interval is 26.5 (24.6, 28.5), and the corresponding figures for $\text{MSE}_{v,3}$ are 45, (41, 49.4), the latter reflecting, as expected, the extra uncertainty contributed by sampling of new records $y_{v,pred}$.

10.3 Bayesian Expectations of MSE

In the Bayesian setting, the expected values of the validating mean squared errors are composed of terms that generate a structure similar to the frequentist counterpart discussed on page 278. The expectations of (10.15) and (10.16) can be obtained as follows. The focus is on single terms of the sum, since the expectation of the sum is equal to the sum of the expectations of each record. From (10.15) for the ith record,

$$\text{E}_{b|y}\left(\text{MSE}_{i,v,2}\right) = \text{E}_{b|y}\left(y_{i,v} - \hat{y}_{i,v}\right)^2 \tag{10.17}$$

where the predictor is $\hat{y}_{i,v} = x'_{i,v}b$, $E_{b|y}$ stands for the expectation over the posterior distribution $[b|y]$ and y includes the vector of training (y_t) and validating (y_v) records. The random variable in (10.17) is $\hat{y}_{i,v}$, a function of the random variable b. Expanding the square and taking expectations yields

$$\begin{aligned}
\text{E}_{b|y}\left(\text{MSE}_{i,v,2}\right) &= y_{i,v}^2 + \text{Var}_{b|y_t}\left(\hat{y}_{i,v}\right) + \left(\text{E}_{b|y_t}\left(\hat{y}_{i,v}\right)\right)^2 - 2y_{i,v}\,\text{E}_{b|y_t}\left(\hat{y}_{i,v}\right) \\
&= \text{Var}_{b|y_t}\left(\hat{y}_{i,v}\right) + \left(y_{i,v} - \text{E}_{b|y_t}\left(\hat{y}_{i,v}\right)\right)^2 \\
&= \text{Var}_{b|y_t}\left(x'_{i,v}b\right) + \left(y_{i,v} - x'_{i,v}\hat{b}\right)^2,
\end{aligned} \tag{10.18}$$

where $\hat{b} = \text{E}(b|y_t)$. The first term is the contribution from the posterior variance of the predictor; the second term is the squared discrepancy between the validating datum and the posterior mean of the predictor.

Using similar algebra, the expectation of single terms in (10.16) is

$$\begin{aligned}
\mathrm{E}_{y_v^*|y}\left(\mathrm{MSE}_{i,v,3}\right) &= \mathrm{Var}_{y_v^*|y_t}\left(y_{i,v}^*\right) + \left(y_{i,v} - \mathrm{E}_{y_v^*|y_t}\left(y_{i,v}^*\right)\right)^2 \\
&= \sigma^2 + \mathrm{Var}_{b|y_t}\left(x_{i,v}'b\right) + \left(y_{i,v} - x_{i,v}'\hat{b}\right)^2, \qquad (10.19)
\end{aligned}$$

where I used $y_{i,v}^*$ instead of $y_{i,v,pred}^*$ to simplify the notation. The random variable in this case is the posterior predicted value of a future validating record $y_{i,v}^*$. The step from the first to the second line uses the decomposition of the posterior predictive variance (10.4b). The first term in (10.19) is the irreducible error: the sampling variance of the future validating datum; the second is the contribution from the variance of the conditional mean of y_v^* under its sampling model (due to posterior uncertainty of b); and the third term, equal to the second term in (10.18) as revealed by (10.4a) is the squared discrepancy between the validating datum y_v and the mean of the posterior predictive distribution of y_v^*.

Let $y_{i,t}^*$ represent the ith draw from the posterior predictive distribution $[\cdot|y_t, x_t]$ involving the training data. The Bayesian expectation of the training mean squared error takes the form

$$\begin{aligned}
\mathrm{E}_{y_t^*|y_t}\left(\mathrm{MSE}_{3,t,i}\right) &= \mathrm{Var}_{y_t^*|y_t}\left(y_{i,t}^*\right) + \left(y_t - \mathrm{E}_{y_t^*|y_t}\left(y_{i,t}^*\right)\right)^2 \\
&= \sigma^2 + \mathrm{Var}_{b|y_t}\left(x_{i,t}'b\right) + \left(y_{i,t} - x_{i,t}'\hat{b}\right)^2. \qquad (10.20)
\end{aligned}$$

This has exactly the same structure as (10.19) with x_v and y_v replaced by x_t and y_t. The important difference between (10.19) and (10.20) is in their third terms. In (10.20), prediction and estimation use the same training data. On the other hand, in (10.19), estimation is based on training data and prediction on validating data. Consequently, the squared term corresponding to the training mean squared error is smaller than the squared term of the validating mean squared error, reflecting overfitting.

A little more specifically, overfitting is reflected in the in-sample correlation between the vectors of data (considered fixed in the Bayesian framework) and the means of the draws from the posterior predictive distributions. The in-sample correlation is a component of the third terms of (10.19) and (10.20). It contributes negatively to these third terms and is larger in the training data than in the validating data.

In summary, the difference between the expected values of training and validating mean squared errors is due to the third terms and constitutes the Bayesian version of the expected optimism.

10.4 Example: Bayesian and Frequentist Measures of Uncertainty

This example illustrates the computation of two measures of validating mean squared error for binary predictions. The first accounts only for uncertainty of the estimated parameters and is labelled MSE_1. The second accounts for both, uncertainty of the estimated parameters and uncertainty due to sampling of single records and is labelled MSE_2.

Bayesian and frequentist approaches are used for inferences; the Bayesian approach is implemented with McMC (Metropolis-Hastings) and the frequentist with classical likelihood using the R-function GLM. The objective is to show how uncertainty can be accounted for using both schools of inference and to make a comparison within the settings of the example.

Binary data (2000 records) are simulated using a logistic model with two covariates, x_1 and x_2. The parameters of the logistic model are $\theta = (\beta_0, \beta_1, \beta_2)$. Specifically for the ith datum, $y_i \sim Br(p_i)$, where

$$\begin{aligned} p_i &= \Pr(y_i = 1|x_{1i}, x_{2i}) \\ &= \frac{\exp(\beta_0 + \beta_1 x_{1i} + \beta_2 x_{2i})}{1 + \exp(\beta_0 + \beta_1 x_{1i} + \beta_2 x_{2i})}. \end{aligned} \tag{10.21}$$

The covariates are simulated from uniform distributions $x_{1i} \sim Un(-1, 1)$, $x_{2i} \sim Un(-1, 1)$ and the three regression parameters are set equal to $\beta_0 = -1.3863$, $\beta_1 = 2$, $\beta_2 = 2$. The intercept β_0 is chosen to generate a proportion of $1's$ in the vicinity of 20%. The 2000 records were randomly divided into two sets of 1000 records each corresponding to training and validating data.

The R-code below generates the binary data.

```
# CODE1002
#########  SIMULATE BINARY RECORDS FROM LOGISTIC MODEL
### AND FIT LOGISTIC MODEL (2 COVARIATES) WITH M-H
#### CREATE TRAINING AND TESTING/VALIDATING DATA
rm(list=ls()) # CLEAR WORKSPACE
set.seed(77111)
require(graphics)
# THE CODE WILL USE THE PACKAGE MVTNORM; IT IS INSTALLED BELOW
#install.packages("mvtnorm", .libPaths()[1])
library(mvtnorm)
library(MASS)
#CHOOSE LENGTH OF CHAIN rep
rep<-6000
result<-matrix(data=NA,nrow=rep,ncol=6)
nindiv <- 2000
x1 <- rep(0,nindiv)
x2 <- rep(0,nindiv)
y <- rep(0,nindiv)
p<-0.2
x1 <- runif(nindiv,-1,1)
x2 <- runif(nindiv,-1,1)
```

```
b_0 <- log(p/(1-p))
b_1 <- 2
b_2 <- 2
z <- b_0 + b_1*x1 + b_2*x2
p1 <- exp(z)/(1+exp(z))
## GENERATE THE BINARY RECORDS
y <- rbinom(length(p1),1,p1)
# CONSTRUCT THE DATA SET
dat1 <- matrix(c(y,x1,x2),nrow = nindiv, ncol = 3)
colnames(dat1) <- c("Y", "X1", "X2")
d<-data.frame(dat1)
#attach(d)
Y <- d$Y
X1 <- d$X1
X2 <- d$X2
set.seed(771)
train=sample(1:length(X1),length(X1)/2)
test=(-train)
y.test=Y[test]
y.train<-Y[train]
length(y.train)
```

```
## [1] 1000
```

The Bayesian approach is implemented using Metropolis-Hastings. The algorithm is very similar to the one used on page 631, with a modification to account for the extra covariate.

The `R`-code to implement the Metropolis-Hastings algorithm is shown below:

```
# CODE1002 (cont)
# CHOOSE TUNING PARAMETER LAMBDA AND COVARIANCE MATRIX C
# OF THE METROPOLIS-HASTINGS ALGORITHM
lambda<-0.015
c <- diag(c(1.5,0.6,0.6))
# INITIALISE THE MEAN OF THE TRIVARIATE DISTRIBUTION
theta<-c(-5,1.0,1.0)

## FUNCTION TO COMPUTE THE LOG-POSTERIOR
logpost <- function(data,theta)
{
  interm <- theta[1] + theta[2]*data$X1 + theta[3]*data$X2
  with(data=data,sum(Y*( interm )-log(1+exp(interm))))
}
#START M-H LOOP
ptm <- proc.time()
accept<-0
for (i in 1:rep)
{
#  print(i)
  #SAMPLE PROPOSAL FOR THETA (Ytheta) FROM N(theta,lamdaC)
  Ytheta<- rmvnorm(1,mean=theta,sigma=lambda*c)
  logalfa<-logpost(d[train,],Ytheta) - logpost(d[train,],theta)
  unif<-runif(1)
  if (unif<exp(logalfa))
  {
    theta[1]<-Ytheta[1]
```

```
    theta[2]<-Ytheta[2]
    theta[3]<-Ytheta[3]
    interm <- theta[1] + theta[2]*X1[test] + theta[3]*X2[test]

    proby1 <- exp(interm)/(1+exp(interm))
    yhat <- rbinom(length(Y[test]),1,proby1)
    yhatBR <- ifelse(proby1 > 0.5, 1, 0)
    accept<-accept+1
  }
  else
  {
    interm <- theta[1] + theta[2]*X1[test] + theta[3]*X2[test]
    proby1 <- exp(interm)/(1+exp(interm))
    yhat <- rbinom(length(Y[test]),1,proby1)
    yhatBR <- ifelse(proby1 > 0.5, 1, 0)
  }
  misclas <- mean((yhat-Y[test])**2)
  misclasBR <- mean((yhatBR-Y[test])**2)
  brier <- mean((Y[test]-proby1)**2)
  vyhat <- var(yhat)
  logscYV <- sum(Y[test]*log(proby1)+(1-Y[test])*log(1-proby1))
  result[i, ]<-c(i,theta[1],theta[2],theta[3],misclas,misclasBR)
}
proc.time()-ptm
```

```
##    user  system elapsed
##    5.77    0.09    5.86
```

```
acceptratio <- accept/rep
# PRINT ACCEPTANCE RATIO OF THE JOINT UPDATING
acceptratio
```

```
## [1] 0.454
```

```
# PRINT THE McMC ESTIMATES OF POSTERIOR MEANS
apply(result[501:rep,5:6],2,mean)
```

```
## [1] 0.2933327 0.2184742
```

```
# PRINT 95% POSTERIOR INTERVALS FOR THE MISCLASSIFICATION RATES
misclasQ <- quantile(result[500:rep,5],c(0.025,0.975))
misclasQ
```

```
##  2.5% 97.5%
## 0.268 0.320
```

```
misclasBRQ <- quantile(result[500:rep,6],c(0.025,0.975))
misclasBRQ
```

```
##  2.5% 97.5%
## 0.212 0.229
```

```
# THE MCMC ESTIMATES OF MEAN OF POSTERIOR DISTRIBUTION OF
# REGRESSION PARAMETERS ARE
meanbetas <- apply(result[500:rep,2:4],2,mean)
meanbetas
```

```
## [1] -1.470091  2.037410  2.063063
```

```
# AND THE 95% POSTERIOR INTERVALS ARE
quantile(result[500:rep,2],c(0.025,0.975))
```

```
##       2.5%      97.5%
## -1.678812 -1.272822
```

```
quantile(result[500:rep,3],c(0.025,0.975))
```

```
##     2.5%    97.5%
## 1.716287 2.389274
```

```
quantile(result[500:rep,4],c(0.025,0.975))
```

```
##     2.5%    97.5%
## 1.688001 2.393371
```

The Monte Carlo estimates of posterior means of β_0, β_1 and β_2 (-1.47, 2.04, 2.06) are in good agreement with the simulated values ($\beta_0 = -1.39$, $\beta_1 = 2$, $\beta_2 = 2$).

The frequentist implementation maximises the logistic loglikelihood

$$\ell(\theta|y_t, x) = \sum_{i=1}^{n_t} \left\{ y_{t,i}\, x_i'\theta - \ln\left(1 + \exp(x_i'\theta)\right) \right\} \tag{10.22}$$

with respect to θ. In (10.22), y_t is the vector of $n_t = 1000$ training records and x_i' is the ith row of the $n_t \times 3$ matrix of covariates x with a column vector of 1's to account for β_0. The R function glm is used to perform the likelihood analysis. Frequentist uncertainty is incorporated generating 1000 bootstrap replications of training/validating splits. The R-code shown below implements the logistic likelihood:

```
# CODE1002 (cont)
# FIT MODEL BY ML
set.seed(771)
nrepl <- 1000
resLik <- matrix(data=NA,nrow=nrepl,ncol=3)
ptm <- proc.time()
```

```
for (i in 1:nrepl){
  train=sample(1:nrow(d),nrow(d)/2)
  f<-glm(Y ~.,data=d[train,],family="binomial")
  #  summary(f)
  #f<-glm(Y ~1,data=d[train,],family="binomial")
  #  summary(f)
  # PRED PROBABILITIES:
  predV <- predict(f,d[-train,],type="response")
  BrierFreq <- mean((predV-d$Y[-train])^2)
  # ASSIGN Y TO ITS CLASS ACCORDING TO BAYES RULE
  yhatBR <- ifelse(predV > 0.5, 1, 0)
  mseBR <- mean((yhatBR-d$Y[-train])^2) # MSE_1
  ## OR SAMPLE Y FROM ITS PREDICTIVE DISTRIBUTION
  ## CONDITIONAL ON ML ESTIMATES - THIS MAKES
  ## IT COMPARABLE TO THE McMC APPROACH
  yhatPpd <- rbinom(length(Y[-train]),1,predV)
  msePpd <- mean((yhatPpd-d$Y[-train])^2) # MSE_2
  resLik[i,] <- c(i,mseBR,msePpd)
}
proc.time()-ptm
```

```
##    user  system elapsed
##    4.44    0.09    4.53
```

```
apply(resLik[,2:3],2,mean)
```

```
## [1] 0.210431 0.288090
```

```
quantile(resLik[,2],c(0.025,0.975))
```

```
##  2.5% 97.5%
## 0.192 0.229
```

```
quantile(resLik[,3],c(0.025,0.975))
```

```
##  2.5% 97.5%
## 0.264 0.313
```

Results are shown in Table 10.1 that includes in the bottom row the predictive performance of the null model with the intercept β_0 as the only parameter of the loglikelihood (10.22).

The uncertainty in θ contributes to MSE_1 whereas both uncertainty in θ and sampling variation of validating records y_v^* contribute to MSE_2. This is reflected in the difference in mean values of MSE_1 and MSE_2.

Table 10.1 Posterior means (for the Bayesian model), means over bootstrap replicates (for the likelihood model) and 95% intervals (in brackets; Monte Carlo estimates of posterior intervals for the Bayesian model; bootstrap frequentist intervals for the likelihood model) of validating mean squared errors MSE_1 and MSE_2, obtained from a Bayesian McMC implementation (Bayesian McMC) and from a frequentist, maximum likelihood implementation (ML), for a logistic model with two predictors. The corresponding results obtained fitting the null model (a model containing only a mean at the level of the logit) are displayed in the row labelled Null model. The null model acts as a benchmark

	MSE_1	MSE_2
Bayesian McMC	0.22	0.29
	(0.21; 0.23)	(0.27; 0.32)
ML	0.21	0.29
	(0.19; 0.23)	(0.26; 0.31)
Null model	0.30	0.42
	(0.28; 0.32)	(0.39; 0.44)

The figures in the table show that the full model does better than the null model on all accounts. Both methods of inference produce very similar results, not only in terms of point estimates but also in terms of measures of uncertainty. This is not a general conclusion. In more complicated hierarchical models, marginalisation of nuisance parameters as part of the standard Bayesian machinery leads to larger (and more appropriate) measures of uncertainty than the standard likelihood approach. This often translates into larger values of MSE and larger uncertainty intervals.

The exercises section on page 562 includes several examples of the computation of mean squared errors, from a frequentist and Bayesian perspective, accounting for different sources of uncertainty.

10.5 Model Checking Using Posterior Predictive Distributions

Predicted data y_{pred} drawn from posterior predictive distributions were used to compute validating mean squared errors, by comparing the average squared difference between predicted (or simulated) data and observed validating data y_v. The MSE is a function of the parameters of the model θ (conditional on training data y_t); its distribution reflects the propagated posterior uncertainty of these parameters and of the sampling uncertainty of the predicted data.

Predicted data can also be used for inferences by checking whether a particular model properly accounts for aspects of the data that may be of scientific relevance. Let M denote the model that is to be questioned and that is presumed to have generated observed data y. The basic idea is to compare a function of y to the same function of predicted (or replicated) data generated under model M. These functions are constructed to address the specific feature of the model under enquiry; they depend on the observed data and often also on the parameters of the model. In the literature they are known as *discrepancy measures* (Gelman et al 1995) and labelled $T(y, \theta_M)$, a function of data and parameters, where θ_M is the parameter vector under model M. The discrepancy measure $T(y, \theta_M)$ based on observed data is compared to $T(y_{pred}, \theta_M)$, based on predicted data y_{pred} generated under model M. If zero is an atypical value in the posterior predictive distribution of $T(y, \theta_M) - T(y_{pred}, \theta_M)$, then model M is making predictions that do not fit that aspect of the data described by the discrepancy measure. Model checking performed in this way is a diagnostic tool for assessing the usefulness of a model for a specific purpose rather than for studying its global fit. Key literature on the subject is Rubin (1984), Gelman et al (1995) and Gelman et al (1996).

An Example with a Genetically Structured Heterogeneous Variance Model

The use of posterior predictive model checking is explained using an example adapted from Sorensen and Waagepetersen (2003). Imagine that unknown to the experimenter, observed data had been generated by the following model. At a first stage, the sampling model is Gaussian of the form

$$y_{ij}|\mu, f_i, \sigma_i^2 \sim N\left(\mu + f_i, \sigma_i^2\right), \quad i = 1, \ldots, n_f; \quad j = 1, \ldots, n. \tag{10.23}$$

In (10.23) y_{ij} is a record, μ is the mean, f_i is a random family effect and there is a structured residual term that gives rise to a different residual variance σ_i^2 for each of the n_f families; there are n offspring per family. The pattern of data generated by this model is quite common in animal breeding, where a family could represent a cohort of half-sibs produced by sires that mate each to n randomly chosen females, which in turn have each one offspring.

The model for the residual variances takes the form

$$\sigma_i^2 = \exp\left(\mu^* + f_i^*\right), \tag{10.24}$$

where μ^* is a parameter in $\mathbb{R}$ and f_i^* is a random family effect acting on the variance. When the elements of f^* are all zero, the residual variance reduces to the usual homogeneous variance $\sigma^2 = \exp(\mu^*)$.

The random variables (f_i, f_i^*), $i = 1, \ldots, n_f$, are *iid* draws from

$$\left(f_i, f_i^*|\sigma_f^2, \sigma_{f^*}^2, \rho\right) \sim N\left[\begin{pmatrix}0\\0\end{pmatrix}, \begin{pmatrix}\sigma_f^2 & \rho\sigma_f\sigma_{f^*}\\ \rho\sigma_f\sigma_{f^*} & \sigma_{f^*}^2\end{pmatrix}\right], \tag{10.25}$$

where σ_f^2 is the variance between half-sibs acting at the level of the mean, σ_{f*}^2 is the variance between half-sibs acting at the level of the variance and ρ is the coefficient of correlation.

The model defined by (10.23), (10.24) and (10.25) postulates that the residual variance has a genetic component because it varies across families. In the half-sib model assumed, variation between half-sib families is entirely genetic and explains 25% of the additive genetic variance.

This model induces a relationship between mean and variance through (10.25). Write

$$\begin{aligned} f_i^* &= \mathrm{E}\left(f_i^*|f_i\right) + \left(f_i^* - \mathrm{E}\left(f_i^*|f_i\right)\right) \\ &= \rho \frac{\sigma_{f*}}{\sigma_f} f_i + u \end{aligned} \tag{10.26}$$

where u is independent of f_i. Then

$$u|\sigma_{f*}, \rho \sim N\left(0, \left(1-\rho^2\right)\sigma_{f*}^2\right). \tag{10.27}$$

Therefore, the distribution $\left[f_i, f_i^*|\sigma_f^2, \sigma_{f*}^2, \rho\right]$ is the same as the distribution

$$\left[f_i, \left(\rho \frac{\sigma_{f*}}{\sigma_f} f_i + u\right)|\sigma_f^2, \sigma_{f*}^2, \rho\right]. \tag{10.28}$$

When $\rho = 1$ the relationship between mean and variance is deterministic; when $\rho = 0$ the variance is genetically determined and homogeneous, independent of the mean. Otherwise, the relationship between mean and variance is controlled by the sign and size of ρ.

The focus of enquiry in this example is to study whether the data provide evidence of a relationship between residual variance and family mean, supporting case 1 or 3.

Taking expectations of (10.23) over the distribution $\left[f_i, f_i^*|\sigma_f^2, \sigma_{f*}^2, \rho\right]$, the marginal distribution of the data is

$$y_{ij}|\mu, \mu^*, \sigma_f^2, \sigma_{f*}^2 \sim N\left(\mu, \sigma_f^2 + \exp\left[\mu^* + \frac{\sigma_{f*}^2}{2}\right]\right), \tag{10.29}$$

using the fact that if $x \sim N\left(\mu, \sigma^2\right)$, $\exp(x)$ is lognormally distributed with mean $E\left[\exp(x)\right] = \exp\left(\mu + \frac{\sigma^2}{2}\right)$.

Consider regressing $(\mu^* + f_i^*)$, the log-residual variance of family i, on $(\mu + f_i)$, the mean of family i. The regression coefficient is

$$\beta = \frac{\text{Cov}_{f_i, f_i^*}\left(\mu^* + f_i^*, \mu + f_i\right)}{\text{Var}_{f_i}\left(f_i\right)} = \rho \frac{\sigma_{f^*}}{\sigma_f}. \tag{10.30}$$

The choice of discrepancy measure described below aimed at establishing an association between family effects and within family variance is motivated by (10.30).

Constructing Discrepancy Measures

The mechanism that generated a particular set of data is seldom known. The closest one may come to knowing this mechanism is when data originate from a carefully designed and controlled experiment.

In the specific situation considered here, we imagine that using field records y, there is interest in investigating a possible genetic source of residual variation inducing a relationship between within family variance and family effect. Before committing resources to the development of the software needed to fit the model defined by (10.23), (10.24) and (10.25), one may start by fitting a model labelled M that assumes homogeneity of residual variance. In a second stage, one can define a discrepancy measure designed to reveal whether the lack of association between family mean and within family variance implied by model M holds. Given model M, replicated or predicted data y_{pred} are generated, $T\left(y_{pred}, \theta_M\right)$ is constructed and compared to $T\left(y, \theta_M\right)$. If zero is an atypical value in the posterior predictive distribution of $T\left(y, \theta_M\right) - T\left(y_{pred}, \theta_M\right)$, then there is justification to extend model M, because model's M assumption of lack of association between family mean and within family variance is not supported.

The starting point is to fit model M of homogeneous variance to data y,

$$y_{ij}|\mu, f_i, \sigma^2, M \sim N\left(\mu + f_i, \sigma^2\right), \quad i = 1, \ldots, n_f; \quad j = 1, \ldots, n, \tag{10.31a}$$

$$f_i|\sigma_f^2 \sim N\left(0, \sigma_f^2\right). \tag{10.31b}$$

The model is implemented using McMC; one may assign improper prior distributions to $\left(\mu, \sigma_f^2, \sigma^2\right)$.

The discrepancy measure to check for an association between family mean and within family variance in the observed data y is constructed as follows:

1. Using observed data first compute for each family

$$S_i\left(y_{ij},\theta_i\right)=\frac{1}{n}\sum\nolimits_{i=j}^{n}\left(y_{ij}-\mu-f_i\right)^2,\quad \theta_i=\left(\mu,f_i\right). \tag{10.32}$$

   The quantity S_i is the average squared residual for family i.
2. Second, fit a simple linear regression of $\log S_i\left(y_{ij},\theta_i\right)$ on f_i, $i=1,\ldots,n_f$, leading to the discrepancy measure

$$T\left(y,\theta\right)=\beta\left(y,\theta\right), \tag{10.33}$$

   where β is the regression coefficient which is a function of data y and θ.
3. Repeat using predicted data generated using the homogeneous variance model

$$\left[y_{pred,ij}|\mu,f_i,\sigma^2\right]\sim N\left(\mu+f_i,\sigma^2\right).$$

   Regress $\log S_i\left(y_{pred,ij},\theta_i\right)$ on f_i, $i=1,\ldots,n_f$, leading to the discrepancy measure

$$T\left(y_{pred},\theta\right)=\beta\left(y_{pred},\theta\right), \tag{10.34}$$

   the regression coefficient based on predicted data.
4. Compute

$$T\left(y,\theta\right)-T\left(y_{pred},\theta\right). \tag{10.35}$$

These steps are conducted using McMC. At round t the algorithm outputs draw $\theta^{[t]}$ and $y_{pred}^{[t]}$ from $\left[\theta,y_{pred}|y,M\right]$ based on model M of homogeneous variance. These draws are used to construct a Monte Carlo estimate of the posterior distribution of the discrepancy measure (10.35). If zero is excluded from the posterior distribution or is very unlikely, then the conjecture of an association between family mean and within family variance in the observed data is supported.

Generating Data Based on the Genetic Heterogeneous Variance Model

Data available for analysis are generated using the model defined in (10.23), (10.24) and (10.25). This model is not known to the researcher.

To create the data, the following set of parameters are used:

1. Number of half-sib families $n_f=400$; number of records per family $n=30$, the same for all families

2. $\mu = 10$; $\mu^* = 3$; $\sigma_f^2 = 5$; $\sigma_{f*}^2 = 1.8$; $\rho = 0.15$. Based on these input parameters, the marginal mean and variance of a datum are expected to be 10 and $\exp(3 + 0.9) = 49.4$, respectively.

The R-code used to generate the data is shown below:

```
# HETVARMODEL MODEL CHECKING
rm(list=ls()) # Clear the workspace
require(graphics)
# GENERATE CORRELATED (FULL-SIB/HALF-SIB) DATA

# INITIALISE PARAMETERS
mu<-10 # MEAN
mu_v <- 3 # MEAN VAR

#VARIANCE BETWEEN FULL-SIBS/HALF-SIBS
vfs <- 5 # mimicking half-sibs
vfs_v <- 1.8 # VARIANCE BETWEEN FULL-SIBS/HALF-SIBS
#             AT THE LEVEL OF THE VAR
rho <- 0.15 # CORRELATION BETWEEN FAMILY EFFECTS  (fs,fs_v)
###           AT THE LEVEL OF MEAN AND VARIANCE
# RESIDUAL VARIANCE (NOT USED)
ves<-exp(log(50))
nf<-400 # NUMBER OF FULL-SIB/HALF-SIB FAMILIES
n<-30 #FULL-SIB/HALF-SIB FAMILY SIZE
N<-nf*n
y<-matrix(data=0,nrow=nf*n,ncol=1)
# z IS COLUMN MATRIX WITH FAMILY ID (ID=1,.,nfs)
z<-matrix(data=0,nrow=nf*n,ncol=1)
# GENERATE nf FULL-SIB/HALF-SIB EFFECTS f
fs<-rnorm(nf,mean=0,sd=sqrt(vfs))
################################################
## GENERATING A FULL-SIB/HALF-SIB STRUCTURE
## VIA z
z <- rep(1:nf,each=n)
###############################################
u <- rnorm(nf,0,sqrt((1-rho^2)*vfs_v))
## SIB EFFECTS AT THE LEVEL OF VARIANCE
fs_v <- (rho*sqrt(vfs_v)/sqrt(vfs))*fs + u
# GENERATE nf*n RESIDUAL EFFECTS
es<-rnorm(nf*n,mean=0,sd=sqrt(exp(mu_v + fs_v[z])))
mean(es)
var(es)
y <- mu + fs[z] + es
f <- as.factor(z)
d <- data.frame(y,f)

################################################
#### USE THE FUNCTION AGGREGATE TO COMPUTE MEANS AND
#### VARIANCES FOR EACH FAMILY
agm <- aggregate(y~f,d,FUN=function(y){mean(y)})
agv <- aggregate(y~f,d,FUN=function(y){var(y)})
```

Detecting a Relationship Between Within Family Variance and Family Mean Using Observed Data

What does a simple analysis based on observed data reveal? Figure 10.2 displays the relationship between the logarithm of the estimated within family variance of the 400 families and the family means based on 30 records per family using the observed (simulated) data (generated under the heterogeneous variance model). A simple least squares regression of log variance within families on family means yields an estimate of the regression coefficient equal to 0.038 with a standard error equal to 0.025. This analysis based on raw averages discloses considerable heterogeneity of residual variance but provides a very tenuous and inconclusive relationship between mean and variance, despite a reasonably sized and well-structured dataset. The question is whether inferences of association between mean and variance are sharper using posterior predictive model checking.

Detecting a Relationship Between Within Family Variance and Family Mean Using Discrepancy Measures

The dataset generated under the genetically heterogeneous variance model defined in (10.23), (10.24) and (10.25) is analysed using a homogeneous variance model. A Gibbs sampling implementation of the homogeneous variance model is described in Chap. 5, page 249. In the current application, the code generates 5000 draws

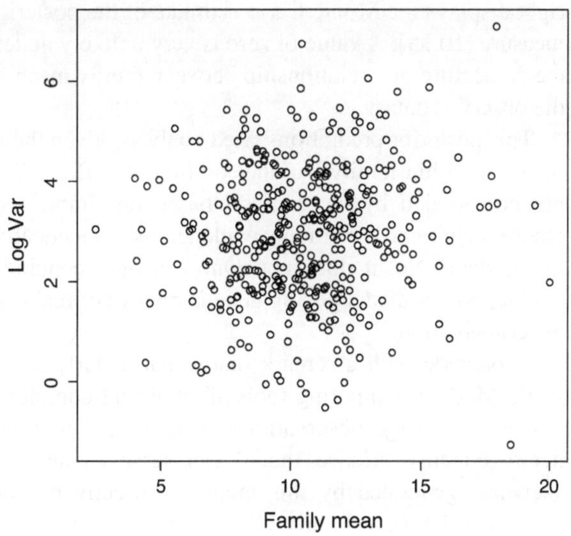

Fig. 10.2 The relationship between the logarithm of the variance within families and the family means induced by the heterogeneous variance model

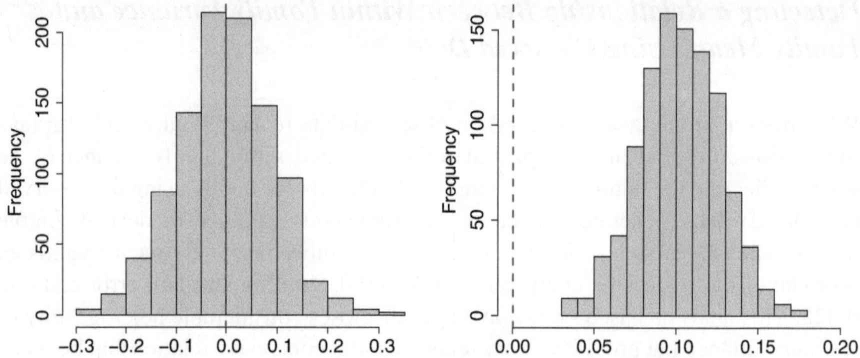

Fig. 10.3 Left: Monte Carlo estimate of posterior distribution of discrepancy measure (10.34) based on predicted data y_{pred} generated using a model of homogeneous variance across families (the reference distribution). Right: Monte Carlo estimate of posterior distribution of discrepancy measure (10.35). The vertical red lines at zero are indicative of a lack of association between family mean and within family variance

from the McMC estimates of the posterior distribution of discrepancies (10.34) and (10.35).

The Monte Carlo estimates of the marginal posterior distributions of the two discrepancy measures (10.34) and (10.35) are shown in Fig. 10.3. The left panel shows that the Monte Carlo estimate of the reference distribution, based on predicted data generated with the homogeneous variance model (discrepancy measure (10.34)), is centred at zero. This is evidence of no signs of association between family mean and within family variance in the replicated data y_{pred}, as expected. The panel on the right displays the Monte Carlo estimate of the posterior distribution of discrepancy measure (10.35). A value of zero is very unlikely under this distribution, supporting the conjecture of a relationship between family mean and within family variance in the observed data y.

This posterior prediction-based analysis, given the model, reveals an association between within family variance and family effects in the observed data that could not be revealed by the analysis based on simple means displayed in Fig. 10.2. Does this provide convincing evidence for a genetically structured residual variance in the data? No, it does not. One can only conclude that the results are not in conflict with a model posing genetic control of residual variance and justify further experimentation.

I conclude with a remark that is particularly relevant in an inferential framework. Modern computing tools allow fitting complex models to investigate subtle structures in large observational datasets that may be scientifically relevant. The heterogeneous variance model can serve as an illustration. The coefficient of skewness generated by this model is directly proportional to ρ, the correlation coefficient between family effects acting on mean and variance (Ros et al 2004). Therefore, if the observed data have a skewed distribution in either direction not necessarily due to genetically structured variance heterogeneity, a model that postulates variance heterogeneity will result in a better fit than the simple alternative assuming homogeneous variance. The analysis would even output an estimate of the non-existent correlation ρ, leading to spurious inference. This makes it strikingly clear that an attempt to understand the state of nature via a statistical analysis of data, particularly observational data, must be regarded only as a first step until more fundamental knowledge becomes available. Model checking using a well-designed discrepancy measure is a contribution to this process.

Identifiability of parameters is also an important consideration. In highly complex models, there is always the pitfall that parameters may be unidentified or very weakly identified and it may not always be possible to check lack of identifiability. Often for convenience, one chooses prior distributions of parameters that are improper (as practised repeatedly in this book!). Such prior assumptions can lead to improper posterior distributions that may go undetected in an McMC environment. This can lead to misleading inferences. An interesting example, initially taken from Carlin and Louis (1996), can be found on page 543 of Sorensen and Gianola (2002) and a useful reference is Natarajan and Kass (2000). The use of proper prior distributions will dispose of the potential problem of posterior impropriety and unidentifiability and, if chosen judiciously, may lead to Bayesian learning. However, the problem of the influence of prior information on posterior inference still needs to be addressed.

It is important to learn as much as possible about the model by experimenting with it before launching a full analysis using modern computational tools.

Chapter 11
Nonparametric Methods: A Selected Overview

Throughout this book a phrase like "assume the data have been generated by the following probability model" has been abundantly used. Indeed, the standard parametric assumption is that observed data represent one realisation from some given probability model and the goal can be to infer the parameters of the model. Alternatively and from a classical frequentist setting, conditionally on estimated parameters, the goal may be to predict future observations. For quantitative responses the parametric model can be written as

$$y_i = m\left(x_i, \theta\right) + e_i, \quad i = 1, 2, \ldots, n, \tag{11.1}$$

where m is some function that relates observations y_i (outputs) to observed covariates x_i (inputs, also referred to as *features*) constituting the dataset $z_i \overset{iid}{=} (y_i, x_i)$, $i = 1, 2, \ldots, n$. The function $m(x_i, \theta) = E(y_i|x_i, \theta)$, a conditional expectation, will be referred to as the *regression function* or the *prediction function*: its estimate uses as input the feature x and produces an output $\widehat{y}$. The residual terms e_i are iid from some distribution and θ is a vector of parameters. The form of m is often simple, as in linear regression, and when the model is approximately correct, inferences about θ are straightforward and efficient. If $\widehat{\theta}$ is an estimate of θ, a predicted value given x_{0i} is $\widehat{y}_{0i} = m\left(x_{0i}, \widehat{\theta}\right)$. For instance, in linear regression, $m\left(x_i, b\right) = x_i'b$, $\widehat{y}_{0i} = x_{0i}'\widehat{b}$, where $\widehat{b}$ is the estimated value of b. For data arising from binary trials, modelling takes place at the level of $\Pr\left(y_i = 1|x_i, \theta\right)$ as in logistic regression.

Nonparametric models, on the other hand, do not commit to a specific form for m and instead regard m as an algorithm that for future x_0 leads to a good predictor of y_0. The prediction algorithm is constructed or trained in the *training set* of observations, and its predictive ability is studied in the *validating or testing set*. Nonparametric theory focuses on the properties of the algorithms, their predictive power, and convergence properties if they are iterative and on factors that affect their accuracy. There is a distinction between *supervised learning* problems, where

D. Sorensen, *Statistical Learning in Genetics*, Statistics for Biology and Health,
https://doi.org/10.1007/978-3-031-35851-7_11

the objective is typically to make a prediction (or a classification in the case of discrete data), and *unsupervised learning*, where the aim is to explore how the data fall into different clusters. The latter is not dealt with here.

Nonparametric methodology offers a great deal of flexibility and can handle model complexities such as a large number of feature variables, complex interactions involving the feature variables and unknown nonlinear relationships between the feature variables and the response. Nonparametric methods have tuning parameters that control their adaptation to the data and therefore the degree of overfitting via the bias-variance trade-off. These tuning parameters can be set by the user or can be estimated from the data.

Prediction algorithms are largely the product of research in fields outside of statistics, especially computer science, and a vast number have been developed. This chapter provides an overview and examples including nonparametric regression methods, kernel methods using basis expansions, a special kind of neural networks known as multilayer perceptrons, decision trees and random forests.

11.1 Local Kernel Smoothing

The first part of this section provides a description of some selected traditional nonparametric methods that possess great flexibility for the estimation of the regression function. The objective is to predict the regression function m at some target value x. The examples and motivation in this first part assume that the response y is a function of a scalar predictor x.

All the nonparametric models described here are of the form $\hat{y} = Hy$, where $\hat{y}$ is the prediction and H is a hat matrix (i.e. page 266), not a function of y.

The Binned Estimator

A convenient starting point is nonparametric estimation of the conditional expectation of a random variable Y, $E(Y|X = x) = m(x)$ using a random sample (training data) $(y_1, x_1), (y_2, x_2), \ldots, (y_n, x_n)$, without assuming a specific form for m. The estimate at the point $X = x$ is the mean of the observations y for which $X_i = x$, $i = 1, 2, \ldots$. An approximate result is to average the observations y associated with X's close to x, such that $|x_i - x| \leq h$ for a small $h > 0$ called the *bandwidth*. Essentially, the X axis is divided into *bins*, and the estimate for the bin is the average of the y's belonging to the bin. The value of the bandwidth determines the width of the bins: large size with more observations produces a smoother fit as a function of X. When $x \pm h$ defines the bin, the estimator of the regression function can be written as

$$\widehat{m}(x) = \frac{\sum_{i=1}^{n} I(|x_i - x| \leq h)\, y_i}{\sum_{i=1}^{n} I(|x_i - x| \leq h)}$$

$$= \sum_{i=1}^{n} w_i(x)\, y_i, \quad w_i(x) = \frac{I(|x_i - x| \leq h)}{\sum_{i=1}^{n} I(|x_i - x| \leq h)}, \tag{11.2}$$

where $I(u)$ is the indicator function equal to 1 if the argument holds and 0 otherwise. For a given x, the sum in the denominator is the number of observations that fall in the corresponding *bin*. Since $\sum_{i=1}^{n} w_i(x) = 1$, (11.2) is a weighted average of the training observations y_i and the ith weight determines the contribution of y_i to the estimator $\widehat{m}(x)$.

The $n \times 1$ column vector of fitted values is

$$\widehat{m} = \left(\widehat{m}(x_1), \widehat{m}(x_2), \ldots, \widehat{m}(x_n)\right)',$$

and it follows that

$$\widehat{m} = Wy, \tag{11.3}$$

where $y = (y_1, y_2, \ldots, y_n)'$ is the training data and W is the $n \times n$ matrix whose ith row contains the weights given to y_i to estimate $\widehat{m}(x_i)$. Expression (11.3) highlights that the estimate is linear in the data. The ith row of W is

$$w(x_i)' = \left[\frac{I(|x_1 - x_i| \leq h)}{\sum_{j=1}^{n} I(|x_j - x_i| \leq h)} \; \frac{I(|x_2 - x_i| \leq h)}{\sum_{j=1}^{n} I(|x_j - x_i| \leq h)} \; \cdots \; \frac{I(|x_n - x_i| \leq h)}{\sum_{j=1}^{n} I(|x_j - x_i| \leq h)} \right]. \tag{11.4}$$

The matrix W is the *hat matrix* encountered in (6.21), page 266.

To illustrate suppose that $n = 7$, $x_i = i/7$, $i = 1, 2, \ldots, 7$ and $h = 1/7$. Then for $i = 1$ and $x = 1/7$, the denominator in (11.2) is 2, the first row of W is

$$\left[\tfrac{1}{2} \, \tfrac{1}{2} \, 0\,0\,0\,0\,0 \right],$$

and $\widehat{m}(x_1 = 1/7) = \frac{1}{2}y_1 + \frac{1}{2}y_2$. For $i = 2$, $x = 2/7$, the denominator in (11.2) is 3, the second row of W is

$$\left[\tfrac{1}{3} \, \tfrac{1}{3} \, \tfrac{1}{3} \, 0\,0\,0\,0 \right],$$

and $\widehat{m}\,(x_2 = 2/7) = \frac{1}{3}y_1 + \frac{1}{3}y_2 + \frac{1}{3}y_3$, and so on. The hat matrix for this example is

$$W = \begin{bmatrix} \frac{1}{2} & \frac{1}{2} & 0 & 0 & 0 & 0 & 0 \\ \frac{1}{3} & \frac{1}{3} & \frac{1}{3} & 0 & 0 & 0 & 0 \\ 0 & \frac{1}{3} & \frac{1}{3} & \frac{1}{3} & 0 & 0 & 0 \\ 0 & 0 & \frac{1}{3} & \frac{1}{3} & \frac{1}{3} & 0 & 0 \\ 0 & 0 & 0 & \frac{1}{3} & \frac{1}{3} & \frac{1}{3} & 0 \\ 0 & 0 & 0 & 0 & \frac{1}{3} & \frac{1}{3} & \frac{1}{3} \\ 0 & 0 & 0 & 0 & 0 & \frac{1}{2} & \frac{1}{2} \end{bmatrix}. \tag{11.5}$$

If instead $h = 2/7$ is used, the hat matrix is

$$W = \begin{bmatrix} \frac{1}{3} & \frac{1}{3} & \frac{1}{3} & 0 & 0 & 0 & 0 \\ \frac{1}{4} & \frac{1}{4} & \frac{1}{4} & \frac{1}{4} & 0 & 0 & 0 \\ \frac{1}{5} & \frac{1}{5} & \frac{1}{5} & \frac{1}{5} & \frac{1}{5} & 0 & 0 \\ 0 & \frac{1}{5} & \frac{1}{5} & \frac{1}{5} & \frac{1}{5} & \frac{1}{5} & 0 \\ 0 & 0 & \frac{1}{5} & \frac{1}{5} & \frac{1}{5} & \frac{1}{5} & \frac{1}{5} \\ 0 & 0 & 0 & \frac{1}{4} & \frac{1}{4} & \frac{1}{4} & \frac{1}{4} \\ 0 & 0 & 0 & 0 & \frac{1}{3} & \frac{1}{3} & \frac{1}{3} \end{bmatrix}. \tag{11.6}$$

Note that as h increases, more observations enter into the bin and smoothing is said to be more "global". Conversely, as h decreases, the neighbourhood size is smaller and the smoothing is more "local".

The R-code below displays an example of a binned estimator. Figure 11.1 shows that as the number of bins increases from 6 (blue line) to 19 (green line), the fit becomes more jagged leading to overfitting. The dots represent the response.

```
# CODE1101
rm(list=ls()) # CLEAR WORKSPACE
set.seed(195021)
x<-seq(from=0, to=2*pi,by=0.2)
f0<-function(x){ 100+sin(2*x)+cos(x/2) }
R2<-2/3
y<-f0(x)+rnorm(n=length(x),sd=sqrt(var(f0(x))*(1-R2)/R2))
z1 <- cut(x,breaks=seq(from=min(x),to=max(x+.01),
                       length=7),right=F)
f1 <- lm(y~z1)
z2 <- cut(x,breaks=seq(from=min(x),to=max(x+.01),
                       length=20),right=F)
f2 <- lm(y~z2)

plot(y~x,main='Binned estimator')
lines(x=x,y=predict(f1),col='blue',lwd=2)
lines(x=x,y=predict(f2),col='green',lwd=2)
legend(5, 103, legend=c("6 bins", "19 bins"),
       col=c("blue", "green"), lty=1:1, cex=0.8)
```

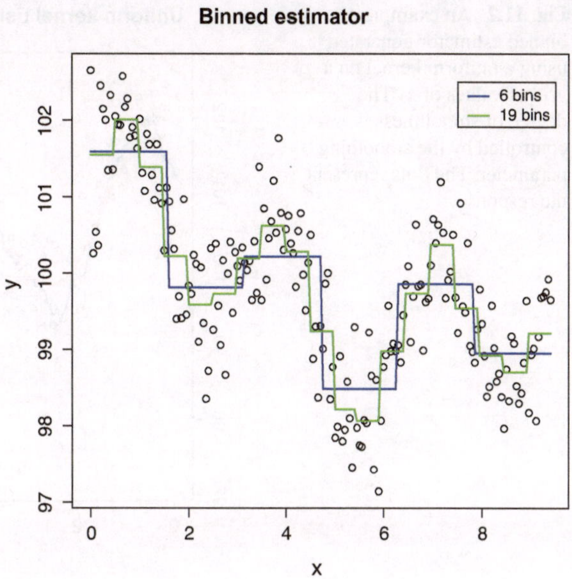

Fig. 11.1 An example of a binned estimator illustrating how the smoothness of the fit is controlled by the number of bins. The dots represent the response

Other Kernel Smoothing Methods

Estimator (11.2) is a step function, discontinuous at each $x = x_i$, due to the weights given by the indicator functions. If the indicator function in (11.2) is replaced by a *kernel function*, $K(\cdot)$, then the estimator can be written as

$$\widehat{m}(x) = \frac{\sum_{i=1}^{n} K\left(\frac{x_i - x}{h}\right) y_i}{\sum_{i=1}^{n} K\left(\frac{x_i - x}{h}\right)}. \tag{11.7}$$

This is the form of a *kernel regression* estimator, known as the Nadaraya-Watson estimator that does not necessarily eliminate the discontinuity unless the kernel function used is continuous. For instance, the uniform density $K(u) = 1/2$ for $u = (x_i - x)/h$ in the support $[-1, 1]$ is a *kernel function* that is a proper probability density function symmetric at zero. The uniform kernel in particular is an alternative representation of the binned estimator (11.2) and suffers from the same discontinuity.

A kernel function satisfies $0 \le K(u) < \infty$, $\int_{-\infty}^{\infty} K(u)\,du = 1$, or equal to a constant, $K(u) = K(-u)$, $\int_{-\infty}^{\infty} uK(u)\,du = 0$ and $\sigma_K^2 = \int_{-\infty}^{\infty} u^2 K(u)\,du < \infty$. A normalised kernel function satisfies $\int_{-\infty}^{\infty} u^2 K(u)\,du = 1$.

The size of h in (11.7) plays a central role in the degree of smoothness of $\widehat{m}(x)$. As h approaches 0, x_i approaches x, $\widehat{m}(x_i)$ approaches y_i and the fitted values go through the observed data. This is an extreme case of overfitting. When h is very large, the computation of $\widehat{m}(x)$ involves all the x_i's and the estimator $\widehat{m}(x)$

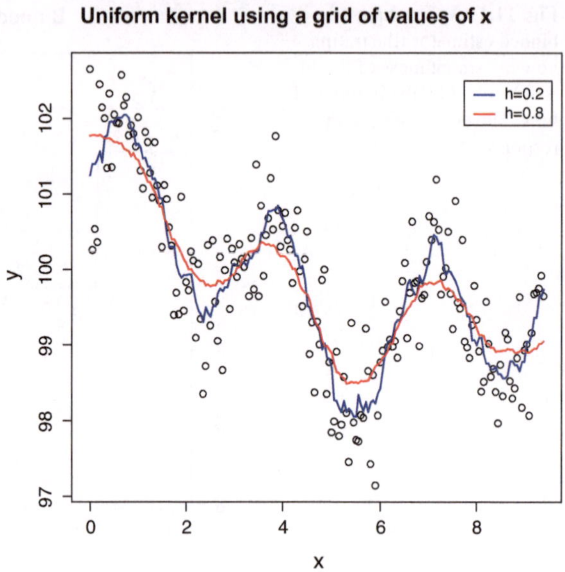

Fig. 11.2 An example of a binned estimator generated using a uniform kernel on a grid of values of x. The degree of smoothness is controlled by the smoothing parameter. The dots represent the response

approaches the mean of the observations for all i, an extreme case of underfitting. The R-code below illustrates this using a uniform kernel on a grid of values of x and the result is displayed in Fig. 11.2.

```
# CODE1102
rm(list=ls()) # CLEAR WORKSPACE
set.seed(195021)
x<-seq(from=0, to=2*pi,by=0.2)
f0<-function(x){ 100+sin(2*x)+cos(x/2) }
R2<-2/3
y<-f0(x)+rnorm(n=length(x),sd=sqrt(var(f0(x))*(1-R2)/R2))
d <- as.matrix(dist(x))
h <- 0.2
d2 <- ifelse(d <= h,1,0)
div <- apply(d2,1,sum)
rx <- d2%*%y/div

h <- 0.8
d2 <- ifelse(d <= h,1,0)
div <- apply(d2,1,sum)
rx <- d2%*%y/div
```

As indicated, the construction of (11.2), either as expressed in terms of indicator functions or in terms of the uniform kernel, places equal weights to those points within a distance h and those outside h are ignored. This generates a discontinuity when the $x's$ are grouped in bins, although a smoother graph results if a grid of values of the $x's$ is chosen instead.

One may wish to place more weights to those observations that are close to x. There are a number of continuous kernel functions that can be used with (11.7) that address these issues and a common one is the Gaussian kernel

$$K\left(\frac{x_i - x}{h}\right) = (2\pi)^{-\frac{1}{2}} \exp\left(-\frac{(x_i - x)^2}{2h^2}\right).$$

The constant term $(2\pi)^{-\frac{1}{2}}$ is typically omitted, and the term in the denominator of the exponential function can be replaced by h. The fitted value at $x_1 = 1/7$ is given by

$$\widehat{m}\left(x_1 = 1/7\right) = \frac{\sum_{i=1}^{7} \exp\left[-\frac{(1/7-x_i)^2}{2h^2}\right] y_i}{\sum_{i=1}^{7} \exp\left[-\frac{(1/7-x_i)^2}{2h^2}\right]} = \sum_{i=1}^{7} w_i\left(x\right) y_i \tag{11.8}$$

that has the form given by (11.7). A prediction for a new scalar value of the covariate $x = x^*$ is

$$\widehat{m}\left(x = x^*\right) = \frac{\sum_{i=1}^{7} \exp\left[-\frac{(x^*-x_i)^2}{2h^2}\right] y_i}{\sum_{i=1}^{7} \exp\left[-\frac{(x^*-x_i)^2}{2h^2}\right]} = \sum_{i=1}^{7} w_i\left(x^*\right) y_i, \tag{11.9}$$

where the weight function is

$$w_i\left(x^*\right) = \frac{\exp\left[-\frac{(x^*-x_i)^2}{2h^2}\right]}{\sum_{i=1}^{7} \exp\left[-\frac{(x*-x_i)^2}{2h^2}\right]}. \tag{11.10}$$

An example of the implementation of a Gaussian kernel applied to the same data used in Figs. 11.1 and 11.2 is shown in the R-code below that produces Fig. 11.3 as output. Compared to the uniform kernel, the Gaussian kernel generates a smoother fit. As shown in the figure, a higher degree of smoothness is associated with larger values of the bandwidth h.

```
# CODE1103
# GAUSSIAN KERNEL REGRESSION
rm(list=ls()) # CLEAR WORKSPACE
set.seed(195021)
# GENERATE DATA
x<-seq(from=0, to=2*pi,by=0.2)
#x<-seq(from=0, to=2*pi,length.out=33)

f0<-function(x){ 100+sin(2*x)+cos(x/2) }
R2<-2/3
y<-f0(x)+rnorm(n=length(x),sd=sqrt(var(f0(x))*(1-R2)/R2))
# CHOOSE h
```

```
h <- 0.25
# CONSTRUCT DISTANCE MATRIX AND GAUSSIAN KERNEL
d <- as.matrix(dist(x))^2
Kh25 <- exp(-(1/(2*h^2))*d)
sc25 <- apply(Kh25,1,sum)
mhgaus25 <- Kh25%*%y/sc25
plot(y~x,main='Gaussian kernel regression')
lines(x,mhgaus25,col="red")
# CHOOSE h
h <- 0.10
Kh10 <- exp(-(1/(2*h^2))*d)
sc10 <- apply(Kh10,1,sum)
mhgaus10 <- Kh10%*%y/sc10
lines(x,mhgaus10,col="blue")
legend(5, 103, legend=c("h=0.25", "h=0.10"),
       col=c("red", "blue"),lty=1:1, lwd=c(1.5,1.5), cex=0.8)
```

Kernel smoothing methods discussed so far suffer from boundary bias. For instance, an estimate for a decreasing function on the left boundary includes points to the right of the boundary, and since the function is decreasing, this creates a downward bias. Bias can also occur in the interior of the function if it has substantial curvature and is aggravated when covariates are multidimensional or unequally spaced. These problems can be alleviated using a generalisation of kernel regression: local polynomial regression.

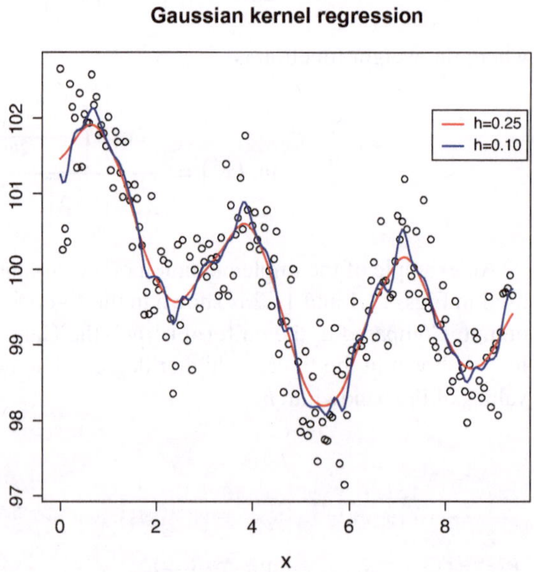

Fig. 11.3 The Gaussian kernel estimator computed with the R-code CODE1103, using two values for the bandwidth parameter

Local Polynomial Regression

The Nadaraya-Watson kernel regression estimator of the conditional expectation $m(x)$ can be framed in terms of a weighted regression. Given the weight function like the one defined in (11.10), choose $\mu = \widehat{m}(x)$ to minimise the weighted sum of squares

$$\sum_{i=1}^{n} w_i(x)(y_i - \mu)^2. \tag{11.11}$$

Setting the derivative with respect to μ equal to zero and solving for μ recovers the kernel estimator

$$\hat{m}(x) = \frac{\sum_i w(x_i) y_i}{\sum_i w(x_i)}. \tag{11.12}$$

This is a weighted regression with a mean only implying the approximation $m(x) \approx \mu$. A generalisation is to replace the local constant μ by a local polynomial of degree p. This polynomial is fitted to each target value x and generates estimates of regression parameters by minimising the weighted sum of squares along the same lines as in (11.11)

$$\sum_{i=1}^{n} w_i(x) \left(y_i - \sum_{j=0}^{p} b_{jx} (x_i - x)^j \right)^2. \tag{11.13}$$

This weighted sum of squares can be written

$$\sum_{i=1}^{n} w_i(x) \left(y_i - \sum_{j=0}^{p} b_{jx} (x_i - x)^j \right)^2 = (y - X_x b_x)' W_x (y - X_x b_x).$$

Minimisation of this expression with respect to b_x gives the standard weighted least squares estimator

$$\widehat{b}_x = \left(X_x' W_x X_x\right)^{-1} X_x' W_x y, \tag{11.14}$$

where

$$X_x = \begin{bmatrix} 1 & (x_1 - x) & \cdots & (x_1 - x)^p \\ \vdots & \vdots & \ddots & \vdots \\ 1 & (x_n - x) & \cdots & (x_n - x)^p \end{bmatrix} \tag{11.15}$$

is an $n \times (p + 1)$ matrix and

$$W_x = \text{diag}\left[w_1(x) \cdots w_n(x)\right] \tag{11.16}$$

is an $n \times n$ diagonal matrix of weights $w_i(x)$. In (11.14) the dependence on the target value x has been made explicit by writing $\hat{b}_x = \left(\hat{b}_{0x}, \hat{b}_{1x}, \ldots, \hat{b}_{px}\right)'$. At the target value $x_i = x$, the terms in $(x_i - x)$ drop out and the fitted value is

$$\begin{aligned} \widehat{m}(x) &= z_1' \left(X_x' W_x X_x\right)^{-1} X_x' W_x y & (11.17) \\ &= H_x y & (11.18) \\ &= \widehat{b}_{0x}, & (11.19) \end{aligned}$$

where z_1 is the $(p + 1) \times 1$ column vector with 1's in the first entry and 0's elsewhere and H_x is a hat vector. It is important to notice that the complete regression function must be computed for each target value, despite the fact that at $x_i = x$, $\hat{m}(x) = \hat{b}_0(x)$. Setting $p = 0$ yields the kernel regression estimator (11.7).

The same result is obtained if (11.13) is replaced by

$$\sum_{i=1}^{n} w_i(x) \left(y_i - \sum_{j=0}^{p} b_{jx} x_i^j\right)^2 = (y - X b_x)' W_x (y - X b_x),$$

where now

$$X = \begin{bmatrix} 1 & x_1 & \cdots & x_1{}^p \\ \vdots & \vdots & \ddots & \vdots \\ 1 & x_n & \cdots & x_n{}^p \end{bmatrix}_{n \times (p+1)}.$$

A predicted value at the target $x^* = \left(x^*, x^{*2}, \ldots x^{*p}\right)$ is

$$\begin{aligned} \widehat{m}\left(x^*\right) &= \left(1, x^*\right) \left(X' W_{x^*} X\right)^{-1} X' W_{x^*} y \\ &= \widehat{b}_{0x^*} + \sum_{j=1}^{p} \widehat{b}_{jx^*} \left(x^*\right)^j, \end{aligned}$$

where $(1, x^*)$ is $(p \times 1) \times 1$ and the weight function W_{x^*} is an $n \times n$ diagonal matrix with ith diagonal element

$$w_i\left(x^*\right) = \frac{\exp\left[-\frac{(x^* - x_i)^2}{2h^2}\right]}{\sum_{i=1}^{n} \exp\left[-\frac{(x* - x_i)^2}{2h^2}\right]}.$$

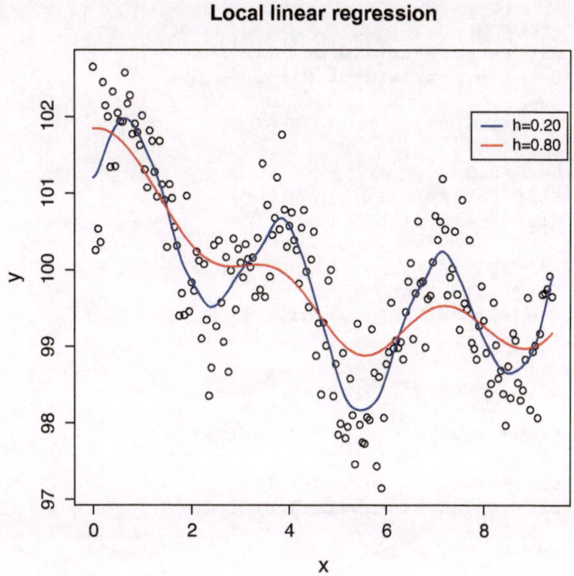

Fig. 11.4 A local linear regression estimator computed with the R-code CODE1104, using two values for the bandwidth parameter

The R-code below implements a local linear regression ($p = 1$) and the output is displayed in Fig. 11.4.

```
# CODE1104
rm(list=ls()) # CLEAR WORKSPACE
set.seed(195021)
# LOCAL LINEAR REGRESSION SETS p = 1
p <- 1
x<-seq(from=0, to=2*pi,by=0.2)
f0<-function(x){ 100+sin(2*x)+cos(x/2) }
R2<-2/3
y<-f0(x)+rnorm(n=length(x),sd=sqrt(var(f0(x))*(1-R2)/R2))

w <- matrix(data=NA,nrow=length(x),ncol=length(x))
X <- matrix(data=NA, nrow=length(y), ncol=(p+1))
W <- matrix(data=NA, nrow=length(y), ncol=length(y))

one <- rep(1,length(y))

X <- cbind(one,x)
Xt <- t(X)
k <- seq(1:length(x))

mhx <- function(k,w,one,X){
  W <- diag(w[k,])
  Xt <- t(X)
  solve(Xt%*%W%*%X,Xt%*%W%*%y)
}

mpred <- function(k,estx,x){
  estx[1,k]+estx[2,k]*x[k]
}
```

```
# CONSTRUCT DISTANCE MATRIX
dst <- as.matrix(dist(x))
d <- as.matrix(dist(x))^2

# CONSTRUCT GAUSSIAN KERNEL
# CHOOSE h
h <- 0.2
Kh25 <- exp(-(1/(2*h^2))*d)
div <- apply(Kh25,1,sum)

# SCALE GAUSSIAN KERNEL: PLACE RESULT IN w
for( i in 1:nrow(Kh25)){
  w[i,] <- Kh25[i,]/div[i]
}

estx <- sapply(k,mhx,w,one,X)

fitx <- sapply(k,mpred,estx,x)
## APPROX PREDICTION FOR A NEW X=5.12:
fitx[which.min(abs(x-5.12))]
fitx[which.min(abs(x-5.0))] ## FIT FOR X=5.0


plot(x,y,main='Local linear regression')
lines(x,fitx,lty=1,col="blue")

h <- 0.5
Kh25 <- exp(-(1/(2*h^2))*d)
div <- apply(Kh25,1,sum)

# SCALE GAUSSIAN KERNEL: PLACE RESULT IN w
for( i in 1:nrow(Kh25)){
  w[i,] <- Kh25[i,]/div[i]
}

estx <- sapply(k,mhx,w,one,X)

fitx <- sapply(k,mpred,estx,x)
## APPROX PREDICTION FOR A NEW X=5.12:
fitx[which.min(abs(x-5.12))]
fitx[which.min(abs(x-5.0))] ## FIT FOR X=5.0

lines(x,fitx,lty=1,col="red")
legend(5, 103, legend=c("h=0.20", "h=0.50"),
       col=c("blue", "red"),lty=1:1, lwd=c(1.5,1.5), cex=0.8)
```

The R-code below computes a prediction based on the local linear regression for a new value of the feature $x = z$.

```
# CODE1105
rm(list=ls()) # CLEAR WORKSPACE
set.seed(195021)
# LOCAL LINEAR REGRESSION SETS p = 1
p <- 1
x<-seq(from=0, to=2*pi,by=0.2)
f0<-function(x){ 100+sin(2*x)+cos(x/2) }
R2<-2/3
y<-f0(x)+rnorm(n=length(x),sd=sqrt(var(f0(x))*(1-R2)/R2))
```

```
z <- 5.12
bz <- matrix(c(1,z),nrow=2)
bzt <- t(bz)
one <- rep(1,length(y))

X <- matrix(data=NA, nrow=length(y), ncol=(p+1))
Wz <- matrix(data=NA, nrow=length(y), ncol=length(y))

X <- cbind(one,x)
Xt <- t(X)
# CHOOSE h
h <- 0.2
d <- x-z
ele <- exp(-(d^2)/(2*h^2))
sm <- sum(ele)
Wz <- diag(ele/sm)
pred_z <- bzt%*%solve(Xt%*%Wz%*%X,Xt%*%Wz%*%y)
cat("x =",z,"\n")
```

```
## x = 5.12
```

```
cat("Prediction =", pred_z,"\n")
```

```
## Prediction = 98.37111
```

Choice of Bandwidth Parameter

Leave-one-out cross-validation can be used to choose the bandwidth parameter h. Since the methods described here are linear smoothers, use can be made of the exact expression (6.58) or of the approximate but more robust generalised cross-validation estimator (6.59). The R-code below computes both for a range of values of h. Figure 11.5 illustrates that for the present data, both estimators produce very similar values of leave-one-out mean squared error (0.64 vs 0.66) and choose a similar amount of smoothing ($h = 0.35$ for (6.58) and $h = 0.325$ for (6.59)).

```
# CODE1106
# COMPUTE HAT MATRIX FOR THE LOCAL LINEAR REGRESSION
rm(list=ls()) # CLEAR WORKSPACE
set.seed(195021)

x<-seq(from=0, to=3*pi,by=0.05)

f0<-function(x){ 100+sin(2*x)+cos(x/2) }
noise <- rnorm(n=length(x),sd=sqrt(var(f0(x))*0.5))
y<- f0(x) + noise

one <- rep(1,length(x))

X <- matrix(data=NA, nrow=length(y), ncol=(2))
W <- matrix(data=NA, nrow=length(y), ncol=length(y))
```

```r
HatMat <- matrix(data=NA, nrow=length(x), ncol=length(x))

h <- seq(from = 0.1, to = 1.5, by = 0.025)
GCV <- rep(0,length(h))
LOOCV <- rep(0,length(h))
form40 <- rep(0,length(h))
dsq <- rep(0,length(h))

tr <- rep(0,length(h))
Trmse <- rep(0,length(h))

j <- seq(1:length(x))

X <- cbind(one,x)
Xt <- t(X)

Hat <- function(j,h,k,x,X){
d <- x - x[j]
ele <- exp(-(d^2)/(2*h[k]^2))
sm <- sum(ele)
w <- diag(ele)/sm
X[j,]%*%solve(Xt%*%w%*%X)%*%Xt%*%w
}

for (k in 1:length(h)) {
  HatMat<- t(sapply(1:length(x),Hat,h,k,x,X))
  predy <- HatMat %*% y
  tr[k] <- sum(diag(HatMat))
  Trmse[k] <- mean((predy - y) ^ 2)
  dsq[k] <- (1 - mean(diag(HatMat))) ^ 2
  GCV[k] <- Trmse[k] / dsq[k] # generalised LOOCV
  omd <- 1 - diag(HatMat)
  LOOCV[k] <- mean(((predy - y) / omd) ^ 2) # LOOCV
}

h[which.min(LOOCV)]
```

```
## [1] 0.35
```

```r
h[which.min(GCV)]
```

```
## [1] 0.325
```

```r
min(LOOCV)
```

```
## [1] 0.660054
```

```r
min(GCV)
```

```
## [1] 0.639314
```

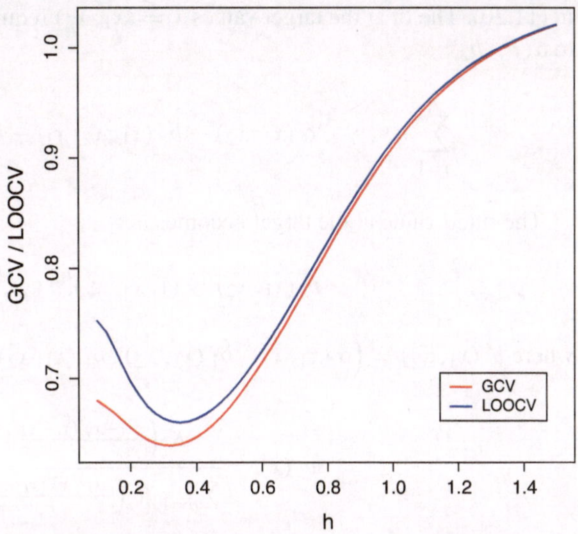

Fig. 11.5 Computation of leave-one-out mean squared error for a range of values of the smoothing parameter h using the generalised cross-validation estimator (6.59) and using estimator (6.58). The former chooses $h = 0.325$ and the latter $h = 0.35$

Extension to Several Dimensions

Kernel smoothing methods such as the Nadaraya-Watson estimator and local regression estimators can be extended to deal beyond the one-dimensional case ($d = 1$) discussed so far. For instance, for $d = 2$ and $p = 1$ (polynomial of degree one for each of the two coordinates, a linear regression), terms of the form $(x_i - x)$ in the kernel (11.7) are replaced by

$$(x_i - x) = (x_{1i} - x_1)\,(x_{2i} - x_2)\,.$$

When a single bandwidth parameter h is used, the kernel reduces to the product of the kernels of each component

$$K\left(\frac{(x_i - x)'\,(x_i - x)}{h}\right) = K_1\left(\frac{x_{1i} - x_1}{h}\right) K_2\left(\frac{x_{2i} - x_2}{h}\right) \tag{11.20}$$

for target values (x_1, x_2). However, a multivariate kernel of the form

$$K\left(\frac{(x_i - x)'\,A\,(x_i - x)}{h}\right),$$

for some positive semidefinite matrix A can also be used. For diagonal A, this would allow to control the influence of certain predictors by assigning different values to h

in (11.20). The fit at the target values $x = (x_1, x_2)$ requires minimising with respect to α, b_1, b_2

$$\sum_{i=1}^{n} w_i \left(y_i - \alpha\left(x_1, x_2\right) - b_1\left(x_1, x_2\right) x_{1i} - b_2\left(x_1, x_2\right) x_{2i}\right)^2 .$$

The fitted value at the target becomes

$$\hat{m}\left(x_1, x_2\right) = \left(1, x_1, x_2\right)' \hat{b}\left(x_1, x_2\right)$$

where $\hat{b}'\left(x_1, x_2\right) = \left(\hat{\alpha}\left(x_1, x_2\right), \hat{b}_1\left(x_1, x_2\right), \hat{b}_1\left(x_1, x_2\right)\right)$ and

$$w_i(x) = \frac{K\left(\frac{(x_i - x)'(x_i - x)}{h}\right)}{\sum_{i=1}^{n} K\left(\frac{(x_i - x)'(x_i - x)}{h}\right)}.$$

Local polynomials become less attractive when the dimension exceeds beyond $d = 2$ or 3. Other nonparametric approaches are required to deal with higher dimensions.

11.2 Kernel Methods Using Basis Expansions

The nonparametric methods discussed so far provide flexibility by fitting a model repeatedly for each target value x. The model could be a simple mean model (polynomial of degree zero: the binned estimator), or polynomials of any degree, such as the simple linear regression, a polynomial of degree one. The contribution that the data features x_i, $i = 1, \ldots, n$ make to the fit at the target x is controlled by a kernel function.

This section introduces kernel methodology where kernel functions expand the original set of features into an implicitly high-dimensional space. The method can involve regularisation, in a similar spirit as in ridge regression, and can capture the effect on the data of complex interactions and nonlinear terms involving the original features, without explicit modelling. Here the treatment is heuristic and based on examples. An authoritative reference is Wahba (1990). The topic is also discussed at length in the books of Bishop (2006) and Hastie et al (2009).

Preliminaries

I review briefly the concepts of *dual representation, basis functions* and *kernel functions* and note how a *dual representation* leads to *kernel functions*. These provide flexible mechanisms to implicitly expand the feature space of a regression model. Kernels are introduced using ridge regression as an example.

The following notation and definitions will be needed.

- The p-norm of a vector $x \in \mathbb{R}^n$ is $\left(|x_1|^p + \cdots + |x_n|^p\right)^{\frac{1}{p}} = \|x\|_p$
- For $p = 1$, the ℓ_1-norm is $\|x\|_1 = |x_1| + \cdots + |x_n|$
- For $p = 2$, the ℓ_2-norm or Euclidean norm or length of vector x is $\|x\|_2 = \left(x_1^2 + \cdots + x_n^2\right)^{\frac{1}{2}} = \langle x, x\rangle^{\frac{1}{2}} = \sqrt{x'x} \geq 0$. The Euclidean norm is sometimes denoted without any subscript: $\|x\|$
- For vectors $x \in \mathbb{R}^n$ and $y \in \mathbb{R}^n$, the squared distance between x and y is

$$\|x - y\|_2^2 = \|x\|_2^2 + \|y\|_2^2 - 2x'y.$$

For instance, for $x = (x_1, x_2)$ and $y = (y_1, y_2)$,

$$\begin{aligned}\|x - y\|_2^2 &= [(x_1 - y_1), (x_2 - y_2)]\,[(x_1 - y_1), (x_2 - y_2)]' \\ &= (x_1 - y_1)^2 + (x_2 - y_2)^2 \\ &= x_1^2 + x_2^2 + y_1^2 + y_2^2 - 2x_1y_1 - 2x_2y_2 \\ &= \|x\|_2^2 + \|y\|_2^2 - 2x'y.\end{aligned}$$

and if x and y are orthogonal, $x'y = 0$.
- A vector whose length is 1 is called the *unit vector*. If a non-zero vector x is divided by its length (or multiplied by $1/\|x\|$), a unit vector results because

$$\|u\| = \frac{1}{\|x\|}\|x\| = 1$$

- The inner product of vectors x and y is a function that can be represented as $\sum_{i=1}^{n} x_i y_i = x'y = y'x = \langle x, y\rangle = \langle y, x\rangle$ (the inner product is symmetrical).
- As noted in the following sections, vectors can be infinitely dimensional. Such infinitely dimensional vector spaces (*Hilbert spaces*) must be endowed with certain structure in order to derive mathematically useful results. This structure guarantees that properties of Hilbert spaces apply also to finite dimensional settings. Technical details are omitted in this overview, and I draw instead on a result based on the *representer theorem* on page 472 that has far-reaching applications. A similar result can be arrived at informally using the concept of dual representation. The general idea is to express the solution of a linear system in terms of kernel functions. A variety of kernel functions can be used. This provides a flexible mechanism to expand the feature space without explicit modelling. In this way, nonlinear associations between the covariates and the records can be explored.

Dual Representation

Linear models admit a *dual representation* where kernel functions arise naturally. For instance, consider a training dataset $S = \{(x_1, y_1), \ldots, (x_n y_n)\}$ where $x_i \in \mathbb{R}^p$ are p-dimensional feature vectors and $y_i \in \mathbb{R}$ are responses. The regression model can be written

$$y = 1\mu + Xb + e, \quad e \sim \left(0, I\sigma^2\right). \tag{11.21}$$

In ridge regression μ and b are obtained as the solution to the minimisation of the cost function

$$J(\mu, b) = (y - 1\mu - Xb)'(y - 1\mu - Xb) + \lambda \|b\|^2, \tag{11.22}$$

where the scalar $\lambda \geq 0$ is a parameter that controls the amount of shrinkage. Differentiation with respect to μ and b gives

$$\begin{bmatrix} 1'1 & 1'X \\ X'X & X'X + \lambda I \end{bmatrix} \begin{bmatrix} \widehat{\mu} \\ \widehat{b} \end{bmatrix} = \begin{bmatrix} 1'y \\ X'y \end{bmatrix} \tag{11.23}$$

that results in the closed-form solution

$$\widehat{b} = \left(X'X + I\lambda\right)^{-1} X'(y - 1\widehat{\mu}). \tag{11.24}$$

A scalar prediction at some target value x (a $p \times 1$ column vector) is equal to

$$m(x) = \hat{\mu} + x'\hat{b}. \tag{11.25}$$

Matrix $X'X + I\lambda$ is of dimension $p \times p$. If $\lambda > 0$ the matrix $X'X + I\lambda$ is positive definite, and therefore the solution $\widehat{b}$ is unique.

Solution (11.24) can be expressed in what is known as the *dual form*. The partial derivative of $J(\mu, b)$ with respect to b results in

$$-X'y + X'1\mu + X'Xb + \lambda b = 0$$

Multiplying by λ^{-1} and rearranging yields

$$\widehat{b} = \lambda^{-1}X'\left(y - 1\widehat{\mu} - X\widehat{b}\right) = X'\widehat{\alpha} = \sum_{i=1}^{n} \widehat{\alpha}_i x_i \tag{11.26}$$

where x_i is the ith column of X' with p entries and the *dual variable* $\hat{\alpha}$ is equal to

$$\begin{aligned}\widehat{\alpha} &= \lambda^{-1}\left(y - 1\widehat{\mu} - X\widehat{b}\right)\\ \Rightarrow \lambda\widehat{\alpha} &= \left(y - 1\widehat{\mu} - XX'\widehat{\alpha}\right)\\ \Rightarrow \widehat{\alpha} &= \left(XX' + I\lambda\right)^{-1}\left(y - 1\widehat{\mu}\right).\end{aligned} \tag{11.27}$$

Expression (11.26) indicates that $\hat{b}$ can be expressed as a linear combination of the training features x_i and $\hat{\alpha}$. The dual variable $\hat{\alpha}$ involves the solution of a linear system of n equations. In contrast, (11.24) requires the solution of a linear system of p equations. The choice between both depends on the relative sizes of n and p. An important observation is that the *dual solution* of the ridge regression (11.27) uses inner products between data points x_i via XX'. Indeed, the ijth element of XX' is

$$\left[XX'\right]_{i,j} = x_i' x_j,$$

where x_i' is the ith row of X and x_j is the jth column of X'. A predicted value at the target x is of the form

$$\begin{aligned}m(x) &= \widehat{\mu} + x'\widehat{b} = \widehat{\mu} + x'\sum_{i=1}^{n}\widehat{\alpha}_i x_i\\ &= \widehat{\mu} + \sum_{i=1}^{n}\widehat{\alpha}_i x' x_i\\ &= \widehat{\mu} + \sum_{i=1}^{n}\widehat{\alpha}_i \langle x, x_i\rangle,\end{aligned} \tag{11.28}$$

again involving the inner product between the training features x_i and the target x. In general, a method will be referred to as *kernelised*, if the training features x appear only inside inner products of functions of x, the *basis functions*. These are dealt with on page 465. The inner product $\langle x, x_i\rangle$ is the simplest example of a *kernel function*, often referred to as *linear kernel.*

Kernelised Predictions

Consider representing the inner product $\langle x, x_i\rangle$ in the last line of (11.28) by $k(x, x_i)$, a *kernel function*. This kernel function takes the form

$$k(x, x_i) = \left(x_1, \ldots, x_p\right)\begin{pmatrix} x_{i1}\\ \vdots \\ x_{ip}\end{pmatrix}.$$

The kernelised prediction formulation of (11.28) at the target x becomes

$$m(x) = \hat{\mu} + \sum_{i=1}^{n} \widehat{\alpha}_i k(x, x_i), \tag{11.29}$$

a linear combination of $k(x, x_1), \ldots, k(x, x_p)$ considered as a function of x. The predicted values for new features x_i', $i = 1, \ldots, m$, can be written as

$$\begin{bmatrix} m(x_1') \\ \vdots \\ m(x_m') \end{bmatrix} = 1\widehat{\mu} + \begin{bmatrix} \widehat{\alpha}_1 k(x_1', x_1) + & \cdots & + \widehat{\alpha}_n k(x_1', x_n) \\ \vdots & \vdots & \vdots \\ \widehat{\alpha}_1 k(x_m', x_1) + & \cdots & + \widehat{\alpha}_n k(x_m', x_n) \end{bmatrix} = 1\widehat{\mu} + K\widehat{\alpha} \tag{11.30}$$

generalising expression (11.29). Matrix $K = XX' = (k(x_i, x_j))_{ij} \in \mathbb{R}^{n \times n}$, which is symmetric and positive semidefinite, is known as the *kernel matrix* or *Gram matrix* for any kernel k in the set $\{x_1, \ldots, x_n\}$. More details on this matrix follow shortly. At this point note that the Gram matrix contains the evaluation of the kernel function on all pairs of feature points x_i. All the information about the data x is contained in K. New predictions require the input of the complete feature points x_i of the training data.

Kernelised Cost Functions

The cost function (11.22) can be written in terms of the $n \times n$ symmetric kernel matrix K. Using (11.26),

$$\begin{aligned} (y - 1\mu - Xb) &= (y - 1\mu - XX'\alpha) \\ &= (y - 1\mu - K\alpha). \end{aligned}$$

Therefore, the residual sum of squares takes the form

$$\|y - 1\mu - Xb\|^2 = \|y - 1\mu - K\alpha\|^2.$$

The penalised term $\lambda \|b\|^2$ can be expressed as

$$\begin{aligned} \lambda \|b\|^2 &= \lambda \|X'\alpha\|^2 \\ &= \lambda\alpha' XX'\alpha \\ &= \lambda\alpha' K\alpha. \end{aligned}$$

Therefore, the cost function (11.22) is equal to the convex and differentiable cost function

$$(y - 1\mu - K\alpha)' (y - 1\mu - K\alpha) + \lambda\alpha' K\alpha. \tag{11.31}$$

Setting the partial derivatives with respect to μ and α equal to zero yields

$$\begin{bmatrix} 1'1 & 1'K \\ K1 & K(K + \lambda I) \end{bmatrix} \begin{bmatrix} \widehat{\mu} \\ \widehat{\alpha} \end{bmatrix} = \begin{bmatrix} 1'y \\ Ky \end{bmatrix}. \tag{11.32}$$

Rearranging the second equation results in $K(K + \lambda I)\alpha = K(y - 1\widehat{\mu})$. One solution is

$$\widehat{\alpha} = (K + \lambda I)^{-1} (y - 1\widehat{\mu}), \tag{11.33}$$

as can be confirmed by replacing (11.33) in the partial derivative of the cost function with respect to α. If K is positive definite, (11.33) is the only solution to the minimisation of (11.31). The vector of n fitted values is given by (11.30)

$$\begin{aligned} \widehat{y} &= 1\widehat{\mu} + K\widehat{\alpha} \\ &= 1\widehat{\mu} + K(K + \lambda I)^{-1} (y - 1\widehat{\mu}). \end{aligned} \tag{11.34}$$

The vector of fitted values $\widehat{y}$ is invariant to whatever solution of $K(K + \lambda I)\alpha = K(y - 1\widehat{\mu})$ is used. The right-hand side of (11.34) can be written as

$$\begin{bmatrix} 1 & K \end{bmatrix} \begin{bmatrix} 1'1 & 1'K \\ K1 & K(K + \lambda I) \end{bmatrix}^{-1} \begin{bmatrix} 1' \\ K \end{bmatrix} y = Hy$$

where H is not a function of y, so this is a linear smoother in the observations.

The subsection concludes with two important messages. First, the algorithm for solving (11.31) involves only the computation of inner products of feature vectors x. All the information about the training data is contained in the matrix K and the output vector y. Second, the kernelised cost function (11.31) with solution (11.33) can be used with a variety of kernel matrices K, providing flexibility. Kernel functions automatically incorporate nonlinear associations between the features x and the records y, without explicit modelling, as discussed in the next subsection.

Nonlinear Feature Mappings Using Kernel Functions

Consider a regression function

$$m(x) = \alpha + bx, \tag{11.35}$$

where b is a scalar parameter and feature $x \in \mathbb{R}$ is also a scalar. One may wish to expand the model to explore nonlinear structures in the training data and, to this end, a *basis function* $\phi(x)$ is used that maps x onto $\phi(x) = (x, x^2, x^3) \in \mathbb{R}^3$, a three-dimensional space. The effect of ϕ is to recode the data from $\{(x_1, y_1), \ldots, (x_n, y_n)\}$ to $\{(\phi(x_1), y_1), \ldots, (\phi(x_n), y_n)\}$. The regression function in the new space is now

$$m(\phi(x)) = \alpha + b_1 x + b_2 x^2 + b_3 x^3 = \alpha + b'\phi(x),$$

where now b is a vector of parameters with three elements. This operation allows the output y to be represented by a nonlinear function of the original features x, but y is still linear in the parameters b in the expanded model. The only difference with (11.35) is that $\phi(x)$ is substituted for x. This linearity in the parameters greatly simplifies the analysis of this class of models. As we shall see, these models lead to convex functions that are relatively easy to optimise. The algorithm that solves (11.31) proceeds in exactly the same manner as in the standard linear model, with x replaced by $\phi(x)$. This requires computation of the inner products $\phi(x)'\phi(x)$.

A more general case is to consider a feature vector $x = (x_1, \ldots, x_p)' \in \mathbb{R}^p$ involving p variables, where the basis function ϕ transforms x into the product of all monomial terms. The notation (x, z) denotes two different values of the feature x, (x_i, x_j) and therefore $x, z \in \mathbb{R}^p$. The basis function is

$$\begin{aligned}\phi(x) &= (x_1 x_1, x_1 x_2, \ldots, x_1 x_n, \ldots, x_p x_p)' \in \mathbb{R}^{p^2}, \\ \phi(z) &= (z_1 z_1, z_1 z_2, \ldots, z_1 z_n, \ldots, z_p z_p)' \in \mathbb{R}^{p^2}.\end{aligned} \tag{11.36}$$

The time required to compute $\phi(x)$ or the inner product $\phi(x)'\phi(z)$ is of order p^2 (because there are p^2 terms and p^2 products), while the time required to compute the inner product in the original feature space $x'z$ is of order p (there are p products). So the complexity of the evaluation of the inner products is proportional to the dimension of the feature space. The function that for all x, z computes the inner products $\phi(x)'\phi(z)$ is a *kernel function* k. Specifically,

$$k(x, z) = \langle \phi(x), \phi(z) \rangle = \phi(x)'\phi(z) = \sum_{i=1}^{p}\sum_{j=1}^{p}(x_i x_j)(z_i z_j). \tag{11.37}$$

The kernel function involves p^2 products, where ϕ is a mapping from X to an inner product feature space F. When ϕ is the identity mapping and the new feature space involves the covariates $x_i \in \mathbb{R}^p$ of the training set, $K = XX'$. The ijth element of XX' is $(XX')_{ij} = \langle x_i, x_j \rangle = x_i' x_j$, where x_i' is the ith row of the incidence matrix $X \in \mathbb{R}^{n \times p}$; this particular kernel is known as the *linear kernel*. The kernel matrix K has dimension $n \times n$.

At face value, it seems that use of the basis function ϕ to explore a higher dimensional space comes at the cost of increased computation. The next section shows that this is not the case.

The Kernel Trick

As indicated above, the computation of inner products involving ϕ in (11.37) can be very demanding when the inner product feature space F is high dimensional. In the case of (11.37), there are p^2 products. As shown below, regardless of the dimension of F, the use of kernel functions allows computation of $\phi(x)'\phi(z)$ using the original features (x, z) without explicitly computing the mapping ϕ. The claim is that the inner product (11.37) involving the basis function (11.36) can be computed using

$$k(x, z) = \phi(x)'\phi(z) = \left(x'z\right)^2 \tag{11.38}$$

requiring p products. To prove the claim, write

$$\begin{aligned} k(x, z) &= \left(x'z\right)^2 \\ &= \left(\sum_{i=1}^{p} x_i z_i\right)\left(\sum_{j=1}^{p} x_j z_j\right) \\ &= \sum_{i=1}^{p}\sum_{j=1}^{p} x_i x_j z_i z_j \\ &= \sum_{i=1}^{p}\sum_{j=1}^{p} \left(x_i x_j\right)\left(z_i z_j\right) \\ &= \phi(x)'\phi(z). \end{aligned} \tag{11.39}$$

The first line in (11.39) requires order p computations involving the inner product of vectors x and z, each with p elements, while the last line requires order p^2. The inner product between the new feature space (high dimensional) has been computed without explicitly evaluating the new feature space, using the original (lower dimensional) feature space. The function that performs this direct computation is the kernel function k, and the operation of swapping the linear kernel with the new kernel is known as the kernel trick.

Choice of Kernel Functions

The use of kernel functions provides flexibility: the same algorithm can be applied replacing one kernel function with another. In addition, very high-dimensional feature spaces involving nonlinear terms among the features and complex interactions can be explored, without their explicit evaluation. The question is, how should a kernel be chosen and what function qualifies as a kernel function?

A function $k(x, z)$ can be used as a kernel function if there exists some ϕ such that $k(x, z) = \phi(x)' \phi(z)$. For a given set of feature vectors $\{x_1, \ldots, x_n\}$, $x_i \in \mathbb{R}^p$, define the symmetric matrix $K \in \mathbb{R}^{n\times n}$ to be the Gram matrix or kernel matrix such that $\left[K_{ij}\right]_{i,j=1,\ldots,n}$ where, $K_{ij} = \langle \phi(z_i), \phi(z_j) \rangle$. Then

$$
\begin{aligned}
z' K z &= \sum_{i=1}^{n} \sum_{j=1}^{n} z_i z_j K_{ij} \\
&= \sum_{i=1}^{n} \sum_{j=1}^{n} z_i z_j \langle \phi(z_i), \phi(z_j) \rangle \\
&= \left\langle \sum_{i=1}^{n} z_i \phi(z_i), \sum_{j=1}^{n} z_j \phi(z_j) \right\rangle \\
&= \left\| \sum_{i=1}^{n} z_i \phi(z_i) \right\|^2 \geq 0
\end{aligned}
$$

and therefore K is a positive semidefinite matrix. This result indicates that the construction of a valid kernel involves first finding a feature mapping ϕ and then computing its inner product to generate K. There is a powerful result known as *Mercer's theorem* that allows going the other way: choose an appropriate K and do not worry about ϕ. An appropriate K is simply any symmetric, positive semidefinite matrix.

A few examples of commonly used kernels are

- the linear kernel $k(x, z) = \langle x, z \rangle$
- the polynomial kernel of degree d, $k(x, z) = (\langle x, z \rangle + 1)^d$
- the Laplacian radial basis kernel $k(x, z) = \exp\left[-\frac{\|x-z\|_1}{2\sigma^2}\right]$
- the Gaussian or radial basis function kernel $k(x, z) = \exp\left[-\frac{\|x-z\|_2^2}{2\sigma^2}\right]$.

(Above, the term $2\sigma^2$ is often replaced by the positive scalar h, the bandwidth parameter).

Any of these kernel functions and many others can be used to minimise (11.31) with solution given by (11.33).

An observation regarding the Gaussian kernel is that, in order to express it as $k(x, z) = \langle \phi(x), \phi(z) \rangle$, the new feature vector $\phi(x)$ must be of infinite dimension. The inner product space of this infinitely dimensional feature vector is referred to as a *reproducing kernel Hilbert space* (RKHS). The finite dimensional inner product space $\mathbb{R}^p$ is a special case. The Gaussian kernel is a square, *positive definite* matrix of dimension $p \times p$, where the original feature space is $\mathbb{R}^p$, despite the fact that $\phi \in \mathbb{R}^\infty$. To illustrate, assume a single scalar feature per response. The basis function $\phi(x)$ corresponding to the Gaussian kernel is an implicitly infinite vector with the form

$$\phi(x) = \exp\left[-\frac{x^2}{2\sigma^2}\right]\left[1, \frac{x}{\sigma\sqrt{1!}}, \frac{x^2}{\sigma^2\sqrt{2!}}, \frac{x^3}{\sigma^3\sqrt{3!}}, \dots\right]'$$

and $\langle \phi(x), \phi(z) \rangle$ is a series that converges to $k(x, z)$. Indeed,

$$\begin{aligned} \langle \phi(x), \phi(z) \rangle &= \sum_{j=0}^{\infty} \frac{1}{j!} \exp\left[-\frac{(x^2+z^2)}{2\sigma^2}\right] \frac{x^j z^j}{(\sigma^2)^j} \\ &= \exp\left[-\frac{(x^2+z^2)}{2\sigma^2}\right] \sum_{j=0}^{\infty} \frac{(xz)^j}{(\sigma^2)^j j!} \\ &= \exp\left[-\frac{(x^2+z^2)}{2\sigma^2}\right] \exp\left[\frac{xz}{\sigma^2}\right] \\ &= \exp\left[\frac{(x-z)^2}{2\sigma^2}\right] = k(x, z) \end{aligned}$$

where the step from the second to third line follows from the characterisation of the exponential function as a Maclaurin series.

Interestingly, the finite dimensional cost function (11.31) with solution (11.33) can be used despite the infinite dimensionality of the inner product space associated with the Gaussian kernel. Positive definite kernel functions give rise to implicitly infinite dimensional feature spaces.

NOTE

The term $2\sigma^2$ in the Laplacian basis kernel and in the Gaussian kernel is the bandwidth parameter and can be replaced by the customary notation h.

Kernel Matrices as Similarity Matrices

Kernel matrices are also known as *similarity matrices* or *covariance functions*. For instance, consider two p dimensional vectors x_i and x_j, where p could represent the number of features, while i and j could represent two individuals. The Gaussian kernel function is

$$k\left(x_i, x_j\right) = \exp\left(-\frac{\left\|x_i - x_j\right\|^2}{\sigma^2}\right),$$

where

$$\left\|x_i - x_j\right\| = \left[\left(x_{i1} - x_{j1}\right)^2 + \left(x_{i2} - x_{j2}\right)^2 + \cdots + \left(x_{ip} - x_{jp}\right)^2\right]^{\frac{1}{2}} \tag{11.40}$$

is the Euclidean distance and σ^2 (or written also as h) is the bandwidth of the kernel. The kernel function $k\left(x_i, x_j\right)$ evaluates to 1 if x_i and x_j are identical and approaches 0 as x_i and x_j become increasingly different. In a genomics context when x represents SNPs, the elements of the kernel matrix are larger for pairs of individuals that are more genetically alike.

The bandwidth parameter controls how fast the Euclidean distance falls with increasing difference between features x_i and x_j. The choice of the bandwidth parameter has important consequences on inference and prediction. Large values of σ^2 lead to entries of the Gaussian kernel approaching 1 resulting in underfitting. Small values of σ^2 generate entries approaching 0. In this case, in a model including a random effect and a residual term, the random effect associated with the kernel matrix becomes confounded with the residual term. The bandwidth parameter can be chosen using cross-validation.

A few remarks are in order.

- Kernel matrices are dense and in large datasets can be computationally challenging to construct.
- Concerning the influence of the bandwidth parameter on predictions based on (11.30): If the elements of K are very small, the predicted values are all approximately equal to $\widehat{\mu}$. Such a kernel model has little or no predictive power.
- On the other hand, if all elements of K are very close to 1, the predicted values are very similar and again, predictive power is compromised.
- Kernels can be used for many types of inputs such as strings, discrete structures, images, time series and more, provided the kernel matrices corresponding to any finite training set are positive semidefinite. They are an extremely flexible tool.
- A new kernel can be generated from existing kernels by addition, element-wise multiplication, or by multiplication by a positive scalar. The requirement is that these operations result in a symmetric, positive semidefinite matrix. This property provides great flexibility. In a genetics context, a new kernel can be constructed from the sum of existing covariance matrices derived from specific parametric

assumptions. The resulting kernel can be used to exploit potential contributions from, for example, non-additive genetic sources of variation or from SNP × environment interactions, despite the fact that the nonparametric model is not explicitly built upon specific mechanistic considerations (Gianola and de los Campos 2008; de los Campos et al 2009).

- The bandwidth parameter of a Gaussian kernel can be selected by cross-validation or can be inferred as in Perez-Elizalde et al (2015).
- As illustrated on page 476, kernels can be viewed as prior information describing covariance structures in the data. The choice of kernel should reflect prior knowledge, but this may be difficult to accomplish. One may want to consider instead a family of kernels weighted by parameters inferred from the data (e.g. de los Campos et al 2010), or attempt to infer the kernel from the data (e.g. Duvenaud et al 2013 and Gianola et al 2020).
- As is clear from (11.40) diagonal elements of the Euclidean distance can become very large when the number of features p is large. In such a situation, the Gaussian kernel matrix is diagonally dominant and the model may predict poorly. Typically, this calls for some form of standardisation at the level of the inputs and perhaps at the level of the Euclidean distance leading to different kernel matrices. This may impinge on the predictive capacity of the model and can be gauged using cross-validation
- Kernel methods are essentially prediction tools. The kernel function expands the original feature space into a new, high-dimensional feature space via inner products of basis functions that are not explicitly constructed. It is therefore not straightforward to measure the importance of a specific feature in the new feature space, contrary to parametrised regression models. However, there are kernels that are constructed in a manner that resemble known parameters of fully parametric models, and this provides a framework for interpretation. An example on page 476 is the linear kernel that takes the form of a genomic relationship matrix, and the smoothing or regularisation parameter is interpreted as a ratio of variance components.

It becomes clear that the linear kernel can be viewed as a similarity matrix if it is transformed into a correlation matrix where the diagonal elements, equal to 1, describe the similarity between the individual with itself.

A genomic relationship matrix is also a linear kernel that can be scaled in a variety of ways. A standard approach is to transform the marker genotype x_{ij} into $w_{ij} \sim (0, 1)$, where i refers to individual (row of X) and j to marker genotype (column of X). The genomic variance under such a model is $p\sigma_b^2$ (this is the marginal variance of $w_i'b$, where vector $b \in \mathbb{R}^p$ has *iid* elements representing p marker effects distributed as $N\left(0, I\sigma_b^2\right)$).

Another approach is to centre x_{ij} and divide by $\left(\sum_{j=1}^{p} \text{Var}\left(x_j\right)\right)^{\frac{1}{2}}$, where $\text{Var}(x_j)$ is the variance of marker j, equal to $2p_j(1 - p_j)$. In practice the variance of marker j is replaced by the empirical variance of the elements of column j of matrix X. With this scaling and centring of x_{ij}, the genomic variance is equal to σ_b^2, independent of the number of markers.

Reducing Infinitely Dimensional Problems to Finite Dimensional Problems

This short tour of kernel methods using basis expansions took ridge regression as starting point from which a number of results were derived. The model is specified in (11.21) with solution (11.23). The dual representation allows to express the solution and predictions in terms of inner products involving the original feature data, as indicated in (11.26) and (11.30). Kernel functions, with associated kernel matrices, are defined as inner products of basis functions. The basis functions expand the original feature space. In this way, one can explore the effects of nonlinear terms of the original feature space, on the observations. Kernel methodology creates a very flexible setup, because the same algorithm to solve the linear system can be applied with different kernel matrices.

A more rigorous approach to arrive at the same result invokes the *representer theorem* (Kimeldorf and Wahba 1971). Consider a training dataset $S = \{(x_1, y_1), \ldots, (x_n y_n)\}$ where $x_i \in \mathbb{R}^p$ are p-dimensional feature vectors, $y_i \in \mathbb{R}$ are responses. Assume a regression function $m(x_i) = \mu + g(x_i)$, where, for example, $g(x_i) = x_i'b$ as in the linear model, but could also accommodate other nonlinear structures. A flexible general formulation for estimating m when it belongs to some possibly infinite dimensional Hilbert space $\mathcal{H}$ of functions is to minimise the cost function

$$J = \min_{m \in \mathcal{H}} \sum_{i=1}^{n} \ell(y_i, m(x_i)) + \lambda \|m\|_{\mathcal{H}}^2 \tag{11.41}$$

where ℓ is some function measuring goodness of fit to data such as $\ell(y, m) = (y - m)^2$, a quadratic function, although other functions (such as the negative of the log-likelihood) can be used, λ is the smoothing or regularisation parameter and the last term $\|m\|_{\mathcal{H}}^2$ is a norm in the space of functions $\mathcal{H}$ generated by a positive definite kernel $K(x_i, x_j)$. Solving for m directly from (11.41) involves an optimisation in a possibly infinite dimension; this is a difficult task.

The representer theorem finds a solution of an equivalent problem in a finite dimension of the form

$$g(x) = \sum_{i=1}^{n} \alpha_i k(x_i, x), \tag{11.42}$$

that requires finding the values of n scalar coefficients α_i (and of the intercept μ). Writing

$$\begin{aligned} m(x_i) &= \mu + g(x_i) \\ &= \mu + \sum_{i=1}^{n} \alpha_i k(x_i, x), \end{aligned} \tag{11.43}$$

this translates into solving

$$\min_{\mu,\alpha} \sum_{i=1}^{n} \ell(y_i, \mu + g(x_i)) + \lambda \|g\|^2 =$$

$$\min_{\mu,\alpha} \sum_{i=1}^{n} \ell\left(y_i, \mu + \sum_{j=1}^{n} \alpha_i k(x_i, x_j)\right) + \lambda \sum_{i=1}^{n}\sum_{j=1}^{n} \alpha_i \alpha_j k(x_i, x_j). \tag{11.44}$$

For instance, in the case of ridge regression, the cost function is

$$J = \sum_{i=1}^{n} (y_i - m(x_i))^2 + \lambda \|g\|^2.$$

Plugging (11.43) into J and writing $\|g\|^2 = \alpha' K \alpha$ yield the finite dimensional optimisation problem of the form

$$J = \|y - 1\mu - K\alpha\|^2 + \lambda \alpha' K \alpha,$$

as in (11.31), where $[K]_{i,j} = k(x_i, x_j)$ is an $n \times n$ Gram matrix. The minimiser over α is (11.33) and the fitted values are given by (11.34); these are obtained by direct application of the representer theorem, without the need for a dual form.

A couple of examples follow, starting with a toy example that highlights details of some of the matrix operations and illustrates how the Gaussian kernel regression can cope with strong nonlinear relationships between the feature x and the output y.

Example: Classical, Kernelised, and Gaussian Kernel Ridge Regression

The R-code below generates a scalar covariate x and a vector of responses y of dimension $N = 100$, nonlinearly associated with x. In a first stage, it performs an analysis using standard ridge scalar regression solving the linear system (11.23). Model (11.21) includes an intercept μ and a regression parameter b, so the left-hand side of the system (11.23) is of dimension 2×2. The fitted values are obtained from

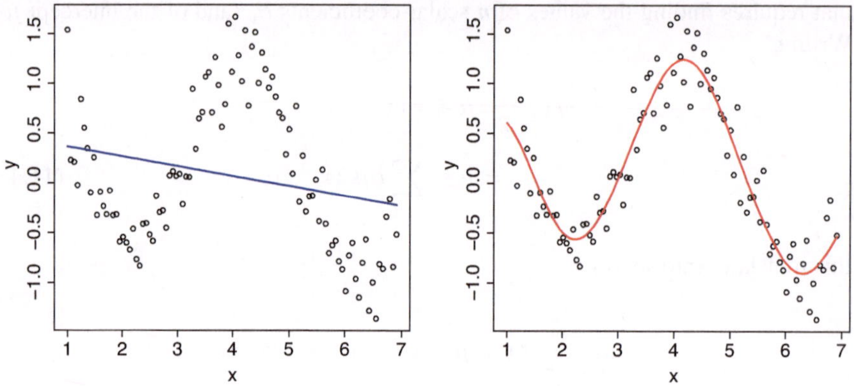

Fig. 11.6 Left: fitted values using a classical linear ridge regression. Right: fitted values using a Gaussian kernel

$X\hat{b}$, where $\hat{b}$ is the solution to (11.23). The estimate of the regression parameter is $\hat{b} = 0.399$ and the code prints the first 10 fitted values. Figure 11.6, left panel, displays the output y as a function of the feature vector x and the fitted values.

```
# CODE1107
rm(list=ls()) # CLEAR WORKSPACE
set.seed(195021)
library(MASS)
N <- 100
# GENERATE DATA
x<-seq(from=1, to=2.2*pi,length=N)
signal <- cos(1.5*x)+ exp(-0.4*x)
noise <- rnorm(N,0,0.25)
y <- signal + noise
lambda <- 0.01
X <- cbind(1,x)
RHS <- crossprod(X,y)
LHS <- crossprod(X)
LHS[-1,-1] <- LHS[-1,-1]+diag(c(rep(1,1)))*lambda # assumes
#           one covariate x
#### Classical LS solution
bh1 <- solve(LHS,RHS)
bh1
```

```
##          [,1]
##    0.46692087
## x -0.09918769
```

```
yhatclassic <- X%*%bh1
yhatclassic[1:5]
```

```
## [1] 0.3677332 0.3618105 0.3558878 0.3499650 0.3440423
```

```
yhatclassic[6:10]
```

```
## [1] 0.3381196 0.3321969 0.3262742 0.3203515 0.3144288
```

The R-code below fits the same model in dual form with $K = XX'$. The left-hand side of the system (11.32) is now 101×101, and the code prints out the first 10 fitted values, which are identical to those generated with the classical representation. The estimates of the intercept and of the regression parameter are the same as in the classical parametrisation.

```
# CODE1107 (cont)
#### kernelised (dual) solution with kernel matrix XX'
K <- x%*%t(x) # LINEAR KERNEL
X <- cbind(1,K)
RHS <- crossprod(X,y)
LHS <- crossprod(X)
LHS[-1,-1] <- LHS[-1,-1]+K*lambda
diag(LHS) <- diag(LHS) + c(0,rep(1e-8,N))
sol <- as.matrix(solve(LHS,RHS))
yhatkernellin <- sol[1]+K%*%sol[-1]
yhatkernellin[1:5]
```

```
## [1] 0.3677332 0.3618105 0.3558878 0.3499650 0.3440423
```

```
yhatkernellin[6:10]
```

```
## [1] 0.3381196 0.3321969 0.3262742 0.3203515 0.3144288
```

```
alfa <- sol[-1]
bhkernellin <- sum(alfa*x)
bhkernellin
```

```
## [1] -0.09918769
```

```
muhatkernellin <- sol[1]
muhatkernellin
```

```
## [1] 0.4669209
```

Finally, I fit the Gaussian kernel regression using the R-code below. The same system of equations (11.32) is solved, where now K represents the Gaussian kernel with $\sigma^2 = h = 0.7$. The fitted values are displayed in Fig. 11.6, right panel. The nonlinear relationship between the covariate and the data is well captured by the Gaussian kernel.

```
# CODE1107(cont)
#######  GAUSSIAN KERNEL SOLUTION #####################
#      CONSTRUCT GAUSSIAN KERNEL
#w <- matrix(data=NA,nrow=length(x),ncol=length(x))
K <- matrix(data=NA,nrow=length(x),ncol=length(x))

d <- as.matrix(dist(x))^2
# CHOOSE h and lambda
h <- 0.7
lambda <- 0.5

K <- exp(-(1/(2*h^2))*d)

X <- cbind(1,K)
RHS <- crossprod(X,y)
LHS <- crossprod(X)
LHS[-1,-1] <- LHS[-1,-1]+K*lambda

diag(LHS) <- diag(LHS)+c(0,rep(1e-8,N))

sol <- solve(LHS,RHS)
fgaus <- sol[1]+K%*%sol[-1]
alfa <- sol[-1]
```

Bayesian View of Kernelised Regression

Consider the derivation of BLUP (best linear unbiased predictor) that can be found in Henderson et al (1959), based on finding the maximiser of the joint density of y (vector of dimension $n \times 1$) and a (vector of dimension $q \times 1$) with respect to b (vector of dimension $p \times 1$) and a, under the assumption of normality. When vector a represents additive genetic values, this probability model is known as the additive genetic model. With known covariances matrices R and G, the logarithm of the joint distribution takes the form (ignoring an additive constant and multiplying by -1)

$$\log\left[p\left(y, a|b, R, G\right)\right] = \log\left[p\left(y|a, b, R\right)\right] + \log\left[p\left(a|G\right)\right]$$
$$\propto \left(y - Xb - Za\right)' R^{-1} \left(y - Xb - Za\right) + a' G^{-1} a. \tag{11.45}$$

Differentiating with respect to b and a leads to the linear system, which arranged in the usual mixed model format is

$$\begin{bmatrix} X'R^{-1}X & X'R^{-1}Z \\ Z'R^{-1}X & Z'R^{-1}X + G^{-1} \end{bmatrix} \begin{bmatrix} \widehat{b} \\ \widehat{a} \end{bmatrix} = \begin{bmatrix} X'R^{-1}y \\ Z'R^{-1}y \end{bmatrix}. \tag{11.46}$$

An alternative interpretation based on a Bayesian argument is to write the joint posterior density $p(b, a|y, G, R)$ as

$$\begin{aligned} p(b, a|y, G, R) &\propto p(y|a, b, R)\, p(a, b|G) \\ &= p(y|a, b, R)\, p(a|G) \end{aligned} \tag{11.47}$$

assuming $p(a, b|G) \propto p(a|G)$. Then $\left[\hat{b}, \hat{a}\right]$ is the mode (and the mean under normality) of the posterior distribution $[b, a|y, G, R]$.

This same structure (11.45) can be applied to the kernel function (replacing vector b with scalar μ)

$$\frac{1}{\sigma_e^2}(y - 1\mu - K\alpha)'(y - 1\mu - K\alpha) + \frac{1}{\sigma_a^2}\alpha' K\alpha \tag{11.48}$$

which implies $\left[y|\mu, \alpha, \sigma_e^2\right] \sim N\left(1\mu + K\alpha, I\sigma_e^2\right)$ and $\left[\alpha|K, \sigma_\alpha^2\right] \sim N\left(0, K^{-1}\sigma_\alpha^2\right)$, where σ_α^2 is the variance captured by the specific kernel adopted. Differentiation with respect to μ and α leads to the linear system

$$\begin{bmatrix} 1'1 & 1'K \\ K'1 & K'K + \lambda K \end{bmatrix} \begin{bmatrix} \hat{\mu} \\ \hat{\alpha} \end{bmatrix} = \begin{bmatrix} 1'y \\ K'y \end{bmatrix}, \tag{11.49}$$

where $\lambda = \sigma_e^2/\sigma_\alpha^2$. This system is identical to (11.32). One can argue that this interpretation brings us back to the parametric setup and offers a framework to infer the regularisation parameter λ from data, using likelihood or Bayesian methods. Parameter λ is expressed as a ratio of the two variance components, i.e., $\lambda = \sigma_e^2/\sigma_\alpha^2$.

Genetic Models Using Kernelised Regressions

The classical additive genetic model assumes that, in (11.45), $G = A\sigma_a^2$ where A is a positive definite $q \times q$ matrix of additive genetic relationships among additive genetic values of q individuals, given a pedigree, and the scalar σ_a^2 is the additive genetic variance of some trait.

In the kernel model (11.48), consider the transformation $g = K\alpha$. Since α is normal with mean zero and variance $K^{-1}\sigma_\alpha^2$, it follows that $g \sim N\left(0, K\sigma_\alpha^2\right)$. The cost function to be minimised along the same lines as in (11.48) is now

$$\frac{1}{\sigma_e^2}(y - 1\mu - g)'(y - 1\mu - g) + \frac{1}{\sigma_\alpha^2}g'K^{-1}g. \tag{11.50}$$

Differentiation with respect to μ and g leads to the linear system

$$\begin{bmatrix} 1'1 & 1' \\ 1 & I + \lambda K^{-1} \end{bmatrix}\begin{bmatrix} \widehat{\mu} \\ \widehat{g} \end{bmatrix} = \begin{bmatrix} 1'y \\ y \end{bmatrix} \tag{11.51}$$

where $\lambda = \sigma_e^2/\sigma_\alpha^2$ is the regularisation parameter. If $K = A$ and $\sigma_\alpha^2 = \sigma_a^2$, the linear system (11.51) becomes identical to that of the classical (infinitesimal) additive genetic model using the numerator relationship matrix as kernel (de los Campos et al 2009). In this case the random variable g represents additive genetic values. Of course the solution for $\widehat{g}$ in (11.51) is identical to the solution for $K\widehat{\alpha}$ in (11.49). To see this, write the second equation in (11.49) as

$$K'(K + \lambda I)\widehat{\alpha} = K'(y - 1\widehat{\mu}).$$

Multiplying both sides by K^{-1} and using $K' = K$ due to symmetry, gives

$$(K + \lambda I)\widehat{\alpha} = (y - 1\widehat{\mu}). \tag{11.52}$$

Similarly from (11.51),

$$\left(I + \lambda K^{-1}\right)\widehat{g} = (y - 1\widehat{\mu})$$

and replacing $\widehat{g}$ by $K\widehat{\alpha}$ recovers (11.52). A similar result applies to the equations for $\widehat{\mu}$.

This development was based on the classical infinitesimal model; a genomic model incorporating a large number of markers leads to the same results. In this case, g are genomic values, σ_a^2 is interpreted as the genomic variance and K as a genomic relationship matrix.

Example: A Bayesian Kernelised Regression

A Bayesian implementation of a RKHS (reproducing kernel Hilbert space) model is applied to the wheat dataset downloaded from BGLR (Perez and de los Campos 2014) consisting of grain yields from $n = 599$ wheat inbred lines genotyped for 1279 genetic markers. A Gaussian kernel with three values of the bandwidth parameter h is used to illustrate the effect on inferences.

The Bayesian RKHS model assumes the following hierarchical structure:

$$y|\mu, g, \sigma_e^2 \sim N\left(1\mu + g, I\sigma_e^2\right),$$

with

$$g|K_h, \sigma_g^2 \sim N\left(0, K_h\sigma_g^2\right),$$

and improper uniform prior distributions for μ, σ_e^2 and σ_g^2. The positive definite $n \times n$ matrix K_h is the Gaussian kernel with bandwidth parameter h.

In order to improve computational efficiency, K is expressed as

$$K = UDU', \tag{11.53}$$

(de los Campos et al 2010) using the eigenvalue decomposition shown on page 91. Details of the Gibbs sampling implementation can be found on page 227 (here, K is positive definite, whereas in the example of page 227, G is singular; this requires a very slight modification of the McMC implementation). In (11.53) U is the $n \times n$ orthogonal matrix of eigenvectors of K and D is an $n \times n$ diagonal matrix with eigenvalues $d_i > 0$, for all i.

A chain of length 25,000 was run and the Monte Carlo estimates of posterior means of σ_g^2, σ_e^2 and $\lambda = \sigma_e^2/\sigma_g^2$ for $h = 0.5$, $h = 1$ and $h = 3$ are shown in Table 11.1.

Inferences about the components of variance are sensitive to the chosen value of the bandwidth parameter h, but the regularisation parameter λ is less affected. The impact on prediction varying h for constant λ is illustrated in an example on page 509.

The R-code to execute the Gibbs sampler is shown below. The lines at the bottom compute the Monte Carlo (sampling) error of features of the posterior samples based on the method of batching.

Table 11.1 McMC-based Bayesian inferences of the components of variance conditional on three values of the bandwidth parameters h for grain yield data of 599 wheat inbred lines from the BGLR package, using a Gaussian kernel RKHS regression. The regularisation parameter λ is the ratio of the variance components σ_e^2 and σ_g^2. The elements in brackets in the fourth column show the 95% posterior intervals for the MC estimates of the posterior distribution of the regularisation parameter λ. The Monte Carlo standard error of the posterior mean of the $\lambda's$ is approximately 0.002

h	σ_g^2	σ_e^2	λ
0.5	1.20	0.33	0.28 (0.17;0.45)
1.0	0.83	0.27	0.33 (0.20;0.52)
3.0	0.72	0.20	0.28 (0.15;0.46)

```
# CODE1108
# BAYESIAN KERNELISED REGRESSION
# WHEAT DATA FROM BGLR


# CODE ASSUMES K IS OF FULL RANK
# THEREFORE IT DOES NOT WORK WITH CENTRED G.
# IT WORKS WITH GAUSSIAN KERNEL


rm(list=ls()) # CLEAR WORKSPACE
set.seed(37111)
library(BGLR)
data(wheat)
### USE BGLR MATRIX X
X <- wheat.X
y<- wheat.Y[,1]
nindiv<-length(y)
nmark<-ncol(X)

#### A GAUSSIAN KERNEL ################

kgaus <- function(X,h){
  X <- scale(X,center=TRUE,scale=FALSE)
  S=sqrt(sum(apply(FUN=var, X=X,MARGIN=2)))
  X <- X/S
  D <- as.matrix(dist(X))^2
  K <- exp(-h*D)
}
############  CHOOSE GAUSSIAN KERNEL ##############
#h <- 0.5
h <- 1
#h <- 3.0

K <- kgaus(X,h)
dim(K)
qr(K)$rank
G<-K
##################################################
# EIGEN DECOMPOSITION OF G
EVD <- eigen(G)
names(EVD)
head(EVD$values)
U <- EVD$vector
```

```
tU<-t(U)
val <- EVD$values
summary(val)
D <- diag(val,nrow=nindiv)
#Dp IS A VECTOR WITH NON-ZERO EIGENVALUES
Dp<-c(val[1:nindiv])
#INITIALISE Ve
Ve<-0.5
#INITIALISE Vg
Vg<-0.5
#INITIALISE k
k<-Ve/Vg
#INITIALISE VECTOR ALFA
alfa<-rep(0,nindiv)
# CHOOSE LENGTH OF GIBBS CHAIN
rep<-25000
#INITIALISE result
result<-matrix(data=NA,nrow=rep,ncol=5)
# START GIBBS CHAIN
ptm <- proc.time()
for (i in 1:rep)
{
  cat(i, "\n",sep="")
  # SAMPLE mu
  avmu<-sum(y-U%*%alfa)/nindiv
  varmu<-Ve/nindiv
  mu<-rnorm(1,mean=avmu,sd=sqrt(varmu))
  # SAMPLE alfa1 (VECTOR OF LENGTH nindiv)
  meanalfa1<-(Dp/(Dp+k))*tU%*%(y-mu)
  varalfa1<-((Dp)/(Dp+k))*Ve
  alfa1<-rnorm(nindiv,meanalfa1,sqrt(varalfa1))
  alfa<-alfa1
  # SAMPLE Vg
  # COMPUTE SCALE
  scVg<-sum(alfa1*alfa1*(1/Dp))
  Vg<-scVg/rchisq(1,nindiv-2)
  #Vg<-0.0001
  # SAMPLE Ve
  # COMPUTE SCALE
  ystar<-y-mu-U%*%alfa
  scVe<-sum(ystar*ystar)
  Ve<-scVe/rchisq(1,nindiv-2)
  k<-Ve/Vg
  ualfa <- U%*%alfa
```

```
  result[i,]<-c(i,mu,Vg,Ve,k)
#  print(result[i,])
}
proc.time()-ptm
apply(result[2000:rep,2:5],2,mean)
##############################################
#CODE FOR THE MC VARIANCE BASED ON BATCHING
y <- result[,5] # READS IN ALL DRAWS STORED IN RESULT
#choose number of batches b
b<-500
batch_size <- length(y)/b
batch_size
x<-matrix(y,ncol=b, byrow=FALSE)
avrb<-apply(x,2,mean)
mcvarb<-var(avrb)/length(avrb)
sqrt(mcvarb)
efchsizebatch<-var(y)/mcvarb
efchsizebatch
#############################################
### PLOT AUTOCORRELATION VERSUS LAG USING
### R-FUNCTION acf
require(graphics)
acf(y) ## AUTOCORRELATION OF McMC DRAWS
acf(avrb) ## AUTOCORRELATION OF BATCH MEANS
```

Kernel Logistic Regression

Section 9.4 introduced penalised logistic regression, where regression coefficients are shrunk as in ridge regression allowing the analysis when $p > n$, where p is the number of features and n is the length of the vector of binary records y. Kernel methods also make use of a penalisation and work instead in an n-dimensional space that can be an advantage when $p \gg n$. Additionally, kernel methods extend the feature space exploring complicated associations among the original features without explicit modelling. Here I describe an application to logistic regression. The starting point is to use $f(x_i) = \mu + x_i'\beta$ in the negative of the log-likelihood (9.12)

$$-\ell(\mu, \beta|y, x) = -\sum_{i=1}^{n} \left\{ y_i \left(f(x_i) \right) - \ln \left[1 + \exp \left(f(x_i) \right) \right] \right\},$$

and use the representer theorem result (11.43) replacing $f(x_i)$ by

$$f(x_i) = \mu + \sum_{j=1}^{n} \alpha_j k(x_i, x_j). \tag{11.54}$$

The next step is to minimise the convex cost function with respect to μ and α_i, $i = 1, 2, \ldots, n$

$$J(\mu, \alpha|\lambda) = -\sum_{i=1}^{n} \{y_i (f(x_i)) - \ln[1 + \exp(f(x_i))]\} + \frac{\lambda}{2} \sum_{i=1}^{n} \sum_{j=1}^{n} \alpha_i \alpha_j k(x_i, x_j). \tag{11.55}$$

Using similar algebra as in Sect. 9.4, the expressions for the first derivatives are

$$\frac{\partial J}{\partial \mu} = -1'(y - \pi), \tag{11.56a}$$

$$\frac{\partial J}{\partial \alpha} = -K(y - \pi) + \lambda K \alpha, \tag{11.56b}$$

where $1'$ is a row vector of n ones, $y = \{y_i\}$ is a column vector with the n responses y_i, $i = 1, \ldots, n$, $\pi = \{\pi_i\}$ is a column vector with the n terms having the form

$$\pi_i = \Pr(y_i = 1|x, \mu, \alpha) = \frac{\exp(f(x_i))}{1 + \exp(f(x_i))},$$

and $K = \{k(x_i, x_j)\}$ is an $n \times n$ Gram matrix whose ijth element is the kernel function $k(x_i, x_j)$. The second derivatives are

$$\frac{\partial^2 J}{(\partial \mu)^2} = \sum_{i=1}^{n} \pi_i (1 - \pi_i) = 1'W1,$$

$$\frac{\partial^2 J}{\partial \alpha \partial \alpha'} = KWK + \lambda K,$$

$$\frac{\partial^2 J}{\partial \alpha \partial \mu} = KW1.$$

The matrix of second derivatives can be written as

$$H = \begin{bmatrix} 1'W1 & 1'WK \\ KW1 & KWK + \lambda K \end{bmatrix},$$

where $W = diag\{\pi_i(1-\pi_i)\}, i = 1, 2, \ldots, n$, an $n \times n$ diagonal matrix. Let

$$\begin{aligned} Z &= [1, K], \text{ of dimension } n \times (n+1), \\ M &= \begin{bmatrix} 0 & 0 \\ 0 & K \end{bmatrix}, \text{ of dimension } (n+1) \times (n+1), \\ \Lambda &= diag\{0, \lambda, \lambda, \ldots, \lambda\}, \text{ of dimension } (n+1) \times (n+1), \\ \theta &= (\mu, \alpha')', \text{ of dimension } (n+1) \times 1. \end{aligned}$$

The matrix of second derivatives can be compactly expressed as

$$H = Z'WZ + \Lambda M$$

and the vector of first derivatives as

$$S(\theta) = -Z'(y-\pi) + \Lambda M\theta.$$

The iterative system of the Newton-Raphson algorithm is

$$\begin{aligned} \theta_{t+1} &= \theta_t - HS(\theta_t) \\ &= \theta_t - \left(Z'WZ + \Lambda M\right)^{-1}\left(-Z'(y-\pi) + \Lambda M\theta_t\right) \\ &= H^{-1}\left(Z'WZ + \Lambda M\right)\theta_t + H^{-1}\left(Z'WW^{-1}(y-\pi) - \Lambda M\theta_t\right) \\ &= H^{-1}Z'W(Z\theta_t) + H^{-1}\Lambda M\theta_t + H^{-1}Z'W\left(W^{-1}(y-\pi)\right) - H^{-1}\Lambda M\theta_t \\ &= H^{-1}Z'W\left(Z\theta_t + W^{-1}(y-\pi)\right) \\ &= H^{-1}Z'Wr, \end{aligned} \tag{11.57}$$

where $H^{-1} = \left(Z'WZ + \Lambda M\right)^{-1}$ and $r = Z\theta_t + W^{-1}(y-\pi)$. This implementation of Newton-Raphson is the *iteratively reweighted ridge regression* algorithm that has a similar structure as (9.17) and at face value requires inversion of the $(n+1) \times (n+1)$ matrix of second derivatives.

In connection with the penalised logistic regression model described on page 377, it was indicated that use of the linear kernel $K = XX'$ generates a reparametrised penalised logistic regression model that operates in the $n+1$-dimensional space. This kernel leads to a dual representation of the original logistic regression model (9.11), which operates in the $p+1$-dimensional space. To see this multiply the second line of (9.17) by $H = \left(Z'WZ + \Lambda\right)$

$$\left(Z'WZ + \Lambda\right)\theta_{t+1} = \left(Z'WZ + \Lambda\right)\theta_t - \left(-Z'(y-\pi) + \Lambda\theta_t\right), \tag{11.58}$$

where $Z = [1, X]$ and $\theta' = \left[\mu, \beta'\right]'$ as in the penalised logistic regression of page 377. This system can be written as

$$\begin{bmatrix} 1'W1 & 1'WX \\ X'W1 & X'WX + I\lambda \end{bmatrix} \begin{bmatrix} \mu_{t+1} \\ \beta_{t+1} \end{bmatrix} = \begin{bmatrix} 1'W1 & 1'WX \\ X'W1 & X'WX + I\lambda \end{bmatrix} \begin{bmatrix} \mu_t \\ \beta_t \end{bmatrix} + \begin{bmatrix} 1'(y-\pi) \\ X'(y-\pi) - \lambda\beta_t \end{bmatrix}.$$

Expanding yields

$$\left(1'W1\right)\mu_{t+1} + \left(1'WX\right)\beta_{t+1} = \left(1'W1\right)\mu_t + \left(1'WX\right)\beta_t + 1'(y-\pi), \tag{11.59}$$

$$\begin{aligned}\left(X'W1\right)\mu_{t+1} + \left(X'WX + I\lambda\right)\beta_{t+1} &= \left(X'W1\right)\mu_t + \left(X'WX + I\lambda\right)\beta_t \\ &\quad + X'(y-\pi) - \lambda\beta_t. \end{aligned} \tag{11.60}$$

Now perform the following:

- Replace $\beta_t = X'\alpha_t$
- Premultiply (11.60) by X and let $K = XX'$

This results in the iterative system

$$\begin{bmatrix} \mu_{t+1} \\ \alpha_{t+1} \end{bmatrix} = \begin{bmatrix} \mu_t \\ \alpha_t \end{bmatrix} + \begin{bmatrix} 1'W1 & 1'WK \\ KW1 & KWK + \lambda K \end{bmatrix}^{-1} \begin{bmatrix} 1'(y-\pi) \\ K(y-\pi) - \lambda K\alpha_t \end{bmatrix}, \tag{11.61}$$

the same as the second line of (11.57). This is a dual representation of the penalised logistic regression model of Sect. 9.4.

Example: Analysis of Binary Observations Using Kernelised Logistic Regression, Penalised Logistic Regression, and Logistic Lasso

The performance of the kernelised logistic regression (KLR) using a Gaussian kernel, the penalised logistic regression (PLR) and logistic lasso (LL) is compared for a particular simulated dataset consisting of 599 binary records. The binary records were simulated using a logistic model. Genetic marker data from 599 wheat inbred lines genotyped for 1279 genetic markers were downloaded from the statistical package BGLR described in Perez and de los Campos (2014). The simulated "true" model on an underlying scale involves 20 loci randomly sampled from these 1279 markers and assigned as QTL. The substitution effects of these QTL were chosen so that the additive genetic variance between lines σ_a^2 on the underlying scale was equal to 1 squared unit. The underlying liability for line i has the linear structure

$$u_i = m + z_i'b + e_i, \quad i = 1, 2, \ldots, 599,$$

where $m = 0.0$, z_i is the column vector for line i of the 20 scaled and standardised QTL genotypes and b is the column vector of the 20 substitution effects. Vector z_i' of dimension 1×20 has (0, 1) elements for each of the two homozygote genotypes of line i and the $e_i's$ are independently distributed standard logistic random variables. The heritability h^2 between lines on the underlying scale $(\sigma_a^2/\left(\sigma_a^2 + \pi^2/3\right))$ is equal to 0.23 and the proportion of 1's in the data is approximately 0.5.

The three operational models used to analyse the data assumed that the liability can be written in terms of the linear structure

$$u_i = \mu + x_i'\beta + \varepsilon_i, \quad i = 1, 2, \ldots, 1, 279,$$

where x_i' is the 1×1279 row vector for line i of the 1279 marker genotypes, β is the vector with the 1279 marker effects and the $\varepsilon_i's$ are independently distributed standard logistic random variables.

The 599 lines were divided into training and validating sets of sizes 299 and 300, respectively. The logistic regression model using the 1279 markers was fitted to the training data using KLR, PLR and LL, and the estimates of the marker effects were used to predict the binary phenotypes of the validating dataset. The package **glmnet** described in Example 7.4 was used to fit the logistic lasso.

The criterion used to evaluate prediction ability was the proportion of misclassifications in the validating data or error rate, given by

$$\frac{1}{N_v}\sum_{i=1}^{N_v}(y_i - \widehat{y}_i)^2 \tag{11.62}$$

where $N_v = 300$ is the number of records in the validating set, y_i is the ith record in the validating set (0 or 1) and $\widehat{y}_i$ is the predicted value computed using (9.3) with $t = 0.5$.

The average proportion of misclassifications (min, max) over 10 random replicates of training/testing samples was 0.33 (0.30, 0.37) for LL, 0.40 (0.38, 0.43) for PLR and 0.33 (0.28, 0.37) for KLR. In principle, LL has the advantage of performing model selection. However, in this particular dataset, the number of covariates not set to zero was highly variable across replicates ranging from 1 to 52 with an average of 14. Therefore, little inferential meaning can be assigned to the choice of covariate.

The R-code to fit the KLR to the wheat dataset using the Newton-Raphson algorithm is shown below:

```
# CODE1109
# SIMULATING BINARY DATA TO BE ANALYSED WITH KLR
#  USES THE X MATRIX FROM WHEAT IN BGLR
rm(list=ls()) # CLEAR WORKSPACE
library(BGLR)
data(wheat)
X <- wheat.X
set.seed(371)
#####################################################
### USE BGLR MATRIX X
nindiv <-nrow(X)
nmark <- ncol(X)
####################################################
nloci<-20
p<-0.5
mu<-log(p/(1-p))

##### GENERATE LIABILITY ###################
va<-1.0 # additive variance of liability
Xc<-matrix(data=NA,nrow=nindiv,ncol= nmark)
# parameter from true model:
be<-matrix(data=0.0,nrow=nmark,ncol=1)
y<-rep(0,nindiv)
cm<-colMeans(X)
### CENTER AND SCALE X #################
for (i in 1:nmark)
{Xc[,i]<- (X[,i]-cm[i])/sd(X[,i])
}
QTLeff<-sqrt(va/nloci)# calculate the QTL effect so that the
# total genetic variance is VA
IDq<-sample(1:nmark,nloci,replace=F) # from the nmark markers,
# choose nloci as QTL
be[IDq]<-QTLeff # the only b's that are not zero are those
# associated with QTL.
########### GENERATE PHENOTYPIC BINARY DATA ###################
xb<-Xc%*%be
pr <- exp(mu+xb)/(1+exp(mu+xb))
y <- rbinom(nindiv,1,pr)
#sum(y)/length(y) # OBSERVED PROPORTION OF 1'S IN SAMPLE
mean(y)
```

```
## [1] 0.5025042
```

```
nitnr <- 10 # NUMBER OF N-R ITERATIONS
nrep <- 10 # NUMBER OF TRAINING / TESTING REPLICATES

#lambda <- 0.0 # ZERO PENALTY !!!!!!!
lambda <- 0.4

newcostvnr <- rep(0,nrep)
res <- matrix(data=NA, nrow=nrep,ncol=9)
resulttvnr <- matrix(data=NA, nrow=nitnr,ncol=8)

msev <- rep(NA,nrep)
msevnr <- rep(NA,nrep)
```

```
#### A GAUSSIAN KERNEL ################
kgaus <- function(X,h){
 X <- scale(X,center=TRUE,scale=FALSE)
 S=sqrt(sum(apply(FUN=var, X=X,MARGIN=2)))
 X <- X/S
 D <- as.matrix(dist(X))^2
 K <- exp(-h*D)
}
############  CHOOSE KERNEL ##############
h <- 0.5
K <- kgaus(Xc,h)
#dim(K)
#qr(K)$rank
##########################################
prob1 <- function(miu,alfa,K){
  pr <- exp(miu+K%*%alfa)/(1+exp(miu+K%*%alfa))
}
cost <- function(miu,alfa,K,y){-sum(y*(miu+K%*%alfa) -
          log(1 + exp(miu+K%*%alfa))) + lambda*crossprod(alfa)}

######### ###########    NEWTON-RAPHSON    ####################
### FIT MODEL TO TRAINING DATA AND TEST IN VALIDATING DATA ###
set.seed(77131111)
ptm <- proc.time()
for (i in 1:nrep) {
#    cat(i, "\n",sep="")
   train=sample(1:nrow(K),floor(0.5*nrow(K)))
   Ktrain <- K[train,train]
   Kval <- K[-train,train]
   ytrain <- y[train]
   yval <- y[-train]

  delta <- diag(c(0,rep(lambda,ncol(Ktrain))))
  M <- cbind(0,Ktrain)
  M <- rbind(0,M)
  ###### START VALUES FOR MIU AND ALFA ################
  miu <- 0.0
  alfa <- rep(0.0, ncol(Ktrain))
  W <- matrix(data = 0,nrow = ncol(Ktrain),ncol = ncol(Ktrain))
  for (j in 1:nitnr) {
    fdmiu <- -sum(ytrain - prob1(miu, alfa, Ktrain))
    fdalfa <-
      -Ktrain %*% (ytrain - prob1(miu, alfa, Ktrain)) +
       lambda * Ktrain %*% alfa
    fd <- matrix(c(fdmiu, fdalfa), nrow = length(alfa) +
                 1, ncol = 1)
    W <-
      diag(c(prob1(miu, alfa, Ktrain) *
      (1 - prob1(miu, alfa, Ktrain))))
    Z <- cbind(1, Ktrain)
    zwz <- t(Z) %*% W %*% Z
    LHS <- zwz + lambda * M
    RHS <- -fd
    sol0 <- matrix(c(miu, alfa), nrow =
             length(alfa) + 1, ncol = 1)
    sol1 <- sol0 + solve(LHS,RHS)
    miu <- sol1[1, 1]
    alfa <- sol1[-1, 1]
    newcostvnr[j] <- cost(miu, alfa, Ktrain, ytrain)
```

```
    resulttvnr[j, ] <- c(j, newcostvnr[j], miu, alfa[1:5])
  }
  probval <- probl(miu, alfa, Kval)
  y_predval <- as.numeric(ifelse(probval > 0.5, 1, 0))
  msev[i] <- mean((y_predval - yval) ^ 2)
  res[i,] <- c(i,j,newcostvnr[j],miu,alfa[1:5])
  }
proc.time()-ptm
```

```
##    user  system elapsed
##    2.11    0.11    0.58
```

```
tail(resulttvnr[,2:6])
```

```
##             [,1]      [,2]       [,3]     [,4]      [,5]
##  [5,] 236.2372 0.1062297 -0.9584729 0.685112 0.8690137
##  [6,] 236.2372 0.1062297 -0.9584729 0.685112 0.8690137
##  [7,] 236.2372 0.1062297 -0.9584729 0.685112 0.8690137
##  [8,] 236.2372 0.1062297 -0.9584729 0.685112 0.8690137
##  [9,] 236.2372 0.1062297 -0.9584729 0.685112 0.8690137
## [10,] 236.2372 0.1062297 -0.9584729 0.685112 0.8690137
```

```
#tail(res)
summary(msev)
```

```
##    Min. 1st Qu.  Median    Mean 3rd Qu.    Max.
##  0.2833  0.3233  0.3283  0.3310  0.3442  0.3700
```

The program outputs the last six iterations of the Newton-Raphson algorithm and replicates 10, for the value of the cost function, μ and the estimates of first three elements of α. The `summary` function at the end displays summary statistics of the misclassification rate over the ten replicates.

11.3 Neural Networks

Neural networks (NN) have a long history (McCulloch and Pitts 1943) and have become the state of the art technique for many machine learning problems involving complex nonlinear data. The subject has been rebranded along its development, and presently it is also known as *machine learning* or *deep learning* and it is being successfully applied in many software areas including object, image and speech processing, robotics, video games and search engines. This section provides a very basic introduction to the subject concentrating exclusively on fully connected NN, also known as multilayer perceptrons and illustrates, via examples, how they can be used for classification and regression. Useful references are the books by Bishop (2006), Goodfellow et al (2016) and Bernard (2021) and a review by LeCun et al (2015), where other types of more modern networks, including convolutional NN, recurrent NN and transformer networks, are discussed. A very comprehensive reference is López et al (2022) that also provides detailed descriptions and implementation of many methods used for genomic prediction with special emphasis on plant breeding. An early Bayesian perspective on the subject can be found in Neal (1996) and a recent neural network Bayesian McMC implementation in Zhao et al (2021).

I motivate the subject by interpreting logistic regression as a special case of a fully connected NN. Multilayer perceptrons (MLP) or fully connected networks are the oldest and classic neural architecture invented in the 1960s. MLP are used today as components of other neural network architectures, such as convolutional neural networks, transformers, Bayesian neural networks, variational autoeconders and others.

Preliminaries: A Logistic Regression

Consider binary observations y and features x that constitute training records $\{(x_1, y_1), \ldots, (x_n, y_n)\}$, $x_i \in \mathbb{R}^p$, $y_i \in \{0, 1\}$ and assume the model adopted is

$$E(y_i|z_i) = \Pr(y_i = 1|z_i) = m(x_i) = \frac{\exp(z_i)}{1 + \exp(z_i)}. \tag{11.63}$$

The linear transformation $z_i = w'x_i + b$ consists of the input (vector x with p features (covariates)), parameters (w (a $(p \times 1)$ vector of unknown regression coefficients)) and b (a scalar intercept). In machine learning the parameters w are known as *weights*, the intercept b is known as the *bias* and the training records are known as *training examples*. Both w and b are unknown.

In Chap. 3 maximum likelihood estimates of the parameters of the logistic model w and b were obtained using the Newton-Raphson and the EM algorithms. Chapter 5 discusses a Bayesian implementation. Here the problem is approached regarding the logistic model as a special case of a NN, and the cost function (negative of the loglikelihood, a convex function) is minimised using a gradient descent algorithm. In what follows, unless otherwise stated, the subscript i identifying the ith training record will be omitted.

Figure 11.7 displays a logistic regression model represented as a neural network. The $x's$ are the $p = 2$ features (covariates). These constitute the first layer. A linear function $z^{(2)}$ of the features and parameters is fed into what is known as *neuron* or *node* (the superscript denotes the layer). The neuron is activated by an *activation function* σ that results in an output of layer 2, $a^{(2)}$, known as an *activation unit*. For

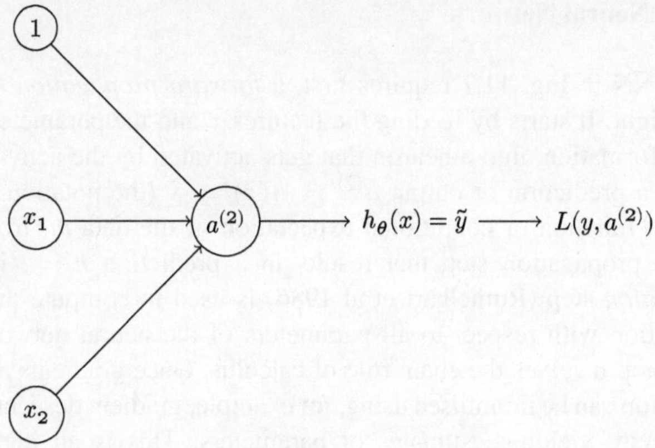

Fig. 11.7 Logistic regression model represented as a *two*-layered neural network for a single training record (y, x_1, x_2). The first layer (outer circles) corresponds to the input: $p = 2$ covariates $x_j, j = 1, 2$, plus 1 for the intercept. A linear transformation of the input, $z^{(2)} = w'x + b$, where $w' = (w_1, w_2)$, is fed into a neuron on the second layer. The neuron is activated by the activation function σ (here, the sigmoid function), producing an output $a^{(2)}$, the activation unit $a^{(2)} = \sigma(z^{(2)})$. In this simple example of a shallow network, the output or prediction $h_\theta(x) = \tilde{y}$ is equal to $a^{(2)}$. This output is evaluated using the loss function L that indicates how well $\tilde{y} = a^{(2)}$ compares with the observation y

the logistic regression, the activation function is the logistic or *sigmoid function* σ (11.63)

$$\sigma\left(z^{(2)}\right) = \frac{\exp\left(z^{(2)}\right)}{1 + \exp\left(z^{(2)}\right)} = \frac{1}{1 + \exp\left(-z^{(2)}\right)}, \tag{11.64}$$

where the linear function $z^{(2)} = w'x + b$. In Fig. 11.7 there are two layers: an input layer (layer 1), where the activation unit is just the input, $a_1^{(1)} = x_1, a_2^{(1)} = x_2$, and a layer 2 that includes a single *activation unit* $a^{(2)} = \sigma\left(z^{(2)}\right) = \tilde{y}$, the output. The output is here a predicted or estimated value of the probability $m(x_i)$ in (11.63). With some abuse of notation, this output will also be denoted by $h(x)$ or $h_\theta(x)$, where θ is a vector of parameters of the NN and x represents the vector of covariates.

In linear models the regression function takes the form $m(x) = w'x$. Kernels introduce nonlinear functions of the features x, via basis functions $\phi(x)$ that could capture more complex structures leading to $m(x) = w'\phi(x)$. Kernel methods require the user to supply the form of the kernel function. In the present representation of logistic regression as a NN, $\phi(x) = \sigma\left(w'x + b\right)$, where w and b are estimated from the training data and σ is a given nonlinear function such as the sigmoid. If σ is the identity function, the neural network reduces to a linear model.

Training the Neural Network

Training the NN in Fig. 11.7 requires first, a *forward propagation* step that runs from left to right. It starts by feeding the features x and the parameters (w, b), via a linear transformation, into a neuron that gets activated by the activation function σ generating a prediction or output $a^{(2)} = h(x) = \tilde{y}$ (the notation distinguishes the regression function or conditional expectation of the data m, from the output of a forward propagation step that results in a prediction $h = \tilde{y}$). Second, a *back propagation* step (Rumelhart et al 1986) is used to compute the gradient of the cost function with respect to all parameters of the neural network. The back propagation step involves the chain rule of calculus. Once gradients are computed, the cost function can be minimised using, for example, gradient descent or stochastic gradient descent, yielding estimates of parameters. This is an iterative process which, on completion, outputs a prediction.

Notation

The simple NN of Fig. 11.7 can be made more complex by increasing the number of layers and the number of neurons per layer. Identification of parameters in such a complex system requires a more involved notation. In general, let

- n_ℓ : number of neurons in layer ℓ. For the input layer, $\ell = 1, n_1 = p$, the number of features in the input datum
- $a_i^{(j)}$: activation unit i (or output produced by the activation function) in layer j. For the input layer $j = 1$, there is no activation function and the notation is interpreted as $a_i^{(1)} = x_i$
- $a^{(j)}$: layer j. This is a column vector with number of elements equal to the number of neurons in layer j
- $W^{(j)}$: matrix of weights (parameters) of the function that controls the mapping from layer j to layer $j + 1$. The number of rows of $W^{(j)}$ is equal to the number of neurons in layer $j + 1$, and the number of columns is equal to the number of neurons in layer j. If a network has n_j neurons in layer j and n_{j+1} neurons in layer $j + 1$, $W^{(j)}$ is of dimension $n_{j+1} \times n_j$
- $b^{(j)}$: column vector of intercept terms of the function that controls the mapping from layer j to layer $j + 1$, with number of elements equal to the number of neurons in layer $j + 1$

Forward Propagation

The logistic regression example displayed in Fig. 11.7 considers a single training datum consisting of the two covariates (x_1, x_2) and the observation y. A scheme of forward propagation is displayed in Fig. 11.8.

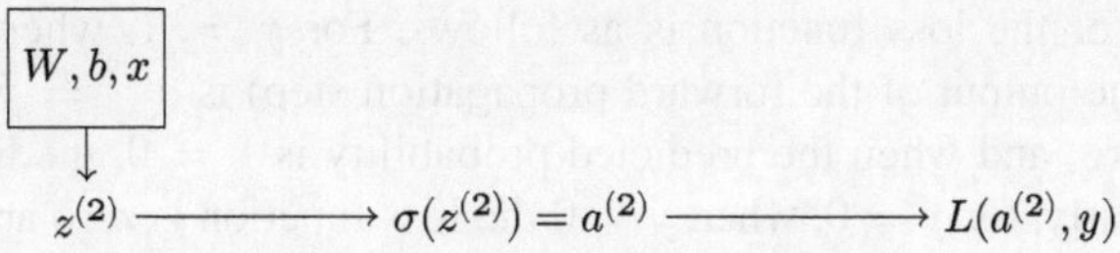

Fig. 11.8 Forward propagation scheme for the logistic regression model interpreted as a neural network. The linear function of parameters W, b and feature vector $x = a^{(1)}$, $z^{(2)}$, is fed into the activation function σ (the logistic function). This activation function outputs the prediction $a^{(2)} = \tilde{y}$. The loss function L quantifies the discrepancy between $a^{(2)}$ and the observed record y

The variable $z^{(2)}$ is a linear function of the input features (the two covariates $x_1 = a_1^{(1)}, x_2 = a_2^{(1)}$) and is fed into the activation function σ, which generates the output $a^{(2)} = \tilde{y}$ (a predicted probability). The *forward propagation* step can be written as

$$z^{(2)} = W_{11}^{(1)} a_1^{(1)} + W_{12}^{(1)} a_2^{(1)} + b^{(1)}, \tag{11.65a}$$

$$a^{(2)} = \sigma\left(z^{(2)}\right) = \tilde{y}. \tag{11.65b}$$

The dimension of $W^{(1)} = (W_{11}^{(1)}, W_{12}^{(1)})$ is 1×2 because this network has two features in layer 1 (i.e., x_1 and x_2) and one single neuron in layer 2 (the output $a^{(2)} = \tilde{y}$). The parameters $W^{(1)}$ and $b^{(1)}$ must be initialised to start the iteration.

Loss Function

The loss function L measures the degree to which the output $\tilde{y}$ matches the observation y (following the machine learning tradition, here *cost function* is a measure involving an entire dataset that may or may not include a regularisation term; *loss function* involves a single datum). The overall objective is to minimise the cost function by minimising the sum of the loss functions for each datum. This requires computation of the gradients $\partial L/\partial W^{(1)}$, $\partial L/\partial b^{(1)}$ that can be accomplished using back propagation. The present simple example provides the background for dealing with more complex situations.

Back propagation is an application of the chain rule of calculus applied in a computationally efficient manner. This efficiency is achieved in part by reusing derivatives computed for higher layers of the network, in the computation of derivatives of the lower layers.

The contribution of the datum (y, x) to the (convex) loss function (i.e., the negative of the loglikelihood) to be minimised with respect to (W, b) follows from the Bernoulli distribution and is

$$L(\tilde{y}, y) = -\left(y \log \tilde{y} + (1 - y) \log (1 - \tilde{y})\right). \tag{11.66}$$

An intuition for the loss function is as follows. For $y = 1$, when the predicted probability (the output of the forward propagation step) is $a^{(2)} = \tilde{y} = 1$, the loss function is zero, and when the predicted probability is $\tilde{y} = 0$, the loss function is infinity. Similarly, for $y = 0$, when $\tilde{y} = 0$ the loss function is zero and when $\tilde{y} = 1$ the loss function is infinity.

Although hidden from the notation, L depends not only on y but also on the input

$$\left(b^{(1)}, W_{11}^{(1)}, W_{12}^{(1)}, a_1^{(1)}, a_2^{(1)}\right).$$

Some of this input, $\left(a_1^{(1)}, a_2^{(1)}\right)$, is given (in this simple example, $a_i^{(1)} = x_i$), but the parameters $\left(b^{(1)}, W_{11}^{(1)}, W_{12}^{(1)}\right)$ are not observed and must be initialised to begin iteration.

Chain Rule

Before deriving the back propagation algorithm for the simple example of Fig. 11.7, I consider a general application of the chain rule of calculus. Figure 11.9 shows a flowchart where input $a \in \mathbb{R}^p$ is fed into a function g that outputs $b \in \mathbb{R}^n$. In turn, b is fed into a function f that outputs $c \in \mathbb{R}^m$. Specifically,

$$\begin{aligned}
a &= (a_1, a_2, \ldots, a_p), \\
b &= g(a) = (b_1, b_2, \ldots, b_n), \\
c &= f(b) = (c_1, c_2, \ldots, c_m).
\end{aligned}$$

The function g performs a mapping from $\mathbb{R}^p$ to $\mathbb{R}^n$ and the function f a mapping from $\mathbb{R}^n$ to $\mathbb{R}^m$. Consider the computation of the change in c_i due to a change in a_j.

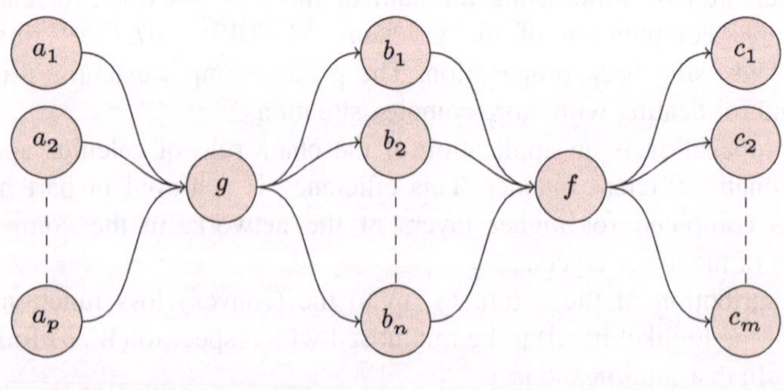

Fig. 11.9 Flowchart where input a is fed into a function g that outputs b. This output is fed into a function f that outputs c

A change in a_j may induce a change in $(b_1, b_2, \ldots, b_n)$, which in turn induces a change in each c_i. The chain rule states that to first order, the net change in c_i due to a change in a_j is given by the sum of the changes induced along each path from a_j to c_i. Letting k denote one of the m paths,

$$\frac{\partial c_i}{\partial a_j} = \sum_{k=1}^{n} \frac{\partial c_i}{\partial b_k} \frac{\partial b_k}{\partial a_j}. \tag{11.67}$$

Back Propagation

The back propagation step computes the gradient of the cost function with respect to the parameters of the NN. Once the gradient is available, the cost function can be minimised using gradient descent.

Minimisation of the cost function involves minimising the sum of the loss functions for each datum (11.66). Using (11.65), this takes the form

$$\begin{aligned}
&\frac{\partial L}{a^{(2)}} : \quad (1 \times 1)\,, \\
\frac{\partial L}{\partial z^{(2)}} &= \frac{\partial L}{\partial a^{(2)}} \frac{\partial a^{(2)}}{\partial z^{(2)}} : \quad (1 \times 1)\,, \\
\frac{\partial L}{\partial W^{(1)}} &= \frac{\partial L}{\partial z^{(2)}} \frac{\partial z^{(2)}}{\partial W^{(1)}} : \quad (2 \times 1)\,, \\
\frac{\partial L}{\partial b^{(1)}} &= \frac{\partial L}{\partial z^{(2)}} \frac{\partial z^{(2)}}{\partial b^{(1)}} : \quad (1 \times 1)\,.
\end{aligned}$$

More specifically,

1. For the output $a^{(2)} = \tilde{y}$ in layer 2, compute

$$\begin{aligned}
\delta^{(2)} = \frac{\partial L}{\partial z^{(2)}} &= \frac{\partial L}{\partial a^{(2)}} \frac{da^{(2)}}{dz^{(2)}} \\
&= \frac{\partial L}{\partial a^{(2)}} \sigma' \left(z^{(2)} \right).
\end{aligned} \tag{11.68}$$

2. The partial derivatives with respect to $W^{(1)}_{11}$, $W^{(1)}_{12}$ are

$$\begin{aligned}
\frac{\partial L}{W^{(1)}_{11}} &= \frac{\partial L}{\partial z^{(2)}} \frac{\partial z^{(2)}}{\partial W^{(1)}_{11}} \\
&= \delta^{(2)} a^{(1)}_1,
\end{aligned} \tag{11.69}$$

$$\frac{\partial L}{W_{12}^{(1)}} = \frac{\partial L}{\partial z^{(2)}} \frac{\partial z^{(2)}}{\partial W_{12}^{(1)}}$$
$$= \delta^{(2)} a_2^{(1)}, \tag{11.70}$$

$$\frac{\partial L}{b^{(1)}} = \frac{\partial L}{\partial z^{(2)}} \frac{\partial z^{(2)}}{\partial b^{(1)}}$$
$$= \delta^{(2)}. \tag{11.71}$$

3. For the sigmoid function in (11.68)

$$\sigma'\left(z^{(2)}\right) = \sigma\left(z^{(2)}\right)\left(1 - \sigma\left(z^{(2)}\right)\right) = a^{(2)}\left(1 - a^{(2)}\right),$$

and

$$\frac{\partial L}{\partial a^{(2)}} = -\frac{y}{a^{(2)}} + \frac{1-y}{1-a^{(2)}}. \tag{11.72}$$

Therefore, substituting in (11.68),

$$\delta^{(2)} = a^{(2)} - y. \tag{11.73}$$

Cost Function

The above computations and the loss function (11.66) pertain to a single training record. For n records the *cost function* is

$$J(\theta) = \frac{1}{n}\sum_{i=1}^{n} L(\tilde{y}_i, y_i) = -\frac{1}{n}\sum_{i=1}^{n}\left(y_i - \log \tilde{y}_i + (1 - y_i)\log(1 - \tilde{y}_i)\right), \tag{11.74}$$

an average of the loss functions over the n records (scaling by $1/n$ does not change the optimisation of J but avoids numerical problems during the computations). In this equation θ is the vector of parameters $\left(W^{(1)}, b^{(1)}\right)$. Minimisation of the cost function requires computation of the gradients with respect to $\left(W^{(1)}, b^{(1)}\right)$. These are the average of the gradients of each datum calculated using (11.69), (11.70) and (11.71). For instance, for $W_{11}^{(1)}$,

$$\frac{\partial J(\theta)}{\partial W_{11}^{(1)}} = \frac{1}{n}\sum_{i=1}^{n} \frac{\partial L(\tilde{y}_i, y_i)}{\partial W_{11}^{(1)}},$$

where now $\partial L(\tilde{y}_i, y_i)/\partial W_{11}^{(1)}$ is the partial derivative of the loss function for the ith record with respect to $W_{11}^{(1)}$.

The general form of a gradient descent update is given by expression (3.57) on page 99. In the present case, once the partial derivatives are available, the gradient descent update is

$$W_{11}^{(1)} := W_{11}^{(1)} - \alpha \frac{\partial J}{W_{11}^{(1)}},$$

$$W_{12}^{(1)} := W_{12}^{(1)} - \alpha \frac{\partial J}{W_{12}^{(1)}},$$

$$b^{(1)} := b^{(1)} - \alpha \frac{\partial J}{b^{(1)}},$$

for a user chosen value of the learning rate α.

A Single Hidden Layer Neural Network

Figure 11.10 shows an example of a neural network with three layers: the input layer showing a single training example, a hidden layer and an output layer with a single output node. The input layer has three nodes ($x_i, i = 1, 2, 3$) plus an extra node 1 for the bias term. The hidden layer has also three nodes representing activation units $\left(a_i^{(2)}, i = 1, 2, 3\right)$, plus a node 1 to account for the bias. Finally, the output layer has a single node $a^{(3)}$, the prediction. For this example, the hidden layer is generated with the sigmoid function, but other activation functions can be used instead, as discussed on page 500.

The single node output layer applied here is appropriate for a regression model where the output is a real number, or for a binary (1/0) outcome. Here the latter is assumed and the sigmoid function (11.64) is used in the generation of the final layer. For a continuous variable, the output is typically generated using the identity function. For a K class classification, the final output can be generated using the *softmax* function

$$a_k^{(L)} = g(z_k) = \frac{\exp(z_k)}{\sum_{k=1}^{k=K} \exp(z_k)}, \tag{11.75}$$

also used for a multilogit model and interpreted as a probability. In this expression L denotes the number of layers in the network, so that $a_k^{(L)}$ is the kth node of the last layer corresponding to class k. The number of nodes in the final layer is K and the sum of terms (11.75) equals 1.

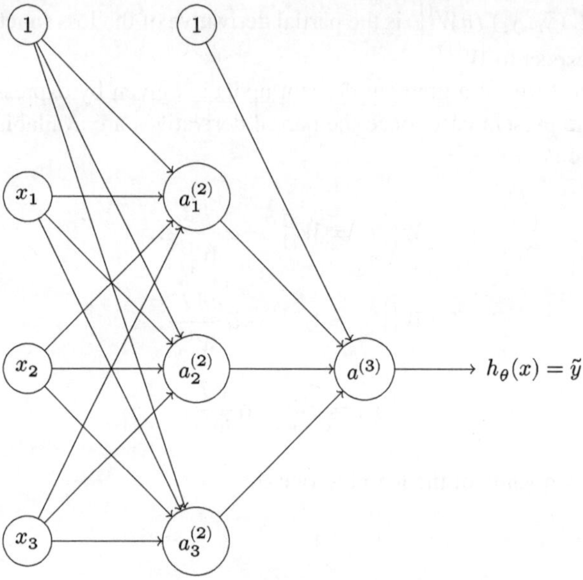

Fig. 11.10 A neural network diagram with three layers: an input layer representing the feature vector with three predictors $x = (x_1, x_2, x_3)$, with 1 to account for the bias term; a hidden layer with three nodes representing activation units $(a_1^{(2)}, a_2^{(2)}, a_3^{(2)})$, with the extra node 1 to account for the bias term and an output layer $a^{(3)} = h_\theta(x) = \tilde{y}$ that yields a prediction

Reverting to the NN of Fig. 11.10 and following the notation defined above, consider the forward propagation step. For a single training datum, the step can be represented by the equations below where input predictors x output the scalar prediction $\tilde{y}$

$$
\begin{aligned}
a^{(1)} &= \left(a_1^{(1)}, a_2^{(1)}, a_3^{(1)}\right)' = (x_1, x_2, x_3)', \\
z_1^{(2)} &= W_{11}^{(1)} a_1^{(1)} + W_{12}^{(1)} a_2^{(1)} + W_{13}^{(1)} a_3^{(1)} + b_1^{(1)}, \\
z_2^{(2)} &= W_{21}^{(1)} a_1^{(1)} + W_{22}^{(1)} a_2^{(1)} + W_{23}^{(1)} a_3^{(1)} + b_2^{(1)}, \\
z_3^{(2)} &= W_{31}^{(1)} a_1^{(1)} + W_{32}^{(1)} a_2^{(1)} + W_{33}^{(1)} a_3^{(1)} + b_3^{(1)}, \\
a_i^{(2)} &= \sigma\left(z_i^{(2)}\right), \quad i = 1, 2, 3, \\
z^{(3)} &= W_{11}^{(2)} a_1^{(2)} + W_{12}^{(2)} a_2^{(2)} + W_{13}^{(2)} a_3^{(2)} + b_1^{(2)}, \\
a^{(3)} &= \sigma\left(z^{(3)}\right) = \tilde{y}.
\end{aligned}
\tag{11.76}
$$

The forward propagation can be written more compactly in vectorised form along the lines in (11.65), yielding

$$\begin{aligned}
&z^{(2)} = W^{(1)}a^{(1)} + b^{(1)}, \qquad z^{(2)} = \left[z_1^{(2)}, z_2^{(2)}, z_3^{(2)}\right]', \\
&W^{(1)} : (3 \times 3), \qquad a^{(1)} : (3 \times 1), \qquad b^{(1)} : (3 \times 1), \\
&a^{(2)} = \sigma\left(z^{(2)}\right), \qquad a^{(2)} : (3 \times 1), \\
&z^{(3)} = W^{(2)}a^{(2)} + b^{(2)}, \\
&W^{(2)} : (1 \times 3), \quad a^{(2)} : (3 \times 1), \quad b^{(2)} : (1 \times 1), \\
&a^{(3)} = \sigma\left(z^{(3)}\right), \qquad a^{(3)} : (1 \times 1).
\end{aligned} \tag{11.77}$$

The logistic regression model was interpreted as a two-layered neural network of the form $f(x) = \Pr(Y = y|x)$. The three-layered neural network represented by the system (11.77) can be written as

$$f(x) = f^{(2)}\left(f^{(1)}(x)\right),$$

with a straightforward extension for a deeper layered network. Above, $f^{(1)}$ is the activation function from the first layer and $f^{(2)}$ is the activation function of the second layer. Each of these functions is indexed by parameters that are learned by the network in order to output/predict a value as close as possible to the observed record. The effectiveness with which this is achieved is measured by the cost function.

From a prediction perspective, closeness between predictions and outputs may be the result of overfitting. This suggests the need for regularisation.

The Cost Function

The cost function (11.74) for the logistic regression is a convex function easy to minimise. In the example there are p parameters and n data points, and when $n \gg p$ there are no issues with the computation. Typically, neural networks are very heavily parametrised and prone to overfitting. One way to avoid overfitting is to include a regularisation term as part of the cost function.

Consider a neural network with m training records $(x_1, y_1), \ldots, (x_m, y_m)$ and assume an output with K classes, so that the output $h_{W,b}(x) \in \mathbb{R}^K$ (with binary outcomes $K = 1$). Let $\left(h_{W,b}\left(x^{(i)}\right)\right)_k$ be the kth output corresponding to datum i. The cost function is

$$J(\theta) = -\frac{1}{m}\left[\sum_{i=1}^{m}\sum_{k=1}^{K} y_k^{(i)} \log\left(h_{W,b}\left(x^{(i)}\right)\right)_k \right.$$
$$\left. + \left(1 - y_k^{(i)}\right)\log\left(1 - \left(h_{W,b}\left(x^{(i)}\right)\right)_k\right)\right]$$
$$+ \frac{\lambda}{2}\sum_{l=1}^{L-1}\sum_{i=1}^{n_l}\sum_{j=1}^{n_{l+1}}\left(W_{ji}^{(l)}\right)^2. \tag{11.78}$$

In this expression, L is the number of layers, n_l is the number of nodes in layer l, n_{l+1} the number of nodes in layer $l + 1$ and λ is the regularisation or tuning parameter. The bias term is not regularised. Expression (11.78) features the quadratic penalisation, but other penalties such as the lasso penalty are also used.

Contrary to what has been the case so far, the cost function (11.78) and of most neural networks are not convex but typically have multiple optima. Global minimisation of such a multidimensional surface is an insurmountable task. Notwithstanding, finding near local optima is not difficult. Recent theoretical and empirical results appear to indicate that, in general, the many local optima found using mini-batch gradient descent or stochastic gradient descent do not constitute an issue in neural network optimisation (LeCun et al 2015).

Minimisation of the cost function requires the gradients $\partial J(\theta)/\partial W_{ji}^{(l)}$, $\partial J(\theta)/\partial b_j^{(l)}$. This is achieved in a computationally efficient manner using back propagation along the same lines as indicated for the simple logistic regression model. Details are a little laborious and are relegated to an Appendix on page 533. The Appendix also discusses vectorisation of back propagation to optimise matrix multiplications and provides an example of back propagation in the presence of multiple connected paths.

Activation Functions

Activation functions are a fundamental building block of neural networks. They turn the neural network model into a highly nonlinear function of the input variables whose parameters are learned as the network is implemented. It is easy to see that use of an identity function (rather than of a nonlinear activation function) in the hidden layer and output layer of the neural network of Fig. 11.10 transforms the neural network into a linear model. Indeed, setting in (11.77)

$$
\begin{aligned}
a^2 = z^{(2)} &= W^{(1)}a^{(1)} + b^{(1)} \\
a^{(3)} &= W^{(2)}a^{(2)} + b^{(2)} \\
&= W^{(2)}\left[W^{(1)}a^{(1)} + b^{(1)}\right] + b^{(2)} \\
&= W^{(2)}W^{(1)}a^{(1)} + \left(W^{(2)}b^{(1)} + b^{(2)}\right) \\
&= \widetilde{W}a^{(1)} + \widetilde{b}, \qquad (11.79)
\end{aligned}
$$

transforms the network into a linear function of the input $a^{(1)}$. This lacks the ability to extract nonlinear patterns from the input data.

The examples discussed so far included as activation functions the sigmoid function, the softmax function and the identity function. The sigmoid function can be used in the hidden layers of a neural network and is used in the outer layer for binary classifiers. It transforms the input to values in $(0, 1)$. For multiclass problems the outer layer uses the softmax function, and for regression the outer layer uses the identity function.

Another activation function is the tanh function or hyperbolic function

$$
\tanh(z) = \frac{\exp(z) - \exp(-z)}{\exp(z) + \exp(-z)},
$$

that outputs values in $(-1, 1)$. The range of values of z generated by this function has the same effect as standardising the output. This often leads to better behaviour of the neural network relative to that of the sigmoid function. The derivative is

$$
\frac{\partial}{\partial z}\tanh(z) = 1 - (\tanh(z))^2 .
$$

One disadvantage of the sigmoid function and of the tanh function is that if z becomes a very large positive number or a very small negative number, the derivatives are very close to zero. This can slow down gradient descent. An activation function that does not suffer from this property is the ReLU (rectified linear unit) function that outputs positive values. It takes the form

$$
\text{ReLU}(z) = \max(0, z) = \begin{cases} 0, & z \leq 0, \\ z, & z > 0, \end{cases}
$$

with derivative

$$
\frac{d\text{ReLU}(z)}{dz} = I(z > 0) .
$$

This derivative does not exist at $z = 0$, but a computer implementation is not sensitive to this since numerically z is never exactly equal to zero. Despite the fact that the derivative is close to zero when $z < 0$, there are enough z values larger than zero so that gradient descent moves at an acceptable pace. A variant is the leaky ReLU

$$\text{leaky ReLU}(z) = \max(cz, z),$$

where c is a small number (e.g. $c = 0.01$) with derivative

$$\frac{\partial \text{leaky ReLU}(z)}{\partial z} = \begin{cases} 1, & z > 0, \\ c, & z < 0, \end{cases}$$

so the derivative, numerically speaking, is never zero.

In modern neural networks, hidden layers often use the ReLU or the leaky ReLU functions rather than the sigmoid function because they often lead to better numerical behaviour and are cheaper to compute. Several other activation functions can be found in the literature but the ones mentioned here are the most commonly used.

Remarks on Fitting Neural Networks

Neural networks are characterised by having many parameters: the weights and biases, the choice of activation functions, the regularisation parameter, the learning rate and the number of hidden layers and neurons per layer. For good performance neural networks require fine tuning of these parameters. This calls for considerable experimentation and expertise. In addition, back propagation, a central operation of the NN, can become very challenging in real-world implementations and is an area with many subtleties. The following is a minimal list of issues that require attention:

- Standardising the features/covariates to mean zero and variance one is particularly advisable when the inputs have different distributions. The standardisation has the effect of speeding up gradient descent and simplifies computations since a common learning rate can be applied.
- Starting values of covariates, parameters and regularisation parameter. The range of values of the covariates has a direct effect on the values of the parameters. Standardisation of the covariates facilitates initialising the parameters; a general advice is to choose starting values at random in the proximity of zero (e.g. random uniform weights over the range $[-0.7, +0.7]$ (Hastie et al 2009), or smaller). Weights (parameters) are typically initialised drawing randomly from Gaussian or uniform distributions. The choice of scale of the initial distribution may require considerable experimentation. One suggestion for initialising $W^{(\ell)}$

is to use $Un(-k, k)$, $k = 1/\sqrt{n_\ell}$, or $N(0, 1) * 1/\sqrt{n_\ell}$ (where n_ℓ is the number of columns of $W^{(\ell)}$) and if a ReLU activation function is used, $N(0, 1) * 2/\sqrt{n_\ell}$ (Hu et al 2021). It is important not to set the initial value of the $W's$ exactly equal to zero. If this is the case, the activation units and the δ terms of each layer take the same values and this symmetry cannot be broken. This creates a redundancy (irrespective of the number of neurons per layer, this would be equivalent to a network with only one neuron per layer since all neurons take the same value) and poor behaviour of the neural network. Judicious choice of starting values for the parameters of a neural network reduces the problem of exploding or vanishing gradients that may arise in deep neural networks (Glorot and Bengio 2010). The value of the regularisation parameter is very influential, with larger values of λ leading to smaller values of the parameters tending to reduce overfitting but to increase bias. An interesting case arises in connection with the use of tanh or sigmoid. Small weights lead to small values of z. Within a fairly broad range of values of z, the relationship between tanh and sigmoid with z is almost linear. In this situation the neural network reduces to a linear model (see (11.79)) and may perform poorly if there are complex nonlinearities in the data

- Number of hidden layers/number of neurons per layer. For any given problem, it is difficult to know in advance how deep (how many layers) a neural network must be. The number of hidden layers is best considered as a hyperparameter to be determined via experimentation and cross-validation. The number of neurons per layer is also a model feature to be determined by experimentation. It is common to use in the order of tens or hundreds of neurons, with this number increasing with the number of features. Shallow (few layers) neural networks would typically require more neurons per layer in order to fit the same function.
- Tuning gradient descent. Depending on the size of the dataset, gradient descent can be applied to the complete data (*batch gradient descent*), or to mini-batches of the data (*mini-batch gradient desdent*). For datasets of the order of say up to 3000 features, batch gradient descent should be a reasonable choice. For larger sizes the features can be divided into mini-batches of size 1024, or 512 features. Within each mini-batch, a single step of gradient descent is computed (as if the whole training set consisted of this unique mini-batch). This requires computation of forward propagation, computation of the cost function, computation of back propagation to obtain the gradients of the cost function with respect to the parameters and finally updating of parameters. This completes the processing of the mini-batch. The calculations are repeated for each mini-batch, until the complete training data have been processed leading to a single pass of the complete training set. The cost function changes for each mini-batch, and the algorithm will never reach a minimum; it will rather fluctuate around a (local) minimum. In contrast with batch gradient descent where a pass through the training data leads to one updating step of the parameters, in mini-batch gradient descent many updating steps are performed, one for each mini-batch. The complete mini-batch algorithm has an outer loop allowing many passes through the complete training data or until satisfactory convergence is reached.

For very large datasets, mini-batch gradient descent is considerably faster than batch gradient descent. The noise associated with mini-batch gradient descent is beneficial in avoiding that the algorithm gets stuck in local minima.

- Gradient checking. During the debugging phase of the development of an algorithm, it is good practice to check the results from the coded gradient with those obtained with numerical derivatives. The latter are easily obtained as follows. Let $\theta \in \mathbb{R}^p$ be the parameters of the neural network and let $J(\theta)$ be the cost function. Then for small ϵ (e.g. $\epsilon = 10^{-4}$), the numerical partial derivatives are approximated by

$$\frac{\partial}{\partial \theta_1} J(\theta) \approx \frac{J\left((\theta_1 + \epsilon), \theta_2, \ldots, \theta_p\right) - J\left((\theta_1 - \epsilon), \theta_2, \ldots, \theta_p\right)}{2\epsilon},$$

$$\vdots$$

$$\frac{\partial}{\partial \theta_p} J(\theta) \approx \frac{J\left(\theta_1, \theta_2, \ldots, (\theta_p + \epsilon)\right) - J\left(\theta_1, \theta_2, \ldots, (\theta_p - \epsilon)\right)}{2\epsilon}.$$

Eyeballing a plot of the exact versus the numerical derivatives should give a first impression (a straight line with slope 1 passing through the origin should be a good indication) or more precisely checking the size of

$$\frac{\left\| \partial\theta_{\text{approx}} - \partial\theta \right\|_2}{\left\| \partial\theta_{\text{approx}} \right\|_2 + \left\| \partial\theta \right\|_2}$$

which for $\epsilon = 10^{-6}$ should be 10^{-6} or 10^{-5}, though this will depend on the number of iterations.

- Software. There is a wealth of software available to fit neural networks. For the simple multilayer perceptron, `neuralnet` proved straightforward to use in the small datasets of the examples below. Other popular choices include `deepnet`, `h20`, and for more ambitious users, Google released the open-source software `TensorFlow` with the friendly user interface `Keras`.

Example: Analysis of Binary Observations Using a Neural Network

The objective is to study the predictive ability of a neural network and to make a comparison with the kernelised logistic regression, the logistic lasso and the penalised logistic regression.

The data consist of 599 binary records simulated with a logistic model using the wheat inbred lines genotyped for 1279 binary genetic markers, as in the example on page 485. The true (simulated) model on the underlying scale involves 20 loci randomly sampled from these 1279 markers and assigned as QTL. The proportion of 1′s in the observations is approximately 50%. A neural network with one hidden layer is implemented in `R` with the vectorised code described on page 537. The neural network is also fitted using the `R` software `neuralnet` as a comparison. The single layered neural network has five neurons in the hidden layer, and the sigmoid function is used in the hidden and output layers. An $\ell - 2$ regularisation term is added to the logistic loss function, and the regularisation parameter is set to 0.002. The learning rate of the gradient descent is set to 0.008, and for each of 20 replicates of training/validating splits, the number of gradient descent iterations is 5000. The predictive ability of the neural network was quantified using the validation MSE (proportion of misclassifications in the validation data, after transforming the predicted probabilities into predicted observations on the 0, 1 scale using Bayes rule, as defined on page 370).

Over the 20 replications, the minimum, median, mean and maximum values of the validating MSE were 0.32, 0.36, 0.36 and 0.40, respectively. The corresponding figures for the training MSE were 0.08, 0.10, 0.10 and 0.15. The analysis was repeated using the `R` software `neuralnet`, and the results for the minimum, median, mean and maximum values of the validation MSE over 20 splits were 0.31, 0.36, 0.36 and 0.41. For the training MSE, the corresponding results were 0.01, 0.03, 0.04 and 0.10.

The predictive performance of the neural work for these data quantified as the average (over 20 replications) validating MSE, was similar to that obtained by the penalised logistic regression (0.40), the kernelised logistic regression (0.33) and the logistic lasso (0.33), as reported on page 485.

Increasing the number of neurons in the hidden layer from 5 to 10 and to 50 deteriorated prediction performance. The same was noted by increasing the number of layers to two and three (with five neurons per layer) using the software `neuralnet`. These were exploratory modifications to the original architecture without aiming at a proper optimisation of the neural network.

The `R`-code to fit the neural network to the simulated wheat data set is shown below. The first code represents the vectorised version, and the second code describes the implementation with `neuralnet`.

```
# CODE1110
# FIT THE NEURAL NETWORK TO SIMULATED BINARY PHENOTYPES
# USING THE WHEAT INBRED LINES WITH THE 1,279 GENETIC MARKERS

rm(list=ls()) # CLEAR WORKSPACE
set.seed(37111)

library(BGLR)
data(wheat)
#####################################################
### USE BGLR MATRIX X

X <- wheat.X
nindiv <-nrow(X)
nmark <- ncol(X)
####################################################
```

```
# NUMBER OF LOCI AFFECTING THE SIMULATED DATA
nloci<-20
p<-0.5
mu<-log(p/(1-p)) # BIAS TERM FOR SIMULATED DATA

##### INITIALISE PARAMETERS AND ALLOCATE MATRICES ##############
va<-1.0 # additive variance of liability
ve<-1.0 # environmental variance
Xc<-matrix(data=NA,nrow=nindiv,ncol= nmark)
be<-matrix(data=0.0,nrow=nmark,ncol=1) # parameter of true model
y<-rep(0,nindiv)
cm<-colMeans(X)
for (i in 1:nmark)
{Xc[,i]<- (X[,i]-cm[i])/sd(X[,i])
}
QTLeff<-sqrt(va/nloci)# QTL effect so that the total
# genetic variance is VA
IDq<-sample(1:nmark,nloci,replace=F) # from the nmark markers,
# choose nloci as QTL
be[IDq]<-QTLeff # the only b's that are not zero are those
# associated with QTL.
########### GENERATE BINARY DATA y ###################
xb<-Xc%*%be
pr <- exp(mu+xb)/(1+exp(mu+xb))
y <- rbinom(nindiv,1,pr)
#####################################################
df = data.frame(cbind(X,y))
m <- length(y)
# BIG X!
X <- t(X)
p <- nrow(X)

Y <- matrix(y,nrow=1,ncol=length(y))

n_1 <- p # NUMBER OF FEATURES IN INPUT DATA
# READ NUMBER OF NEURONS IN LAYER 2
n_2 <- 5
# READ NUMBER OF NEURONS IN LAYER 3
n_3 <- 1
########################################
### FUNCTIONS:
# SIGMOID FUNCTION
sigm <- function(par){
  1/(1+exp(-par))
}
# COST FUNCTION EXCLUDING REGULARISATION TERM
cost <- function(A,Y){-(tcrossprod(Y,log(A))+
    tcrossprod((1-Y),(log(1-A))))/m}
########################################

# READ REGULARISATION PARAMETER delta

delta <- 0.02

# READ GD LEARNING RATE (HERE LABELLED gamma)
gamma <- 0.08

# eps: range of initial values of elements of W_1: (-eps,eps)
eps <- 0.85
```

```
# READ NUMBER OF GRADIENT DESCENT ITERATIONS
nit <- 5000
# READ NUMBER OF TRAINING / VALIDATING REPS
nitval <- 20

resultval <- matrix(data=NA, nrow=nitval,ncol=3)
result <- matrix(data=NA, nrow=nit,ncol=8+n_2)

ptm<-proc.time()

for (j in 1:nitval) {
  # INITIALISE MATRIX OF WEIGHT W_1 (n_2 x n_1), n_1=p=rows of X
  # INITIALISE MATRIX OF WEIGHT W_2 (n_3 x n_2)
  W_1 <- matrix(nrow=n_2,ncol=n_1,runif(n_2*n_1,-eps,eps))
  W_2 <- matrix(nrow=n_3,ncol=n_2,runif(n_3*n_2,-eps,eps))
  train=sample(1:ncol(X),floor(0.5*ncol(X)))
Yt <- matrix(Y[train],nrow=1,ncol=length(Y[train]))
Yv <- matrix(Y[-train],nrow=1,ncol=length(Y[-train]))
Xt <- X[ ,train]
Xv <- X[,-train]
b_1 <- matrix(0,nrow=n_2,ncol=ncol(Xt))
b_2 <- matrix(0,nrow=n_3,ncol=ncol(Xt))
  for (i in 1:nit) {
    cat("j=",j, " ","i=",i, "\n", sep = "")
    # FORWARD PROPAGATION
    Z_2 <- W_1 %*% Xt + b_1
    A_2 <- sigm(Z_2) # SIGMOID FUNCTION
#     A_2 <- pmax(Z_2,0.01*Z_2) # Leaky ReLU function
    Z_3 <- W_2 %*% A_2 + b_2
    A_3 <- sigm(Z_3)
#   A_3 <- Z_3 # USE THE IDENTITY FUNCTION FOR CONTINUOUS DATA
    # BACK PROPAGATION
    DZ_3 <- A_3 - Yt
    DW_2 <- (DZ_3 %*% t(A_2) / m) + delta * W_2
    Db_2 <- mean(DZ_3)
    DZ_2 <- t(W_2) %*% DZ_3 * A_2 * (1 - A_2) # sigmoid function
    # Leaky ReLU function:
#   DZ_2 <- t(W_2)%*%DZ_3 * ifelse(Z_2 > 0,1,0.01)

    DW_1 <- (DZ_2 %*% t(Xt) / m) + delta * W_1
    Db_1 <- apply(DZ_2, 1, mean)
    # GRADIENT DESCENT ON TRAINING DATA Xt
    W_1 <- W_1 - gamma * DW_1
    W_2 <- W_2 - gamma * DW_2
    b_1 <- b_1 - gamma * Db_1
    b_2 <- b_2 - gamma * Db_2
# BELOW: ADD PENALTY TERM TO THE LOSS FUNCTION
    newcost <- cost(A_3, Yt) + (delta/2)*(sum(W_2^2)+sum(W_1^2))
    result[i,] <- c(i,newcost,DW_2[1:5],W_1[1],W_2)

  }
  ytrain <- as.numeric(ifelse(A_3 > 0.5, 1, 0))
  msetrain <- mean((ytrain - Yt) ^ 2)
#  print(table(yh, Y))
  # VALIDATION STAGE WITH DATA Xv
  # A LITTLE TWIST: IN LINES 400-404 b_1 & b_2 MUST BE ADJUSTED
  # BECAUSE/IF # TRAINING RECORDS < # VALIDATING RECORDS !!!!!!
  a <- floor(0.5*ncol(X))
  b <- 0.5*ncol(X)
```

```
  bind <- function(b){cbind(b,b[,1])}
  b_1 <- if(a<b) {bind(b_1)}
  b_2 <- if(a<b) {bind(b_2)}
  Z_2 <- W_1 %*% Xv + b_1
  A_2 <- sigm(Z_2)
  Z_3 <- W_2 %*% A_2 + b_2
  A_3 <- sigm(Z_3)
  yval <- as.numeric(ifelse(A_3 > 0.5, 1, 0))
  mseval <- mean((yval - Yv) ^ 2)
  resultval[j, ] <- c(j, msetrain,mseval)

}
proc.time() - ptm
print(table(yval,Yv))
plot(resultval[,2],type="l",ylim=c(min(resultval[,2]),
max(resultval[,3])))
lines(resultval[,3],col="red")
summary(resultval[,3])
```

R-code to fit the neural network using `neuralnet` is shown below:

```
# CODE1110 (cont)
# FIT \texttt{neuralnet} ON TRAINING DATA AND
# TEST ON VALIDATING DATA
# REPEAT nrepnn TIMES
set.seed(371111)
library(neuralnet)

nrepnn <- 20

resmsenn <- matrix(data=NA,nrow=nrepnn,ncol=3)

for (i in 1:nrepnn){
  cat(i, "\n",sep="")
  train=sample(1:ncol(X),floor(0.5*ncol(X))) # ASSUMES X
#   HAS BEEN TRANSPOSED!
# SETTING BELOW hidden = c(5,6) FITS TWO LAYERS OF 5 NEURONS
# IN LAYER 1 AND 6 NEURONS IN LAYER 2.  hidden =c(5) FITS
# A SINGLE LAYER WITH 5 NEURONS
  nn=neuralnet(y~.,data=df[train,],hidden=c(5),linear.output=
  FALSE,act.fct="logistic")
  pnv <- as.numeric(predict(nn,df[-train,]))
  predv <- ifelse(pnv > 0.5, 1, 0)
  msev <- mean((pnv-Y[-train])^2)
  pnt <- as.numeric(predict(nn,df[train,]))
  predt <- ifelse(pnt > 0.5, 1, 0)
  mset <- mean((pnt-Y[train])^2)
  resmsenn[i,] <- c(i,msev,mset)
}
plot(resmsenn[,2],type="l",ylim=c(min(resmsenn[,3]),
max(resmsenn[,2])))
lines(resmsenn[,3],col="red")
summary(resmsenn[,2])
```

Example: Prediction Using a Neural Network, a RKHS Regression and a Genomic BLUP Model

In this second example, prediction ability is evaluated for the neural network and for the RKHS (reproducing kernel Hilbert space) regression with a Gaussian kernel. A genomic BLUP model is also fitted, which is a special case of the RKHS regression with a linear kernel appropriately scaled. In contrast with the previous two models, the genomic BLUP model captures only linear relationships of the feature space.

Grain yields (centred and standardised to unit variance) from the 599 inbred lines were downloaded from the BGLR package and, as was the case in the previous example, 1279 binary genetic markers were available as input features for each line. The neural network architecture was as in the previous example, with the only difference that the output neuron was generated using the identity function since grain yield data are continuous. The cost function for this example has terms of the form $(y - a^{(3)})^2$ rather than the logistic likelihood employed for the binary records. The expression for $z^{(3)}$ on page 537 is unchanged.

The analysis of grain yield using the RKHS regression was based on the model described in the example on page 478. The Bayesian RKHS model used $\lambda = 0.3$ and three values of the bandwidth parameter h were entertained. The best prediction result (reported in Table 11.2) was obtained using $h = 1$.

The prediction ability of the models was quantified using the validating MSE (average sum of squared differences between observed and predicted validating records) and the correlation between observed and predicted validating phenotypes, over 20 replicates of training/validating splits. For the RKHS regression, predictions were obtained from expression (11.30), and details are disclosed in the R-code below. Results in the form of *Minimum Mean* and *Maximum* over the 20 replicates, for the validating MSE and for the correlation between observed and predicted validating phenotypes, are displayed in Table 11.2.

For these data the RKHS regression with the Gaussian kernel was a slightly better prediction machine than the neural network and the BLUP model, both of which performed similarly. The RKHS regression model was also implemented using $\lambda = 0.3$, $h = 0.5$ and $\lambda = 0.3$, $h = 3.0$. The mean validation MSE and correlations

Table 11.2 Grain yield of 599 inbred lines for wheat data from the BGLR package. Prediction ability quantified as the *minimum*, *mean* and *maximum* over 20 random training/validating splits (50% of each) for the validation MSE (X; e.g. for the first entry, $X = 0.72$) and for the correlation between observed and predicted validating phenotypes (Y; e.g. for the first entry, $Y = 0.36$) for the neural network model (NN), the RKHS regression model (RKHS) and the genomic BLUP model. The results in the table are in the form X/Y

	Minimum	Mean	Maximum
NN	0.72/0.36	0.83/0.43	0.91/0.51
RKHS	0.68/0.42	0.78/0.49	0.90/0.56
BLUP	0.70/0.35	0.83/0.43	0.93/0.51

between observed and predicted validation phenotypes for these two RKHS variants were (0.79; 0.47) and (0.86; 0.42), respectively.

The R-code used to fit the RKHS model is shown below:

```
# CODE1111
# TRAINING AND TESTING OF GRAIN YIELD -
# WHEAT INBRED LINES FROM BGLR PACKAGE
rm(list=ls()) # CLEAR WORKSPACE
set.seed(37111)
library(BGLR)
data(wheat)
#####################################################
### USE BGLR MATRIX X
x <- wheat.X
y<- wheat.Y[,4]
#### A GAUSSIAN KERNEL ################

kgaus <- function(X,h){
  X <- scale(X,center=TRUE,scale=FALSE)
  S=sqrt(sum(apply(FUN=var, X=x,MARGIN=2)))
  X <- X/S
  D <- as.matrix(dist(X))^2
  K <- exp(-h*D)
}
#############################################
##### A LINEAR KERNEL #################
klinl <- function(X){
  X=scale(X,center=TRUE,scale=FALSE)
  S=sqrt(sum(apply(FUN=var, X=X,MARGIN=2)))
  X=X/S
  K=tcrossprod(X)
}
############  CHOOSE GAUSSIAN KERNEL ##############

#h <- 0.5
h <- 1
# <- 3

K <- kgaus(x,h)
#dim(K)
#qr(K)$rank

############## CHOOSE LINEAR KERNEL ###############
#K <- klinl(x)
#dim(K)
#qr(K)$rank
####################################################
# READ NUMBER OF TRAINING / VALIDATING SPLITS nitval
nitval <- 20
# READ REGULARISATION PARAMETER lambda
lambda <- 0.3

result <- matrix(data=NA, nrow=nitval,ncol=5)

ptm <- proc.time()
for (j in 1:nitval) {
  cat(j, "\n",sep="")
  train = sample(1:nrow(x), floor(0.5 * nrow(x)))
  xt <- x[train, ]
```

```
  yt <- y[train]
  xv <- x[-train, ]
  yv <- y[-train]
  YHATt <- matrix(nrow = length(yt), ncol = 1)
  YHATv <- matrix(nrow = length(yv), ncol = 1)
  Ktrain <- K[train, train]
  Kval <- K[-train, train]
  Xt <- cbind(1, Ktrain)
  RHSt <- crossprod(Xt, yt)
  LHSt <- crossprod(Xt)
  LHSt[-1, -1] <- LHSt[-1, -1] + Ktrain * lambda
# diag(LHSt) <- diag(LHSt) + c(0, rep(1e-8,ncol(Xt)-1))
  solt <- solve(LHSt, RHSt)
  YHATt <- solt[1] + Ktrain %*% solt[-1]
  YHATv <- solt[1] + Kval %*% solt[-1]
  mset <- mean((YHATt - yt) ^ 2)
  msev <- mean((YHATv - yv) ^ 2)
  cort <- cor(YHATt, yt)
  corv <- cor(YHATv, yv)
  result[j, ] <- c(j, mset, msev, cort, corv)
}
proc.time() - ptm
summary(result[,3])
summary(result[,5])
```

11.4 Classification and Regression Trees

This section describes an approach to modelling where the feature space is partitioned into regions within which the responses are relatively homogeneous. The partitioning of the feature space follows a set of binary splitting rules that gives rise to a decision tree. Decision trees can be used for continuous as well as for categorical variables. In the former they are known as regression trees and in the latter as classification trees. The tree has terminal nodes or leaves that arise when a stopping criterion for further partitioning has been met. Finally, a prediction associated with a particular terminal node is obtained based on the mean or mode of the training output in the terminal node, in the case of regression trees, or, based on the most frequent class, in the case of classification tress. The method lends itself to a graphical representation that possesses pedagogical quality and ease of interpretation.

This section shows how to build a tree and how to apply it with the South African Heart Disease Data (Rousseauw et al 1983) used also in Hastie et al (2009), where a more detailed account can be found. The data is a subset of a larger dataset and includes 462 males in a high-risk heart disease region of the Western Cape in South Africa. There are $p = 9$ sources of information (features or covariates or inputs) available on each patient: systolic blood pressure (sbp), cumulative tobacco (tobacco), low-density lipoprotein cholesterol (ldl), adiposity, family history of heart disease (famhist), type A behaviour (type-a), obesity, alcohol and age. The output or response is coronary heart disease (chd). Adiposity is a measure of percent body

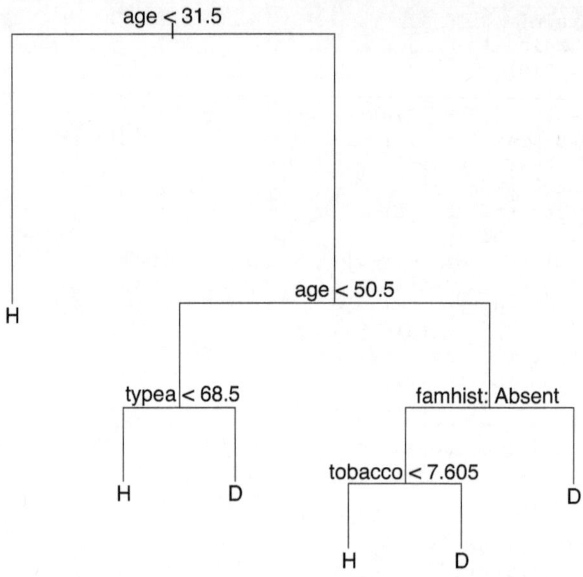

Fig. 11.11 Classification tree fit to the South African Heart Disease data. The goal is to predict the health status of a patient ("Healthy" or "Diseased, "H" and "D", respectively), based on nine features or covariates. The split at the top of the tree results in two branches: one on the left for age < 31.5 years and one on the right for age > 31.5 years. The left split does not undergo further subdivision and the group is classified as "Healthy". The split on the right undergoes a further split whereby those with age < 50.5 years are moved to the left and those over 50.5 years to the right. The split on the left has a further split governed by type a score whereby those < 68.5 are moved to the left and are classified as "Healthy" and those scoring > 68.5 are moved to the right and are classified as "Diseased". The final tree has six terminal nodes or leaves with a certain number of observations in each terminal node (not shown in the figure). The predictions "D" or "H" at each terminal node depend on the proportion of observations falling into each category

fat, whereas obesity is measured as body mass index (bmi). Type-A refers to a behaviour pattern characterised by an excessive competitive drive, impatience and anger/hostility. Coronary heart disease (chd) is a classification variable that takes values "Diseased" and "Healthy". Out of the 462 patients, 160 are classified as "Diseased".

Figure 11.11 shows the output of a classification tree fitted to these data. The tree arises from questions associated with the inputs/covariates. These questions are known as *splits*. Starting at the top of the tree, the first split assigns observations to the left if age is less than 31.5 years. These observations do not undergo further splits and constitute a *terminal node* or *leaf*. Observations whose age is older than 31.5 years are moved to the right. This branch undergoes a further split, again based on age, where the threshold now is 50.5 years; those younger define a left branch and those older a right branch. The left branch has a split according to scoring for type-a behaviour. Those scoring less than 68.5 are moved to the left and constitute another terminal node; those scoring higher than 68.5 are moved to the right and give rise to a third terminal node. Observations whose age is older than 50.5 years undergo a split due to family history, (absent/present), a categorical variable. The left branch includes observations that do not show family history for chd, while the right branch includes those observations that do. These constitute a terminal node. The branch on the left has a further split according to tobacco: less than an accumulated amount of 7.605 go to the left and give rise to a terminal node. Those with a value larger than 7.605 comprise another terminal node. The tree has 6 terminal nodes.

The shape of the final tree is governed by the splits. The data consist of p variables or features (in the case of the heart data, $p = 9$) and a response that is $(x_i, y_i), i = 1, 2, \ldots, n$, with $x_i = \left(x_{i1}, x_{i2}, \ldots, x_{ip}\right)$. For each split, the algorithm must decide which of the p variables to choose and then must perform a search to find the *optimal* point to split (the first split in the tree of Fig. 11.11 is based on the feature age and the split occurs at 31.5 years). One must also decide how deep the tree must be, that is, when to stop splitting and thereby declaring the records from this last split as belonging to a terminal node. The predictions for each terminal node shown in Fig. 11.11 arise by counting the number of D's and H's among the observations in the terminal node and choosing as the predicted value the most frequent.

Among the 462 observations, the proportion "Diseased" is 0.346. Consider the first decision at the top of the tree (actually, the root, since the tree is upside down). For each of the nine covariates, the algorithm finds the binary partition that optimises one of two criteria (*Gini index* or *deviance*, described at the end of this subsection) that measure the *impurity* among the response in the binary partition (if all the observations in the partition belong to the same class, the impurity is zero and it is maximum when 50% belong to each class). The aim is ideally to find a partition of the covariate within which the responses have the same label. For the tree in Fig. 11.11, the choice falls on age. There are 49 "age classes" among the 462 records and the age 31.5 years minimises the *impurity*. At this point there are 117 observations whose age is less than 31.5 years, and among these, 10 are classified as "Diseased" resulting in a proportion of "Healthy" equal to 0.915 and in a prediction equal to H. This constitutes the first terminal node of the tree. After this split the algorithm works with those observations on the right, and a new split is found by the same procedure. The same covariate, age, now at the point 50.5 years is chosen as the one minimising impurity. Those records whose age is less than 50.5 years are moved to the left. Among these records, 119 are "Healthy" and 54 are "Diseased" resulting in a prediction at this stage of H.

A characteristic of decision trees is that optimality is defined locally for each split, one at a time, and the algorithm searches for all possible splitting points for each covariate. The process is repeated until a stopping criterion is reached (based on the number of observations in a splitting region, or on the reduction of impurity). A compromise must be found between growing very deep trees, leading to overfitting, or shallow trees that do not capture enough of the variability of the response. The common practice is first to grow a deep tree stopping when the number of observations in a node reaches a minimum (5, say). This tree will typically overfit the data. This is then followed by tree *pruning* which requires cross-validation: the bottom branches of the tree are removed until the cross-validation error increases. The tree in Fig. 11.11 is a result of this pruning procedure. The original tree had 15 terminal nodes and included seven of the nine predictors. The pruned tree has only six terminal nodes and four predictors. Both trees were fitted to the complete set of 462 observations. The misclassification rate of these trees using the complete data was 0.21 and 0.23 for the complete and pruned tree, respectively.

Regression trees follow the same rationale as the classification trees, except that the residual sum of squares is used instead of the Gini index or the deviance as criteria to split the nodes or to prune the tree. The response y now is quantitative and, initially, the residual sum of squares before the first split is $\sum_{i=1}^{n}(y_i - \overline{y})^2$ where $\overline{y}$ is the mean of the n responses. For a first partition around a splitting point s, the n_l observations falling on the left of s have mean $\widehat{\mu}_l$ and those n_r falling on the right of s, $\widehat{\mu}_r$. These means constitute the predicted values for the first partition. The residual sum of squares for this node for this value of s is reduced to

$$\sum_{i:x_{ij}\leq s}(y_i - \widehat{\mu}_l)^2 + \sum_{i:x_{ij}>s}(y_i - \widehat{\mu}_r)^2 . \tag{11.80}$$

The algorithm searches the covariate j and the split point s that minimises Eq. (11.80). For each branch a new split is carried out to refine predictions. Once a final tree has been constructed, there will be a predicted value attached to each terminal node given by the node's mean. For a new set of covariates for a new individual, a predicted value is obtained by starting at the top of the tree and following the splits downwards until the terminal node with its predicted value is reached. For a regression tree, the new individual is assigned the prediction given by that terminal node's mean.

The Gini Index and the Deviance

Let R_j denote the jth node and let n_{jk} denote the number of observations of class k, $k = 1, 2, \ldots, K$ in node j. The estimate of the probability of observing class k in node j is the proportion of observations y that fall in class k

$$\widehat{p}_{jk} = \frac{1}{n_{jk}} \sum_{i:x_i\in R_j} I\,(y_i = k)\,,$$

and the class assigned to node j is the class with the largest estimated probability, arg max $\widehat{p}_{jk}$. The *Gini index* associated with node j is

$$G_j = \sum_{k=1}^{K} \widehat{p}_{jk}\,(1 - \widehat{p}_{jk})\,. \tag{11.81}$$

G takes on small values when the estimated probabilities are extreme and reaches a maximum when estimated probabilities are intermediate.

An alternative to the Gini index is the *deviance* or *entropy* given by

$$D_j = -\sum_{k=1}^{K} \widehat{p}_{jk} \log \widehat{p}_{jk}. \tag{11.82}$$

Since $\lim_{p->0} p \log p = 0$, we take $\widehat{p}_{jk} \log \widehat{p}_{jk} = 0$ whenever $\widehat{p}_{jk} = 0$. Therefore, entropy is at a minimum value of 0 when one of the $\widehat{p}_{jk} = 1$ and all other equal to 0, and it attains a maximum when all $\widehat{p}_{jk}$ in node j are equal and $\widehat{p}_{jk} = 1/K$ for all k.

As shown in the examples that follow, for classification trees, the function `tree` in `R` uses $2n_j D_j = -2\sum_{k=1}^{K} n_{jk} \log \widehat{p}_{jk}$ instead of (11.82), where n_j is the number of responses in node j.

Evaluating Prediction Performance with Cross-Validation

The misclassification rates of the full and pruned trees quoted above (21% and 23%, respectively) were estimated using the complete data. This is likely to underestimate the misclassification rate when evaluated on new data. Here the predicted ability of the classification tree is investigated further using cross-validation. The approach is as follows:

1. Split the data into training and validating sets (50% of the observations in each)
2. Construct a full tree using the training data
3. Compute the misclassification rate for this full tree in the training and validating data
4. Prune the full tree
5. Compute the misclassification rate of this pruned tree in the validating data

For a particular replicate, the full tree from step 2 has 25 terminal nodes and includes 7 of the 9 variables (dropped *famhist* and *alcohol*). The training and validating misclassification rates of this tree are 14% and 38% respectively. The misclassification rate of the null model obtained by classifying all the observations in the complete data as healthy is 34.6% (equal to the proportion of diseased individuals). The result is not encouraging.

The pruned tree from step 4 has three terminal nodes and includes only *age* and *ldl*, a remarkable simplification from the full tree. The training and validating misclassification rates of this pruned tree are 24% and 34.2% respectively, better than the full tree but not an improvement over the performance of the null model.

Example: Analysis of the Heart Data Using a Classification Tree

The code below reads the South African Heart Disease dataset and executes the function `tree` in R. The presentation follows closely the format in James et al (2017), pages 323–327. We look a little more closely into the output of the `tree` function and of other related functions that reveal some of the mechanics of the constructed trees.

```
# CODE1112
# READ SOUTH AFRICAN HEART DISEASE DATA
rm(list=ls()) # CLEAR WORKSPACE
library(sda)
library(tree)
library(glmnet)
library(randomForest)
library(loon.data) # MUST INSTALL PACKAGE loon.data
data("SAheart")
sahd <- SAheart
sahd$chd <- factor(as.numeric(sahd$chd),levels=c(2,1),
                   labels=c("D","H"))
#length(which(sahd$chd=="H")) # NO HEART DISEASE
#length(which(sahd$chd=="D")) # HEART DISEASE
##############################################################
# FUNCTION "ACCURACY":
accuracy = function(actual, predicted) {
  mean(actual == predicted)
}
# FIT TREE TO TRAINING AND VALIDATING HEART DATA
replicate <- 1
result <- matrix(data=NA, nrow=replicate,ncol=2)
# REPLICATION
for(i in 1:replicate){
  # SPLIT DATA INTO TRAINING / VALIDATING SET
  set.seed(31)
  train=sample(1:nrow(sahd),nrow(sahd)/2)
  treetrain <- sahd[train,]
  treevalid <- sahd[-train,]
  # FIT TREE TO THE TRAINING DATA
  treetr <- tree(sahd$chd[train] ~. ,data=treetrain)
  trtrpred <- predict(treetr,treetrain,type="class")
  trvapred <- predict(treetr,treevalid,type="class")
  table(predicted = trvapred, actual = sahd$chd[-train] )
  table(predicted = trtrpred, actual = sahd$chd[train] )
  acval <- accuracy(actual=sahd$chd[-train],predicted=trvapred)
  actst <- accuracy(actual=sahd$chd[train],predicted=trtrpred)
  result[i,] <- c(1-acval,1-actst)
}
summary(treetr)
```

```
##
## Classification tree:
## tree(formula = sahd$chd[train] ~ ., data = treetrain)
## Variables actually used in tree construction:
## [1] age, typea, tobacco, obesity, ldl, sbp,
## [7] adiposity
## Number of terminal nodes:  25
```

```
## Residual mean deviance:  0.5449 = 112.2 / 206
## Misclassification error rate: 0.1342 = 31 / 231
```

```
result
```

```
##              [,1]      [,2]
## [1,] 0.3809524 0.1385281
```

The code reads the 462 records and randomly samples two sets of 231 records that constitute the training and validating data. The `tree` function is fit to the training data (*treetrain*) and creates the object `treetr`. The `summary` function displays information about the resulting tree. I discuss this function below in connection with the pruned tree. At this stage we note that the resulting tree has 25 terminal nodes and includes 7 out of the 9 features of the dataset. The reported misclassification error rate of 13% corresponds to the training data; it is therefore likely to be underestimated. This is confirmed applying the `predict` function using as arguments the training and validating data. The misclassification error rate using the validating data is 38% and using the training data slightly under 14% (showing a mild discrepancy with the result of a little over 13% reported in the `summary` function).

The next step is to investigate whether a pruned tree leads to a better prediction performance. The code below fits the function `cv.tree` using as one of the arguments the object generated by fitting the `tree` to the training data. It uses cross-validation in order to find the size of the tree that produces the smallest misclassification rate. Size here refers to the number of terminal nodes.

```
# CODE1112 (cont)
# PRUNE TREE FROM TRAINING RUN
set.seed(33)
treetr_cv <- cv.tree(treetr, FUN = prune.misclass)
#treetr_cv
# index of tree with minimum error
min_idx = which.min(treetr_cv$dev)
#min_idx
# number of terminal nodes in that tree
treetr_cv$size[min_idx]
```

```
## [1] 3
```

```
# misclassification rate of each tree
#treetr_cv$dev / length(train)
# IT APPEARS THAT A TREE WITH treetr_cv$size[min_idx]
# TERMINAL NODES HAS THE SMALLER MISCLASSIFICATION
# EXECUTE PRUNE.MISCLASS: TREE WITH LOWEST C-V ERROR RATE
# SET best = TO NR TERMINAL NODES FROM THIS BEST TREE
tree_prune<-prune.misclass(treetr,best=treetr_cv$size[min_idx])
#summary(tree_prune)
```

```
# OBTAIN PREDICTIONS USING THIS PRUNED TREE
tree_prune_trn <- predict(tree_prune, sahd[train,],type="class")
table(predict = tree_prune_trn, actual = sahd$chd[train] )
```

```
##          actual
## predict   D   H
##         D 35  10
##         H 45 141
```

```
1-accuracy (predict = tree_prune_trn, actual = sahd$chd[train])
```

```
## [1] 0.2380952
```

```
tree_prune_val<-predict(tree_prune,sahd[-train,],type="class")
table(predict = tree_prune_val, actual = sahd$chd[-train] )
```

```
##          actual
## predict   D   H
##         D 20  19
##         H 60 132
```

```
1-accuracy (predict = tree_prune_val, actual = sahd$chd[-train])
```

```
## [1] 0.3419913
```

```
tree_prune
```

```
## node), split, n, deviance, yval, (yprob)
##       * denotes terminal node
##
## 1) root 231 298.10 H ( 0.3463 0.6537 )
##   2) age < 51.5 149 155.00 H ( 0.2148 0.7852 ) *
##   3) age > 51.5 82 111.30 D ( 0.5854 0.4146 )
##     6) ldl < 4.92 37  47.97 H ( 0.3514 0.6486 ) *
##     7) ldl > 4.92 45  47.67 D ( 0.7778 0.2222 ) *
```

The training and validating misclassification rates ($1 - accuracy$) of this pruned tree (that has three terminal nodes only) are 24% and 34.2% respectively. The function `table` displays the number of misclassifications in the off-diagonals and the correctly classified records in the diagonals. For instance, in the case of the *training* data, the first table, the number of misclassified observations is $10+45 = 55$, and the number of correctly classified observations is $35+141 = 176$, out of a total of 231 observations.

Turning attention to the output from the pruned tree and typing the `tree` object `tree_prune` disclose details about each branch of the tree. The first branch of the tree (the *split* at $age < 51.5$ years; corresponds to the second *node*) has $n =$ 149 observations. The *deviance* for the split is 155.00 and the predicted value is $yval = H$, because the proportion of H's, 0.7852, is larger than the proportion of

D's, 0.2148. This split is a *terminal node* as indicated by the asterisk $*$. As a check on the reported proportions, compute:

```
# CODE1112 (cont)
# A LITTLE CALCULATION: PROPORTION OF H's IN LEFT BRANCH
# OF THE FIRST SPLIT OF THE PRUNED TREE
trtrlt <- treetrain[which(treetrain$age < 51.5),]
dim(trtrlt)
```

```
## [1] 149  10
```

```
trtrltH <- trtrlt[which(trtrlt$chd == "H"),]
dim(trtrltH)[1]/dim(trtrlt)[1]
```

```
## [1] 0.7852349
```

The reported *deviance* for the split equal to 155.00 results from the following computation. The starting point is the entropy (11.82); for $j = 2$, the second node,

$$D_2 = -(0.7852 \times log(0.7852) + 0.2148 \times log(0.2148))$$
$$= 0.5202$$

and the printed *deviance* is $2(149)(0.5202) = 155.02$.

It is also possible to obtain the deviance of the pruned tree by executing:

```
summary(tree_prune)
```

```
##
## Classification tree:
## snip.tree(tree = treetr, nodes = c(6L, 7L, 2L))
## Variables actually used in tree construction:
## [1] "age" "ldl"
## Number of terminal nodes:  3
## Residual mean deviance:  1.099 = 250.7 / 228
## Misclassification error rate: 0.2381 = 55 / 231
```

The numerator of the printed *residual mean deviance* is

$$2\sum_j n_j D_j = -2 \times 149 \times (0.7852 \times \log(0.7852) + 0.2148 \times \log(0.2148)) +$$
$$-2 \times 37 \times (0.6486 \times \log(0.6486) + 0.3514 \times \log(0.3514)) +$$
$$-2 \times 45 \times (0.2222 \times \log(0.2222) + 0.7778 \times \log(0.7778))$$
$$= 250.68,$$

where D_j is given by expression (11.82), n_j is the number of observations in terminal node j and there are three terminal nodes in the pruned tree. The *residual mean deviance* is obtained by dividing the numerator by the number of observations in the data minus the number of terminal nodes. In the example, this gives $250.68/(231 - 3) = 1.099$.

Variability of Estimated Trees

Figure 11.12 shows the number of features and the validating misclassification error rate versus replicate number for 10 pruned trees obtained by resampling the complete data to construct ten training/ validating sets of data, fitting a tree to the training data in each replicate, pruning the tree obtained from the training data and obtaining predictions on the validating data. The number of features included in the final pruned tree ranged from 8 in replicate 1, with MSE = 0.35, to 1 in replicate 6, with a MSE = 0.32. The smallest MSE = 0.29 was observed in replicates 7 and 9, with 5 and 2 features, respectively. The figures display large variation, particularly in the number of features included in the final pruned tree.

There are two sources of variability at play here. One is the data. The number of records, the relative contribution of each covariate and the correlation structure among covariates make it difficult to find the "best" predictive model. More than one combination of covariates leads to similar predictive ability. The other important source of variability is methodological: trees are known to be non-robust, such that small changes in the data cause large changes in the final estimated tree. The large variability of decision trees makes them relatively poor prediction machines. This limitation is in contrast to other attractive properties such as ease of interpretation,

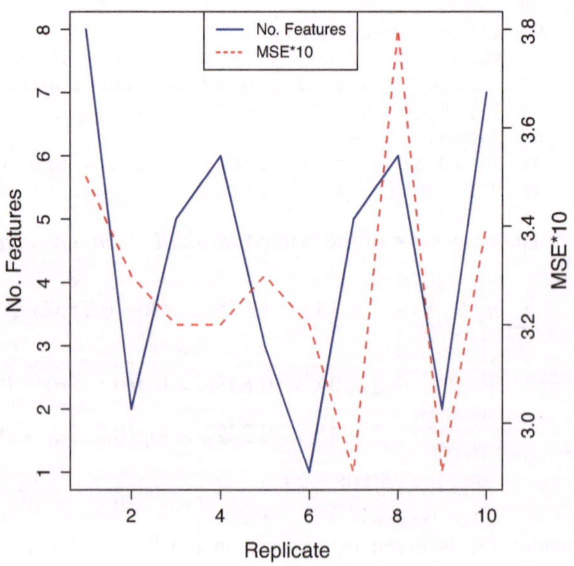

Fig. 11.12 Left y-axis: number of features included in the pruned tree (in blue). Right y-axis: misclassification error (MSE $*10$) (dashed red), versus replicate number. The smallest MSE = 0.29 over the 10 replicates was for replicates 7 and 9. The pruned trees for these replicates included 5 and 2 features, respectively. The smallest number of features is 1 (*age* $<$ 49.5 years) for replicate 6, with a $MSE = 0.32$

coupled with a highly pedagogic graphical display. As indicated in the next section, the non-robustness issue can be improved upon by constructing many trees and averaging the final predictions.

11.5 Bagging and Random Forests

As indicated in (6.51), the variance of the predictor is one of the three terms that influence the expected validation prediction error. Bootstrap aggregation or *bagging* (Breiman 1996) improves prediction ability by reducing this variance. This is achieved by constructing a prediction averaged over several bootstrap samples whose variance is smaller than the variance of an estimate based on a single sample.

The idea of bagging is illustrated here with decisions trees, but it can be used with other approaches. The algorithm is quite straightforward: create B bootstrap samples of training data each of size n_t (sampling the n_t rows of the training data (y_i, x_i), $i = 1, 2, \ldots, n_t$ with replacement n_t times) and grow a tree from each. Let $\widehat{f}_b(x_0)$ be the prediction from the bth bootstrap sample at the prediction point x_0 and let $\widehat{f}_B(x_0)$ be the prediction obtained using the B samples. In the case of regression trees, $\widehat{f}_B$ is a simple average over the B samples, and for classification trees $\widehat{f}_B$ is based on a majority vote: the most frequent class is chosen as the label for $\widehat{f}_B$. Each tree is grown deep and not pruned. This ensures low bias, and by averaging the B predictions, the large variance of each tree is substantially reduced. Specifically, we have

$$
\begin{aligned}
\mathrm{Var}\left(\widehat{f}_B(x_0)\right) &= \mathrm{Var}\left[\frac{1}{B}\sum_{b=1}^{B}\widehat{f}_b(x_0)\right] \\
&= \frac{1}{B^2}\left[B\ \mathrm{Var}\left(\widehat{f}_b(x_0)\right) + B(B-1)\ \mathrm{Cov}\left(\widehat{f}_b(x_0), \widehat{f}_{b}'(x_0)\right)\right] \\
&= \frac{(1-\rho)}{B}\ \mathrm{Var}\left(\widehat{f}_b(x_0)\right) + \rho\ \mathrm{Var}\left(\widehat{f}_b(x_0)\right) \qquad (11.83)
\end{aligned}
$$

where ρ is the pairwise correlation between predictors from the different trees. With large number of bootstrap samples B, the first term in the right-hand side is reduced, while the second depends on the correlation between the predictor functions.

The initial use of bagging with decision trees is based on inclusion of all the p features each time a split is considered. The trees fitted to the bootstrap samples constructed in this way tend to be correlated because the same feature is likely to be chosen in the different trees, particularly at the top of the tree. In a *random forest* (Breiman 2001), only $m < p$ features are randomly selected (without replacement) as candidates for splitting and a different set of m features is selected at each split. This has the effect of reducing the correlation between the trees and the second term in (11.83). Typical values for m are $p/3$ for regression trees and $\sqrt{p}$ for classification. The essentials of a random forest construction are that each tree is based on a random subset of the observations (the bootstrapping step) and each split is based on a random subset of candidate features.

Out-of-Bag Error Estimates

The use of bagging provides an opportunity to study the prediction ability of a model without resorting to cross-validation. Consider data $z_i = (y_i, x_i)$, $i = 1, 2, \ldots, n$. The probability of selecting a particular observation in each of the B bootstrap samples is $1/n$, and the probability of not selecting it is $(1 - 1/n)$. When n random samples are taken with replacement to form the bootstrap sample, the probability of not selecting the observation in the n samples is $(1 - 1/n)^n$, which in the limit when $n \to \infty$, is equal to $\exp(-1) \approx 0.37$. Therefore, the observation is absent in approximately one third of the B bootstrap samples, on average (the number of bootstrap samples with the missing observation is binomially distributed, with parameters B and $\exp(-1)$). In random forests, *out-of-bag* error estimates for classification or regression are obtained by constructing random forest predictors of each observation y_i, by averaging the predictors from trees grown in which z_i is absent. The error estimates obtained in this manner agree well with those obtained using cross-validation.

Variable Importance

Although the primary objective of random forests is prediction, it may be of interest to obtain an overall summary of the importance of each predictor or feature. There are two popular measures of importance of the predictor variables. One is based on recording the decrease of the residual sum of squares, in the case of regression trees, or of the Gini index in the case of classification trees, each time a particular predictor is used in a split. Averaging this decrease over the B trees reveals the important predictors as those leading to large decreases. The other approach addresses the contribution of a feature to prediction more directly: it considers the decrease in prediction accuracy when the feature is randomly permuted in the out-of-bag samples, thereby removing its association with the response. The rationale is that a permutation is supposed to mimic the absence of the predictor variable from the model. When the permuted predictor, together with the remaining non-permuted predictors, is used to predict the response with these out-of-bag observations, the prediction accuracy (proportion of observations classified correctly) decreases if the permuted predictor is associated with the response. In regression problems, classification rate is replaced by mean squared error. The difference between the two prediction accuracies (permuted/non-permuted) associated with a particular feature averaged over the B trees constitutes its variable importance score.

Example: Analysis of the Heart Data Using a Random Forest

The R-code below fits a random forest to the heart data using the function `randomForest`. In order to look more closely into the output, the argument `mtry` that specifies the number of features randomly selected in each split is set to a single value (3 in the example; setting mtry= 9 chooses bagging).

```
# CODE1113
# RANDOM FOREST WITH HEART DATA
#rm(list=ls()) # CLEAR WORKSPACE
library(sda)
library(tree)
library(glmnet)
library(randomForest)
# sahd <- read.table(
# "http://www-stat.stanford.edu/~tibs/ElemStatLearn/
# datasets/SAheart.data", sep=",",head=T,row.names=1)
# FUNCTION ACCURACY:
accuracy = function(actual, predicted) {
  mean(actual == predicted)
}

# CHANGE RESPONSE VARIABLE AND FAMHIST TO FACTOR
#sahd$chd<-as.factor(ifelse(sahd$chd == 1,
#                     "Disease","Healthy"))
#sahd$famhist <- as.factor(sahd$famhist)

#head(sahd)
#str(sahd)
sumd <- data.frame()
nrep <- 1
reslt <- rep(NA,nrep)
mtry <- c(1,3,5,9)
mtry <- 3
set.seed(2)
for (m in mtry){
    cat("mtry ",m,"\n",sep="")
  for ( rep in 1:nrep ) {
    cat("Replicate ",rep,"\n",sep="")
    train=sample(1:nrow(sahd),nrow(sahd)/2)
    rf <- randomForest(sahd$chd ~.,data=sahd,
                       subset=train,mtry=m,importance=TRUE)
    predict <- predict(rf,sahd[-train,])
    observed <- sahd$chd[-train]
    t <- table(observed,predict)
    print(t)
    reslt[rep]<-1-accuracy(predicted=predict,actual=observed)
  }
  sumd <- rbind(
    sumd,c(m,min(reslt),mean(reslt),median(reslt),max(reslt),
           var(reslt)))
}
```

```
## mtry 3
## Replicate 1
##         predict
```

```
## observed   D   H
##          D  30  47
##          H  25 129
```

```
names(sumd) <- c("mtry","min","mean","median","max","var")
sumd[3]
```

```
##         mean
## 1 0.3116883
```

The data are divided into a training and a validating set (each of 231 observations), the random forest is fitted to the training set and the resulting validating mean squared error is MSE ≈ 0.31 (misclassification rate: $(47+25)/(30+47+25+129)$). This is a little improvement over the validating MSE of the pruned tree, with MSE $= 0.34$, reported on page 518.

When `randomForest` is executed on the complete data, details of the output can be displayed typing the object `rf` (not shown).

```
# CODE1113 (cont)
library(randomForest)
rf <- randomForest(sahd$chd ~.,data=sahd,mtry=3,importance=TRUE)
rf$mtry
```

```
## [1] 3
```

```
rf$ntree
```

```
## [1] 500
```

```
rf$confusion
```

```
##    D   H class.error
## D 68  92   0.5750000
## H 54 248   0.1788079
```

Particular elements of the output file can be extracted. The output indicates that 3 variables are used per split (`mtry`) and that by default $B = 500$ trees are constructed (`ntree`). The estimate of the out-of-bag error rate computed from the complete data ($(92 + 54)/(92 + 54 + 248 + 68) \approx 0.32$) is in good agreement with the misclassification error rate using cross-validation reported above (31%).

The importance of each feature can be obtained using the `importance` function:

```
# CODE1113 (cont)
importance(rf,type=1)
```

```
##              MeanDecreaseAccuracy
## sbp                      3.869507
## tobacco                 13.272608
## ldl                      4.080764
## adiposity                4.032976
## famhist                  7.679644
## typea                    1.714666
## obesity                 -3.309007
## alcohol                 -3.556895
## age                     17.041426
```

The output indicates that on average over the 500 trees, *age*, *tobacco* and *family history* are the three most important features affecting prediction accuracy.

Variability of Random Forests

Variation over repeated samples of training/testing data can be obtained using the code above by modification of the parameter `nrep`. The performance of random forest for different choices of the number of features sampled per split can be studied setting `mtry <- c(1,3,5,9)`. The code above was run setting `nrep=100`, and the prediction accuracy (misclassification rate) is shown in Table 11.3.

For this particular dataset, the results are similar across scenarios with a slight advantage for the choice `mtry=1`.

Example: An Analysis Involving Genetic Epistatic Interactions

An attractive property of decision trees and random forests is their potential ability to capture interactions involving covariables. This is examined here with a stylised example in the context of gene discovery and genomic prediction.

A matrix of genotypic markers (of order *number of individuals* (sample size)} $\times$ *number of markers* $= 1000$) is generated from a binomial distribution $Bi\ (n = 2, p = 0.5)$. These marker data are part of the operational models used for

Table 11.3 Prediction accuracy of random forests over 100 samples of training/testing splits and for 4 different numbers of features sampled out of 9 per split (`mtry`; `mtry=9` indicates *bagging*). The figures in the table display the minimum, mean and maximum prediction accuracy over 100 replicates

mtry	min	mean	max
1	0.22	0.30	0.35
3	0.27	0.32	0.36
5	0.26	0.32	0.40
9	0.26	0.32	0.38

inferences and prediction. Three sample sizes are studied: 1000, 5000 and 10,000 observations.

Observations are simulated using two models that differ in their genetic architectures. In the first model, the *additive model*, 10 marker genotypes are randomly sampled from the 1000 genotypic markers and designated as QTL. The 10 QTL are assigned additive genetic effects equal to 0.22σ each, where $\sigma = \sqrt{5}$ is the standard deviation of the phenotypic distribution. The 10 QTL combine additively to determine genetic values. This model generates an additive genetic variance equal to 1.25 squared units. Setting the environmental variance equal to $(5 - 1.25)$ results in a heritability or proportion of variance of the observations captured by the linear regression on QTL genotypes equal to 0.25.

In the second model, the *interaction model*, the genetic values are the result of the following process. First, one locus contributing to additive genetic variation is randomly chosen among the 10 QTL. This additive locus contributes an additive genetic variance equal to 0.125 squared units. Second, among the 45 possible combinations of two loci taken from the 10 QTL, 10 are randomly sampled to generate additive × additive epistatic interactions. The interaction effects are all set equal to 0.44σ where, as before, $\sigma = \sqrt{5}$. When gene frequencies are intermediate and loci are in linkage equilibrium, the epistatic part of the interaction model constructed in this example does not generate additive genetic variation; the only source of additive genetic variation stems from the single additive locus. The genetic variance arising from the interacting genotypes is of additive × additive type only and equal to 2.5 squared units. As in the first model, the phenotypic variance is equal to 5 squared units. Details of the interaction model are described below in the **NOTE: The epistatic model**.

Three operational models used for analysis of the data are compared. The first two are fully parametric models and include the lasso and a Bayesian spike and slab mixture described on page 321. In both, the conditional mean of the data is assumed to be linearly related to the 1000 marker genotypes. The third is a nonparametric model: a random forest. The three models are fitted to data generated using both genetic architectures (the *additive* and the *interaction models*) and to the 3 sample sizes.

The two parametric models assume a multiple linear regression of the response y on the 1000 marker genotypes. It follows that, for the additive model, the 10 QTL are a subset of the marker panel of the operational model. For the interaction model, only the single QTL is a subset of the marker panel. The interaction terms that define the remaining proportion of the true genotypic values are not members of the operational models.

The models do not intend to mimic any particular trait. Rather, the intention is to examine the following conjectures:

1. with a purely additive genetic model, genomic operational models that postulate a linear regression of response on marker genotypes (encoded additively) should perform as well as or better than random forests, in terms of prediction ability and in terms of QTL detection.

2. In the presence of genetic interactions between QTL, random forests should outperform linear models, both in prediction ability and in QTL detection because linear models do not account for interactions.

The justification for point 1 is that operational models that postulate a linear regression of data on additively encoded marker genotypes are in close agreement with the true model. In such a setup, the linear models should be difficult to outperform. While the exact location of QTL is unknown, their detection depends on size of the effects and on the amount of data. In the case of point 2, the linear models are at a disadvantage: there are no interaction terms in the operational models. On the other hand, the building process of a decision tree (which is an integral part of the random forests) is designed to capture interactions provided signals are strong enough.

The `R`-code below generates the data for 10,000 individuals and for the *interaction model*. The *additive model* is retrieved by minor changes of the code.

```
# CODE1114
# GENERATING AN EPISTATIC MODEL AND FIT RANDOM FOREST,
# LASSO, BAYESIAN MIXTURE
rm(list=ls()) # CLEAR WORKSPACE
set.seed(303371)
nindiv<-10000
nmark <- 1000
nqtl <- 10
nintqtl <- 10
mu_y<-0
Xq<-matrix(data=NA,nrow= nindiv,ncol= nqtl)
#######  GENETIC MARKERS Xm  #########################
Xm<-matrix(nrow= nindiv,ncol= nmark,
           rbinom(nindiv*nmark,size=2,p=.5)-1)
# from the nmark markers, choose nqtl as QTL:
IDq<-sample(1:nmark,nqtl,replace=F)
Xq <- Xm[,IDq] # QTL GENOTYPIC MATRIX
# INTERACTION GENOTYPIC MATRIX:
Xi<-matrix(data=NA,nrow= nindiv,ncol= nintqtl)
b <- rep(0,nmark+nintqtl)
nr <- ncol(Xq)
i1 <- combn(nr,2)
i2 <- sample(ncol(i1),nintqtl,replace=FALSE)
i3 <- as.matrix(i1[,i2])
for (i in 1:nintqtl){
  Xi[,i] <- Xq[,i3[1,i]]*Xq[,i3[2,i]]+1
}
# GENERATE GENOTYPIC VALUES g
b[IDq] <- 0.5
```

```
# ########################
# BELOW: ZERO OUT length(idzero) ADDITIVE EFFECTS
idzero <- sample(IDq,floor(0.9*nqtl),replace=FALSE)
b[idzero] <- 0
# ###########################
lb <- nmark+1
ub <- nmark+nintqtl
b[lb:ub] <- 1.0
gi <-Xi%*%b[lb:ub]
ga <- Xq%*%b[IDq]
g <- ga + gi
va <- var(ga)
vi <- var(gi)
vg <- var(g)
y <- ga+gi+rnorm(nindiv,0,sqrt(5-va-vi))
vy <- var(y)
her_a <- va/vy
her_i <- vi/vy
V <- vy*(1-her_a-her_i) # CONDITIONAL VARIANCE
cov(ga,gi)
```

```
##              [,1]
## [1,] -0.0001627963
```

```
va
```

```
##           [,1]
## [1,] 0.1254941
```

```
vi
```

```
##          [,1]
## [1,] 2.553642
```

```
vy
```

```
##          [,1]
## [1,] 4.922617
```

The results for the *additive Model* are displayed in Table 11.4. In terms of prediction ability quantified by the validating mean squared error (MSE), there is little difference between the lasso and the Bayesian mixture although the latter is consistently superior. Both outperformed the random forest. In terms of detection of causal loci, the proportion of true and false positives in the discovery set differs markedly among methods and population sizes (see the table legend for the

Table 11.4 *Additive model.* Validating mean square error (MSE), true positive calls (TP) and false positive calls (FP) among flagged loci in the discovery set (TP+FP) using lasso, the Bayesian mixture and the random forest for three sample sizes (n). The discovery set was arbitrarily defined as follows. For lasso, the set of regression estimates not set equal to zero. For the Bayesian mixture, the loci whose McMC estimate of the probability of belonging to the non-zero mixture component is larger than 0.5. For random forest, the 10 loci that had the highest importance score. NULL MODEL refers to the validation MSE of predictions based on the mean of the training observations, a benchmark

	MSE			TP / FP		
n	1000	5000	10,000	1000	5000	10,000
LASSO	4.59	3.98	3.96	9 / 10	10 / 17	10 / 33
MIXTURE	4.12	3.89	3.71	7 / 0	10 / 0	10 / 0
R FOREST	5.25	4.53	4.32	4 / 6	9 / 1	10 / 0
NULL MODEL	5.55	5.03	5.03			

Table 11.5 *Interaction Model.* Validating mean square error (MSE), true positive calls (TP) and false positive calls (FP) among flagged loci in the discovery set (TP+FP) using lasso, the Bayesian mixture and the random forest for three sets of number of individuals (n). The discovery set was arbitrarily defined as follows. For lasso, the set of regression estimates not set equal to zero. For the Bayesian mixture, the loci whose McMC estimate of the probability of belonging to the non-zero mixture component is larger than 0.5. For random forest, the 10 loci that had the highest importance score. As point of reference, NULL MODEL refers to the validation MSE of predictions based on the mean of the training observations, a benchmark

	MSE			TP / FP		
n	1000	5000	10,000	1000	5000	10,000
LASSO	4.47	5.04	4.89	1 / 15	1 / 2	1 / 0
MIXTURE	4.46	5.03	4.87	1 / 1	1 / 0	1 / 0
R FOREST	4.61	4.95	4.69	1 / 9	4 / 6	8 / 2
NULL MODEL	4.61	5.16	5.01			

definition of discovery set). With $n = 1000$ observations the random forest and the Bayesian mixture detect four and seven of the ten causal loci, respectively. The random forest flags six false positives and the Bayesian mixture none. Lasso on the other hand detects nine of the ten causal loci and generates ten false positives. With the two largest population sizes, the three methods successfully detect the ten causal loci (with the exception of the random forest that detects nine out of the ten with one false positive when fitted to $n = 5000$ observations). However, lasso, in contrast with the other two methods, flags a considerable number of false positive calls with the larger population sizes.

The results for the *interaction Model* are displayed in Table 11.5. The prediction ability of lasso and Bayesian mixture is slightly better to that of the random forest for $n = 1000$, but the reverse is true for the two larger sample sizes.

When $n = 1000$, all three methods detect the single additive QTL. Lasso and random forest produce a substantial amount of false positive calls, but the Bayesian mixture produces only 1. When $n = 5000$ or 10,000, lasso and the Bayesian mixture detect the single additive QTL. On the other hand, the random forest detects not only the single additive QTL but also flags 4 and 8 of the 10 genotypes involved in epistatic interactions, for sample sizes $n = 5000$ or 10,000, respectively. The number of false positive results in the random forest discovery set is 6 and 2, for $n = 5000$ and $n = 10{,}000$, respectively.

We conclude the example with a few remarks:

- The analysis confirms both conjectures in terms of prediction ability. When it comes to identification of causal loci involved in epistatic interactions, the conjecture holds provided $n > p$.
- When the relationship between signal and noise is unfavourable for detection of loci involved in epistatic interactions, a possible strategy is to fit a random forest locally, in specific chromosome segments, searching for loci that are not captured by a regression based on linear terms.
- A common way of searching for interactions is, first, to fit a regression based on linear terms and, second, using the detected marker loci fit for the quadratic terms. This strategy does not work when the QTL emit signals only through epistatic combinations.
- The random forest, as implemented in the example, can potentially capture individual QTL that are involved in epistatic interactions but does not flag the interaction itself. A simple discovery strategy, particularly when the number of single loci discovered is not large relative to n, is to include these individual loci in a second step constructing all possible interaction combinations using, for example, simple least squares regression, or alternatively, a Bayesian mixture.

NOTE: The Epistatic Model

The epistatic model used in the example induces additive genetic variation through the single additive QTL. The 10 epistatic loci generate additive × additive epistatic variance only. This is stylised and holds provided that loci are in linkage equilibrium and gene frequencies are all intermediate.

The details of the model for a specific interaction genotype are as follows. Each biallelic locus with genotypes A_1A_1, A_1A_2 and A_2A_2 is coded as $X_{11} = 1$, $X_{12} = X_{21} = 0$ and $X_{22} = -1$, respectively. Then the quadratic term is obtained as $X_{ij}X_{kl} + 1$, $i, j, k, l = 1, 2$. This generates the following genotypic codes for the additive × additive epistatic model:

	A_1A_1	A_1A_2	A_2A_2
B_1B_1	2	1	0
B_1B_2	1	1	1
B_2B_2	0	1	2

The R-code below provides an example with 10^6 individuals in the sample and a single additive × additive genotype.

```
# CODE1115
###       GENETIC MODELS
rm(list=ls()) # CLEAR WORKSPACE
set.seed(303371)
nindiv<-1000000
nqtl <- 2
nintqtl <- 1
ba <- rep(1,nqtl)
bi <- rep(1,nintqtl)
mu_y<-0
# INTERACTION GENOTYPIC MATRIX
Xi<-matrix(data=NA,nrow= nindiv,ncol= nintqtl)
Xq<-matrix(nrow= nindiv,ncol= nqtl,
    rbinom(nindiv*nqtl,size=2,p=.5)-1) # LINEAR TERMS
nr <- ncol(Xq)
i1 <- combn(nr,2)
i2 <- sample(ncol(i1),nintqtl,replace=FALSE)
i3 <- as.matrix(i1[,i2])
# CONSTRUCT INTERACTION GENOTYPE
for (i in 1:nintqtl){
  Xi[,i] <- Xq[,i3[1,i]]*Xq[,i3[2,i]]+1
}
gi <-Xi%*%bi # INTERACTION GENETIC VALUES
ga <- Xq%*%ba # ADDITIVE GENETIC VALUES
g <- ga + gi # TOTAL GENETIC VALUES
va <- var(ga)
vi <- var(gi)
vg <- var(g)
y_i <- gi+rnorm(nindiv,0,sqrt(5-vi))
y_ai <- ga + gi+rnorm(nindiv,0,sqrt(5-va-vi))
vy <- var(y_i)
cor(Xq)
```

```
##              [,1]         [,2]
## [1,] 1.0000000000 0.0003593763
## [2,] 0.0003593763 1.0000000000
```

```
vy
```

```
##          [,1]
## [1,] 4.982423
```

```
vi
```

```
##              [,1]
## [1,] 0.2497092
```

```
va
```

```
##              [,1]
## [1,] 0.9998697
```

```
vg
```

```
##             [,1]
## [1,] 1.249443
```

```
# FIT LINEAR REGRESSION OF
# INTERACTION GENETIC VALUES ON QTL LOCI
fa <- lm(gi ~ Xq)
# CONFIRM THAT THE MODEL DOES NOT CAPTURE ANY
# (ADDITIVE GENETIC) VARIATION
res <- summary(fa)
res$r.squared
```

```
## [1] 9.507938e-07
```

```
res$coefficients[,1:3]
```

```
##                   Estimate   Std. Error      t value
## (Intercept)  1.0001798275 0.0004997101 2001.5199827
## Xq1         -0.0005506352 0.0007067582   -0.7790999
## Xq2          0.0004147291 0.0007069803    0.5866204
```

The 2×2 correlation matrix of the QTL genotypes indicates virtually no linkage disequilibrium. The output of the summary indicates that the linear regression of epistatic genetic values on the two QTL genotypes fails to capture epistatic variation and, in agreement with this, the two linear regression coefficients do not differ from zero. In other words, for these data the additive genetic variance is zero. This is no longer the case if matrix Xq of QTL genotypes is generated from a binomial distribution with $n = 2$ and $p \neq 0.5$. The subject is further explored by, for example, Hill et al (2008) and Mäki-Tanila and Hill (2014).

11.6 Appendix

This appendix provides the details of back propagation for the single hidden layer neural network described on page 497. It is shown how matrix multiplications can be optimised computationally by means of vectorisation. The last subsection provides an example of minimisation of a cost function, when the architecture of the NN requires computation of partial derivatives following more than a single path.

Minimisation of the Cost Function: Back Propagation

Finding local minima of (11.78) requires the gradients $\partial J(\theta)/\partial W_{ji}^{(l)}, \partial J(\theta)/\partial b_j^{(l)}$. For the neural network of Fig. 11.10, $K = 1$, $L = 3$, $n_1 = 3$, $n_2 = 3$, $n_3 = 1$, and the predicted value for the ith datum is $h_{W,b}(x^{(i)}) = a_i^{(3)}$. Then (11.78) reduces to

$$J(\theta) = -\frac{1}{m}\left[\sum_{i=1}^{m} y^{(i)} \log a_i^{(3)} + \left(1 - y^{(i)}\right)\log\left(1 - a_i^{(3)}\right)\right] + \frac{\lambda}{2}\sum_{l=1}^{2}\sum_{i=1}^{n_l}\sum_{j=1}^{n_{l+1}}\left(W_{ji}^{(l)}\right)^2. \tag{11.84}$$

In this neural network, as indicated in (11.77), the dimension of the matrix of parameters is

$$\begin{aligned} W^{(1)} &= (3 \times 3), \\ W^{(2)} &= (1 \times 3), \\ b^{(1)} &= (3 \times 1), \\ b^{(2)} &= (1 \times 1). \end{aligned}$$

There are a total of 16 parameters including four bias terms.

The back propagation for this neural network is derived in detail. The different steps pertain to Fig. 11.13 that displays a flowchart of the operations and make

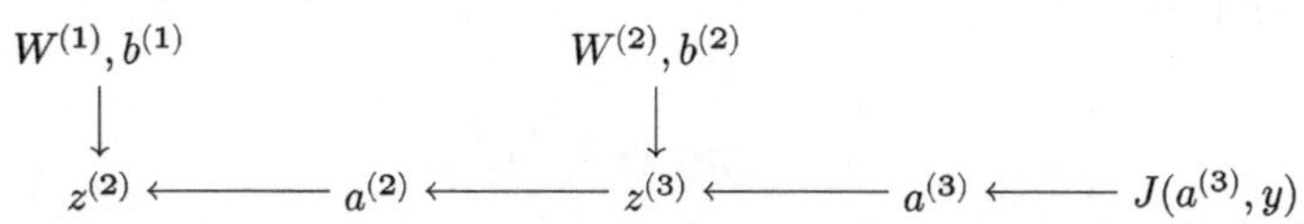

Fig. 11.13 Scheme of back propagation for the neural network displayed in Fig. 11.10, moving from right to left. Forward propagation moves from left to right. Back propagation computes partial derivatives of the cost function J with respect to $W^{(2)}$, $b^{(2)}$, $W^{(1)}$, $b^{(1)}$. This involves 16 parameters

reference to Eqs. (11.76). In a first stage, I start with a single record/example and the computations involve only the first term on the right-hand side of (11.84) (partial derivatives for the regularisation term are of the form $\lambda W_{ji}^{(\ell)}$).

- Step 1. Start from the last layer (layer 3) and compute $\delta^{(3)}$ in two steps

$$\delta^{(3)} = \frac{\partial J}{\partial z^{(3)}} = \frac{\partial J}{\partial a^{(3)}}\frac{\partial a^{(3)}}{\partial z^{(3)}}$$
$$= -\frac{\partial\left[y\log a^{(3)} + (1-y)\log\left(1-a^{(3)}\right)\right]}{\partial a^{(3)}}\frac{\partial a^{(3)}}{\partial z^{(3)}}.$$

From the first term in (11.84), dropping the superscript (i) denoting the record

$$\frac{\partial J}{\partial a^{(3)}} = -\frac{y}{a^{(3)}} + \frac{(1-y)}{\left(1-a^{(3)}\right)},$$
$$\frac{\partial a^{(3)}}{\partial z^{(3)}} = a^{(3)}\left(1-a^{(3)}\right),$$
$$\Rightarrow \delta^{(3)} = a^{(3)} - y.$$

These partial derivatives take the same form as (11.72) and (11.73).

- Step 2. Compute partial derivatives of J with respect to $W^{(2)} = \left[W_{11}^{(2)}, W_{12}^{(2)}, W_{13}^{(2)}\right]$, $b^{(2)}$,

$$\frac{\partial J}{\partial W^{(2)\prime}} = \frac{\partial J}{\partial z^{(3)}}\frac{\partial z^{(3)}}{\partial W^{(2)\prime}},$$
$$\frac{\partial z^{(3)}}{\partial W_{11}^{(2)}} = a_1^{(2)},$$
$$\frac{\partial z^{(3)}}{\partial W_{12}^{(2)}} = a_2^{(2)},$$
$$\frac{\partial z^{(3)}}{\partial W_{13}^{(2)}} = a_3^{(2)},$$

and therefore

$$\frac{\partial J}{\partial W^{(2)}} = \delta^{(3)} a^{(2)}, \tag{11.85}$$

where $\delta^{(3)} = \partial J/\partial z^{(3)}$ is a scalar and $a^{(2)} = \left[a_1^{(2)}, a_2^{(2)}, a_3^{(2)}\right]$ is a (3×1) column vector. Above, $W^{(2)}$ is a matrix with a single row and three columns consistent with the definition given for $W^{(j)}$. We need also

$$\frac{\partial J}{\partial b^{(2)}} = \frac{\partial J}{z^{(3)}}\frac{\partial z^{(3)}}{\partial b^{(2)}} = \delta^{(3)} \tag{11.86}$$

because $\partial z^{(3)}/\partial b^{(2)} = 1$.

- Step 3. Compute $\delta^{(2)}$ in two steps. The first step is

$$\delta^{(2)} = \frac{\partial J}{\partial z^{(2)}} = \frac{\partial J}{\partial a^{(2)}}\frac{\partial a^{(2)}}{\partial z^{(2)}},$$

$$\frac{\partial J}{\partial a^{(2)}} = \frac{\partial J}{\partial z^{(3)}}\frac{\partial z^{(3)}}{\partial a^{(2)}},$$

$$\frac{\partial z^{(3)}}{\partial a_1^{(2)}} = W_{11}^{(2)},$$

$$\frac{\partial z^{(3)}}{\partial a_2^{(2)}} = W_{12}^{(2)},$$

$$\frac{\partial z^{(3)}}{\partial a_3^{(2)}} = W_{13}^{(2)}.$$

The second step is

$$\frac{\partial a^{(2)}}{z^{(2)}} = \sigma'\left(z^{(2)}\right) = \begin{bmatrix} a_1^{(2)}\left(1 - a_1^{(2)}\right) \\ a_2^{(2)}\left(1 - a_2^{(2)}\right) \\ a_3^{(2)}\left(1 - a_3^{(2)}\right) \end{bmatrix}$$

$$= a^{(2)} \cdot \left(1 - a^{(2)}\right),$$

where $\cdot$ denotes the Hadamard (or elementwise) product and 1 is a column vector with three elements. Finally,

$$\delta^{(2)} = W^{(2)\prime}\delta^{(3)} \cdot \sigma'\left(z^{(2)}\right).$$

This can be written compactly as follows:

$$\delta^{(2)} = \frac{\partial J}{\partial z^{(2)}} = \begin{bmatrix} \delta_1^{(2)} \\ \delta_2^{(2)} \\ \delta_3^{(2)} \end{bmatrix} = \delta^{(3)} \begin{bmatrix} W_{11}^{(2)} \\ W_{12}^{(2)} \\ W_{13}^{(2)} \end{bmatrix} \cdot \begin{bmatrix} a_1^{(2)}\left(1 - a_1^{(2)}\right) \\ a_2^{(2)}\left(1 - a_2^{(2)}\right) \\ a_3^{(2)}\left(1 - a_3^{(2)}\right) \end{bmatrix}.$$

- Step 4. Compute partial derivatives of J with respect to $W^{(1)}$, $b^{(1)}$.

$$\frac{\partial J}{\partial W_{11}^{(1)}} = \frac{\partial J}{\partial z_1^{(2)}} \frac{\partial z_1^{(2)}}{\partial W_{11}^{(1)}} = \delta_1^{(2)} a_1^{(1)},$$

$$\frac{\partial J}{\partial W_{12}^{(1)}} = \frac{\partial J}{\partial z_1^{(2)}} \frac{\partial z_1^{(2)}}{\partial W_{12}^{(1)}} = \delta_1^{(2)} a_2^{(1)},$$

$$\vdots$$

$$\frac{\partial J}{\partial W_{33}^{(1)}} = \frac{\partial J}{\partial z_3^{(2)}} \frac{\partial z_3^{(2)}}{\partial W_{33}^{(1)}} = \delta_3^{(2)} a_3^{(1)},$$

and

$$\frac{\partial J}{\partial b_1^{(1)}} = \frac{\partial J}{\partial z_1^{(2)}} \frac{\partial z_1^{(2)}}{\partial b^{(1)}} = \delta_1^{(2)},$$

$$\frac{\partial J}{\partial b_2^{(1)}} = \frac{\partial J}{\partial z_2^{(2)}} \frac{\partial z_2^{(2)}}{\partial b_2^{(1)}} = \delta_1^{(2)},$$

$$\frac{\partial J}{\partial b_3^{(1)}} = \frac{\partial J}{\partial z_3^{(2)}} \frac{\partial_3^{(2)}}{\partial b_3^{(1)}} = \delta_3^{(2)}.$$

For the ijth element of W of layer ℓ and for the ith bias term b of layer ℓ, we have

$$\frac{\partial J}{\partial W_{ij}^{(\ell)}} = \delta_i^{(\ell+1)} a_j^{(\ell)},$$

$$\frac{\partial J}{\partial b_i^{(\ell)}} = \delta_i^{(\ell+1)}.$$

In vectorised form, for any matrix $W^{(\ell)}$, vector $b^{(\ell)}$ in layer ℓ and activation function g, the back propagation for a single record, in standard multilayer neural networks, reduces to the following three equations:

$$\begin{aligned}
\delta^{(\ell)} &= \left(W^{(\ell)'} \delta^{(\ell+1)}\right) \cdot g'\left(z^{(\ell)}\right), \\
\frac{\partial J}{\partial W^{(\ell)}} &= \delta^{(\ell+1)} a^{(\ell)'} + \lambda W^{(\ell)}, \\
\frac{\partial J}{\partial b^{(\ell)}} &= \delta^{(\ell+1)}.
\end{aligned} \tag{11.87}$$

For instance,

$$\frac{\partial J}{\partial W^{(1)}} = \begin{bmatrix} \delta_1^{(2)} \\ \delta_2^{(2)} \\ \delta_3^{(2)} \end{bmatrix} \begin{bmatrix} a_1^{(1)} a_2^{(1)} a_3^{(1)} \end{bmatrix} + \lambda W^{(1)}$$

$$= \begin{bmatrix} \delta_1^{(2)} a_1^{(1)} & \delta_1^{(2)} a_2^{(1)} & \delta_1^{(2)} a_3^{(1)} \\ \delta_2^{(2)} a_1^{(1)} & \delta_2^{(2)} a_2^{(1)} & \delta_2^{(2)} a_3^{(1)} \\ \delta_3^{(2)} a_1^{(1)} & \delta_3^{(2)} a_2^{(1)} & \delta_3^{(2)} a_3^{(1)} \end{bmatrix} + \lambda W^{(1)},$$

and in particular

$$\frac{\partial J}{\partial W_{23}^{(1)}} = \delta_2^{(2)} a_3^{(1)} + \lambda W_{23}^{(1)}.$$

With the gradients available from (11.87) and averaging the gradients over the m training records, implementation of gradient descent to find a local minimum of J requires the updating steps

$$W^{(\ell)} : W^{(\ell)} - \alpha \frac{\partial J}{W^{(\ell)}},$$

$$b^{(\ell)} : b^{(\ell)} - \alpha \frac{\partial J}{\partial b^{(\ell)}},$$

where gradients are interpreted as averages over the m training records.

Vectorising Forward and Back Propagation for the Complete Training Data

Much of the material in this section is guided by Andrew Ng's course on Machine Learning, Stanford University.

Most neural network computations involve matrix multiplications. This can be exploited in order to improve computational efficiency by means of vectorisation.

The system of equations (11.77) for forward propagation and (11.87) for back propagation involves vectorised forms for a single-input datum. Therefore, the complete propagation requires a loop over all the training records. Such a loop is computationally inefficient and can be avoided by vectorising over the complete set of training records.

The complete vectorisation of forward propagation is achieved as follows. Assume that the m training records are $\left(x^{(1)}, y^{(1)}\right), \ldots, \left(x^{(m)}, y^{(m)}\right)$, where $x^{(i)} \in \mathbb{R}^p$ and $y^{(i)} \in \mathbb{R}$. First, stack the input feature vectors column-wise and generate the matrix X of dimension $p \times m$,

$$X = \left[x^{(1)} \; x^{(2)} \; \cdots \; x^{(m)} \right]$$

where each $x^{(i)}$ is a column vector with p elements. I illustrate with the neural network of Fig. 11.10, with $p = n_1 = 3, n_2 = 3, n_3 = 1$ and X is $n_1 \times m$. Then the complete forward propagation involves computation of

1. $Z^{(2)} = W^{(1)}X + b^{(1)}$, $Z^{(2)} : n_2 \times m = 3 \times m$; $W^{(1)} : n_2 \times n_1 = 3 \times 3$; $X : n_1 \times m = 3 \times m$; $b^{(1)} : n_2 \times m = 3 \times m$
2. $A^{(2)} = \sigma\left(Z^{(2)}\right)$, $A^{(2)} : n_2 \times m = 3 \times m$
3. $Z^{(3)} = W^{(2)}A^{(2)} + b^{(2)}$, $Z^{(3)} : n_3 \times m = 1 \times m$; $W^{(2)} : n_3 \times n_2 = 1 \times 3$; $b^{(2)} : n_3 \times m = 1 \times m$
4. $A^{(3)} = \sigma\left(Z^{(3)}\right)$, $A^{(3)} : n_3 \times m = 1 \times m$

In step 1, $b^{(1)}$ is the original 3×1 column vector with the single column repeated m times. Similarly in step 3, $b^{(2)}$ is the original scalar, repeated column-wise m times. Computation of the 4 steps generates the forward propagation for the complete training data. The activation function σ in steps 2 and 4 can be arbitrary, but in the present example, it is the sigmoid function.

To compute back propagation in vectorised form, define

- $\Delta Z^{(\ell)}$: vectorised version of $\delta^{(\ell)} = \partial J / \partial z^{(\ell)}$
- $\Delta W^{(\ell)}$: vectorised version of $\partial J / \partial W^{(\ell)}$
- $\Delta b^{(\ell)}$: vectorised version of $\partial J / \partial b^{(\ell)}$
- $Y = \left(y^{(1)}, y^{(2)}, \ldots, y^{(m)}\right)$: $1 \times m$ vector of observed records

The complete back propagation involves computation of

1. $\Delta Z^{(3)} = A^{(3)} - Y$, $\Delta Z^{(3)} : 1 \times m$
2. $\Delta W^{(2)} = \frac{1}{m}\Delta Z^{(3)} A^{(2)'} + \lambda W^{(2)}$, $\Delta W^{(2)} : 1 \times 3$; $A^{(2)'} : m \times 3$; $W^{(2)} : 1 \times 3$
3. $\Delta b^{(2)} = \frac{1}{m}\Delta Z^{(3)} 1$, $\Delta b^{(2)} : 1 \times 1$ and 1 is a $m \times 1$ column vector of $1's$ ($\Delta b^{(2)}$ is the result of averaging the elements of the row vector $\Delta Z^{(3)}$ over the number of training records)
4. $\Delta Z^{(2)} = W^{(2)'} \Delta Z^{(3)} \cdot \sigma'\left(Z^{(2)}\right)$, $\Delta Z^{(2)} : 3 \times m$; $W^{(2)'} : 3 \times 1$; $\Delta Z^{(3)} : 1 \times m$; $\sigma'\left(Z^{(2)}\right) : 3 \times m$
5. $\Delta W^{(1)} = \frac{1}{m}\Delta Z^{(2)} X' + \lambda W^{(1)}$, $\Delta W^{(1)} : 3 \times 3$; $\Delta Z^{(2)} : 3 \times m$; $X' : m \times 3$; $\lambda W^{(1)} : 3 \times 3$
6. $\Delta b^{(1)} = \frac{1}{m}\Delta Z^{(2)} 1$, $\Delta b^{(1)} : 3 \times 1$; 1 is a $m \times 1$ column vector of $1's$ ($\Delta b^{(1)}$ is the result of averaging the elements of each of the three rows of matrix $\Delta Z^{(2)}$ over the number of columns).

Computation of the six steps generate the back propagation for the complete training data, and as in the case of forward propagation, the activation function in

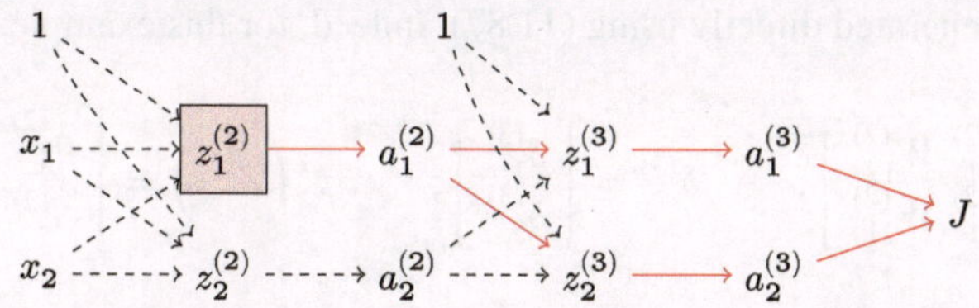

Fig. 11.14 A three-layer neural network with 2 output classes, $a_1^{(3)}$, $a_2^{(3)}$. The computation of the partial derivative of the cost function J with respect to $z_1^{(2)}$ must trace two paths (marked in red), since a change in $z_1^{(2)}$ has an effect on J via $z_1^{(3)}$ and via $z_2^{(3)}$

step 4 is arbitrary. The result in step 1 is specific for the sigmoid function used in the output layer.

Back Propagation with Multiple Paths

Neither of the two examples of back propagation discussed so far involved the computation of partial derivatives following more than one path along the lines indicated in (11.67). To illustrate a case where this is necessary, consider the 3-layer neural network displayed in Fig. 11.14 with two output classes rather than the single output class neural network of Fig. 11.10 (another example could involve a neural network with a single output neuron and more than 3 layers).

The forward propagation for a single data point for this network generates

$$\begin{aligned} z_i^{(2)} &= W_{i1}^{(1)} x_1 + W_{i2}^{(1)} x_2 + b_i^{(1)}, \quad i = 1, 2, \\ z_i^{(3)} &= W_{i1}^{(2)} a_1^{(2)} + W_{i2}^{(1)} a_2^{(2)} + b_i^{(2)}, \quad i = 1, 2, \\ a_i^{(2)} &= \sigma\left(z_i^{(2)}\right) \quad i = 1, 2, \\ a_i^{(3)} &= \sigma\left(z_i^{(3)}\right) \quad i = 1, 2. \end{aligned}$$

Consider the computation of $\partial J \big/ \partial z_1^{(2)} = \delta_1^{(2)}$. This requires

$$\begin{aligned} \frac{\partial J}{\partial z_1^{(2)}} &= \frac{\partial J}{\partial a_1^{(2)}} \frac{\partial a_1^{(2)}}{\partial z_1^{(2)}} \\ &= \left[\frac{\partial J}{\partial z_1^{(3)}} \frac{\partial z_1^{(3)}}{\partial a_1^{(2)}} + \frac{\partial J}{\partial z_2^{(3)}} \frac{\partial z_2^{(3)}}{\partial a_1^{(2)}} \right] \frac{\partial a_1^{(2)}}{\partial z_1^{(2)}} \\ &= \left[\delta_1^{(3)} W_{11}^{(2)} + \delta_2^{(3)} W_{21}^{(2)} \right] a_1^{(2)} \left(1 - a_1^{(2)}\right). \end{aligned} \tag{11.88}$$

This can be generated directly using (11.87). Indeed, for this example, $\ell = 2$,

$$W^{(2)} = \begin{bmatrix} W_{11}^{(2)} & W_{12}^{(2)} \\ W_{21}^{(2)} & W_{21}^{(2)} \end{bmatrix}; \quad \delta^{(3)} = \begin{bmatrix} \delta_1^{(3)} \\ \delta_2^{(3)} \end{bmatrix}; \quad g'\left(Z^{(2)}\right) = \begin{bmatrix} a_1^{(2)}\left(1 - a_1^{(2)}\right) \\ a_2^{(2)}\left(1 - a_2^{(2)}\right) \end{bmatrix}$$

and substituting in

$$\delta^{(l)} = \left(W^{(\ell)'}\delta^{(\ell+1)}\right) \cdot g'\left(z^{(\ell)}\right)$$

yields

$$\delta^{(2)} = \frac{\partial J}{\partial z^{(2)}} = \begin{bmatrix} \delta_1^{(2)} \\ \delta_2^{(2)} \end{bmatrix} = \begin{bmatrix} \delta_1^{(3)} W_{11}^{(2)} + \delta_2^{(3)} W_{21}^{(2)} \\ \delta_1^{(3)} W_{12}^{(2)} + \delta_2^{(3)} W_{22}^{(2)} \end{bmatrix} \cdot \begin{bmatrix} a_1^{(2)}\left(1 - a_1^{(2)}\right) \\ a_2^{(2)}\left(1 - a_2^{(2)}\right) \end{bmatrix}.$$

Part III
Exercises and Solutions

Chapter 12
Exercises

12.1 Likelihood Exercises I

Exercise 1

In a binomial experiment with n trials and probability of success θ, x successes are observed. The setup could represent a trial designed to estimate the proportion of individuals in a population that suffer from a particular disease. A sample of n randomly selected individuals is taken and the number diseased, x, is recorded:

(a) Write down the probability mass function of the data
(b) Write down the likelihood function and the loglikelihood function of θ.
(c) Derive the maximum likelihood estimator of θ. Assume $n = 10$ and $x = 8$. What is the maximum likelihood estimate of θ?
(d) Derive a 95% confidence interval for θ using the asymptotic approximation to the ML estimator of θ.

Exercise 2

In the above experiment, use the transformation (known as the logit or logodds):

$$\beta = g(\theta) = \ln\left(\frac{\theta}{1-\theta}\right). \tag{12.1}$$

The inverse transformation is

$$\theta = g^{-1}(\beta) = \frac{\exp(\beta)}{1+\exp(\beta)}. \tag{12.2}$$

D. Sorensen, *Statistical Learning in Genetics*, Statistics for Biology and Health,
https://doi.org/10.1007/978-3-031-35851-7_12

(a) Write the likelihood based on β.
(b) Plot the likelihood based on β and the likelihood based on θ and compare both.
(c) Derive the ML estimator and the ML estimate of β based on the likelihood (a).
(d) What is the 95% confidence interval for β?
(e) Finally, transform the confidence interval obtained in (d) back in terms of θ. How does this confidence interval compare with that obtained in (1d)? Which do you believe is more reliable?

Exercise 3

Suppose you observe the following $n = 10$ iid records each from $N\left(\mu, \sigma^2\right)$:

$$0.88; 1.07; 1.27; 1.54; 1.91; 2.27; 3.84; 4.50; 4.64; 5.41,$$

$$\sum_{i=1}^{n} y_i = 27.33; \sum_{i=1}^{n} (y_i - \widehat{\mu})^2 = 25.8052 \text{ where } \hat{\mu} = \frac{1}{n} \sum_{i=1}^{n} y_i, n = 10.$$

(a) Write down the likelihood and the loglikelihood of $\left(\mu, \sigma^2\right)$.
(b) Derive the ML estimators and the ML estimates of μ and σ^2.
(c) Derive the observed information matrix $I(\mu, \sigma^2)$ and evaluate it at the ML estimates.
(d) Derive the asymptotic variance of $\left(\hat{\mu}, \hat{\sigma}^2\right)$ based on 3c.
(e) Derive Fisher's (expected) information of $\left(\mu, \sigma^2\right)$ and evaluate it at the ML estimates.
(f) Find the asymptotic variance of $\left(\hat{\mu}, \widehat{\sigma}^2\right)$ based on Fisher's (expected) information obtained in 3e.
(g) According to asymptotic theory, $\widehat{\sigma}^2$ is normally distributed with mean equal to σ^2 and variance given by the result you derived in 3f. What is the exact (small sample) distribution of $\widehat{\sigma}^2$ and what is its mean and variance?
(h) Compute an exact (small sample) 95% confidence interval for σ^2 and an approximate 95% confidence interval based on asymptotic results.

Exercise 4

Consider the following hypothetical data collected from the field:

$$x_1, x_2, \ldots, x_m, x_{m+1}, \ldots, x_n,$$

$$y_1, y_2, \ldots, y_m.$$

There are m bivariate observations $\{(x_i, y_i), i = 1, 2, \ldots, m\}$ and $(n - m)$ univariate records. However sampling is not at random: the structure of the data mimics records on first and second lactation, where only individuals with the highest first lactation records $(x_1, \ldots, x_m)$ were allowed to produce a second lactation $(y_1, \ldots, y_m)$.

Using these data, the analyst wishes to estimate various parameters associated with milk production, such as the mean and variance of first and second lactation records, their correlation and the regression coefficient of second lactation records on first lactation records. The objective of the example is to show how the construction of the "correct" likelihood model avoids the potential problem associated with the non-random sampling mechanism that generated the data.

Let $\theta = (\mu_y, \mu_x, \sigma_{xy}, \sigma_{yy}, \sigma_{xx})$. Ignoring the selection mechanism and the $n-m$ univariate records $x_{m+1}, \ldots, x_n$, one can assume that each of the m pairs of records is bivariate normal:

$$(y_i, x_i)|\theta \sim N\left(\begin{bmatrix} \mu_y \\ \mu_x \end{bmatrix}, \begin{bmatrix} \sigma_{yy} = 4 & \sigma_{yx} = 2.4 \\ \sigma_{yx} = 2.4 & \sigma_{xx} = 4 \end{bmatrix}\right), \tag{12.3}$$

$$i = 1, 2, \ldots, m,$$

with $\mu_y = 10$, $\mu_x = 15$ and the pairs (y_i, x_i) are iid. A standard result is that, given (12.3),

$$y_i | x_i, \theta \sim N\left(\mathrm{E}(y_i|x_i, \theta), \mathrm{Var}(y_i|x_i, \theta)\right), \tag{12.4}$$

where

$$\begin{aligned} \mathrm{E}(y_i|x_i, \theta) &= \mu_y + \frac{\sigma_{yx}}{\sigma_{xx}}(x_i - \mu_x) \\ &= 10 + 0.6\,(x_i - 15) \end{aligned}$$

and

$$\begin{aligned} \mathrm{Var}(y_i|x_i, \theta) &= \sigma_{yy} - \sigma_{yx}(\sigma_{xx})^{-1}\sigma_{yx} \\ &= 4 - 2.4^2/4. \end{aligned}$$

For the m pair of records, the joint pdf is

$$p\left(x, y|\mu_y, \theta\right) = \left(2\pi\sigma_{yy}\sigma_{xx}\left(1-\rho^2\right)\right)^{-\frac{m}{2}} \tag{12.5}$$

$$\exp\left[-\frac{1}{2\left(1-\rho^2\right)}\right. \tag{12.6}$$

$$\left.\left(\frac{\sum_{i=1}^m (x_i-\mu_x)^2}{\sigma_{xx}} + \frac{\sum_{i=1}^m \left(y_i-\mu_y\right)^2}{\sigma_{yy}} - 2\rho\frac{\sum_{i=1}^m (x_i-\mu_x)\left(y_i-\mu_y\right)}{(\sigma_{xx})^{\frac{1}{2}}\left(\sigma_{yy}\right)^{\frac{1}{2}}}\right)\right], \tag{12.7}$$

where the coefficient of correlation $\rho = \sigma_{xy}/(\sigma_{xx}\sigma_{yy})^{0.5}$. Ignoring the sampling mechanism that gave rise to the data, the likelihood is proportional to (12.7) and the closed form likelihood estimators are given by

$$\widehat{\mu}_x = \frac{1}{m}\sum_{i=1}^{n} x_i, \qquad \widehat{\mu}_y = \frac{1}{m}\sum_{i=1}^{m} y_i, \tag{12.8a}$$

$$\widehat{\sigma}_{xx} = \frac{1}{m}\sum_{i=1}^{m} (x_i - \widehat{\mu}_x)^2, \qquad \widehat{\sigma}_{yy} = \frac{1}{m}\sum_{i=1}^{m} \left(y_i - \widehat{\mu}_y\right)^2, \tag{12.8b}$$

$$\widehat{\rho} = \frac{\widehat{\sigma}_{xy}}{\sqrt{\widehat{\sigma}_{xx}\widehat{\sigma}_{yy}}}, \qquad \widehat{\sigma}_{xy} = \frac{1}{m}\sum_{i=1}^{m} \left(y_i - \widehat{\mu}_y\right)(x_i - \widehat{\mu}_x). \tag{12.8c}$$

The R-code below generates bivariate observations where some of the records in one of the variables are missing. The strategy makes use of the factorisation:

$$p(x, y) = p(y|x)\, p(x).$$

```
rm(list=ls()) # Clear the workspace
set.seed(772231)
n <- 5000 # Number of records generated for Y_1
p <- 0.20 # Proportion to be selected
m <- round(p*n) # Number of records selected
y1 <- rnorm(n,mean=10,sd=2) # Generate random variable Y_1
y1_sort <-sort(y1,decreasing=TRUE) # Sort Y_1
y1s <- y1_sort[1:m] # The highest m Y_1 records are kept
#length(y1) # Number of records in Y_1
#length(y1s) # Number of records in the selected subset of Y_1
# Generate Y_2
y2<-rnorm(length(y1s),mean=15+0.6*(y1s-10),sd=sqrt(4-(2.4^2)/4))
```

(a) Using only the fraction of the data that has no missing offspring (i.e. the m pairs (y_i, x_i)), estimate the parameters with estimators in (12.8). From the same subset of data, estimate also the regression of y on x, using

$$\widehat{b}_{yx} = \frac{\widehat{\sigma}_{xy}}{\widehat{\sigma}_{xx}}. \tag{12.9}$$

The regression is a parameter that can be directly derived from the set of parameters in likelihood (12.7) or from (12.3). That is

$$b_{yx} = \frac{\sigma_{xy}}{\sigma_{xx}} = 0.6.$$

Similarly, the correlation coefficient is

$$\rho = \frac{\sigma_{xy}}{\sigma_{xx}\sigma_{yy}} = 0.6. \tag{12.10}$$

How do the estimates obtained using this approach compare with the parameters defined in (12.3)? An eyeball evaluation is adequate at this stage.

(b) Write down the correct likelihood and loglikelihood accounting for the observed pattern of data based on parametrisation $(\mu_x, \sigma_{xx}, \beta_0, \beta_1, \sigma_{y.x})$.

(c) Using the likelihood in (b), obtain ML estimators and ML estimates of $\phi = (\mu_x, \sigma_{xx}, \beta_0, \beta_1, \sigma_{y.x})$ by analytical maximisation of the loglikelihood. Backtransform these estimates in terms of $\theta = (\mu_y, \mu_x, \sigma_{xy}, \sigma_{yy}, \sigma_{xx})$. How do the estimates obtained using this second approach compare with the true value of the parameters defined in (12.3)?

Exercise 5

This is a theoretical exercise designed to train the skills necessary to construct and to maximise a likelihood function in the absence of random sampling. An exponential distribution is used often applied to study waiting times until the next event. For example, in a clinical trial, one may be interested in studying how a new drug affects time until a relapse of the disease. At the end of the trial, some patients may not have shown signs of relapse; this gives rise to censoring.

In the simple exponential sampling model used here, expectations can be obtained in closed form. The censoring mechanism is as follows: each observation is compared to a censoring or truncation point c (the termination time of the clinical trial). If it is smaller than or equal to c, it is registered as such; if it is larger than c (no relapse observed), it is set equal to c. Since selection depends on the censored records, it is not ignorable and must be incorporated in the likelihood in order to draw inferences correctly. To set the background, the first part of the example deals with standard inferences based on a random sample from an exponential distribution in the absence of censoring.

1. A random sample of size n is obtained from *iid* draws from an exponential distribution. The pdf of one observation is

$$p(y_i|b) = b \exp(-by_i), \quad b, y_i > 0.$$

The parameter b is known as the rate, the focus of inference.

It can readily be verified that

$$\int_0^\infty p(y_i|b)\, dy_i = 1.$$

The mean and variance of Y_i are

$$\mathrm{E}(Y_i|b) = \frac{1}{b}$$

and

$$\text{Var}(Y_i|b) = \frac{1}{b^2},$$

respectively (here I distinguish the random variable Y from its realised value y). The expression for the mean indicates that mean relapse time is inversely proportionately to the rate, as expected.

The pdf associated with the n independent observations is

$$p(y|b) = \prod_{i=1}^{n} p(y_i|b)$$
$$= b^n \exp\left(-b \sum_{i=1}^{n} y_i\right). \tag{12.11}$$

(a) Write the loglikelihood
(b) Write the score
(c) Derive the maximum likelihood estimator of b
(d) Derive Fisher's information from $-\mathrm{E}\left[l''(b|Y)\right]$

2. Assume that after the random sample of size n has been drawn, $(n-r)$ observations larger than a known point c are censored. Clearly, $r \leq n$. There are r observed records which happen to be smaller than c, and $(n-r)$ records whose values are equal to c.

The data-generating mechanism can be written as follows. To derive the pdf of a single observation, define a latent variable X_i whose density is $p(x_i|b) = b\exp(-bx_i)$, $b, x_i > 0$ and let the data be the observed values of the random variable:

$$Y_i = \begin{cases} c & \text{if } X_i \geq c \\ X_i & \text{if } X_i < c. \end{cases}$$

Note that

$$\Pr(X_i \geq c) = 1 - \int_0^c p(x_i|b)\,dx_i$$
$$= \exp(-bc)$$
$$= 1 - F(c),$$

$$\Pr(X_i < c) = \int_0^c p(x_i|b)\,dx_i$$
$$= 1 - \exp(-bc)$$
$$= F(c).$$

Then the pdf of Y_i is

$$\begin{aligned} p_{Y_i}(y_i|b) &= p_{X_i}(y_i|Y_i<c, b)\Pr(Y_i<c|b) + \Pr(Y_i=c|b, Y_i=c)\Pr(Y_i=c|b) \\ &= I(y_i<c)\, p_{X_i}(y_i|b) + I(y_i=c)\Pr(Y_i=c|b) \\ &= I(y_i<c)\, b\exp(-by_i) + I(y_i=c)\exp(-bc), \end{aligned} \tag{12.12}$$

and due to independence, $p(y|b) = \prod_{i=1}^{n} p(y_i|b)$. Expression (12.12) is a proper pdf. Indeed,

$$\begin{aligned} \int p(y_i|b)\, dy_i &= \int I(y_i<c)\, b\exp(-by_i)\, dy_i + \exp(-bc) \\ &= 1 - \exp(-bc) + \exp(-bc) = 1. \end{aligned}$$

(a) Write the loglikelihood
(b) Write the score (loglikelihood)
(c) Derive the maximum likelihood estimator of b
(d) Derive Fisher's information from $-\mathrm{E}\left[l''(b|Y)\right]$

12.2 Likelihood Exercises II

Exercise 1

The R-code below generates 30 binary records mimicking the presence-1/absence-0 of disease among subjects exposed to (scaled) levels of a drug, the covariate. The analyst could be interested in learning how the probability of the disease is affected by the levels of the drug:

```
rm(list=ls()) # Clear the workspace
set.seed(12371)
# CREATE BINARY DATA
mu <- -2
beta <- 0.7
cov <- rnorm(30,2,3) # GENERATE THE COVARIATE
xb <- cov*beta
p1 <- pnorm(mu+xb) # PROBABILITIES ACCORDING TO PROBIT MODEL
#p1 <- rbeta(30,2,2)
dat1 <- cbind(rbinom(30,1,p1),round(cov,digits=0)) # CREATE DATA
colnames(dat1) <- c("Y", "X")
d <- data.frame(dat1)
head(d)
```

```
##   Y X
## 1 1 7
## 2 1 6
## 3 1 3
## 4 0 1
## 5 1 4
## 6 0 0
```

In an initial analysis, one can study the following approximate relationship between the probability of disease and the level of the drug. Let

$$\theta_i = \Pr(Y_i = 1|\mu, \beta, x_i),$$

be the probability that the ith individual shows the disease, given the covariate and parameters μ and β:

(a) Sort the covariate x and then create *five* groups where group 1 includes the *six* individuals exposed to the lowest levels of the drug and group 5 includes individuals exposed to the highest levels of the drug. Plot the proportion of individuals that show the disease in each group versus the mean level of the group. Does the relationship look approximately linear?

(b) Fit a model of the form

$$Y_i = \mu + \beta x_i + e_i, \tag{12.13}$$

and estimate μ and β by least squares. In this expression Y_i is a proxy for θ_i and represents the number of cases that show the disease in group i, divided by the number of cases in group i (i.e. *six* cases) and x_i is the average level of the drug in group i over the *six* cases. What is the probability that a future individual will show the disease when exposed to a scaled drug level equal to −3, 1 or 9?

(c) Fit the logistic regression:

$$\text{logit}(\theta_i) = \ln\left(\frac{\theta_i}{1-\theta_i}\right) = \mu + \beta x_i. \tag{12.14}$$

From this expression, solving for the probability θ_i (the probability that $Y_i = 1$)

$$\text{logit}^{-1}(\theta_i) = \theta_i = \frac{\exp(\mu + \beta x_i)}{1 + \exp(\mu + \beta x_i)}. \tag{12.15}$$

The parametrisation used is

$$Y_i = \begin{cases} 1 \text{ if } u_i > 0, \\ 0 \text{ if } u_i < 0, \end{cases}$$

$$u_i = \mu + \beta x_i + e_i,$$

where the independent errors e_i follow the standard logistic distribution. Therefore,

$$\begin{aligned}
\Pr(Y_i = 1|\mu, \beta, x_i) &= \Pr(u_i > 0|\mu, \beta, x_i) \\
&= \Pr(\mu + \beta x_i + e_i > 0|\mu, \beta, x_i) \\
&= \Pr(e_i > -\mu - \beta x_i|\mu, \beta, x_i) \\
&= \Pr(e_i < \mu + \beta x_i|\mu, \beta, x_i) \\
&= \int_{-\infty}^{\mu+\beta x_i} p(e_i)\, de_i \\
&= \int_{-\infty}^{\mu+\beta x_i} \frac{\exp(e_i)}{[1+\exp(e_i)]^2} de_i \\
&= \frac{\exp(\mu + \beta x_i)}{1 + \exp(\mu + \beta x_i)} \qquad (12.16)
\end{aligned}$$

as in (12.15), indicating that in the binary case, the logit and the threshold are equivalent models.

With the present parametrisation, the joint density of the data is

$$p(y|\mu, \beta, x) = \prod_{i=1}^{30} \left[\frac{\exp(\mu + \beta x_i)}{1 + \exp(\mu + \beta x_i)}\right]^{y_i} \left[\frac{1}{1 + \exp(\mu + \beta x_i)}\right]^{1-y_i}. \qquad (12.17)$$

(c1) Write the loglikelihood of the parameters

(c2) Calculate first and second derivatives of the loglikelihood

(c3) Fit the model (12.14) using Newton-Raphson and obtain ML estimates of μ and of β.

(c4) What is the asymptotic variance covariance matrix of the ML estimators of μ and of β?

(c5) What is the probability that a future individual will show the disease when exposed to a scaled drug level equal to -3, 1 or 9? Compare with what you obtained in 1b.

Exercise 2

Fit the probit regression model to the previous data using the EM algorithm.

Exercise 3

Table 12.1, taken from Gelman et al 1995, shows data from a bioassay experiment. Four different doses of a drug (measured as $\ln(g/ml)$) are administered to 20 animals, 5 in each dose, and the number of deaths is recorded. The objective of the experiment is to study the toxicity of the drug and how it varies with dose.

Let y_i represent the number of deaths observed out of n_i with dose level x_i. It will be assumed that y_i is binomial $Bi\ (n_i, \theta_i)$, where θ_i is modelled with the logistic regression:

$$\text{logit}\,(\theta_i) = \ln\left(\frac{\theta_i}{1-\theta_i}\right) = \beta_0 + \beta_1 x_i, \quad i = 1, \ldots, 4. \tag{12.18}$$

The inverse function yields:

$$\text{logit}^{-1}\,(\theta_i) = \theta_i = \frac{\exp\,(\beta_0 + \beta_1 x_i)}{1 + \exp\,(\beta_0 + \beta_1 x_i)}. \tag{12.19}$$

The likelihood takes the form

$$L\,(\beta_0, \beta_1 | n, x) \propto \prod\nolimits_{i=1}^{4} \left[\frac{\exp\,(\beta_0 + \beta_1 x_i)}{1 + \exp\,(\beta_0 + \beta_1 x_i)}\right]^{y_i} \left[1 - \frac{\exp\,(\beta_0 + \beta_1 x_i)}{1 + \exp\,(\beta_0 + \beta_1 x_i)}\right]^{n_i - y_i} \tag{12.20}$$

where $n = (n_1, \ldots, n_4)'$, $x = (x_1, \ldots, x_4)'$. The loglikelihood, after a little simplification, reduces to

$$\ell\,(\beta_0, \beta_1 | n, x) = \sum_{i=1}^{4} \{y_i\,(\beta_0 + \beta_1 x_i) - n_i \ln\,(1 + \exp(\beta_0 + \beta_1 x_i))\}\,. \tag{12.21}$$

(i) Obtain ML estimates of β_0 and β_1 for the logit model using Newton-Raphson.
(ii) Obtain ML estimates of β_0 and β_1 for the probit model using the EM algorithm.
(iii) Suppose that within each dose, the five animals happen to be full-sibs. What modifications to the loglikelihood (12.20) are necessary to account for the correlated structure of the data?

Table 12.1 Data from a bioassay experiment

Dose	Deaths	Sample size
−0.86	0	5
−0.30	1	5
−0.05	3	5
0.73	5	5

Example 4

The R-code below generates genomic data: 2000 nominally unrelated individuals (meaning distantly related) with 20,000 genetic markers available for each individual. Modern datasets include of the order of 200,000 individuals and 2 million marker genotypes per individual. This example scales down by factor 100. The simulated data assumes that all the genetic markers have an effect on the trait of the same magnitude.

In a preliminary analysis, one could be interested in fitting a classic genomic model to estimate the genomic variance.

Assuming that records have zero mean, the genomic model is

$$y|g, \sigma_e^2 \sim N\left(g, I\sigma_e^2\right), \tag{12.22a}$$

$$g|W, \sigma_g^2 \sim SN\left(0, G\sigma_g^2\right), \tag{12.22b}$$

$$G = \frac{1}{m}WW', \tag{12.22c}$$

$$W = \left\{W_{ij}\right\}. \tag{12.22d}$$

In these expressions, m is the number of SNPs, and W_{ij} is the label for the jth maker in individual i, $(i = 1, \ldots n;\ j = 1, \ldots, m;\ m > n)$:

$$W_{ij} = \frac{X_{ij} - \mathrm{E}\left(X_{ij}\right)}{\sqrt{\mathrm{Var}\left(X_{ij}\right)}}, \quad X_{ij} = 0, 1, 2.$$

The eigenvalue decomposition of WW' is

$$\begin{aligned} WW' &= U\Delta U' \\ &= \sum\nolimits_{i=1}^{n} \lambda_i U_i U_i^{'}, \end{aligned}$$

where $U = [U_1, U_2, \ldots, U_n]$, of order $n \times n$ is the matrix of eigenvectors of WW', U_j is the jth column (dimension $n \times 1$) and Δ is a diagonal matrix with elements equal to the eigenvalues $\lambda_1, \lambda_2, \ldots, \lambda_n$ associated with the n eigenvectors.

The loglikelihood, up to an additive constant, is

$$\ln p\left(k, \sigma_e^2|y, W\right) = -\frac{1}{2}\left\{n \ln \sigma_e^2 + \sum_{i=1}^{n} \ln\left(\lambda_i k + 1\right) + \frac{1}{\sigma_e^2}\sum_{i=1}^{n} \frac{\widetilde{y}_i^2}{\lambda_i k + 1}\right\}, \tag{12.23}$$

where $\tilde{y}_i$ is the ith element of the $n \times 1$ column vector $\tilde{y} = U'y$.

(i) Obtain ML estimates of σ_e^2 and k using Newton-Raphson.
(ii) Obtain ML estimates using the R-function OPTIM and compare with the Newton-Raphson computations.
(iii) The loglikelihood (12.23) is parametrised in terms of σ_e^2 and $k = \sigma_g^2/\sigma_e^2$. Obtain the ML estimate of σ_g^2 and its asymptotic variance from the Newton-Raphson output in (i).
(iv) Obtain ML estimates of σ_e^2, k and σ_g^2 using the EM algorithm.

```
# LIKELIHOOD PROBLEMS II QUESTION 4
# DATA BASED ON GENOMIC MODEL; OBTAIN SVD OF WW' (1/m)
rm(list=ls()) # CLEAR WORKSPACE
set.seed(1327)
nindiv<-2000
nmark<-20000
nt <- nindiv*nmark
X<-matrix(nrow=nindiv,ncol=nmark,rbinom(n=nt,size=2,p=.5))
stdev <- matrix(data=NA,nrow= nmark,ncol=1)
W <- matrix(data=NA,nrow= nindiv,ncol=nmark)
U <- matrix(data=NA,nrow= nindiv,ncol= nindiv)
G<-matrix(data=NA,nrow= nindiv,ncol= nindiv)
cm <- colMeans(X)
# MATRIX OF STANDARDISED MARKER GENOTYPE CODES
for (i in 1:nmark)
{
  W[,i] <-( X[,i]-cm[i]) / sd(X[,i])
}
# COULD USE INSTEAD:
# W <- scale(X)
#qr(X)$rank
#qr(W)$rank
# GENERATE nindiv GENOMIC VALUES N(0,(1/nmark)WW'*10); Vg=10
g <- (1/sqrt(nmark))*W%*%rnorm(nmark,mean=0,sd=sqrt(10))
# GENERATE nindiv PHENOTYPES WITH MEAN 0, VAR=10+15,
# HERITABILITY=10/(10+15)=0.4
#PARAMETER k = Vg/Ve = 10/15 =0.67
y <- g+rnorm(nindiv,mean=0,sd=sqrt(15))
# GENOMIC RELATIONSHIP MATRIX G
G <- (1/nmark)*W%*%t(W)
# SVD OF G
EVD <- eigen(G)
names(EVD)
head(EVD$values)
U <- EVD$vector
val <- EVD$values
val[length(y)] <-0
D <- diag(val,nrow=nindiv)
ytilde <- t(U)%*%y
dim(ytilde)
#END OF GENERATION OF DATA
```

12.3 Bayes Exercises I

Exercise 1

You are given the following data y drawn from a normal distribution:

$$y = (45.83,\ 50.37,\ 50.06,\ 51.59,\ 48.43,\ 52.75,\ 42.92,\ 48.57,\ 46.18,\ 50.20)'.$$

Assuming the likelihood is proportional to

$$y|\mu, \sigma^2 \sim N\left(1\mu, I\sigma^2\right),$$

that μ and σ^2 are a priori independent and their prior distributions are

$$[\mu] \propto constant, \tag{12.24a}$$

$$[\sigma^2] \propto \frac{1}{\sigma^2}, \tag{12.24b}$$

then the posterior distribution can be written

$$p\left(\mu, \sigma^2|y\right) \propto p\left(y|\mu, \sigma^2\right) p\left(\sigma^2\right) p\left(\mu\right)$$

$$\propto \exp\left[-\frac{\sum_{i=1}^{n}(y_i - \mu)^2}{2\sigma^2}\right]\left(\sigma^2\right)^{-\left(\frac{n}{2}+1\right)}. \tag{12.25}$$

(i) Derive analytically the marginal distributions $[\mu|y]$ and $[\sigma^2|y]$.
(ii) From these marginal distributions, obtain $\mathrm{E}[\mu|y]$, $\mathrm{Var}[\mu|y]$, $\mathrm{E}[\sigma^2|y]$, $\mathrm{Var}\left[\sigma^2|y\right]$ and the mode of $[\sigma^2|y]$.
(iii) Compute 95% posterior intervals for μ and σ^2.
(iv) Derive the asymptotic posterior distribution $[\mu, \sigma^2|y]$
(v) Obtain the 95% posterior intervals for μ and σ^2 based on the asymptotic posterior distribution $[\mu, \sigma^2|y]$.

Exercise 2

Using data y in Exercise 1, write a computer programme to implement the above model with:

(i) A Metropolis-Hastings, single-site updating of μ and σ^2. From the output obtain Monte Carlo estimates of $\mathrm{E}[\mu|y]$, $\mathrm{Var}[\mu|y]$, $\mathrm{E}[\sigma^2|y]$, $\mathrm{Var}\left[\sigma^2|y\right]$ and 95% posterior intervals for μ and σ^2.
(ii) A Metropolis-Hastings, joint updating of μ and σ^2. From the output obtain Monte Carlo estimates of $\mathrm{E}[\mu|y]$, $\mathrm{Var}[\mu|y]$, $\mathrm{E}[\sigma^2|y]$, $\mathrm{Var}\left[\sigma^2|y\right]$ and 95% posterior intervals for μ and σ^2.
(iii) For the Metropolis-Hastings single-site updating algorithm, use two sets of tuning parameters. In the first set, try $k_\mu = 0.19$, $k_{\sigma^2} = 1$. In the second set, try $k_\mu = 19$, $k_{\sigma^2} = 9$. Compute the Monte Carlo variance, the effective chain length (effective sample size) and the integrated autocorrelation for μ and for σ^2 for both strategies. What do you observe? Can you provide an explanation?

Exercise 3

Use the vector of records y from Exercise 1 and write a computer programme to implement the model using Gibbs sampling:

(i) From the output of the Gibbs sampler, compute Monte Carlo estimates of $\mathrm{E}[\mu|y]$, $\mathrm{E}[\sigma^2|y]$ and 95% posterior intervals for μ and σ^2. Calculate also Monte Carlo variances of the estimator of $\mathrm{E}[\mu|y]$ and $\mathrm{E}[\sigma^2|y]$.
(ii) Compute the effective chain length and the integrated autocorrelation for μ and for σ^2.

12.4 Bayes Exercises II

Exercise 1

Consider the data generated in Likelihood Exercises II, Exercise 1, on page 549. As before, let

$$\theta_i = \Pr\left(Y_i = 1|\mu, \beta, x_i\right),$$

be the probability that the ith individual shows the disease, given the covariate and parameters μ and β.

The logistic regression model is

$$\operatorname{logit}(\theta_i) = \ln\left(\frac{\theta_i}{1-\theta_i}\right) = \mu + \beta x_i. \tag{12.26}$$

Solving for θ_i gives

$$\theta_i = \text{logit}^{-1}(\theta_i) = \frac{\exp(\mu + \beta x_i)}{1 + \exp(\mu + \beta x_i)}.$$

This model can be interpreted directly as a non-linear model for the probability θ_i or also, indirectly, in terms of an unobserved latent variable (or underlying variable) u_i. In this second formulation, the discrete observation Y_i is associated with the latent variable u_i as follows:

$$Y_i = \begin{cases} 1 \text{ if } u_i > 0 \\ 0 \text{ if } u_i < 0 \end{cases}, \text{ and}$$
$$u_i = \mu + \beta x_i + e_i,$$

where the $e's$ are iid residuals from a logistic probability distribution. Therefore in this case,

$$\begin{aligned}\theta_i &= \Pr(Y_i = 1|\mu, \beta, x_i) = \Pr(u_i > 0|\mu, \beta, x_i) \\ &= \frac{\exp(\mu + \beta x_i)}{1 + \exp(\mu + \beta x_i)},\end{aligned} \tag{12.27}$$

as in (12.16). The likelihood is proportional to

$$p(\mu, \beta|y, x) = \prod_{i=1}^{30} \left[\frac{\exp(\mu + \beta x_i)}{1 + \exp(\mu + \beta x_i)}\right]^{y_i} \left[\frac{1}{1 + \exp(\mu + \beta x_i)}\right]^{1-y_i} \tag{12.28}$$

and assuming a uniform prior for (μ, β), (12.28) is also proportional to the posterior distribution of (μ, β).

(i) Write a programme to implement this logistic model using the Metropolis-Hastings algorithm with joint updating for (μ, β).
(ii) Compute MC estimators of the mean, variance and 95% posterior intervals of $[\mu|y]$ and $[\beta|y]$.
(iii) Compute MC standard errors and effective chain lengths of $\hat{\text{E}}(\mu|y)$ and $\hat{\text{E}}(\beta|y)$.
(iv) Compute a MC estimate of $\Pr(Y = 1|x = 3, y)$.

Exercise 2

Fit the probit regression model to the previous data using the Metropolis-Hastings algorithm with joint updating for (μ, β):

(i) Write a programme to implement this probit model using the Metropolis-Hastings algorithm with joint updating of (μ, β).
(ii) Compute MC estimators of the mean, variance and 95% posterior intervals of $[\mu|y]$ and $[\beta|y]$.
(iii) Compute MC standard errors and effective chain lengths of $\hat{\mathrm{E}}(\mu|y)$ and $\hat{\mathrm{E}}(\beta|y)$.
(iv) Compute the MC estimate of $\Pr(Y = 1|x = 3, y)$. How does this estimate differ from the one computed in Exercise 1, iv ?

Exercise 3

Fit the probit regression model to the previous data using a Gibbs sampling algorithm and data augmentation:

(i) Write a programme to implement this probit model using the Gibbs sampling algorithm.
(ii) Compute MC estimators of the mean, variance and 95% posterior intervals of $[\mu|y]$ and $[\beta|y]$.
(iii) Compute MC standard errors and effective chain lengths of $\hat{E}(\mu|y)$ and $\hat{E}(\beta|y)$.

Exercise 4

The `R`-code below generates correlated binary data based on a probit threshold model. A full-sib family structure is assumed: nfs full-sib families with fs full-sibs per family. At the level of liability, the (true) hierarchical model is

$$u_{ij} = f_i + e_{ij}, \ , i = 1, \ldots, nfs, \ j = 1, \ldots, fs,$$

$$f_i|\sigma_f^2 \overset{iid}{\sim} N\left(0, \sigma_f^2\right),$$

$$e_{ij} \overset{iid}{\sim} N(0, 1).$$

```
#SINGLE-SITE GIBBS - CORRELATED PROBIT MODEL
# DOES NOT USE THE SVD OF ZZ'
rm(list=ls()) # Clear the workspace
set.seed(7713)

require(graphics)
# GENERATE CORRELATED (FULL-SIBS) BINARY DATA  (THRESHOLD MODEL)
# THE CODE WILL USE THE PACKAGE MVTNORM; IT IS INSTALLED BELOW
#install.packages("MCMCpack", .libPaths()[1])
#install.packages("mvtnorm", .libPaths()[1])
library(mvtnorm)
#library(MCMCpack)
# INITIALISE PARAMETERS
#p0<-0.2
p0 <- 0.5
mu <- qnorm(p0)
iccfs<-0.25 #INTRACLASS CORRELATION FS
# VARIANCE BETWEEN FAMILIES: iccfs /(1- iccfs)
# PHENOTYPIC VARIANCE: 1/(1-iccfs)
nfs<- 400 # NUMBER OF FULL-SIB FAMILIES
fs<-3 #FULL-SIB FAMILY SIZE
y<-matrix(data=0,nrow=fs*nfs,ncol=1)
x<-matrix(data=0,nrow=fs*nfs,ncol=1)
# GENERATE NFS FULL-SIB EFFECTS f
f<-rnorm(nfs,mean=0,sd=sqrt(iccfs/(1-iccfs)))
##########################################################
####  GENERATE BINARY RECORDS Y
f<-rnorm(nfs,mean=0,sd=sqrt(iccfs/(1-iccfs)))
p <- pnorm(mu+f)
y <- rbinom(nfs*fs,1,rep(p,each=fs))
w <- rep(1:nfs,each=fs)
d<-data.frame(w,y)
family <- w
family <- as.factor(family)
Z<-model.matrix(~0+family)
head(d)
```

```
##   w y
## 1 1 1
## 2 1 1
## 3 1 0
## 4 2 0
## 5 2 0
## 6 2 0
```

The Bayesian model used to analyse the data assumes

$$u_{ij} = \mu + f_i + e_{ij}, \ , i = 1, \ldots, nfs, \ j = 1, \ldots, fs,$$

$$f_i | \sigma_f^2 \overset{iid}{\sim} N\left(0, \sigma_f^2\right),$$

$$e_{ij} \overset{iid}{\sim} N(0, 1).$$

The prior distributions for μ and σ_f^2 are assumed to be proportional to constants. In matrix notation the model for the vector of liabilities is

$$u = X\mu + Zf + e$$

where X (a column vector of $1's$) and Z are observed incidence matrices.

(i) Write a computer programme to draw inferences about μ and σ_f^2 using a single-site updating Gibbs sampling algorithm and data augmentation.
(ii) Compute MC estimators of the mean and 95% posterior intervals of $[\mu|y]$, $\left[\sigma_f^2|y\right]$ and $\left[h^2|y\right]$.
(iii) Compute MC standard errors and effective chain lengths for $\hat{E}(\mu|y)$, $\hat{E}\left(\sigma_f^2|y\right)$ and $\hat{E}\left(h^2|y\right)$.

Exercise 5

The R-code below generates genomic data: 500 nominally unrelated individuals with 1000 genetic markers available for each individual. The genomic model is

$$y|\mu, g, \sigma_e^2 \sim N\left(1\mu + g, I\sigma_e^2\right), \tag{12.29a}$$

$$g|W, \sigma_g^2 \sim SN\left(0, G\sigma_g^2\right), \tag{12.29b}$$

$$G = \frac{1}{m}WW', \tag{12.29c}$$

$$W = \left\{W_{ij}\right\}. \tag{12.29d}$$

In these expressions, m is the number of SNPs, and W_{ij} is the label for the jth marker genotype of individual i, $(i = 1, \ldots n;\ j = 1, \ldots, m;\ m > n)$:

$$W_{ij} = \frac{X_{ij} - \mathrm{E}\left(X_{ij}\right)}{\sqrt{\mathrm{Var}\left(X_{ij}\right)}}, \quad X_{ij} = 0, 1, 2.$$

The eigenvalue decomposition of G is

$$\begin{aligned} G &= U\Delta U' \\ &= \sum\nolimits_{i=1}^{n} \lambda_i U_i U_i', \end{aligned}$$

where $U = [U_1, U_2, \ldots, U_n]$, of order $n \times n$ is the matrix of eigenvectors of G, U_j is the jth column (dimension $n \times 1$) and Δ is a diagonal matrix with elements equal to the eigenvalues $\lambda_1, \lambda_2, \ldots, \lambda_n$ associated with the n eigenvectors.

The loglikelihood, up to an additive constant, is (see expression (3.48) on page 95)

$$\ln p\left(\mu, k, \sigma_e^2|y, W\right) = -\frac{1}{2}\left\{n \ln \sigma_e^2 + \sum_{i=1}^{n} \ln\left(\lambda_i k + 1\right)\right.$$

$$\left. + \frac{1}{\sigma_e^2}\left(y - 1\mu\right)' U \left(\Delta k + I\right)^{-1} U' \left(y - 1\mu\right)\right\}, \tag{12.30}$$

where $k = \sigma_g^2/\sigma_e^2$.

```
# BAYES PROBLEMS II. GENOMIC MODEL
# DATA BASED ON GENOMIC MODEL; OBTAIN SVD OF WW'(1/m)
rm(list=ls()) # CLEAR WORKSPACE
nindiv<-500
nmark<-1000
nt <- nindiv*nmark
X<-matrix(nrow=nindiv,ncol=nmark,rbinom(n=nt,size=2,p=.5))
stdev <- matrix(data=NA,nrow= nmark,ncol=1)
W <- matrix(data=NA,nrow= nindiv,ncol=nmark)
U <- matrix(data=NA,nrow= nindiv,ncol= nindiv)
G<-matrix(data=NA,nrow= nindiv,ncol= nindiv)
cm <- colMeans(X)
#CHOOSE VALUE FOR GENOMIC VARIANCE vgs
vgs<-10
#CHOOSE VALUE FOR ENVIRONMENTAL VARIANCE ves
ves<-25
# CREATE MATRIX OF STANDARDISED MARKER GENOTYPE CODES
for (i in 1:nmark)
{
  W[,i] <-( X[,i]-cm[i]) / sd(X[,i])
}
# CAN USE INSTEAD:
# W <- scale(X)
# GENERATE nindiv GENOMIC VALUES FROM N(0,(1/nmark)WW'*vgs,)
# nmark MARKER VALUES: REALISATIONS FROM N(0,I sqrt(vgs/nmark))
g <- (1/sqrt(nmark))*W%*%rnorm(nmark,mean=0,sd=sqrt(vgs))
# GENERATE nindiv PHENOTYPES WITH MEAN 0, VAR=vgs+ves,
#  HERITABILITY=vgs/(vgs+ves)
e<- rnorm(nindiv,mean=0,sd=sqrt(ves))
y <- g+ e
# GENOMIC RELATIONSHIP MATRIX G
G <- (1/nmark)*W%*%t(W)
# SVD OF G
EVD <- eigen(G)
#names(EVD)
#head(EVD$values)
U <- EVD$vector
tU<-t(U)
val <- EVD$values
val[length(y)] <-0
D <- diag(val,nrow=nindiv)
# Dp IS A VECTOR WITH NON-ZERO EIGENVALUES
Dp<-c(val[1:nindiv-1])
```

(i) Write a computer programme to draw inferences about μ, σ_e^2 and σ_g^2 using a single-site updating Gibbs sampling algorithm.

(ii) Compute MC estimators of the mean and 95% posterior intervals of $[\mu|y]$, $\left[\sigma_g^2|y\right]$, $\left[\sigma_e^2|y\right]$ and $\left[h^2|y\right]$.

(iii) Compute MC standard errors and effective chain lengths for $\widehat{\mathrm{E}}\,(\mu|y)$, $\widehat{\mathrm{E}}\left(\sigma_g^2|y\right)$, $\widehat{\mathrm{E}}\left(\sigma_e^2|y\right)$ and $\widehat{\mathrm{E}}\left(h^2|y\right)$.

(iv) Obtain ML estimates of μ, σ_e^2 and σ_g^2 using the `R`-function `OPTIM` to compare with the Bayesian results. Obtain the asymptotic covariance matrix and compare the 95% frequentist interval with the 95% Bayesian posterior interval of σ_g^2.

12.5 Prediction Exercises

The expectations of the validating mean squared errors on pages 278 and 428 were obtained by simple expansion of the squared term of the MSE_v. The same results can be obtained by expressing the MSE as a quadratic form. It is quite simple to check whether the quadratic form is chi-square distributed, and therefore an analytical form of the complete distribution of MSE becomes available. The result rests on the assumption of normality.

The following result is used (used also on page 584):

- If the random vector $x \sim N(\mu, V)$, then the random variable $x'Ax \sim \chi^2(r(A), \lambda)$, a chi-square distribution with $r(A)$ degrees of freedom and non-centrality parameter λ, if AV is idempotent, where $r(A)$ denotes the rank of matrix A. The non-centrality parameter λ is equal to

$$\lambda = \mu' A\mu.$$

The mean and variance are

$$\mathrm{E}\left(x'Ax\right) = \mu'A\mu + \mathrm{tr}\,(AV)\,, \tag{12.31a}$$

$$\mathrm{Var}\left(x'Ax\right) = 4\mu'AVA\mu + 2\,\mathrm{tr}\,(AV)^2\,. \tag{12.31b}$$

In the special case when $\mu = 0$, then $\mathrm{E}\left(x'Ax\right) = \mathrm{tr}\,(AV)$ and $\mathrm{Var}\left(x'Ax\right) = 2\,\mathrm{tr}\,(AV)^2 = 2\,\mathrm{tr}\,(AV)$. Typically, $r(A) < r(V)$ and $\mathrm{tr}\,(AV) = r(AV) = r(A)$. Then for $\mu = 0$, $\mathrm{E}(x'Ax) = r(A)$ and $\mathrm{Var}(x'Ax) = 2r(A)$.

- For a symmetric matrix A and a random vector $x \sim (\mu, V)$, the quadratic form $x'Ax$ has mean given by

$$\mathrm{E}\left(x'Ax\right) = \mu'A\mu + \mathrm{tr}\,(AV)\,. \tag{12.32}$$

If x is normally distributed

$$\mathrm{Var}\left(x'Ax\right) = 4\mu' AVA\mu + 2\,\mathrm{tr}\,(AV)^2\,.$$

In the above expressions, expectations are taken with respect to $[x|u, V]$.

Exercise 1

A classical problem in quantitative genetics is to predict a future phenotypic value or an unobserved genetic value, given available information. The starting point is to construct a model of the joint distribution of observables and unobservables. To be specific, consider a **training-validating scenario** where the training data consist of n_1 observed phenotypic records y_1 and the scalar to be predicted could be either a hold-out phenotypic record y_2 or an unobserved genetic value g_2. A hierarchical genetic model assumes

$$\begin{bmatrix} y_1 \\ y_2 \end{bmatrix} = \begin{bmatrix} 1\mu_1 \\ \mu_2 \end{bmatrix} + \begin{bmatrix} g_1 \\ g_2 \end{bmatrix} + \begin{bmatrix} e_1 \\ e_2 \end{bmatrix}, \tag{12.33}$$

where y_1 is the vector of observed phenotypic records in the training set with n_1 elements. The scalar y_2 is a hold-out validating datum, μ_1 and μ_2 are unobserved means, and g_1 (a column vector with n_1 elements) and g_2 (a scalar) are normally distributed unobserved (additive) genetic values

$$\begin{bmatrix} g_1 \\ g_2 \end{bmatrix} \sigma_g^2 \sim N\left[\begin{pmatrix} 0 \\ 0 \end{pmatrix}, \begin{pmatrix} G_{11} & G_{12} \\ G_{21} & G_{22} \end{pmatrix} \sigma_g^2\right], \tag{12.34}$$

where σ_g^2 is the additive genetic variance (here, assumed known). In (12.34) G_{11} is the matrix of dimension $n_1 \times n_1$ of expected additive genetic relationships in the training data constructed on the basis of a given pedigree. The dimensions of the remaining blocks are

$$\begin{aligned} G_{12} &: n_1 \times 1, \\ G_{21} &: 1 \times n_1, \\ G_{22} &: 1 \times 1. \end{aligned}$$

The off-diagonal block G_{12}, a column vector, specifies the coefficient of additive genetic relationships between the n_1 individuals in the training data and the individual in the validating data. The scalar G_{22} is the coefficient of additive genetic relationship of the individual in the validating data with itself, equal to 1, in the absence of inbreeding. Residual terms are assumed to follow the independent normal distributions:

$$\begin{bmatrix} e_1 \\ e_2 \end{bmatrix} \Bigg| \sigma_e^2 \sim N\left[\begin{pmatrix} 0 \\ 0 \end{pmatrix}, \begin{pmatrix} I & 0 \\ 0 & 1 \end{pmatrix}\sigma_e^2\right],$$

where σ_e^2 is a residual variance $\text{Var}(y_i|\mu, g_i)$ (assumed known) and I is the $n_1 \times n_1$ identity matrix.

The purpose of the exercise is to obtain predictors of y_2 and of g_2, given the observable y_1, the training data. Assume that the predictors take the form of the conditional means:

$$\widehat{y}_2 = \text{E}(y_2|y_1),$$
$$\widehat{g}_2 = \text{E}(g_2|y_1).$$

1. Derive $\text{Var}(\widehat{y}_2)$, $\text{Cov}(y_2, \widehat{y}_2)$, $\text{Var}(\widehat{g}_2)$, $\text{Cov}(g_2, \widehat{g}_2)$.
2. Derive the prediction error variances $\text{Var}(y_2 - \widehat{y}_2)$, $\text{Var}(g_2 - \widehat{g}_2)$
3. Derive the squared correlations $R^2(y_2, \widehat{y}_2)$, $R^2(g_2, \widehat{g}_2)$, where

$$R^2(x, y) = \frac{[\text{Cov}(x, y)]^2}{\text{Var}(x)\,\text{Var}(y)}$$

   Show that

$$R^2(y_2, \widehat{y}_2) = h^2 R^2(g_2, \widehat{g}_2), \qquad h^2 = \frac{\text{Var}(g_2)}{\text{Var}(y_2)}$$

4. Argue that as n_1 increases, $R^2(g_2, \widehat{g}_2)$ approaches 1 and $R^2(y_2, \widehat{y}_2)$ approaches h^2.
5. Using the hold-out single data point y_2, derive the expected value of the validating mean squared error:

$$\text{E}(\text{MSE}_v) = \text{E}_{y_1 y_2}(y_2 - \widehat{y}_2)^2$$

Exercise 2

The components of the expected validating mean squared error MSE_v were discussed on page 277. From a classical frequentist point of view, when the expectation is taken over the distribution of training and validating data, three components contribute to the expected validating mean squared error as indicated in (6.51) on page 279.

Let y denote training data, y_0 validating data and let $\hat{y}_0 = x_0'\hat{\theta}$ denote a frequentist point prediction of the validating datum y_0, where x_0' is a row vector of p covariates and $\hat{\theta}$ is some estimate based on the training data of parameter θ that indexes the linear model. Then, given x_0

$$\begin{aligned} \mathrm{E}_{yy_0}\left(\mathrm{MSE}_v\right) &= \mathrm{E}_{yy_0}\left(y_0 - \hat{y}_0\right)^2 \\ &= \mathrm{E}_{y_0}\left(y_0^2\right) + \mathrm{E}_y\left(\hat{y}_0^2\right) - 2\,\mathrm{E}_{yy_0}\left(y_0\,\hat{y}_0\right). \end{aligned} \quad (12.35)$$

1. Obtain an analytical expression for (12.35) by calculating the expectations of its three terms and, in so doing, reproduce result (6.51). Expectations are taken over the distributions of training and validating data.
2. Adopt a Bayesian perspective. Conceptually, this involves drawing θ^* from $[\theta|y]$ and given θ^*, drawing a predicted validating datum $\hat{y}_0^*$ from $\left[\hat{y}_0|\theta^*, y, x_0\right]$ that has the same distribution as the validating datum $\left[y_0|\theta^*, y, x_0\right]$. The draws $\left(\theta^*, \hat{y}_0^*\right)$ are extractions from $\left[\theta, \hat{y}_0|y, x_o\right]$, while the margins are draws from $[\theta|y]$ and from $\left[\hat{y}_0|y, x_0\right]$. The latter is the posterior predictive distribution of the Bayesian predictor. The mean (dropping the conditioning on x_0 to avoid cluttering the notation) is

$$\mathrm{E}\left(\hat{y}_0|y\right) = \mathrm{E}_{\theta|y}\left[\mathrm{E}\left(\hat{y}_0|\theta, y\right)\right] = \mathrm{E}_{\theta|y}\left(x_0'\theta\right)$$

and the variance

$$\mathrm{Var}\left(\hat{y}_0|y\right) = \sigma^2 + \mathrm{Var}_{\theta|y}\left(x_0'\theta\right)$$

where σ^2 represents sampling uncertainty of the draws from $\left[\hat{y}_0|\theta^*, y, x_0\right]$, as explained in (10.4) on page 420. Obtain an analytical expression for the Bayesian expected validating mean squared error:

$$\begin{aligned} \mathrm{E}\left(\mathrm{MSE}_v\,|y, y_0\right) &= \mathrm{E}\left(y_0 - \hat{y}_0|y, y_0\right)^2 \\ &= y_0^2 + \mathrm{E}\left(\hat{y}_0^2|y\right) - 2y_0\,\mathrm{E}\left(\hat{y}_0|y\right). \end{aligned}$$

that involves expectations over the posterior predictive distribution of $\hat{y}_0$.

Exercise 3

Let the $n \times 1$ vector of observations y be a realisation from the probability model:

$$y|\mu \sim N\left(1\mu, I\sigma^2\right). \quad (12.36)$$

In what follows σ^2 is assumed known and μ is the only unknown parameter.

Partition $y' = \left(y_t', y_v'\right)$ into training and validating sets of lengths n_t, n_v. Let

$$\widehat{\mu} = \frac{1'y_t}{n_t} \tag{12.37}$$

be the ML estimate of μ based on the training records, where 1 is a column vector of 1′s with n_t elements. The sampling distribution of the ML estimator is

$$\hat{\mu} \sim N\left(\mu, \frac{\sigma^2}{n_t}\right). \tag{12.38}$$

A vector of predicted validating records $\hat{y}_v$ is

$$\hat{y}_v = 1\hat{\mu}$$

and the validating MSE_v is

$$\text{MSE}_v = \frac{1}{n_v}\left(\hat{y}_v - y_v\right)'\left(\hat{y}_v - y_v\right).$$

1. Derive the expected value of MSE_v under the following three scenarios:
   (a) Case 1: the training data y_t are treated as fixed and the validating data y_v as random. Here the MSE_v quantifies the ability of the model to predict new records accounting for their sampling variation, without accounting for sampling variation of $\hat{\mu}$.
   (b) Case 2: the training data y_t are treated as random and the validating data y_v as fixed. Here the MSE_v quantifies the ability of the model to predict an average validating datum, accounting for sampling variation of $\hat{\mu}$.
2. Case 3: Derive the expected value of MSE_v and of MSE_t when training data y_t and validating data y_v are treated as random. Here the MSE_v quantifies the ability of the model to predict a new record accounting for its sampling variation and also for sampling variation of $\hat{\mu}$.
   Obtain an expression for $\text{E}(\text{MSE}_v) - \text{E}(\text{MSE}_t)$. Formulate any insights from this calculation, perhaps with help from a glance at expression (6.68).

Exercise 4

A Bayesian perspective is adopted extending model (12.36) by assigning an improper uniform prior distribution to μ. Bayesian inferences are conditional on the data, and therefore y_t and y_v are treated as fixed observed quantities:

1. Derive the expected value of the validating mean squared error under the following three scenarios:

   (a) Compute the mean of the posterior distribution of $[\mu|y_t]$ that takes the same form as the ML estimator (12.37); label this mean $\widehat{\mu}$. Generate validating predictions y_v^* drawing from $\left[y_v|\hat{\mu}, y_t\right]$. Define the validating mean squared error:

$$\text{MSE}_v = \frac{1}{n_v}\left(y_v^* - y_v\right)'\left(y_v^* - y_v\right). \tag{12.39}$$

   Show that MSE_v has a scaled chi-square distribution.

   This approach does not account for the posterior uncertainty of μ but accounts for sampling uncertainty of $y_v^\star$.

   (b) Account for the uncertainty about μ that is fully captured by its posterior distribution. Given the likelihood (12.36) and an improper uniform prior distribution for μ, the posterior distribution is

$$\mu|y_t \sim N\left(\hat{\mu}, \frac{\sigma^2}{n_t}\right).$$

   Construct the vector of predicted validating records

$$\hat{y}_v = 1_{n_v}\mu$$

   and using these, calculate

$$\text{MSE}_v = \frac{1}{n_v}\left(\hat{y}_v - y_v\right)'\left(\hat{y}_v - y_v\right) \tag{12.40}$$

   and its expectation.

   (c) Account for the uncertainty about μ and for sampling uncertainty of new validating data $y_v^\star$ (the predictors), fitting the hierarchical Bayesian model:

$$\mu|y_t \sim N\left(\hat{\mu}, \frac{\sigma^2}{n_t}\right), \tag{12.41a}$$

$$y_v|\mu, y_t \sim N\left(1\mu, I\sigma^2\right), \tag{12.41b}$$

   where $\widehat{\mu}$ is the posterior mean; obtain an analytical expression for $\text{E}(\text{MSE}_v)$.

   A Monte Carlo approach based on the method of composition is to draw the pairs $\left(\mu_1^*, y_{v1}^*\right), \ldots, \left(\mu_N^*, y_{vN}^*\right)$, where μ_i^* is a draw from the distribution $[\mu|y_t]$ and y_{vi}^* is a draw from the distribution $\left[y_v|\mu_i^*, y_t\right]$. The quantities $y_{v1}^*, \ldots, y_{vN}^*$ are an *iid* sample from the posterior predictive distribution $[y_v|y_t]$ and can be used to compute the validating mean squared error (12.39).

2. Discuss the results obtained using a, b and c. Provide an interpretation of the distribution of the MSE_v under the Bayesian and frequentist perspectives.

Exercise 5

This exercise is in the same spirit as the previous one; the model is a little more parametrised.

The R-code below generates n individuals each genotyped for p covariates (genetic markers). Among these genetic markers nqtl are defined as causal genotypes. The size of gene substitution effects of these causal loci is chosen to generate an additive genetic variance equal to 10 squared units, and the heritability of the continuous trait is set equal to 0.5. The data vector y with n elements is divided into a training set y_t and a validating set y_v, each of size $n/2$. Let W represent the centred and scaled matrix X, where $X = \{X_{ij}\}$ is an $n/2 \times p$ observed matrix with genotypic codes X_{ij} equal to $0, 1, 2$ according to the number of the arbitrarily chosen allele of individual i and marker j.

The operational statistical model is as follows:

$$y_t|\mu, b, \sigma_e^2 \sim N\left(1\mu + Wb, I\sigma_e^2\right), \tag{12.42a}$$

$$b|\sigma_b^2 \sim N\left(0, I\sigma_b^2\right), \tag{12.42b}$$

where y_t is the vector of training records of length $n/2$, μ is an unobserved mean, b is a vector of unknown genetic marker effects of length p, 1 is a vector of $1'$s of length $n/2$, σ_e^2 is the residual variance and σ_b^2 reflects prior uncertainty for each element of b. In other words, σ_b^2 is the prior variance of the effect of one marker, the same for all p markers. These two variance components are assumed known:

```
# PREDICTION EXERCISE 5
rm(list=ls()) # CLEAR WORKSPACE
set.seed(123)
nindiv<-100
nmark <- 50
nt <- nindiv*nmark
# NUMBER QTL
nqtl <- 50

# GENERATE MARKER MATRIX FROM BINOMIAL DISTRIBUTION
X<-matrix(nrow= nindiv,ncol= nmark,rbinom(n=nt,size=2,p=.5))
##########################################################
# CHOOSE VALUE FOR MEAN mu AND GENOMIC VARIANCE vgs
mu <- 10
vgs<-10
# CHOOSE VALUE FOR ENVIRONMENTAL VARIANCE ves
ves<-20
her <- vgs/(vgs+ves)
```

```
btrue<-matrix(data=0.0,nrow=nmark,ncol=1) # parameter from
#      true model
IDq<-sample(1:nmark,nqtl,replace=F) # from nmark markers, choose
#      nqtl as QTL
QTLeff<-sqrt(vgs/nqtl)# QTL effect so that the total genetic
#      variance is VA
btrue[IDq]<-QTLeff # QTL b's are not zero
W <- matrix(data=NA,nrow= nindiv,ncol=nmark)
cm <- colMeans(X)
# CREATE MATRIX OF STANDARDISED MARKER GENOTYPE CODES
for (i in 1:nmark)
{
  W[,i] <-( X[,i]-cm[i]) / sd(X[,i])
}
# can use more compactly:
# W <- scale(X)
# GENERATE nindiv PHENOTYPES
e<- rnorm(nindiv,mean=0,sd=sqrt(ves))
y <- mu + W%*%btrue+ e
k <- (ves/vgs)*nmark # ratio of residual to genomic variance
#      Vb = vgs/nmark
train <- sample(1:nrow(W),floor(0.5*nrow(W)))
Xt <- W[train,]
yt <- y[train]
Xv <- W[-train,]
yv <- y[-train]
Zt <- cbind(1,Xt)
Zv <- cbind(1,Xv)
######################
##   ridge regression coefficient matrix, rhs & solution solt
RHSt <- crossprod(Zt,yt)
LHSt <- crossprod(Zt)
LHSt[-1,-1] <- LHSt[-1,-1]+diag(k,nrow=nrow(LHSt)-1)
solt <- solve(LHSt,RHSt)
# PREDICTION, CONDITIONAL ON ESTIMATED PARAMETERS (solt)
predval <- Zv%*%solt # VALIDATING
predtrain <- Zt%*%solt # TRAINING
```

A standard ridge regression is fitted to the training data y_t and assuming known dispersion parameters, the posterior mean of $\left[\mu, b|y_t, \sigma_e^2, \sigma_b^2\right]$ satisfies the linear system

$$\begin{bmatrix} 1'1 & 1'W_t \\ W_t'1 & W_t'W_t + I\lambda \end{bmatrix} \begin{bmatrix} \widehat{\mu} \\ \widehat{b} \end{bmatrix} = \begin{bmatrix} 1'y_t \\ W'y_t \end{bmatrix}, \tag{12.43}$$

where $\lambda = \frac{\sigma_e^2}{\sigma_b^2}$ and matrix W has been appropriately partitioned into training and validating blocks:

$$W = \begin{bmatrix} W_t \\ W_v \end{bmatrix},$$

each of dimension $n/2 \times p$.

In a Bayesian setting, predictors y_v^* of validating records are drawn from the posterior predictive distribution $[y_v|y_t]$, and the validating mean squared error is computed with the usual expression:

$$\text{MSE}_v = \frac{1}{n_v}\left(y_v^* - y_v\right)'\left(y_v^* - y_v\right), \tag{12.44}$$

1. Derive the exact distribution of MSE_v (12.44) conditional on the solutions $\left(\hat{\mu}, \hat{b}\right)$ from (12.43). In this case, y_v^* is a draw from $\left[y_v|y_t, \hat{\mu}, \hat{b}\right]$. Write a computer programme to generate a Monte Carlo estimate of the posterior distribution of MSE_v, drawing validating predictions $y_v^\star$ from $\left[y_v|y_t, \hat{\mu}, \hat{b}\right]$. Compare with the analytic results.
2. Derive an analytic expression for the expected value of (12.44) based on the hierarchical Bayesian model (12.42), using the formula for the expected value of a quadratic form (12.32). This provides a description of MSE based on a single point: its mean. Here, as in 1. above, account is taken of the contribution from sampling variation of new records, but in contrast with 1., account is also taken of the posterior uncertainty of $[\mu, b]$.
3. Allowing for uncertainty in (μ, b), write a code that uses the method of composition to obtain draws $\left(y_v^*, \mu^*, b^*\right)$, by sampling repeatedly from

$$\left(\mu^*, b^*\right) \sim [\mu, b|y_t],$$
$$y_v^* \sim \left[y_v|\mu^*, b^*, y_t\right],$$

from which the validating mean squared error (12.44) is calculated. This generates a second MC estimate of the posterior distribution of MSE_v. In contrast with 2. above, this provides a complete description of the marginal posterior distribution of MSE. Compute the Monte Carlo estimates of the mean and variance and compare the mean with the exact result obtained in 2.

Exercise 6

The exercise involves estimation of validating mean squared error using training data only. This entails obtaining an expression for the expected optimism followed by use of expression (6.69).

Consider the hierarchical linear model:

$$y|b, f, \sigma^2 \sim N\left(Xb + Zf, I\sigma^2\right), \tag{12.45a}$$

$$f|\sigma_f^2 \sim N\left(0, I\sigma_f^2\right). \tag{12.45b}$$

Above, y is the vector of observed training data with n elements, X and Z are observed incidence matrices of dimension $(n \times p)$, and $(n \times n_f)$, b is a vector of unobserved p elements representing fixed effects and f is a vector of unobserved *iid* random effects with n_f elements. Assume that there are n_o records in each of the n_f random effects, so that the total number of records is $n = n_f n_o$. A typical example arises when data cluster in full-sib families, where f is a vector of random family effects (here, assumed to be unrelated across families) and there are n_o offspring per family. Given known variance components σ^2 and σ_f^2, BLUP of f and BLUE of b are obtained solving the linear system:

$$\begin{bmatrix} X'X & X'Z \\ Z'X & Z'Z + Ik \end{bmatrix} \begin{bmatrix} \widehat{b} \\ \widehat{f} \end{bmatrix} = \begin{bmatrix} X'y \\ Z'y \end{bmatrix}, \quad k = \frac{\sigma^2}{\sigma_f^2}. \tag{12.46}$$

The vector of predictors (fitted values, linear in y) is

$$\hat{y} = W\hat{\theta}$$

where $W = [X\ Z]$, $\hat{\theta}' = \left[\hat{b}', \hat{f}'\right]$ and $\hat{\theta}$ is the solution to the linear system

$$\left[W'W + \Sigma\right]\hat{\theta} = W'y$$

with

$$\Sigma = \begin{bmatrix} 0 & 0 \\ 0 & Ik \end{bmatrix}.$$

The R-code below generates data based on the model, with $n_f = 500$ full-sib families, $n_o = 2$ full-sibs per family and $n = n_f n_o = 1000$ records. The vector of fixed effects b has $p = 2$ elements, here mimicking 2 breeds. The variance components σ^2 and σ_f^2 are set equal to 5 and 1 squared units, respectively, leading to a heritability equal to 1/3.

The code below constructs and solves the mixed model equations:

```
#FULL-SIB CONTINUOUS DATA
rm(list=ls()) # Clear the workspace
set.seed(123771)
ptm<-proc.time()
require(graphics)
# INITIALISE PARAMETERS
mus<-10 # MEAN
vfs<-1 #VARIANCE BETWEEN FULL-SIBS
#vfs<-0.5 #VARIANCE BETWEEN FULL-SIBS
#vfs <- 0.1
# RESIDUAL VARIANCE
```

```
ves<-5
k <- ves/vfs
nf<-500 # NUMBER OF FULL-SIB FAMILIES
n<-2 # FULL-SIB FAMILY SIZE
nb <- 2 # NUMBER OF BREEDS
N<-nf*n # TOTAL NUMBER OF RECORDS
y<-matrix(data=0,nrow=nf*n,ncol=1)
z<-matrix(data=0,nrow=nf*n,ncol=1)
# GENERATE nf FULL-SIB EFFECTS fs
fs<-rnorm(nf,mean=0,sd=sqrt(vfs))
# BREED EFFECTS
br <- rep(0,nb)
br[1] <- 5
br[2] <- 8
# GENERATE nf*n RESIDUAL EFFECTS
es<-rnorm(nf*n,mean=0,sd=sqrt(ves))
##################################################
## GENERATING A FULL-SIB STRUCTURE
b <- rep(1:nb,each=N/2)
z <- rep(1:nf,each=n)
y <- br[b] + fs[z] + es
d <- data.frame(y,z)
##################################################
d<-data.frame(y,z)
# GENERATE INCIDENCE MATRICES X & Z
family <- z
breed <- b
family <- as.factor(family)
breed <- as.factor(breed)
X<-model.matrix(~0+breed)
Z<-model.matrix(~0+family)
W <- cbind(X,Z)
LHS <- crossprod(W) # LHS OF MME
LHS[-(1:2),-(1:2)] <- LHS[-(1:2),-(1:2)]+
  diag(k,nrow=nrow(LHS)-2)
RHS <- crossprod(W,y) # RHS OF MME
SOL <- solve(LHS,RHS) # SOLUTION
```

1. Derive the analytical form for the expected optimism:

$$\frac{2}{n}\sum_{i=1}^{n}\mathrm{Cov}\left(y_i,\widehat{y}_i\right). \tag{12.47}$$

Data are correlated within the training and validating data due to the full-sib structure. However training and validating data are independent because it is assumed that different *iid* families are assigned to both. Therefore the expression for expected optimism given in (6.68) is valid.

2. Write a computer programme to obtain a Monte Carlo estimate of (12.47) using a parametric bootstrap and compare with the exact result obtained in 1.

Chapter 13
Solution to Exercises

13.1 Likelihood Exercises I

Exercise 1

(a) The probability mass function of the data x is

$$p(x|n, \theta) = \binom{n}{x} \theta^x (1-\theta)^{n-x}, \quad x = 0, 1, \ldots, n \tag{13.1}$$

where

$$\binom{n}{x} = \frac{n!}{(n-x)!x!}$$

is the binomial coefficient. In (13.1) the random variable is x, for fixed values of n and θ.

(b) The likelihood function is proportional to (13.1) and can be written

$$L(\theta|n, x) \propto \theta^x (1-\theta)^{n-x}, \quad 0 < \theta < 1. \tag{13.2}$$

which is a function of θ, for fixed values of n and x.

The loglikelihood is obtained by taking the natural logarithm of (13.2):

$$\ell(\theta|n, x) = x \ln \theta + (n-x) \ln (1-\theta), \quad 0 < \theta < 1. \tag{13.3}$$

(c) The ML estimator of θ is obtained by maximising (13.3). Taking the first derivative of $\ell(\theta|n, x)$ with respect to θ and setting the resulting expression

D. Sorensen, *Statistical Learning in Genetics*, Statistics for Biology and Health,
https://doi.org/10.1007/978-3-031-35851-7_13

equal to zero yields:

$$\frac{\partial \ell\left(\theta|n, x\right)}{\partial \theta} = \frac{x}{\theta} - \frac{n-x}{1-\theta} = 0. \tag{13.4}$$

The ML estimator of θ is

$$\widehat{\theta} = \frac{x}{n}. \tag{13.5}$$

With $n = 10$ and $x = 8$, the ML estimate is

$$\widehat{\theta} = \frac{8}{10} = 0.8.$$

(d) First-order asymptotic theory asserts that

$$\widehat{\theta} \sim N\left(\theta, i\left(\theta\right)^{-1}\right), \tag{13.6}$$

where Fisher's expected information is

$$\begin{aligned}
i\left(\theta\right) &= -\operatorname{E}_x\left[\frac{\partial^2 \ell\left(\theta|n, x\right)}{\left(\partial \theta\right)^2}\right] \\
&= -Ex_x\left[-\frac{x}{\theta^2} - \frac{n-x}{(1-\theta)^2}\right] \\
&= \frac{n\theta}{\theta^2} + \frac{n-n\theta}{(1-\theta)^2} \\
&= \frac{n}{\theta\left(1-\theta\right)}.
\end{aligned}$$

The asymptotic variance of the ML estimator is obtained using

$$i\left(\widehat{\theta}\right)^{-1} = \frac{\widehat{\theta}\left(1-\widehat{\theta}\right)}{n} = 0.016.$$

The asymptotic 95% confidence interval for θ is

$$\Pr\left(0.8 - 1.96 \times \sqrt{0.016} < \theta < 0.8 + 1.96 \times \sqrt{0.016}\right) = 0.95,$$

equal to

$$\Pr\left(0.55 < \theta < 1.05\right) = 0.95. \tag{13.7}$$

In repeated sampling, the random interval (13.7) includes θ with probability 95%. The asymptotic confidence interval (13.7) includes values of θ outside its bounds.

Exercise 2

(a) The transformation is

$$\beta = g(\theta) = \ln\left(\frac{\theta}{1-\theta}\right). \tag{13.8}$$

with inverse transformation

$$\theta = g^{-1}(\beta) = \frac{\exp(\beta)}{1+\exp(\beta)}. \tag{13.9}$$

Under the transformation (13.8), the likelihood is

$$\begin{aligned} L(\theta|n,x) = L\left(g^{-1}(\beta)|n,x\right) &\propto \left[\frac{\exp(\beta)}{1+\exp(\beta)}\right]^x \left(1-\frac{\exp(\beta)}{1+\exp(\beta)}\right)^{n-x}, \\ &= \left[\frac{\exp(\beta)}{1+\exp(\beta)}\right]^x \left[\frac{1}{1+\exp(\beta)}\right]^{n-x} \quad -\infty < \beta < \infty. \end{aligned} \tag{13.10}$$

(b) The result of using the transformed parameter (13.8) translates into a more symmetric likelihood function, as displayed in Fig. 13.1. This in turn has consequences for the quality of inferences.

(c) There are two ways of deriving the MLE of β. The simplest is using the invariance property of the MLE. The MLE of β is

$$\hat{\beta} = \ln\left(\frac{\hat{\theta}}{1-\hat{\theta}}\right) = \ln\frac{0.8}{0.2} = 1.386. \tag{13.11}$$

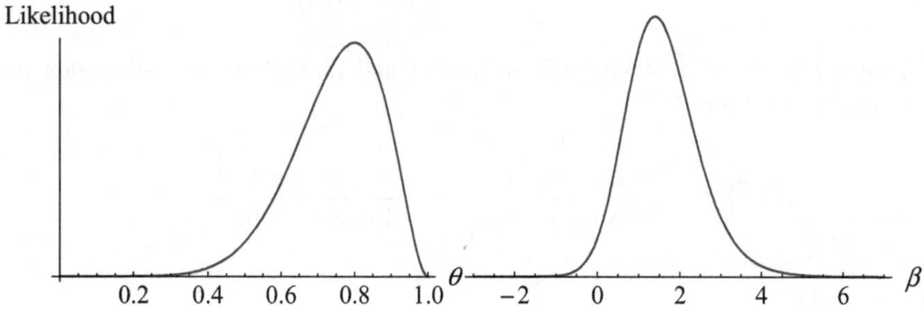

Fig. 13.1 Left panel: plot of $L(\theta|n,x)$ (13.2). Right panel: plot of $L\left(g^{-1}(\beta)\right)$ (13.10)

The asymptotic variance of $\hat{\beta}$ is

$$\begin{aligned}\operatorname{Var}\left(\hat{\beta}\right) &= \operatorname{Var}\left(\hat{\theta}\right)\left(\frac{dg(\theta)}{d\theta}\right)\Bigg|^{2}_{\theta=\hat{\theta}} \\ &= 0.016\left(\frac{1}{\hat{\theta}\left(1-\hat{\theta}\right)}\right)^{2} \\ &= 0.625. \end{aligned} \tag{13.12}$$

A longer calculation requires maximising (13.10) with respect to β. It is easier to work with the loglikelihood. A simple calculation shows that the loglikelihood is

$$\ell\left(g^{-1}(\beta)\,|n,x\right) = x\beta - n\ln\left[1+\exp(\beta)\right]. \tag{13.13}$$

Then,

$$\frac{\partial}{\partial\beta}\ell\left(g^{-1}(\beta)\,|n,x\right) = x - \frac{n\exp(\beta)}{1+\exp(\beta)},$$

whose root is

$$\widehat{\beta} = -\ln\left(\frac{n-x}{x}\right) = 1.386$$

as in (13.11).

To compute the asymptotic variance of $\widehat{\beta}$, take second derivatives of the loglikelihood with respect to β. This gives:

$$\begin{aligned}\frac{\partial^2}{(\partial\beta)^2}\ell\left(g^{-1}(\beta)\,|n,x\right) &= \frac{\partial}{\partial\beta}\left[x - \frac{n\exp(\beta)}{1+\exp(\beta)}\right] \\ &= -\frac{n\exp(\beta)}{(1+\exp(\beta))^2}.\end{aligned}$$

Evaluated at $\beta = \hat{\beta}$, taking the reciprocal and multiplying by -1 yields the asymptotic variance:

$$\operatorname{Var}\left(\widehat{\beta}\right) = \frac{(1+\exp(\beta))^2}{n\exp(\beta)}\Bigg|_{\beta=\widehat{\beta}} = \frac{(1+\exp(1.386))^2}{10\exp(1.386)} = 0.625$$

as in (13.12).

(d) The asymptotic 95% confidence interval (CI) for β is

$$\Pr\left(1.386 - 1.96 \times \sqrt{0.625} < \beta < 1.386 + 1.96 \times \sqrt{0.625}\right) = 0.95,$$

equal to

$$\Pr(-0.16 < \beta < 2.94) = 0.95. \tag{13.14}$$

(e) Using (13.9), interval (13.14) can be transformed back in terms of θ:

$$\Pr\left(g^{-1}(-0.16) < g^{-1}(\beta) < g^{-1}(2.94)\right)$$

yielding

$$\Pr(0.46 < \theta < 0.95). \tag{13.15}$$

This interval based on the more regular likelihood function (13.10) is more reliable than that obtained with the original likelihood function (13.2). Interval (13.15) includes values of θ within the permissible parameter space.

Note

There is a (pure) likelihood-based confidence interval. Useful references from two advocates of the method are Edwards (1992) and Royall (1997), and a tutorial is given by Meeker and Escobar (1995). In this approach to inference, the likelihood function is regarded as conveying a complete measure of uncertainty about the parameter of interest, without the need to invoke sampling distributions (like in inferences based on a posterior distribution). In the present example of the binomial model with $x = 8$ successes out of $n = 10$ trials, the value of the loglikelihood (13.3) evaluated at $\theta = \hat{\theta}$ is equal to -5.004024. In the words of Edwards (1992), this is the *maximum support*, at $\hat{\theta} = 0.8$. The pure likelihood approach considers the values of θ which lie within, for example, two units of support of the best supported value (the choice of "two" units is arbitrary, like the choice of two units of standard deviations of the "best" estimate in conventional analyses). Then the MLE and the "confidence interval" that arises from such an exercise in the present example are 0.8 (0.494; 0.965), quite close to the value reported in (13.15).

The `R`-code below evaluates first, the loglikelihood when the parameter is replaced by its MLE. Second, the code executes the function `OPTIM` twice, with judicious choice of the parameter `lower` and `upper`, generating the two values of θ that define the lower and upper limits within two units of support of the best supported value.

In multiparameter settings this approach is less transparent, in contrast with Bayesian methods that handle multidimensional parameters in a natural way via marginalisation. In general, this is a vast subject with considerable statistical and philosophical controversy:

```
rm(list=ls()) # Clear the workspace
set.seed(123771)
n <- 10
x <- 8
loglik<-function(n,x,theta)
{
  llik <- (x*log(theta)+(n-x)*log(1-theta))
}
LL <- loglik(10,8,0.8) # loglikelihood at ML of theta
LL
```

```
## [1] -5.004024
```

```
# The value "-7" below, is the (approximate) value of
# the loglikelihood two units of support from its value
# at the MLE (which is approx, -5)
fn <- function(theta){
 (-7-(8*log(theta)+(10-8)*log(1-theta)))^2
}
# EXECUTE OPTIM TWICE, ADJUSTING LOWER AND UPPER LIMITS
# TO GENERATE THE TWO ROOTS
lower<-optim(0.4,fn,method="Brent",lower=0.01,upper=0.8)$par
upper<-optim(0.4,fn,method="Brent",lower=0.8,upper=0.999)$par
# LOWER BOUND OF CI
lower
```

```
## [1] 0.4943429
```

```
# UPPER BOUND OF CI
upper
```

```
## [1] 0.9652074
```

As a closing comment, from a Bayesian perspective, (13.3) is the logarithm of the beta density $Be(a, b)$, with $a = x + 1$ and $b = n - x + 1$. Therefore this "pure likelihood" approach in this unidimensional example leads to the same inferences about θ as those drawn from the beta posterior $Be(x + 1, n - x + 1)$. To illustrate, the 95 % posterior interval for θ from $Be(x + 1, n - x + 1)$ using `R` is obtained executing `qbeta(c(0.025,0.975),9,3)` that yields (0.482, 0.940).

Exercise 3

(a) First write the model using the more general formulation:

$$y = Xb + e,$$

where X is a column vector of ones of length $n = 10$, b is a vector with a single element equal to μ and the random vector $e \sim N\left(0, I\sigma^2\right)$. With this notation the likelihood is proportional to

$$L\left(b, \sigma^2|y\right) \propto \left(2\pi\sigma^2\right)^{-\frac{n}{2}} \exp\left[-\frac{1}{2\sigma^2}(y - Xb)'(y - Xb)\right],$$

and the loglikelihood, up to an additive constant is equal to

$$\ell\left(b, \sigma^2|y\right) = -\frac{n}{2}\ln\left(\sigma^2\right) - \frac{1}{2\sigma^2}(y - Xb)'(y - Xb). \tag{13.16}$$

(b) Derivation of the ML estimators of b and of σ^2 requires a joint maximisation of (13.16) with respect to b and σ^2. The partial derivative with respect to b is

$$\begin{aligned}\frac{\partial}{\partial b}\ell\left(b, \sigma^2|y\right) &= \frac{\partial}{\partial b}\left[-\frac{1}{2\sigma^2}\left(y'y - 2y'Xb + b'X'Xb\right)\right] \\ &= \frac{1}{2\sigma^2}\left(2X'y - 2X'Xb\right) \\ &= \frac{1}{\sigma^2}\left(X'y - X'Xb\right).\end{aligned}$$

Setting this equal to zero and solving for b yields the ML estimator:

$$\widehat{b} = \left(X'X\right)^{-1}X'y. \tag{13.17}$$

The partial derivative of (13.16) with respect to σ^2 is

$$\begin{aligned}\frac{\partial}{\partial\sigma^2}\ell\left(b, \sigma^2|y\right) &= \frac{\partial}{\partial\sigma^2}\left[-\frac{n}{2}\ln\left(\sigma^2\right) - \frac{1}{2\sigma^2}(y - Xb)'(y - Xb)\right] \\ &= -\frac{n}{2\sigma^2} + \frac{1}{2\sigma^4}(y - Xb)'(y - Xb).\end{aligned}$$

On setting this equation equal to zero and multiplying out by $2\sigma^2$ yields the MLE of σ^2:

$$\widehat{\sigma}^2 = \frac{\left(y - X\widehat{b}\right)'\left(y - X\widehat{b}\right)}{n} \tag{13.18}$$

where $X\hat{b} = \widehat{y}$ is the predicted or fitted value of y.

In the special case of this simple model with the $n \times 1$ matrix $X = 1$ and $b = \mu$, the MLE (13.17) is

$$\begin{aligned}\widehat{\mu} &= \left(1'1\right)^{-1} 1'y \\ &= \frac{1}{10}\sum_{i=1}^{10} y_i = \overline{y} = 2.733\end{aligned}$$

and (13.18) is

$$\begin{aligned}\widehat{\sigma}^2 &= \frac{(y - 1\widehat{\mu})'(y - 1\widehat{\mu})}{n} \\ &= \frac{\sum_{i=1}^{10}(y_i - \widehat{\mu})^2}{10} = 2.581.\end{aligned}$$

(c) Let $\theta' = \left(b, \sigma^2\right)$. The observed information is equal to minus the Hessian :

$$I\left(b, \sigma^2\right) = -\frac{\partial^2 \ell\left(b, \sigma^2|y\right)}{\partial\theta\partial\theta'}. \tag{13.19}$$

The Hessian requires

$$\frac{\partial^2 \ell\left(b, \sigma^2|y\right)}{\partial b \partial b'}, \frac{\partial^2 \ell\left(b, \sigma^2|y\right)}{\partial b \partial \sigma^2}, \frac{\partial^2 \ell\left(b, \sigma^2|y\right)}{\left(\partial\sigma^2\right)^2}.$$

These second derivatives are

$$\begin{aligned}\frac{\partial^2 \ell\left(b, \sigma^2|y\right)}{\partial b \partial b'} &= \frac{\partial\left[\frac{1}{\sigma^2}\left(X'y - X'Xb\right)\right]}{\partial b} = -\frac{X'X}{\sigma^2}, \\ \frac{\partial^2 \ell\left(b, \sigma^2|y\right)}{\partial b \partial \sigma^2} &= \frac{\partial\left[\frac{1}{\sigma^2}\left(X'y - X'Xb\right)\right]}{\partial \sigma^2} = -\frac{X'\left(y - Xb\right)}{\sigma^2},\end{aligned}$$

$$\frac{\partial^2 \ell\left(b, \sigma^2|y\right)}{\left(\partial\sigma^2\right)^2} = \frac{\partial\left[-\frac{n}{2\sigma^2} + \frac{1}{2\sigma^4}(y - Xb)'(y - Xb)\right]}{\partial\sigma^2}$$

$$= \frac{n}{2}\left(\sigma^2\right)^{-2} - (y - Xb)'(y - Xb)\left(\sigma^2\right)^{-3}.$$

Therefore the observed information matrix is

$$I(b, \sigma^2) = \begin{bmatrix} \frac{X'X}{\sigma^2} & \frac{X'(y-Xb)}{\sigma^2} \\ -\frac{(y-Xb)'X}{\sigma^2} & -\frac{n}{2(\sigma^2)^2} + \frac{(y-Xb)'(y-Xb)}{(\sigma^2)^3} \end{bmatrix}$$

$$= \begin{bmatrix} \frac{n}{\sigma^2} & \frac{\sum_{i=1}^n (y_i-\mu)}{\sigma^2} \\ \frac{\sum_{i=1}^n (y_i-\mu)}{\sigma^2} & -\frac{n}{2(\sigma^2)^2} + \frac{\sum_{i=1}^n (y_i-\mu)^2}{(\sigma^2)^3} \end{bmatrix}, \tag{13.20}$$

which evaluated at the ML estimates $\hat{\mu} = 2.733$ and $\hat{\sigma}^2 = 2.581$ is equal to

$$I(\hat{\mu}, \hat{\sigma}^2) = \begin{bmatrix} 3.874 & 0 \\ 0 & 0.7503 \end{bmatrix}.$$

(d) The asymptotic variance of $(\hat{\mu}, \hat{\sigma}^2)$ based on (13.19) is

$$\text{Var}\left(\widehat{\mu}, \widehat{\sigma^2}\right) = I(\widehat{\mu}, \widehat{\sigma^2})^{-1}$$

$$= \begin{bmatrix} 3.874^{-1} & 0 \\ 0 & 0.7503^{-1} \end{bmatrix}$$

$$= \begin{bmatrix} 0.258 & 0 \\ 0 & 1.333 \end{bmatrix}. \tag{13.21}$$

(e) The expected information is

$$i\left(b, \sigma^2\right) = \mathrm{E}_y\left[-\frac{\partial^2 \ell\left(b, \sigma^2|y\right)}{\partial\theta\partial\theta'}\right],$$

which requires the expectation of each of the *four* elements of (13.20). These are

$$\mathrm{E}_y\left(\frac{n}{\sigma^2}\right) = \frac{n}{\sigma^2},$$

$$\mathrm{E}_y\left(\frac{\sum_{i=1}^n (y_i - \mu)}{\sigma^2}\right) = 0,$$

$$\mathrm{E}_y\left(-\frac{n}{2\left(\sigma^2\right)^2}+\frac{\sum_{i=1}^n\left(y_i-\mu\right)^2}{\left(\sigma^2\right)^3}\right)=-\frac{n}{2\left(\sigma^2\right)^2}+\frac{n\sigma^2}{\left(\sigma^2\right)^3}$$
$$=\frac{n}{2\left(\sigma^2\right)^2}.$$

Therefore,

$$i\left(\hat{\mu},\hat{\sigma}^2\right)=\begin{bmatrix}\frac{n}{\hat{\sigma}^2} & 0\\ 0 & \frac{n}{2\left(\hat{\sigma}^2\right)^2}\end{bmatrix}$$
$$=\begin{bmatrix}3.874 & 0\\ 0 & 0.7503\end{bmatrix},$$

the same as (13.21).

(f) The asymptotic variance of $\left(\hat{\mu},\hat{\sigma}^2\right)$ based on Fisher's (expected) information is the same as (13.21). The form of the expression is

$$\mathrm{Var}\left(\hat{\mu},\hat{\sigma}^2\right)=\begin{bmatrix}\frac{\hat{\sigma}^2}{n} & 0\\ 0 & \frac{2\left(\hat{\sigma}^2\right)^2}{n}\end{bmatrix}. \tag{13.22}$$

(g) The following results are useful:

(i) If vector x is $N(\mu, V)$, then $x'Ax$ is $\chi^2\left(r(A), \mu'A\mu\right)$, a chi-square distribution with $r(A)$ degrees of freedom and non-centrality parameter $\mu'A\mu$, if AV is idempotent. An idempotent matrix is a matrix which, when multiplied by itself, yields itself. If K is idempotent, $K^2=K$ and $rank\,(K)=r\,(K)=\mathrm{tr}\,(K)$.

(ii) Write $\frac{\sum_{i=1}^n(y_i-\mu)^2}{\sigma^2}=\frac{1}{\sigma^2}(y-1\mu)'(y-1\mu)$. Given $y\sim N\left(1\mu, I\sigma^2\right)$, then

$$\mathrm{E}\,(y-1\mu)=0;\quad V=\mathrm{Var}\,(y-1\mu)=I\sigma^2;\quad A=\frac{1}{\sigma^2}I$$

and $AV=I$ which is idempotent. It follows that $\frac{1}{\sigma^2}(y-1\mu)'(y-1\mu)\sim\chi^2[n,0]$ because $r\left(\frac{1}{\sigma^2}I\right)=n$. The mean and variance are

$$\mathrm{E}\left(\frac{1}{\sigma^2}(y-1\mu)'(y-1\mu)\right)=n$$
$$\mathrm{Var}\left(\frac{1}{\sigma^2}(y-1\mu)'(y-1\mu)\right)=2n.$$

Given the linear model $y = Xb + e$, with $e \sim N\left(0, I\sigma^2\right)$, the MLE of b is

$$\widehat{b} = \left(X'X\right)^{-1} X'y.$$

and the MLE of σ^2 is

$$\widehat{\sigma}^2 = \frac{\left(y - X\widehat{b}\right)'\left(y - X\widehat{b}\right)}{n}.$$

Write

$$\begin{aligned} y - X\widehat{b} &= y - X\left(X'X\right)^{-1} X'y \\ &= \left(I - X\left(X'X\right)^{-1} X'\right) y. \end{aligned}$$

Therefore,

$$\begin{aligned} \left(y - X\widehat{b}\right)'\left(y - X\widehat{b}\right) &= y'\left(I - X\left(X'X\right)^{-1} X'\right) y \\ &= y'Py \end{aligned}$$

where P is idempotent. Then

$$\frac{1}{\sigma^2}\left(y - X\widehat{b}\right)'\left(y - X\widehat{b}\right) = \frac{1}{\sigma^2} y'Py.$$

Since $V = \mathrm{Var}(y) = I\sigma^2$ and $A = \frac{1}{\sigma^2}P$, $AV = P$, and idempotent matrix . Therefore,

$$\begin{aligned} \frac{1}{\sigma^2}\left(y - X\widehat{b}\right)'\left(y - X\widehat{b}\right) &\sim \chi^2\left(r(P), b'X'PXb\frac{1}{2\sigma^2}\right) \\ &= \chi^2\left(n - r(X)\right), \end{aligned} \tag{13.23}$$

and the exact sampling distribution of the MLE of σ^2 is

$$\frac{\left(y - X\widehat{b}\right)'\left(y - X\widehat{b}\right)}{n} \sim \frac{\sigma^2}{n}\chi^2\left(n - r(X)\right), \tag{13.24}$$

which is proportional to a chi-square distribution. The second line follows because $PX = 0$ and $r(P) = r\left(I - X\left(X'X\right)^{-1} X'\right) = r(I) - r\left(X\left(X'X\right)^{-1} X'\right) = n - r(X)$. In view of (13.24), since $X = 1$ (a vector of ones whose rank is 1) and $b = \mu$, a scalar, the expected value and variance of $\widehat{\sigma}^2$ are

$$\mathrm{E}\left[\frac{\left(y-X\widehat{b}\right)'\left(y-X\widehat{b}\right)}{n}\right]=\frac{\sigma^2}{n}\,(n-1)\,,$$

$$\mathrm{Var}\left[\frac{\left(y-X\widehat{b}\right)'\left(y-X\widehat{b}\right)}{n}\right]=\left(\frac{\sigma^2}{n}\right)^2 2\,(n-1)\,. \tag{13.25}$$

Therefore the MLE of σ^2 is biased. The small sample variance of $\widehat{\sigma}^2$ using (13.25) evaluated at $\sigma^2=\widehat{\sigma}^2$ is

$$2\left(\frac{\widehat{\sigma}^2}{n}\right)^2(n-1)=\left(\frac{2.581}{10}\right)^2 2\,(9)=1.199,$$

a little smaller than the asymptotic value $2\left(\hat{\sigma}^2\right)^2\Big/ n$ reported in (13.22).

Exercise 4

The R-code that generates the data is reproduced below:

```
rm(list=ls()) # Clear the workspace
set.seed(772231)
n <- 5000 # Number of records generated for Y_1
p <- 0.20 # Proportion to be selected
m <- round(p*n) # Number of records selected
y1 <- rnorm(n,mean=10,sd=2) # Generate random variable Y_1
y1_sort <-sort(y1,decreasing=TRUE) # Sort Y_1
y1s <- y1_sort[1:m] # The highest m Y_1 records are kept
#length(y1) # Number of records in Y_1
#length(y1s) # Number of records in the selected subset of Y_1
# Generate Y_2
y2<-rnorm(length(y1s),mean=15+0.6*(y1s-10),sd=sqrt(4-(2.4^2)/4))
```

Let $\theta=(\mu_y,\mu_x,\sigma_{xy},\sigma_{yy},\sigma_{xx})$. If the selection mechanism is (incorrectly) ignored and only the m bivariate records are considered, these can be assumed to be a realisation from

$$(y_i,x_i)|\theta\sim N\left(\begin{bmatrix}\mu_y\\ \mu_x\end{bmatrix},\begin{bmatrix}\sigma_{yy}=4 & \sigma_{yx}=2.4\\ \sigma_{yx}=2.4 & \sigma_{xx}=4\end{bmatrix}\right), \tag{13.26}$$

$$i=1,2,\ldots,m,$$

with $\mu_y=10$, $\mu_x=15$. The closed forms of the ML estimators are

$$\widehat{\mu}_x=\frac{1}{m}\sum_{i=1}^{m}x_i,\qquad \widehat{\mu}_y=\frac{1}{m}\sum_{i=1}^{m}y_i, \tag{13.27a}$$

$$\widehat{\sigma}_{xx} = \frac{1}{m}\sum_{i=1}^{m}(x_i - \widehat{\mu}_x)^2, \qquad \widehat{\sigma}_{yy} = \frac{1}{m}\sum_{i=1}^{m}\left(y_i - \widehat{\mu}_y\right)^2, \qquad (13.27b)$$

$$\widehat{\rho} = \frac{\widehat{\sigma}_{xy}}{\sqrt{\widehat{\sigma}_{xx}\widehat{\sigma}_{yy}}}, \qquad \widehat{\sigma}_{xy} = \frac{1}{m}\sum_{i=1}^{m}\left(y_i - \widehat{\mu}_y\right)(x_i - \widehat{\mu}_x). \qquad (13.27c)$$

(a) The ML estimates based on (13.27) can be computed with the R-code below:

```
m <- length(y2)
n <- length(y1)
round(mean(y1s),digits=2)
```

```
## [1] 12.8
```

```
round(mean(y2),digits=2)
```

```
## [1] 16.56
```

```
round(mean((y1s-mean(y1s))^2),digits=2)
```

```
## [1] 0.91
```

```
round(mean((y2-mean(y2))^2),digits=2)
```

```
## [1] 2.85
```

```
round((cov(y1s,y2)*(m-1)/m),digits=2)
```

```
## [1] 0.54
```

```
round(cov(y1s,y2)/(sqrt(var(y1s)*var(y2))),digits=2)
```

```
## [1] 0.34
```

```
round(cov(y1s,y2)/var(y1s),digits=2)
```

```
## [1] 0.59
```

The ML estimates (13.27) are

$$\begin{aligned}
\hat{\mu}_x &= 12.8, \\
\hat{\mu}_y &= 16.56, \\
\hat{\sigma}_{xx} &= 0.91, \\
\hat{\sigma}_{yy} &= 2.85, \\
\hat{\sigma}_{xy} &= 0.54, \\
\hat{\rho} &= 0.34, \\
\hat{b}_{yx} &= 0.59.
\end{aligned}$$

With the exception of $\hat{b}_{yx}$, these estimates are in very poor agreement with the true parameter values defined in (13.26).

If the individuals that produced a second lactation had been randomly selected, this approach using a subset of the data would have performed adequately. The topic of inferences under selection is a delicate one; important references are Rubin (1976) and Little and Rubin (1987).

(b) The density function of the **complete** data under parametrisation $\phi = (\mu_x, \sigma_{xx}, \beta_0, \beta_1, \sigma_{y.x})$ is

$$\begin{aligned}
p(x, y|\phi) &= \prod_{i=1}^{m} p(x_i, y_i|\phi) \prod_{i=1+m}^{n} p(x_i|\phi) \\
&= \left[\prod_{i=1}^{m} p(y_i|x_i, \phi)\, p(x_i|\phi)\right]\left[\prod_{i=1+m}^{n} p(x_i|\phi)\right] \\
&= \prod_{i=1}^{n} p(x_i|\mu_x, \sigma_{xx}) \prod_{i=1}^{m} p\left(y_i|x_i, \beta_0 + \beta_1 x_i, \sigma_{y.x}\right) \\
&= (2\pi\sigma_{xx})^{-\frac{n}{2}} \exp\left[-\frac{\sum_{i=1}^{n}(x_i - \mu_x)^2}{2\sigma_{xx}}\right] \left(2\pi\sigma_{y.x}\right)^{-\frac{m}{2}} \\
&\quad \exp\left[-\frac{\sum_{i=1}^{m}(y_i - \beta_0 - \beta_1 x_i)^2}{2\sigma_{y.x}}\right]
\end{aligned} \tag{13.28}$$

where

$$\sigma_{y.x} = \sigma_{yy} - \frac{\left(\sigma_{xy}\right)^2}{\sigma_{xx}}, \tag{13.29}$$

is the variance of the conditional distribution of Y given X. The likelihood is proportional to (13.28) and the loglikelihood is

$$\ell\left(\mu_x, \sigma_{xx}, \beta_0, \beta_1, \sigma_{y.x} | x, y\right) = -\frac{n}{2} \ln \sigma_{xx} - \frac{\sum_{i=1}^{n} (x_i - \mu_x)^2}{2\sigma_{xx}} - \frac{m}{2} \ln \sigma_{y.x} - \frac{\sum_{i=1}^{m} (y_i - \beta_0 - \beta_1 x_i)^2}{2\sigma_{y.x}}. \quad (13.30)$$

(c) The first two terms in the right-hand side of (13.30) are the loglikelihood of the $N\left(\mu, \sigma^2\right)$ linear model. The ML estimates of μ and σ^2 are given for the general case in (13.17) and (13.18), which in this example with $X = 1$ (a vector of ones) and $b = \mu_x$ reduce to

$$\widehat{\mu}_x = \frac{\sum_{i=1}^{n} x_i}{n}, \quad (13.31a)$$

$$\widehat{\sigma}_{xx} = \frac{\sum_{i=1}^{n} (x_i - \widehat{\mu})^2}{n}. \quad (13.31b)$$

To obtain the MLE of $\sigma_{y.x}$, β_0 and β_1, differentiate (13.30) with respect to $\sigma_{y.x}$, β_0 and β_1. After a little simplification, the three equations with the three unknowns are

$$-m + \frac{\sum_{i=1}^{m} (y_i - \beta_0 - \beta_1 x_i)^2}{\sigma_{y.x}} = 0,$$

$$\sum_{i=1}^{m} (y_i - \beta_0 - \beta_1 x_i) = 0,$$

$$\sum_{i=1}^{m} x_i (y_i - \beta_0 - \beta_1 x_i) = 0.$$

From the second equation

$$\widehat{\beta}_0 = \overline{y} - \widehat{\beta}_1 \overline{x}^*, \quad (13.32)$$

where

$$\overline{y} = \frac{\sum_{i=1}^{m} y_i}{m},$$

$$\overline{x}^* = \frac{\sum_{i=1}^{m} x_i}{m},$$

In these expressions, $\overline{x}^*$ is the mean of the selected first lactation records. From the third equation

$$\widehat{\beta}_1 = \frac{\sum_{i=1}^m x_i y_i - m\overline{x}^*\,\overline{y}}{\sum_{i=1}^m x_i^2 - m\overline{x}^{*2}} = \frac{S_{xy}}{S_{xx}}. \tag{13.33}$$

Finally from the first equation,

$$\widehat{\sigma}_{y.x} = \frac{\sum_{i=1}^m \left(y_i - \widehat{\beta}_0 - \widehat{\beta}_1 x_i\right)^2}{m}. \tag{13.34}$$

Substituting (13.32) and (13.33) in (13.34) leads to the alternative form (see NOTE below):

$$m\,\widehat{\sigma}_{y.x} = S_{yy} - \frac{S_{xy}^2}{S_{xx}} \tag{13.35}$$

which is more in line with the parameter (13.29), where

$$S_{yy} = \sum_{i=1}^m (y_i - \overline{y})^2 .$$

Note

Substituting (13.32) in (13.34) gives:

$$\begin{aligned}
\sum_{i=1}^m \left(y_i - \widehat{\beta}_0 - \widehat{\beta}_1 x_i\right)^2 &= \sum_{i=1}^m \left(y_i - \left(\overline{y} - \widehat{\beta}_1\overline{x}^*\right) - \widehat{\beta}_1 x_i\right)^2 \\
&= \sum_{i=1}^m \left((y_i - \overline{y}) - \widehat{\beta}_1\left(x_i - \overline{x}^*\right)\right)^2 \\
&= \sum_{i=1}^m (y_i - \overline{y})^2 - 2\widehat{\beta}_1 \sum_{i=1}^m (y_i - \overline{y})\left(x_i - \overline{x}^*\right) + \widehat{\beta}_1^2 \sum_{i=1}^m \left(x_i - \overline{x}^*\right)^2 \\
&= S_{yy} - 2\widehat{\beta}_1 S_{xy} + \widehat{\beta}_1^2 S_{xx} \\
&= S_{yy} - \frac{S_{xy}^2}{S_{xx}},
\end{aligned}$$

where the last line is obtained by replacing $\hat{\beta}_1 = S_{xy}/S_{xx}$ in the 4th line.

These MLEs of $\phi = (\mu_x, \sigma_{xx}, \beta_0, \beta_1, \sigma_{y.x})$ can be expressed in terms of the MLEs of $\theta = (\mu_y, \mu_x, \sigma_{xy}, \sigma_{yy}, \sigma_{xx})$. First, MLEs of μ_x and σ_{xx} are displayed in (13.31). The remaining MLEs of the parameters in ϕ are obtained as follows. Recall that

$$\beta_0 = \mu_y - \beta_1 \mu_x = 1,$$

$$\beta_1 = \frac{\sigma_{xy}}{\sigma_{xx}} = 0.6,$$

$$\sigma_{y.x} = \sigma_{yy} - \frac{(\sigma_{xy})^2}{\sigma_{xx}} = 2.56.$$

Then from the first equation

$$\widehat{\mu}_y = \widehat{\beta}_0 + \widehat{\beta}_1 \widehat{\mu}_x = 14.9.$$

From the second equation

$$\widehat{\sigma}_{xy} = \widehat{\beta}_1 \widehat{\sigma}_{xx} = 2.36,$$

and from the third

$$\widehat{\sigma}_{yy} = \widehat{\sigma}_{y.x} + \frac{(\widehat{\sigma}_{xy})^2}{\widehat{\sigma}_{xx}} = \widehat{\sigma}_{y.x} + \widehat{\beta}_1^2 \widehat{\sigma}_{xx} = 4.25.$$

The R-code that performs the computations is shown below:

```
sxy <- cov(y1s,y2)/var(y1s)
syy <- sum((y2-mean(y2))^2)
sxx <- mean((y1-mean(y1))^2)
beta1hat <- cov(y1s,y2)/var(y1s)
sigmaxxhat <- mean((y1-mean(y1))^2)
sigmaxyhat <- beta1hat*sigmaxxhat
beta0hat <- mean(y2)-beta1hat*mean(y1s)
muxhat <- mean(y1)
muyhat <- beta0hat+beta1hat*muxhat
sigmay.xhat <- (syy-sxy^2/sxx)/length(y2)
sigmayyhat <- sigmay.xhat +( (sigmaxyhat^2)/sigmaxxhat)
rhohat <- sigmaxyhat/sqrt(sigmaxxhat*sigmayyhat)
byxhat <- cov(y1s,y2)/var(y1s)
```

and generates the ML estimates:

$$\hat{\mu}_x = 10,$$

$$\hat{\mu}_y = 14.9,$$

$$
\begin{aligned}
\hat{\sigma}_{xx} &= 3.97, \\
\hat{\sigma}_{yy} &= 4.25, \\
\hat{\sigma}_{xy} &= 2.36, \\
\hat{\rho} &= 0.57, \\
\hat{b}_{yx} &= 0.59.
\end{aligned}
$$

These estimates are all in good agreement with the true value of the parameters defined in (13.26).

Exercise 5

1. Uncensored model:

   (a) The loglikelihood is obtained taking logarithms of (12.11):

   $$l\left(b|y\right) = n \ln b - b \sum_{i=1}^{n} y_i. \tag{13.36}$$

   (b) The score is

   $$l'\left(b|y\right) = \frac{n}{b} - \sum_{i=1}^{n} y_i. \tag{13.37}$$

   (c) The maximum likelihood estimator is obtained setting the derivative with respect to b equal to zero and solving for b. This results in

   $$\widehat{b} = \frac{n}{\sum_{i=1}^{n} y_i}. \tag{13.38}$$

   (d) Fisher's information is computed from

   $$l''\left(b|y\right) = -\frac{n}{b^2},$$

   that yields

   $$
   \begin{aligned}
   I\left(b\right) &= -\mathrm{E}\left[l''\left(b|Y\right)\right] \\
   &= \frac{n}{b^2}.
   \end{aligned} \tag{13.39}
   $$

2. Censored model:

(b) From (12.12), the contribution to the loglikelihood from one observation is

$$l\left(b|y_i\right) = I\left(y_i < c\right)\left(\ln b - by_i\right) - I\left(y_i = c\right)bc. \tag{13.40}$$

(c) Associated first and second derivatives are

$$l'\left(b|y_i\right) = I\left(y_i < c\right)\left(\frac{1}{b} - y_i\right) - I\left(y_i = c\right)c \tag{13.41}$$

and

$$l''\left(b|y_i\right) = -I\left(y_i < c\right)\frac{1}{b^2}. \tag{13.42}$$

The score based on all the data is

$$\begin{aligned}\sum_{i=1}^{n} l'\left(b|y_i\right) &= \sum_{i=1}^{n}\left\{I\left(y_i < c\right)\left(\frac{1}{b} - y_i\right) - I\left(y_i = c\right)c\right\} \\ &= \frac{r}{b} - \sum_{i=1}^{r} y_i - (n-r)\,c. \end{aligned} \tag{13.43}$$

(d) Setting to zero and solving leads to MLE of b :

$$\hat{b} = \frac{r}{\sum_{i=1}^{r} y_i + (n-r)\,c}. \tag{13.44}$$

In the absence of censoring, $r = n$ and (13.44) is equal to (13.38).

a. Information is computed using (13.42):

$$\begin{aligned} I\left(b\right) &= -\,\mathrm{E}\left\{\sum_{i=1}^{n}\left[l''\left(b|Y_i\right)\right]\right\} \\ &= \mathrm{E}\left\{\sum_{i=1}^{n}\left[I\left(Y_i < c\right)\frac{1}{b^2}\right]\right\} \\ &= \sum_{i=1}^{n}\frac{1}{b^2}\Pr\left(Y_i < c\right) \\ &= \frac{n}{b^2} - \frac{n\exp\left(-bc\right)}{b^2} \\ &= \frac{n}{b^2}\Pr\left(Y_i < c\right). \end{aligned} \tag{13.45}$$

In the absence of censoring $c = \infty$, $\Pr(Y_i < c) = 1$ and (13.45) is equal to (13.39). Also, $n \Pr(Y_i < c) = \mathrm{E}(r)$ (the number of non-censored records, r, is binomially distributed, i.e. $r \sim Bi\,(n, \Pr(Y_i < c))$).

Censoring reduces the amount of information in the data to infer b.

13.2 Likelihood Exercises II

Exercise 1

(a)

(b) Figure 13.2 depicts the relationship between Y (the rough measure of θ) and X, the covariate (average drug level). A linear fit seems like a good approximation. Here, Y is an 5×1 vector with the averages of the 5 groups of the 0/1 observations, and X is an 5×1 vector with the average drug level of the 5 groups. The linear fit is based on the model $E(Y_i|x_i) = \mu + \beta x_i$.

The least squares estimates of μ and β are

$$\hat{\beta} = \frac{\hat{\mathrm{Cov}}\,(X, Y)}{\hat{\mathrm{Var}}\,(X)} = \frac{1.07}{8.35} = 0.13$$

$$\hat{\mu} = \overline{X} - \hat{\beta}\,\overline{Y} = 0.47 - \widehat{\beta}\,2.33 = 0.17.$$

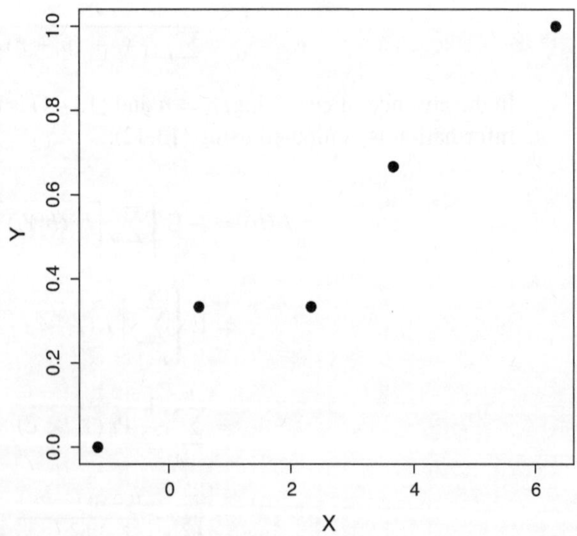

Fig. 13.2 Plot of Y versus X—the data collected in five groups with six observations in each group

In these expressions, $\widehat{\text{Cov}}(X, Y)$ is the sample covariance between the elements of vectors X and Y, $\widehat{\text{Var}}(X)$ is the sample variance among the *five* observations in X, $\overline{X}$ is the mean of the *five* observations in X and $\overline{Y}$ is the mean of the *five* observations in Y.

According to the linear approximation, the estimated probabilities that future individuals will show the disease when exposed to a drug level equal to −3, 1, 9 are

$$\widehat{\mu} + \widehat{\beta}(-3) = -0.22$$
$$\widehat{\mu} + \widehat{\beta}1 = 0.30$$
$$\widehat{\mu} + \widehat{\beta}9 = 1.32.$$

With the exception of the second, the probabilities do not lie within the interval (0, 1).

c1. Taking logarithms of the likelihood results in the loglikelihood

$$\begin{aligned}\ell(\mu, \beta|y, x) &= \sum_{i=1}^{30}\left\{y_i \ln\left[\frac{\exp(\mu+\beta x_i)}{1+\exp(\mu+\beta x_i)}\right]\right.\\ &\quad \left. +(1-y_i)\ln\left[\frac{1}{1+\exp(\mu+\beta x_i)}\right]\right\}\\ &= \sum_{i=1}^{30}\{y_i(\mu+\beta x_i) - \ln(1+\exp(\mu+\beta x_i))\}. \end{aligned} \quad (13.46)$$

c2. The first derivatives of $\ell(\mu, \beta|y, x)$ with respect to μ and β are

$$\frac{\partial \ell(\mu, \beta|y, x)}{\partial \mu} = \sum_{i=1}^{30}\left(y_i - \frac{\exp(\mu+\beta x_i)}{1+\exp(\mu+\beta x_i)}\right),$$

$$\frac{\partial \ell(\mu, \beta|y, x)}{\partial \beta} = \sum_{i=1}^{30}\left(x_i y_i - \frac{x_i\exp(\mu+\beta x_i)}{1+\exp(\mu+\beta x_i)}\right).$$

The second derivatives are

$$\frac{\partial^2 \ell(\mu, \beta|y, x)}{(\partial \mu)^2} = -\sum_{i=1}^{30}\frac{\exp(\mu+\beta x_i)}{(1+\exp(\mu+\beta x_i))^2},$$

$$\frac{\partial^2 \ell(\mu, \beta|y, x)}{(\partial \beta)^2} = -\sum_{i=1}^{30}\frac{x_i^2\exp(\mu+\beta x_i)}{(1+\exp(\mu+\beta x_i))^2},$$

$$\frac{\partial^2 \ell(\mu, \beta|y, x)}{\partial \beta \partial \mu} = -\sum_{i=1}^{30}\frac{x_i\exp(\mu+\beta x_i)}{(1+\exp(\mu+\beta x_i))^2}.$$

c3. The iterative system based on Newton-Raphson is

$$\begin{bmatrix}\mu\\ \beta\end{bmatrix}_{t+1}=\begin{bmatrix}\mu\\ \beta\end{bmatrix}_{t}+\begin{bmatrix}\sum_{i=1}^{30}\frac{\exp(\mu+\beta x_i)}{(1+\exp(\mu+\beta x_i))^2} & \sum_{i=1}^{30}\frac{x_i\exp(\mu+\beta x_i)}{(1+\exp(\mu+\beta x_i))^2}\\ \sum_{i=1}^{30}\frac{x_i\exp(\mu+\beta x_i)}{(1+\exp(\mu+\beta x_i))^2} & \sum_{i=1}^{30}\frac{x_i^2\exp(\mu+\beta x_i)}{(1+\exp(\mu+\beta x_i))^2}\end{bmatrix}^{-1}_{\substack{\mu=\mu_t\\ \beta=\beta_t}}$$

$$\begin{bmatrix}\sum_{i=1}^{30}\left(y_i-\frac{\exp(\mu+\beta x_i)}{1+\exp(\mu+\beta x_i)}\right)\\ \sum_{i=1}^{30}\left(x_i y_i-\frac{x_i\exp(\mu+\beta x_i)}{1+\exp(\mu+\beta x_i)}\right)\end{bmatrix}_{\substack{\mu=\mu_t\\ \beta=\beta_t}}. \tag{13.47}$$

The following `R-code` implements the Newton-Raphson algorithm:

```
# CODE1301
rm(list=ls()) # Clear the workspace
set.seed(12371)
# CREATE BINARY DATA
mu <- -2
beta <- 0.7
cov <- rnorm(30,2,3) # GENERATE THE COVARIATE
xb <- cov*beta
p1 <- pnorm(mu+xb) # PROBABILITIES ACCORDING TO PROBIT MODEL
# CREATE DATA:
dat1 <- cbind(rbinom(30,1,p1),round(cov,digits=0))
colnames(dat1) <- c("Y", "X")
### END OF GENERATION OF DATA
nit <- 10 # NUMBER OF N-R ITERATIONS
miu <- matrix(data=NA, nrow=nit+1,ncol=1)
beta <- matrix(data=NA, nrow=nit+1,ncol=1)
resultnr <- matrix(data=NA,nrow=nit,ncol=3)
# START VALUES FOR MIU AND BETA
miu [1]<- 0.17
beta[1] <- 0.13
for (i in 1:nit)
{
  vc11 <- - sum(exp(miu[i]+beta[i]*dat1[,2])/((1+exp(miu[i]+
                              beta[i]*dat1[,2]))^2))
  vc22 <- - sum(dat1[,2]^2*exp(miu[i]+ beta[i]*dat1[,2])/
                    ((1+exp(miu[i]+ beta[i]*dat1[,2]))^2))

  vc12 <- - sum(dat1[,2]*exp(miu[i]+ beta[i]*dat1[,2])/
                    ((1+exp(miu[i]+beta[i]*dat1[,2]))^2))

  vcmat <- matrix(c(vc11,vc12,vc12,vc22),nrow=2,ncol=2)
  vcmatinv <- solve(vcmat)
  fd1 <- sum((dat1[,1]-(exp(miu[i]+ beta[i]*dat1[,2]))/
                  (1+exp(miu[i]+beta[i]*dat1[,2]))))

  fd2 <- sum(((dat1[,1]*dat1[,2])-(dat1[,2]*exp(miu[i]+
            beta[i]*dat1[,2]))/(1+exp(miu[i]+beta[i]*dat1[,2]))))

  fd <- matrix(c(fd1,fd2),nrow=2,ncol=1)
  sol0 <- matrix(c(miu[i], beta[i]),nrow=2,ncol=1)
  sol1 <- sol0+(-vcmatinv%*%fd)
  miu[i+1] <-sol1[1]
```

```
  beta[i+1] <- sol1[2]
  resultnr[i,] <- c(i,sol1[1],sol1[2])
}
resultnr
```

```
##         [,1]      [,2]      [,3]
##  [1,]     1 -1.366182 0.5567651
##  [2,]     2 -2.014602 0.7949203
##  [3,]     3 -2.278267 0.8957553
##  [4,]     4 -2.314792 0.9096833
##  [5,]     5 -2.315397 0.9099107
##  [6,]     6 -2.315397 0.9099107
##  [7,]     7 -2.315397 0.9099107
##  [8,]     8 -2.315397 0.9099107
##  [9,]     9 -2.315397 0.9099107
## [10,]    10 -2.315397 0.9099107
```

```
# ASYMPTOTIC COVARIANCE MATRIX
-vcmatinv
```

```
##            [,1]       [,2]
## [1,]  0.8641485 -0.2427031
## [2,] -0.2427031  0.1003475
```

```
## COMPUTE PROBABILITIES THAT Y=1, GIVEN X = -3, 1, 9
p1<-exp(miu[i+1]+beta[i+1]*(-3))/(1+exp(miu[i+1]+
                                        beta[i+1]*(-3)))
p1
```

```
## [1] 0.006399413
```

```
p2<-exp(miu[i+1]+beta[i+1]*1)/(1+ exp(miu[i+1]+beta[i+1]*1))
p2
```

```
## [1] 0.196947
```

```
p3<-exp(miu[i+1]+beta[i+1]*9)/(1+ exp(miu[i+1]+beta[i+1]*9))
p3
```

```
## [1] 0.9971957
```

After a few iterations, the Newton-Raphson method converges to $\hat{\mu} = -2.32$ and $\hat{\beta} = 0.91$

c4. The asymptotic covariance matrix of the ML estimators is shown above. The 95% confidence interval for β based on the asymptotic variance is

$$\Pr\left(0.91 - 1.96 \times \sqrt{0.100} < \beta < 0.91 + 1.96 \times \sqrt{0.100}\right) = 0.95,$$

equal to

$$\Pr(0.29 < \beta < 1.53) = 0.95. \quad (13.48)$$

c5. The three probabilities (also displayed at the bottom of the code) are

$$\Pr\left(Y = 1|\hat{\mu}, \hat{\beta}, x = -1\right) = 0.006,$$
$$\Pr\left(Y = 1|\hat{\mu}, \hat{\beta}, x = 3\right) = 0.20,$$
$$\Pr\left(Y = 1|\hat{\mu}, \hat{\beta}, x = 9\right) = 0.99,$$

all within the allowed range (0, 1).

As a comparison, the following code uses the R function OPTIM to maximise the loglikelihood and to obtain asymptotic variances (OPTIM minimises the cost function defined as the negative of the loglikelihood). The bottom part of the code presents the simplest solution to this problem using the standard R function glm:

```
# CODE1301 (cont)
### USE THE R-FUNCTION OPTIM TO COMPARE WITH THIS PROGRAMME
dat <- data.frame(dat1)
logl <- function(data,par)
{
  with(data,-sum(Y*(par[1]+par[2]*X)-
                log(1+exp(par[1]+par[2]*X))))
}
result<-optim(par=c(-3.5,0.05),logl,data=dat,
              hessian=TRUE,method="BFGS")
# IF METHOD IS NOT INCLUDED IN THE CALL, OPTIM USES
#  THE NELDER-MEAD ALGORITHM
# result <- optim(par=c(-3.5,0.05),logl,data=dat,hessian=TRUE,
#  method="BFGS", control=list(trace=1,REPORT=1))
# THIS CALL INCLUDES  control=list(trace=1,REPORT=1) WHICH
#  PROVIDES THE PROGRESS OF THE ITERATION
result$par
```

```
## [1] -2.3154794  0.9099291
```

```
solve(result$hessian)
```

```
##             [,1]       [,2]
## [1,]  0.8641949 -0.2427165
## [2,] -0.2427165  0.1003514
```

```
###############################################
########### USE GLM in R
d <- data.frame(dat1)
logreg <- glm(d$Y~d$X, data=d, family=binomial(link="logit"))
resglm <- summary(logreg)
pred <- predict(logreg,d,type="response")
```

```
logscore <- sum(d$Y*log(pred)+(1-d$Y)*log(1-pred)) # LogLik of
#           full model evaluated at ML estimates
-2*logscore # DEVIANCE FULL MODEL
```

```
## [1] 22.81623
```

```
resglm$deviance
```

```
## [1] 22.81623
```

```
resglm$coefficients[,1:3]
```

```
##                Estimate Std. Error   z value
## (Intercept) -2.3153968  0.9295501 -2.490879
## d$X          0.9099107  0.3167592  2.872563
```

```
# NULL MODEL
logregnull <- glm(d$Y~1,data=d,family=binomial(link="logit"))
resglmnull <- summary(logregnull)
prednull <- predict(logregnull,d,type="response")
logscorenull <- sum(d$Y*log(prednull)+(1-d$Y)*log(1-prednull))
# LogLik of null model evaluated at ML estimate
-2*logscorenull # DEVIANCE NULL MODEL
```

```
## [1] 41.4554
```

```
resglmnull$deviance
```

```
## [1] 41.4554
```

```
resglmnull$coefficients[1:3]
```

```
## [1] -0.1335314  0.3659621 -0.3648776
```

Exercise 2

The model is paramatrised as

$$\begin{aligned}\Pr(Y_i = 1|x_i, \beta) &= \Pr(u_i > 0|x_i, \beta)\\ &= \Phi\left(x_i'\beta\right)\end{aligned} \tag{13.49}$$

and

$$\Pr(Y_i = 0|x_i, \beta) = 1 - \Phi\left(x_i'\beta\right). \tag{13.50}$$

The conditional expectations are

$$\mathrm{E}\left(u_i|\beta, x_i, y_i = 1\right) = x_i'\beta + \frac{\phi\left(x_i'\beta\right)}{\Phi\left(x_i'\beta\right)},$$

$$\mathrm{E}\left(u_i|\beta, x_i, y_i = 0\right) = x_i'\beta - \frac{\phi\left(x_i'\beta\right)}{1 - \Phi\left(x_i'\beta\right)}.$$

The likelihood is

$$L\left(\beta|x, y\right) \propto \prod_{i=1}^{N}\left[\left(\Phi\left(x_i'\beta\right)\right)^{y_i}\left(1 - \Phi\left(x_i'\beta\right)\right)^{1-y_i}\right]$$

and the loglikelihood

$$\ell\left(\beta|x, y\right) = \sum_{i=1}^{N}\left[y_i \ln\Phi\left(x_i'\beta\right) + \left(1 - y_i\right)\ln\left(1 - \Phi\left(x_i'\beta\right)\right)\right].$$

The R-code below implements the EM algorithm using the probit model. The result is in good agreement with that obtained with the R function OPTIM shown at the bottom:

```
# CODE1302
# LikIIQ2 EM algorithm with probit model
rm(list=ls()) # Clear the workspace
set.seed(12371)
# CREATE BINARY DATA
mu <- -2
beta <- 0.7
cov <- rnorm(30,2,3) # GENERATE THE COVARIATE
xb <- cov*beta
p1 <- pnorm(mu+xb) # PROBABILITIES ACCORDING TO PROBIT MODEL
#p1 <- rbeta(30,2,2)
dat1 <- cbind(rbinom(30,1,p1),round(cov,digits=0)) # CREATE DATA
colnames(dat1) <- c("Y", "X")
d <- data.frame(dat1)
attach(d)
m <- cbind(1,X)
mt <- t(m)
mtm <- mt%*%m
miu <- -1
beta <- 1
# FUNCTION expu THAT COMPUTES CONDITIONAL EXPECTATIONS
# HERE WE ASSUME THAT Y=1 IF u>0
expu <- function(data,par,i)
{
  miu <- par[1]
  beta <- par[2]
  if(Y[i]==1)
  {
  with(data,miu+beta*X[i]+
          (dnorm(miu+beta*X[i])/pnorm(miu+beta*X[i])))
  } else
```

```
  {
  with(data,miu+beta*X[i]-
        (dnorm(miu+beta*X[i])/(1-pnorm(miu+beta*X[i]))))
  }
}
## FUNCTION loglik TO COMPUTE THE LOG-LIKELIHOOD
loglik <- function(data,par)
{
  miu <-par[1]
  beta <- par[2]
  with(data,sum(Y*log(pnorm(miu+beta*X))+
                  (1-Y)*log(1-pnorm(miu+beta*X))))
}
iter <- 100
euvec <- matrix(data=NA, nrow=30,ncol=1)
sol <- matrix(data=NA, nrow=iter, ncol=2)
llik <- matrix(data=NA,nrow=iter,ncol=1)
result <- matrix(data=NA,nrow=iter,ncol=3)
# PLACE 30 CONDITIONAL EXPECTATIONS IN euvec AND ITERATE
for (i in 1:iter)
{
  for (j in 1:length(X))
  {
    euvec[j] <- expu(d,c(miu,beta),j)
  }
  sol[i,] <- t(solve(mtm)%*%mt%*%euvec)
  miu <- sol[i,1]
  beta <- sol[i,2]
  llik[i] <-loglik(d,c(miu,beta))
  result[i,] <- c(miu,beta,llik[i])
}
# FINAL ITERATES OF THE EM ALGORITHM
tail(result)
```

```
##             [,1]      [,2]      [,3]
##  [95,] -1.386253 0.5462922 -11.22912
##  [96,] -1.386253 0.5462921 -11.22912
##  [97,] -1.386253 0.5462921 -11.22912
##  [98,] -1.386253 0.5462920 -11.22912
##  [99,] -1.386253 0.5462920 -11.22912
## [100,] -1.386253 0.5462919 -11.22912
```

```
#############################################################
### USE THE R-FUNCTION OPTIM TO COMPARE WITH THIS PROGRAMME
logl <- function(data,par)
{
  miu <- par[1]
  beta <- par[2]
  with(data,-sum(Y*log(pnorm(miu+beta*X))+
              (1-Y)*log(1-pnorm(miu+beta*X))))
}
result1<-optim(par=c(-1,1),logl,data=d,hessian=TRUE,
              method="BFGS")
# ML ESTIMATES USING OPTIM
result1$par
```

```
## [1] -1.386252  0.546292
```

Exercise 3

(i) To implement the Newton-Raphson algorithm, first and second derivatives of the loglikelihood are needed. These are

$$\frac{\partial \ell(\beta_0, \beta_1|n, x)}{\partial \beta_0} = \sum_{i=1}^{30} \left(y_i - \frac{n_i \exp(\beta_0 + \beta_1 x_i)}{1 + \exp(\beta_0 + \beta_1 x_i)} \right),$$

$$\frac{\partial \ell(\beta_0, \beta_1|n, x)}{\partial \beta} = \sum_{i=1}^{30} \left(x_i y_i - \frac{n_i x_i \exp(\beta_0 + \beta_1 x_i)}{1 + \exp(\beta_0 + \beta_1 x_i)} \right).$$

and

$$\frac{\partial^2 \ell(\beta_0, \beta_1|n, x)}{(\partial \beta_0)^2} = -\sum_{i=1}^{30} \frac{n_i \exp(\beta_0 + \beta_1 x_i)}{(1 + \exp(\beta_0 + \beta_1 x_i))^2},$$

$$\frac{\partial^2 \ell(\beta_0, \beta_1|n, x)}{(\partial \beta)^2} = -\sum_{i=1}^{30} \frac{n_i x_i^2 \exp(\beta_0 + \beta_1 x_i)}{(1 + \exp(\beta_0 + \beta_1 x_i))^2},$$

$$\frac{\partial^2 \ell(\beta_0, \beta_1|n, x)}{\partial \beta \partial \beta_0} = -\sum_{i=1}^{30} \frac{n_i x_i \exp(\beta_0 + \beta_1 x_i)}{(1 + \exp(\beta_0 + \beta_1 x_i))^2}.$$

Therefore the iterative system based on Newton-Raphson is

$$\begin{bmatrix} \beta_0 \\ \beta \end{bmatrix}_{t+1} = \begin{bmatrix} \beta_0 \\ \beta \end{bmatrix}_t + \begin{bmatrix} \sum_{i=1}^{30} \frac{n_i \exp(\beta_0+\beta_1 x_i)}{(1+\exp(\beta_0+\beta_1 x_i))^2} & \sum_{i=1}^{30} \frac{n_i x_i \exp(\beta_0+\beta_1 x_i)}{(1+\exp(\beta_0+\beta_1 x_i))^2} \\ \sum_{i=1}^{30} \frac{n_i x_i \exp(\beta_0+\beta_1 x_i)}{(1+\exp(\beta_0+\beta_1 x_i))^2} & \sum_{i=1}^{30} \frac{x_i^2 \exp(\beta_0+\beta_1 x_i)}{(1+\exp(\beta_0+\beta_1 x_i))^2} \end{bmatrix}^{-1}_{\substack{\beta_0=\beta_{0t} \\ \beta=\beta_t}}$$

$$\begin{bmatrix} \sum_{i=1}^{30} \left(y_i - \frac{n_i \exp(\beta_0+\beta_1 x_i)}{1+\exp(\beta_0+\beta_1 x_i)} \right) \\ \sum_{i=1}^{30} \left(x_i y_i - \frac{n_i x_i \exp(\beta_0+\beta_1 x_i)}{1+\exp(\beta_0+\beta_1 x_i)} \right) \end{bmatrix}_{\substack{\beta_0=\beta_{0t} \\ \beta=\beta_t}}. \tag{13.51}$$

To obtain start values for the iteration, rough estimates of β_0 and β_1 can be obtained from $\theta_i \approx y_i/n_i$. Then,

$$\text{logit}(\theta_i) = \ln\left(\frac{\theta_i}{1 - \theta_i}\right) = \beta_0 + \beta_1 x_i, \quad i = 1, \ldots, 4. \tag{13.52}$$

and

$$\text{logit}(y_i/n_i) = z = \beta_0 + \beta_1 x_i + e_i, \quad i = 1, \ldots, 4. \tag{13.53}$$

The logits of 0 and 1 are not defined, so for the purpose of this approximation, change $y_1 = 0.5$ and $y_4 = 4.5$. Writing (13.53) as $z = Xb + e$, the least squares estimates are

$$\widehat{\beta_0} = 0.1; \quad \widehat{\beta_1} = 2.9.$$

The R-code below fits the likelihood using Newton-Raphson. As before, the function OPTIM is included as a check.

```
# CODE1303
## NEWTON-RAPHSON IMPLEMENTATION OF LIKELIHOOD PROBLEM II,
rm(list=ls()) # Clear the workspace
set.seed(771133)
## BINOMIAL DATA SET (Y "successes" out of n=5 "trials")
dat1 <-
matrix(c(0,-0.86,5,1,-0.3,5,3,-0.05,5,5,0.73,5),
       nrow=4,ncol=3,byrow=T)
colnames(dat1) <-c("Y","X","n")
dat1
```

```
##      Y     X n
## [1,] 0 -0.86 5
## [2,] 1 -0.30 5
## [3,] 3 -0.05 5
## [4,] 5  0.73 5
```

```
nit <- 10
miu <- matrix(data=NA, nrow=nit+1,ncol=1)
beta <- matrix(data=NA, nrow=nit+1,ncol=1)
resultnr <- matrix(data=NA, nrow=nit,ncol=2)

miu [1]<- 0.1
beta[1] <- 2.9
for (i in 1:nit)
{
  vc11 <- - sum(5*exp(miu[i]+beta[i]*dat1[,2])/
                ((1+exp(miu[i]+ beta[i]*dat1[,2]))^2))
  vc22 <- - sum(5*dat1[,2]^2*exp(miu[i]+ beta[i]*dat1[,2])/
                ((1+exp(miu[i]+ beta[i]*dat1[,2]))^2))
  vc12 <- - sum(5*dat1[,2]*exp(miu[i]+ beta[i]*dat1[,2])/
                ((1+exp(miu[i]+ beta[i]*dat1[,2]))^2))
  vcmat <- matrix(c(vc11,vc12,vc12,vc22),nrow=2,ncol=2)
  vcmatinv <- solve(vcmat)
  fd1 <- sum((dat1[,1]-(5*exp(miu[i]+ beta[i]*dat1[,2]))/
              (1+exp(miu[i]+ beta[i]*dat1[,2]))))
  fd2 <- sum(((dat1[,1]*dat1[,2])-(5*dat1[,2]*exp(miu[i]+
        beta[i]*dat1[,2]))/(1+exp(miu[i]+ beta[i]*dat1[,2]))))

  fd <- matrix(c(fd1,fd2),nrow=2,ncol=1)
  sol0 <- matrix(c(miu[i], beta[i]),nrow=2,ncol=1)
  sol1 <- sol0+(-vcmatinv%*%fd)
  miu[i+1] <-sol1[1]
  beta[i+1] <- sol1[2]
  resultnr[i,] <- c(miu[i+1],beta[i+1])
}
resultnr
```

```
##          [,1]     [,2]
##  [1,] 0.3048018 4.425389
##  [2,] 0.5560914 5.914432
##  [3,] 0.7610883 7.169775
##  [4,] 0.8394160 7.696086
##  [5,] 0.8465271 7.748401
##  [6,] 0.8465802 7.748817
##  [7,] 0.8465802 7.748817
##  [8,] 0.8465802 7.748817
##  [9,] 0.8465802 7.748817
## [10,] 0.8465802 7.748817
```

```
-vcmatinv
```

```
##          [,1]      [,2]
## [1,] 1.038535  3.545987
## [2,] 3.545987 23.743865
```

```
### USE THE R-FUNCTION OPTIM TO COMPARE WITH THIS PROGRAMME
dat <- data.frame(dat1)
logl <- function(data,par)
{
  with(data,-sum(Y*(par[1]+par[2]*X)-
                5*log(1+exp(par[1]+par[2]*X))))
}
result <- optim(par=c(0.1,2.9),logl,data=dat,
               hessian=TRUE,method="BFGS")
# IF METHOD NOT INCLUDED, OPTIM USES NELDER-MEAD ALGORITHM
# THIS CALL INCLUDES control=list(trace=1,REPORT=1):
# PROVIDES THE PROGRESS OF THE ITERATION
result$par
```

```
## [1] 0.8468243 7.7496258
```

```
solve(result$hessian)
```

```
##          [,1]      [,2]
## [1,] 1.038684  3.546793
## [2,] 3.546793 23.748757
```

(ii) The EM algorithm is constructed assuming the probit model :

$$\begin{aligned}\Pr(Y_i = 1|x_i, \beta) &= \Pr(u_i > 0|x_i, \beta)\\ &= \Pr\left(x_i'\beta + e_i > 0|x_i, \beta\right)\\ &= \Pr\left(e_i > -x_i'\beta|x_i, \beta\right)\\ &= \Phi\left(x_i'\beta\right),\end{aligned}$$

$$\Pr(Y_i = 0|x_i, \beta) = 1 - \Phi\left(x_i'\beta\right),$$

where $e_i \sim N(0, 1)$ and $\Phi(z)$ is the cdf of the standard normal distribution evaluated at z. The binomial likelihood is proportional to

$$L(\beta|n, N, x) \propto \prod_{i=1}^{C} \left[\Phi\left(x_i'\beta\right)\right]^{n_i} \left[1 - \Phi\left(x_i'\beta\right)\right]^{N_i - n_i}$$

and the loglikelihood is

$$\ell(\beta|n, N, x) = \sum_{i=1}^{C} n_i \ln \Phi\left(x_i'\beta\right) + (N_i - n_i) \ln\left(1 - \Phi\left(x_i'\beta\right)\right)$$

where C is the number of levels of the covariate (in the example, $C = 4$).

After 950 iterations the system converges to $\widehat{\beta_0} = 0.484$ and $\widehat{\beta_1} = 4.459$. At this point the loglikelihood is equal to -5.869818. The R-code below fits the likelihood using the EM algorithm based on the iterative system (3.93) on page 116. The R-function OPTIM is included as a comparison. It provides the same estimates and, in addition, an estimate of the asymptotic covariance matrix. This is

$$\text{Var}\left(\widehat{\beta_0}, \widehat{\beta_1}\right) = \begin{bmatrix} 0.434 & 1.575 \\ 1.575 & 9.733 \end{bmatrix}.$$

Based on $\text{Var}\left(\widehat{\beta_0}, \widehat{\beta_1}\right)$, the 95% confidence intervals for β_0 and β_1 are

$$\Pr(-0.807 < \beta_0 < 1.775) = 0.95,$$

$$\Pr(-1.655 < \beta_0 < 10.574) = 0.95,$$

reflecting large uncertainty in the inference.

```
# CODE1304
## EM PROBIT MODEL; LIKELIHOOD PROBLEMS II, QUESTION 3ii
rm(list=ls()) # Clear the workspace
set.seed(12371)
### BINOMIAL DATA (Y successes out of n=5 trials)
dat1 <-
matrix(c(0,-0.86,5,1,-0.3,5,3,-0.05,5,5,0.73,5),
       nrow=4,ncol=3,byrow=T)
colnames(dat1) <-c("n","X","N")
dat1
```

```
##      n     X N
## [1,] 0 -0.86 5
## [2,] 1 -0.30 5
## [3,] 3 -0.05 5
## [4,] 5  0.73 5
```

```
nit <- 1000
miu <- 0.1
beta <- 2.9
dat <- data.frame(dat1)
col1 <- matrix(rep(1:1,length(dat1[,1])),
               nrow=length(dat1[,1]),ncol=1)
m <- matrix(c(col1,dat1[,2]),nrow=length(dat1[,1]),ncol=2)
mt <- t(m)
dN <- diag(dat$N)
dn <- diag(dat$n)
dd <- dN-dn
## FUNCTION loglik TO COMPUTE THE LOG-LIKELIHOOD
loglik <- function(data,par)
{
  miu <-par[1]
  beta <- par[2]
  with(data, sum(n*log(pnorm(miu+beta*X))+
                 (N-n)*log(1-pnorm(miu+beta*X))))
}
# CONSTRUCT FUNCTIONS expu0 and expu1 THAT COMPUTE
#   CONDITIONAL EXPECTATIONS
# ASSUME: Y=1 IF u>0; THEREFORE Pr[Y=1]=F (CDF of std. normal)
expu1 <- function(data,par,i)
{
  miu <- par[1]
  beta <- par[2]
  with(data,(miu+beta*X[i]+
      dnorm(miu+beta*X[i])/pnorm(miu+beta*X[i])))
}
expu0 <- function(data,par,i)
{
  miu <- par[1]
  beta <- par[2]
  with(data,(miu+beta*X[i]-
      dnorm(miu+beta*X[i])/(1-pnorm(miu+beta*X[i]))))
}
sol <- matrix(data=NA, nrow=nit, ncol=2)
e1 <- matrix(data=NA,nrow=length(dat1[,1]),ncol=1)
e0 <- matrix(data=NA,nrow=length(dat1[,1]),ncol=1)
llik <- matrix(data=NA,nrow=nit,ncol=1)
result <- matrix(data=NA,nrow=nit,ncol=3)
for (i in 1:nit)
{
  for (j in 1:length(dat1[,1]))
  {
    e1[j] <- expu1(data=dat,c(miu,beta),j)
    e0[j] <- expu0(data=dat,c(miu,beta),j)
  }
  sol[i,] <- t(solve(mt%*%dN%*%m)%*%(mt%*%dn%*%e1+mt%*%dd%*%e0))
  miu <- sol[i,1]
  beta <- sol[i,2]
  llik[i] <-loglik(dat,c(miu,beta))
  result[i,] <- c(miu,beta,llik[i])
}
tail(result)
```

```
##                [,1]    [,2]      [,3]
## [995,] 0.4839674 4.45866 -5.869818
## [996,] 0.4839674 4.45866 -5.869818
```

```
## [997,] 0.4839674 4.45866 -5.869818
## [998,] 0.4839674 4.45866 -5.869818
## [999,] 0.4839674 4.45866 -5.869818
## [1000,] 0.4839674 4.45866 -5.869818
```

```
### USE THE R-FUNCTION OPTIM TO COMPARE WITH THIS PROGRAMME
logl <- function(data,par)
{
  miu <- par[1]
  beta <- par[2]
  with(data,-sum(n*log(pnorm(miu+beta*X))+
         (N-n)*log(1-pnorm(miu+beta*X))))
}
result1 <- optim(par=c(0.1,2.9),logl,data=dat,
                 hessian=TRUE,method= "BFGS")
result2 <- optim(par=c(0.1,2.9),logl,data=dat,
                 hessian=TRUE)
result1$par
```

```
## [1] 0.4839727 4.4587022
```

```
solve(result1$hessian)
```

```
##             [,1]      [,2]
## [1,] 0.4343711 1.575394
## [2,] 1.5753937 9.732975
```

Figure 13.3 displays the trajectory of the EM iterates for μ. The corresponding plot for β shows a similar pattern.

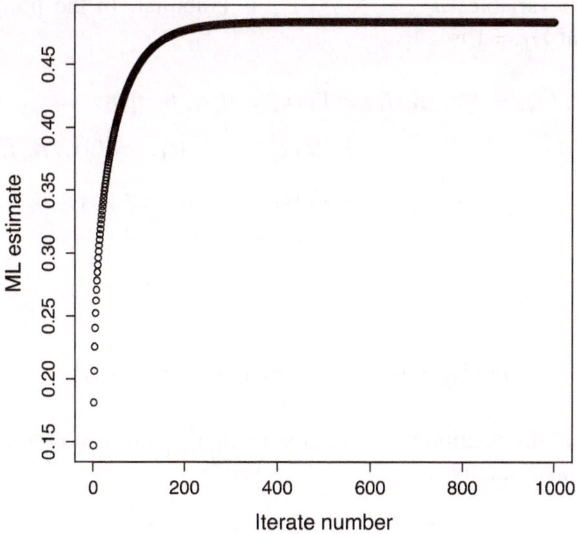

Fig. 13.3 Trajectory of the EM iterates for μ

(ii) Data (number of deaths) are assumed to be binomially distributed. This requires that the binary observations contributing to the number of deaths are independent. This will not be the case if the data consist of related individuals: in the presence of genetic variation or of effects of common environment records of relatives are correlated.

When observations within a class consist of full-sibs, the liability for a given individual (assume the probit model for the example) can be written

$$u_{ij} = x_i'\beta + f_i + e_{ij}, \ldots j = 1, \ldots, N_i \tag{13.54}$$

where now f_i is the effect of the ith full-sib family common to the N_i records that belong to covariate level i and $e_{ij} \overset{iid}{\sim} N(0, 1)$. Assume that $f_i \overset{iid}{\sim} N\left(0, \sigma_f^2\right)$, where σ_f^2 is the covariance between full-sibs. Then,

$$Cov\left(u_{ij}, u_{ik}\right) = \sigma_f^2.$$

Conditioning on f_i, $[u_i|x_i, \beta, f_i] \overset{iid}{\sim} N\left(x_i'\beta + f_i, 1\right)$ and

$$\begin{aligned}
\Pr\left(Y_{ij} = 1, Y_{ik} = 1|x_i, \beta, f_i\right) &= \Pr\left(u_{ij} > 0, u_{ik} > 0|x_i, \beta, f_i\right) \\
&= \Pr\left(u_{ij} > 0|x_i, \beta, f_i\right) \Pr\left(u_{ik} > 0|x_i, \beta, f_i\right) \\
&= \Pr\left(Y_{ij} = 1|x_i, \beta, f_i\right) \Pr\left(Y_{ik} = 1|x_i, \beta, f_i\right).
\end{aligned}$$

In view of the conditional independence of Y_{ij}, given f_i, the sum of independent binary random variables $n_i = \sum_{j=1}^{N_i} Y_{ij}$ is binomial. In the probit model, the probability that $Y_{ij} = 1$ is

$$\begin{aligned}
\Pr\left(Y_{ij} = 1|x_i, \beta, f_i\right) &= \Pr\left(u_i > 0|x_i, \beta, f_i\right) \\
&= \Pr\left(x_i'\beta + f_i + e_{ij} > 0|x_i, \beta, f_i\right) \\
&= \Pr\left(e_{ij} > -x_i'\beta - f_i|x_i, \beta, f_i\right) \\
&= \Phi\left(x_i'\beta + f_i\right)
\end{aligned}$$

and

$$\Pr\left(Y_{ij} = 0|x_i, \beta, f_i\right) = 1 - \Phi\left(x_i'\beta + f_i\right).$$

The conditional distribution of the data $n = \{n_i\}_{i=1}^C$ given vector $f = \{f_i\}_{i=1}^C$ is binomial and takes the form:

$$\Pr\left(n|x_i, \beta, f\right) = \prod_{i=1}^{C} \left[\Phi\left(x_i'\beta + f_i\right)\right]^{n_i} \left[1 - \Phi\left(x_i'\beta + f_i\right)\right]^{N_i - n_i}$$

and the marginal distribution of the data is obtained averaging over the distribution $\left[f|\sigma_f^2\right]$

$$\Pr(n|x_i,\beta) = \int \prod_{i=1}^{C} \left[\Phi\left(x_i'\beta + f_i\right)\right]^{n_i} \left[1 - \Phi\left(x_i'\beta + f_i\right)\right]^{N_i - n_i} p\left(f|\sigma_f^2\right) df$$

$$= \prod_{i=1}^{C} \int \left[\Phi\left(x_i'\beta + f_i\right)\right]^{n_i} \left[1 - \Phi\left(x_i'\beta + f_i\right)\right]^{N_i - n_i} p\left(f_i|\sigma_f^2\right) df_i \qquad (13.55)$$

where the equality in the second line holds if the full-sib families are independent. The likelihood is proportional to (13.55) whose construction requires the solution of univariate integrals. In such univariate cases, numerical methods such as Gaussian quadrature work well, and the problem does not pose a computational obstacle. A benchmark paper pre McMC is Anderson and Aitkin (1985).

Exercise 4

The first R-code below generates genomic data (assumes genomic variance is 10 and residual variance is 15):

```
# CODE1305
# LIKELIHOOD PROBLEMS II QUESTION 4
# DATA BASED ON GENOMIC MODEL AND OBTAIN THE SVD OF WW'(1/m)
rm(list=ls()) # CLEAR WORKSPACE
set.seed(1327)
nindiv<-2000
nmark<-20000
nt<-nindiv*nmark
X<-matrix(nrow= nindiv,ncol= nmark,rbinom(n=nt,size=2,p=.5))
stdev <- matrix(data=NA,nrow= nmark,ncol=1)
W <- matrix(data=NA,nrow= nindiv,ncol=nmark)
U <- matrix(data=NA,nrow= nindiv,ncol= nindiv)
G<-matrix(data=NA,nrow= nindiv,ncol= nindiv)
cm <- colMeans(X)
# CREATE MATRIX OF STANDARDISED MARKER GENOTYPE CODES
for (i in 1:nmark)
{
  W[,i] <-( X[,i]-cm[i]) / sd(X[,i])
}
# COULD USE INSTEAD:
# W <- scale(X)
#qr(X)$rank
#qr(W)$rank
#nmark MARKER VALUES: REALISATIONS FROM N(0,I sqrt(0.001))
g <- (1/sqrt(nmark))*W%*%rnorm(nmark,mean=0,sd=sqrt(10))
# GENERATE nindiv PHENOTYPES WITH MEAN 0, VAR=10+15,
# GENOMIC HERITABILITY=10/(10+15)=0.4
#PARAMETER k = Vg/Ve = 10/15 =0.67
y <- g+rnorm(nindiv,mean=0,sd=sqrt(15))
# GENOMIC RELATIONSHIP MATRIX G
G <- (1/nmark)*W%*%t(W)
```

```
# SVD OF G
EVD <- eigen(G)
names(EVD)
```

```
## [1] "values"  "vectors"
```

```
head(EVD$values)
```

```
## [1] 1.731427 1.724233 1.722874 1.715446 1.710233 1.707764
```

```
U <- EVD$vector
val <- EVD$values
val[length(y)] <-0
D <- diag(val,nrow=nindiv)
ytilde <- t(U)%*%y
dim(ytilde)
```

```
## [1] 2000    1
```

```
#END OF GENERATION OF DATA
```

(i) The R-code below performs the Newton-Raphson computations and the top part includes a call to OPTIM as a check:

```
#################################################################
# CODE1305(cont)
# FUNCTIONS loglik AND logliktransf TO COMPUTE LOG-LIKELIHOODS
loglik <- function(data,par)
{
  ve <- par[1]
  k <- par[2]
  ll <- -0.5*(length(ytilde)*log(ve)+sum(log(val*k+1))+
              (1/ve)*sum(ytilde^2/(val*k+1)))
  return(-ll)
}
# FUNCTION logliktransf TO COMPUTE TRANSFORMED LOG-LIKELIHOOD
logliktransf <- function(data,par)
{
nue <- par[1]
nug <- par[2]
lltransf<--0.5*(length(ytilde)*nue+sum(log(val*exp(nug-nue)+1))+
              (1/exp(nue))*sum(ytilde^2/(val*exp(nug-nue)+1)))

  return(-lltransf)
}
################################################################
# FUNCTION OTPIM TO COMPARE WITH RESULTS TO COME
result1 <-optim(par=c(5,0.5),loglik,data =ytilde,hessian=TRUE)
result1$par
```

```
## [1] 18.2720286  0.2991165
```

```
# OBTAIN ASYMPTOTIC VARIANCES BY INVERSION OF THE -HESSIAN
solve(result1$hessian)
```

```
##            [,1]        [,2]
## [1,]  5.6228947 -0.38289962
## [2,] -0.3828996  0.02772035
```

```
# USE OPTIM TO MAXIMIZE TRANSFORMED LOG-LIKELIHOOD
result2 <-
optim(par=c(exp(5),exp(0.5)),logliktransf,
      data=ytilde,hessian=TRUE)
result2$par
```

```
## [1] 2.901324 1.714862
```

```
solve(result2$hessian)
```

```
##             [,1]        [,2]
## [1,]  0.01619379 -0.04998708
## [2,] -0.04998708  0.17211488
```

```
###############################################################
# NEWTON-RAPHSON COMPUTATIONS
nit <- 20
resultnr<-matrix(data=NA,nrow=nit,ncol=3)
llike<-matrix(data=NA,nrow=nit,ncol=1)
ve <- matrix(data=NA, nrow=nit+1,ncol=1)
k <- matrix(data=NA, nrow=nit+1,ncol=1)
ve[1] <- 7
k[1] <- 0.4
llike[1] <- -loglik(ytilde,c(ve[1],k[1]))
for (i in 1:nit)
{
  vc11 <- - 0.5*((2/ve[i]^3)*sum(ytilde^2/(1+k[i]*val))-
            length(ytilde)/ve[i]^2)
  vc12 <- -0.5*(1/ve[i]^2)*sum(val*ytilde^2/(1+k[i]*val)^2)
  vc22 <- -0.5*((1/ve[i])*sum(2*val^2*ytilde^2/(1+k[i]*val)^3)-
            sum(val^2/(1+k[i]*val)^2))
  vcmat <- matrix(c(vc11,vc12,vc12,vc22),nrow=2,ncol=2)
  vcmatinv <- solve(vcmat)
  fd1 <- -0.5*((length(ytilde)/ve[i])-
        (1/ve[i]^2)*sum(ytilde^2/(1+k[i]*val)))
  fd2<--0.5*(sum(val/(1+k[i]*val))-
        (1/ve[i])*sum(val*ytilde^2/((1+k[i]*val)^2)))
  fd <- matrix(c(fd1,fd2),nrow=2,ncol=1)
  sol0 <- matrix(c(ve[i],k[i]),nrow=2,ncol=1)
  sol1 <- sol0 - (vcmatinv%*%fd)
  ve[i+1] <- sol1[1]
  k[i+1] <- sol1[2]
  llike[i+1] <- loglik(ytilde,c(ve[i+1],k[i+1]))
  resultnr[i,] <- c(sol1[1],sol1[2],llike[i])
}
#PRINT RESULTS
tail(resultnr)
```

```
##          [,1]      [,2]     [,3]
## [15,] 18.2763 0.2986934 4164.218
## [16,] 18.2763 0.2986934 4164.218
## [17,] 18.2763 0.2986934 4164.218
## [18,] 18.2763 0.2986934 4164.218
## [19,] 18.2763 0.2986934 4164.218
## [20,] 18.2763 0.2986934 4164.218
```

```
-vcmatinv
```

```
##            [,1]        [,2]
## [1,]  5.6031133 -0.38126712
## [2,] -0.3812671  0.02758818
```

(i) After 8-10 iterations the Newton-Raphson algorithm converges to $\widehat{\sigma}_e^2 = 18.28$ and $\widehat{k} = 0.30$. The asymptotic variance-covariance matrix is

$$\text{Var}\left(\widehat{\sigma}_e^2, \widehat{k}\right) = \begin{bmatrix} 5.603 & -0.381 \\ -0.381 & 0.028 \end{bmatrix}.$$

(ii) The values obtained using OPTIM are $\widehat{\sigma}_e^2 = 18.27$ and $\widehat{k} = 0.30$ with asymptotic variances:

$$\text{Var}\left(\widehat{\sigma}_e^2, \widehat{k}\right) = \begin{bmatrix} 5.622 & -0.383 \\ -0.383 & 0.028 \end{bmatrix}.$$

(iii)

The ML estimate of $\sigma_g^2 = k\sigma_e^2$ is $\widehat{k\sigma_e^2} = 5.48$. The asymptotic variance is

$$\text{Var}\left(g\left(\widehat{\theta}\right)\right) = \left.\frac{\partial \lambda}{\partial \theta'}\right|_{\theta=\widehat{\theta}} \text{Var}\left(\widehat{\theta}\right) \left.\frac{\partial \lambda'}{\partial \theta}\right|_{\theta=\widehat{\theta}}. \tag{13.56}$$

In this expression, the original parameters are $\theta' = (\sigma_e^2, k)$ and we are interested in

$$\lambda = g(\theta) = \left(\sigma_e^2, k\sigma_e^2\right)',$$

where $k\sigma_e^2 = \sigma_g^2$. Now,

$$\frac{\partial \lambda}{\partial \theta'} = \begin{bmatrix} \frac{\partial \sigma_e^2}{\partial \sigma_e^2} & \frac{\partial \sigma_e^2}{\partial k} \\ \frac{\partial k\sigma_e^2}{\partial \sigma_e^2} & \frac{\partial k\sigma_e^2}{\partial k} \end{bmatrix} = \begin{bmatrix} 1 & 0 \\ k & \sigma_e^2 \end{bmatrix}.$$

Therefore (13.56) is

$$\text{Var}\left(\widehat{\sigma}_e^2, \widehat{k\sigma}_e^2\right) = \begin{bmatrix} 1 & 0 \\ 0.30 & 18.28 \end{bmatrix} \begin{bmatrix} 5.603 & -0.381 \\ -0.381 & 0.028 \end{bmatrix} \begin{bmatrix} 1 & 0.30 \\ 0 & 18.28 \end{bmatrix}$$

$$= \begin{bmatrix} 5.603 & -5.266 \\ -5.266 & 5.693 \end{bmatrix}$$

and $\text{Var}\left(\hat{\sigma}_g^2\right) = 5.693$.

(iv) The R-code below performs the EM computations based on the iterative system (3.140):

```
# CODE1305 (cont)
# EM COMPUTATIONS
emiter<-1000
vgem<-5
veem<-10
kem<-vgem/veem
resultem<-matrix(data=NA,nrow=emiter,ncol=4)
for (i in 1:emiter)
{
expalfa<-kem*(val/(val*kem+1))*ytilde
tol<-0.00001
trdiv<-vgem*sum(1/(val[val>tol]*kem+1))
trv<- vgem*sum(val/(val*kem+1))
vgem<-((kem^2)*sum(ytilde^2*val/(val*kem+1)^2)+trdiv)/(nindiv-1)
veem<-(sum(y^2)-2*sum(expalfa*ytilde)+sum(expalfa^2)+trv)/nindiv
kem<-vgem/veem
loglike<-loglik(ytilde,c(veem,kem))
resultem[i,]<-c(vgem,veem,kem,-loglike)
}
tail(resultem)
```

```
##             [,1]     [,2]      [,3]      [,4]
##   [995,] 5.464388 18.27113 0.2990722 -4164.218
##   [996,] 5.464361 18.27116 0.2990703 -4164.218
##   [997,] 5.464335 18.27118 0.2990684 -4164.218
##   [998,] 5.464308 18.27121 0.2990666 -4164.218
##   [999,] 5.464282 18.27124 0.2990647 -4164.218
##  [1000,] 5.464256 18.27126 0.2990629 -4164.218
```

After approximately 950 iterations, the EM algorithm converges to the same estimates obtained using Newton-Raphson (to 2 decimal places). These are $\hat{\sigma}_g^2 = 5.46$, $\widehat{\sigma}_e^2 = 18.27$ and $\hat{k} = 0.30$.

13.3 Bayes Exercises I

Exercise 1

Start by reading in the data:

```
y <- c(45.83,50.37,50.06,51.59,48.43,52.75,42.92,48.57,
       46.18,50.20)
ybar <- mean(y)
ssq <- var(y)
```

The sample mean $\bar{y}$ and the sample variance $S^2 = \frac{\sum_{i=1}^{n}(y_i-\bar{y})^2}{n-1}$ are 48.69 and 8.859, respectively, with sample size $n = 10$:

(i) Derivation of the density of $[\mu|y]$ requires integration of (12.25) with respect to σ^2. Using

$$\sum_{i=1}^{n} (y_i - \mu)^2 = (n-1)\,S^2 + n\,(\bar{y} - \mu)^2\,,$$

where

$$S^2 = \frac{\sum_{i=1}^{n} (y_i - \bar{y})^2}{n-1},$$

leads to

$$p\left(\mu, \sigma^2|y\right) \propto \left(\sigma^2\right)^{-\left(\frac{n}{2}+1\right)} \exp\left(-\frac{1}{2\sigma^2}\left[(n-1)\,S^2 + n\,(\bar{y}-\mu)^2\right]\right). \tag{13.57}$$

Let $\alpha = n/2$. Then

$$p\,(\mu|y) = \int_0^\infty K\left(\sigma^2\right)^{-\alpha-1} \exp\left(-\frac{1}{2\sigma^2}\left[(n-1)\,S^2 + n\,(\bar{y}-\mu)^2\right]\right) d\sigma^2, \tag{13.58}$$

where K is a constant of integration. Let

$$A = (n-1)\,S^2 + n\,(\bar{y}-\mu)^2\,,$$

$$x = \frac{A}{2\sigma^2},$$

$$d\sigma^2 = -\frac{A}{2}x^{-2}dx.$$

Substituting in (13.58)

$$
\begin{aligned}
p(\mu|y) &= -\frac{A}{2}\int_0^\infty K\left(\frac{A}{2x}\right)^{-\alpha-1}\exp(-x)\,x^{-2}dx \\
&= -\left(\frac{A}{2}\right)\left(\frac{A}{2}\right)^{-\alpha-1}\int_0^\infty K\left(\frac{1}{x}\right)^{-\alpha-1}\exp(-x)\,x^{-2}dx \\
&\propto A^{-\alpha}\int_0^\infty x^{\alpha-1}\exp(-x)\,dx \\
&\propto A^{-\alpha}
\end{aligned}
$$

where the last line results from the fact that the integral, known as the Gamma function is

$$\int_0^\infty x^{\alpha-1}\exp(-x)\,dx = \Gamma(\alpha), \quad \alpha > 0,$$

a constant with respect to μ. Then,

$$
\begin{aligned}
p(\mu|y) &\propto \left((n-1)S^2 + n(\overline{y}-\mu)^2\right)^{-\frac{n}{2}} \\
&= \left((n-1)S^2 + n(\overline{y}-\mu)^2\right)^{-\left(\frac{v+1}{2}\right)},
\end{aligned}
$$

where $v = n - 1$. Finally,

$$p(\mu|y) \propto \left(1 + \frac{n(\overline{y}-\mu)^2}{vS^2}\right)^{-\left(\frac{v+1}{2}\right)}, \tag{13.59}$$

that is the kernel of the density of a $t-$distribution with $v = n - 1$ degrees of freedom, mean $\overline{y}$ and scale S^2/n; that is

$$[\mu|y] \sim t\left(n-1, \overline{y}, \frac{S^2}{n}\right). \tag{13.60}$$

The mean and variance are

$$
\begin{aligned}
E(\mu|y) &= \overline{y}, \\
\text{Var}(\mu|y) &= \frac{S^2}{n}\frac{n-1}{n-3}.
\end{aligned}
$$

Derivation of the density of $\left[\sigma^2|y\right]$ requires the following integration of (12.25) with respect to μ:

$$\begin{aligned}
p\left(\sigma^2|y\right) &\propto \int_{-\infty}^{\infty} \left(\sigma^2\right)^{-\left(\frac{n}{2}+1\right)} \exp\left(-\frac{1}{2\sigma^2}\left[(n-1)\,S^2 + n\,(\overline{y}-\mu)^2\right]\right) d\mu \\
&= \left(\sigma^2\right)^{-\left(\frac{n}{2}+1\right)} \exp\left(-\frac{1}{2\sigma^2}(n-1)\,S^2\right) \int_{-\infty}^{\infty} \\
&\quad \times \exp\left(-\frac{1}{2\sigma^2} n\,(\overline{y}-\mu)^2\right) d\mu \\
&= \left(\sigma^2\right)^{-\left(\frac{n}{2}+1\right)} \exp\left(-\frac{1}{2\sigma^2}(n-1)\,S^2\right) \left(2\pi\frac{\sigma^2}{n}\right)^{\frac{1}{2}} \\
&\propto \left(\sigma^2\right)^{-\left(\frac{n}{2}+1\right)} \left(\sigma^2\right)^{\frac{1}{2}} \exp\left(-\frac{1}{2\sigma^2}(n-1)\,S^2\right) \\
&= \left(\sigma^2\right)^{-\left(\frac{n-1}{2}+1\right)} \exp\left(-\frac{1}{2\sigma^2}(n-1)\,S^2\right), \qquad (13.61)
\end{aligned}$$

which is the kernel of the density of a scaled inverted chi-square distribution with $(n-1)$ degrees of freedom and scale S^2. Therefore,

$$\left[\sigma^2|y\right] \sim \chi^{-2}\left(n-1, S^2\right). \qquad (13.62)$$

(ii) The expected value and variance of the t−distribution (13.60) are given by

$$\begin{aligned}
\mathrm{E}\,(\mu|y) &= \overline{y} = 48.69, \\
\mathrm{Var}\,(\mu|y) &= \frac{n-1}{n-3}\frac{S^2}{n} = \frac{9}{7}\frac{8.859}{10} = 1.139.
\end{aligned}$$

The modal value is equal to the mean.

The mean, variance and mode of the scale inverted chi-square distribution (13.62) are

$$\begin{aligned}
\mathrm{E}\left(\sigma^2|y\right) &= \frac{n-1}{n-3}S^2 = 11.39, \\
\mathrm{Var}\left(\sigma^2|y\right) &= \frac{2(n-1)^2}{(n-3)^2\,(n-5)}S^4 = \frac{162}{49\times 5}8.86^2 = 51.91, \\
\arg\max\left[\sigma^2|y\right] &= \frac{n-1}{n+1}S^2 = \frac{9}{11}8.86 = 7.25.
\end{aligned}$$

(iii) To compute confidence intervals for the t–distribution, it is easiest to transform μ to the standard t–distribution as follows.

Let

$$t = f(\mu) = \frac{\mu - \overline{y}}{S / \sqrt{n}};$$

$$\mu = f^{-1}(t) = \frac{S}{\sqrt{n}} t + \overline{y};$$

$$\frac{df^{-1}(t)}{t} = S / \sqrt{n}\ ; \text{ not a function of } t;$$

$$\Rightarrow p(t) \propto \left(1 + \frac{n \left(\overline{y} - \frac{S}{\sqrt{n}} t - \overline{y} \right)^2}{v S^2} \right)^{-\left(\frac{v+1}{2}\right)}$$

$$= \left(1 + \frac{t^2}{v} \right)^{-\left(\frac{v+1}{2}\right)}$$

which is the kernel of the density of a standard t–distribution $t(v)$, with $v = n - 1$. Then using the R-function `qt(c(.025, .975), df=9)`, the higher and lower bounds are

$$h = 2.262157,$$
$$\ell = -2.262157$$

which in the original scale (obtained as $\left(h \times \frac{S}{\sqrt{n}}\right) + \overline{y}$ and $\left(\ell \times \frac{S}{\sqrt{n}}\right) + \overline{y}$) results in the 95% posterior interval

$$\Pr(46.56 < \mu < 50.82 | y) = 0.95. \tag{13.63}$$

To compute the 95% posterior interval of σ^2, it seems easy to use the function `igamma` in R. First recall that the relationship between the scaled inverse chi-square and the inverse gamma distributions is

$$\chi^{-2}\left(v, S^2\right) = IG\left(\frac{v}{2}, \frac{v S^2}{2}\right).$$

Therefore in terms of (13.62) requires working with

$$IG\left(\frac{n-1}{2}, \frac{(n-1) S^2}{2}\right).$$

The R function qigamma is part of the pscl package that must be downloaded and installed. The R-code below does this and computes the 95% posterior interval:

```
#install.packages("pscl", .libPaths()[1])
library(pscl)
a <- (length(y)-1)/2
b <- (length(y)-1)*var(y)/2
qigamma(c(0.025,0.975),a,b)
```

```
## [1]  4.191167 29.524482
```

The 95% posterior interval is

$$\Pr\left(4.19 < \sigma^2 < 29.52|y\right) = 0.95. \tag{13.64}$$

(iv) According to (first order) asymptotic theory, the posterior distribution of $[\mu, \sigma^2|y]$ is

$$\left[\mu, \sigma^2|y\right] \sim N\left(\begin{bmatrix} \hat{\mu} \\ \hat{\sigma}^2 \end{bmatrix}, -\begin{bmatrix} \left(\frac{\partial^2 p(\mu,\sigma^2|y)}{(\partial\mu)^2}\right) & \left(\frac{\partial^2 p(\mu,\sigma^2|y)}{\partial\mu\partial\sigma^2}\right) \\ \left(\frac{\partial^2 p(\mu,\sigma^2|y)}{\partial\mu\partial\sigma^2}\right) & \left(\frac{\partial^2 p(\mu,\sigma^2|y)}{(\partial\sigma^2)^2}\right) \end{bmatrix}^{-1}\right), \tag{13.65}$$

where $\hat{\mu}$ and $\hat{\sigma}^2$ are the mode of $\left[\mu, \sigma^2|y\right]$.

The density of the joint posterior distribution was given in (13.57):

$$p\left(\mu, \sigma^2|y\right) \propto \left(\sigma^2\right)^{-\left(\frac{n}{2}+1\right)} \exp\left(-\frac{1}{2\sigma^2}\left[(n-1)\,S^2 + n\,(\overline{y}-\mu)^2\right]\right),$$

and the logposterior density, up to an additive constant, is

$$\ln p\left(\mu, \sigma^2|y\right) = -\left(\frac{n}{2}+1\right)\ln\sigma^2 - \frac{1}{2\sigma^2}\left[(n-1)\,S^2 + n\,(\overline{y}-\mu)^2\right]. \tag{13.66}$$

First derivatives are

$$\frac{\partial \ln p\left(\mu, \sigma^2|y\right)}{\partial\mu} = \frac{n\,(\overline{y}-\mu)}{\sigma^2},$$

$$\frac{\partial \ln p\left(\mu, \sigma^2|y\right)}{\partial\sigma^2} = -\frac{n+2}{2\sigma^2} + \frac{(n-1)\,S^2 + n\,(\overline{y}-\mu)^2}{2\sigma^4}.$$

Setting these equations equal to zero and solving for μ and σ^2 gives:

$$\begin{bmatrix} \widehat{\mu} \\ \widehat{\sigma}^2 \end{bmatrix} = \begin{bmatrix} \overline{y} \\ \frac{n-1}{n+2} S^2 \end{bmatrix} = \begin{bmatrix} \overline{y} \\ \frac{\sum (y_i - \overline{y})^2}{n+2} \end{bmatrix}. \tag{13.67}$$

Second derivatives are

$$\frac{\partial^2 p\left(\mu, \sigma^2 | y\right)}{(\partial \mu)^2} = -\frac{n}{\sigma^2},$$

$$\frac{\partial^2 p\left(\mu, \sigma^2 | y\right)}{\left(\partial \sigma^2\right)^2} = \frac{n+2}{2\sigma^4} - \frac{(n-1)\, S^2 + n\left(\overline{y} - \mu\right)^2}{\sigma^6},$$

$$\frac{\partial^2 p\left(\mu, \sigma^2 | y\right)}{\partial \mu \partial \sigma^2} = -\frac{n\left(\overline{y} - \mu\right)}{\sigma^4}.$$

Evaluated at the mode, these second derivatives are (and substituting $(n-1)\, S^2 = (n+2)\, \widehat{\sigma}^2$)

$$\left.\frac{\partial^2 p\left(\mu, \sigma^2 | y\right)}{(\partial \mu)^2}\right|_{\mu=\widehat{\mu}, \sigma^2=\widehat{\sigma}^2} = -\frac{n}{\widehat{\sigma}^2},$$

$$\left.\frac{\partial^2 p\left(\mu, \sigma^2 | y\right)}{\left(\partial \sigma^2\right)^2}\right|_{\mu=\widehat{\mu}, \sigma^2=\widehat{\sigma}^2} = \frac{n+2}{2\widehat{\sigma}^4} - \frac{n+2}{\widehat{\sigma}^4} = -\frac{n+2}{2\widehat{\sigma}^4},$$

$$\left.\frac{\partial^2 p\left(\mu, \sigma^2 | y\right)}{\partial \mu \partial \sigma^2}\right|_{\mu=\widehat{\mu}, \sigma^2=\widehat{\sigma}^2} = 0.$$

Therefore the asymptotic approximation to $\left[\mu, \sigma^2 | y\right]$ is

$$\left[\mu, \sigma^2 | y\right] \sim N\left(\begin{bmatrix} \overline{y} \\ \frac{\sum (y_i - \overline{y})^2}{n+2} \end{bmatrix}, \begin{bmatrix} \frac{\widehat{\sigma}^2}{n} & 0 \\ 0 & \frac{2\widehat{\sigma}^4}{n+2} \end{bmatrix} \right) \tag{13.68}$$

Another approximation based on the likelihood function is

$$[\theta | y] \sim N\left(\widehat{\theta}, \left. I\left(\theta\right)^{-1} \right|_{\theta=\widehat{\theta}} \right) \tag{13.69}$$

where $\widehat{\theta}$ is the ML estimator and $I(\theta)$ is the observed information. I derive this approximation and compare it to (13.68). Ignoring the prior, write

$$
\begin{aligned}
p\left(\mu, \sigma^2|y\right) &\propto p\left(y|\mu, \sigma^2\right) \\
&\propto \exp\left[-\frac{\sum\limits_{i=1}^{n} (y_i - \mu)^2}{2\sigma^2}\right]\left(\sigma^2\right)^{-\left(\frac{n}{2}\right)} \\
&= \exp\left[-\frac{(n-1)\,S^2 + n\left(\overline{y} - \mu\right)^2}{2\sigma^2}\right]\left(\sigma^2\right)^{-\left(\frac{n}{2}\right)},
\end{aligned}
$$

and the logposterior is

$$
\ln p\left(\mu, \sigma^2|y\right) = -\left(\frac{n}{2}\right)\ln \sigma^2 - \frac{1}{2\sigma^2}\left[(n-1)\,S^2 + n\left(\overline{y} - \mu\right)^2\right].
$$

First derivatives are

$$
\begin{aligned}
\frac{\partial \ln p\left(\mu, \sigma^2|y\right)}{\partial \mu} &= \frac{n\left(\overline{y} - \mu\right)}{\sigma^2}, \\
\frac{\partial \ln p\left(\mu, \sigma^2|y\right)}{\partial \sigma^2} &= -\frac{n}{2\sigma^2} + \frac{(n-1)\,S^2 + n\left(\overline{y} - \mu\right)^2}{2\sigma^4}.
\end{aligned}
$$

Setting these equations equal to zero and solving for μ and σ^2 gives

$$
\begin{bmatrix} \hat{\mu} \\ \hat{\sigma}^2 \end{bmatrix} = \begin{bmatrix} \overline{y} \\ \frac{n-1}{n} S^2 \end{bmatrix} = \begin{bmatrix} \overline{y} \\ \frac{\sum (y_i - \overline{y})^2}{n} \end{bmatrix}.
$$

Second derivatives are

$$
\begin{aligned}
\frac{\partial^2 p\left(\mu, \sigma^2|y\right)}{(\partial \mu)^2} &= -\frac{n}{\sigma^2}, \\
\frac{\partial^2 p\left(\mu, \sigma^2|y\right)}{\left(\partial \sigma^2\right)^2} &= \frac{n}{2\sigma^4} - \frac{(n-1)\,S^2 + n\left(\overline{y} - \mu\right)^2}{\sigma^6}, \\
\frac{\partial^2 p\left(\mu, \sigma^2|y\right)}{\partial \mu \partial \sigma^2} &= -\frac{n\left(\overline{y} - \mu\right)}{\sigma^4}.
\end{aligned}
$$

Evaluated at the mode, these second derivatives are (substituting $(n-1)\,S^2 = n\widehat{\sigma}^2$) are

$$\left.\frac{\partial^2 p\left(\mu, \sigma^2|y\right)}{(\partial\mu)^2}\right|_{\mu=\hat{\mu},\sigma^2=\hat{\sigma}^2} = -\frac{n}{\hat{\sigma}^2},$$

$$\left.\frac{\partial^2 p\left(\mu, \sigma^2|y\right)}{\left(\partial\sigma^2\right)^2}\right|_{\mu=\hat{\mu},\sigma^2=\hat{\sigma}^2} = \frac{n}{2\hat{\sigma}^4} - \frac{n}{\hat{\sigma}^4} = -\frac{n}{2\hat{\sigma}^4},$$

$$\left.\frac{\partial^2 p\left(\mu, \sigma^2|y\right)}{\partial\mu\partial\sigma^2}\right|_{\mu=\hat{\mu},\sigma^2=\hat{\sigma}^2} = 0.$$

Therefore the asymptotic approximation to $\left[\mu, \sigma^2|y\right]$ is now

$$\left[\mu, \sigma^2|y\right] \sim N\left(\begin{bmatrix} \overline{y} \\ \frac{\sum(y_i-\overline{y})^2}{n} \end{bmatrix}, \begin{bmatrix} \frac{\hat{\sigma}^2}{n} & 0 \\ 0 & \frac{2\hat{\sigma}^4}{n} \end{bmatrix}\right) \tag{13.70}$$

As n increases, (13.68) and (13.70) approach the same value.

(v) The approximate 95% posterior intervals for μ and σ^2 based on (13.68) can be obtained using the R functions :

```
qnorm(c(0.025,0.975),48.69,sqrt(6.644/10))
```

```
## [1] 47.09242 50.28758
```

```
qnorm(c(0.025,0.975),6.644,sqrt(7.357))
```

```
## [1]  1.327835 11.960165
```

that yield

$$\Pr(47.09 < \mu < 50.28|y) = 0.95$$
$$\Pr\left(1.33 < \sigma^2 < 11.96|y\right) = 0.95.$$

The approximate 95% posterior intervals for μ and σ^2 based on (13.70) are obtained using the R functions:

```
qnorm(c(0.025,0.975),48.69,sqrt(7.973/10))
```

```
## [1] 46.93992 50.44008
```

```
qnorm(c(0.025,0.975),7.973,sqrt(12.714))
```

```
## [1]  0.9844159 14.9615841
```

that yield

$$\Pr(46.94 < \mu < 50.44|y) = 0.95,$$
$$\Pr\left(0.98 < \sigma^2 < 14.96|y\right) = 0.95.$$

Comparison with the exact results (13.63) and (13.64) shows that the interval for μ is well approximated by the normal distribution, but this is not the case for σ^2. Clearly a sample size of $n = 10$ is not big enough for the normal approximation to hold in the case of σ^2.

Exercise 2

(i) The R-code below implements the Bayesian model with the Metropolis-Hastings single-site updating algorithm:

```
# CODE1306
#METROPOLIS-HASTINGS SINGLE-SITE UPDATING - USES ACF
rm(list=ls()) # Clear the workspace
set.seed(123456)
require(graphics)
y<-c(45.83,50.37,50.06,51.59,48.43,52.75,42.92,48.57,
     46.18,50.20)
# SET LENGTH OF CHAIN rep
rep<-10000
result<-matrix(data=NA,nrow=rep,ncol=3)

mu<- 15
v<- 10
#CHOOSE TUNING PARAMETERS kmu AND kv
#kmu<-0.19
#kv<-0.1
kmu<-19
kv<-9
acceptv<-0
acceptmu<-0
ptm <- proc.time()
for (i in 1:rep)
{
```

```
  #UPDATING THE VARIANCE
  logYv<-rnorm(1,mean=log(v),sd=sqrt(kv))
  logalfav <-sum((y-mu)^2)/(2*v)-sum((y-mu)^2)/
    (2*exp(logYv))+(length(y)/2)*(log(v)-logYv)
  unif<-runif(1)
  if (unif<exp(logalfav))
  {
    v<-exp(logYv)
    acceptv<-acceptv+1
  }
  #UPDATING THE MEAN
  Ymu<-rnorm(1,mean=mu,sd=sqrt(kmu))
  logalfamu<- sum((y-mu)^2)/(2*v)-sum((y-Ymu)^2)/(2*v)
  unif<-runif(1)
  if (unif<exp(logalfamu))
  {
    mu<-Ymu
    acceptmu<-acceptmu+1
    result[i, ]<-c(i,mu,v)
  }
  else  {
    result[i, ]<-c(i,mu,v)
  }
}
proc.time()-ptm
```

```
##    user  system elapsed
##    0.15    0.00    0.19
```

```
acceptratiomu<-acceptmu/rep
acceptratiov<-acceptv/rep
# DISCARD THE FIRST 1500 DRAWS
mu<-result[1501:rep,2]
v<-result[1501:rep,3]
meanmus<-mean(mu)
meanmus
```

```
## [1] 48.67492
```

```
varmus<-var(mu)
varmus
```

```
## [1] 1.315822
```

```
cimus<-quantile(mu,c(0.025,0.975))
cimus
```

```
##     2.5%    97.5%
## 46.35049 50.93522
```

```
meanvs<-mean(v)
meanvs
```

```
## [1] 12.15146
```

```
varvs<- var(v)
varvs
```

```
## [1] 66.8197
```

```
civs<-quantile(v,c(0.025,0.975))
civs
```

```
##     2.5%    97.5%
##  4.29519 32.92001
```

The single-site updating algorithm generates the following output. The chain length is 10, 000, and the first 1, 500 samples are discarded:

$$\begin{aligned}
\hat{\mathrm{E}}\,(\mu|y) &= 48.67 \\
\widehat{\mathrm{Var}}\,(\mu|y) &= 1.32 \\
\hat{\Pr}\,(46.35 < \mu < 50.94|y) &= 0.95 \\
\hat{\mathrm{E}}\left(\sigma^2|y\right) &= 12.15 \\
\widehat{\mathrm{Var}}\left(\sigma^2|y\right) &= 66.82 \\
\hat{\Pr}\left(4.3 < \sigma^2 < 32.92|y\right) &= 0.95.
\end{aligned}$$

(ii) The R-code below implements the Bayesian model with the Metropolis-Hastings joint updating algorithm:

```
# CODE1307
#METROPOLIS-HASTINGS JOINT UPDATING
rm(list=ls()) # Clear the workspace
set.seed(123456)
require(graphics)
```

```r
y<-c(45.83,50.37,50.06,51.59,48.43,52.75,42.92,48.57,
     46.18,50.20)
# SET LENGTH OF CHAIN rep
rep<-10000
result<-matrix(data=NA,nrow=rep,ncol=3)
#INITIALISE mu AND v
mu<-15
v<-10
#CHOOSE TUNING PARAMETERS kmu AND kv
kmu<-0.15
kv<-0.08
accept<-0
ptm <- proc.time()
for (i in 1:rep)
{
  Ymu<-rnorm(1,mean=mu,sd=sqrt(kmu))
  logYv<-rnorm(1,mean=log(v),sd=sqrt(kv))
  logalfa <-sum((y-mu)^2)/(2*v)-sum((y-Ymu)^2)/
    (2*exp(logYv))+(length(y)/2)*(log(v)-logYv)
  unif<-runif(1)
  if (unif<exp(logalfa))
  {
    mu<-Ymu
    v<-exp(logYv)
    result[i, ]<-c(i,mu,v)
    accept<-accept+1
  }
  else
  {
    result[i, ]<-c(i,mu,v)
  }
}
proc.time()-ptm
```

```
##    user  system elapsed
##    0.13    0.00    0.13
```

```r
acceptratio<-accept/rep
mu<-result[1501:rep,2]
v<-result[1501:rep,3]
meanmuj<-mean(mu)
meanmuj
```

```
## [1] 48.62203
```

```
varmuj<-var(mu)
varmuj
```

```
## [1] 1.175362
```

```
cimuj<-quantile(mu,c(0.025,0.975))
cimuj
```

```
##     2.5%    97.5%
## 46.34522 50.69246
```

```
meanvj<-mean(v)
meanvj
```

```
## [1] 11.91832
```

```
varvj<- var(v)
varvj
```

```
## [1] 55.73139
```

```
civj<-quantile(v,c(0.025,0.975))
civj
```

```
##      2.5%     97.5%
##  4.242654 30.864260
```

The joint updating algorithm produces the following output. The chain length is 10, 000, and the first 1, 500 samples are discarded:

$$\begin{aligned}
\hat{\mathrm{E}}(\mu|y) &= 48.62 \\
\widehat{\mathrm{Var}}(\mu|y) &= 1.18 \\
\hat{\Pr}(46.35 < \mu < 50.69|y) &= 0.95 \\
\hat{\mathrm{E}}\left(\sigma^2|y\right) &= 11.92 \\
\widehat{\mathrm{Var}}\left(\sigma^2|y\right) &= 55.73 \\
\hat{\Pr}\left(4.24 < \sigma^2 < 30.86|y\right) &= 0.95.
\end{aligned}$$

A glance at these figures confirms that they are in good agreement with the exact results.

(iii)

For $k_\mu = 0.19$, $k_{\sigma^2} = 0.1$, the following results are obtained:

(a) acceptance ratios for μ and σ^2 are 0.87 and 0.47, respectively;
(b) The MC variances for μ and σ^2 are 0.0095 and 0.63, respectively;
(c) effective chain length for μ and σ^2 are 160 and 194, respectively;
(d) integrated autocorrelations for μ and σ^2 are 53 and 43, respectively.

For $k_\mu = 19$, $k_{\sigma^2} = 9$, the following results are obtained:

(a) acceptance ratios for μ and σ^2 are 0.29 and 0.19, respectively;
(b) The MC variances for μ and σ^2 are 0.00097 and 0.093, respectively;
(c) effective chain length for μ and σ^2 are 1357 and 720, respectively;
(d) integrated autocorrelations for μ and σ^2 are 6.3 and 11.8, respectively.

When the tuning parameters (variances) are very small, the proposed values differ very little from the previous values and are accepted with high probability; this generates a highly correlated chain. When the tuning parameters are larger, proposed values differ, on average, relatively more from previous values, are accepted with smaller probability and the degree of autocorrelations is relatively smaller. This is also illustrated in Fig. 13.4 that displays autocorrelations versus lag. The left panel corresponds to $k_\mu = 0.19$, $k_{\sigma^2} = 0.1$ that generates high autocorrelations decreasing at a low rate with lag. The right panel corresponds to $k_\mu = 19$, $k_{\sigma^2} = 9$; the autocorrelations fall at a higher rate.

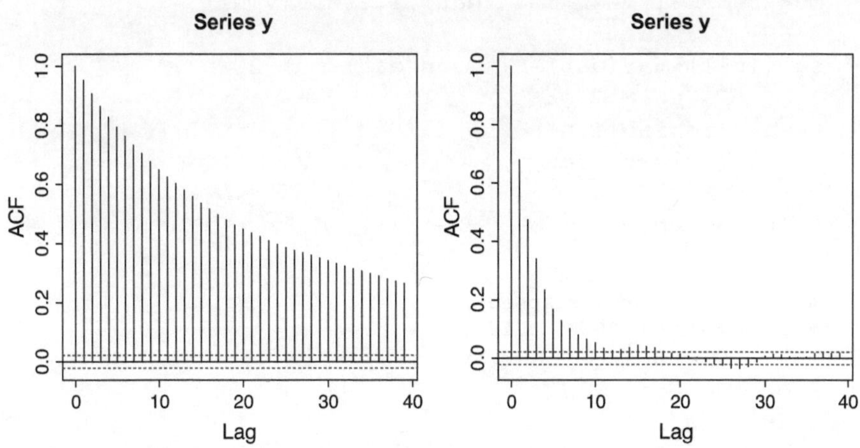

Fig. 13.4 Plots of autocorrelations versus lag of draws from $[\mu|y]$ using $k_\mu = 0.19$, $k_{\sigma^2} = 0.1$ (left panel) and $k_\mu = 19$, $k_{\sigma^2} = 9$ (right panel)

Exercise 3

(i) The R- code below implements the Bayesian model with the single-site Gibbs sampling algorithm:

```
# CODE1308
#GIBBS SAMPLING ALGORITHM
rm(list=ls()) # Clear the workspace
set.seed(12345)
require(graphics)
dat<-c(45.83,50.37,50.06,51.59,48.43,52.75,42.92,48.57,
       46.18,50.20)
# SET LENGTH OF CHAIN rep
rep<-10000
result<-matrix(data=NA,nrow=rep,ncol=3)
#INITIALISE mu AND v
#mu<-mean(dat)
#v<-var(dat)
mu<-1
v<-2
n<-length(dat)
# START GIBBS CHAIN
ptm <- proc.time()
for (i in 1:rep)
{
  # GENERATE MIU
  mu<-rnorm(1,mean=mean(dat),sd=sqrt(v/n))
  # COMPUTE SCALE
  s<-((n-1)*var(dat)+n*(mean(dat)-mu)^2)/n
  # GENERATE V
  v<-n*s/rchisq(1,n)
  result[i,]<-c(i,mu,v)
}
proc.time()-ptm
```

```
##    user  system elapsed
##    0.45    0.00    0.46
```

```
# END OF GIBBS CHAIN
#mu<-result[1000:rep,2]
mu<-result[,2]

#v<-result[1000:rep,3]
v<-result[,3]
```

```
meanmu<-mean(mu)
meanmu
```

```
## [1] 48.70177
```

```
varmu<-var(mu)
varmu
```

```
## [1] 1.132319
```

```
cimu<-quantile(mu,c(0.025,0.975))
cimu
```

```
##     2.5%    97.5%
## 46.59150 50.84902
```

```
meanv<-mean(v)
meanv
```

```
## [1] 11.35799
```

```
varv<- var(v)
varv
```

```
## [1] 48.91507
```

```
civ<-quantile(v,c(0.025,0.975))
civ
```

```
##      2.5%     97.5%
##  4.184609 29.596819
```

The Gibbs sampling algorithm yields the following estimates:

$$
\begin{aligned}
\hat{\mathrm{E}}\,(\mu|y) &= 48.7 \\
\widehat{\mathrm{Var}}\,(\mu|y) &= 1.13 \\
\hat{\mathrm{E}}\left(\sigma^2|y\right) &= 11.36 \\
\widehat{\mathrm{Var}}\left(\sigma^2|y\right) &= 48.92 \\
\hat{\Pr}\,(46.59 < \mu < 50.85|y) &= 0.95 \\
\hat{\Pr}\left(4.18 < \sigma^2 < 29.6|y\right) &= 0.95,
\end{aligned}
$$

in good agreement with those obtained with the Metropolis-Hastings algorithms. The Monte Carlo variances are

$$\widehat{\mathrm{Var}}\left[\widehat{\mathrm{E}}\left(\mu|y\right)\right] = 0.0001,$$
$$\widehat{\mathrm{Var}}\left[\widehat{\mathrm{E}}\left(\sigma^2|y\right)\right] = 0.0070.$$

(ii) The effective chain lengths for μ and σ^2 are $9,931$ and $7,512$, respectively.

The integrated autocorrelations μ and σ^2 are 1.0 and 1.33, respectively. The R-code that computes these quantities is shown below:

```
# CODE1308 (cont)
#POST-McMC ANALYSIS
# CODE FOR THE MC VARIANCE BASED ON GEYER

# CHOOSE MU OR V AND PLACE IN VECTOR Y
y<-result[,2] # READS IN mu
#y <- result[,3] # READS IN v
ns <- length(y)
svar<-var(y)*(ns-1)/ns
tau<-1
tausum<-0
for(i in 0:ns)
{
  gamaj<-0.0
  gamak<-0.0
  j<-2*i
  k<-(2*i)+1
  lag1<-j
  lag2<-k
  #USE THE R-FUNCTION ACF TO COMPUTE AUTOCORRELATIONS
  cov1<-acf(y,type="covariance",lag.max=lag1,plot=FALSE)
  cov2<-acf(y,type="covariance",lag.max=lag2,plot=FALSE)
  gamaj<-cov1$acf[lag1+1]
  gamak<-cov2$acf[lag2+1]
  tau<-gamaj+gamak
  if(tau<0)
  {
    break
  }
  tausum<-tausum+tau
}

varch<- -svar+2*tausum
mcvar<-varch/ns
mcvar
```

```
## [1] 0.0001140094
```

```
efchsize<-svar/mcvar
efchsize
```

```
## [1] 9930.804
```

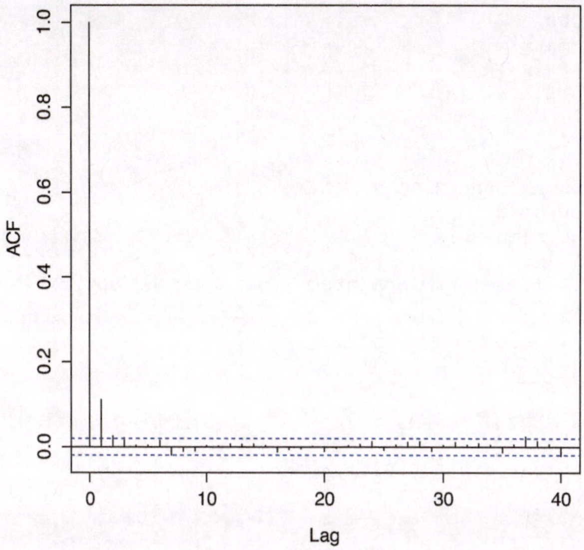

Fig. 13.5 Plots of autocorrelations versus lag of draws from $[\sigma^2|y]$ using the Gibbs sampler

```
integrautoc<-varch/svar
integrautoc
```

```
## [1] 1.006968
```

For this toy model and dataset, the draws from the McMC chain generated by the Gibbs sampler have very small autocorrelation. This is illustrated in Fig. 13.5.

13.4 Bayes Exercises II

Exercise 1

(i) The R-code below fits the Bayesian logistic regression model implemented with a Metropolis-Hastings algorithm with joint updating of (μ, β):

```
# CODE1309
#BINARY DATA - M-H JOINT UPDATING - LOGISTIC MODEL
rm(list=ls()) # Clear the workspace
set.seed(123)
#require(graphics)
# THE CODE WILL USE THE PACKAGE MVTNORM; IT IS INSTALLED BELOW
#install.packages("mvtnorm", .libPaths()[1])
library(mvtnorm)
#CHOOSE LENGTH OF CHAIN rep
```

```
rep<-10000
result<-matrix(data=NA,nrow=rep,ncol=4)
###############################################
# CREATE BINARY DATA
mu <- -2
beta <- 0.7
nrec <- 30
cov <- rnorm(nrec,2,3) # GENERATE THE COVARIATE
xb <- cov*beta
p1 <- pnorm(mu+xb) # PROBABILITIES ACCORDING TO PROBIT MODEL
# CREATE DATA:
d <- data.frame(Y=rbinom(nrec,1,p1),X=round(cov,digits=0))
mean(d$Y)
```

```
## [1] 0.3
```

```
# CHOOSE TUNING PARAMETER LAMBDA AND COVARIANCE MATRIX C
lambda<-1
c<-matrix(c(1,0.0,0.0,0.1),nrow=2,ncol=2,byrow=T)

# INITIALISE THE MEAN OF THE BIVARIATE DISTRIBUTION
theta<-c(-2,1)
accept<-0
## FUNCTION TO COMPUTE THE LOG-POSTERIOR
logpost <- function(data,par)
{
  theta[1] <-par[1]
  theta[2] <- par[2]
  with(data=d,sum(Y*(theta[1]+theta[2]*X)-log(1+exp(theta[1]+
                                   theta[2]*X))))
}
#START MH LOOP
ptm <- proc.time()
for (i in 1:rep)
{
  #SAMPLE PROPOSAL FOR THETA (Ytheta) FROM N(theta,lamdaC)
  Ytheta<- rmvnorm(1,mean=theta,sigma=lambda*c)
  logalfa<-logpost(d,c(Ytheta[1],Ytheta[2])) -
        logpost(d,c(theta[1],theta[2]))
  unif<-runif(1)
  if (unif<exp(logalfa))
  {
    theta[1]<-Ytheta[1]
    theta[2]<-Ytheta[2]
    result[i, ]<-c(i,theta[1],theta[2],exp(theta[1]+3*theta[2])/
                    (1+exp(theta[1]+3*theta[2])))
    accept<-accept+1
  }
  else
  {
    result[i, ]<-c(i,theta[1],theta[2],exp(theta[1]+3*theta[2])/
                    (1+exp(theta[1]+3*theta[2])))
  }
}
proc.time()-ptm
```

```
##    user  system elapsed
##    2.56    0.03    2.59
```

```
acceptratio<-accept/rep
acceptratio
```

```
## [1] 0.4313
```

```
meanmu<-mean(result[,2])
meanmu
```

```
## [1] -3.996897
```

```
varmu<-var(result[,2])
varmu
```

```
## [1] 1.986835
```

```
cimu<-quantile(result[,2],c(0.025,0.975))
cimu
```

```
##      2.5%     97.5%
## -7.311877 -1.717778
```

```
meanbeta<-mean(result[,3])
meanbeta
```

```
## [1] 1.168351
```

```
varbeta<- var(result[,3])
varbeta
```

```
## [1] 0.1899689
```

```
cibeta<-quantile(result[,3],c(0.025,0.975))
cibeta
```

```
##      2.5%     97.5%
## 0.4968853 2.1923000
```

```
meanprob <- mean(result[,4])
meanprob
```

```
## [1] 0.3901256
```

```
ciprob <- quantile(result[,4],c(0.025,0.975))
ciprob
```

```
##      2.5%     97.5%
## 0.1421850 0.6921409
```

(ii) The following features of the posterior distribution are obtained from the output:

$$
\begin{aligned}
\hat{\mathrm{E}}\,(\mu|y) &= -3.997,\\
\hat{\mathrm{E}}\,(\beta|y) &= 1.168,\\
\widehat{\mathrm{Var}}\,(\mu|y) &= 1.987,\\
\widehat{\mathrm{Var}}\,(\beta|y) &= 0.19,\\
\hat{\Pr}\,(-7.312 < \mu < -1.718|y) &= 0.95,\\
\hat{\Pr}\,(0.497 < \beta < 2.192|y) &= 0.95,\\
\overline{\hat{\Pr}}\,(Y = 1|x = 3, y) &= 0.39,
\end{aligned}
$$

where in the last line, $\overline{\hat{\Pr}}\,(Y = 1|x = 3, y)$ is the mean of the McMC draws from $\Pr\,(Y = 1|x = 3, y)$. The McMC estimate of the 95% posterior interval of the distribution of $\Pr(Y = 1|x = 3, y)$ is (0.142; 0.692). Given x, this probability is only a function of (μ, β), and the confidence interval reflects the posterior uncertainty associated with (μ, β).

Figure 13.6 displays histograms of MC estimates of the marginal posterior distributions $[\mu|y]$ (left panel) and of $[\beta|y]$ (right panel). Both histograms reveal considerable asymmetry.

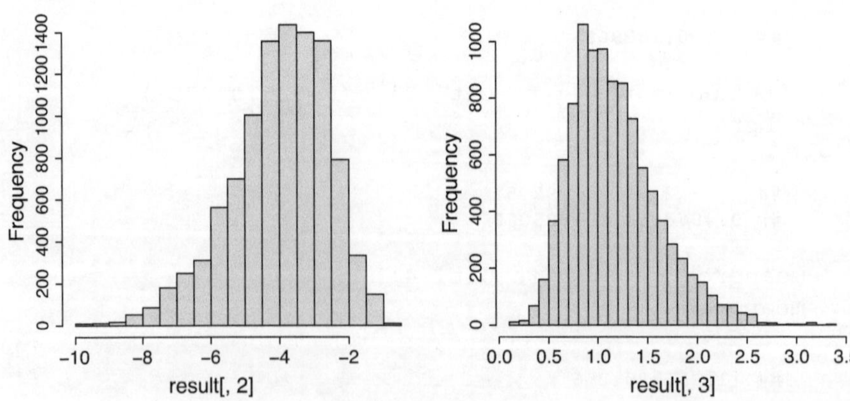

Fig. 13.6 Histograms of MC estimates of $[\mu|y]$ (left panel) and of $[\beta|y]$ (right panel)

It is interesting to compare the output from the Bayesian implementation with that derived from a maximum likelihood analysis. The R-function below fits the OPTIM function :

```
# CODE1309 (cont)
### USE THE R-FUNCTION OPTIM TO COMPARE WITH THIS PROGRAMME
## FUNCTION TO COMPUTE THE NEGATIVE OF THE LOG-POSTERIOR
logpostoptim <- function(data,par)
{
  theta[1] <-par[1]
  theta[2] <- par[2]
  with(data=d,-sum(Y*(theta[1]+theta[2]*X)-log(1+exp(theta[1]+
                                   theta[2]*X))))
}
result1 <- optim(par=c(-3,1),logpostoptim,
                 data=d,hessian=TRUE,method="BFGS")
result1$par
```

```
## [1] -3.2918405  0.9354898
```

```
solve(result1$hessian)
```

```
##             [,1]       [,2]
## [1,]  1.3069454 -0.3367133
## [2,] -0.3367133  0.1197200
```

The ML estimates of uncertainty are based on asymptotic results that do not quite hold in the small sample setting of this example. There is considerable difference between the uncertainty reported by both methods of inference.

The ML estimates of μ and β are -3.292 and 0.935, respectively. These differ from the McMC estimates of posterior means due, in part, to the asymmetry referred to above that results in different values for mean and mode. Further, the likelihood approach relies on a joint maximisation of the loglikelihood function, whereas marginal inferences are reported with the Bayesian method. One can bring both approaches a little closer by computing the modal value of the estimates of the marginal posterior distributions $[\mu|y]$ and $[\beta|y]$. The R-code below uses the function modeest and calculates these modal values using a kernel estimator:

```
# CODE1309 (cont)
# computing MODE with package modeest
# and via density estimation
#install.packages("modeest", .libPaths()[1])
y<- result[,2]
x<-result[,3]
library(modeest)
myDensity<-density(y)
modeEstmu <- mlv(y,method = "kernel", kernel = "gaussian")
modeEstmu
```

```
## [1] -3.498042
```

```
modeEstbeta <- mlv(x,method = "kernel", kernel = "gaussian")
modeEstbeta
```

```
## [1] 0.9147319
```

The estimate of the posterior mode of $[\mu|y]$ is seen to be -3.498 and of $[\beta|y]$, 0.915, more in line with the results reported from the likelihood approach.

(iii) The R-code below calculates MC variances, effective chain lengths and integrated autocorrelations:

```
# CODE1309 (cont)
#POST-McMC ANALYSIS
# CODE FOR THE MC VARIANCE BASED ON GEYER
ns<-rep
# CHOOSE MU OR V AND PLACE IN VECTOR Y
y<- result[,2]
#y<-result[,3]
svar<-var(y)*(ns-1)/ns
tau<-1
tausum<-0
for(i in 0:ns)
{
  gamaj<-0.0
  gamak<-0.0
  j<-2*i
  k<-(2*i)+1
  lag1<-j
  lag2<-k
  #USE THE R-FUNCTION ACF TO COMPUTE AUTOCORRELATIONS
  cov1<-acf(y,type="covariance",lag.max=lag1,plot=FALSE)
  cov2<-acf(y,type="covariance",lag.max=lag2,plot=FALSE)
  gamaj<-cov1$acf[lag1+1]
  gamak<-cov2$acf[lag2+1]
  tau<-gamaj+gamak
  if(tau<0)
  {
    break
  }
  tausum<-tausum+tau
}
varch<- -svar+2*tausum
mcvar<-varch/ns
mcvar
```

```
## [1] 0.009202506
```

```
efchsize<-svar/mcvar
efchsize
```

```
## [1] 215.88
```

```
integrautoc<-varch/svar
integrautoc
```

```
## [1] 46.32203
```

The MC variance of the estimated mean of $[\mu|y]$, the effective chain length and integrated autocorrelation of the draws from $[\mu|y]$ obtained from the output are 0.0092, 215.88 and 46.32, respectively. For $[\beta|y]$ the respective values are 0.00088, 213.57 and 46.82, respectively. These figures indicate that the McMC samples display a relatively high degree of autocorrelation.

Exercise 2

(i) The R-code below fits the probit regression model with the Metropolis-Hastings joint updating algorithm:

```
# CODE1310
#BINARY DATA - METROPOLIS-HASTINGS JOINT UPDATING - PROBIT MODEL
rm(list=ls()) # Clear the workspace
set.seed(123)
#require(graphics)
# CODE USES PACKAGE MVTNORM; IT IS INSTALLED BELOW
#install.packages("mvtnorm", .libPaths()[1])
library(mvtnorm)
#CHOOSE LENGTH OF CHAIN rep
rep<-10000
result<-matrix(data=NA,nrow=rep,ncol=4)
##################################################
# CREATE BINARY DATA BASED ON PROBIT MODEL
mu <- -2
beta <- 0.7
nrec <- 30
cov <- rnorm(nrec,2,3) # GENERATE THE COVARIATE
xb <- cov*beta
p1 <- pnorm(mu+xb) # PROBABILITIES ACCORDING TO PROBIT MODEL
# CREATE DATA:
d <- data.frame(Y=rbinom(nrec,1,p1),X=round(cov,digits=0))
mean(d$Y)
```

```
## [1] 0.3
```

```
################################################################
# CHOOSE TUNING PARAMETER LAMBDA AND COVARIANCE MATRIX C
lambda<-0.25
c<-matrix(c(1,0.0,0.0,0.1),nrow=2,ncol=2,byrow=T)

# INITIALISE THE MEAN OF THE BIVARIATE DISTRIBUTION
theta<-c(-2,1)

accept<-0
## FUNCTION TO COMPUTE THE LOG-POSTERIOR
```

```
logpost <- function(data,par)
{
  theta[1] <-par[1]
  theta[2] <- par[2]
  with(data=d,sum(Y*log(pnorm(theta[1]+theta[2]*X))+
       (1-Y)*log(1.000001-pnorm(theta[1]+theta[2]*X))))
}
#START MH LOOP
ptm <- proc.time()
for (i in 1:rep)
{
  #SAMPLE PROPOSAL FOR THETA (Ytheta) FROM N(theta,lamdaC)
  Ytheta<- rmvnorm(1,mean=theta,sigma=lambda*c)
  lp1<- logpost(d,c(Ytheta[1],Ytheta[2]))
  lp2<- logpost(d,c(theta[1],theta[2]))
  logalfa<-lp1-lp2
  unif<-runif(1)
  if (unif<exp(logalfa))
  {
    theta[1]<-Ytheta[1]
    theta[2]<-Ytheta[2]
    result[i, ]<-c(i,theta[1],theta[2],
                pnorm(theta[1]+3*theta[2]))
    accept<-accept+1
  }
  else
  {
    result[i, ]<-c(i,theta[1],theta[2],
                pnorm(theta[1]+3*theta[2]))
  }
}
proc.time()-ptm
```

```
##    user  system elapsed
##    2.73    0.06    2.80
```

```
acceptratio<-accept/rep
meanmu<-mean(result[,2])
meanmu
```

```
## [1] -2.230753
```

```
varmu<-var(result[,2])
varmu
```

```
## [1] 0.4690703
```

```
cimu<-quantile(result[,2],c(0.025,0.975))
cimu
```

```
##      2.5%     97.5%
## -3.696849 -1.062495
```

```
meanbeta<-mean(result[,3])
meanbeta
```

```
## [1] 0.6450377
```

```
varbeta<- var(result[,3])
varbeta
```

```
## [1] 0.04395467
```

```
cibeta<-quantile(result[,3],c(0.025,0.975))
cibeta
```

```
##      2.5%     97.5%
## 0.2879808 1.0886044
```

```
meanprob <- mean(result[,4])
meanprob
```

```
## [1] 0.3900007
```

```
ciprob <- quantile(result[,4],c(0.025,0.975))
ciprob
```

```
##      2.5%     97.5%
## 0.1653514 0.6497098
```

(iii) The following features of the posterior distribution are obtained from the output:

$$\begin{aligned}
\hat{\mathrm{E}}\,(\mu|y) &= -2.231,\\
\hat{\mathrm{E}}\,(\beta|y) &= 0.645,\\
\widehat{\mathrm{Var}}\,(\mu|y) &= 0.469,\\
\widehat{\mathrm{Var}}\,(\beta|y) &= 0.044,\\
\hat{\Pr}\,(-3.697 < \mu < -1.062|y) &= 0.95,\\
\hat{\Pr}\,(0.288 < \beta < 1.089|y) &= 0.95,\\
\overline{\hat{\Pr}}\,(Y = 1|x = 3, y) &= 0.39,
\end{aligned}$$

where in the last line, $\overline{\hat{\Pr}}\,(Y = 1|x = 3, y)$ is the mean of the McMC draws from $\Pr\,(Y = 1|x = 3, y)$. The McMC estimate of the 95% posterior interval of the distribution of $\Pr(Y = 1|x = 3, y)$ is (0.165; 0.65). Given x this probability is only a function of (μ, β) and the confidence interval reflects the posterior uncertainty associated with (μ, β). Despite the difference in estimates of the parameters between the logit and the probit models, the estimates of the posterior probability $\Pr(Y = 1|x = 3, y)$ are in good agreement.

(iii) The MC variance of the mean of $[\mu|y]$, the effective chain length and integrated autocorrelation of the draws from $[\mu|y]$ are 0.0018, 264.0 and 37.9, respectively. For $[\beta|y]$ these figures are 0.00017, 264.92 and 37.75, respectively.

Exercise 3

(i) The R-code below generates the same 30 binary records used in the previous examples and fits a probit model implemented with a single-site updating Gibbs sampling algorithm:

```
# CODE1311
# BINARY DATA; GIBBS WITH DATA AUGMENTATION PROBIT MODEL
# LIABILITIES SAMPLED IN ONE GO AFTER JA  pg 241
rm(list=ls()) # Clear the workspace
set.seed(123)
#require(graphics)
# THE CODE WILL USE THE PACKAGE MVTNORM; IT IS INSTALLED BELOW
#install.packages("mvtnorm", .libPaths()[1])
library(mvtnorm)
#CHOOSE LENGTH OF CHAIN rep
rep<-10000
result<-matrix(data=NA,nrow=rep,ncol=3)
##################################################
# CREATE BINARY DATA
mu <- -2
beta <- 0.7
nrec <- 30
cov <- rnorm(nrec,2,3) # GENERATE THE COVARIATE
xb <- cov*beta
p1 <- pnorm(mu+xb) # COMPUTE PROBABILITIES FOR PROBIT MODEL
#p1 <- rbeta(30,2,2)
# CREATE DATA:
d <- data.frame(Y=rbinom(nrec,1,p1),X=round(cov,digits=0))
mean(d$Y)
```

```
## [1] 0.3
```

```
##############################################################
One<-rep(1,nrec)
X<-matrix(c(One,d[,2]),nrow=nrec,ncol=2)
Y<-matrix(d[,1],nrow=length(d[,1]),ncol=1)
#INITIALISE THETA (THE VECTOR WITH MU AND BETA)
```

```
theta<-matrix(data=0,nrow=2,ncol=1)
# INITIALISE VECTOR u
u<-rep(0,nrec)
#COMPUTE XTX INVERSE
xtxinv<-solve(t(X)%*%X)
#START GIBBS LOOP
ptm <- proc.time()
for (i in 1:rep)
{
  thetahat<-xtxinv%*%t(X)%*%u
  #SAMPLE THETA FROM MVN(thetahat,xtxinv)
  theta<- t(rmvnorm(1,mean=thetahat,sigma= xtxinv))
  # SAMPLE LIABILITIES u_j FROM A TN(beta+Uf,1)

  av<-X%*%theta # MEAN OF UNTRUNCATED NORMAL(mu+U[j,]alfa,1)
  cutoff<-pnorm(-av)
  interm<-(cutoff*(1-Y)+(1-cutoff)*Y)*runif(length(Y))+cutoff*Y
  interm[interm==1]<-0.999
  u<-qnorm(interm)+av
  # END SAMPLING LIABILITIES u
  result[i,]<-c(i,t(theta))
}
proc.time() - ptm
```

```
##    user  system elapsed
##    2.64    0.09    2.73
```

```
meanmu<-mean(result[,2])
meanmu
```

```
## [1] -2.177854
```

```
varmu<-var(result[,2])
varmu
```

```
## [1] 0.4747578
```

```
cimu<-quantile(result[,2],c(0.025,0.975))
cimu
```

```
##      2.5%     97.5%
## -3.728438 -1.014262
```

```
meanbeta<-mean(result[,3])
meanbeta
```

```
## [1] 0.6290715
```

```
varbeta<- var(result[,3])
varbeta
```

```
## [1] 0.04445811
```

```
cibeta<-quantile(result[,3],c(0.025,0.975))
cibeta
```

```
##       2.5%     97.5%
## 0.2846762 1.1175002
```

(ii) The following features of the posterior distribution are obtained from the output:

$$
\begin{aligned}
\hat{\mathrm{E}}\,(\mu|y) &= -2.178, \\
\hat{\mathrm{E}}\,(\beta|y) &= 0.629, \\
\widehat{\mathrm{Var}}\,(\mu|y) &= 0.475, \\
\widehat{\mathrm{Var}}\,(\beta|y) &= 0.044, \\
\hat{\mathrm{Pr}}\,(-3.728 < \mu < -1.014|y) &= 0.95, \\
\hat{\mathrm{Pr}}\,(0.285 < \beta < 1.118|y) &= 0.95,
\end{aligned}
$$

that can be seen to be in good agreement with the output in Exercise 13.4.

(iii) The MC variance of the estimated mean of $[\mu|y]$, the effective chain length and integrated autocorrelation of the draws from $[\mu|y]$ are 0.0014, 340.8 and 29.3, respectively. For $[\beta|y]$ these figures are 0.00013, 336.3 and 29.7, respectively. A comparison with Exercise 13.4 indicates that the Gibbs sampler generated slightly less correlated draws from $[\mu, \beta|y]$ than the Metropolis-Hastings algorithm, as implemented in this example. The latter can be tuned to improve its behaviour; this was not attempted here.

Exercise 4

(i) The R-code below generates binary data from 500 full-sib families each with 2 full-sibs and fits a probit threshold model implemented via a Gibbs sampling algorithm using a single chain of length 2, 000.

```
# CODE1312
#SINGLE-SITE GIBBS - CORRELATED PROBIT MODEL
# DOES NOT USE THE SVD OF ZZ'
```

```
rm(list=ls()) # Clear the workspace
set.seed(7713)

require(graphics)
# GENERATE CORRELATED (FULL-SIBS) BINARY DATA (THRESHOLD MODEL)
#install.packages("MCMCpack", .libPaths()[1])
#install.packages("mvtnorm", .libPaths()[1])
library(mvtnorm)
#library(MCMCpack)
# INITIALISE PARAMETERS
p0 <- 0.5
mu <- qnorm(p0)
iccfs<-0.15 #INTRACLASS CORRELATION FS
# VARIANCE BETWEEN FAMILIES: iccfs /(1- iccfs)
# PHENOTYPIC VARIANCE: 1/(1-iccfs)
nfs<-500 # NUMBER OF FULL-SIB FAMILIES
fs<-2 #FULL-SIB FAMILY SIZE
# SET DATA Y= 0
y<-matrix(data=0,nrow=fs*nfs,ncol=1)
# x IS COLUMN MATRIX WITH FAMILY ID (ID=1,.,nfs)
x<-matrix(data=0,nrow=fs*nfs,ncol=1)
# GENERATE NFS FULL-SIB EFFECTS f
f<-rnorm(nfs,mean=0,sd=sqrt(iccfs/(1-iccfs)))
# SET COUNTER c EQUAL TO ZERO
# (c = nfs*fs IS EQUAL TO THE LENGTH OF BINARY DATA VECTOR y)
c<-0
##########################################################
####  GENERATE BINARY RECORDS Y
f<-rnorm(nfs,mean=0,sd=sqrt(iccfs/(1-iccfs)))
p <- pnorm(mu+f)
y <- rbinom(nfs*fs,1,rep(p,each=fs))
w <- rep(1:nfs,each=fs)
d<-data.frame(w,y)
family <- w
family <- as.factor(family)
Z<-model.matrix(~0+family)
# WITH INDEPENDENT FAMLIES Z'Z IS DIAGONAL
ztz<-rep(fs,nfs)
#CHOOSE LENGTH OF CHAIN rep
rep<-2000
result<-matrix(data=NA,nrow=rep,ncol=3)
# INITIALISE LIABILITY VECTOR u
u<-rep(0,fs*nfs)
#INITIALISE THE VECTOR OF FAMILIY EFFECTS fam
fam<-rep(0,nfs)
# INITIALISE BETWEEN FAMILY VARIANCE COMPONENT vf
vf<-0.2
# INITIALISE THE MEAN (HERE BETA, A SCALAR)
beta<-0
#START GIBBS LOOP
ptm<-proc.time()
for (i in 1:rep)
{
  zfam <- Z%*%fam
  # SAMPLE BETA
  betahat<-sum(u-zfam)/(fs*nfs)
  beta<-rnorm(1,mean=betahat,sd=sqrt(1/(fs*nfs)))
  # SAMPLE LIABILITIES u FROM A TN(beta+Zf,1)
  av<-beta+ zfam # MEAN OF UNTRUNCATED NORMAL(mu+Zf,1)
```

```
  prob<-pnorm(-av)
  interm<-( prob *(1-y)+(1- prob)*y)*runif(fs*nfs)+ prob*y
  interm[interm==1]<-0.999
  u<-qnorm(interm)+av
  # END SAMPLING LIABILITIES u
  # SAMPLE FAMILY EFFECTS fam
  varfam<-1/(fs+(1/vf))
  fammean<-varfam*(t(Z)%*%(u-beta))
  fam<-rnorm(nfs,mean=fammean, sd=sqrt(varfam))
  #SAMPLE vf
  #COMPUTE SCALE
  ftf<-sum(fam*fam)
  vf<-ftf/rchisq(1,nfs-2)
  result[i,]<-c(i,beta,vf)
}
proc.time()-ptm
```

```
##    user  system elapsed
##   25.33    0.64    6.52
```

```
# CONSTRUCT THE DERIVED PARAMETER "HERITABILITY" herit
herit <- 2*result[,3]/(result[,3]+1)
meanbeta <- mean(result[,2])
meanbeta
```

```
## [1] -0.05565647
```

```
meanvf <- mean(result[,3])
meanvf
```

```
## [1] 0.377655
```

```
meanher <- mean(herit)
meanher
```

```
## [1] 0.5408147
```

```
cibeta <- quantile(result[,2],c(0.025,0.975))
cibeta
```

```
##        2.5%       97.5%
## -0.15785926  0.03703906
```

```
civf <- quantile(result[,3],c(0.025,0.975))
civf
```

```
##      2.5%     97.5%
## 0.2110779 0.5870883
```

```
ciher <- quantile(herit,c(0.025,0.975))
ciher
```

```
##      2.5%     97.5%
## 0.3485786 0.7398306
```

(ii) The code outputs the following results:

$$\hat{\mathrm{E}}\,(\mu|y) = -0.06,$$

$$\hat{\mathrm{E}}\left(\sigma_f^2|y\right) = 0.38,$$

$$\hat{\mathrm{E}}\left(h^2|y\right) = 0.54,$$

$$\hat{\Pr}\,(-0.158 < \mu < 0.037|y) = 0.95,$$

$$\hat{\Pr}\left(0.211 < \sigma_f^2 < 0.587|y\right) = 0.95,$$

$$\hat{\Pr}\left(0.349 < h^2 < 0.74|y\right) = 0.95.$$

The parameter h^2, the heritability on the underlying scale, is constructed from the output of the McMC, using σ_f^2, the variance component between families and the variance of the residual term of the liability, equal to 1: $h^2 = 2\sigma_f^2/(\sigma_f^2 + 1)$.

(iii) Execution of the R-code on page 636 generates the following results.

The MC variances of the estimate of the mean of $[\mu|y]$, the mean of $[\sigma_f^2|y]$ and the mean of $[h^2|y]$ are 5.74×10^{-6}, 2.3×10^{-4} and 2.6×10^{-4}, respectively.

The effective chain lengths of $[\mu|y]$, $[\sigma_f^2|y]$ and $[h^2|y]$ are 448.2, 42.2 and 40.2, respectively.

The integrated autocorrelations of $[\mu|y]$, $[\sigma_f^2|y]$ and $[h^2|y]$ are 4.46, 47.4 and 49.8, respectively.

Due to the considerable degree of autocorrelation among samples, particularly for $[\sigma_f^2|y]$ and $[h^2|y]$, a chain of length 2,000 results in effective chain lengths equivalent to approximately 40 independent draws.

Exercise 5

(i) The R-code below fits the genomic model using a single-site Gibbs sampling algorithm:

```
# CODE1313
# BAYES PROBLEMS II Exercise 5. GENOMIC MODEL
```

```
# DATA BASED ON GENOMIC MODEL; OBTAIN SVD OF WW'(1/m)
rm(list=ls()) # CLEAR WORKSPACE
set.seed(12345)
nindiv<-500
nmark<-1000
nt <- nindiv*nmark

# GENERATE MARKER MATRIX FROM BINOMIAL DISTRIBUTION
X<-matrix(nrow= nindiv,ncol= nmark,rbinom(n=nt,size=2,p=.5))
stdev <- matrix(data=NA,nrow= nmark,ncol=1)
W <- matrix(data=NA,nrow= nindiv,ncol=nmark)
U <- matrix(data=NA,nrow= nindiv,ncol= nindiv)
G<-matrix(data=NA,nrow= nindiv,ncol= nindiv)
cm <- colMeans(X)
#CHOOSE VALUE FOR GENOMIC VARIANCE vgs
vgs<-10
#CHOOSE VALUE FOR ENVIRONMENTAL VARIANCE ves
ves<-25
# CREATE MATRIX OF STANDARDISED MARKER GENOTYPE CODES
for (i in 1:nmark)
{
  W[,i] <-( X[,i]-cm[i]) / sd(X[,i])
}
# GENERATE nindiv GENOMIC VAL FROM N(0,(1/nmark)WW'*vgs)
#NOTE: MARKER VALUES ARE DRAWS FROM N(0,I sqrt(vgs/nmark))
g <- (1/sqrt(nmark))*W%*%rnorm(nmark,mean=0,sd=sqrt(vgs))
# GENERATE nindiv PHENOTYPES WITH MEAN 0,
#  VAR=vgs+ves, HERITABILITY=vgs/(vgs+ves)
e<- rnorm(nindiv,mean=0,sd=sqrt(ves))
y <- g+ e
# GENOMIC RELATIONSHIP MATRIX G
G <- (1/nmark)*W%*%t(W)
# SVD OF G
EVD <- eigen(G)
names(EVD)
```

```
## [1] "values"  "vectors"
```

```
#head(EVD$values)
U <- EVD$vector
tU<-t(U)
val <- EVD$values
val[length(y)] <-0
D <- diag(val,nrow=nindiv)
#Dp IS A VECTOR WITH NON-ZERO EIGENVALUES
Dp<-c(val[1:nindiv-1])
#INITIALISE Ve
Ve<-5
#INITIALISE Vg
Vg<-5
#INITIALISE k
k<-Ve/Vg
#INITIALISE VECTOR ALFA
alfa<-rep(0,nindiv)
# CHOOSE LENGTH OF GIBBS CHAIN
rep<-4000
#INITIALISE result
result<-matrix(data=NA,nrow=rep,ncol=7)
```

```
# START GIBBS CHAIN
ptm<-proc.time()

for (i in 1:rep)
{
 # print (i)
  # SAMPLE mu
  avmu<-sum(y-U%*%alfa)/nindiv
  varmu<-Ve/nindiv
  mu<-rnorm(1,mean=avmu,sd=sqrt(varmu))
  #mu<-0
  # SAMPLE alfa1 (VECTOR OF LENGTH nindiv-1)
  meanalfa1<-(Dp/(Dp+k))*tU[1:nindiv-1,]%*%(y-mu)
  varalfa1<-((Dp)/(Dp+k))*Ve
  alfa1<-rnorm(nindiv-1,meanalfa1,sqrt(varalfa1))
  alfa<-c(alfa1,0)
  # SAMPLE Vg
  # COMPUTE SCALE
  scVg<-sum(alfa1*alfa1*(1/Dp))
  Vg<-scVg/rchisq(1,nindiv-3)
  #Vg<-0.0001
  # SAMPLE Ve
  # COMPUTE SCALE
  ystar<-y-mu-U%*%alfa
  scVe<-sum(ystar*ystar)
  Ve<-scVe/rchisq(1,nindiv-2)
  #Ve<-25
  k<-Ve/Vg
  result[i,]<-c(i,mu,Vg,Ve,Vg/(Vg+Ve),1/k,mean(alfa*alfa))
#  print(result[i,])
}
proc.time()-ptm
```

```
##    user  system elapsed
##   28.87    0.67    7.39
```

```
# FUNCTION LOGLIK TO CONSTRUCT THE LOGLIKELIHOOD
# TO COMPARE THE BAYESIAN RESULTS WITH OPTIM
#NOTE: k IN THE LOGLIKELIHOOD IS Vg/Ve
loglik<-function(data,par)
{
  mu<-par[1]
  ve<-par[2]
  k<-par[3]
  dkiinv<-diag(1/((val*k)+1),nrow=nindiv,ncol=nindiv)
  ll<- -0.5*(length(y)*log(ve)+sum(log((val*k)+1))+
             (1/ve)*t(y-mu)%*%U%*%dkiinv%*%tU%*%(y-mu))
  return(-ll)
}
result1<-optim(par=c(0.5,12,2),loglik,data=y,hessian=TRUE)
result1$par
```

```
## [1] -0.02250109 28.28775687  0.32911418
```

```
vgml <- result1$par[2]*result1$par[3]
solve(result1$hessian)
```

```
##                [,1]          [,2]          [,3]
## [1,]  5.657552e-02  0.0003634452 -1.420182e-05
## [2,]  3.634452e-04 13.2109992444 -5.162269e-01
## [3,] -1.420182e-05 -0.5162269466  2.661989e-02
```

```
meanmu <- mean(result[,2])
meanmu
```

```
## [1] -0.02252344
```

```
#apply(result[,2:6],2,mean)
cimu <- quantile(result[,2],c(0.025,0.975))
cimu
```

```
##       2.5%      97.5%
## -0.4810972  0.4374214
```

```
meanvg <- mean(result[,3])
meanvg
```

```
## [1] 10.11903
```

```
civg <- quantile(result[,3],c(0.025,0.975))
civg
```

```
##      2.5%     97.5%
##  3.210285 16.971752
```

```
meanve <- mean(result[,4])
meanve
```

```
## [1] 28.27113
```

```
cive <- quantile(result[,4],c(0.025,0.975))
cive
```

```
##     2.5%    97.5%
## 22.05225 35.56460
```

```
meanher <- mean(result[,5])
meanher
```

```
## [1] 0.2624171
```

```
ciher <- quantile(result[,5],c(0.025,0.975))
ciher
```

```
##       2.5%      97.5%
## 0.08389514 0.41867397
```

```
meank <- mean(result[,6])
meank
```

```
## [1] 0.3749238
```

```
cik <- quantile(result[,6],c(0.025,0.975))
cik
```

```
##       2.5%      97.5%
## 0.09157809 0.72020512
```

```
meanalfasq <- mean(result[,7])
meanalfasq
```

```
## [1] 9.928053
```

```
cialfasq <- quantile(result[,7],c(0.025,0.975))
cialfasq
```

```
##     2.5%    97.5%
##  3.17666 15.95473
```

(ii) The code outputs the following quantities:

$$\hat{\mathrm{E}}(\mu|y) = -0.02,$$
$$\hat{\mathrm{E}}\left(\sigma_g^2|y\right) = 10.12,$$
$$\hat{\mathrm{E}}\left(\sigma_e^2|y\right) = 28.27,$$
$$\hat{\mathrm{E}}\left(h^2|y\right) = 0.26,$$
$$\hat{\Pr}(-0.481 < \mu < 0.437|y) = 0.95,$$
$$\hat{\Pr}\left(3.21 < \sigma_g^2 < 16.972|y\right) = 0.95,$$
$$\hat{\Pr}\left(22.052 < \sigma_e^2 < 35.565|y\right) = 0.95,$$
$$\hat{\Pr}\left(0.084 < h^2 < 0.419|y\right) = 0.95.$$

The last lines at the bottom of the code (`meanalfasq` and `cialfasq`) show the Monte Carlo estimate of the mean and posterior interval of the genomic variance defined in (5.53) on page 234.

(iii) Execution of the `R`-code on page 636 yields MC variances of $\hat{E}(\mu|y)$, $\hat{E}\left(\sigma_g^2|y\right)$, $\hat{E}\left(\sigma_e^2\right)$ and $\hat{E}\left(h^2|y\right)$. These are 1.47×10^{-5}, 0.23, 0.14 and 1.41×10^{-4}, respectively. The associated effective chain lengths are 3, 758, 55, 84 and 54, revealing a high degree of autocorrelation for the draws from $\sigma_e^2|y$, $\sigma_g^2|y$ and from the posterior distribution $h^2|y$. A longer chain length than the one used here would result in less noisy estimates of MC estimates of posterior means. For example, executing the code setting the length of the Gibbs chain to 40, 000 results in a MC variance of $\hat{E}\left(\sigma_g^2|y\right)$ equal to 0.04 and an effective chain length of 326, an increase of factor 6 for the latter. This leads to more precise estimates of features of posterior distributions.

(iv) The likelihood is parametrised in terms of $\theta' = (\mu, \sigma_e^2, k)$. In order to derive the ML estimate of σ_g^2, one proceeds as follows. Let $\lambda' = g(\theta)' = (\mu, \sigma_e^2, k\sigma_e^2)$, where $k\sigma_e^2 = \sigma_g^2$. Using the transformation invariance property (see page 56)

$$\widehat{\sigma}_g^2 = \widehat{k\sigma}_e^2 = 9.31.$$

The asymptotic variance of λ is

$$\operatorname{Var}\left(\widehat{\lambda}\right) = \left.\frac{\partial \lambda}{\partial \theta'}\right|_{\theta=\widehat{\theta}} \operatorname{Var}\left(\widehat{\theta}\right) \left.\frac{\partial \lambda'}{\partial \theta}\right|_{\theta=\widehat{\theta}}. \tag{13.71}$$

In this expression (see page 54)

$$\frac{\partial \lambda'}{\partial \theta} = \begin{bmatrix} \frac{\partial \mu}{\partial \mu} & \frac{\partial \sigma_e^2}{\partial \mu} & \frac{\partial k\sigma_e^2}{\partial \mu} \\ \frac{\partial \mu}{\partial \sigma_e^2} & \frac{\partial \sigma_e^2}{\partial \sigma_e^2} & \frac{\partial k\sigma_e^2}{\partial \sigma_e^2} \\ \frac{\partial \mu}{\partial k} & \frac{\partial \sigma_e^2}{\partial k} & \frac{\partial k\sigma_e^2}{\partial k} \end{bmatrix} = \begin{bmatrix} 1 & 0 & 0 \\ 0 & 1 & k \\ 0 & 0 & \sigma_e^2 \end{bmatrix}.$$

When evaluated at the ML estimates, this gives:

$$\left.\frac{\partial \lambda'}{\partial \theta}\right|_{\theta=\widehat{\theta}} = \begin{bmatrix} 1 & 0 & 0 \\ 0 & 1 & 0.3291142 \\ 0 & 0 & 28.2877569 \end{bmatrix}.$$

Using $\text{Var}(\widehat{\theta})$ from the `OPTIM` output, (13.71) is

$$\text{Var}\left(\widehat{\mu}, \widehat{\sigma}_e^2, \widehat{\sigma}_g^2\right) = \begin{bmatrix} 0.00566 & 0.00036 & -0.00028 \\ 0.00036 & 13.21010 & -10.25614 \\ -0.00028 & -10.25614 & 13.12319 \end{bmatrix}.$$

The frequentist 95% confidence interval for σ_g^2 is

$$\begin{aligned} &\Pr\left(9.31 - 1.96\sqrt{13.12319} < \sigma_g^2 < 9.31 + 1.96\sqrt{13.12319}\right) \\ &= \Pr\left(2.21 < \sigma_g^2 < 16.41\right) = 0.95 \end{aligned}$$

that is numerically in reasonable agreement with the McMC estimate of the Bayesian posterior interval. The chain of length 40, 000 leads to the 95% Bayesian posterior interval $\widehat{\Pr}\left(2.25 < \sigma_g^2 < 16.92|y\right) = 0.95$, quite in line with the frequentist confidence interval.

13.5 Prediction Exercises

Exercise 1

The following results are useful:

$$E(\widehat{y}_2) = E[E(y_2|y_1)] = E(y_2), \tag{13.72}$$

and

$$\begin{aligned} \text{Cov}(\widehat{y}_2, y_2) &= E(\widehat{y}_2 y_2) - E(\widehat{y}_2) E(y_2), \\ E(\widehat{y}_2 y_2) &= E[E(\widehat{y}_2 y_2|y_1)] \\ &= E[\widehat{y}_2 E(y_2|y_1)] \\ &= E\left(\widehat{y}_2^2\right). \end{aligned} \tag{13.73}$$

Using (13.72) in (13.73)

$$\text{Cov}(\widehat{y}_2, y_2) = \text{Var}(\widehat{y}_2). \tag{13.74}$$

Similar algebra yields

$$\text{Cov}(\widehat{g}_2, g_2) = \text{Var}(\widehat{g}_2).$$

1. Given the model, the variance in the training data is

$$\mathrm{Var}(y_1) = \sigma_g^2 (G_{11} + I\lambda) = \sigma_g^2 T, \qquad \lambda = \frac{\sigma_e^2}{\sigma_g^2}, \quad T = (G_{11} + I\lambda).$$

$$\widehat{y}_2 = \mathrm{E}(y_2|y_1) = \mu_2 + G_{21} T^{-1} (y_1 - 1\mu_1).$$

Then,

$$\begin{aligned}\mathrm{Var}(\widehat{y}_2) &= G_{21} T^{-1} \left(T\sigma_g^2\right) T^{-1} G_{12} \\ &= G_{21} T^{-1} G_{12} \sigma_g^2,\end{aligned}$$

Also, given the model,

$$\widehat{g}_2 = \mathrm{E}(g_2|y_1) = G_{21} T^{-1} (y_1 - 1\mu_1).$$

Then

$$\begin{aligned}\mathrm{Var}(\widehat{g}_2) &= G_{21} T^{-1} G_{12} \sigma_g^2 \\ &= \mathrm{Var}(\widehat{y}_2).\end{aligned}$$

2. The prediction error variances are

$$\mathrm{Var}(y_2 - \widehat{y}_2) = \mathrm{Var}(y_2) - \mathrm{Var}(\widehat{y}_2) = \mathrm{E}[\mathrm{Var}(y_2|y_1)] = \mathrm{Var}(y_2|y_1),$$

Similarly,

$$\mathrm{Var}(g_2 - \widehat{g}_2) = \mathrm{Var}(g_2) - \mathrm{Var}(\widehat{g}_2) = \mathrm{E}[\mathrm{Var}(g_2|y_1)] = \mathrm{Var}(g_2|y_1),$$

where the last equalities in the right-hand sides hold in the Gaussian setup because the conditional variances do not depend on y.
3. The squared correlations or prediction accuracies are

$$\begin{aligned}R^2(g_2, \widehat{g}_2) &= \frac{[\mathrm{Cov}(g_2, \widehat{g}_2)]^2}{\mathrm{Var}(g_2)\,\mathrm{Var}(\widehat{g}_2)} \\ &= \frac{\mathrm{Var}(\widehat{g}_2)}{\mathrm{Var}(g_2)},\end{aligned}$$

equal to the proportion of the additive genetic variance captured by the linear predictor $\widehat{g}_2$. Similarly

$$
\begin{aligned}
R^2(y_2, \widehat{y}_2) &= \frac{[\mathrm{Cov}(y_2, \widehat{y}_2)]^2}{\mathrm{Var}(y_2)\,\mathrm{Var}(\widehat{y}_2)} \\
&= \frac{\mathrm{Var}(\widehat{y}_2)}{\mathrm{Var}(y_2)} \\
&= \frac{\mathrm{Var}(\widehat{g}_2)}{\mathrm{Var}(y_2)} \frac{\mathrm{Var}(g_2)}{\mathrm{Var}(g_2)} \\
&= h^2 R^2(g_2, \widehat{g}_2) .
\end{aligned}
\tag{13.75}
$$

4. $R^2(g_2, \widehat{g}_2)$ can be written as

$$
\begin{aligned}
R^2(g_2, \widehat{g}_2) &= \frac{\mathrm{Var}(\widehat{g}_2)}{\mathrm{Var}(g_2)} \\
&= \frac{\mathrm{Var}(g_2) - \mathrm{E}[\mathrm{Var}(g_2|y_1)]}{\mathrm{Var}(g_2)}.
\end{aligned}
$$

As the size of the training data n_1 increases, given that the data structure is such that this leads to a steady increase of information about the scalar g_2, the posterior variance $\mathrm{Var}(g_2|y_1)$ decreases and $R^2(g_2, \widehat{g}_2)$ will increase towards 1. Therefore in (13.75), the upper limit of $R^2(y_2, \widehat{y}_2)$ is equal to h^2.

5.

$$
\begin{aligned}
\mathrm{E}(\mathrm{MSE}_v) &= \mathrm{E}_{y_1 y_2}(y_2 - \widehat{y}_2)^2 \\
&= \mathrm{E}_{y_1 y_2}\left(y_2^2 + \widehat{y}_2^2 - 2 y_2 \widehat{y}_2\right) \\
&= \mathrm{E}_{y_2}\left(y_2^2\right) + \mathrm{E}_{y_1}\left(\widehat{y}_2^2\right) - 2\,\mathrm{E}_{y_1 y_2}(y_2 \widehat{y}_2) \\
&= \mathrm{Var}(y_2) + [\mathrm{E}(y_2)]^2 + \mathrm{Var}(\widehat{y}_2) + [\mathrm{E}(\widehat{y}_2)]^2 \\
&\quad -2\,[\mathrm{Cov}(y_2, \widehat{y}_2) + \mathrm{E}(y_2)\,\mathrm{E}(\widehat{y}_2)] \\
&= \mathrm{Var}(y_2) + \mathrm{Var}(\widehat{y}_2) + [\mathrm{E}(y_2) - \mathrm{E}(\widehat{y}_2)]^2 - 2\,\mathrm{Cov}(y_2, \widehat{y}_2) \\
&= \mathrm{Var}(y_2) - \mathrm{Var}(\widehat{y}_2)
\end{aligned}
$$

because $\mathrm{E}(y_2) = \mathrm{E}(\widehat{y}_2)$ and $\mathrm{Cov}(y_2, \widehat{y}_2) = \mathrm{Var}(\widehat{y}_2)$. This expression is equal to $\mathrm{Var}(y_2 - \widehat{y}_2)$, the prediction error variance. The equality holds when the predictor is based on the conditional mean. Indeed, by definition of variance

$$\begin{aligned}\operatorname{Var}(y_2 - \widehat{y}_2) &= \mathrm{E}\left[(y_2 - \widehat{y}_2)^2\right] - [\mathrm{E}(y_2 - \widehat{y}_2)]^2 \\ &= \mathrm{E}(\mathrm{MSE}) - [\mathrm{E}(y_2) - \mathrm{E}(\widehat{y}_2)]^2 \\ &= \mathrm{E}(\mathrm{MSE})\end{aligned}$$

when $\mathrm{E}(y_2) = \mathrm{E}(\widehat{y}_2)$.

The assumption throughout is that expectations are computed over the true model. For example, the conditional mean is

$$\widehat{y}_2 = \mathrm{E}(y_2|y_1) = \int y_2 p(y_2|y_1)\, dy_2$$

and this requires knowledge of $[y_2|y_1]$, not only its form but also of the parameters that index this distribution.

Exercise 2

1. The frequentist expectation of the validating mean squared error (12.35), given x_0, is

$$\begin{aligned}\mathrm{E}_{yy_0}(\mathrm{MSE}_v) &= \mathrm{E}_{y_0}\left(y_0^2\right) + \mathrm{E}_y\left(\hat{y}_0^2\right) - 2\,\mathrm{E}_{y_0}(y_0)\,\mathrm{E}_y(\hat{y}_0) \\ &= \operatorname{Var}(y_0) + (\mathrm{E}(y_0))^2 + \operatorname{Var}(\hat{y}_0) + (\mathrm{E}(\hat{y}_0))^2 \\ &\quad -2\,\mathrm{E}_{y_0}(y_0)\,\mathrm{E}_y(\hat{y}_0) \\ &= \operatorname{Var}(y_0) + \operatorname{Var}(\hat{y}_0) + (\mathrm{E}(y_0 - \hat{y}_0))^2\end{aligned}$$

as in (6.51). Due to independence, $\mathrm{E}_{yy_0}(y_0\,\hat{y}_0) = \mathrm{E}_{y_0}(y_0)\,\mathrm{E}_y(\hat{y}_0)$. The first term represents sampling variation of a new record; the second accounts for variation of the predictor over replications of training data, and the third is an average squared bias.
2. The algebra of the Bayesian calculation (given x_0) is based on taking expectations over the posterior predictive distribution of $\hat{y}_0$, $[\hat{y}_0|y]$:

$$\begin{aligned}\mathrm{E}(\mathrm{MSE}_v\,|y, y_0) &= y_0^2 + \mathrm{E}\left(\hat{y}_0^2|y\right) - 2y_0\,\mathrm{E}(\hat{y}_0|y) \\ &= y_0^2 + \sigma^2 + \operatorname{Var}_{\theta|y}(x_0'\theta) + (\mathrm{E}(\hat{y}_0|y))^2 \\ &\quad -2y_0\,\mathrm{E}(\hat{y}_0|y) \\ &= \sigma^2 + \operatorname{Var}_{\theta|y}(x_0'\theta) + (y_0 - \mathrm{E}(\hat{y}_0|y))^2. \quad (13.76)\end{aligned}$$

The first term accounts for sampling variation of a new validating datum; the second represents the propagated posterior uncertainty of parameter θ to the conditional mean of the predictor $\hat{y}_0$, given θ (the sharper the posterior distribution of θ, the smaller the size of this term); the third is the squared discrepancy between the observed validating datum y_0 and the expected value of the predictor, taken over its posterior predictive distribution (see (10.4a)). In the second line, I used (see (10.4b))

$$\begin{aligned}
\mathrm{Var}\left(\hat{y}_0|y\right) &= \mathrm{E}\left(\hat{y}_0^2|y\right) - \left(\mathrm{E}\left(\hat{y}_0|y\right)\right)^2 \\
&= \sigma^2 + \mathrm{Var}_{\theta|y}\left(x_0'\theta\right), \\
\mathrm{E}\left(\hat{y}_0^2|y\right) &= \sigma^2 + \mathrm{Var}_{\theta|y}\left(x_0'\theta\right) + \left(\mathrm{E}\left(\hat{y}_0|y\right)\right)^2.
\end{aligned}$$

On the other hand, if instead of predicting single records, a Bayesian predictor $\hat{y}_0 = x_0'\theta^*$ is constructed drawing θ^* from the posterior distribution $[\theta|y]$, then, taking expectations over $[\theta|y]$

$$\begin{aligned}
\mathrm{E}\left(\mathrm{MSE}_v\,|y, y_0\right) &= y_0^2 + \mathrm{E}\left(\left(x_0'\theta\right)^2|y\right) - 2y_0\,\mathrm{E}\left(x_0'\theta|y\right) \\
&= \mathrm{Var}_{\theta|y}\left(x_0'\theta\right) + \left(y_0 - \mathrm{E}\left(x_0'\theta|y\right)\right)^2,
\end{aligned}$$

which is equal to the last two terms in (13.76). This is so because in the linear model, with $\hat{y}_0 = x_0'\theta$

$$\mathrm{E}_{y_0|y}\left(\hat{y}_0|y\right) = \mathrm{E}_{\theta|y}\left(x_0'\theta|y\right)$$

(see (10.4a)). The expectation of MSE_v in (13.76) has an extra term accounting for the sampling variance of the draw of the predictor from $\left[\hat{y}_0|\theta, y\right]$.

Exercise 3

Case 1. When training data y_t are treated as fixed, $\mathrm{E}(\hat{y}_v - y_v) = \hat{y}_v - \mathrm{E}(y_v)$, $\mathrm{Var}(\hat{y}_v - y_v) = \mathrm{Var}(y_v)$ and

$$\begin{aligned}
\mathrm{E}\left(\hat{y}_v - y_v\right) &= 1\widehat{\mu} - 1\mu. \\
\mathrm{Var}\left(\hat{y}_v - y_v\right) &= V = I\sigma^2.
\end{aligned}$$

Let $A = \frac{1}{\sigma^2} I$. Then $AV = I$, an $n_v \times n_v$ identity matrix (idempotent). It follows that

$$\frac{1}{\sigma^2}\left(\hat{y}_v - y_v\right)'\left(\hat{y}_v - y_v\right) \sim \chi^2\left(r\left(A\right), \lambda\right).$$

Since $r(A) = n_v$, $\lambda = \frac{1}{\sigma^2}\left(1\widehat{\mu} - 1\mu\right)'\left(1\widehat{\mu} - 1\mu\right)$, then,

$$\text{MSE}_v = \frac{\sigma^2}{n_v}\chi^2\left(n_v, \lambda\right).$$

Using (12.31a)

$$\begin{aligned} \text{E}\left(\text{MSE}_v\right) &= \frac{\sigma^2}{n_v}\left(n_v + \frac{1}{\sigma^2}\left(1\widehat{\mu} - 1\mu\right)'\left(1\widehat{\mu} - 1\mu\right)\right) \\ &= \sigma^2 + \left(\hat{\mu} - \mu\right)^2. \end{aligned} \tag{13.77}$$

The first term in the right-hand side of (13.77) represents sampling uncertainty of y_v, and the second represents the squared difference between the predictor and the expected value of the predictand.

Using (12.31b) the variance is

$$\text{Var}\left(\text{MSE}_v\right) = \left(\frac{\sigma^2}{n_v}\right)^2\left(2n_v + \frac{4}{\sigma^2}\left(1\widehat{\mu} - 1\mu\right)'\left(1\widehat{\mu} - 1\mu\right)\right). \tag{13.78}$$

Case 2. When training data are treated as random and validating data as fixed

$$\begin{aligned} \text{E}\left(\hat{y}_v - y_v\right) &= 1_{n_v}\mu - y_v \\ \text{Var}\left(\hat{y}_v - y_v\right) &= 1_{n_v}1'_{n_v}\frac{\sigma^2}{n_t}, \end{aligned}$$

where 1_{n_v} is a column vector of 1′s with n_v elements.

Consider the quadratic form:

$$\frac{1}{\sigma^2}\left(\hat{y}_v - y_v\right)'\left(\widehat{y}_v - y_v\right) = \left(\hat{y}_v - y_v\right)' A\left(\hat{y}_v - y_v\right)$$

where $A = I\frac{1}{\sigma^2}$. Then AV is idempotent, as shown below:

$$\begin{aligned} AVAV &= \left(11'\frac{1}{n_t}\right)\left(11'\frac{1}{n_t}\right) \\ &= \left(11'\frac{1}{n_t}\right). \end{aligned}$$

It follows that

$$\frac{1}{\sigma^2}\left(\widehat{y}_v - y_v\right)'\left(\widehat{y}_v - y_v\right) \sim \chi^2\left(r\left(A\right), \frac{1}{\sigma^2}\left(1\mu - y_v\right)'\left(1\mu - y_v\right),\right)$$

a chi-square distribution with $r(A) = n_v$ degrees of freedom and non-centrality parameter $\frac{1}{\sigma^2}(1\mu - y_v)'(1\mu - y_v)$. Then the sample validating mean squared error is a draw from

$$\text{MSE}_v = \frac{\sigma^2}{n_v}\chi^2\left(r\left(A\right), \frac{1}{\sigma^2}\left(1\mu - y_v\right)'\left(1\mu - y_v\right)\right)$$

and has expected value

$$\begin{aligned}
\text{E}\left(\text{MSE}_v\right) &= \frac{\sigma^2}{n_v}\left[\frac{n_v}{n_t} + \frac{1}{\sigma^2}\left(1\mu - y_v\right)'\left(1\mu - y_v\right)\right] \\
&= \frac{\sigma^2}{n_t} + \frac{1}{n_v}\sum_{i=1}^{n_v}\left(y_{v_i} - \mu\right)^2.
\end{aligned}$$

The first term represents sampling variance of $\hat{\mu}$ and the second is the average squared difference between the validating records and the expected value of their predictions.

Case 3. When both the training and the validating data are treated as random, the mean squared error is not proportional to a chi-square distribution. However, its expected value can be computed using (12.32).

In this case

$$\begin{aligned}
\text{E}\left(\hat{y}_v - y_v\right) &= \text{E}\left(1_{n_v}\hat{\mu} - 1_{n_v}\mu\right) \\
&= \left(1_{n_v}\mu - 1_{n_v}\mu\right) = 0, \\
\text{Var}\left(\hat{y}_v - y_v\right) &= 1_{n_v}1'_{n_v}\frac{\sigma^2}{n_t} + I_{n_v}\sigma^2.
\end{aligned}$$

Then,

$$\begin{aligned}
\text{E}\left(\text{MSE}_v\right) &= \frac{1}{n_v}\text{tr}\left[\text{Var}\left(\hat{y}_v - y_v\right)\right] \\
&= \frac{\sigma^2}{n_v}\left(\frac{n_v}{n_t} + n_v\right) \\
&= \sigma^2\left(\frac{1}{n_t} + 1\right). \qquad (13.79)
\end{aligned}$$

The first term accounts for the sampling variance of the ML estimator and the second for the sampling variance of the validating records. There is no third term because the expectation of $\left(\hat{y}_v - y_v\right)$ over the distribution of (y_v, y_t) is zero.

To compute $\mathrm{E}\left(\mathrm{MSE}_t\right)$, I use

$$\begin{aligned}\widehat{y}_t &= 1_{n_t}\left(1_{n_t}1'_{n_t}\right)^{-1}1_{n_t}y_t \\ &= Hy_t\end{aligned}$$

where H is idempotent. Then with $H1_{n_t} = 1_{n_t}$

$$\mathrm{E}\left(\widehat{y}_t - y_y\right) = 0.$$

The variance is

$$\begin{aligned}\mathrm{Var}\left(Hy_t - y_y\right) &= H\sigma^2 + I\sigma^2 - 2H\sigma^2 \\ &= \left(I_{n_t} - H\right)\sigma^2.\end{aligned}$$

Therefore,

$$\begin{aligned}\mathrm{E}\left(\mathrm{MSE}_t\right) &= \frac{\sigma^2}{n_t}\,\mathrm{tr}\left(I_{n_t} - H\right) \\ &= \frac{\sigma^2}{n_t}\left(n_t - 1\right).\end{aligned}$$

The average difference between validating and training mean squared errors

$$\begin{aligned}\mathrm{E}\left(\mathrm{MSE}_v\right) - \mathrm{E}\left(\mathrm{MSE}_t\right) &= \sigma^2\left(\frac{1}{n_t} + 1\right) - \frac{\sigma^2}{n_t}\left(n_t - 1\right) \\ &= 2\frac{\sigma^2}{n_t}.\end{aligned}$$

This is the expected optimism (see (6.68)) as coined by Efron (1986), the extent by which, the MSE based on training data underestimates the validating MSE, on average.

Exercise 4

1.

(a) In the first scenario, assuming y_v^* is a draw from $y_v|\hat{\mu}, y_t \sim N\left(1\hat{\mu}, I\sigma^2\right)$

$$\mathrm{E}\left(y_v^* - y_v\right) = 1\hat{\mu} - y_v.$$

Let

$$z = \frac{1}{\sigma}\left(y_v^* - y_v\right).$$

Then,

$$z|y_v \sim N\left(\frac{1}{\sigma}\left(1\hat{\mu} - y_v\right), I\right)$$

and therefore,

$$z'z|y_v \sim \chi^2\left(n_v, \lambda\right),$$

where the non-centrality parameter is

$$\lambda = \mathrm{E}\left(z|y_v\right)' \mathrm{E}\left(z|y_v\right) = \frac{1}{\sigma^2}\left(1\hat{\mu} - y_v\right)'\left(1\hat{\mu} - y_v\right).$$

In terms of z, the validating MSE_v is

$$\mathrm{MSE}_v = \frac{\sigma^2}{n_v} z'z$$

and therefore, using (12.31),

$$\mathrm{E}\left(\mathrm{MSE}_v\right) = \frac{\sigma^2}{n_v}\left(n_v + \frac{1}{\sigma^2}\left(1\hat{\mu} - y_v\right)'\left(1\hat{\mu} - y_v\right)\right), \tag{13.80a}$$

$$\mathrm{Var}\left(\mathrm{MSE}_v\right) = \left(\frac{\sigma^2}{n_v}\right)^2\left(2n_v + \frac{4}{\sigma^2}\left(1\hat{\mu} - y_v\right)'\left(1\hat{\mu} - y_v\right)\right). \tag{13.80b}$$

The first term in the right-hand side of (13.80a) accounts for the sampling variance associated with the generation of predicted validating records $y_v^\star$.

(b) In this second scenario, the posterior distribution of μ is

$$\mu|y_t \sim N\left(\hat{\mu}, \frac{\sigma^2}{n_t}\right),$$

where in this model, the posterior mean $\hat{\mu}$ takes the same form as the ML estimator. The vector of predicted validating records is

$$\widehat{y}_v = 1_{n_v}\mu.$$

The computation of MSE_v requires

$$\begin{aligned}\text{E}\left(\hat{y}_v|y_t\right) &= \text{E}_{\mu|y_t}\left[\text{E}\left(\hat{y}_v|\mu\right)\right] \\ &= \text{E}_{\mu|y_t}\left[1_{n_v}\mu\right] = 1_{n_v}\hat{\mu}\end{aligned}$$

and

$$\begin{aligned}\text{Var}\left(\hat{y}_v|y_t\right) &= \text{E}_{\mu|y_t}\left[\text{Var}\left(\hat{y}_v|\mu\right)\right] + \text{Var}_{\mu|y_t}\left[E\left(\hat{y}_v|\mu\right)\right] \\ &= \text{Var}_{\mu|y_t}\left(1_{n_v}\mu\right) = 1_{n_v}1'_{n_v}\frac{\sigma^2}{n_t}.\end{aligned}$$

Therefore,

$$\begin{aligned}\text{E}\left(\hat{y}_v - y_v\right) &= 1_{n_v}\hat{\mu} - y_v, \\ \text{Var}\left(\hat{y}_v - y_v\right) &= 1_{n_v}1'_{n_v}\frac{\sigma^2}{n_t},\end{aligned}$$

and

$$\begin{aligned}\text{E}\left(\text{MSE}_v\right) &= \frac{1}{n_v}\text{E}\left[\left(\widehat{y}_v - y_v\right)'\left(\widehat{y}_v - y_v\right)\right] \\ &= \frac{1}{n_v}\left[\frac{\sigma^2}{n_t}\text{tr}\left(1'_{n_v}1_{n_v}\right) + \left(1_{n_v}\widehat{\mu} - y_v\right)'\left(1_{n_v}\widehat{\mu} - y_v\right)\right] \\ &= \frac{\sigma^2}{n_t} + \frac{1}{n_v}\left(1_{n_v}\widehat{\mu} - y_v\right)'\left(1_{n_v}\widehat{\mu} - y_v\right).\end{aligned} \tag{13.81}$$

The first term on the right-hand side is the contribution from the posterior uncertainty of μ, and the second term is the average squared discrepancy between the validating observations and the validating mean predictions.

(c) The third scenario is based on the hierarchical model defined in (12.41). Under this model, the validating mean squared error does not reduce to a chi-square variable. An analytic expression for the expected value of MSE_v is derived using the formula for the expectation of a quadratic form (12.32). Under the hierarchy defined in (12.41), dropping the subscript n_v in 1_{n_v}

$$\begin{aligned}\text{E}\left(y_v|y_t\right) &= \text{E}_{\mu|y_t}\left[\text{E}\left(y_v|\mu, y_t\right)\right] \\ &= \text{E}_{\mu|y_t}\left(1\mu\right) = 1\hat{\mu}\end{aligned}$$

and

$$\begin{aligned}\text{Var}(y_v|y_t) &= \text{E}[\text{Var}(y_v|\mu, y_t)] + \text{Var}[E(y_v|\mu, y_t)] \\ &= I\sigma^2 + 11'\frac{\sigma^2}{n_t}.\end{aligned}$$

Then,

$$\text{E}\left(y_v^\star - y_v\right) = \left(1\hat{\mu} - y_v\right),$$

$$\text{Var}\left(y_v^\star - y_v\right) = I\sigma^2 + 11'\frac{\sigma^2}{n_t},$$

where the random variable $y_v^\star$, the vector of predicted validating records, is drawn from the posterior predictive distribution $[y_v|y_t]$. Since $A = I$, using (12.32), it follows that the expected value of the validating mean squared error is

$$\begin{aligned}\text{E}(\text{MSE}_v) &= \frac{1}{n_v}\left[(1\widehat{\mu} - y_v)'(1\widehat{\mu} - y_v) + \sigma^2\left(\text{tr}\left(I_{n_v} + \frac{1}{n_t}\text{tr}(11')\right)\right)\right] \\ &= \sigma^2 + \frac{\sigma^2}{n_t} + \frac{1}{n_v}(1\widehat{\mu} - y_v)'(1\widehat{\mu} - y_v). \qquad (13.82)\end{aligned}$$

2. The difference in MSE_v between (13.80) and (13.82) is the term $\frac{\sigma^2}{n_t}$ that reflects posterior uncertainty of μ.

   As indicated in (13.79), the frequentist expectation of MSE_v taken over the distribution of training and validating data generates a term that accounts for the sampling uncertainty of $\hat{\mu}$ (the term $\frac{\sigma^2}{n_t}$). The estimator is unbiased and therefore there is no third term contributing the $\text{E}(\text{MSE}_v)$.

   Exercises 1 and 2 illustrate that the expectation of MSE derived either from frequentist or from Bayesian perspectives contain similar terms. The distribution over which the expectation is computed determines the breakdown of E(MSE). In the frequentist approach, expectations are taken over the training data, or the validating data, or both. The Bayesian counterpart conditions on training and validating data; expectations are computed over the posterior distribution of parameters, or over the conditional distribution of simulated records, given parameters. These are draws from the same distribution as the validating records. Alternatively and to account for both sources of uncertainty, one resorts to the hierarchical model (12.41), and expectations are taken over the posterior predictive distribution of predicted values of validating data.

Exercise 5

All computations are conditional on the variance components that are assumed known.

1. The exact distribution of MSE_v when validating predictions y_v^* are drawn from the distribution $[y_v|\hat{\mu}, \hat{b}, y_t]$ can be obtained as follows. Let

$$z = \frac{1}{\sigma_e}\left(y_v^* - y_v\right),$$

$$\Rightarrow z|\hat{\mu}, \hat{b}, y_v \sim N\left(\frac{1}{\sigma_e}\left(1_v\widehat{\mu} + W_v\widehat{b} - y_v\right), I\right).$$

Then,

$$z'z \sim \chi^2\left(n_v, \lambda\right),$$

because $A = I\sigma_e^{-2}$, $V = I\sigma_e^2$ and $AV = I$, an idempotent matrix. The non-centrality parameter is

$$\lambda = E\left(z|\hat{\mu}, \hat{b}, y_v\right)' E\left(z|\hat{\mu}, \hat{b}, y_v\right)$$
$$= \frac{1}{\sigma_e^2}\left(1_v\widehat{\mu} + W_v\widehat{b} - y_v\right)'\left(1_v\widehat{\mu} + W_v\widehat{b} - y_v\right).$$

In terms of z, the validating mean squared error is

$$\text{MSE}_v = \frac{\sigma_e^2}{n_v}z'z.$$

Using (12.31), it follows that

$$\text{E}\left(\text{MSE}_v\right) =$$
$$= \left(\frac{\sigma_e^2}{n_v}\right)\left(n_v + \frac{1}{\sigma_e^2}\left(1_v\widehat{\mu} + W_v\widehat{b} - y_v\right)'\left(1_v\widehat{\mu} + W_v\widehat{b} - y_v\right)\right) \tag{13.83a}$$

$$\text{Var}\left(\text{MSE}_v\right) =$$
$$= \left(\frac{\sigma_e^2}{n_v}\right)^2\left(2n_v + \frac{4}{\sigma_e^2}\left(1_v\widehat{\mu} + W_v\widehat{b} - y_v\right)'\left(1_v\widehat{\mu} + W_v\widehat{b} - y_v\right)\right). \tag{13.83b}$$

The R-code below generates n individuals each genotyped for p covariates (genetic markers). Among these genetic markers, `nqtl` are chosen as causal genotypes. The gene substitution effects of these causal loci are chosen to generate an additive genetic variance equal to 10 squared units, and the heritability of the continuous trait is set equal to 0.5. The data y are divided into a training y_t and a validating set y_v, each of size $n/2$. Let W represent the centred and scaled matrix X, where $X = \{X_{ij}\}$ is an $n/2 \times p$ observed matrix with genotypic codes X_{ij} equal to 0, 1, 2 according to the number of the arbitrarily chosen allele of individual i and marker j.

The operational model is defined in terms of the following distributions:

$$y_t|\mu, b, \sigma_e^2 \sim N\left(1\mu + Wb, I\sigma_e^2\right), \tag{13.84a}$$

$$b|\sigma_b^2 \sim N\left(0, I\sigma_b^2\right), \tag{13.84b}$$

where y_t is a vector of records of length $n/2$, μ is an unobserved mean, b is a vector of unknown genetic marker effects of length p, 1 is a vector of 1's of length $n/2$, σ_e^2 is the residual variance and σ_b^2 reflects prior uncertainty for each element of b. The two variance components are assumed known:

```
# CODE1314
# PREDICTION EXERCISE 5
rm(list=ls()) # CLEAR WORKSPACE
set.seed(123)
nindiv<-100
nmark <- 500
nt <- nindiv*nmark
# NUMBER QTL
nqtl <- 50

# GENERATE MARKER MATRIX FROM BINOMIAL DISTRIBUTION
X<-matrix(nrow=nindiv,ncol=nmark,rbinom(n=nt,size=2,p=.5))
##########################################################
# CHOOSE VALUE FOR MEAN mu AND GENOMIC VARIANCE vgs
mu <- 10
vgs<-10
# CHOOSE VALUE FOR ENVIRONMENTAL VARIANCE ves
ves<-20
her <- vgs/(vgs+ves)

btrue<-matrix(data=0.0,nrow=nmark,ncol=1) # parameter from
#             true model
IDq<-sample(1:nmark,nqtl,replace=F) # from the nmark markers,
#             choose nqtl as QTL
QTLeff<-sqrt(vgs/nqtl)# calculate the QTL effect so that the
#             total genetic variance is VA
btrue[IDq]<-QTLeff # the only b's that are not zero are those
#            associated with QTL.
W <- matrix(data=NA,nrow= nindiv,ncol=nmark)
cm <- colMeans(X)
# CREATE MATRIX OF STANDARDISED MARKER GENOTYPE CODES
for (i in 1:nmark)
{
  W[,i] <-( X[,i]-cm[i]) / sd(X[,i])
}
# more efficiently, could use:
# W <- scale(X)
# GENERATE nindiv PHENOTYPES WITH MEAN 0, VAR=vgs+ves,
```

```
#            HERITABILITY=vgs/(vgs+ves)
e<- rnorm(nindiv,mean=0,sd=sqrt(ves))
y <- mu + W%*%btrue+ e
k <- (ves/vgs)*nmark # ratio of residual to
#            genomic variance Vb = vgs/nmark
train <- sample(1:nrow(W),floor(0.5*nrow(W)))
Xt <- W[train,]
yt <- y[train]
Xv <- W[-train,]
yv <- y[-train]
Zt <- cbind(1,Xt)
Zv <- cbind(1,Xv)
######################
# ridge regression coefficient matrix, rhs & solution solt
RHSt <- crossprod(Zt,yt)
LHSt <- crossprod(Zt)
LHSt[-1,-1] <- LHSt[-1,-1]+diag(k,nrow=nrow(LHSt)-1)
solt <- solve(LHSt,RHSt)
# PREDICTION, CONDITIONAL ON ESTIMATED PARAMETERS (solt)
predval <- Zv%*%solt # VALIDATING
predtrain <- Zt%*%solt # TRAINING
```

The R-code below generates a MC estimate of the posterior distribution of the MSE_v, where the validating predictions are drawn from the distribution $[y_v|\hat{\mu}, \hat{b}, y_t]$.

```
# CODE1314 (cont)
# COMPUTE SAMPLING DISTRIBUTION OF MSE, CONDITIONAL ON
#              (mu_hat,b_hat) AND VARIANCES
rep <- 10000
res1 <- matrix(data=NA, nrow=rep,ncol=1)

meany <- predval
vary <- diag(ves,nrow=length(yv))
ptm <- proc.time()
for (i in 1:rep){
  yrep <- rnorm(length(yv),meany,sqrt(ves))
  mse1 <- mean((yrep-yv)^2)
  ztz <- (1/ves)*sum((yrep-yv)^2)
  res1[i,] <- mse1
}
proc.time()-ptm
```

```
##    user  system elapsed
##    0.70    0.01    0.21
```

```
meanmsev <- apply(res1,2,mean)
meanmsev
```

```
## [1] 52.34182
```

```
varmsev <- apply(res1,2,var)
varmsev
```

```
## [1] 67.52094
```

```
ncp <- sum((yv-meany)^2)/ves
expmse <- (ves/length(yv))*(length(yv)+ncp)
expmse
```

```
## [1] 52.35257
```

```
varmse <- (2* length(yv) + 4*ncp)*(ves/length(yv))^2
varmse
```

```
## [1] 67.76411
```

```
expQF <- ves + (mean((yv-meany)^2))
expQF
```

```
## [1] 52.35257
```

The MC estimates of the mean and variance are 52.342 and 67.521. These agree well with the exact results, 52.353 and 67.764 obtained from (13.83). The last line of the code computes the expected value of MSE_v using the formula for the expectation of a quadratic form, which, of course, agrees with the mean of the scaled chi-square distribution.

2. The validating mean squared error arising from this model does not reduce to a chi-square variable. However an analytic expression for the expected value of MSE_v based on the expectation of a quadratic form can be derived as follows. First express the hierarchical model as

$$\theta^*|y_t \sim N\left(\widehat{\theta}, C_t^{-1}\sigma_e^2\right),$$

$$y_v^*|\theta^*, y_t \sim N\left(Z_v\theta^*, I\sigma_e^2\right),$$

where

$$\theta = (\mu, b),$$

$$Z_v = (1\ W_v).$$

The solution to the linear system, $\hat{\theta}$, is

$$C_t\widehat{\theta} = Z_t'y_t,$$

and

$$C_t = \begin{bmatrix} 1'1 & 1'W_t \\ W_t'1 & W_t'W_t + Ik \end{bmatrix}, \quad k = \frac{\sigma_e^2}{\sigma_b^2}.$$

Then

$$\mathrm{E}\left(y_v^*|y_t\right) = \mathrm{E}_{\theta^*|y_t}\left[\mathrm{E}\left(y_v^*|\theta^*, y_t\right)\right] = \mathrm{E}_{\theta^*|y_t}\left(Z_v\theta^*\right) = Z_v\widehat{\theta},$$

and

$$\begin{aligned}\mathrm{Var}\left(y_v^*|y_t\right) &= \mathrm{E}_{\theta^*|y_t}\left[\mathrm{Var}\left(y_v^*|\theta^*, y_t\right)\right] + \mathrm{Var}_{\theta^*|y_t}\left[\mathrm{E}\left(y_v^*|\theta^*, y_t\right)\right] \\ &= I\sigma_e^2 + Z_v C_t^{-1} Z_v' \sigma_e^2.\end{aligned}$$

Therefore, conditional on the complete data $y' = \left(y_t', y_v'\right)$,

$$\begin{aligned}\mathrm{E}\left(\mathrm{MSE}_v\right) &= \frac{1}{n_v}\mathrm{E}\left[\left(y_v^* - y_v\right)'\left(y_v^* - y_v\right)\right] \\ &= \frac{1}{n_v}\left[\mathrm{tr}\left(\sigma_e^2\left(I + Z_v C_t^{-1} Z_v'\right)\right) + \left(Z_v\widehat{\theta} - y_v\right)'\left(Z_v\widehat{\theta} - y_v\right)\right] \\ &= \sigma_e^2 + \frac{\sigma_e^2}{n_v}\mathrm{tr}\left(Z_v C_t^{-1} Z_v'\right) + \frac{1}{n_v}\left(Z_v\widehat{\theta} - y_v\right)'\left(Z_v\widehat{\theta} - y_v\right)\end{aligned} \quad (13.85)$$

The first term is the contribution from sampling variation of validating predictions, the second reflects the propagated posterior uncertainty of θ, and the third is an average squared discrepancy between the observed validating records and the predictions.

3. The R-code below applies the method of composition to obtain draws from the posterior predictive distribution of y_v and constructs the validating mean squared error based on these draws:

```
# CODE1314 (cont)
# METHOD OF COMPOSITION:
# (ACCOUNTING FOR UNKNOWN LOCATION PARAMETERS)
# 1. USING TRAINING DATA Yt, SAMPLE THETA* ~ THETA|Yt
# 2. SAMPLE VALIDATING DATA Yv* ~ Yv|THETA*
# 3. COMPUTE VALIDATION MSEv = MEAN((Yv*-Yv)^2)
# 4. GOTO 1 UNTIL ENOUGH SAMPLES
rep <- 1000
res2 <- matrix(data=NA, nrow=rep,ncol=2)
theta <- solt
Cinv <- solve(LHSt)
```

```
ch <- chol(Cinv*ves)
varcov <- Cinv*ves
ptm <- proc.time()
for (i in 1:rep){
#  print(i)
  theta <- solt + t(ch)%*%rnorm(length(theta),0,1)
# DRAWS FROM THE VALIDATING DATA:
  ystarval <- rnorm(length(yv),Zv%*%theta,sqrt(ves))
# DRAWS FROM THE TRAINING DATA:
  ystartrain <- rnorm(length(yt),Zt%*%theta,sqrt(ves))

  mse2val <- mean((ystarval-yv)^2) # VALIDATION MSE
  mse2train <- mean((ystartrain-yt)^2) # TRAINING MSE
  res2[i,] <- c(mse2val,mse2train)
}
proc.time()-ptm
```

```
##    user  system elapsed
##    6.51    0.15    1.50
```

```
# hist(res2[,1])
apply(res2,2,mean)
```

```
## [1] 62.99294 43.39716
```

```
meanmse2val <- mean(res2[,1])
varmse2val <-  var(res2[,1])
interm <- Zv%*%Cinv%*%t(Zv)
expQF <- ves + (ves*sum(diag(interm)))/length(yv) +
        mean((predval-yv)^2)
expQF
```

```
## [1] 62.65027
```

The MC estimate of the mean validating mean squared error 62.993 agrees well with the theoretical result 62.65 obtained from (13.85). The posterior uncertainty of $[\theta|y_t]$ propagates onto the distribution of the validating mean squared error. As a consequence, the mean validating mean squared error is larger than the one obtained when the calculation is conditional on $\theta = \hat{\theta}$. The posterior variance of MSE_v, equal to 116.464, is also considerably larger.

The attraction of the Monte Carlo approach is that it generates an estimate of the complete marginal posterior distribution of MSE.

Exercise 6

1. The linear predictor is

$$\widehat{y} = W\left[W'W + \Sigma\right]^{-1} W'y$$
$$= Hy.$$

Contrary to the non-hierarchical case the "hat" matrix H is not idempotent:

The expected optimism is

$$\frac{2}{n}\operatorname{tr}\left[\operatorname{Cov}\left(y, \widehat{y}'\right)\right] = \frac{2}{n}\operatorname{tr}\left(\operatorname{Cov}\left(y, \widehat{y}'\right)\right)$$
$$= \frac{2}{n}\operatorname{tr}\left(\operatorname{Cov}\left(y, y'H'\right)\right)$$
$$\frac{2}{n}\operatorname{tr}\left[\operatorname{Cov}\left(y, y'W\left[W'W + \Sigma\right]^{-1} W'\right)\right]$$
$$= \frac{2}{n}\operatorname{tr}\left[VW\left[W'W + \Sigma\right]^{-1} W'\right] \tag{13.86}$$

where $V = \operatorname{Var}(y) = ZZ'\sigma_f^2 + I\sigma^2 = \sigma^2\left(ZZ'\frac{\sigma_f^2}{\sigma^2} + I\right) = \sigma^2\widetilde{V}$. An additive genetic model without non-genetic sources of covariation between full-sibs imposes the constraint $0 < \sigma_f^2 \leq \sigma^2$.

The R-code below generates the full-sib family data, constructs the mixed model equations, the predictor (fitted values), and obtains an expression for expected optimism. The bottom part of the code generates a Monte Carlo estimate of expected optimism:

```
# CODE1315
#FULL-SIB CONTINUOUS DATA
rm(list=ls()) # Clear the workspace
set.seed(123771)
ptm<-proc.time()
require(graphics)
# INITIALISE PARAMETERS
mus<-10 # MEAN
vfs<-1 #VARIANCE BETWEEN FULL-SIBS
#vfs<-0.5 #VARIANCE BETWEEN FULL-SIBS
#vfs <- 0.1
# RESIDUAL VARIANCE
ves<-5
k <- ves/vfs
nf<-500 # NUMBER OF FULL-SIB FAMILIES
n<-2 # FULL-SIB FAMILY SIZE
nb <- 2 # NUMBER OF BREEDS
N<-nf*n # TOTAL NUMBER OF RECORDS
y<-matrix(data=0,nrow=nf*n,ncol=1)
z<-matrix(data=0,nrow=nf*n,ncol=1)
```

```
# GENERATE nf FULL-SIB EFFECTS fs
fs<-rnorm(nf,mean=0,sd=sqrt(vfs))
# BREED EFFECTS
br <- rep(0,nb)
br[1] <- 5
br[2] <- 8
# GENERATE nf*n RESIDUAL EFFECTS
es<-rnorm(nf*n,mean=0,sd=sqrt(ves))
##################################################
## GENERATING A FULL-SIB STRUCTURE
b <- rep(1:nb,each=N/2)
z <- rep(1:nf,each=n)
y <- br[b] + fs[z] + es
d <- data.frame(y,z)
##################################################
d<-data.frame(y,z)
# GENERATE INCIDENCE MATRICES X & Z
family <- z
breed <- b
family <- as.factor(family)
breed <- as.factor(breed)
X<-model.matrix(~0+breed)
Z<-model.matrix(~0+family)
W <- cbind(X,Z)
LHS <- crossprod(W) # LHS OF MME
LHS[-(1:2),-(1:2)]<-LHS[-(1:2),-(1:2)]+diag(k,nrow=nrow(LHS)-2)
RHS <- crossprod(W,y) # RHS OF MME
SOL <- solve(LHS,RHS) # SOLUTION
HAT <- W%*%solve(LHS)%*%t(W)
V <- Z%*%t(Z)*vfs + diag(ves,nrow=length(y))
COVyyhat <- sum(diag(V%*%t(HAT)))
lambda <- 1/k
Vtilde <- (Z%*%t(Z)*lambda + diag(1,nrow=length(y)))
df <- sum(diag(Vtilde%*%HAT))
yhat <- HAT%*%y
MSEt <- mean((y-yhat)^2)
MSEt
```

```
## [1] 4.754883
```

```
optim1 <- 2*COVyyhat/length(y)
optim1
```

```
## [1] 2.02
```

```
MSEv <- MSEt + optim1
MSEv
```

```
## [1] 6.774883
```

The analytical results yield an expression for expected optimism (parameter `optim1` in the bottom of the code) based on (13.86) equal to 2.02. The sample training mean squared error is

$$\text{MSE}_t = \frac{1}{n}\sum_{i=1}^{N}(y_i - \widehat{y}_i)^2$$

equal to 4.75. Therefore the estimate of the validating mean square error is

$$\text{MSE}_v = \text{MSE}_t + \frac{2}{n}\,\text{tr}\left(\text{Cov}\left(y, \widehat{y}'\right)\right)$$

equal to 6.77 (parameter MSE_v at the bottom of the code).

2. The R-code below generates a Monte Carlo estimate of expected optimism using a parametric bootstrap:

```
# CODE1315 (cont)
# MONTE CARLO ESTIMATE OF OPTIMISM
## SIMULATE DATA & STORE Y, YHAT
rep <- 1000
gemY <- matrix(data=NA,nrow=rep,ncol=length(y))
gemYHAT <- matrix(data=NA,nrow=rep,ncol=length(y))

br[1] <- SOL[1]
br[2] <- SOL[2]
for (i in 1:rep){
  fs<-rnorm(nf,mean=0,sd=sqrt(vfs))
  es<-rnorm(nf*n,mean=0,sd=sqrt(ves))
  y <- br[b] + fs[z] + es
  RHS <- crossprod(W,y)
  SOL <- solve(LHS,RHS)
  yhat <- W%*%SOL
  gemY[i,] <- y
  gemYHAT[i,] <- yhat
}
sumcov <- 0
## COMPUTE SUM(COV(Y,YHAT))
for (i in 1:length(y)){
  sumcov <- sumcov + cov(gemY[,i],gemYHAT[,i])
}
#######################################
##  A MORE EFFICIENT AND LESS TRANSPARENT CODE IS
# sumcov <-
# sum(sapply(1:length(y),FUN=function(i)
# {cov(gemY[,i],gemYHAT[,i])}))
#######################################
sumcov
```

```
## [1] 1010.313
```

```
optim2 <- 2*sumcov/length(y)
optim2
```

```
## [1] 2.020626
```

The Monte Carlo estimate of expected optimism, conditional on the model (parameter `optim2` in the bottom of the code), based on $1,000$ replications of data, is equal to 2.0206. This compares well with the analytical result.

Degrees of Freedom

An expression for the model's *degrees of freedom* (or *effective number of parameters*) is

$$df = \frac{1}{\sigma^2} \operatorname{tr}\left(\operatorname{Cov}\left(y, \widehat{y}'\right)\right), \tag{13.87}$$

(Hastie et al 2009) that for the present setup gives

$$df = \operatorname{tr}\left[\widetilde{V} W \left[W'W + \Sigma\right]^{-1} W'\right].$$

Setting $\sigma_f^2 = 1$ and $\sigma^2 = 5$ results in an estimate of the model's degrees of freedom or effective number of parameters based on (13.87) equal 202 (parameter `df` in the code).

This estimate depends on the ratio $\sigma_f^2 \big/ \sigma^2$. To illustrate, setting σ_f^2 to a very small number in the code above (say, $\sigma_f^2 = 10^{-5}$) produces an estimate of the model's degrees of freedom of 2.00 (rounded off to two decimal places). This is equal to the number of breeds. A simple way to understand this result is by studying the linear system (12.46). The model for a scalar record can be written

$$y_{ijk} = b_i + f_{ij} + e_{ijk},$$

where b_i is the fixed effect of breed i, f_{ij} is the random effect of family j nested in breed i and e_{ijk} is the residual associated with record k of family j in breed i. A closer look at the mixed model equations (12.46) reveals that the equation for the ijth family is

$$n_o \widehat{b}_i + \left(n_o + \frac{\sigma_f^2}{\sigma^2}\right) \widehat{f}_{ij} = y_{ij.}$$

where $y_{ij.}$ is the sum of the records belonging to the ijth family. A little algebra shows that for the data structure of the example

$$\widehat{f}_{ij} = \left(y_{ij.} - n_o\widehat{b}_i\right) - \frac{(n_o - 1)\,\sigma_f^2 + \sigma^2}{n_o\sigma_f^2 + \sigma^2}\left(y_{ij.} - n_o\widehat{b}_i\right).$$

When $\sigma_f^2 \rightarrow 0$, $\widehat{f}_{ij} \rightarrow 0$ and when $\sigma_f^2 \rightarrow \sigma^2$ (its maximum value, given the constraint),

$$\widehat{f}_{ij} \rightarrow \frac{1}{(n_o + 1)}\left(y_{ij.} - n_o\widehat{b}_i\right).$$

The first case, $\sigma_f^2 \rightarrow 0$, sets family effects equal to zero, and the model's degrees of freedom correspond to the number of breeds, equal to 2.

Imagine that instead of parametrising the model as in (12.45), the random family effects are integrated out yielding the alternative form:

$$y|b \sim N\left(Xb, \sigma^2\widetilde{V}\right).$$

The generalised least squares estimator of b is

$$\widehat{b} = \left(X'\widetilde{V}^{-1}X\right)^{-1} X'\widetilde{V}^{-1}y.$$

The vector of fitted values is $\widehat{y} = X\widehat{b}$ and the degrees of freedom of the integrated parametrisation are

$$\begin{aligned}
df &= \frac{1}{\sigma^2}\operatorname{tr}\left[\operatorname{Cov}\left(y, \widehat{y}'\right)\right] \\
&= \frac{1}{\sigma^2}\operatorname{tr}\left[\operatorname{Cov}\left(y, y'\widetilde{V}^{-1}X\left(X'\widetilde{V}^{-1}X\right)^{-1}X'\right)\right] \\
&= \frac{1}{\sigma^2}\operatorname{tr}\left[\sigma^2\widetilde{V}\widetilde{V}^{-1}X\left(X'\widetilde{V}^{-1}X\right)^{-1}X'\right] \\
&= \operatorname{tr}\left[X'X\left(X'\widetilde{V}^{-1}X\right)^{-1}\right].
\end{aligned}$$

Setting $\sigma_f^2 = 1$ and $\sigma^2 = 5$ results in an estimate of the model's degrees of freedom equal to 2.8, which differs from the estimate 202 obtained with the alternative parametrisation. Different parametrisations of the model result in different measures of its complexity, as pointed out by Spiegelhalter et al (2002). On the other hand, when $\sigma_f^2 \rightarrow 0$, $\widetilde{V} \rightarrow I$ and the model's degree of freedom approaches $\operatorname{tr}\left[X'X\left(X'X\right)^{-1}\right] = 2$, the number of breeds, as with the previous parametrisation.

The concept of degrees of freedom has not been discussed in the book and is not pursued further. The example illustrates that the topic deserves considerable reflection (see Janson et al 2015). The definition used here is based on the optimism of the training mean squared error as an estimate of validating error and is due to Efron (1986). Spiegelhalter et al (2002, 2014) discuss the subject and propose an alternative expression. Further elaborations of the concept can be found in Gelman et al (2014).

References

Aitken AC (1934) Note on selection from a multivariate normal population. Proc Edinb Math Soc 4:106–110

Albert J (2009) Bayesian Computation with R. Springer, Berlin

Anderson DA, Aitkin M (1985) Variance component models with binary response: interviewer variability. J R Stat Soc Ser B 47:203–210

Anderson TW (1957) Maximum likelihood estimation for the multivariate normal distribution when some observations are missing. J Am Stat Assoc 52:200–203

Benjamini Y, Hochberg Y (1995) Controlling false discovery rate: A practical and powerful approach to multiple testing. J R Stat Soc Ser B Methodol 57:289–300

Benjamini Y, Yekutieli Y (2001) The control of false discovery rate in multiple testing under dependency. Ann Stat 29:1165–1188

Benjamini Y, Krieger AM, Yekutieli D (2006) Adaptive linear step-up procedures that control the false discovery rate. Biometrika 93:491–507

Bernard E (2021) Introduction to machine learning. Wolfram Media, Champaign

Bernardo JM, Smith AFM (1994) Bayesian Theory. Wiley, London

Berndt E, Hall B, Hall R, Hausman J (1974) Estimation and inference in nonlinear structural models. Ann Econ Soc Meas 3:653–665

Bishop CM (2006) Pattern recognition and machine learning. Springer, Berlin

Bottou L (2012) Stochastic gradient descent tricks. In: Montavon G, Orr GB, Müller KR (eds) Neural networks: tricks of the trade. Springer, Berlin, pp 421–436. chap 18

Box GEP, Tiao GC (1973) Bayesian inference in statistical analysis. Wiley, London

Boyle EA, Li YI, Pritchard JK (2017) An expanded view of complex trits: from polygenic to omnigenic. Cell 169:1177–1186

Breiman L (1996) Bagging predictors. Mach Learn 24:123–140

Breiman L (2001) Random forests. Mach Learn 45:5–32

Breiman L, Friedman JH, Olshen RA, Stone CJ (1984) Classification and regression trees. Chapman and Hall, London

Brown PJ (1977) Centering and scaling in ridge regression. Technometics 19:35–36

Brzyski D, Peterson CB, Sobczyk P, Candes E, Bogdan M, Sabatti C (2017) Controlling the rate of GWAS false discoveries. Genetics 205:61–75

Bulmer MG (1971) The effect of selection on genetic variability. Am Nat 105:201–211

Bulmer MG (1980) The mathematical theory of quantitative genetics. Oxford University Press, Oxford

Carlin BP, Louis TA (1996) Bayes and empirical Bayes methods for data analysis. Chapman and Hall, London

D. Sorensen, *Statistical Learning in Genetics*, Statistics for Biology and Health,
https://doi.org/10.1007/978-3-031-35851-7

Casella G, Berger RL (1990) Statistical inference. Brooks–Cole, Belmont
Chen MH, Shao QM, Ibrahim JG (2000) Monte Carlo methods in Bayesian computation. Springer–Verlag, Berlin
Cowles MK, Carlin BP (1996) Markov chain Monte Carlo convergence diagnostics: a comparative review. J Am Stat Assoc 91:883–904
Curnow RN (1961) The estimation of repeatability and heritability from records subject to culling. Biometrics 17:553–566
Dahlquist B Å, Björck Å (1974) Numerical methods. Prentice-Hall, Englewood Cliffs
de los Campos G, Gianola D, Rosa GJM (2009) Reproducing kernel Hilbert spaces regression: a general framework for genetic evaluation. J Anim Sci 87:1883–1887
de los Campos G, Gianola D, Rosa GJM, Weigel KA, Crossa J (2010) Semi-parametric genomic-enabled prediction of genetic values using reproducing kernel Hilbert spaces methods. Genet Res 92:295–308
de los Campos G, Hickey J, Pong-Wong R, Daetwyler HD, Calus MPL (2013a) Whole genome regression and prediction methods applied to plant and animal breeding. Genetics 193:327–345. https://doi.org/10.1534/genetics.112.143313
de los Campos G, Vazquez AI, Fernando R, Klimentidis YC, Sorensen D (2013b) Prediction of complex human traits using the genomic best linear unbiased predictor. PLoS Genet 9(3), e1003608
de los Campos G, Sorensen D, Gianola D (2015) Genomic heritability: what is it? PLOS Genet 11(5), e1005048
de los Campos G, Grueneberg A, Funkhouser S, Perez-Rodriguez P, Samaddar A (2022) Fine mapping and accurate prediction of complex traits using Bayesian Variable Selection models applied to biobank-size data. Eur J Hum Genet. https://doi.org/10.1038/s41431-022-01135-5
Dempster AP, Laird NM, Rubin DB (1977) Maximum likelihood from incomplete data via de EM algorithm (with discussion). J R Stat Soc Ser B 39:1–38
Devroye L (1986) Non-uniform random variate generation. Springer–Verlag, Berlin
Diggle PJ (2011) Estimating prevalence using an imperfect test. Epidemiol Res Int 2011. https://doi.org/10.1155/2011/608719
Duvenaud D, Lloyd JR, Grosse R, Tenenbaum JB, Ghahramani Z (2013) Structure discovery in nonparametric regression through compositional kernel search. ArXiv:1302.4922
Edwards AWF (1992) Likelihood. The John Hopkins University Press, Baltimore
Efron B (1986) How biased is the apparent rate of a prediction rule? J Am Stat Assoc 81:461–470
Efron B (2010) Large-scale inference. Cambridge University Press, Cambridge
Efron B (2020) Prediction, estimation, and attribution. J Am Stat Assoc 115:636–655
Efron B, Hastie T (2016) Computer age statistical inference. Cambridge University Press, Cambridge
Efron B, Tibshirani R (2002) Microarrays, empirical Bayes methods and false discovery rates. Genet Epidemiol 1:70–86
Efron B, Tibshirani R, Storey JD, Tusher V (2001) Empirical Bayes analysis of a microarray experiment. J Am Stat Assoc 96:1151–1160
Falconer DS (1965) The inheritance of liability to certain diseases, estimated from the incidence among relatives. Ann Hum Genet 29:51–76
Falconer DS, Mackay TFC (1996) Introduction to quantitative genetics. Longman, New York
Fisher RA (1922) On the mathematical foundations of theoretical statistics. Philos Trans R Soc Lond A 222:309–368
Friedman J, Hastie T, Höfling H, Tibshirani R (2007) Pathwise coordinate optimization. Ann Appl Stat 1(2):302–332
Friedman J, Hastie T, Tibshirani R (2009) Glmnet: Lasso and elastic-net regularized generalized linear models. R package version 1.1-4
Friedman J, Hastie T, Tibshirani R (2010) Regularization paths for generalized linear models via coordinate descent. J Stat Softw 33(1):1–22
Galton F (1885) Regression towards mediocrity in hereditary stature. J Anthropol Insti 15:246–263

García-Cortés LA, Sorensen D (1996) On a multivariate implementation of the Gibbs sampler. Genet Sel Evol 28:121–126

Gelfand AE (1996) Model determination using sampling-based methods. In: Gilks WR, Richardson S, Spiegelhalter DJ (eds) Markov chain Monte Carlo in practice. Chapman and Hall, London, pp 145–161

Gelfand AE, Dey DK, Chang H (1992) Model determination using predictive distributions with implementation via sampling-based methods. In: Bernardo JM, Berger JO, Dawid AP, Smith AFM (eds) Bayesian statistics, vol. 4. Oxford University Press, Oxford, pp 147–167

Gelman A, Carlin JB, Stern HS, Rubin DB (1995) Bayesian data analysis. Chapman and Hall, London

Gelman A, Meng XL, Stern H (1996) Posterior predictive assessment of model fitness via realized discrepancies (with discussion). Stat Sin 6:733–807

Gelman A, Hwang J, Vehtari A (2014) Understanding predictive information criteria for Bayesian models. Stat Comput 24:997–1016

George EI, McCulloch RE (1993) Variable selection via Gibbs sampling. J Am Stat Assoc 8:881–889

Geyer CJ (1992) Practical Markov chain Monte Carlo. Stat Sci 7:473–511

Gianola D, de los Campos G (2008) Inferring genetic values for quantitative traits non-parametrically. Genet Res 90:525–540

Gianola D, de los Campos G, Hill WG, Manfredi E, Fernando R (2009) Additive genetic variability and the Bayesian alphabet. Genetics 183:347–363

Gianola D, Fernando R, Schön CC (2020) Inferring trait-specific similarity among individuals from molecular markers and phenotypes with Bayesian regression. Theor Popul Biol 132:47–59

Glorot X, Bengio Y (2010) Understanding the difficulty of training deep feedforward neural networks. J Mach Learn Res Proc Track 9:249–256

Goodfellow I, Bengio Y, Courville A (2016) Deep learning. MIT Press, Cambridge

Green P (1995) Reversible jump MCMC computation and Bayesian model determination. Biometrika 82:711–732

Habier D, Fernando R, Deckers J (2007) The impact of genetic relationship information on genome-assisted breeding values. Genetics 177:2389–2397

Habier D, Fernando R, Kizilkaya K, Garrik DJ (2011) Extension of the Bayesian alphabet for genomic selection. BMC Bioinf 12(186):1–2

Hald A (1998) A History of mathematical statistics, from 1750 to 1930. Wiley, London

Halldorsson BV, Eggertsson HP et al (2022) The sequences of 150,119 genomes in the UK Biobank. Nature 607:732–740

Harville DA (1977) Maximum likelihood approaches to variance component estimation and to related problems. J Am Stat Assoc 72:320–340

Hastie T, Qian J (2016) Glmnet vignette. https://web.stanford.edu/~hastie/Papers/Glmnet_Vignette.pdf

Hastie T, Tibshirani R, Friedman J (2009) The elements of statistical learning. Springer, New York, p 745

Hastings WK (1970) Monte Carlo sampling methods using Markov chains and their application. Biometrika 57:97–109

Hayes B, Price J, Chamberlain AJ, Bowman PJ, Goddard ME (2010) Genetic architecture of complex traits and accuracy of genomic prediction: coat colour, milk-fat percentage, and type in Holstein cattle as contrasting model traits. PLoS Genet 6(9), e1001139

Henderson CR (1953) Estimation of variance and covariance components. Biometrics 9:226–252

Henderson CR (1975) Best linear unbiased estimation and prediction under a selection model. Biometrics 31:423–447

Henderson CR, Kempthorne O, Searle SR, Von Krosigk CN (1959) Estimation of environmental and genetic trends from records subject to culling. Biometrics 15:192–218

Hill WG, Weir BS (2011) Variation in actual relationship as a consequence of Mendelian sampling and linkage. Genet Res 93:47–64

Hill WG, Goddard ME, Visscher PM (2008) Data and theory point to mainly additive genetic variance for complex traits. PLoS Genet 4:e1000008

Hoerl AE, Kennard RW (1970a) Ridge regression: Applications to nonorthogonal problems. Technometrics 12:69–82

Hoerl AE, Kennard RW (1970b) Ridge regression: biased estimation for nonorthogonal problems. Technometrics 12:55–67

Hu Z, Zhang J, Ge Y (2021) Handling vanishing gradient problem using artificial derivative. IEEE Access 9:22,371–22,377

James G, Witten D, Hastie T, Tibshirani R (2017) An introduction to statistical learning. Springer, Berlin

Janson L, Fithian W, Hastie TJ (2015) Effective degrees of freedom: a flawed metaphor. Biometrika 102:479–485

Kass RE, Carlin BP, Gelman A, Neal RM (1998) Markov chain Monte Carlo in practice: a roundtable discussion. Am Stat 52:93–100

Kimeldorf G, Wahba G (1971) Some results on Tchebycheffian spline functions. J Math Anal Appl 33:82–95

LeCun Y, Bengio Y, Hinton G (2015) Deep learning. Nature 521:436–444

Lehmann EL, Casella G (1998) Theory of point estimation. Springer–Verlag, Berlin

Lindley DV, Smith AFM (1972) Bayesian estimates for the linear model. J R Stat Soc Ser B 34:1–41

Little RJA, Rubin DB (1987) Statistical analysis with missing data. Wiley, London

Liu JS, Wong HW, Kong A (1994) Covariance structure of the Gibbs sampler with applications to the comparisons of estimators and augmentation schemes. Biometrika 81:27–40

López OAM, López AM, Crossa J (2022) Mulitivariate statistical machine learning methods for genomic prediction. Springer, Berlin

Lush JL, Shrode RR (1950) Changes in milk production with age and milking frequency. J Dairy Sci 33:338–357

Mäki-Tanila A, Hill WG (2014) Influence of gene interaction on complex trait variation with multilocus models. Genetics 198:355–367

Mallows CL (1973) Some comments on *Cp*. Technometrics 15:661–675

Mardia KV, Kent JT, Bibby JM (1979) Multivariate analysis. Academic Press, New York

McCulloch WS, Pitts W (1943) A logical calculus of the ideas immanent in nervous activity. Bull Math Biophys 5:115–133

McLachlan GJ, Krishnan T (1997) The EM algorithm and extensions. Wiley, London

Meeker WQ, Escobar LA (1995) Teaching about approximate confidence regions based on maximum likelihood estimation. Am Stat 49:48–53

Meng X (2018) Statistical paradises and paradoxes in big data (i): law of large populations, big data paradox, and the 2016 presidential election. Ann Appl Stat 12:685–726

Metropolis N, Rosenbluth AW, Rosenbluth MN, Teller AH, Teller E (1953) Equations of state calculations by fast computing machines. J Chem Phys 21:1087–1092

Meuwissen THE, Hayes BJ, Goddard ME (2001) Prediction of total genetic value using genome-wide dense marker maps. Genetics 157:1819–1829

Mitchell AJ, Beauchamp JJ (1988) Bayesian variable selection in linear regression (with discussion). J Am Stat Assoc 83:1023–1036

Müller P, Parmigiani G, Rice K (2007) FDR and Bayesian multiple comparison rules. In: Bernardo JM, Bayarri MJ, Berger JO, Dawid AP, Heckerman D, Smith AFM, West M (eds) Bayesian statistics, vol 8, Oxford University Press, Oxford, pp 349–370

Natarajan R, Kass RE (2000) Reference Bayesian methods for generalized linear models. J Am Stat Assoc 95:227–237

Neal RM (1996) Bayesian learning for neural networks. Springer-Verlag, Berlin

Park MY, Hastie T (2008) Penalized logistic regression for detecting gene interactions. Biostatistics 9:30–50

Park T, Casella G (2008) The Bayesian LASSO. J Am Stat Assoc 103:681–686

Patterson HD, Thompson R (1971) Recovery of inter-block information when block sizes are unequal. Biometrika 58:545–554

Patxot M, Banos TD, Kousathanas A, Orliac EJ, Ojavee SE, Moser G, Holloway A, Sidorenko J, Kutalik Z, Mägi R, Visscher PM, Rönnegård L, Robinson MR (2021) Probabilistic inference of the genetic architecture underlying functional enrichment of complex traits. Nat Commun 12:6972

Pearson K (1894) Contributions to the mathematical theory of evolution. Philos Trans R Soc Lond Ser A 185:71–110

Pearson K (1903) Mathematical contributions to the theory of evolution. XI. On the influence of natural selection on the variability and correlation of organs. Philos Trans R Soc Lond Ser A 200:1–66

Perez P, de los Campos G (2014) Genome-wide regression and prediction with the BGLR statistical package. Genetics 198:483–495

Perez-Elizalde S, Cuevas J, Perez-Rodriguez P, Crossa J (2015) Selection of the bandwidth parameter in a Bayesian regression model for genomic-enabled prediction. J Agric Biol Environ Stat 20:512–532

Peskun PH (1973) Optimum Monte Carlo sampling using Markov chains. Biometrika 60:607–612

Prentice RL, Pyke R (1979) Logistic disease incidence models and case-control studies. Biometrika 66:403–411

Priestley MB (1981) Spectral analysis and time series. Academic Press, New York

Ripley B (1987) Stochastic simulation. Wiley, London

Ros M, Sorensen D, Waagepetersen R, Dupont-Nivet M, SanCristobal M, Bonnet JC, Mallard J (2004) Evidence for genetic control of adult weight plasticity in the snail *Helix aspersa*. Genetics 168:2089–2097

Rousseauw J, du Plessis J, Benade A, Jordaan P, Kotze J, Jooste P (1983) Coronary risk factor screening in three rural communities. S Afr Med J 64:430–436

Royall R (1997) Statistical Evidence. Chapman and Hall, London

Rubin DB (1976) Inference and missing data. Biometrika 63:581–592

Rubin DB (1984) Bayesianly justifiable and relevant frequency calculations for the applied statistician. Ann Stat 12:1151–1172

Rubin DB (2002) Statistical analysis with missing data. Wiley, London

Rumelhart D, Hinton G, Williams R (1986) Learning representations by back-propagation. Nature 323:533–536

San Cristobal-Gaudy M, Elsen JM, Bodin L, Chevalet C (1998) Prediction of the response to a selection for canalisation of a continuous trait in animal breeding. Genet Sel Evol 30:423–451

Searle SR (1971) Linear models. Wiley, London

Seber GAF, Lee AJ (2003) Linear regression analysis. Wiley, London

Singh D, Febbo PG, Ross K, Jackson DG, Manola J, Ladd C, Tamayo P, Renshaw AA, D'Amico AV, Richie JP, Lander ES, Loda M, Kantoff PW, Golub TR, Sellers WR (2002) Gene expression correlates of clinical prostate cancer behavior. Cancer Cell 1:203–209

So HC, Kwan JSH, Cherny SS, Sham PC (2011) Risk prediction of complex diseases from family history and known susceptibility loci, with applications for cancer screaning. Am J Hum Genet 88:548–565

Sorensen D, Gianola D (2002) Likelihood, Bayesian, and MCMC methods in quantitative genetics. Springer-Verlag, Berlin, pp 740. Reprinted with corrections, 2006

Sorensen D, Waagepetersen R (2003) Normal linear models with genetically structured residual variance heterogeneity: a case study. Genet Res 82:207–222

Sorensen D, Fernando RL, Gianola D (2001) Inferring the trajectory of genetic variance in the course of artificial selection. Genet Res 77:83–94

Speed D, Hermani G, Johnson MR, Balding DJ (2012) Improved heritability estimation from genome-wide SNPs. Am J Hum Genet 91:1011–1021

Speed D, Cai N, Johnson S M R Nejentsev, Balding DJ (2017) Reevaluation of SNP heritability in complex human traits. Nat Genet 49:986–992

Speed D, Kaphle A, Balding DJ (2021) SNP-based heritability and selection: improved models and new results. BioEssays. https://doi.org/10.1002/bies.202100170

Spiegelhalter DJ, Best NG, Carlin BP, van der Linde A (2002) Bayesian measures of model complexity and fit (with discussion). J R Stat Soc Ser B 64:583–639

Spiegelhalter DJ, Best NG, Carlin BP, van der Linde A (2014) The deviance information criterion: 12 years on. J R Stat Soc Ser B 76:485–493

Storey JD (2002) A direct approach to false discovery rates. J R Stat Soc Ser B 64:479–498

Storey JD (2003) The positive false discovery rate: a Bayesian interpretation and the $q-$value. Ann Stat 31:2013–2035

Storey JD, Bass AJ (2021) Bioconductor's qvalue package. https://www.bioconductor.org/packages/devel/bioc/vignettes/qvalue/inst/doc/qvalue.pdf

Storey JD, Tibshirani R (2007) Statistical significance for genomewide studies. Proc Natl Acad Sci 100:9440–9445

Sun X, Qu L, Garrick DJ, Dekkers JCM, Fernando RL (2012) A fast EM algorithm for Bayes-like prediction of genomic breeding values. Plos One https://doi.org/10.1371/journal.pone.0049157

Swendsen R, Wang J (1987) Non-universal critical dynamics in Monte Carlo simulations. Phys Rev Lett 58:86–88

Tanner MA (1996) Tools for statistical inference. Springer–Verlag, Berlin

Tanner MA, Wong W (1987) The calculation of posterior distributions by data augmentation. J Am Stat Assoc 82:528–550

Tibshirani R (1996) Regression shrinkage and selection via the LASSO. J R Stat Soc Ser B 58:267–288

VanRaden PM (2008) Efficient methods to compute genomic predictions. J Dairy Sci 91:4414–4423

Visscher PM (2008) Sizing up human height variation. Nat Genet 40:489–490

Waagepetersen R, Sorensen D (2001) A tutorial on reversible jump MCMC with a view towards applications in QTL-mapping. Int Stat Rev 69:49–61

Wahba G (1990) Spline models for observational data. SIAM, Philadelphia

Wainschtein P, Jain D et al (2022) Assessing the contribution of rare variants to complex trait heritability from whole-genome sequence data. Nat Genet 54:263–273

Wei GCG, Tanner MA (1990) A Monte Carlo implementation of the EM algorithm and the poor man's data augmentation algorithm. J Am Stat Assoc 85:699–704

Weir B, Cockerham CC, Reynolds J (1980) The effects of linkage and linkage disequilibrium on the covariance of noninbred relatives. Heredity 45:351–359

Weir BS (1996) Genetic data analysis II. Sinauer Associates

Weir BS, Cockerham CC (1977) Two-locus theory in quantitative genetics. In: Pollak E, Kempthorne O, Bailey TB (eds) Proceedings of the international conference on quantitative genetics. The Iowa State University Press, Ames, pp 247–269

Whittemore AS (2007) A Bayesian false discovery rate for multiple testing. J Appl Stat 34:1–9

Wright S (1934) An analysis of variability in number of digits in an inbred strain of guinea pigs. Genetics 19:506–536

Yang J, Benyamin B, McEvoy BP, Gordon S, Henders AK, Nyholt DR, Madden PA, Heath AC, Martin NG, Montgomery GW, Goddard ME, Visscher PM (2010) Common SNP's explain a large proportion of the heritability for human height. Nat Genet 42:565–569

Yang J, Ferreira T, Morris AP, Medland SE, Genetic Investigation of Anthropometric Traits (GIANT) Consortium DGR, Meta-analysis (DIAGRAM) Consortium PAF Madden, Heath AC, Martin GN, Montgomery GW, Weedon MN, Loos RJ, Frayling TM, McCarthy MI, Hirschhorn JN, Goddard ME, Visscher PM (2012) Conditional and joint multiple-SNP analysis of GWAS summary statistics identifies additional variants influencing complex traits. Nat Genet 44:369–378

Yekutieli D, Benjamini Y (1999) Resampling-based false discovery rate controlling multiple tests procedures for correlated test statistics. J Stat Plann Inference 82:171–196

Yengo L, Vedantam S, Marouli E et al (2022) A saturated map of common genetic variants associated with human height. Nature 610:704–712

Zhao T, Fernando R, Cheng H (2021) Interpretable artificial neural networks incorporating Bayesian alphabet models for genome-wide prediction and association studies. G3. https://doi.org/10.1093/g3journal/jkab228

Zou H, Hastie T (2005) Regularization and variable selection via the elastic net. J R Stat Soc Ser B 67:301–320

Author Index

D. Sorensen, *Statistical Learning in Genetics*, Statistics for Biology and Health,
https://doi.org/10.1007/978-3-031-35851-7

Subject Index

D. Sorensen, *Statistical Learning in Genetics*, Statistics for Biology and Health,
https://doi.org/10.1007/978-3-031-35851-7